Collection Systems 2017 Conference

Collections in the Heartland – Collecting Today for a Cleaner Tomorrow

Omaha, Nebraska, USA
21 - 24 May 2017

ISBN: 978-1-5108-6316-3

Printed from e-media with permission by:

Curran Associates, Inc.
57 Morehouse Lane
Red Hook, NY 12571

Some format issues inherent in the e-media version may also appear in this print version.

Printed by Curran Associates, Inc. (2018)

Water Environment Federation (WEF)
601 Wythe Street
Alexandria, Virginia 22314
USA

Phone: 1-800-666-0206
Fax: 1-703-684-1545

csc@wef.org

Additional copies of this publication are available from:

Curran Associates, Inc.
57 Morehouse Lane
Red Hook, NY 12571 USA
Phone: 845-758-0400
Fax: 845-758-2633
Email: curran@proceedings.com
Web: www.proceedings.com

Water Environment Federation
the water quality people™

Collection Systems 2017 Conference

Collections in the Heartland – Collecting Today for a Cleaner Tomorrow

May 21-24, 2017
CenturyLink Center
Omaha, Nebraska, USA

NEBRASKA WATER ENVIRONMENT ASSOCIATION

This conference is held by the Water Environment Federation in cooperation with the Nebraska Water Environment Association.

TECHNICAL PROGRAM

Session 1: Operation & Maintenance

Monday, May 22, 2017
10:45 AM – 11:45 AM

Moderators: Sean W. Fitzgerald, Eric Lienhard, Hazen & Sawyer

TECHNICAL PROGRAM

Session 2: Invited Session I: Missouri and Kansas Case Studies

Monday, May 22, 2017
1:30 PM – 4:45 PM

Moderators: Srinivas Vallabhaneni, MWH Global; Christy Willson, Horner & Shifrin

TECHNICAL PROGRAM

Session 3: Sanitary Sewer Overflow

Monday, May 22, 2017
1:30 PM – 4:45 PM

Moderators: Gregory R. Heath, AECOM; Diane Sumego, Black & Veatch

TECHNICAL PROGRAM

Session 4: Infiltration and Inflow

Monday, May 22, 2017
1:30 PM – 4:45 PM

Moderators: Phillip Hubbard, Hampton Roads Sanitation District; Carl C. Chan, CDM Smith

3:00 PM Networking Break

3:45 PM Facilitated Discussion:
Panelists and Specific Topics TBD

TECHNICAL PROGRAM

Session 5: Trenchless

Tuesday, May 23, 2017
8:30 AM – 11:45 AM

Moderators: David Servidio, Hatch Mott MacDonald; Daniel J. Murray, US Environmental Protection Agency

TECHNICAL PROGRAM

Session 6: Pump Stations & Forcemain

Tuesday, May 23, 2017
8:30 AM – 11:45 AM

Moderators: Jennifer Steffens, Pure Technoligies; David Bennett, Freese & Nichols

TECHNICAL PROGRAM

Session 6: Pump Stations & Forcemain (Continued)

Tuesday, May 23, 2017
8:30 AM – 11:45 AM

TECHNICAL PROGRAM

Session 7: Modeling

Tuesday, May 23, 2017
8:30 AM – 11:45 AM

Moderators: Timothy Ruggaber, Emnet LLC; Gregory R. Heath, AECOM

TECHNICAL PROGRAM

Session 7: Modeling (Continued)

Tuesday, May 23, 2017
8:30 AM – 11:45 AM

Alternate 1 **Riverview Lift Station: A Hydraulic Balancing Act .....N/A**
Weston Engel, Brian Clow, Maya Turner, Ryan Rossell, Carollo Engineers; Jacob Hansen, City of Omaha

Alternate 2 **SSO 700 Integrated Watershed Action Plan: Selection of A Robust Data Set to Support the Successful Continuous Calibration of a Collection System Model .....N/A**
Victoria Berry, Susan Moisio, CH2M; Andrew Spurgeon, MSDGC

TECHNICAL PROGRAM

Session 8: Planning

Tuesday, May 23, 2017
1:30 PM – 3:00 PM

Moderators: Christy Willson, Horner & Shifrin; Gary T. Moore, St. Louis Metropolitan Sewer District

TECHNICAL PROGRAM

Session 9: Intelligent Networks

Tuesday, May 23, 2017
1:30 PM – 3:00 PM

Moderators: Daniel J. Murray, US Environmental Protection Agency; Steve Donovan, Brown and Caldwell

09A	**1:30 PM**	**Into the Information Age: Application of Internet of things for Sewer Maintenance Optimization .....499** Richard Haimann, Haimann Engineering; Tamsen McNarie, San Antonio Water System; Greg Quist, Keith Lewinger, SmartCover Systems
09B	**2:00 PM**	**Intelligent Platform Approach Enables Breakthrough Results: Case Studies from the US and UK Wastewater and Water Markets .....517** Pete Percivalle, Srini Vallabhaneni, John Abrera, Kevin Johnson, MWH Global now part of Stantec
09C	**2:30 PM**	**Big Pipes Plus Low Flow Equals Cost Savings -- Startup and Lessons Learned .....528** Jon Schommer, EmNet; Oluwale McFoy, Buffalo Sewer Authority; Michael Quinn, GHD; Timothy Ruggaber, EmNet
	Alternate	**Using Hydraulic Model Based Optimization to Help Determine the Best Solution to Accommodate Future Growth in the Mill Creek Watershed .....N/A** Ryan Eisele, HDR Inc.; Stuart Lord, Johnson County Wastewater; Joel Wilson, WCS; Sasa Tomic, HDR Inc.

TECHNICAL PROGRAM

Session 10: Design / Construction

Tuesday, May 23, 2017
1:30 PM – 3:00 PM

Moderators: Fernando F. Gonzalez, City of Los Angeles Industrial Waste Mgmt Div; Pono Hanson, Brown and Caldwell

TECHNICAL PROGRAM

Session 11: Combined Sewer Overflows

Wednesday, May 24, 2017
8:30 AM – 11:45 AM

Moderators: David Servidio, Hatch Mott MacDonald; Gary T. Moore, St. Louis Metropolitan Sewer District

Session 11: Combined Sewer Overflow (Continued)

Wednesday, May 24, 2017
8:30 AM – 11:45 AM

TECHNICAL PROGRAM

Session 12: Asset Management

Wednesday, May 24, 2017
8:30 AM – 11:45 AM

Moderators: Diane Sumego, Black & Veatch; Luis R. Leon

TECHNICAL PROGRAM

Session 12: Asset Management (Continued)

Wednesday, May 24, 2017
8:30 AM – 11:45 AM

12F	**11:15 AM**	**Managing The Service Life of a Brick & Cast in Place Concrete Sewer.....811** Bryon Livingston, Black & Veatch; Andy Clements, City of St. Joseph
	Alternate	**Large Scale, Priority-driven Field Data Collection and Rehabilitation: How Dekalb County Department of Watershed Management Managed Multiple Assessment Contractors to Target and Coordinate Sewer Rehabilitation .....817** Darren Eastall, Dekalb County

TECHNICAL PROGRAM

Session 13: Invited Session II: Nebraska and Iowa Case Studies

Wednesday, May 24, 2017
8:30 AM – 11:45 AM

Moderators: Jimmy Stewart, Compliance EnviroSystems; Kevin Vander Tuig

13A	**8:30 AM**	**Update on the City of Omaha's CSO Long-term Control Plan .....830** Thomas Heinemann, CH2M; James Theiler, City of Omaha
13B	**9:00 AM**	**Tapping into Omaha Parks with Green Infrastructure .....854** Andy Szatko, City of Omaha Stormwater Program
13C	**9:30 AM**	**Enhanced Affordability Analysis of Combined Sewer Overflow Long-Term Control Plans - Omaha's CSO! Program .....862** Jason Mumm, Hawksley Consulting; Jim Theiler, City of Omaha; Andy Baker, Carol Malesky, Hawklsey Consulting
	10:00 AM	**Networking Break**
13D	**10:15 AM**	**CSO Program Combines Non-Traditional Separation Implementation with Innovative CSO Disinfection .....869** Jay Brady, Stanley Consultants, Inc.
13E	**10:45 AM**	**Parallel Master Planning and Implementation to Address Significant Stormwater Needs ....888** Michael Butterfield, HDR; Sandy Pumphrey, City of Cedar Rapids; David Dechant, HDR
13F	**11:15 AM**	**Davenport, Iowa 14th District Diversion Study** Karol Giokas, RJN Group 899
	Alternate	**Man-Entry Inspections of Large Sewers Using New, Old and Creative Technologies .....N/A** John Schroeder, Drew Richards, CDM Smith

Implementing Baltimore County's Sewer Line Rapid Assessment Tool (SL-RAT) Program

Paul Sayan, PE, Louis Berger
Thomas Kiefer, PE, Baltimore County DPW

ABSTRACT

To improve inspection production rates and reduce staff/financial costs, the Bureau of Utilities implemented in 2015 their Sewer Line Rapid Assessment Tool (SL-RAT) inspection program. The SL-RAT tool transmits an acoustic signal through a sewer and, based on the analysis of the signal, determines if there are any restrictions that may result in surcharge conditions and/or overflows/basement backups. Based on the SL-RAT inspection data, the Bureau of Utilities has determined that 50 percent of the inspected sewers do not require cleaning; thereby, resulting in significant time and cost savings.

KEYWORDS

Utility management, condition assessment, sewer maintenance, acoustic testing

INTRODUCTION

The Baltimore County Department of Public Works, Bureau of Utilities is responsible for maintaining the County's sewer collection system. One of the County's sewer maintenance levels of service is to clean the entire collection system, which consists of approximately 2,100 miles of sewer, every 7 years; however, like any municipality, the Bureau of Utilities has limited staff and equipment resources. With the realization that hiring more staff and/or purchasing more equipment to maintain the County's level of service is unrealistic, the Bureau of Utilities has focused on maximizing resource efficiencies. For sewer cleaning work, maximizing efficiencies has resulted in a paradigm shift for the Bureau to inspect sewers and clean, when needed.

Between December 2012 and May 2014 Baltimore County completed three pilot studies to assess the Sewer Line Rapid Assessment Tool (SL-RAT), manufactured by Infosense, Inc. Since CCTV inspection is the most reliable method to determine sewer cleanliness; estimated percent blockages in the sewer were determined by reviewing the CCTV inspections and the percent blockage value was compared to the corresponding SL-RAT score. Blockage was defined as any defect such as debris deposition, root intrusion, grease deposit or pipe sag which reduces a sewer's available hydraulic capacity. No pre-cleaning work was completed prior to the CCTV inspections and the CCTV inspections were completed by field crews that were not involved with the SL-RAT inspections.

Phase I Pilot

SL-RAT and CCTV inspections were performed on 78 individual sewer segments. After initial review of the SL-RAT scores, Louis Berger (on behalf of the County) contacted InfoSense, the SL-RAT manufacturer, to discuss the preliminary findings. Based on their review, InfoSense noted several data discrepancies as detailed below:

1. Duplicate records
2. Incorrect inputted sewer lengths (compared to the County GIS information and GIS data recorded by the SL-RAT equipment)
3. Incorrect manhole IDs

After deleting the duplicate records and addressing the data discrepancies, InfoSense provided to Louis Berger the corrected SL-RAT scores. The analyses and findings were based on the corrected SL-RAT scores for 78 tested sewers. While the SL-RAT was designed to minimize physical or technical ability, the data discrepancies identified in this study indicated that some level of post-construction quality control (QC) review is necessary.

The SL-RAT and CCTV inspections were not performed on the same day; therefore, the hydraulic conditions at the time of the SL-RAT inspection may have differed from the conditions at the time of the CCTV inspection. However, both inspections were performed during dry weather and it is assumed that dry weather hydraulic conditions did not significantly vary. Table 1 below summarizes the expected correlation between the SL-RAT score and the percent blockage.

Table 1: Summary of SL-RAT/Estimate Percent Blockage Correlation

SL-RAT Score	Estimated Percent Blockage[1]
1	90
2	80
3	70
4	60
5	50
6	40
7	30
8	20
9	10
10	0

[1] Based on review of CCTV video inspections

The blue rows in Figure 1 below represent the average estimated percent blockage for each SL-RAT score between 0 and 10. As the SL-RAT score decreases, the estimated percent blockage should increase, as indicated by the red line. The R^2 value of the linear regression (shown as the dashed white line) is an indication of a correlation between the SL-RAT score and estimated percent blockage. An R^2 value closer to 1.00 indicates a strong correlation. The low R^2 value shown below (0.0072) indicates that there is no correlation between the SL-RAT score and the estimated percent blockage.

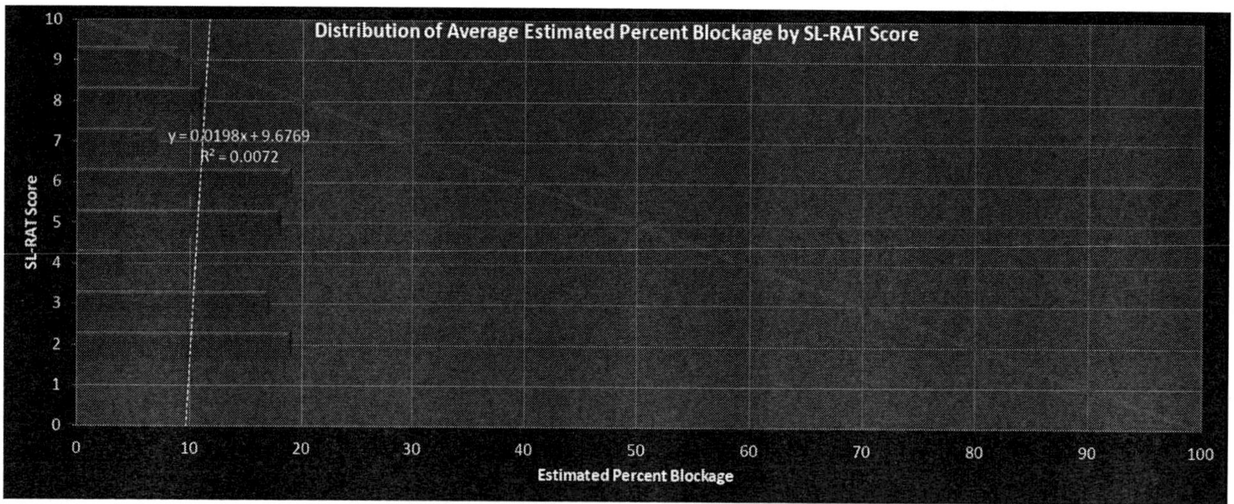

Figure 1: Summary Phase I Pilot SL-RAT/Estimated Blockage Correlation

The analyses indicate that the SL-RAT will, in general, conservatively estimated the percent blockage with a high degree of precision. Because of this conservative estimate, some sewers with low SL-RAT scores may have minor blockages, which may result in cleaning recommendations for sewers that do not require cleaning.

The lack of any correlation between the SL-RAT score and the estimated percent blockage could have resulted by inspecting the incorrect sewer; however, this was not confirmed.

Phase II Pilot

The County completed a second pilot using the SL-RAT technology to confirm the findings from the initial SL-RAT study and to provide County staff with an opportunity to use the SL-RAT equipment. The sewers included in the second study area were identified from:

1. A list of sewers with known structural defects
2. Sewer service area with known sewer debris issues

SL-RAT and CCTV inspection were performed on 23 sewer segments.

The Phase 2 analysis indicated that the SL-RAT conservatively estimated the percent blockage with a high degree of precision, thus confirming the findings from the initial study. There were no sewer segments with a high SL-RAT score (i.e. minor/no blockage) and high estimated percent blockages.

The low R^2 value shown in Figure 2 indicates insignificant correlation between the SL-RAT score and estimated blockage, thus confirming the findings of the initial study; however, the R^2 value was significantly higher than the Phase I pilot findings. It was suspected that the higher R^2 value resulted from better data quality and that there may be a correlation between the SL-RAT score and estimated blockage.

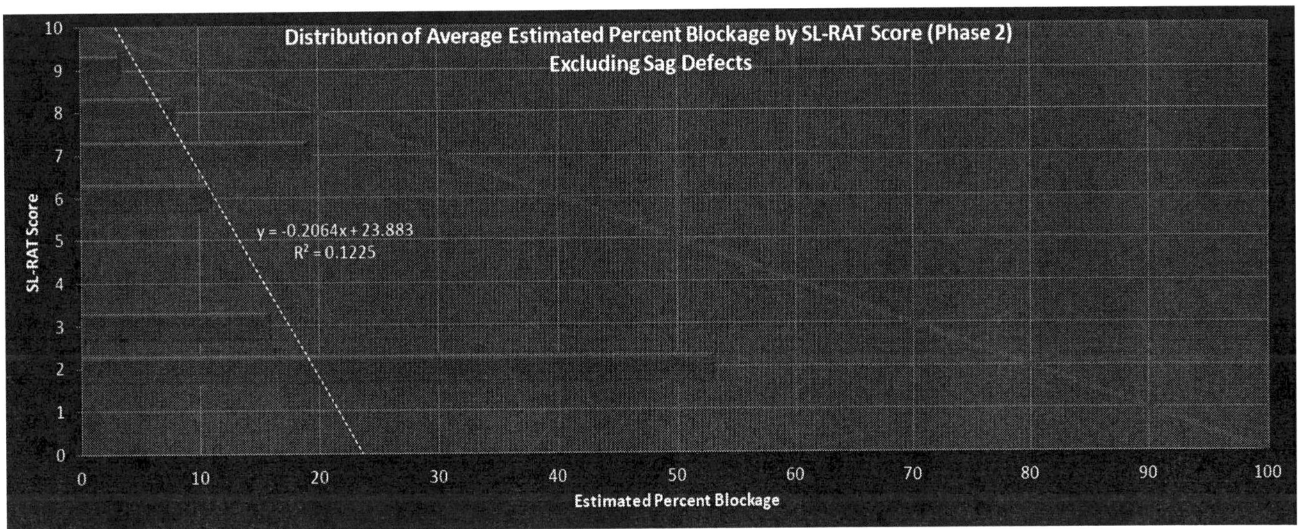

Figure 2: Summary Phase II Pilot SL-RAT/Estimated Blockage Correlation

Phase III Pilot

The County completed a third phase of the pilot study to confirm the findings from either the initial and Phase 2 SL-RAT studies. Inspections were completed by County field crews between October 2013 and May 2014. The sewers included in the third study area were selected from a list of sewers with known operational defect. A total of 55 sewer segments were inspected by CCTV and SL-RAT.

The Phase 3 analysis indicated that the SL-RAT conservatively estimated the percent blockage with a high degree of precision, thus confirming the Phase 1 and 2 findings. Figure 3 below shows the average estimated percent blockage for each SL-RAT score. The R^2 value indicates a moderate correlation between the SL-RAT score and estimated blockage. The R^2 value was significantly higher than either the Phase I or Phase II values and was suspected to be a result of improved data quality, familiarity with the equipment and indicative of a true correlation between the SL-RAT score and estimated blockage.

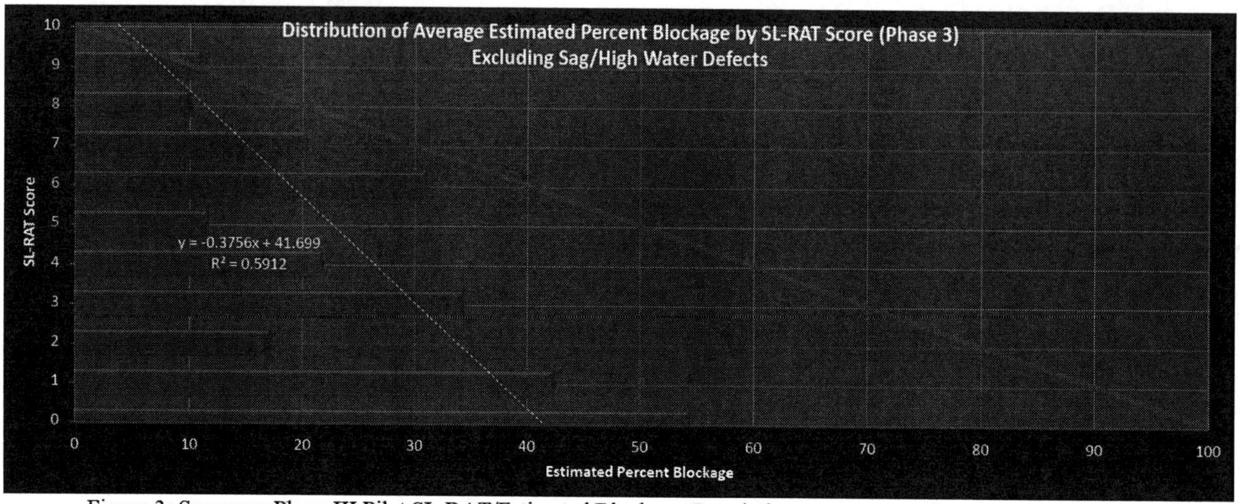

Figure 3: Summary Phase III Pilot SL-RAT/Estimated Blockage Correlation

Based on the Phase II and III pilot studies, the County developed and implemented a SL-RAT inspection program to identify and prioritize sewer cleaning work.

SL-RAT Implementation

The County's SL-RAT inspection program began in early 2015. One SL-RAT unit was purchased and a part-time two-man field crew was trained to inspect the sewers. Based on internal discussions, SL-RAT scores with corresponding sewer cleaning activities were identified as shown in Table 2 below.

Table 2: Summary of SL-RAT Follow-Up Activities

SL-RAT Score	Follow-Up Activity
0-3	Clean and CCTV Inspection
4-6	Clean
7-10	None

Figures 4 below is a summary of the SL-RAT scores that have been recorded since the program's inception. Based on the SL-RAT inspections, approximately 56% of the sewers that have been inspected, did not need to be cleaned and/or inspected, which has greatly reduced work and improved staff/equipment resource efficiencies.

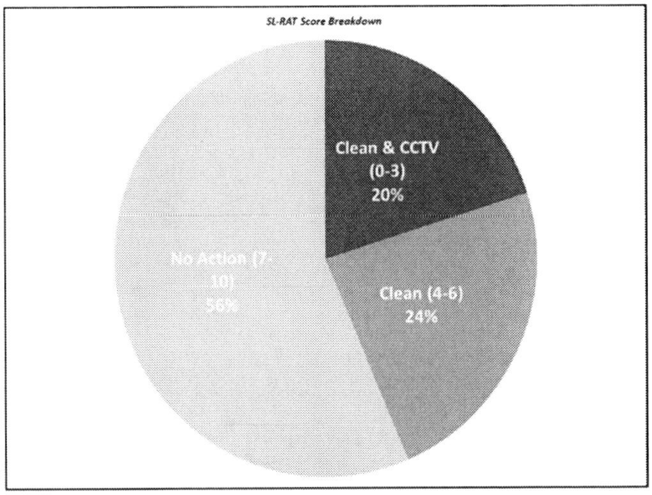

Figure 4: Summary of SL-RAT Scores

As shown in Figure 5 below, the two-man SL-RAT inspection crew average between 2,500 and 3,000 linear feet of inspections per day, which is significantly lower than the 10,000 linear feet per day production rates that were completed for the pilot studies. Based on staff interviews, it is believed that the high production rates completed for the pilot studies was a result of:

1. Significant supervisor oversight, which is not possible to sustain for every-day activities
2. Short SL-RAT inspection duration (i.e. two week pilot study compared to inspections completed every day)
3. Dedicated inspection crew. As shown in Figure 5, the SL-RAT inspection are performed approximately 13 days per month. The part-time SL-RAT inspection crew is assigned other operational/maintenance activities that require dedicated time each month.
4. Quality of inspection staff skills

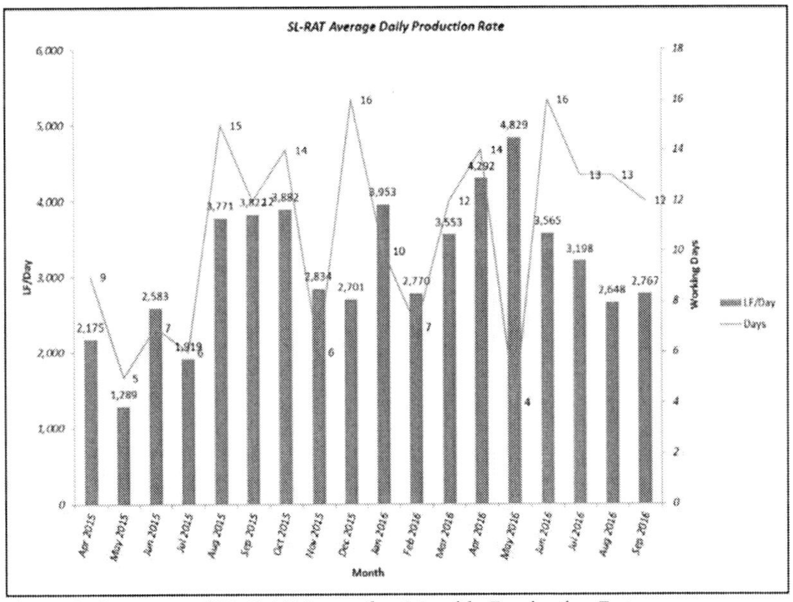

Figure 5: Summary of SL-RAT Inspection Monthly Production Rate

SL-RAT Program Improvements

The County has continuously reviewed the SL-RAT data since program implementation. As shown in Figure 6 below, nine dry-weather sewer overflows/basement backups have occurred from a sewer where a SL-RAT inspection had previously been completed; however, after further review, two of the overflows/backups where not a result of a mainline sewer maintenance issue.

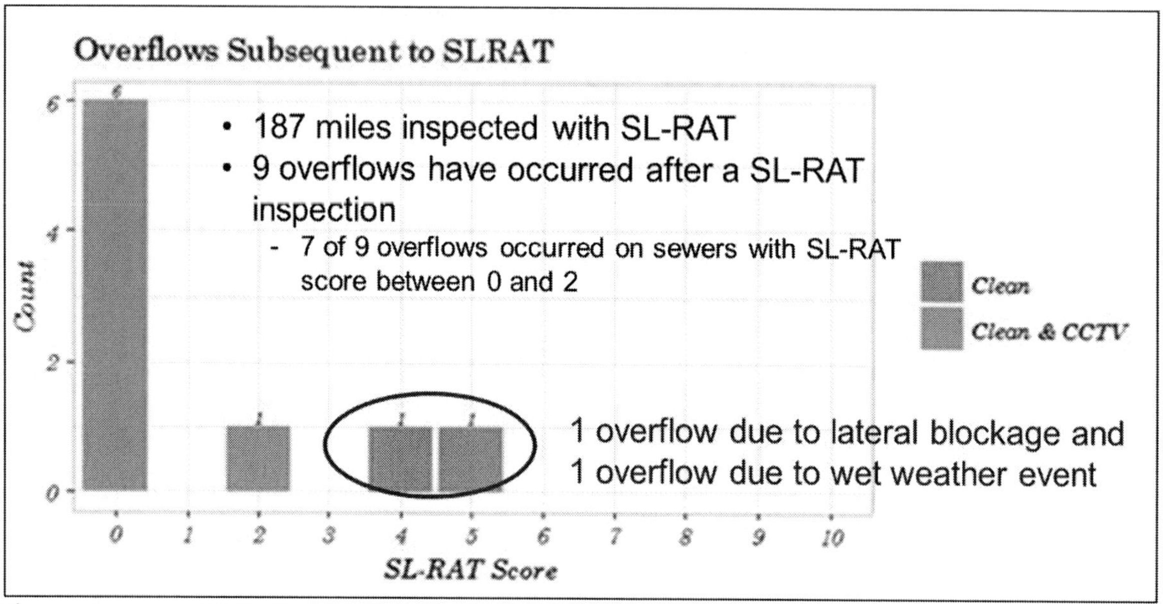

Figure 6: Summary of SSOs/Basement Backups post SL-RAT Inspection

Based on the analysis, the County revised the criteria for follow-up SL-RAT work as shown below.

Table 3: Summary of Current SL-RAT Follow-Up Activities

SL-RAT Score	Follow-Up Activity
0-3	Clean and CCTV Inspection
4-6	None
7-10	None

As summarized in Figure 7 below, approximately 20 percent of the SL-RAT inspections result in any follow-up sewer cleaning/inspection work using the revised criteria.

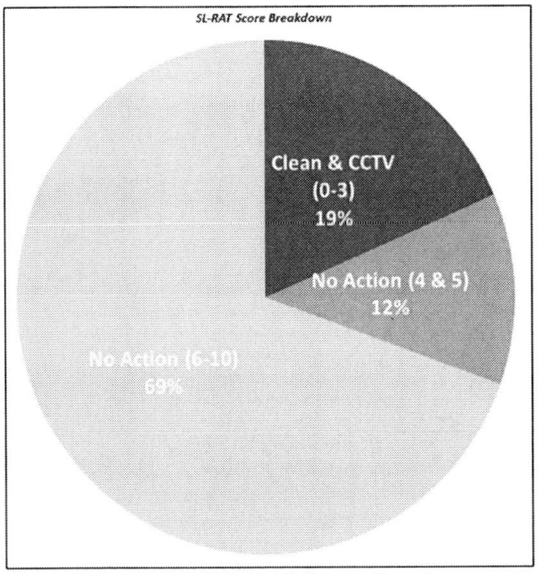

Figure 7: Summary of SL-RAT Scores (based on current follow-up work criteria)

SUMMARY AND CONCLUSIONS

To maintain the Baltimore County's sewer overflow/basement backup level of service and resource limitations, the County's Bureau of Utilities has shifted from a mindset of clean all sewers to inspect sewers and clean where needed. The shift has meant finding and using new technologies that improve sewer inspection/assessment efficiency and production rate, which resulted in creating and implementing the County's Sewer Line Rapid Assessment Tool (SL-RAT) program.

The SL-RAT inspection program has significantly reduced and prioritized sewer cleaning activities. Since the SL-RAT program began, the number of dry weather overflows and basement backups has greatly decreased (reference Figures 8 and 9, respectively), which is partly attributed to the SL-RAT program.

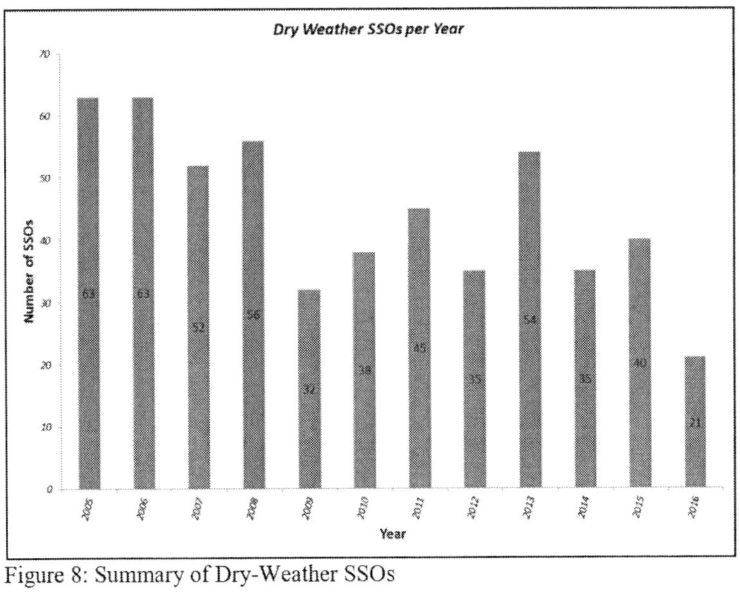

Figure 8: Summary of Dry-Weather SSOs

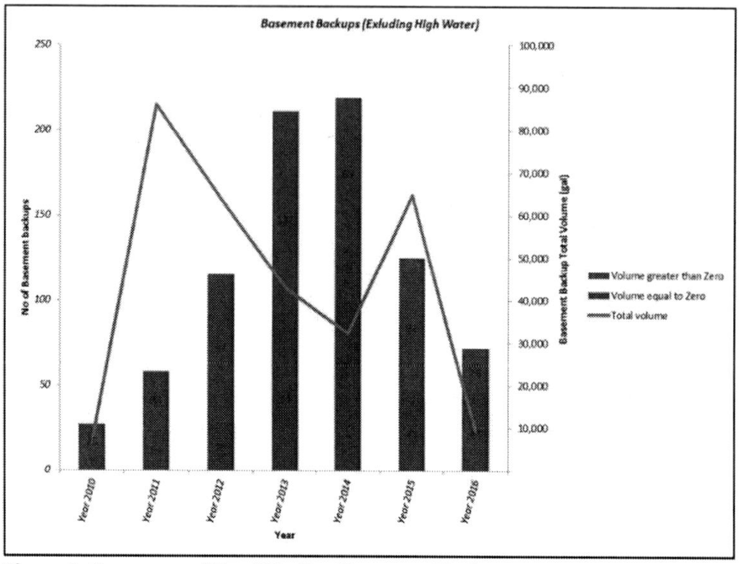

Figure 9: Summary of Dry-Weather Basement Backups

The County believes that the SL-RAT program is maximizing each sewer's operational/maintenance service life, which is helping the County better manage their staff and equipment resources. Based on the SL-RAT program's success, the County is expanding the program by:

1. Adding a second SL-RAT unit and crew
2. Implementing the SL-RAT inspections to inspect the sewers that require frequent and recurring cleaning
3. Adding SL-RAT line items to the County's sewer cleaning/inspection on-call contracts

The SL-RAT program has helped effect operational management paradigm change at the County. Traditionally, the cleaning of the wastewater collection system was based on time – when was the system and each individual asset last cleaned and when is it due for cleaning? Today, system cleaning is based on performance – does the most recent inspection indicate that cleaning is required? This paradigm shift allows the County to maximize the system's operational and maintenance service life and improve resource management so that the County can continue providing and improving their level of service.

Multi-Sensor Inspection

Author Affiliation

Dennis Sullivan
Gonzalez Companies, LLC (formerly with the City of O'Fallon, IL)

Terry Adams
Gonzalez Companies, LLC

ABSTRACT

The City of O'Fallon, IL required inspections of their North and South Sanitary Sewer Trunk Mains which were placed between 1968 and 1974. The North Main was constructed with lined corrugated metal pipe (LCMP), and the South Main was constructed using reinforced concrete pipe (RCP). Records of installation were minimal, and there were limited records of inspection or condition assessment of the piping since placement.

A combined Laser (LIDAR-Light Detection and Ranging), Sonar and High Definition-Closed Circuit Television (HDCCT) inspection of the wastewater piping was selected to provide results that would allow assessment of the pipes' interior conditions and potential for failure – structural or corrosive breakthrough.

The conclusions were that the interior of the LCMP, with the exception of a single segment (manhole to manhole), were much better than anticipated. Deformation of the metal pipe segments or individual pipe sections were minimal as well. Where liner material was missing point repairs of individual pipe sections were needed in the near term. Remaining useful life (RUL) of the LCMP was generally 30 to 50+ years based on the analysis of the interior of the pipe segments. However, many segments of the RCP trunk main examined showed significant loss of interior wall thickness and severe deterioration of the reinforcing metal cage. Resulting in a RUL of those segments of less than 10 years.

KEYWORDS

Lined corrugated metal pipe (LCMP), reinforced concrete pipe (RCP), laser (LIDAR- Light Detection and Ranging), sonar, high definition (HD), closed circuit television (CCT), deformation, wall thickness, remaining useful life (RUL), maximum wall loss (MWL), allowable wall loss (AWL), Deterioration Rate per Year (DRY), minimum allowable wall thickness (MAWT)

INTRODUCTION

The City of O'Fallon, IL's north and south interceptors convey sewage flow from the City and the new Memorial Hospital East in the Village of Shiloh, IL to the City's wastewater treatment plant (WWTP) located east of O'Fallon on Rieder Road adjacent to Silver Creek, a tributary to the Kaskaskia River. The north interceptor is approximately 7.2 kilometers (4.5 miles) in length with pipe ranging in diameter from 46 to 76 cm (18" to 30"). The south interceptor is approximately 6.6 kilometers (4.1 miles) in length with pipe ranging in diameter from 30 to 53 cm (12" to 21"). The interceptors combine in a 122 cm (48") diameter pipe for approximately 0.5 kilometers (0.3 miles) before discharging into the WWTP's terminal lift station.

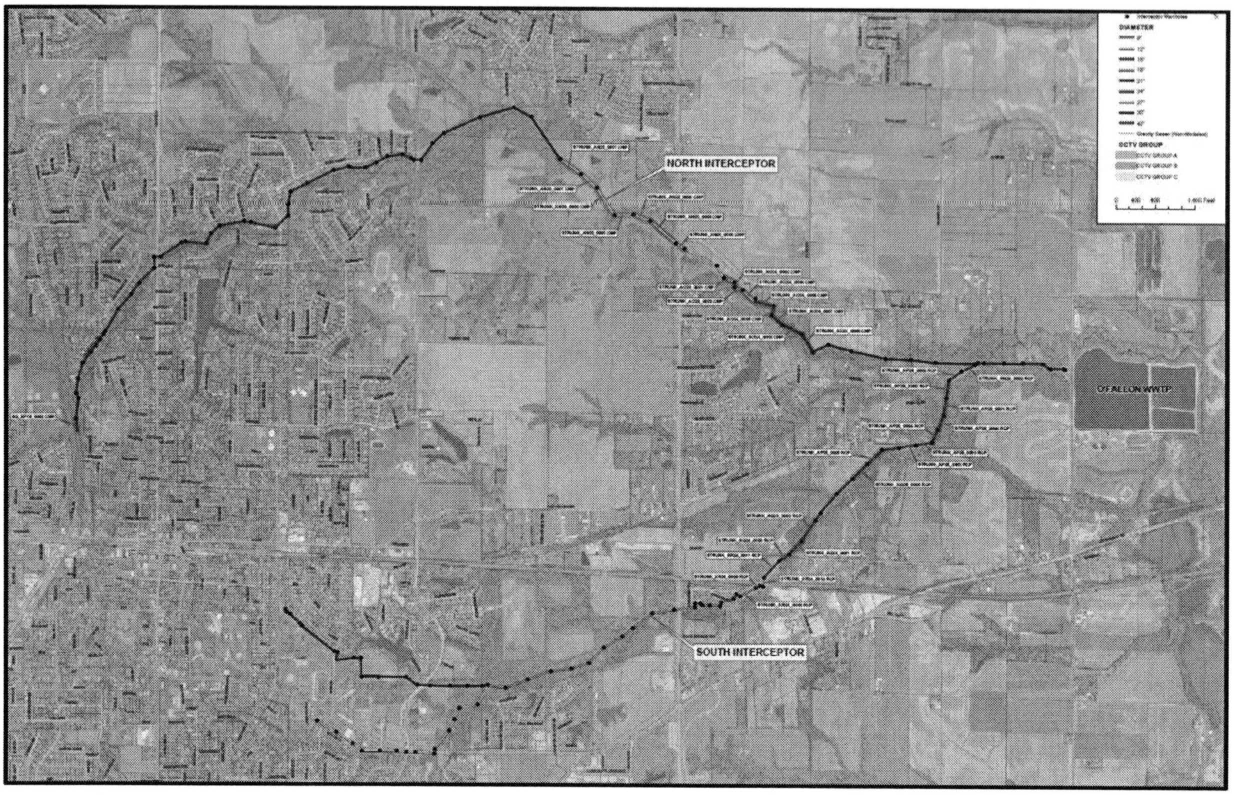

City of O'Fallon North and South Sewer Trunk Mains

The trunk mains were placed between 1968 and 1974, and constructed with lined corrugated metal pipe (LCMP) (see diagrams below) along the Northern Trunk Main route and reinforced concrete pipe (RCP) along the Southern Trunk Main route. Little information about the mains had been recorded since that period, other than visual inspections of the routes for the potential of main failure and resultant pollution of the nearby streams. As the City of O'Fallon had experienced main failures in 2004 and 2005 due to erosion of creek bank areas, Gonzalez Companies (including former City of O'Fallon employees) was involved in an effort to locate manholes along the trunk mains and the true path of the mains which had been overgrown with vegetation associated with the stream reaches that the mains were placed along. The outcome of that surface inspection was that the manholes and mains were not always as shown on the design drawings. The installers had deviated from the planned route and those changes were not reflected in any available as-built drawings. Additionally, the condition of some manholes inspected at that time needed immediate repair and the condition of others would need to be addressed in the near term. However, manholes intentionally buried in farm fields were not inspected. The concern about the condition of the LCMP main was heightened as WWTP operators had recently reported that they believed portions of the pipe lining (which before the inspection were thought to be lined with a latex or plastic coating) had been arriving at the WWTP through the mains.

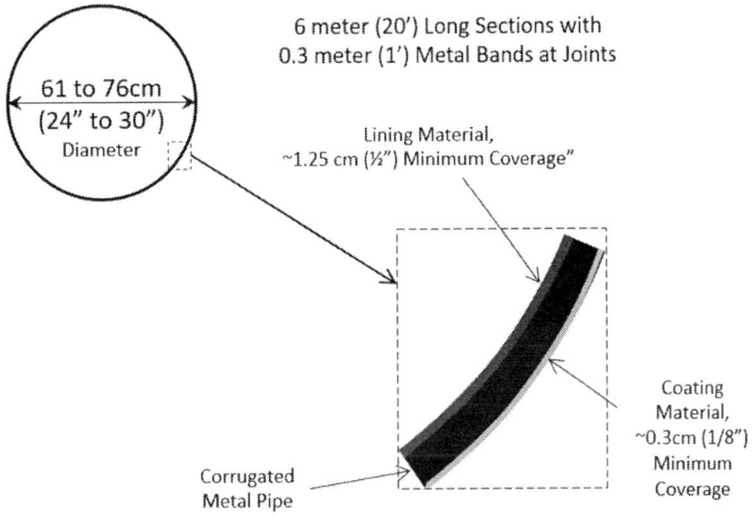

North Trunk Main's Lined Corrugated Metal Pipe

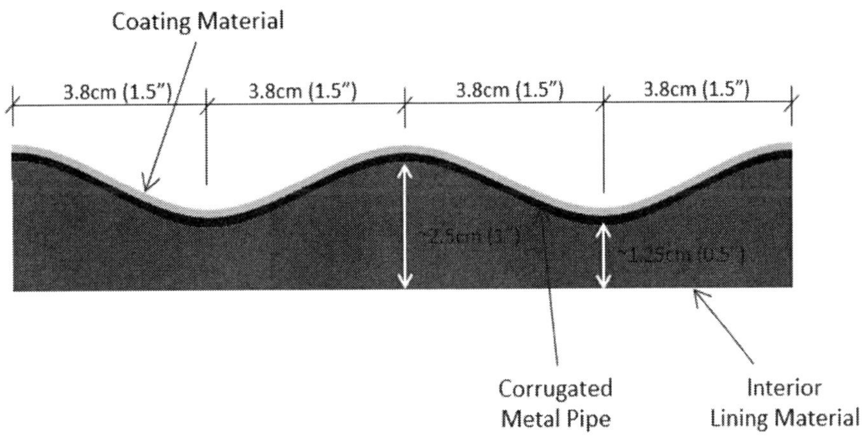

Coating Material appears to be comprised of a thin tack coat of asphalt on the pipe, then a paper-like material and a top coat of asphaltic material approximately 3 mm (1/8") thick.

Lining Material appears to be comprised of a thin tack coat of asphalt on the pipe, then a paper-like material with a thick, protective layer of asphalt coating the interior of the pipe to make the interior surface of the pipe smooth.

Lined Corrugated Pipe Construction Detail

Then failure of two segments of the RCP constructed main (South Trunk Main) not affected by stream bank erosion in 2013 suggested that the condition of that pipe needed to be a focus of further inspection at a minimum.

Gonzalez Companies had become familiar with the work of Ace Pipe Cleaning, of Kansas City, and their multi-sensor inspection services in the City of Fort Worth ICAP program. As a result, Gonzalez Companies was chosen in late 2013 to accomplish an evaluation of trunk main segments along both the north and south routes.

Gonzalez Companies sub-contracted with Ace Pipe Cleaning to provide a multi-sensor analysis of the interceptor or trunk mains and the cleaning of the mains as required to provide access for Ace Pipe Cleaning. Additionally, White Rock Consultants, familiar with the techniques used during extensive inspections in the Dallas-Ft Worth area, was brought in to review the O'Fallon inspections. However, inspection was not without its trials as the vegetation growth associated with 40+ years of inattention to the route of the mains made it difficult to get into position for insertion and retrieval of data recording equipment. To provide access, the City acquired the service of local earth moving contractor to fall and clear trees, as well as establish temporary roads to manhole locations and excavate areas of farm fields where manholes were buried in the original construction.

METHODOLOGY

While the equipment used in the combined Laser (LIDAR), Sonar and High Definition-Closed Circuit Television (HDCCT) inspection is sophisticated, the application in the field is not. A pickup truck and enclosed trailer carries all the equipment needed to a manhole location. Once on site, a series of small floats carried the equipment through the pipe segments from manhole to manhole (see photo below). The floats are tethered by nylon rope at the insertion and retrieval manholes, and are propelled by the flow of the wastewater through the pipe. Antennae are placed in the insertion and retrieval manholes to collect data transmitted by radio signals from the sensory equipment atop the boats as they pass through the pipe segments. That data is collected on a laptop computer for future manipulation.

During the analysis of a pipe segment, a technician can monitor the progress of the equipment through the pipe via the HDCCT picture and telemetry showing the distance the equipment has traveled from the insertion manhole. Field manipulation of the data recorded is minimal. Time to collect the data in the field for a segment of pipe from insertion into a manhole to retrieval of the equipment at the end of the segment for a total pipe length of 100 meters (328') to 122 meters (400') is less than 3 minutes. For straight runs of piping 200+ meters (650') or better of pipe can be analyzed in one insertion and retrieval, allowing passage through two segments of pipes if an intermediate manhole is particularly difficult to access with equipment.

Manpower utilized at the site, once a pathway to a manhole is cleared and the pipe has been cleaned to allow passage of the combined LIDAR, Sonar and High Definition-Closed Circuit Television (HDCCT), is three technicians, unless pipe flow constriction is needed. One technician monitors data collection on the laptop to ensure it is complete. A technician is needed at the insertion manhole to man the tether rope and the data retrieval antenna, and the remaining tech is needed at the retrieval manhole to man the lead tether there and the data retrieval antenna. Data retrieved is then manipulated in an office environment to provide a final report.

The final report includes a Flat Graph™ which is produced to topographically display the pipe segment and highlight its radial differences from the shape it should be in (see below). The bottom of the pipe is at the 6 o'clock position (invert), with 12 o'clock being the top of the pipe (crown).

Small Floats Used to Transport Sensor Equipment in Trunk Mains

Truck Used to Transport Field Equipment

Vegetation Clearing Effort

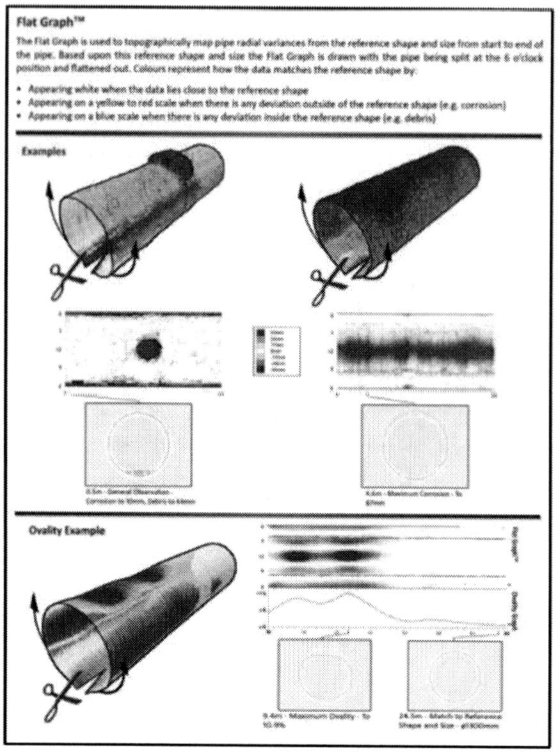

The Flat Graph™ is produced by LIDAR for surfaces above the water line in the pipe with flow, and the by sonar for the surfaces of the pipe submerged. Color on the graph shows how the data matches the defined shape of the pipe (circular, oval, etc.)

Sonar profiling when coupled with LIDAR data provides a Debris Graph with water height and debris height.

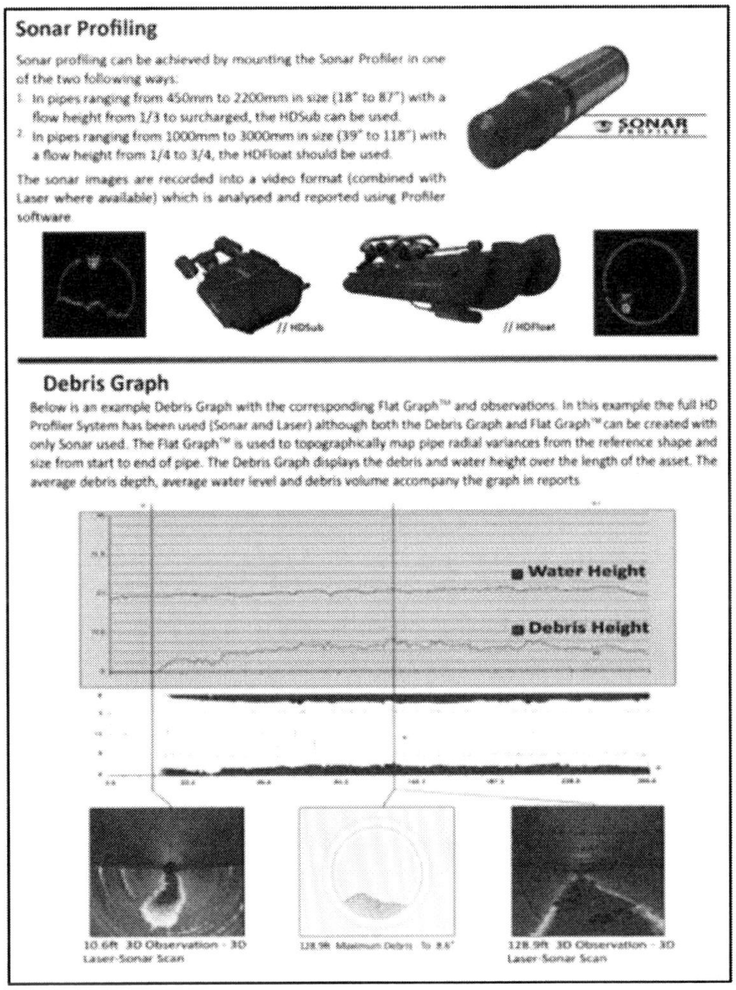

By using the data received from the LIDAR and sonar as well as the HDCCTV an overview via Ace Pipe Cleaning's Profiler software can be assembled.

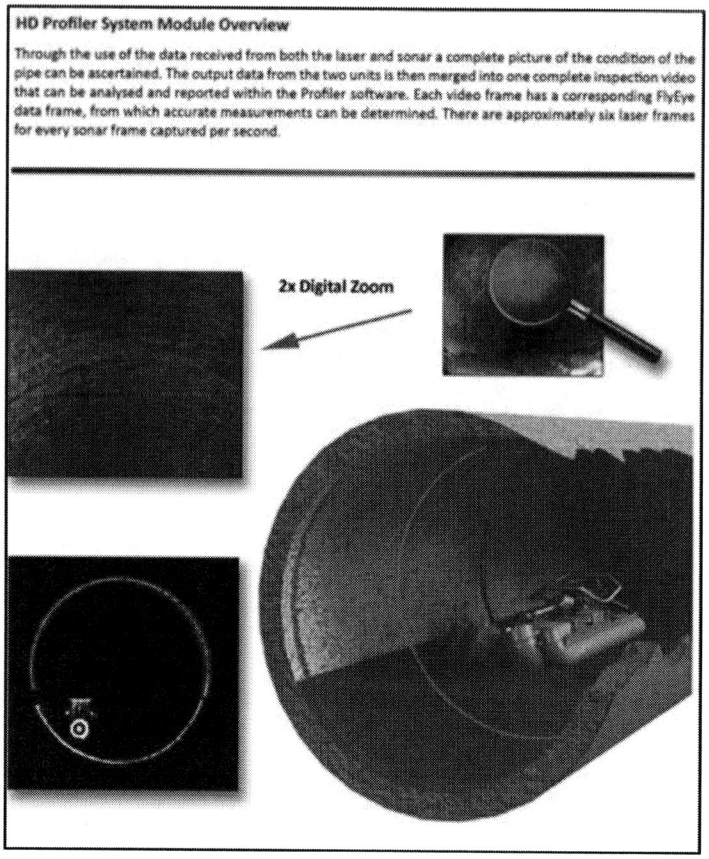

This data can then be portrayed in a Flat Graph™ Summary, which by segment is a bit difficult to interpret. However, when Flat Graph™ Summaries for all pipe segments analyzed are shown together, a visual comparison can show which segments are the most problematic.

In addition to the graphs produced, data is provided in a tabular format (see below):

PROFILER

FLAT GRAPH SUMMARY

Flat Graphs for all the segments that are inspected by Ace Pipe Cleaning are listed together providing a quick reference. In this case, listed on Page 9 of

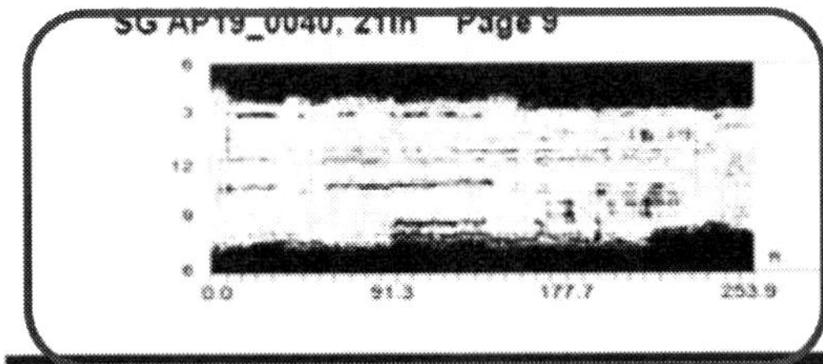

This page is from the O'Fallon inspection. Ace Pipe Cleaning reviewers of the HDCCTV pictures mistook the images of the LCMP as PVC due to the white slime on the interior walls of the pipe with the lining's depth masking the

OVERVIEW - SG AP19_0040

Asset Number	SG AP19_0040		
Upstream MH	SM_AP19_0054	Material	Polyvinyl chloride
Downstream MH	SM_AP19_0053	Size & Shape	21 in Circle
Survey Direction	Downstream	Match to Reference	135.9 ft

Distance	Observation	Comment
0 ft	**Beginning of Inspection**	**SM_AP19_0054**
0	Note	Water on lens affects laser data intermittently
0	Note	Unable to use Ovality calculation due to water on the lens
0	General Observation	Measurement outside reference to 0.4"
50	General Observation	Measurement outside reference to 0.5"
50.4	Maximum Measurement Outside Reference	To 0.5"
100	General Observation	Measurement outside reference to 0.5"
135.9	Match to Reference Shape and Size	ø21.0"
150	General Observation	Measurement outside reference to 0.2"
199.9	General Observation	Measurement outside reference to 0.2"
249.9	General Observation	Measurement outside reference to 0.5"
253.8 ft	**End of Inspection**	**SM_AP19_0053**

The report produced by Ace Pipe Cleaning also includes still pictures of the inside of the pipe at the locations commented upon in the Overview table.

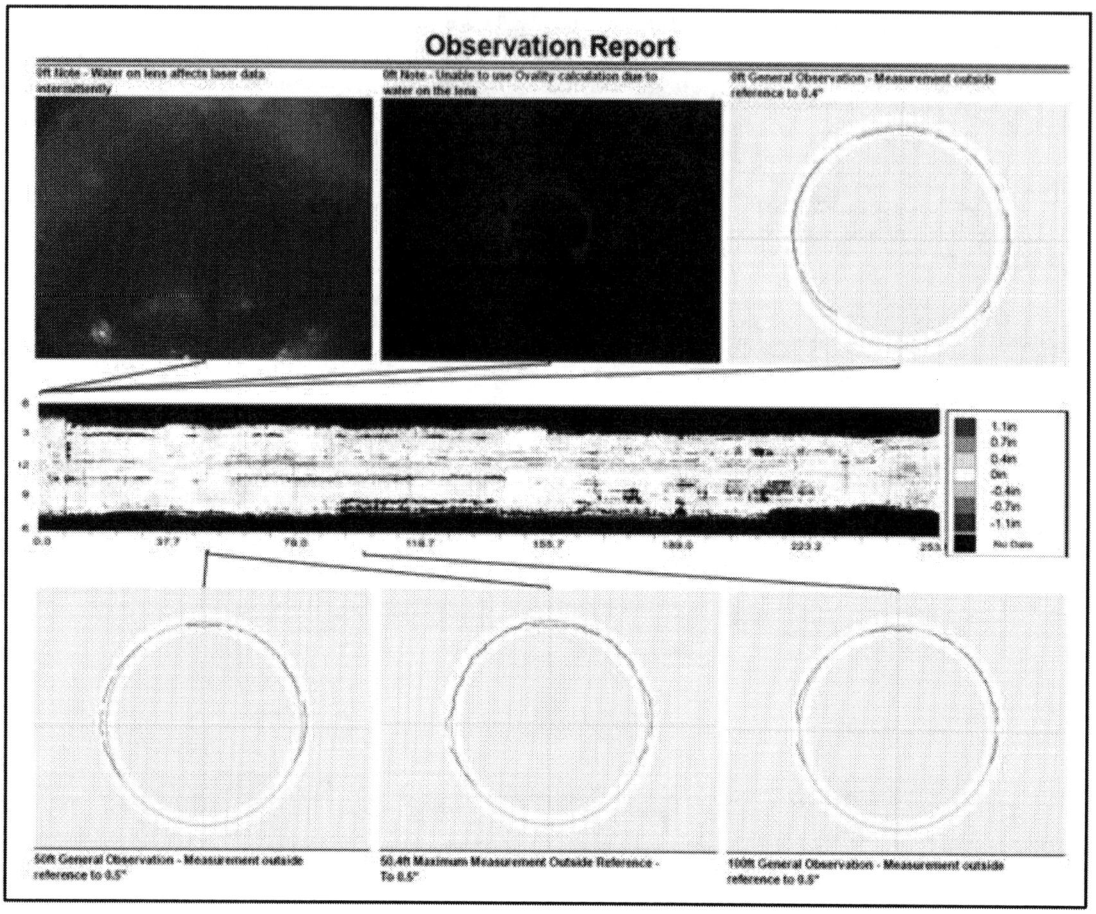

Along with the written report on pipe segments, the HDCCTV files are provided electronically which provides endless opportunity to compare future inspections to this one.

RESULTS

North Trunk Main (LCMP)

The LCMP segments were found in generally good condition considering they were about 50 years old. Some deformation was measured, but the deformation did not exceed 10%. Many of the segments had some liner defects, but none to the extent they would need repair or replacement now. Several LCMP segments had root intrusion and infiltration at the joints.

One segment was found to contain a section damaged by a data cable cross-bore. Also, in the same segment, a large section of the liner is missing and the corrugated metal looks to be corroding.

South Trunk Main (RCP)

Most of the concrete pipe segments were found to have measured wall loss of about ½ inch or greater. Two segments were found to have measured wall loss greater than 1.5 inches. Additionally, one segment had significant wall loss. Subsequent to the profiling of the pipe and before the report was issued, we did have failure of the reinforced concrete pipe along the south trunk main. The pipe was in such poor condition that several sticks or sections of the pipe were destroyed trying to find structurally sound pipe at which to place a rubber gasketing or banding material around an old piece of pipe to form a watertight seal between the old and new pipe added in the repair.

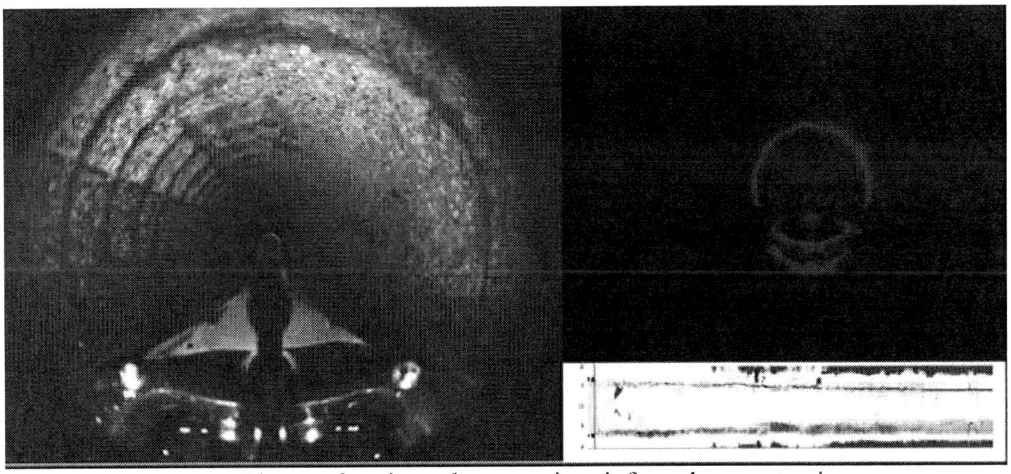

Picture showing rebar cage in reinforced concrete pipe.

DISCUSSION

The following was written by Rod Thornhill, PE, 2016, of White Rock Consultants for use by Gonzalez Companies on the findings of the Ace Pipe Cleaning inspection of the O'Fallon Trunk Mains:

> With concrete pipes, LIDAR data provides the ability to determine the location of the pipe wall (surface) in relation to the pipe's axis at the time of the inspection anywhere along the pipe's inside wall. Pipe research using the original specifications and drawings provides the best available information on the original geometry of the pipe including inside diameter and wall thickness.
>
> The difference in the original inside pipe wall (surface) location and the laser measured pipe wall location during inspection provides the data for the wall loss. By subtracting the wall loss from the original pipe thickness, the amount of remaining wall thickness can be calculated. Therefore, it is easy to see that reliable estimates of original wall geometry are as important as laser measurement.

The O'Fallon project only inspected reinforced concrete pipes of 44 and 53cm (18" and 21") in internal diameter. The assumption was made the pipes were ASTM C76 Wall B specification based on the research into the original specifications. That ASTM standard defines the 44cm (18") pipe have a minimum wall thickness of 6.4cm (2.5") and the 53cm (21") pipe with a minimum wall thickness of 7cm (2.75").

The point where the maximum wall loss (MWL) occurs in a segment is used as a basis for estimating RUL, since that is considered the weakest "link" in the segment and the most likely place where failure will occur in the RCP. Since the amount of wall loss will vary throughout a pipe segment (from manhole to manhole), the RUL will also vary. The variability of wall loss may often be a significant factor in whether full or only partial replacement of a pipe segment is needed.

The maximum wall loss is divided by the estimated age of the pipe when the inspection was conducted. The result is an estimated Deterioration Rate per Year (DRY). When wall loss in concrete pipe is accurately measured with a tool like laser inspection, the RUL is the amount of time before the remaining wall thickness is so thin the risk of failure is unacceptable. The recommended minimum allowable wall thickness (MAWT) for this project is 1.9cm (0.75"). To calculate the remaining useful life, we first calculate the minimum remaining wall thickness based on the MWL. Then we subtract from the minimum remaining wall thickness, 1.9cm (0.75"). The result is the amount of allowable wall loss (AWL) or that which can be tolerated before the wall thickness reaches 1.9cm (0.75"). The AWL by the DRY results in the Remaining Useful Life (RUL). The RUL is based on the condition of the pipe at the time of inspection. By adding the RUL to the date of inspection provides the estimated calendar date (Fail Date) that a pipe segment will reach the minimum allowable wall thickness (MAWT). Generally, pipes with an estimated Fail Date of less than 10 years are recommended for renovation.

The Lined CMP (LCMP) unlike concrete pipes with measured wall loss from laser inspection, the assignment of RUL for LCMP is subjective. The only benchmark data available is the condition of the pipe segments at the time of inspection. In this case, the primary failure mechanisms anticipated based on the multi-sensor inspection of the pipes are expected to be the loss of liner material failure resulting in accelerated corrosion of the pipe core and deformation of the pipe over time. After nearly 50 years in the ground, the liner in most of the LCMP segments was in at least fair condition, with one exception. That one segment was assigned an RUL of 30 years, but it was not recommended for replacement. Also, none of the LCMP segments had excessive measured deformation. Therefore, all of the Lined CMP segments with the exception of the one discussed above were assigned an RUL of 50 years.

However, as the multi-sensor inspection of the pipes cannot provide information on the *exterior surface of the pipe or the condition of the banding material at the joints of individual sections of pipe, point repairs of the pipe segments may be required in the future, inside the 50-year prediction. Visual surveys for root intrusion may be an indicator for such repairs in the future at the joints or where the exterior protection of the pipe was compromised during construction or as in the case of the cable cross bore mentioned above, later in the life of the pipe.*

CONCLUSIONS

The Ace Pipe Cleaning inspection provides quality data on the interior of the RCP segments, and the interior of the pipe segments of the LCMP. However, it cannot assess the condition of the interior nor the exterior of the steel coupling bands that connect the segments of the LCMP.

Referring to the diagram below, at the butt-joint of two pipe sections, the coupling band ties the sections together and holds the rubber gasket material that keeps the sewage in the pipe and groundwater out. In a physical examination of the LCMP and the bands removed during remediation of pipe uncovered by a meandering stream reach, some of the bands showed break-throughs or pin-holes (see Picture #8 from the O'Fallon Report below) where wastewater and groundwater caused corrosion on the bands. The coupling bands do not have asphaltic protection on either the interior or exterior surfaces as the LCMP sections do. The bands have only a thin asbestos based sheeting applied to their interior and exterior surfaces. It is assumed that the sheeting applied to the bands and the pipe sections is asbestos containing, as that was a common practice at the time of the pipe's manufacture.

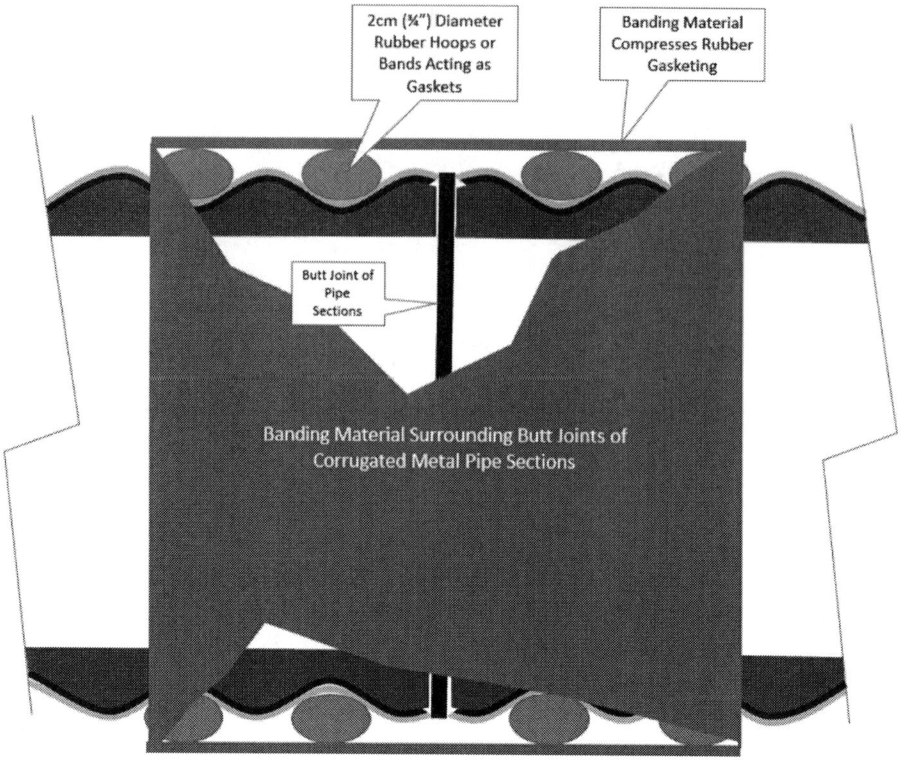

Butt Joint Detail of Pipe Sections

Picture #8 - Hole in banding material. Bands are not protected by inner or outer coatings of asphalt.

Corrosion Breakthrough in Banding Material

Therefore, beyond the concern of the LCMP pipe sections losing their interior asphaltic protection layer and becoming vulnerable to the corrosive nature of wastewater, the unprotected interiors of the coupling band material may be a bigger concern. Penetration of the LCMP at the joints by roots may be the best indication of a problem with the banding material. An examination of the gasketing material removed during the creek bank failure found the material to still resilient, rebounding from compression and regaining a round shape.

Due to the inherent weakness of these joints in the LCMP, periodic inspection of the LCMP trunk main segments, at minimum HDCCTV, is required to look for root invasion at the joints which would be an indication of band material failure.

Periodic HDCCTV inspection of the RCP segments is needed as well to check on reinforcing gage appearance, an indication of wall thickness loss and eventual structure failure of pipe segments. The Ace Pipe Cleaning inspection revealed misaligned joints in the RCP and failure of gasket material allowing root invasion which was removed during the inspection process.

REFERENCES

Judge, R. Patrick, PE and Sullivan, Dennis, PE; 2016, O'Fallon, Illinois Sanitary Sewer Trunk Main Ace Pipe Cleaning HDCCTV & Capacity Analysis; Gonzalez Companies, LLC

Renton, Tim, BE (Hons) and Jameson, Bruce; 2014; O'Fallon Laser Profiling, Part A; Ace Pipe Cleaning

Thornhill, Rod, PE, 2016, O'Fallon, Illinois Ace Pipe Cleaning Multi-Sensor Inspection Project Data Review and Recommendation, Technical Memorandum; White Rock Consultants

Performance Measurement Helps Manage Your CMOM Program

Steve Donovan
Brown and Caldwell
25 Merchant Street, Suite 145
Cincinnati, Ohio 45246

ABSTRACT

Many Utilities have developed full or partial Capacity, Management, Operations and Maintenance (CMOM) Program Plans, typically in response to regulatory compliance requirements, such as Consent Orders, Administrative Consent Decrees or Permit requirements. But how do they prove that the improvement initiatives that have been committed to are meeting the stated goals of the Plan and satisfying Regulatory expectations?

Performance Measurement.

Many utilities face issues and angst when they realize they need to develop a performance measurement program to defend their regulatory compliance.

Obviously, under a CMOM Program, the ultimate measure is the reduction in Sanitary Sewer Overflows (SSOs) and building backups due to system failures. Very few, if any, CMOM Program Plans are accepted without some definition of Key Performance Indicators (KPIs) being assigned to the Program elements.

Most utilities have tracked and reported various performance measures internally, but have not looked at them beyond just numbers, with the goal of using them for punitive actions. However, the goal is to use the analysis of the KPIs against defined targets to adjust management strategies to improve effectiveness and/or remove constraints that hinder achievement of the goals, not just to be submitted in the monthly or annual reports that no one ever looks at.

Effective Measures and targets should be SMART:

Specific to the goals of the organization or the regulatory requirements,
Measurable to an appropriate degree of accuracy (integrated into tracking tools),
Achievable and controllable, not to frustrate those who perform the work,
Relevant to meeting the goals of reducing costs and risks, improving customer service, and
Time-bound to establish temporal milestones.

The display of the KPIs can be initially as simple as being shown through spreadsheets, but more sophisticated tools and dashboards can be generated for more advance Performance Measurement programs. Several Computerized Maintenance Management Programs (CMMS) programs now have built-in capabilities of dashboard displays of the data that has been configured into the Work Order modules.

Utilities should be careful embarking on benchmarking performance against industry data compilations such as presented in the AWWA's Benchmarking Performance Indicators for Water and Wastewater Utilities: Survey Data and Analyses Report which is a voluntary submittal rather than a blind survey. It is suspected that in recent years only the higher performing utilities have responded to these requests. Utilities should perform internal benchmarking against past internal performance until a time when they have a more mature program.

KEYWORDS

CMOM, performance measures, key performance indicators (KPI), organizational optimization, preventive maintenance, benchmarking, computerized maintenance management system (CMMS), data dashboard

INTRODUCTION

Many utilities have undertaken or about to develop a Capacity, Management, Operations and Maintenance (CMOM) program to comply with regulatory requirements or to improve how their collection system is run. Most utilities are performing some elements of a CMOM program (e.g. hotspot cleaning, etc.) however the program is not comprehensive and is typically performed in a reactive manner. Some have been prescribed by regulatory agencies a defined level of performance (e.g. clean on an X-year cycle) or have developed methods of establishing what their level of service should be.

Regardless, an organization cannot adjust management strategies without first understanding what has been accomplish, or can be accomplished by their staff.
Tracking and analysis of key performance indicators is essential to keeping tabs on system performance, production rates and various factors that can constrain attainment of the organization's desired outcomes or established goals.

If a regulatory action requires the utility to "conduct routine preventive operation and maintenance activities for its sewers" to minimize overflows and building backups, then tracking of the number of overflows and building backup occurrences must be reviewed and a determination of how much and which type of preventive maintenance is needed to help reduce those occurrences.

However, many performance measures evolve from demonstrating effectiveness to exposing efficiencies. In that context, one must differentiate between the two – Effectiveness is "producing a result that is wanted" and Efficiency is "producing results without wasting materials, energy or time."

CMOM BACKGROUND

The Capacity, Management, Operation and Maintenance (CMOM) language under the proposed Sanitary Sewer Overflow (SSO) rule was developed to provide utilities with guidance on how to manage and operate their system to reduce overflows. Though the SSO rule has yet to be adopted on a national basis, many states and USEPA regions have been requiring wastewater agencies to implement the elements of a CMOM Program. Additionally, many utilities have realized that implementation of a CMOM Program will result in improved operation, increased productivity and provide economic benefits.

Even though the guidance was developed to minimize the impact of SSOs (and more recently backups into buildings - WIBs) the processes and principles can be applied to storm water, combined sewer and water distribution systems, especially since, in many cities and counties, the same staff are responsible for management, operations and maintenance of each type of system. Additionally, in numerous cases, a CMOM Program becomes the first step in developing a comprehensive Asset Management Program that may engage the entire organization, not just the wastewater agency.

Many utilities have already implemented many elements of the guidance and merely need to provide some formality to their Program, typically through documentation and tracking of performance measures, to realize the improvements or adjustments that are needed. The analysis of the appropriate metric data will allow the staff to evaluate productivity, performance and adjust management strategies to resolve constraints that may hinder CMOM Program success and regulatory compliance.

A CMOM Program is a continuous improvement process, makes good business sense and should not be a daunting endeavor. The utility should understand that the development of an improvement plan is only the beginning and it needs to be revisited and revised based on the analysis of performance.

Collectively, the activities undertaken through a CMOM Program can result in many rewards. Increased productivity, improved operations and maintenance (O&M) practices and reduced system failures will increase the level of customer service and reduce the negative effects on the environment.

Beyond the General Standard, drafted in 2001, to *"Properly operate and maintain, at all times, the parts of the collection system that the permittee owns or over which it has operational control"*, several of the elements of a CMOM program to meet that Standard include *"Identify program goals consistent with the general standards"*, *"Address appropriate measures and activities necessary to meet the performance standards"* through the *"use of timely, relevant information"*, and *"Monitor program implementation and measure its effectiveness"*.

Additionally, the CMOM Guidance recommends periodic audits of the programs, and performance measures tell a great story when updating the CMOM program.

PERFORMANCE MEASURE DEVELOPMENT

Obviously, under a CMOM Program, the ultimate measure is the reduction in Sanitary Sewer Overflows (SSOs) and building backups due to system failures. Very few, if any, CMOM Program Plans are accepted without some definition of Key Performance Indicators (KPIs) being assigned to the Program elements.

Many industry publications provide example Key Performance Indicators a utility can track. Many are common industry standards, such as number of SSOs, feet cleaned, feet CCTV'd, etc.

However, identifying the appropriate measures for a particular utility is dependent on a number of factors.
Who is the audience?
What has caused problems in the past?
How will the information be used?
What is to be proved, defended, justified?
Who in the organization will be analyzing the measures?

Though the ultimate goal of the organization may boil down to one target, such as zero SSOs, there are number of measures that are required for analysis that feed into that goal. Throughout the process, the measures of interest will vary depending on the tier within the organization. The elected officials may just what to know the number of customer complaints responded to within an hour or the number of SSOs. Whereas the Manager may want to see those measures, and SSOs by cause, as well as a handful of others to satisfy the reporting requirements of the regulatory agency. However, it takes many more measures to manage and inform the operations and maintenance activities throughout the organization to meet that goal.

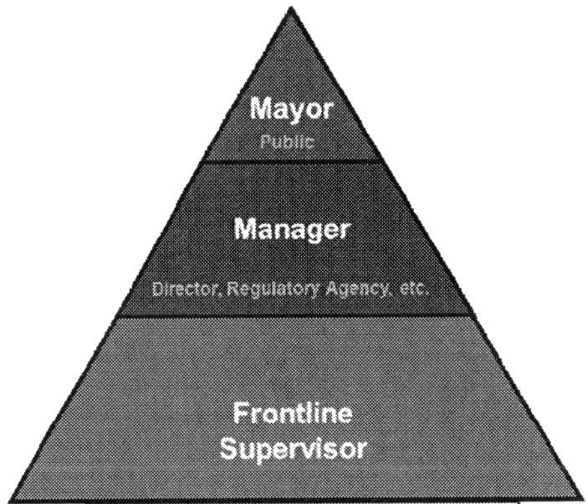

Figure 1 – Organizational Tiers

On one project, we facilitated the development of a series of measures in discussions with front line supervisors, knowing that the strategic goal was to reduce the number of SSOs. With input from eleven (11) functional work groups within the organization, about 100 unique performance measures were identified that conveyed the information they felt was most important to them to manage their work group, both in effectiveness and efficiency. That may sound like a lot of measures, but that amounted to about an average of nine (9) per supervisor and only of total of seven (7) were fed into the system that was reported to the elected officials. The added benefit from these discussions was that the measures were selected by the supervisors which ensured buy-in throughout the organization

Often specific measures are dictated to the utility under the regulatory action (e.g. Consent Decree, Administrative Order on Consent, etc.). Regulatory agencies have been dictating specific time intervals for specific operations and maintenance activities, particularly in cases where utilities do not have a sufficient data set of multiple measures to help negotiate a more appropriate schedule.

However, most regulatory actions include language where the intervals can be adjusted based on defensible data. But that may take upwards of 10 years to complete multiple cycles and to acquire an adequate data set.

Many Utilities have tracked and reported various performance measures internally, but have not looked at them beyond just the numbers, with the goal of using them for punitive actions. However, the goal should be to use the analysis of the KPIs against defined targets to adjust management strategies to improve effectiveness and/or remove constraints that hinder achievement of the goals, not just to be submitted in the monthly or annual reports that no one ever reviews. The best measures are ones that can be used to manage stuff, not just track stuff – however, in some cases, indicators just need to be tracked. Too often utilities focus on proving what they are doing, rather than managing what they are doing.

TARGET DEVELOPMENT

To better understand what the measures are indicating, the should be compared to defined targets. However, if performance measurement is a new concept within the organization, it might be better to allow the staff to get used to tracking accurate KPIs before springing the specter of targets on them.

Nonetheless, effective measures and targets should bc SMART:

Specific to the goals of the organization or the regulatory requirements,
Measurable to an appropriate degree of accuracy (integrated into tracking tools),
Achievable and controllable, not to frustrate those who perform the work,
Relevant to meeting the goals of reducing costs and risks, improving customer service, and
Time-bound to establish temporal milestones.

The initial target setting efforts for production measures is typically performed through the review of historical performance in the recent past. By taking into consideration staffing, technology or equipment changes, and identifying obvious areas for higher efficiency, achievable targets can be initially set.

Establishing temporal milestones is very important and must be recognized through the analysis phase. Monthly, quarterly or annual milestones for measurement should be established based on a reasonable expectation when and how that information can be useful. It does not make sense to set daily targets for cleaning foot, for example, in that there is so many factors that can affect a daily production rate. However, they obviously should be tracked on an ongoing basis and trended to ensure there is not a rash of failures occurring that could be stopped well before the reporting milestone.

When setting targets, seasonal factors that could affect scheduling must also be accounted for. For example, stormwater pump station maintenance is typically scheduled in the summer when it is expected that long periods of dry weather will occur and the station will not be needed. An annual target averaged over twelve (12) months will show unmet goals over a significant portion of the year.

KEY PERFORMANCE INDICATOR ANALYSIS

Analysis of the measures versus targets provides insight to assist in changes to management strategies, adjustment of priorities and re-deployment of resources to achieve the goals of the program.

For a high-level analysis, each measure should be evaluated in context of the establish milestone, whether that is daily, weekly, monthly or annually. In some cases, it can be very misleading to perform an analysis on an earlier interval than has been established for that specific measure, unless other measures are taken into consideration. Meaning, for example, if you have a target of 100,000 feet of cleaning per month, and the production is evaluated in the first week and is not on target, it needs be determined if it is a scheduling issue, where more crews are scheduled for that activity in the follow weeks or if there is a reason production is off track, such as lack of available resources or the occurrences of a holiday or all day training session.

For measures that are reliant on efficient practices, once the targets are consistently achieved, the targets should be elevated until the utility finds the "sweet spot" of what can be achieved. However, some measures should always be 100% of the target such as routine inspections, but still need to be tracked as documentation that the work is being conducted and may provide some level of affirmative defense should an unforeseen circumstance results in a failure. This is proof of due diligence and if the target is not met, a huge "red flag" will be identified that needs to be addressed.

Sometimes the analysis does not need to be tied to a goal, but rather to be used for identifying trends. Trends that may indicated the utility is approaching a need for additional resources or that is a pre-indicator which maintenance strategies need to be adjusted.

Additional measures may be identified during the analysis to better inform and identify constraints that need to be remedied. During the project discussed earlier, we heard "We could not achieve our target because the trucks were in for repair" or "We could not reach are target because it's hunting season". The remedy was to add vehicle up-time and staff availability as additional measures. Upon further analysis, it was identified that fleet needed to prioritize vehicle repairs and that smaller crew sizes could be used for less complex jobs.

The bottom line is to ask, "Is what we are doing working? If not, why and what can we do about it?"

Tracking and Display Platforms

Often it is much easier to perform an analysis of data if it can be visualized. Bar charts, pipe charts, spider graphs, dashboard "gauges" are very useful in seeing the numbers as visuals for comparison. Spatial visualization can be reasonably achieved through a Geographical Information System (GIS) map.

The display of the KPIs can be initially as simple as being shown through spreadsheets, but more sophisticated tools and dashboards can be generated for more advance performance measurement programs. Several Computerized Maintenance Management System (CMMS) programs now have built-in capabilities of dashboard displays of the data that has been configured into the Work Order modules.

Graphs can be generated to support the analysis of the measures and to assist in identifying trends. With the systematic collection of performance measures into a central location, graphs are relatively easy to produce and provide managers and supervisors with the tools needed to effectively manage the work load in near real-time. If the data collection is clumsy or too time consuming, the inconvenience could far out-weigh the benefit and lead to erroneous data being collected.

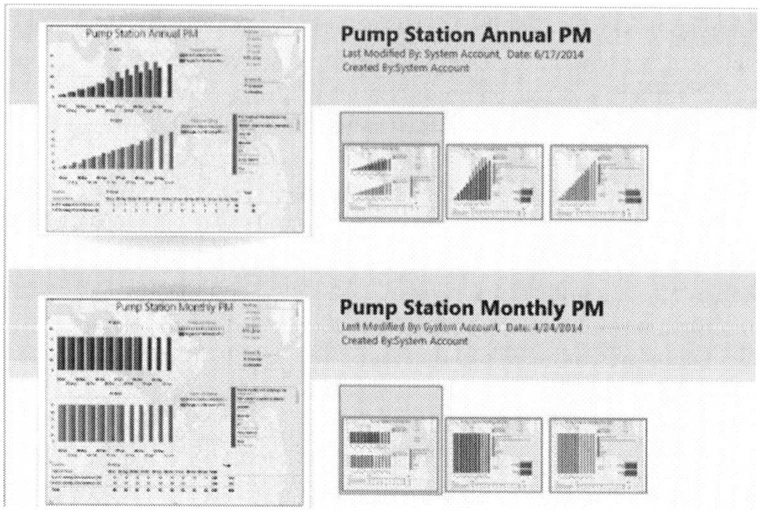

Figure 2. Example Data Dashboard

An example of analysis from one utility shows that 38 sewersheds had been significantly completed under the preventive maintenance program through the end of 2009. The utility proactively cleaned more than 756 miles of sewers (36-inches and smaller) through this program and an estimated 1200 cubic yards of debris has been removed from the system which contributed to the reduction in building backups.

In figure 3, the inset graphs represent the annual number of WIBs (in red) and the number of residents that had participated in the backflow prevention program, within the completed sewersheds.

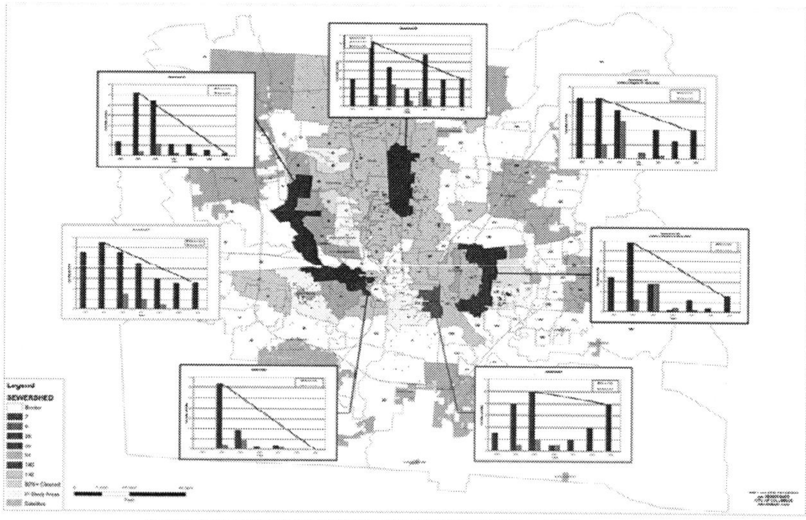

Figure 3. Example of GIS display

As a result of the preventive maintenance program, over 60 percent of the cleaning footage was performed in a proactive manner, with the monthly average footage increasing 40,000 feet over what was cleaned four years earlier. Additionally, the average emergency cleaning dropped to about 2,300 feet a month, which was slightly over one percent of the total footage cleaned.

Benchmarking

Utilities should be careful embarking on benchmarking performance against industry data compilations such as presented in the AWWA's Benchmarking Performance Indicators for Water and Wastewater Utilities: Survey Data and Analyses Report which relies on voluntary submittal of performance rather than being a blind survey. It is suspected that in recent years only the higher performing utilities have responded to these requests.

Too often Utilities look at these surveys as a launching off point for where they want to be or think they should be. Very rarely a similarly sized utility, with a similar geo-political environment and who have the same responsibilities can be found for a fair comparison.

This is especially of concern if a utility that is just embarking on a CMOM program is attempting to benchmark against another utility with well-established CMOM program that has been under development for numerous years. It can be reasonably expected that utility did not start out with a high level of service either.

Utilities should perform internal benchmarking against past internal performance until a time when they have a more mature program. This is a continuous improvement process where progress should be shown over the previous period of record for a specific data set. Management needs to looks at all the factors that are affecting performance and the resulting outcomes to determine what is right for their organization and meets the expectations of the ratepayers.

However, it may be advantageous for a utility who is trying to abate regulatory action or provide some other type of justification, to benchmark against industry survey data, assuming they are already performing at a high level in comparison to the benchmark numbers.

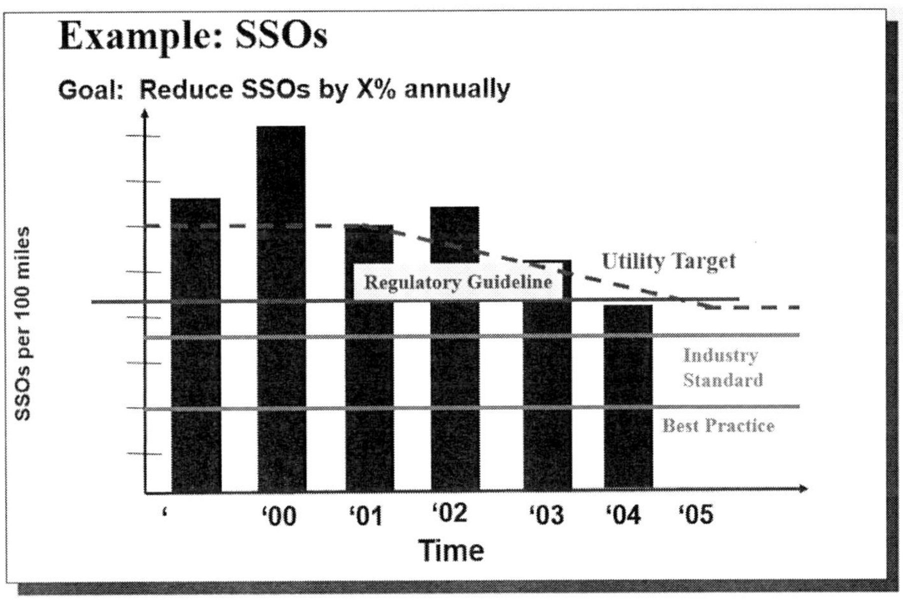

Figure 4. Example of SSO analysis

CONCLUSION

Many utilities have developed full or partial Capacity, Management, Operations and Maintenance (CMOM) Programs, typically in response to regulatory compliance requirements, such as Consent Orders, Administrative Consent Decrees or Permit requirements.

To demonstrate compliance with the regulatory requirements, a variety of measures need to be tracked and adjusted over time to tell the real story.

Adjusting management strategies for operations and maintenance activities, based on data, is the most effect way to address the causes of failures in the system.

The best measures are ones that can be used to manage operations and maintenance, not just for tracking numbers. If the review of the measures versus target is only used to punish staff, miss-information is sure to follow. Accountability for true performance and accurate data is essential.

Improved performance should be recognized, celebrated and rewarded to sustain buy-in to the process.

The performance measure process is a continuous improvement process, as well as an integral part of on-going operations and maintenance activities within a wastewater collection system.

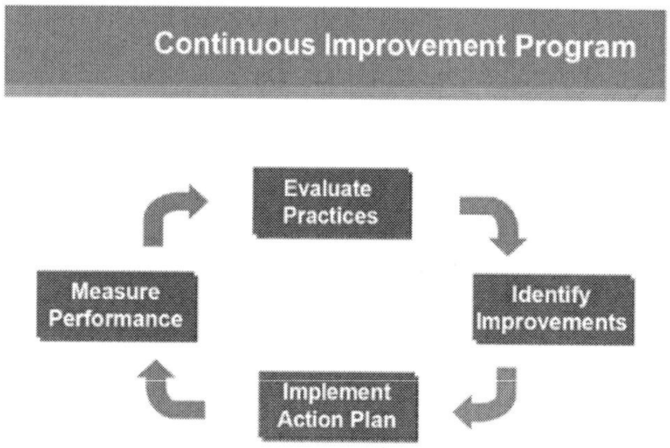

Figure 5. Continuous Improvement Program

Springfield – Greene County Integrated Plan: A Citizen Focused Approach

Errin Kemper[1], **Trent Stober**[2]

[1] City of Springfield, Missouri
Email: EKemper@springfieldmo.gov
[2] HDR Engineering
Email: trent.stober@hdrinc.com

ABSTRACT

The leaders from the City of Springfield, Greene County and City Utilities of Springfield (integrated Plan Partners) have embarked on developing an Integrated Plan to address the regulatory challenges facing these wastewater, stormwater, drinking water, and power utilities. Balancing and prioritizing water quality improvements are challenging in Integrated Municipal Stormwater and Wastewater Planning due to the dissimilar nature of stormwater and wastewater alternatives. The Integrated Plan Partners have moved forward to develop and use a Multiple Criteria Decision Analysis (MCDA) process to identify the most signficant pollution sources that impact the regional environment. The MCDA process is data-driven and aligned with community priorities established through extensive stakeholder engagement. The City elected to pilot HDR's Sustainable Return on Investment (SROI) methodology to assess its use as a prioritization tool for Springfield and Greene County, Missouri's nationally-recognized Integrated Plan. Using best-practice cost-benefit analysis techniques, SROI quantifies and monetizes a broad array of triple-bottom-line - environmental, social, and economic - impacts typically assessed only qualitatively, allowing decision makers to transparently compare diverse wastewater/stormwater projects on an 'apples to apples' basis.

KEYWORDS: integrated planning, sustainability, triple-bottom-line, wet weather, sanitary sewer overflows, stormwater, nutrients

INTRODUCTION

Like many cities and counties across the nation, Springfield and Greene County, Missouri is challenged by increasingly stringent environmental regulations on several fronts. From stormwater and wastewater to air quality and drinking water, as regulations continue to evolve, our community is required to devote more of its limited resources to comply. The US Environmental Protection Agency (Stoner and Giles, 2012) realizes the struggles many communities face. To provide a path forward, USEPA released its "Integrated Municipal Stormwater and Wastewater Planning Approach Framework", which emphasized a commitment to work with states and communities to integrate environmental goals.

In response to this opportunity, leaders from the City of Springfield, Greene County and City Utilities of Springfield (Integrated Planning or IP Partners) developed a local approach to Integrated Planning titled "A Citizen Focused Approach." This holistic approach proposes to use local knowledge to examine environmental resources related to wastewater and stormwater as well as solid waste, drinking water, and air quality. The planning approach has received written endorsement from both state and federal regulators. The Springfield – Greene County community is now developing an Integrated Plan that will ensure our vital natural resources are protected in a manner that is affordable to the community.

Integrated Plan Guiding Principles:
The Integrated Planning Partners developed an Integrated Plan Approach based largely on the USEPA Framework (Stoner and Giles, 2012). At the heart of this proposal are six guiding principles:

- Affordability – Ensure that the plan is affordable to the community's citizens.
- Effectiveness – Ensure that the plan addresses environmental issues in an efficient and effective manner.
- Fairness – Ensure that the plan results in all segments of the community being treated equally and fairly.
- Attainability – Ensure that the plan outlines actions that can reasonably be accomplished within the community's financial capability.
- Measurability – Ensure that the plan includes performance measures that track progress over time.
- Adaptability – For the plan to be effective, we must be able to adjust and improve our plan based upon our experiences and results.

Furthermore, this approach will develop local solutions using community expertise and values to determine how best to improve our environment while still making the solutions affordable to our citizens. What makes the Springfield-Greene County approach unique is that it proposes to include, not only stormwater and wastewater, as the USEPA guidance suggests, **but also includes resources related to solid waste, drinking water, and air quality.**

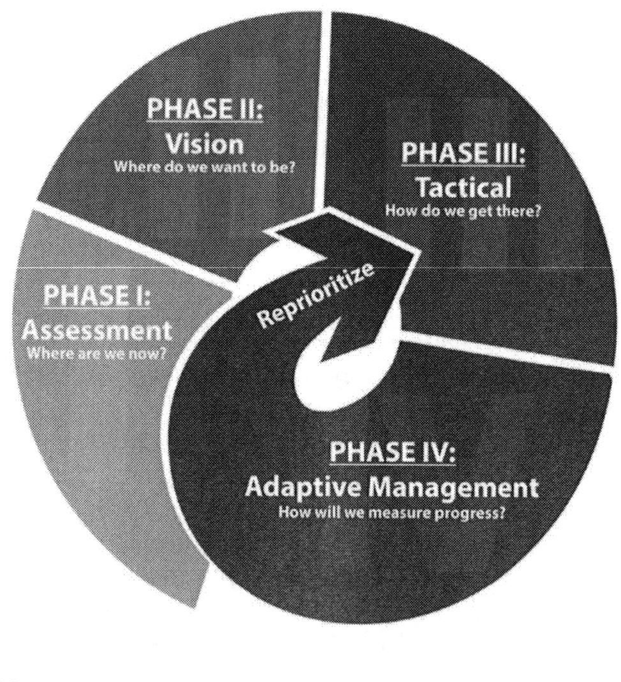

Figure 1: Springfield-Greene County Integrated Planning Approach

Integrated Planning Approach:
Development and implementation of the Springfield-Greene County Integrated Plan includes a four-phased approach as displayed in Figure 1 and covers each of the elements outlined within the USEPA Integrated Plan Framework.

Phase I – Assessment Phase:
The Assessment Phase answers the question "Where are we now?". During this phase, local stakeholder groups are gathering data to assess the current status of our environmental resources. The Assessment Phase will yield a comprehensive database of environmental information, including hydrologic and water quality

data, critical site information (e.g., legacy sites), and any other information used to assess the health of our environment and indicate any potential causes for impairments. This action will not only provide an overview of current conditions but will also establish a baseline for tracking future progress.

Phase II – Vision Phase:
The second phase of our approach is the Vision Phase and answers the question "Where do we want to be?". Our Integrated Plan will have achieved success when ***community resources are directed towards managing environmental issues using the most effective solutions to address the most significant problems in a way that is affordable to our citizens***.

In addition:

- We are in compliance with Federal and State regulations while addressing the specific needs of our community.
- We have the ability to address water, air, and solid waste issues holistically allowing both our community and the regulators to operate more efficiently.
- We have a community culture that understands and supports the goal of high-quality environmental resources and supports these efforts through stakeholder involvement. Our community has a high level of trust that resources are being used to address environmental issues efficiently and effectively.
- Our community has a clear understanding of how funding and other resources will be used to improve environmental quality.
- Our community realizes a competitive advantage toward growth and economic development and an increase in quality of life as a consequence of this plan.

Phase III – Tactical Phase:
The Tactical Phase answers the question "How will we get there?". During this phase, technical committees and stakeholder groups will prioritize our community's environmental needs based on four key elements:

Element 1 - Prioritize the Sources of Pollution:
Identify and prioritize the most significant **Sources of Pollution**: Using a Multiple-Criteria Decision Analysis (MCDA) toolset developed specifically for the Integrated Plan, we are able to take a structured look at how different pollutants impact the natural environment and the relative significance of each source. The partnering agencies selected HDR to lead a team of national experts who will develop the following components related to prioritizing pollutant sources:

- Assess pollutant sources related to the relative magnitude in which our water, air, and land resources are impacted by different pollutants.
- Develop a tool that will be used to rank alternatives from a series of multiple conflicting criteria. This MCDA tool will be used to rank alternatives representing the pollutant sources impacting our community. It is expected that the results of this analysis will be based on the relationships described above. HDR will also develop the selection methodology most appropriate for this type of analysis. An initial version of the MCDA toolset has been developed and will serve as a starting point for the analysis.
- Collect and summarize the existing data and research necessary to feed the decision making process. A strong emphasis will be placed on the use of local data and resources

to ensure that the final recommendations reflect the needs of the region. With the volume of information available, it is likely that this task will include a phased approach where initial estimates are further refined as data and research is uncovered. It is anticipated that some level of water/air quality modeling will eventually be needed to help establish the relationship between pollutants and their source.

- Develop a data gap analysis to determine what additional sampling and analytical data is needed to supplement and further refine the existing research. This additional data will be used to further refine the analysis.
- Facilitate input from local stakeholders and regulators to gain consensus on the parameters (e.g., pathogen indicators, nutrients) used to drive the priority decisions. There is an existing Integrated Planning Technical Committee made up of staff from each partnering organization. We will be engaged with this group, other local experts, and regulatory agencies to reach a level of agreement on the parameters used.

Element 2 - Prioritize the Most Effective Solutions:
Identify and prioritize the most **Effective Solutions:** Using the HDR's Sustainable Return on Investment (SROI) approach, the Integrated Plan Partners are able to evaluate the environmental, economic, and social costs and benefits for many of the solutions considered by our community. By finding the triple-bottom-line, we can ensure that the most effective solutions are being targeted toward the most serious challenges.

Element 3 – Capture Community Priorities:
Capture our **Community's Priorities**: We realize the importance of protecting our natural resources and the ways in which our community is unique. A citizen-based Environmental Priorities Task Force has been assembled and with input from this group, our Integrated Plan will define the issues that our community is focused on. By proactively addressing the issues that our citizens find important, rather than simply reacting to the latest regulation, we will build trust and support for our programs.

Element 4 – Assess Financial Capability:
Assess our community's **Financial Capability**: Our community will work together to evaluate financial capability and take a candid look at how community resources should be allocated toward environmental stewardship, using the most recent measures and approaches to assess financial capability.

Phase IV – Adaptive Management:
The fourth phase of our approach is the Adaptive Management phase. We realize that a true Integrated Plan will never be complete. As we achieve success and learn more, the target will continue to move. In this phase, we will continue to refine our analysis, check the effectiveness of our solutions, and constantly reprioritize.

MCDA OVERVIEW

A MCDA toolset will be used to address the first of the four key elements within Phase III. At its core, MCDA is a tool for solving complex problems that are characterized as a choice among alternatives (NRLI 2016). MCDA supports group decision making as it promotes consideration and discussion of trade-offs among alternatives. In effect, MCDA facilitates the critical thinking process in an open and transparent manner. For the IP, a MCDA will be used to identify and prioritize the most significant sources of pollution. After the most significant pollution sources have been identified, the IP partner agencies can focus IP efforts on evaluating the most effective solutions to reduce the impacts of those sources on the regional environment. This latter step will follow the MCDA and fall outside of this effort.

The five basic components of MCDA are as follows (NRLI 2016):

1. **Goal** - The goal is defined by here by Element 1 of Phase III of the IP, which is to prioritize the most significant sources of pollution in the Springfield-Greene County region.
2. **Decision maker or group of decision makers with opinions** - The decision maker for the MCDA includes citizens and leaders from the IP Partners.
3. **Decision alternatives** - During this initial evaluation, the decision alternatives are defined by 15 pollutant sources, which are grouped by air, land or water.
4. **Evaluation criteria** – Evaluation criteria represent the interests of the decision makers. They are defined here by the community's priorities and the pollutants or conditions that impact those priorities. For example, safe drinking water is a community priority and waterborne pathogens potentially impact that priority.
5. **Outcomes or consequences associated with alternatives** – Outcomes are defined here by the ratings and scores for each pollutant source as determined by the evaluation criteria.

MCDA METHODOLOGY

Work completed as part of the Springfield-Greene County Environmental Priorities Task Force Report (Task Force report) addresses some of the key elements of the MCDA (Shockey 2015). As noted above, the MCDA goal and decision makers have already been determined. Additionally, environmental issues and priorities established within the Task Force report help define both the decision alternatives and subjective elements of the evaluation criteria. What the Task Force report lacks, with respect to the MCDA, is a decision framework linking pollutant sources to community priorities and an objective means of scoring and ranking these sources.

A key element in MCDA development for the IP is the convening of an Expert Panel. This multi-disciplinary team of national experts will help develop the MCDA structure, selection of the evaluation criteria, the weightings of individual pollutants and conditions, and ratings of the magnitude of impacts by pollutant sources. The intent is to leverage the expert's professional judgment to address data gaps, where possible.

In an effort to finalize an MCDA decision framework, HDR conducted an MCDA workshop with the IP Partner in March 2016. The MCDA model hierarchy that resulted from this workshop is presented within this memorandum, which is structured as follows:

- **Pollutant sources** – This section includes a finalized list of pollutant sources in the model.
- **Evaluation criteria** - Evaluation criteria include both impacts from the pollutant sources and indicators used to assess those impacts.
- **Initial decision framework** – The initial decision framework provides the link between pollutant sources and impacts as well as model weightings.

- **Rating and Scoring Pollutant Sources** – This section describes how the decision framework will be used to rate and score pollutant sources.

POLLUTION SOURCES

A key task in developing the MCDA model was to identify 15 pollution sources for the initial evaluation; each of which represents an MCDA alternative. An important consideration of this task was to identify sources representative of air, water and land. Other considerations in the selection of pollution sources included the ability to control the source, regulatory emphasis and impact. After consultation with key City personnel and based on feedback from the March 28, 2016 MCDA workshop, the following 15 pollutant sources were identified.

Water Pollution Sources

Agricultural Runoff – Agricultural runoff is water leaving farm fields because of precipitation runoff, melted snow, or excess irrigation. As runoff moves across the land, pollutants can be mobilized and transported into streams, ponds, and lakes. Agricultural runoff can include pollution from soil erosion, feeding operations, tillage, animal waste (e.g., horses, cattle and poultry), and fertilizer. Overgrazing leads to high runoff volumes, increased erosion and water quality impacts. Livestock waste deposited in or adjacent to waterways also leads to increased pollutant loading from agricultural runoff. Livestock direct access to streams for watering can also degrade riparian and instream habitat.

Failing On-Site Wastewater Systems – Where central sewer service is not available, onsite (or decentralized) wastewater treatment systems are used to treat wastewater from a home or business and disperse it on the property where it is generated. Frequently referred to as septic systems, when functioning properly, onsite systems prevent human contact with sewage, and prevent contamination of surface and groundwater. Failing onsite systems allow the sewage to leave the property and can be a contributor of bacteria and nutrient contamination to surface water and ground water. Factors that affect the proper functioning of onsite systems include the site and soil conditions, design, installation, operation and maintenance. EPA has estimated that 70% of all septic systems fail at some point in their operation.

Industrial Runoff – Stormwater runoff is water that originates from precipitation and travels across the land rather than seeping into the ground. Stormwater will pickup pollutants from the land surface and transport them to waterways. Industrial stormwater runoff comes from industrial sites regulated under the National Pollutant Discharge Elimination System (NPDES) industrial stormwater program. These sites are regulated due to having industrial activities and materials exposed to stormwater. Common industrial pollutants include equipment deposits (oil, grease, and metals), dust deposits, chemicals, organic waste and outdoor materials storage.

Land Disturbance Runoff – As stormwater flows over a construction site, it can pick up pollutants like sediment, debris, and chemicals and transport them to nearby storm sewer systems or directly into rivers, and lakes. The NPDES stormwater program requires permits for discharges from construction activities that disturb one or more acres, and discharges from smaller sites that are part of a larger common plan of development or sale.

Permitted Wastewater Discharges – Permitted direct dischargers include facilities designed and operated to effectively treat municipal and industrial wastewater. Permitted direct dischargers include publically owned wastewater treatment plants, industrial facilities and power plants. Discharges includes treated cooling, industrial and municipal wastewater. Effluent from wastewater treatment facilities is discharged to nearby streams. Treatment levels are based upon technology and water quality-based requirements. Water quality-based effluent limitations are tied to water quality standards, which are established to protect designated uses such as aquatic life protection and swimming. Extensive pollutant removal occurs at the facilities; however, a certain amount of pollutants remains within treatment plant effluent. Common pollutants that are allowed to be discharged by current EPA rules include small amounts of metals, nutrients, bacteria, suspended solids and organics. The regulatory pollutant limits and the number of pollutants monitored change over time as the aquatic impacts are better understood.

Sanitary Sewer Exfiltration – Sanitary sewer exfiltration occurs when untreated sewage is discharged from a sanitary sewer into the surrounding geology. Exfiltration may occur due to cracks and defects in pipe, manhole defects, defective laterals and other sources within a sanitary sewer system. Exfiltration can increase during dry weather periods as a result of the regional groundwater table lowering.

Sanitary Sewer Overflows – Sanitary Sewer Overflows (SSOs) occur when untreated sewage is discharged from a sanitary sewer into a building or the surface water environment prior to reaching wastewater treatment facilities. The discharge comes from manholes, lift stations, emergency relief outlets, and other sources within a sanitary sewer system. Sanitary sewer overflows are typically caused by wet weather events, blockages (tree roots, etc.,), power outages and vandalism. When caused by rainfall it is also known as a wet weather SSO.

Stream Bank Erosion – Streambank erosion refers to the removal of soil, rock and vegetation from the streambank. Streambank erosion is a natural process, but the rate at which it occurs is often increased by human activities such as urbanization and agriculture. Acceleration of this natural process leads to a disproportionate sediment supply, stream channel instability, and habitat loss. Activities that contribute to increased streambank erosion include the urbanization of watersheds, loss of riparian forests or buffers (e.g., removal of trees/vegetation, mowing/agriculture to the edge of waterway), overgrazing, and trampling from livestock (e.g., cattle) and other animals (e.g., geese).

Urban Runoff – Stormwater runoff is surface water that originates from precipitation events such as snow, ice melt, and more commonly direct rainfall events that travels across the land rather than seeping into the ground. The runoff discharge comes from rooftops, lawns, parking lots, roads, restaurants, golf courses, parks and driveways. Stormwater will mobilize pollutants from the land surface and transport pollutants to waterways. In addition, urbanization generally increases runoff and reduces groundwater recharge. Common urban pollutants include herbicides, pesticides, excess fertilizer, yard clippings, trash, and pet waste.

Land Pollution Sources

Improper Disposal & Dumping – Improper disposal and dumping includes discarded sources of trash, furniture, appliances, household chemicals, yard waste, electronics, tires and other waste streams that are not recycled or disposed at a licensed facility. When discarded on land it may impact wildlife, aesthetics and lead to a reduction in property values. A portion of the debris may be washed into the storm sewer system causing blockages and accumulation of trash in water quality treatment facilities. A portion of the debris maybe washed into our streams and impact terrestrial and aquatic habitats. Sinkholes also are used for improper disposal which may place the debris in a direct connection to groundwater.

Legacy Contaminated Sites – Legacy pollution refers to pollution that remains from historical activities. Legacy pollution is often associated with historic unlined landfills, former salvage operations, brownfields, hazardous waste dumping sites, rail yards, leaking underground storage tanks, mine tailings and former manufacturing sites. Collectively referred to as legacy contaminated sites, such areas pose environmental and health risks from a variety of pollutants (e.g., toxic organic compounds, lead, and other toxics) and exposure pathways (e.g., air, groundwater, soil).

Significant Contaminated Sites – Significant contaminated sites are primarily those sites regulated under Superfund or Comprehensive Environmental Response, Compensation, and Liability Act of 1980 (CERCLA) is a federal law designed to clean up sites contaminated with hazardous substances and pollutants. The National Priorities List (NPL) is the list of hazardous waste sites eligible for long-term cleanup financed under the federal Superfund Program.

Air Pollution Sources

Mobile – Mobile sources include cars, trucks, buses, non-road engines, equipment and vehicles. Mobile sources are responsible for emissions of green house gases, air toxics and precursor emissions which react to form secondary pollutants. Examples of mobile source air toxics include benzene, 1,3-butadiene, formaldehyde, acetaldehyde, acrolein, polycyclic organic matter (POM), naphthalene, and diesel particulate matter.

Residential Wood Burning – Pollutants from wood stoves and fireplaces include fine particulates, nitrogen oxides, sulfur oxides, carbon monoxide, volatile organic compounds, dioxins, and furans. Pollution from wood burning is a particular concern in the winter when cold, stagnant air and temperature inversions limit air movement. Smoke from wood burning is generated primarily by incomplete combustion, which can be caused by a number of different factors related to the wood burning device efficiency. With proper burning techniques and well-seasoned wood, emissions can be significantly reduced, even in older appliances.

Stationary – There are a variety of industrial, manufacturing and power generating plants that are stationary sources of air pollution within Greene County. The largest producers in the County include power generation, manufacturing and automotive repair related industries. EPA's National Air Toxics Assessments indicates hydrogen fluoride, toluene, and sulfuric acid dominate the toxic air releases in the Springfield metropolitan area. Other air pollution releases identified on EPA's National Emissions Inventory (NEI) include fine particulates, carbon dioxide, ozone and certain other chemicals generally related to smog. Stationary sources generally not related to industrial, manufacturing or power generation such as dirt roads and residential burning are not included within this source description.

EVALUATION CRITERIA

The MCDA evaluation criteria represent the interests of the stakeholders and are defined here by impacts from the pollutant sources and indicators used to assess those impacts.

Impacts

Impacts were selected based on desired outcomes identified in the Task Force report (Figure 2). Desired outcomes are tiered in the Task Force report such that Tier I is the most important (Shockey 2015). In some instances, impacts identified for the MCDA, which are described below, represent a combination of outcomes identified in the Task Force report.

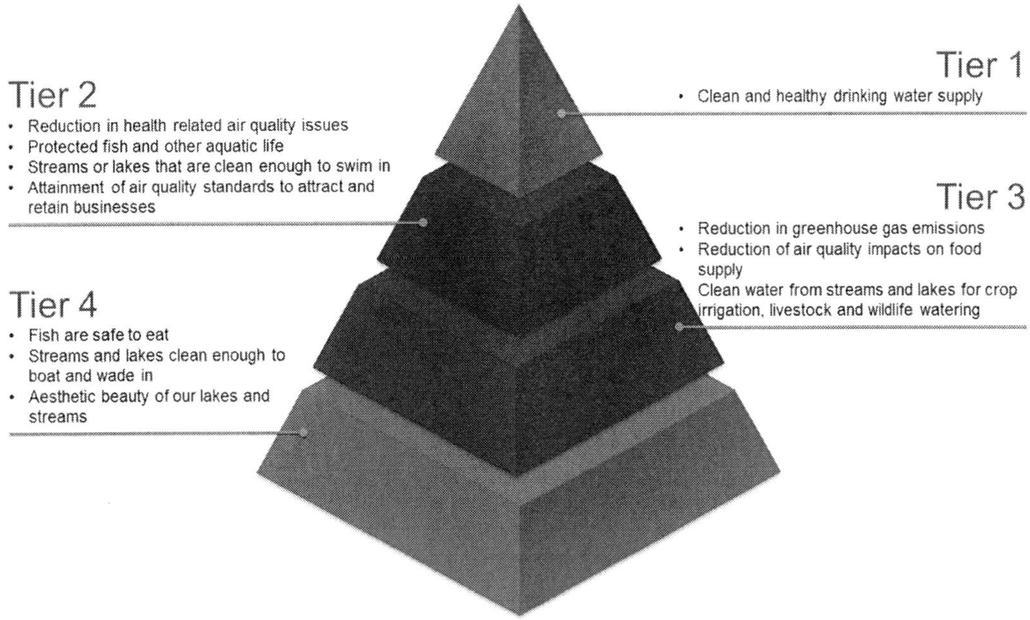

Figure 2. Overall Priorities Identified in the Springfield-Greene County Environmental Priorities Task Force Report

Climate Change

Climate change refers to any substantial change in measures of climate lasting for an extended period of time such as major changes in temperature, precipitation, or wind patterns. Global warming is one, but an important, aspect of climate change and refers to an average increase in the temperature of the atmosphere near the Earth's surface. Scientists believe that increases in greenhouse gases associated with human activities are contributing to climate change and global warming. Anthropogenic sources of greenhouse gases include the burning of fossil fuels to generate electricity, heat and cool buildings, and power vehicles. The major greenhouse gases emitted into the atmosphere are carbon dioxide, methane, nitrous oxide, and fluorinated gases. The reduction of greenhouse gases was identified as a Tier III priority in the Task Force report.

Aquatic Life Impacts

Aquatic life impacts refers to the altering or impairment of fish and other aquatic life in steams, rivers, ponds and lakes due to pollutants and habitat modification. Missouri regulations at 10 CSR 20-7.031 provide for aquatic life protections through the designation of aquatic habitat uses. Warm water habitat protections apply to most all surface waters in the Springfield-Greene County region, which are defined in state regulations as waters in which naturally-occurring water quality and habitat conditions allow the maintenance of a wide variety of warm water biota. The protection of fish and other aquatic life was identified as a Tier II priority in the Task Force report.

Agricultural Impacts

Agricultural impacts refers to impacts on food supply from the reduction of air quality. These impacts may be due to climate change and ozone pollution. Both these issues were identified as a Tier III priority in the Task Force report.

Ability to Attract and Retain Local Businesses

The Task Force report identified that attainment of air quality standards to attract and retain businesses as a Tier II priority. The Task Force report noted that potential nonattainment could limit the types of businesses attracted to the region or place additional restrictions on existing businesses. The Task Force report also included an overall goal of protecting the environmental to attract/retain business and maintain a high quality of life.

Water Quality Aesthetics
Aesthetics of streams and lakes was identified as a Tier IV priority in the Task Force report. The Task Force report also called for improving the aesthetics of Wilsons Creek noting that there is an important trail system in this watershed and it is positioned upstream of important recreational uses. Pollutants linked to aesthetic issues in the Task Force report included sediment, trash and bank erosion.

Primary and Secondary Contact Recreation
Streams or lakes that are clean enough to swim in (primary recreation) and boat and wade in (secondary recreation) were identified as Tier II and IV priorities, respectively, in the Task Force report. Specifically, the Task Force report prioritized protecting water from pollution in the Lower James River, Upper James River, Sac River, and Little Sac River in areas where people swim. These rivers are classified for both Whole Body Contact Recreation (WBCR) – Class A and Secondary Contact Recreation (SCR) designated use protections. WBCR Class A protections apply to waters that have been established by the property owner as public swimming areas welcoming access by the public for swimming purposes and waters with documented existing whole body contact recreational use(s) by the public. WBCR and SCR waters are protected in state regulations by *E. coli* criteria.

Clean Drinking Water
Clean and healthy drinking water was identified as the most important priority in the Task Force report. Drinking water sources targeted for protection in the Task Force report included McDaniel Lake, Fellows Lake, Upper James River, Fulbright Spring Recharge Area, and the Upper Little Sac. CU pulls its raw water supply from these sources plus Stockton Lake and deep groundwater wells. Additionally, tens of thousands of residents in Greene County rely on groundwater for their drinking water. Potential threats to the drinking water sources identified in the Task Force report include nutrients and increased sediment load.

Fish Consumption Advisories
Fish that are safe to eat was identified as a Tier IV priority in the Task Force report. Each year, the Missouri Department of Health and Senior Services (DHSS) create a fish consumption advisory based on an evaluation of contaminants in Missouri sport-caught fish. The advisory is based on annual fish-tissue studies by the Missouri Department of Conservation (MDC) and the Missouri Department of Natural Resources (MDNR) at various Missouri lakes, ponds, rivers, and streams (DHSS 2016). In the Springfield-Greene County region, the 2016 fish advisory includes Lake Springfield due to elevated levels of polychlorinated biphenyls (PCBs) being found in catfish. There is also currently a statewide advisory for mercury.

Health Impacts from Air Pollution
Air pollution can contribute to a variety of human health issues, affecting a number of different systems and organs. Health related issues range from minor upper respiratory irritation to chronic respiratory and heart disease, lung cancer, acute respiratory infections in children and chronic bronchitis in adults, aggravating pre-existing heart and lunch disease, or asthmatic attacks. To protect for health related issues, the Clean Air Act requires EPA to set National Air Quality Standards (NAAQS) for six common air pollutants also known as "criteria pollutants". These include particle pollution, photochemical oxidants and ground-level ozone, carbon monoxide, sulfur oxides, nitrogen oxides, and lead. Man-made sources of air pollutants can

include emissions from automobiles, factories, power plants, construction equipment, small businesses, and open burning. The reduction in health related air quality issues was identified as a Tier II priority in the Task Force report.

Indicators

The indicators, which are analogous to sub-criteria, provide an objective means for assessing impacts and linking impacts to pollutant sources. Indicators are pollutants (e.g., toxins) or conditions (e.g., aquatic habitat) that are critically linked to the Impacts defined by the Task Force report. The Expert Panel will be used to select indicators for each Impact identified through the Task Force.

DECISION FRAMEWORK

A draft decision framework was developed to explicitly link pollutant sources with evaluation criteria specified above. The decision framework also weights impacts and indicators on a scale of 0 to 1, with a total sum of 1. Impact weights are based on community priorities and indicator weights are based on technical considerations and feedback from the IP partners and the Expert Panel.

Ratings and Scoring of Pollution Sources

The MCDA draft decision framework presented above is a tool for rating and scoring pollutant sources. Ratings should first be applied to all links between pollutant sources and indicators. These links are assigned a rating on a scale of 0 to 3. A rating of 3 indicates the pollutant source significantly contributes to high levels of the indicator. Conversely, a rating of 1 indicates the pollutant source is minimal with respect to the indicator and associated impacts. A rating of 2 falls in the middle. Alternatively, a scale of 0 to 5 may be used depending upon input from the Expert Panel. A multi-dimensional rating matrix will be used to assign ratings to each pollution source for each indicator and associated community priority (see Table 1). Ratings will be based on consideration of multiple factors aligned that help characterize the significance of each source based on existing data, where available, best professional judgment, and feedback from the Expert Panel. Weightings of each factor may be qualitatively or quantitatively applied to aid in setting the overall rating for each source and indicator. An example of the rating process is provided as Table 2.

Table 1. Preliminary Pollution Source Multi-Dimensional Rating Matrix

Rating	Pollution Source Significance	Watershed Scale	Relative Contribution	Hydrologic Condition	Risk of Pollutant Species
3	Major	Impacts all regional watersheds	Contributes 50-100% of Pollutant	Impacts range of hydrologic conditions	Pollutant species associated with source poses high risk to Community Priority
2	Moderate	Impacts only Community Priority watersheds	Contributes 25-50% of Pollutant	Impacts primarily dry weather conditions only	Pollutant species associated with source poses medium risk to Community Priority
1	Minor	Only impacts one minor watershed	Contributes 10-25% of Pollutant	Impacts primarily wet weather conditions only	Pollutant species associated with source poses low risk to Community Priority
0	Insignificant	No Impact	No Impact	No Impact	No Impact

Table 2. Example of Pollution Source Rating for Exfiltration Impacts on *E. coli* as related to Primary and Secondary Recreation

Rating	Pollution Source Significance	Watershed Scale	Relative Contribution	Hydrologic Condition	Risk of Pollutant Species
3	Major	Impacts all regional watersheds	**Contributes 50-100% of Pollutant**	Impacts range of hydrologic conditions	**Pollutant species associated with source poses high risk to Community Priority**
2	Moderate	**Impacts only Community Priority watersheds**	Contributes 25-50% of Pollutant	**Impacts primarily dry weather conditions only**	Pollutant species associated with source poses medium risk to Community Priority
1	Minor	Only impacts one minor watershed	Contributes 10-25% of Pollutant	Impacts primarily wet weather conditions only	Pollutant species associated with source poses low risk to Community Priority
0	Insignificant	No Impact	No Impact	No Impact	No Impact

Overall Rating: 3

Scoring is a function of ratings and weights. Each rating is multiplied by the impact and the indicator weight. The products are summed for each pollutant source for a maximum score of 3. Pollutant sources can then be ranked to determine the most significant sources of pollution. However, this process is intended to be iterative and prompt discussion. Refinement and re-scoring may be necessary before there is a final determination on pollutant source rankings. This example is only being provided for illustrative purposes. The final MCDA structure, selection of indicators, weightings, and scoring of pollution sources is expected to change through the Expert Panel process.

MCDA RESULTS

The Expert Panel was convened on January 17 and 18, 2017. Expert Panel participants included Bob Pitt, Ph.D., Professor Emeritus University of Alabama, Jon Jones, Wright Water Engineers, Jane Clary, Wright Water Engineers, Dave Clark, HDR, Chance Lauderdale, HDR, and Tom Dupuis, HDR. MCDA indicators were selected and weights were assigned. Pollution source impact ratings were also assigned for each indicator. The Expert Panel and IP Partners also established data gaps within the available body of environmental data for the region. Final results will be released after additional analyses and consensus among the Expert Panel and IP Partners is established.

SUSTAINABLE RETURN ON INVESTMENT OVERVIEW

The City retained HDR to conduct a pilot economic assessment and provide decision support using HDR's SROI analytical framework. For the City, SROI provides an objective, transparent and defensible economic business case approach to assess the costs and benefits pertaining to three Integrated Planning Opportunities to help the City evaluate and prioritize the competing investments the City is facing within its Clean Water Act obligations.

The SROI process draws from standard economic benefit-cost analysis (BCA) methods and the best available data to systematically calculate and compare the benefits and costs of project options. The process addresses sustainability goals and outcomes from a triple bottom-line perspective – i.e. the full range of environmental, social, and economic impacts. Economic impacts account for the capital, operating and maintenance costs of the project. In this case, environmental and social benefits include the value of water quality improvements as well as changes in air pollution, greenhouse gas (GHG) emissions, land value, and other factors. The process includes:

- Comprehensive identification of impacts,
- Formulation of data into a draft model,
- Workshop-based discussion of evidence, and
- Accounting for risk and uncertainty in key drivers of outcomes.

Issues related to sustainability, sustainable communities, and sustainable development is being actively discussed across a broad sector of our society. Sustainable development is typically defined as the pattern of development that meets the needs of the present without compromising the ability of future generations to meet their own needs. Sustainable development combines the financial considerations of development with broader socio-economic concerns including environmental stewardship, human health and equity issues, social well-being, and the social implications of decisions.

While the importance of these issues is widely recognized, organizations are challenged when they try to integrate sustainability considerations into their investment and operating decisions. Traditional financial evaluation tools used to assess an investment project, such as Business Case Analysis or Life-Cycle Cost Analysis (LCCA), rely exclusively on financial impacts. These traditional tools have two primary drawbacks:

1. An inability to accurately quantify the non-cash benefits and costs accruing to both the organization in question and to society as a whole resulting from a specific investment (sustainable benefits and costs).
2. A failure to adequately incorporate the element of risk and uncertainty.

The SROI process is both analogous to and an improvement upon triple bottom-line oriented MCDA methods that are more commonly used to support decisions in the water and wastewater sectors. SROI and MCDA both account for the financial, environmental, and social outcomes; however, SROI adopts principles of economic analysis to evaluate these outcomes (e.g., BCA). The important difference here is that MCDA is a more ad hoc method that permits consideration of a wide range of scoring criteria, even potentially overlapping drivers (e.g., energy costs and pollution emissions). Overlapping criteria can lead to over-estimated benefits. A second difference is that while MCDA applies subjective weights to decision criteria so that scores for each criterion can be totaled, SROI's benefit categories are measured in monetary terms using the best available objective data. This feature of SROI enables benefits to be based on objective research and generate results that are directly comparable to costs in metrics such as a *Net Present Value* or *Benefit-Cost Ratio*. Finally, SROI directly accounts for data gaps using simulation and probabilistic modeling of uncertainty – a process that generates results for decision makers which indicate the likelihood that benefits exceed costs, for all categories of triple bottom-line benefits.

In the SROI evaluation process, costs and benefits are measured relative to a baseline. Typically a baseline represents the status quo economic, environmental and social conditions, also known as "the do nothing alternative." Best-practices in economic analysis would also dictate that the costs and benefits should be presented in a way to maximize their consistency or comparability. A monetized estimate is one that either occurs naturally in dollars (e.g., increased costs by a business to purchase equipment needed to comply with a regulation) or has been converted into dollars using a specified methodology (e.g., the number of avoided environmental impacts multiplied by individuals' estimated willingness-to-pay to avoid them). Presentation of monetized benefits and costs is preferred where acceptable estimates are possible. The key analytical principles employed in developing the SROI analysis are listed as follows:

- Assessing benefits, including economic, social and environmental categories, using appropriate metrics that are directly relevant and measurable;
- Accounting for any potential double-counting of benefits;
- Accounting for life-cycle costs associated with building, operating and maintaining the facilities; and
- Converting benefits to equivalent monetary values following best-practices.

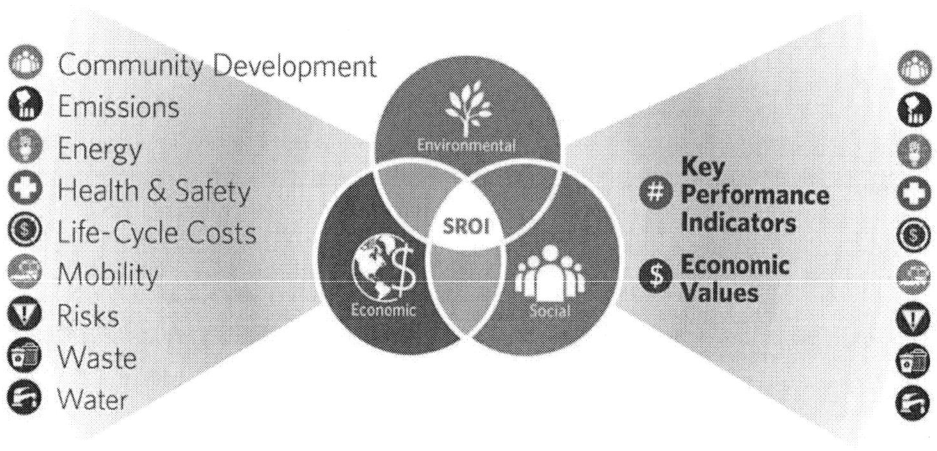

Figure 3: SROI Concept

At this stage, the analysis acts as a pilot project to highlight the role SROI can play in the decision making process at the City using real investments. While the analysis is rigorous and employs best-practices in economics and financial modelling, for the most part, detailed technical engineering analysis has been completed at a preliminary stage by HDR's engineering team. The technical engineering assessment produced the inputs necessary for the SROI evaluation, including capital cost and annual operations and maintenance estimates, flow and pollutant loading estimates, water quality modeling, recreational users counts, and energy impact modeling, amongst others. The technical evaluations are included in this report, alongside the economic analysis. The outcomes from this SROI analysis will be used to narrow down the list of Integrated Planning Opportunities the City will pursue with further analysis.

HDR utilized a multi-disciplined team of engineers and economists to produce the analysis, and collaborated with key City staff and stakeholders throughout the process. The list of three broad Integrated Planning Opportunities investment alternatives (seven in total) analyzed for the City. The opportunities include stormwater detention basin retrofits, polycyclic aromatic hydrocarbon reduction policies, enhanced nutrient removal at the Southwest Treatment Plant, and differing levels of sanitary sewer overflow control. Each of the nine alternatives was compared to the base case of status-quo, which is forward-looking.

SROI METHODOLOGY

The best possible information is needed to make rational investment decisions that properly reflect value-for-money comparisons among different capital project alternatives. This includes the need to compare competing projects on a 'common language' basis – whereby costs and benefits, including those environmental and social impacts, are converted into monetary values to estimate the overall impacts in comparable financial terms. Cost-benefit analysis is typically conducted as part of a regulatory impact analysis for proposed regulatory changes; for example, Executive Orders 13563 and 12866 require Federal agencies to provide to the public and to OMB an analysis of the anticipated consequences of significant regulatory actions. This includes an assessment and (to the extent feasible) a quantification and monetization of the benefits and costs anticipated from the proposed action.

Costs and benefits must be measured relative to a baseline; best-practices establishes that the baseline should be the best assessment of the way the world would look absent the proposed investment/regulation. Typically, the baseline should start with the world before the investment/action taken and applied equally to benefits and costs. For Springfield, this baseline is the status quo for each of the independent opportunities analyzed. Best-practices in economic analysis would also dictate that the costs and benefits should be presented in a way to maximize their consistency or comparability. A monetized estimate is one that either occurs naturally in dollars (e.g., increased costs by a business to purchase equipment needed to comply with a regulation) or has been converted into dollars using some specified methodology (e.g., the number of avoided health effects multiplied by individuals' estimated willingness-to-pay to avoid them). Presentation of monetized benefits and costs is preferred where acceptable estimates are possible. The key analytical principles employed in developing the cost-benefit analysis are listed as follows:

- Assessing benefits using appropriate metrics that are directly relevant and measurable;
- Accounting for any potential double-counting of benefits;
- Accounting for life-cycle costs associated with building, operating and maintaining the facilities;
- Converting benefits to equivalent monetary values following best-practices.

Most professionals in the Architecture, Engineering, and Construction industry rely on basic financial tools to quantify the first costs of sustainable projects and to show the financial return on investment of sustainable strategy benefits. The first instance where these traditional methods often fall short is in the accurate quantification of the benefits that accrue to society as a whole as a result of these decisions. The second instance where traditional tools fall short is that they fail to incorporate the element of risk. Sustainable project decisions require the forecasting of future costs and benefits and these are subject to uncertainty, which is typically not captured by conventional methods.

Since traditional analytical methods fall short in accurately quantifying all positive and negative externalities, HDR developed the SROI process. SROI is an enhanced form of cost-benefit analysis - a systematic process for calculating and comparing benefits and costs of a project or policy, and is generally conducted to justify an investment or compare projects. The SROI process accounts for a project's triple-bottom-line – its full range of environmental, social and economic impacts. The process builds on best practices in cost-benefit and financial analysis methodologies, complemented by advanced risk analysis and stakeholder elicitation.

In this analysis, actual financial costs and benefits incurred by the City are accounted for, in addition to the monetized value of various social and environmental impacts to account for the triple bottom line. In other words, through SROI, the City and HDR were able to quantify and monetize those impacts typically assessed only qualitatively, allowing for an apples-to-apples comparison of various stormwater and wastewater investments. Economic viability is determined on the basis of whether the long-term benefits exceed the costs and whether the rate of return is adequate.

HDR's SROI analysis involves four distinct steps:

STEP 1	Develop Structure and Logic Diagrams	Map out economic, social and environmental variables and graphically illustrate the calculations required.
STEP 2	Assign Monetary Values and Risk Ranges	Measure the impacts of the project by assigning monetary values and probability distributions, where possible, to each variable.
STEP 3	Develop Consensus among Stakeholders	Bring stakeholders together and develop consensus on model and the assigned values.
STEP 4	Simulate and Quantify Outcomes	Compute SROI metrics such as Net Present Value, Benefit-Cost Ratio, Discounted Payback Period, etc.

Risk analysis and Monte Carlo simulation techniques were used to account for uncertainty in both the input values and model parameters. Projections were expressed as probability distributions (a range of possible outcomes and the probability of each outcome). Finally, each element was developed or converted into monetary values to estimate the overall impacts in comparable financial terms.

Our analysis produced results on both a financial and a sustainable basis using several of the most recognized evaluation metrics. For example:

Net Present Value (NPV)	The net value that a project generates to the public benefit, calculated as the sum of the present value of future benefits less the present value of the project's costs.
Benefit-Cost Ratio (BCR)	The overall "value for money" of a project, expressed as the ratio of the benefits of a project relative to its costs, with both expressed in present-value monetary terms.

SROI originated from a Commitment to Action by HDR to develop a new generation of public decision support metrics for the Clinton Global Initiative (CGI) in 2007. SROI was developed with input from Columbia University's Graduate School of International Public Affairs and launched at the 2009 CGI annual meeting. Since then, the SROI process has been used by HDR to evaluate the monetary value of sustainability programs and projects with a combined value of over $20 billion. It has been used by corporations and all levels of government.

The methodology for the various benefits and costs is presented graphically in the form of a flow chart called a "structure and logic model" (see Figure 4). Such models provide a graphical illustration of how the various inputs combine to determine the benefit or cost evaluated. They are intended to provide a transparent record of how each benefit and cost is calculated.

In this analysis, actual financial costs and benefits incurred by the City are accounted for (excluding accounting-related financial impacts such as amortization/depreciation schedules and taxes), in addition to the monetized value of various social and environmental impacts to account for the triple bottom line. In this case, the analysis included the upfront capital expenditures and ongoing operations and maintenance (O&M) cost impacts, avoided costs related to the various opportunities, and any revenue impacts to the City. The analysis also monetized non-cash benefits and costs including those related to water quality impacts; greenhouse gases and criteria air contaminants production (from changes in electricity consumption and chemicals usage); biosolids and chemicals trucking-related impacts (accidents, noise, congestion, vehicle O&M, pavement O&M, and emissions); public and stakeholder relations impacts; tree impacts; Total Maximum Daily Load costs, regulatory compliance costs, and private property flooding cost impacts.

HDR used methodologies developed by McClelland (1974) and Van Houtven et al. (2007) for estimating water quality benefits. Using these approaches, a 10-point water quality index (WQI) was developed to support the application of the SROI tool, which was used for assessing the monetary benefits of different water quality improvement opportunities in Springfield area. The WQI provides a single metric that expresses overall water quality based on several parameters and is critical in evaluating water quality improvements as it allows for uniform comparisons between different opportunities, pollutants, and waterbodies.

The anticipated change in the WQI was estimated for each water quality improvement opportunity and the associated receiving waters. Receiving waters considered as part of this evaluation were categorized as follows:

- Springfield urban streams
- Wilsons Creek below the Southwest Treatment Plant (SWTP)
- James River below Wilsons Creek
- Lake Springfield
- Table Rock Lake

The number of users and non-users also factors into the SROI's tool for assessing monetary benefits associated with each water quality improvement opportunity as described by Van Houtven et al. (2007). Therefore, the number of users and non-users were estimated for the five receiving waters under evaluation.

Figure 4: Sample S&L Diagram for tree planting within existing grass detention ponds and the associated water quality improvements

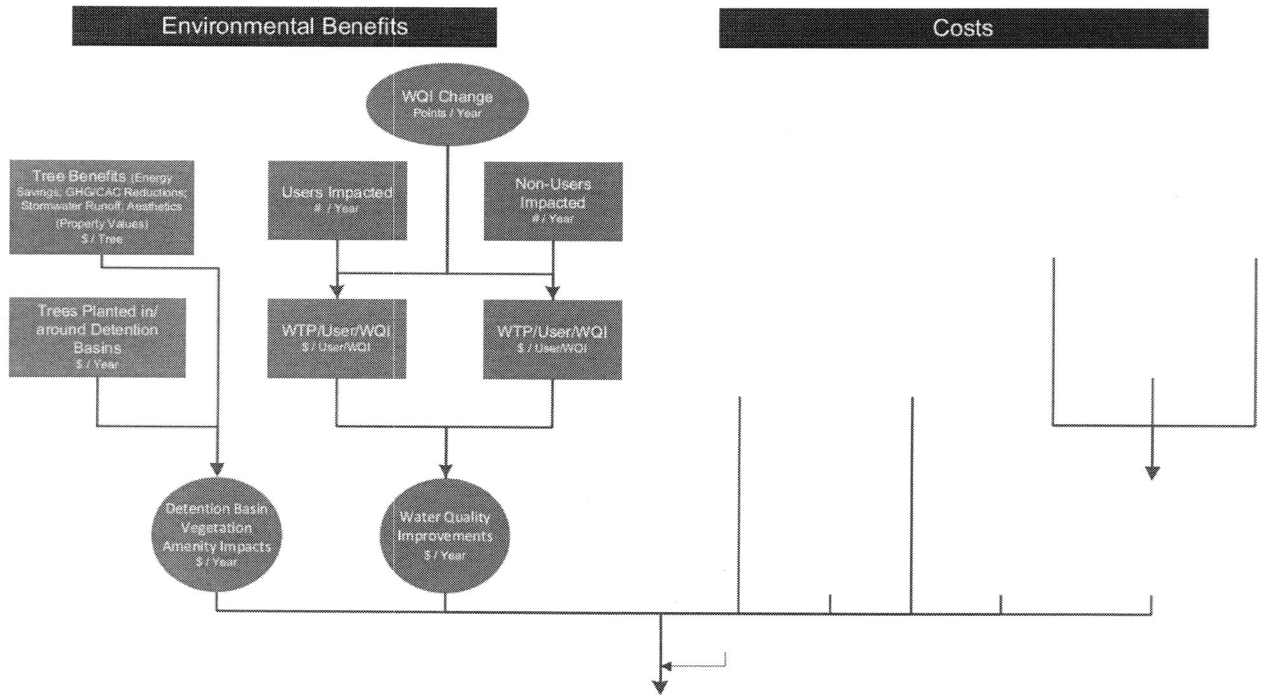

SROI RESULTS

The SROI process has been implemented to evaluate a range of Integrated Planning Opportunities. This report highlights the results of several types of project opportunities – each of which includes one or more design and implementation alternatives. These opportunities and alternatives include consideration of:

- **Stormwater Detention Basin Retrofits:** This opportunity includes two separate evaluations of detention basin retrofit alternatives – simple outlet modification to improve stormwater treatment, and enhanced vegetation within the basins for hydrologic and water quality benefits.
- **Polycyclic Aromatic Hydrocarbon (PAH) Reduction Measures:** The evaluation of this opportunity includes two alternatives – as a coal tar sealant reduction policy and physical excavation of accumulated PAHs within the stormwater drainage system.
- **Enhanced Nutrient Removal at Southwest Wastewater Plant (SWTP):** This opportunity is based upon providing enhanced nutrient removal with target effluent quality of 0.1 mg/l TP and 3 mg/l TN at the SWTP.
- **Sanitary Sewer Overflow (SSO) Controls:** This opportunity includes evaluation of four different SSO control levels of service (LOS).

Overall, nine total alternatives among four types of projects. This preliminary assessment of project options and corresponding SROI analysis will be used to refine further the list of Integrated Planning Opportunities that the City will consider for further analysis and future implementation.

Analysis of project impacts requires an analysis of baseline conditions, and changed conditions because of the project, across each of the benefit categories. Data on these changed conditions are generated by engineering data and expert judgment. A multi-disciplined team of engineers, key City staff, local stakeholders, and economists have been involved to produce data and review analytical methods. Some of the key technical engineering inputs include: capital cost and annual operations and maintenance estimates, flow and pollutant loading estimates, water quality modeling, recreational users counts, and energy impact modeling, among others.

Benefits and Costs by Opportunity Alternative

Opportunity alternatives accrue different combinations of environmental and social benefits depending on their characteristics. Table 3 indicates (with a shaded cell) how benefit categories apply differently to project alternatives. All of the projects generate environmental water quality benefits. Some projects generate changes in GHG emissions from energy use in hauling biosolids or treatment facility electricity use. Note that CAC emissions are included as a social impact because the basis for estimating impacts comes from changes in human health conditions. Other social impacts, as discussed above, relate to impacts of trucks on the road and a variety of impacts on private property.

Table 3: Matrix of Benefit Categories by Project Alternative

Benefit	Extended Detention	Extended Enhanced Detention	PAH Reduction - Policy	PAH Reduction – Policy, Excavation	Enhanced Nutrient Removal at SWTP	SSO Reduction: I/I Control	SSO Reduction: LOS 1 YR	SSO Reduction: LOS 2 YR	SSO Reduction: LOS 5 YR
Environmental									
Water Quality Impacts	■	■	■	■	■	■	■	■	■
GHG Emissions Impacts - Increased Electricity Use at SWTP					■				
GHG Emissions Impacts - Chemical Use at SWTP					■				
GHG Emissions Impacts - Reduced Electricity - SSO Reduction						■	■	■	■
GHG Emissions Impacts from Trucking (Chemicals/Biosolids)					■				
Tree Impacts Around Detention Basins		■							
Social									
CAC Emissions Impacts - Increased Electricity Use at SWTP					■				
CAC Emissions Impacts - Reduced Electricity - SSO Reduction						■	■	■	■
CAC Emissions Impacts from Trucking (Chemicals/Biosolids)					■				
Accident Impacts - Trucking (Chemicals/Biosolids)					■				
Pavement Damage Impacts - Trucking (Chemicals/Biosolids)					■				
Congestion Impacts - Trucking (Chemicals/Biosolids)					■				
Noise Impacts - Trucking (Chemicals/Biosolids)					■				
Private Property Impacts - Reduced Flooding, SSO Reduction						■	■	■	■
Capital Costs									
Capital Expenditure Impacts (Including Replacement Costs)	■	■		■	■	■	■	■	■
Residual Value of Capital Costs (Including Replacement Costs)	■	■			■	■	■	■	■
Operating and Maintenance (O&M) Costs									
O&M Impacts	■	■			■				
Sinkhole Repair Expenditure Impacts	■	■							
O&M Cost Impacts at SWTP/NWTP from SSO Reduction						■	■	■	■
Vehicle Operating Cost Impacts - Increased Biosolids Hauling					■				
Public and Stakeholder Relations Impacts	■	■	■	■		■	■	■	■
Sealant Cost Impacts to Private Property			■	■					
TMDL Impairment Cost Impacts (Administrative/Legal Only)					■	■	■	■	■
Avoided Permit Violation Costs (Regulatory Compliance Costs)						■	■	■	■

The analysis of projects can be viewed from a variety of perspectives including cost-only and total value, in which water quality improvement and other outcomes are incorporated. In all cases, the measures of comparison in this SROI analysis are in monetary units the analytical methods for estimating monetary value of benefits is discussed in the previous chapter. A cost-only perspective would rank projects based on the total lifecycle cost, irrespective of the type and degree of water quality improvements.

The results in **Error! Reference source not found.** provide the estimates of present value costs and benefits for each of the project opportunities. The results indicate that there is a wide variety of results within the four opportunities and nine alternatives. Each investment generates net social & environmental benefits with varying magnitudes. In some cases, the monetized value of water quality improvements are not enough to offset the implementation costs, the value of these improvements are still significant. These results clearly demonstrate the usefulness of the SROI process for evaluation of dissimilar stormwater and wastewater alternatives that carry significantly different costs and benefits.

A summary of the financial lifecycle costs:

- There are varying degrees of magnitudes of net lifecycle impacts to each alternative.
- A lower cost option entails $7.7M in detention basin retrofits
- SSO options including alternative designs for LOS 1, 2, and 5 years all exceed $200M. While there are operational cost savings, these savings do not offset the construction costs.

From a triple bottom-line perspective, the 16 categories of impacts are each valued based on methods described above and generated the following results:

- Each project produces positive net social/environmental benefits over the study period.
- The magnitudes of the impacts vary greatly – from $57.6M for nutrient removal at the wastewater treatment plant to $9.8M for detention basin retrofits.
- The predominant driver of benefits in each case is the value of improved water quality. The benefits range from $59.8M for enhanced nutrient removal and $9.8M for detention basin retrofits.
- The sanitary sewer alternatives show diminishing returns to the LOS – each subsequent LOS shows increasing benefits; however, costs are increasing at a greater rate, reducing the BCR at each subsequent level. SSO with I/I control has the highest BCR, while SSO with 5 year LOS has the lowest among the SSO projects.

Table 4: Summary of Results – Integrated Planning Opportunities Alternatives Analysis ($2014)

SROI Analysis Results: Present Value 27-Year Study Results from 2014-2040 (2014 Base Year, 2015 Capital Investment, 2016 Ongoing Cash flows)	Option 1: Extended Detention	Option 2: Extended Enhanced Detention	Option 3: PAH Reduction – Policy	Option 4: PAH Reduction – Policy, Excavation	Option 5: Enhanced Nutrient Removal at SWTP	Option 6: SSO Reduction: I/I Control	Option 7: SSO Reduction: LOS 1 YR	Option 8: SSO Reduction: LOS 2 YR	Option 9: SSO Reduction: LOS 5 YR
Environmental Benefits									
Water Quality Impacts	$9.8	$10.0	$15.5	$18.8	$59.8	$16.4	$17.6	$18.1	$18.3
GHG Emissions Impacts - Increased Electricity Use at SWTP					($1.1)				
GHG Emissions Impacts - Chemical Use at SWTP					($0.2)				
GHG Emissions Impacts - Reduced Electricity - SSO Reduction						$1.1	$1.1	$1.1	$1.1
GHG Emissions Impacts from Trucking (Chemicals/Biosolids)					($0.1)				
Tree Impacts Around Detention Basins		$0.1							
Social Benefits									
CAC Emissions Impacts - Increased Electricity Use at SWTP					($0.3)				
CAC Emissions Impacts - Reduced Electricity - SSO Reduction						$0.3	$0.3	$0.3	$0.3
CAC Emissions Impacts from Trucking (Chemicals/Biosolids)					($0.1)				
Accident Impacts - Trucking (Chemicals/Biosolids)					($0.0)				
Pavement Damage Impacts - Trucking (Chemicals/Biosolids)					($0.3)				
Congestion Impacts - Trucking (Chemicals/Biosolids)					($0.2)				
Noise Impacts - Trucking (Chemicals/Biosolids)					($0.0)				
Private Property Impacts - Reduced Flooding, SSO Reduction						$0.3	$0.8	$0.9	$0.9

SROI Analysis Results: Present Value 27-Year Study Results from 2014-2040 (2014 Base Year, 2015 Capital Investment, 2016 Ongoing Cash flows)	Option 1: Extended Detention	Option 2: Extended Enhanced Detention	Option 3: PAH Reduction – Policy	Option 4: PAH Reduction – Policy, Excavation	Option 5: Enhanced Nutrient Removal at SWTP	Option 6: SSO Reduction: I/I Control	Option 7: SSO Reduction: LOS 1 YR	Option 8: SSO Reduction: LOS 2 YR	Option 9: SSO Reduction: LOS 5 YR
Capital Costs									
Capital Expenditure Impacts (Including Replacement Costs)	($2.0)	($14.4)		($9.3)	($34.5)	($40.5)	($359.3)	($416.7)	($495.1)
Residual Value of Capital Costs (Including Replacement Costs)	$0.2	$1.1			$12.8	$11.3	$105.2	$121.9	$144.1
Operating and Maintenance Costs									
O&M Impacts	($5.6)	($25.9)			($48.5)				
Sinkhole Repair Expenditure Impacts	($0.1)	($0.2)							
O&M Cost Impacts at SWTP/NWTP from SSO Reduction						$6.4	$6.4	$6.4	$6.4
Vehicle Operating Cost Impacts - Increased Biosolids Hauling					($1.0)				
Public and Stakeholder Relations Impacts	($0.2)	($0.2)	($0.1)	($0.1)		$0.4	$1.0	$1.2	$1.3
Sealant Cost Impacts to Private Property			($14.7)	($14.7)					
TMDL Impairment Cost Impacts (Administrative/Legal Only)			$1.0	$1.0	$0.2	$1.0	$1.0	$1.0	$1.0
Avoided Permit Violation Costs (Regulatory Compliance Costs)						$1.7	$4.6	$5.3	$5.6
Summary Results									
Financial Lifecycle Cost	($7.7)	($39.5)	($13.7)	($23.1)	($71.0)	($19.7)	($240.9)	($280.9)	($337.0)
Total Social, Environmental Benefits	$9.8	$10.2	$15.5	$18.8	$57.6	$18.0	$19.6	$20.2	$20.5
Total Value - All Costs and Benefits	$2.1	($29.3)	$1.8	($4.3)	($13.4)	($1.7)	($221.3)	($260.7)	($316.5)
Benefit-Cost Ratio	1.28	0.26	1.13	0.81	0.81	0.91	0.08	0.07	0.06

The SROI results also include risk-based outputs using Monte Carlo simulation techniques to capture the likelihood of generating positive benefits compared to costs. Box plots of the BCRs for each opportunity and alternative is displayed as Figure 5. These plots illustrate the differences in the potential that benefits could outweigh costs (BCR > 1) with some probability, after accounting for a wide range of uncertainties in cost and benefit drivers. The analysis suggests that *Alt. 1 - Extended Detention*, *Alt. 3 - PAH Reduction - Policy*, and *Alt. 6 - SSO Reduction: I/I Control* provide the highest potential for yielding positive triple bottom line outcomes. Other options indicate that costs are highly likely to outweigh benefits. These comparisons provide valuable insights into prioritization of projects and programs through the Integrated Planning process.

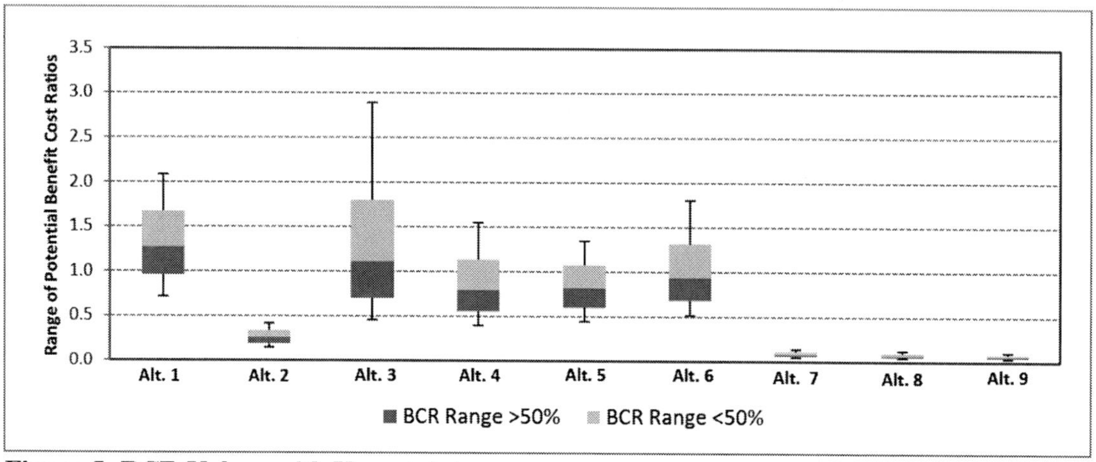

Figure 5: BCR Values with Uncertainty

CONCLUSIONS

The IP Partners have taken great strides in identifying the most significant pollution sources that impact the regional environment through the MCDA process. Final MCDA results will be available within the next few months after additional analyses and consensus from Expert Panel participants is established.

The SROI pilot study opportunities vary widely in terms of financial costs and social and environmental benefits. All opportunities generate net social and environmental benefits with water quality improvements comprising the greatest benefits. While these estimated benefits are significant with the greatest exceeding $60 million, most alternatives have net costs with financial costs exceeding social and environmental benefits. The SROI results suggest two base alternatives for the stormwater basin retrofits and coal tar sealant reduction policy opportunities yield benefits that outweigh costs. In addition, some alternatives nearly result in break even with benefit-cost ratios approaching or exceeding 0.9. There are also significant probabilities that these alternatives will generate net benefits. Additional analyses to fill data gaps would provide a greater understanding of triple-bottom-line costs and benefits and reduce uncertainty. Lastly and most importantly, the SROI pilot study has demonstrated the utility of this process for the integrated planning by weighing all costs and benefits to balance and prioritize dissimilar investments.

REFERENCES

McClelland, N. (1974) Water quality index application in the Kansas River Basin. Prepared for US Environmental Protection Agency-Region 7. EPA-907/9-74-001.

Missouri Department of Health and Senior Services (DHSS). (2016). 2016 Missouri Fish Advisory – A Guide to Eating Missouri Fish.

Natural Resources Leadership Institute (NRLI). (2016). Multi-Criteria Decision Analysis. Retrieved August 10, 2016, from https://www.ncsu.edu/nrli/decision-making/MCDA.php.

Shockey Consulting Services, LLC (Shockey). (2015). Integrated Plan for the Environment – Environmental Priorities Task Force Report. Prepared for the City of Springfield, Greene County and Springfield City Utilities. February 3, 2015.

Stoner, N. and Giles, C. (2012) Integrated Municipal Stormwater and Wastewater Planning Framework. Memorandum from US Environmental Protection Agency Regional Administrators and Regional Permit and Enforcement Division Directors, June 5, 2012.

Van Houtven, G., Powers, J. and Pattanayak, S. (2007) Valuing water quality improvements in the United States using meta-analysis: Is the glass half-full or half-empty for national policy analysis? Resource and Energy Economics, 29, pp. 206-228.

State of the Rates - 2016 Missouri Wastewater and Water Rate Survey

Philip Walsack, Missouri Public Utility Alliance

ABSTRACT

The Missouri Water Environment Association and many other industry advocacy groups emphatically suggest that the Nation, as well as Missouri, is falling behind the infrastructure replacement curve. Municipalities around the country are fretting about utility rates, publically-elected officials dodged the rate issue during the election season, while the public is getting increasingly hostile over proposed rate increases. The Missouri State Legislature passed House Bill 1251 in 2012 which specifies that new NPDES permit conditions and enforcement actions must be "affordable". The State law further states that the affordability "determinations shall be based upon empirical data".

KEYWORDS

Water Rates, Wastewater Rates, Sewer Rates, Rate Survey, Median Household Income, Affordability

INTRODUCTION

This following professional endeavor describes the findings of the largest wastewater and water rate survey ever conducted in American history. It compares the 2016 rate data to the median household income values of each municipality. This presentation will enable the attendees to gain an understanding of the complexity nature of affordability. A comparison between the 2016 rate survey and the previous biennial survey was made. Important factors that are paramount when addressing affordability will also be made, including: community size and growth; geography; and infrastructure conditions.

2016 RATE SURVEY

The Missouri Public Utility Alliance (MPUA) conducted a water and wastewater rate survey (Survey) in 2016. This Survey is a continuation of biennial efforts that were conducted in 2010, 2012, and 2014.

The 2016 Survey was a massive undertaking. Our records show that it is the most complete and extensive rate survey in the United States. While this Survey lists communities with a population greater than 100 people, the survey does, in fact, contain data from the very, very small communities. Their data is not included in the public version of the Survey because it lacks appropriateness when attempting to compare it to larger utilities. The hardest utilities to gather rate data from are the very small, those with populations between 100 and 1,000 people. The Survey covers most of the 489 communities in this size grouping. Overall, this survey covers 93% of all Missouri communities over 100 people. Of the 838 communities in Missouri that have "public utilities", 783 are covered in the rate survey. For all six size groupings greater than 1,000 people, MPUA lacked only one response.

Each line of the database contains the municipality name and municipality's legal status (village, town, or city). The following data points were gleaned from the U.S. Census Bureau: 2014 population;

population change from 2010 to 2014; and the 2014 median household income (2014 American Community Survey (ACS) 5-year estimate [ID # B19013]).

The heart of the database is the monthly water and wastewater rate. Rates for each community were researched. A few Missouri communities have their rates posted on the internet. Many communities responded to an emailed request. Those non-respondents were called by telephone to ascertain their rates. This enables the data to be viewed as a "snapshot in time". The survey was begun in early January and completed by March 15. Each individual municipality's rate scenario was then calculated to reflect an individual residential water / wastewater bill. It was assumed that all residential customers consumed 5,000 gallons of potable water when the "monthly bill" was computed. This monthly bill value was entered into the database. The database's yearly value is the monthly bill multiplied by 12 months. The yearly water and wastewater bill value was then divided by each city's Median Household Income (MHI) value. The resulting fraction (expressed as a percentage) is the water and wastewater bill expressed in terms of a municipality's MHI. The following table is the summary of our findings.

MPUA's 2016 Rate Survey

Community Size Group (Population)	Number of Communities in Size Group	Surveyed Communities in Size Group	Water Rate (as % of MHI)	Wastewater Rate (as % of MHI)
100 - 999	489	433	1.40%	1.28%
1,000 - 3,499	174	173	1.30%	1.28%
3,500 - 5,999	60	60	1.09%	1.08%
6,000 - 9,999	37	37	0.92%	0.88%
10,000 - 49,999	65	65	0.82%	0.93%
Greater than 50,000	13	13	0.90%	1.17%

The data is compiled in an Excel ™ spreadsheet. The rate survey is grouped by community size for and then alphabetized for quick reference. Rates from private, for-profit utilities were not included in the Survey nor were county public water and/or wastewater districts included.

For most municipal size groupings, effective rates have been on the rise in the past two years. The effective rate considers both MHI increases or decreases and the "actual dollar value" of the rates (in 2016 dollars). The 100 to 999 population size category (with 489 municipalities) and the large cities (with greater than 50,000 people) showed the highest effective rate increases. For the smallest size grouping, effective rates rose 0.08% of the MHI for water and 0.28% of the MHI for wastewater over the past two years. For the biggest 13 cities, effective rates rose 0.13% of the MHI for water and 0.26% of the MHI for wastewater over the same time period.

While this state-wide database is very large (in terms of number of utilities surveyed), it has shortcomings when attempts are made to use it as a comparative tool between utilities. A direct comparison between two municipalities' utility bills, expressed in terms of a percentage of the MHI, can be problematic. The database is ill-equipped to compare cities with large population differences. Generally, large cities have lower rates than smaller cities because they can capitalize on "economies of scale". It is for this reason that cities are grouped by population size.

For water utilities, the type of treatment plays a role in rates. One city may have a surface water treatment plant, while another may have a high-yielding groundwater well. The fiscal needs differ between these two cities. Likewise, from the wastewater perspective, one small city may have an advanced wastewater treatment plant, while another city has a lagoon system. Again, this creates different financial needs for each utility. A city's geographic location can also play a role in rates. For wastewater utilities, a small receiving stream may have more stringent effluent requirements when compared to a municipality located along a major river.

There are several managerial and political factors that must be considered. Each utility funds its operations and maintenance (O&M) activities differently. One utility may have an aggressive O&M program and replace assets before they fail. A neighboring community may employ a "run-to-failure" management style. These policies create utility rate differences. This management philosophy can be established by either the utility manager, the governing body, or both in combination.

Some communities have financed large infrastructure projects by borrowing funds. The repayment of loans has a large impact to existing and future rates. Utility rates can increase steeply to accommodate such payments.

The survey creates interesting opportunities for analysis. Imagine for a moment the asset conditions of a water utility that is charging a rate of less than $10.00 / month for 5,000 gallons of potable water. This equals $0.002 per gallon. There are three such water utilities in Missouri. There are nineteen water utilities that serve 5,000 gallons of water for less than $15.00. Consider the condition of a wastewater utility that provides monthly service for less than $10.00. Seven such utilities exist in Missouri. There are thirty-seven wastewater providers that service for less than $15 monthly. Knowing that this is the National Sewer Collection Systems Conference, let your imagination run wild as you consider what a city's sewer collection system looks like if the utility charges less than $0.50 per day for service.

MPUA's 2016 water and wastewater rate survey is a definitive source of information and sets the cornerstone for rate/affordability discussions. Care must be taken if municipalities or regulatory agencies attempt to use the database as a comparative tool. Rigorous consideration of a myriad of factors must be made before conclusions between different utilities' rates can be determined.

As the national conversation about utility affordability continues, the use of the sewer rate as a percentage of MHI is getting increased attention. There are many in the industry who disapprove of this type of affordability analysis". However, in the absence of other affordability indicators, MHI percentages are a great place to begin the public policy debate about affordability and the long-term impacts to residential customer rates.

St. Louis MSD CSO Volume Reduction Green Infrastructure Program

Melantha D. Norton, P.E.[1], Gary T. Moore, P.E.[2]

[1]Civil Engineer, Metropolitan St. Louis Sewer District, St. Louis, Missouri.
Email: mnorton@stlmsd.com
[2] Program Manager, Metropolitan St. Louis Sewer District, St. Louis, Missouri.
Email:gtmore@stlmsd.com

ABSTRACT

The Metropolitan St. Louis Sewer District (MSD) is a regional Special Sewer District that provides Wastewater and Stormwater services to the City of St. Louis and most of St. Louis County, Missouri. The service area includes a Combined Sewer System (CSS) in the City and the older portions of the County. In 2011, MSD completed a Combined Sewer Overflow Long-Term Control Plan (LTCP), which was formalized in a Consent Decree (CD) with the U.S. Environmental Protection Agency (EPA) in April, 2012. MSD's Combined Sewer Overflow (CSO) control plan includes a $100 million CSO Volume Reduction Green Infrastructure Program in the areas of the City of St. Louis that flow toward the Mississippi River CSOs. MSD spent five years conducting a $3 million CSO Volume Reduction Green Infrastructure Pilot Program, culminating in a CD required final report completed in December 2015. The report explains what work was completed in the GI Pilot Program and the results, and outlines MSD's plan for full implementation of the CSO Volume Reduction Green Infrastructure Program control measure. This presentation will briefly review the GI Pilot Program, with the main focus to discuss the findings of the GI Pilot Program and the plan for full-implementation of the $100 million CSO Volume Reduction Green Infrastructure Program. Included will be major components of the program, the expected reduction in CSO volume and use of an adaptive management approach to manage the CSO Volume Reduction Green Infrastructure Program through 2034.

Implementation of green infrastructure for CSO volume reduction is one of many ways green infrastructure can be strategized for use in urban areas. As a separate entity from the City of St. Louis, MSD's plan highlights the important role that collaboration must have in the implementation of the CSO Volume Reduction Green Infrastructure Program. The plan for full implementation of the CSO Volume Reduction Green Infrastructure Program was developed with critical consideration of public-private partnerships, support of planned use areas and encouraging development and redevelopment in the City of St. Louis. Found to be effective in the GI Pilot Program, these same strategies are applicable to other communities planning and implementing green infrastructure programs.

Keywords: Green Infrastructure, Combined Sewer Overflows, Consent Decree, Long Term Control Plan, Adaptive Management, Public-Private Partnerships

INTRODUCTION

MSD is a regional sewer district, formed in 1954, that provides wastewater and stormwater service to approximately 1.3 million people in a 525 square-mile service area encompassing the independent City of St. Louis and approximately 80% of St. Louis County, Missouri. As of 2013, the collection system owned by MSD consists of 9,578 miles of pipe, making it the fourth largest in the United States. Most of MSD's customers are served by Separate Sanitary and Storm Sewers. However, approximately 75 square miles of the City of St. Louis and adjoining St. Louis County are served by a CSS containing approximately 1,806 miles of sewer.

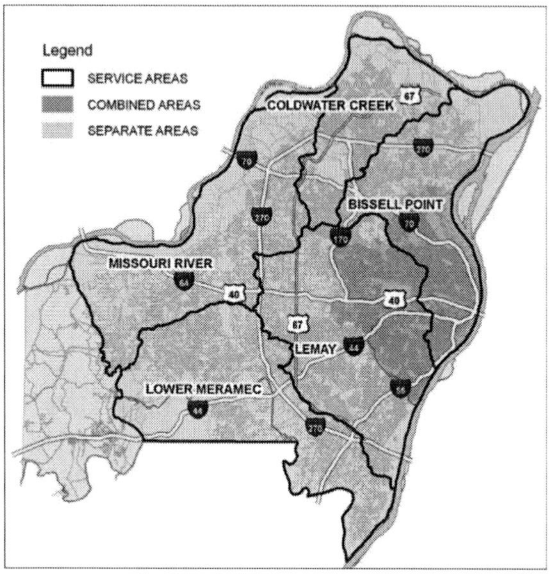

Figure 1-1: MSD's Service Area and Combined Sewer Area

MSD's total area is divided into five major service areas: Bissell Point, Coldwater Creek, Lemay, Lower Meramec, and Missouri River. The Combined Sewer area is located within the Bissell Point and Lemay service areas, as shown in Figure 1-1.

During dry weather, the capacity of the CSS is sufficient so that wastewater is conveyed to MSD's wastewater treatment plants. During heavy rainfall, the combination of stormwater and wastewater may exceed the capacity of the Combined Sewer System. The excess flow, (i.e., the CSO), is discharged directly to the Mississippi River or to one of the river's tributary streams through permitted outfall pipes.

In 1994, EPA issued a CSO Control Policy which required agencies with CSOs to prepare a Long-Term Control Plan (LTCP). The LTCP had to describe how the agency would: 1) ensure that if CSOs occur, they are only as a result of wet weather; 2) bring all wet weather CSO discharge points into compliance with the technology-based and water quality-based requirements of the Clean Water Act; and 3) minimize the impacts of CSOs on water quality, aquatic biota, and human health.

MSD prepared and submitted its original CSO LTCP to Missouri Department of Natural Resources (MDNR) in June 1999. In 2007, the State of Missouri and EPA filed a lawsuit against

MSD regarding overflows. The Missouri Coalition for the Environment later joined the lawsuit as an intervener. Following discussions aimed at resolving conflicts between state and federal law/regulation, EPA and MDNR requested that MSD update its original CSO LTCP. In 2011, MSD prepared an updated CSO LTCP report to describe the development and selection of MSD's plan for controlling CSOs. Along with several components, MSD's control plan included system wide controls and watershed specific controls, including green infrastructure (GI), illicit connect controls, inflow and infiltration (I/I) controls, sewer separation, and storage tunnel controls. This control plan schedules a 23-year investment in the implementation of the selected CSO controls. June 2011 MDNR approved MSD's CSO LTCP.

In August 2011, the Department of Justice filed a settlement, or Consent Decree (CD), requiring MSD to spend a minimum $4.7 billion over the following 23 years to address the issue of overflows and other sewer system improvements. The CD agreement between MSD, EPA, and the Missouri Coalition for the Environment has been accepted by the United States District Court, has been entered by the Court, and went into effect on April 2012.This agreement includes a $100 million commitment to the use of GI as control measure for CSO volume reduction to the Mississippi River.

With a $100 million commitment to green infrastructure in the CD, MSD developed a CSO Volume Reduction GI Program (GI Program) with an overall objective to identify and implement projects and programs that will significantly reduce CSO volumes and pollutants loading to the Mississippi River. At the start of the GI Program, MSD developed a 5 year GI Pilot Program, which would be used to inform the development of the full implementation plan for the program. The GI Pilot Program gave MSD the opportunity to plan for anticipated regulatory, logistical, and financial aspects of the projects among the multiple stakeholders. At the conclusion of the GI Pilot Program, the full-scale program was developed for the full implementation of the GI Program through 2034.

GI can supplement redevelopment efforts, add green space to cities, increase recreational opportunities, increase property values, enhance urban quality of life, and improve human health. The GI Program focuses on a Combined Sewer area within St. Louis County and the City of St. Louis. These same areas represent some of the most economically-distressed portions of the St. Louis community.

This paper will provide an overview of the GI Pilot Program, with the focus to discuss the findings of the GI Pilot Program and the Plan for full-implementation of the $100 million GI Program through 2034. The findings and lessons learned during the GI Pilot Program and applied to the full-scale GI Program will be outlined; along with the CSO volume reduction expected and accomplishments of the full-scale GI Program in the first year of implementation.

GI PROGRAM AREA

During the LTCP development, a cost benefit analysis and water quality benefits for different levels of control was developed. The Mississippi River is not as accessible to the public as urban streams, and CSOs are a much smaller portion of the total river flow; thus the river is less impacted by CSOs than the urban tributary streams. GI in the Bissell Point Watershed, which mostly drains to the Mississippi River, has the additional benefit of partnerships and opportunities to implement green infrastructure. This would directly benefit the city of St. Louis. In addition, EPA was interested in a green component to the Plan. It became clear early on that there was an opportunity to take advantage of the additional community benefits of GI in a part of the City with greater opportunity for redevelopment and creation of green space.

During public outreach sessions, stakeholders voiced a preference that gray infrastructure was focused on urban streams, which are the locations where people live and play. These measures would have measureable water quality benefits and benefits downstream to the Mississippi River. Spending on gray infrastructure in tributaries will achieve significant flow reductions to the Mississippi River as well. Stakeholders voiced the preference for GI in areas tributary to the Mississippi River. These areas are also considered locations with high redevelopment potential. Additional gray infrastructure in the areas tributary to the Mississippi River would yield small benefits to the river, as a relatively small percent of its flow is from CSO volume or stormwater runoff flows. The majority of the Mississippi River flow comes from base flow contributed by tributary streams.

With consideration to public input, water quality benefits, and cost benefit analysis, the Bissell Point Watershed was designated as the GI Program Area. MSD realized that understanding how GI can be best used in this area was critical. In 2008, MSD commissioned a preliminary study of GI in the Bissell Point Watershed to evaluate the potential for GI to reduce discharges from 11 target CSOs, representing approximately 90% of the total average annual CSO volume from the Bissell Point CSS. The study was used to identify GI techniques that could be applicable in the main Bissell Point CSO drainage areas. The study also identified potential retrofit opportunities and their potential to reduce CSO volumes and peak rates.

The study found that the GI techniques that are likely to be most applicable in MSD's CSS areas include green roofs, bioretention, green streets, green parking retrofits, rain barrels, and site-scale & neighborhood-scale stormwater retrofitting. Because of the soil types in the CSS areas, rapid infiltration techniques were not recommended as an effective GI practice. This study has been used to advise the utilization of GI in the Bissell Point Watershed and the development of MSD's GI Program.

GI PILOT PROGRAM IMPLEMENTATION

MSD committed to spending $3 million in the GI Pilot Program. The actual total expenditure was considerably higher in order to maximize construction of GI. The experience gained from the GI Pilot Program was used to inform the full-scale implementation of the $100 million CSO GI Program, update the goal of the GI Program in terms of CSO discharge volumes and establish performance criteria for measuring progress toward the GI Program goals. The major components of the GI Pilot Program included construction and monitoring of site-scale & neighborhood-scale projects, demolitions, and public education & outreach.

Site-Scale & Neighborhood-Scale Projects

During the GI Pilot Program, MSD constructed a total of 35 GI site-scale and neighborhood-scale facilities, as summarized in Table 1-1.

Table 1-1: GI Pilot Program Facility Type Summary

Facility Type	Green Infrastructure Practice	# of Facilities
Site-scale	Planter Box	13
Site-scale	Rain Garden	3
Site-scale	Amended Soil[a]	13
Neighborhood-scale	Bioretention	4
Neighborhood-scale	Bioretention with Detention	2

[a]Amended soil was done across multiple parcels at five locations; each parcel is considered a "facility" for the purpose of this table.

Site-scale facilities are GI facilities that are designed to capture runoff from a single parcel, typically from the roof or other impervious area on the property. Like all privately owned stormwater management facilities in the Separate Sewer Area regulated by the MS4 Program by MSD, these facilities are constructed within a "reserve area". The reserve areas specifies the minimum area need for the GI facility on the site, and is used to ensure the GI facility is sustained. The reserve area is recorded with the property, as part of a Maintenance Agreement, assigning maintenance responsibility to the current owner of the property.

Neighborhood-scale facilities are GI facilities designed to capture runoff from multiple lots as well as adjacent roadways and alleys. MSD acquired the properties for these projects through an agreement with St. Louis Land Reutilization Authority (LRA) and owns and maintains these facilities. The facilities are designed for a future redeveloped condition in the drainage, or tributary area. Therefore, if redevelopment occurs in that area, MSD will consider the stormwater management provided by the neighborhood-scale facility in reviewing the requirements for the development. Besides being an amenity for the existing neighborhood, these facilities were planned to provide incentive for redevelopment due to at least part of the stormwater management being provided by the facility.

Demolition Program

In addition to site-scale and neighborhood-scale GI facilities, demolitions for the purpose of removing impervious surface within the GI Program area and securing land to construct Pilot retrofit projects were a key component of the GI Pilot Program. Since MSD is not part of the City of St. Louis or St. Louis County government organizations, it does not own street right-of-way, parks or other public spaces. Collaboration to address MSDs limited access to property in the GI Program Area for construction of GI facilities became imperative. Therefore, MSD partnered with the LRA which receives the title to all tax delinquent properties not sold at the Sheriff's sale. It also receives titles to properties through donations. The LRA maintains, markets, and sells these properties and performs land assemblage for future development. The LRA owns thousands of parcels throughout the City of St. Louis. In addition, the LRA has an established program for demolition of unsafe buildings on their properties.

Through the partnership agreement with LRA, MSD provided $1.5 million to LRA for their demolition program. In exchange, MSD was able to use these lots, or others as agreed, to construct GI demonstration projects. MSD's partnership with the LRA for implementation of the GI Pilot Program was critical to resolving the logistical and political aspects of locating and constructing demonstration projects.

The $1.5 million in MSD funding for LRA's demolition program resulted in the demolition of 221 buildings on 219 LRA owned parcels. The demolitions not only improved safety, but removed 9.4 acres of impervious surface in the GI Program Area; much of it was directly connected to the Combined Sewer System. In addition to an immediate reduction in the connected impervious surface, MSD was able to build GI facilities on or record deed restrictions (development agreements) against those properties in order to provide lasting control for stormwater volume reduction. The development agreement stipulates that MSD review of development plans is required, future downspouts for roof drainage will not be re-connected to the combined sewer system, sets a maximum area of impervious surface, and specifies a minimum reserve area needed for future MSD funded GI facility to be built when the lot is re-developed.

Public Education & Outreach

A critical component of the GI Pilot Program was the education and outreach efforts. Unlike most gray infrastructure practices, GI practices are very visible facilities in the community. For this reason, community acceptance and support is vital to the success of any GI program. Additionally, to understand other potential community benefits, input from the community is important. MSD reached out to various sectors of the community through a variety of means, including face-to-face meetings, workshops, presentations, website, blogs, social media, press and a targeted public survey. Other efforts included coordination with local government, neighborhood meetings, and educational signs and materials, including ownership and maintenance brochures.

As part of the overall Consent Decree outreach efforts, MSD formed the Stakeholder Engagement Committee (SEC) made up of 51 representatives from economic, religious, community, political, engineering, not-for-profit, and higher education organizations from the St. Louis region. The committee meets quarterly and provides input on all aspects of the Consent Decree efforts, including the GI Program. They also distribute information to their respective communities.

GI PILOT PROGRAM LESSONS LEARNED

One of the major goals of the GI Pilot Program was to inform the planning for implementation of the full GI Program. Lessons learned were considered during the development of the full GI Program Plan. Some of the key lessons learned were in the area of the demolition program, project site selection, GI design and construction, and the importance of public education & outreach.

Demolitions & Project Site Selection

During the Pilot, the parcels with structures to be demolished were selected by LRA, and the demolition list had already been prioritized. MSD was limited to these parcel locations and had to work with parcels scattered throughout the city of St. Louis. Ideally for stormwater collection, parcels that are clustered together, contiguous, or located at low points, would be the best case scenario. MSD had to work within the constraints and priorities of the partnering organization, LRA, to make the best use of various types of GI technologies for both the neighborhood-scale or site-scale projects.

Of the demolition sites, there were only a few locations where the parcels with demolitions parcels are contiguous. Site investigations revealed that several parcels were located between dilapidated houses. It is anticipated that these buildings will eventually be torn down. Due to lack of working room on these narrow parcels, there is a strong probability that if a GI facility was built, it would be significantly disturbed during the demolition of the neighboring parcels. Therefore, the GI that could be built would be subject to impacts from future activities in the block.

During the site selection process for the site-scale projects, an alternative was investigated to disconnect and re-direct neighboring downspouts to a GI facility on the neighboring vacant LRA-owned property. However, in many cases the roof gutters and downspouts on the neighboring house were missing. Replacing gutters on a private residence and re-directing the flow across a property was determined to be problematic. Therefore, existing structure conditions provided limited ability to direct downspouts to green infrastructure. This was considered in the site selection process.

GI Design and Construction Lessons Learned

The final costs of the neighborhood-scale projects varied from 3% to 361% higher than the conceptual cost. The highest change was the smallest facility on Geraldine Avenue. The original plan for that facility did not include a monitoring manhole, as well as a water source which was added to other facilities. If the Geraldine Rain Garden is taken out, the average is a 20% increase in final cost over conceptual cost.

Part of this differential in cost was due to providing a water source at the facilities. Also, the original cost estimating data was associated with construction of stormwater best management practices (BMPs) constructed as part of an overall development typically in green fields. Conversely, MSD constructed the GI Pilot Program facilities individually in a retrofit situation, which is essentially more costly.

In addition to costs, lessons were learned related to GI facility design. The design of the forebay transition area in the bioretention basins raised several questions. On one hand the inflow velocity must be reduced enough for the sediment to drop out; on the other hand, the forebay needs to drain in a reasonable amount of time. Any system that is designed to drain slowly through narrow openings is also prone to clogging. From the GI Pilot Program, a technique that was adopted is the "boulder berm" where water is able to seep between the spaces between the boulders. The natural boulders also complement the plantings and add to the aesthetic rock features incorporated in the design. Early projects experienced some clogging by leaves and mulch if the rocks were too close together. MSD modified the specifications to include a minimum spacing of one inch between the boulders, in addition to the maximum spacing of three inches. Another design feature adopted from the GI Pilot Program was ensuring the elevation in the basin drops after the forebay to ensure flow through the berm.

Excavation and grading of the facilities was complicated by the presence of buried rubble due to previously demolished buildings. The typical standard for demolition by LRA's contractors was followed, which allows for the basement floors and walls to be broken up, and non-hazardous rubble to be used to fill the basement. Only a shallow layer of "clean" fill is required at the top. To accommodate for this in the design and bidding of the projects, the geotechnical investigation used open trench exploration, as well as review of historic "fire marshal maps" to estimate the extents of the building rubble, which was included in the construction specifications. Additionally, the construction specifications required that building rubble encountered during excavation for the bioretention facilities be removed to a depth of two (2) feet below the proposed bottom of the bioretention facility. Geotextile fabric was used to "bridge" over remaining rubble, to obtain smooth and workable grading surfaces. The amounts of rubble excavated were extensive at each site.

A similar situation existed with the amended soil projects. Within the first eight inches of soil, broken pieces of bricks and concrete were common. All of this led to higher excavation costs and higher total project costs.

Public Education & Outreach Lessons Learned

The success of any GI program relies on the practices being accepted by the community. Knowledge and acceptance can unveil new opportunities for incorporating GI into the landscape, and allows a GI Program to build on existing energy to promote sustainable practices into new development and redevelopment practices. LRA was found to be a valuable resource for communication with the City of St. Louis Board of Alderman (BOA) and other City of St. Louis departments. Using the City of St. Louis networks including the LRA and the Neighborhood Improvement Specialists, was an effective way to learn about community meetings and possibly get on the agenda. In order to reach the community in the GI Program Area, considerable time in face-to-face meetings with aldermen and women, community group meetings, individual meetings with interested parties is required to get the most inclusion possible into the GI Program.

Educating private homeowners about GI facilities on their individual properties is a time consuming task with mixed results, and will need to be repeated periodically as property changes hands. Of the strategies attempted, one-on-one time with experts, such as Master Gardeners, in plant design and maintenance, was one of the most effective strategies.

Monitoring Lessons Learned

Several of the monitors utilized a flume for flow measurements, which were susceptible to quick buildup of debris in the flumes and behind the weir, which disrupts the flow elevation. This resulted in unusable data during the initial monitoring set up. To rectify this situation, inspection of the equipment was completed within 24 hours after a storm event. In addition, inspections were made more frequently to ensure that good data would be obtained for use in determination of the effectiveness of the GI facilities.

Monitoring equipment in storm pipes was found to be vulnerable to damage by vermin like raccoons, and could potentially attract mosquitos due to holding water for extended periods.

With consideration of these issues and gathering usable data, MSD also used controlled tests with fire hydrant flow to validate flow monitoring data.

EARLY ACTION PROGRAM

Early Action Grant Program

The CD agreement and the LTCP allowed MSD the opportunity for early implementation of high priority projects before the GI Program Plan was finalized and approved by the EPA. While the core of the GI Pilot Program was focused on stormwater retrofitting in partnership with the City of St. Louis, MSD sought opportunities to build partnerships with other municipalities, schools, community development organizations, and private developers within the GI Program Area. These partnerships provide an opportunity to incorporate GI into ongoing programs and redevelopment projects. One of the main challenges to encouraging implementation of GI on privately-owned land stems from funding and financing barriers (Brown, 2014). To help private land owners overcome the funding and financial barriers and to encourage partnerships and GI practices, MSD implemented the CSO Early Action Grant Program (Early Action Program).

The Early Action Program was developed to reimburse accepted applicants for approved GI elements incorporated on projects within the GI Program Area. From 2012 to 2015, MSD accepted four rounds of applications. From January 2013 to January 2016, MSD submitted 34 projects to the EPA for approval as Early Action projects, committing over $20 million.

MSD's success in the design and implementation of the GI Pilot Program encouraged the submittal of an additional six MSD-acquired Neighborhood-scale facilities to EPA as Early Action Program projects. The project areas were selected from the remainder of the acceptable sites from MSD's site selection effort under the GI Pilot Program. These projects have been constructed and are owned and maintained by MSD.

Rainscaping Small Grants

The GI Pilot Program and Early Action Program generated a lot of inquiries by the general public for installing GI on their individual properties. These small projects did not warrant the level of design and rigorous paperwork required for the Early Action Program projects. Therefore, MSD proposed to EPA in 2013, to spend $250,000 under the Early Action Program to develop a "Rainscaping Small Grants" program, primarily for the purpose of education and outreach. The term "Rainscaping" was developed by MSD's Public Information Office as another name for green infrastructure, so that the public would be able to visualize the effect of green infrastructure on the landscape. Grants up to $3,000 were made available during a "model" round with area churches and schools to help work out the logistics and promote the Rainscaping Small Grants program. The first full round was conducted in the spring of 2015 and the second full round in spring 2016.

The main objective for the Small Grants Program is to encourage the use of simple rainscaping techniques by individual landowners to build capacity for community acceptance. Additionally, this program helps educate the public to understand how stormwater impacts the CSS, and how the community can be part of the solution through the use of rainscaping techniques on their properties. To assist MSD with administrating the grant program, MSD partnered with the

Missouri Botanical Garden (MBG), who conducts a similar grant program in the Deer Creek Watershed in St. Louis County.

GI PILOT PROGRAM MONITORING & RESULTS

The GI Pilot Program required MSD to evaluate project effectiveness by monitoring selected GI projects. For MSD, this meant monitoring stormwater runoff volumes from the project sites and computing estimates of runoff volume reduction based upon hydrologic modeling of the drainage area to each project. The focus of the monitoring was solely on the hydrologic benefit, specifically runoff volume reduction of the GI element. The monitoring program was able to successfully collect data to inform MSD of expected runoff volume reduction from the various GI practices. The data collected during the GI Pilot Program was used to help predict the expected runoff volume reduction value of future GI projects.

The monitoring program was able to successfully collect data to inform MSD of expected runoff volume reduction from the various GI practices. Monitoring results are summarized in Table 1-2. From 2013 to 2015, MSD conducted monitoring of the outflow at seven GI Pilot Program project sites. GI practices at the sites included bioretention, amended soils, permeable pavement, rain gardens, and planter boxes.

Table 1-2: GI Pilot Program Monitoring Summary

GI Practice	Rate of Runoff Volume Reduction (%)
Neighborhood-Scale Bioretention[c]	96%
Site-Scale Rain Garden	95%
Amended Soils[a]	N/A
Permeable Interlocking Concrete Pavement (PICP) Blocks[c]	92%
Site-Scale Planter Box[b]	55 - 65%

[a] Due to the limited quantity of available data, MSD was not able to make prediction of the potential runoff volume reduction for amended soil practices.

[b] Data from the site-scale planter boxes was collected and evaluated by the Department of Civil Engineering at Southern Illinois University Edwardsville (SIUE). Findings of the SIUE study have been published in the Proceedings of the World Environmental and Water Resources Congress in 2014 and 2015 (Bloorchian, et al., 2014 and Bloorchian, et al., 2015).

[c] Statistical models were created for the bioretention sites and the permeable pavement project from the monitoring data to estimate runoff reduction

Generally, for each GI practice which yielded monitoring results for runoff volume reduction, it was found that performance is very good and higher than typically reported in other literature (e.g. Davis 2008, Hunt et. al. 2006, Li et. al. 2009). Recognizing variability of soil conditions and changes in performance over the life cycle of the GI practices, MSD decided to use an annual runoff volume reduction during the typical year of 70%. MSD also performed hydrant tests which flooded the bioretention basins with a volume greater than any of the observed events with volume reduction ranging from approximately 70% to 100% of the volume. Because MSD had monitoring data suggesting that all practices monitored can perform at levels much greater than 70%, using it is considered a reasonably conservative value that can be uniformly applied across GI Program Area.

Runoff Retention

The GI Program directly increases runoff retention for the drainage areas managed. The captured runoff is partially or completely infiltrated to the ground, permanently removing it from entering the CSS. This has the effect of reducing the discharge volume of CSOs. MSD has experience implementing runoff volume reducing Best Management Practices (BMPs) and GI in the Separate Storm Sewer System areas and select CSS areas since 2006. MSD has rules and regulations governing the design standards of such BMPs. The current design standard, which is intended to treat the water quality storm, or the 90% non-exceedance 24 hour storm depth of 1.14 inches, was used in the GI Pilot Program. This design standard targets the removal of suspended sediments and is essential to water quality within the Separate Storm Sewer systems. The current design standard is intended to mimic natural hydrology and minimize increases in annual runoff volume from new development.

In the final Plan and GI Pilot Program Report it is concluded that the design of GI for the 90th percentile storm, or water quality storm, is not an essential priority for CSS areas as low flows will be treated at a wastewater treatment facility, thus meeting water quality standards. With the overall goal of CSO discharge volume reduction as the focus, the largest benefit to the CSS will come from maximizing drainage area managed. The particular design storm depth is of secondary importance. Because urban environments have limited available space for GI facilities, in order to maximize area managed without exceeding the GI footprint allotted to a particular site, it may be desirable for the design storm depth to be reduced. Runoff volume retention during the more frequent smaller wet weather events could have significant benefits for reducing CSO discharge volume. The Plan supports reviewing GI design standards within the Combined Sewer areas such as to permit alternative designs to maximize drainage and impervious area managed for consideration over the GI Program timeline.

Peak Flow Control

GI also has the ability to limit the rate at which flow can enter the sewer system. A reduction in peak flow has the benefit of reducing the required conveyance capacity of a sewer and can reduce the frequency and discharge volume of CSOs. This function can be realized at all GI facilities and is an important benefit which may be significant even without runoff volume reduction.

Relating Runoff Reduction to CSO Volume Reduction

One significant purpose and outcome of the GI Pilot Program is the development of a technique to connect GI practices to the CSO LTCP collection system model. To do this MSD staff used the collection system model developed for use in the LTCP to compute the effect of GI on the CSO discharge volume during the typical year. The model represents the actual geometric data and hydrologic parameters at the time of calibration in 2006 and represents the most current and comprehensive model of the collection system. MSD staff identified within the model 12 geographical regions served by CSS draining to the Mississippi River which are both hydraulically and hydrologically independent of each other, meaning the systems had no cross-connections in the sewers or intersecting overland flow paths. For each of the 12 regions, not including Lemay North, the runoff and CSO volumes during the typical year rainfall were related to a model land cover parameter. Relating runoff and CSO volume to a model land cover parameter assisted MSD in connecting GI practices to the collection system model similar as with other communities (EPA 2014).

MSD used the XP-SWMM model to perform a set of systematic numerical simulations which were used to define relationships between stormwater runoff depth, percent of effective impervious area, and CSO volume. Combining these relationships with observed runoff volume reduction from GI, MSD derived a scaling parameter which estimates CSO volume reduction per acre managed by GI. The parameters, presented in Table 1-3, have been termed an efficiency parameter by MSD because it can be interpreted as representing the effectiveness of GI practices in reducing CSO discharge volume for a particular CSO Region. The efficiency parameter represents the CSO volume (in million gallons) reduced per acre of drainage area in the region that is managed by GI facility or acre of impervious area removed.

The equation developed from the analysis linked GI area managed to CSO discharge through runoff reduction as shown in Equation 1-1. Where Am is the total managed area of the particular CSO region, and ε is the CSO discharge volume reduced in million gallons per acre managed by GI practice.

$$Vr = -\varepsilon * A_m \quad \textbf{(Equation 1-1)}$$

Table 1-3: Efficiency Parameter for CSO Regions

CSO Region	Efficiency Parameter (MG/Acre)	Normalized Efficiency Parameter
Central	0.320	1.385
Gimblin	0.271	1.176
Southern Arsenal	0.250	1.084
Utah	0.247	1.071
Louisa, Trudeau, Barton	0.238	1.033
Entire Mississippi River CSO Drainage	0.231	1.000
Dakota	0.208	0.903
Rocky Branch, Salisbury	0.198	0.860
Harlem Creek	0.186	0.806
Humboldt, Baden	0.182	0.788
Ferry	0.135	0.585
Prairie	0.129	0.561
Biddle, Chambers South Benton	0.126	0.548

As presented in the final Plan, this approach developed by MSD is considered conservative in two ways. First, it assumes a mixed land use which is neither completely impervious nor completely grass for the area draining to a GI facility. MSD believes that a large portion of projects completed will take runoff from primarily impervious surfaces, and thus the net impact on CSO discharge volume reduction would be greater than predicted using the described approach. Second, it does not take advantage of the peak flow reduction inherent in GI facilities.

MSD believes that this benefit will be significant, but accounting for it given the uncertainty in both facility type and location in the future is difficult. Because of the uncertainty in estimating the impact of peak flow reduction on the CSO outfalls, MSD chose to only take credit for runoff volume reduction in GI facilities in the final report for the GI Pilot Program.

Predicted CSO Volume Reduction from GI Pilot Program and Early Action Program
Using Equation 1-1, the CSO volume reduction was calculated for the GI Pilot Program projects and the Early Action Program projects. Table 1-4 summarizes the CSO volume reduction calculations.

Table 1-4: Calculated CSO Volume Reduction for GI Practices

GI Practice	Calculated CSO Volume Reduction
GI Pilot Program	4 MG
Early Action Program (MSD-owned)	1.6 MG
Early Action Program (non MSD-owned)	1.4 MG[a]

[a]CSO volume reduction calculated for projects fully completed and reported to MSD as of September 2015.

GREEN INFRASTRUCTURE FULL PROGRAM PLAN COMPONENTS

The full GI Program period began in 2016, with approval of the Plan by EPA, and will be completed by June 30, 2034. Including the $3 million used to fund the Pilot, MSD has committed to spend a total of $100 million on GI under the GI Program. The full GI Program plan is composed of four major components:
1.) Indirect Projects
2.) Direct Projects
3.) Demolitions
4.) Small Grants

Indirect Projects
Indirect projects are projects that MSD provides partial or full payment for the design and construction of GI components that have CSO volume reduction benefits. The property owner will maintain these GI practices as required through a maintenance agreement with MSD. These are planned to be realized through public-private partnership projects via MSDs Rainscaping Grant Program for large-scale projects.

The Early Action Program set a framework for opportunities to build public-private partnerships for the incorporation of GI in future private development and redevelopment projects. While the cost of the public-private partnership projects was higher than the MSD Pilot Projects, these projects are still considered a preferred and valuable option for investing in green infrastructure. Being completed in conjunction with development, MSD believes this can increase the longevity of the green infrastructure as it will have higher visibility as part of the development's landscape. Investing in green infrastructure where development is occurring is in line with the triple bottom line of sustainability and green infrastructure. The private-public partnerships allow MSD to take advantage of current opportunities and provides for the social and economic needs of the community.

MSD will continue to offer reimbursement for specified GI elements that are eligible for funding using two methods. The first method is through a process streamlined as part of MSD's private development review process. MSD will facilitate the application process with developers as they submit development projects which incorporate GI within the GI Program Area. As projects are submitted, MSD will communicate to the developer and project owner the intent of the GI Program and offer to reimburse them for the GI elements of their project that will reduce CSO volume. This program reimburses up to a max rate of $180,000 per acre of total tributary area to the approved volume reducing GI features.

The second method used for establishing public-private partnerships for indirect projects is through an application process similar to that of the Early Action Program. The application period will be open as funds are available at least on an annual basis. The intent of the application process is to take advantage of partnerships which might not be available otherwise such as with non-profit organizations and community organizations which would not have funding for a project without the partnership. Additionally, MSD believes more innovative technologies will be used through the application process than might be used in a more traditional private development submittal. The application process will use the project ranking and benefit criteria developed for the Early Action Program and will reimburse up to 100% of GI costs for approved projects.

Direct Projects

Direct projects (e.g. neighborhood-scale bioretention projects) are those implemented by MSD staff and/or consultants on real estate obtained and owned by MSD. Direct projects are anticipated to be an important part of the Plan to reach the Program goals for spending the $100

Figure 1-2: Neighborhood-Scale Bioretention Basin - Clinton Street and Geraldine Avenue Rain Gardens

million before the critical milestone date. MSD will work with the City of St. Louis and LRA on opportunities for purchasing land feasible for GI projects. MSD also plans to review LRA vacant lots for these opportunities.

MSD has 46 existing facilities in the Mississippi River CSO regions, including pump stations, a wastewater treatment facility and an office building. Flow management at some MSD facility buildings and sites can reduce the amount of storm water entering the CSS with GI at these locations. While there are 46 existing facilities, it is not practical to implement GI practices at the majority of these locations. The MSD headquarters is considered to be the best opportunity for incorporating GI at an existing MSD-owned facility. To date, MSD has not designed or constructed projects incorporating GI practices at MSD facilities within the GI Program area.

Detention basin retrofitting, where beneficial, is a feasible option as a green practice for reduction of CSO volume. Basins designed as dry extended detention basins promote infiltration of a portion of the volume into the soil, achieving a runoff volume reduction of nearly 30% in some dry basins (Fernandez, 2012). MSD owns a number of detention basins constructed under a flood mitigation program called the Cityshed Mitigation Program. The Cityshed Mitigation Program areas can also present excellent opportunities for GI to be co-located, particularly when basins cannot be most effectively constructed on the property. There is an opportunity to incorporate GI elements into projects, such as converting buyout properties to green space or installing bioretention on those properties found to be unsuitable for flood control structures. In addition, detention basins can incorporate designs that naturalize the basins. These project types, within the GI Program Area, can result in CSO discharge volume reduction and contribute to CSO volume reduction goals.

Demolitions

Figure 1-3: Example of LRA vacant buildings for demolition

In partnership with the City of St. Louis Planning Department, St. Louis Land Clearance for Redevelopment Authority (LCRA), and the Missouri Department of Conservation (MDC), MSD will strategically select vacant LRA buildings for demolitions. Demolitions can provide desirable locations for GI practices and can yield long-term volume reduction when maintained as green space or when runoff from future development is treated through other GI practices. These parcels will be maintained as permanent green space, or will require a deed recorded agreement to provide for future impervious surfaces to be treated, as in the GI Pilot Program. MSD has agreed to limit the total expenditure on demolition in this category to 15% of the $100 million program providing for an additional $13.5 million in the Full Program Plan.

The City of St. Louis intends to turn its many vacant parcels from a liability into one of its greatest assets. Also, from a health and safety perspective for residents, the City of St. Louis plans to tear down and/or redevelop hazardous vacant buildings, determine the optimal density for sustainable neighborhoods, and strategically consolidate vacant and open space for infill housing. Many plans exist and more are under development. The funds provided by MSD help further put the plans to work through implementation.

Small Grants Program

The Rainscaping Small Grants Program encourages the use of simple rainscaping techniques by individual landowners. The program's focus is to help educate the public on how stormwater impacts the CSS and associated issues. It will continue to engage the community and help them understand how they can be a part of the solution through the use of rainscaping techniques on their properties.

The Rainscaping Small Grants Program provides workshops and more in-depth education on the rainscaping topic. This level of outreach and education is important for the long term sustainability of the program to ensure the practices are acceptable to the community for implementation and then maintained as an amenity over time. The full program commits to continuing this program up to a maximum expenditure of $4 Million over the course of the GI Program.

PROGRAM PROJECTED TIMELINE & EXPENDITURES

For planning purposes, MSD has outlined specific near term actions within a 5-year timeframe, and plans to manage the balance of the GI Program over the long term through adaptive management. During the near term period, MSD has secured funding for the GI Program at the rate of $5 million per year through the completion of a 4-year Rate Commission process for fiscal year 2017-2020, whereby the public has provided input for the rates to be set by MSD to collect revenue for the wastewater program.

Adaptive Management

MSD understands that with a 23-year program, things will change over time due to the duration, for example governmental changes within the City, grant program participation or MSD funds. In the long term period, MSD will continue to secure and promote the funding through the Rate Commission process to complete the $100 million GI Program. The planned expenditures and time schedule for implementation were developed using knowledge gained from the Pilot and accounting for anticipated opportunities. Nevertheless, the planned project expenditures and time frame, as shown in Figure 1-4, are subject to change if MSD finds that opportunities are different than anticipated or prompted by other adaptive management changes.

As the GI Grant Program is one of the primary mechanisms to reach the GI Program goals, a major part of MSD's adaptive management strategy is to evaluate the structure of the public-private partnerships. MSD will consider changes to the program if it is found that participation is not consistent with long term goals and performance criteria. In the event that milestone goals are not met, MSD will focus on planning more direct projects and also explore new opportunities to acquire land.

Measure	Planned Expenditure (Million Dollars)*	Time Frame for Implementation*
		2012 - 2034
Pilot Program	$3	
Indirect Projects		
Grant Program	$40 to $60	
Targeted Partnerships	$10 to $20	
Direct Projects		
Neighborhood-Scale	$10 to $15	
Cityshed Mitigation	$5 to $15	
MSD Facilities	≤ $5	
Demolitions	≤ $15	
Education and Outreach		
Small Grants Program	≤ $4	
Cityshed Mitigation Program	$230	

*Planned Expenditures and Time Frames are subject to available opportunities and adaptive management changes

Figure 1-4: Initial CSO GI Program Plan Planned Expenditures and Time Frame

CSO GI PROGRAM PERFORMANCE GOALS

Including the 33 acres accomplished during the GI Pilot Program, MSD expects to manage a total of 405-500 acres with GI. With an estimated 50% impervious area within the GI Program Area, MSD expects this to include a minimum of 202 to 250 acres of impervious area. Utilizing the expected performance of GI from the hydrologic/ hydraulic model, between 90-141 MG of CSO discharge volume are expected to be removed during the typical year. MSD is committed to meeting a minimum of 405 acres managed and $100 million in eligible expenses by June 30, 2034. MSD will report expenditures and acres managed on an annual basis to the USEPA.

GI FULL PROGRAM – YEAR 2016-2017 ACCOMPLISHMENTS

Urban Greening Program

The demolition program for the full GI Program has been branded to the public as the Urban Greening Program (UGP). Planning for demolitions in partnership with LCRA, MDC, and the St. Louis Planning Division began in early 2016. It is important to MSD, from a socio-economically perspective, that the demolition dollars be used to support current federal initiatives and local city planning and revitalization efforts. The $13.5 million allocated for this program is being used to directly provide the funds that will achieve the strategies envisioned during the planning with the developed partnership.

MSD spent Year 1 of the full program coordinating and continues to coordinate with the City of St. Louis, LCRA, and MDC to identify vacancy clusters for demolitions sites. This approach is focused on avoiding a "patchwork" approach to the demolition locations and focuses on clustered demolitions to create permanent green space or large areas for potential redevelopment. With concerted effort from MDC and the St. Louis Planning Division, vacancy clusters have been created by aggregating a group of vacant parcels that are separated by no more than 20 feet.

Vacant parcels in the clusters include vacant lots with vacant buildings owned by LRA, lots owned by MSD, privately-owned vacant lots and lots with vacant buildings. Priority scores were then given to the vacancy clusters, scoring on cluster size, hydrology, proximity to MSD trunk sewers, roads and green corridor vicinity. These clusters were then analyzed to create a list of top ranking vacancy clusters by looking at if the cluster met three criteria: 1) is within the GI Program Area 2) is 2 acres or larger and 3) is in the top 35th percentile of priority scores city wide. Consideration was also given to the cluster not containing a historic building. Using these criteria over 1,300 parcel lots have been identified for demolition locations.

MSD held a press event announcing the UGP, during the demotion of the first 11 demolitions in early 2017. The UGP is projected to produce at least 50 acres of permeable ground, reducing stormwater runoff in the combined sewer system, ultimately reducing sewer overflows to the Mississippi River.

Rainscaping Large Scale Grant Program

Rainscaping Large Scale Grant Program – Annual Call for Applications

The fifth round of large scale grants now referred to as the Rainscaping Large Scale Grant Program – Annual Call for Applications was completed. The previous four rounds were under the Early Action Grant program title. In this round, MSD awarded 9 grants totaling just over $1.4 million. This brought the total commitment to $22,600,000 approved for grant funding.

Rainscaping Large Scale Grant Program – Development Review

MSD successfully developed and implemented a process and internal procedures for a streamlined reimbursement program within MSD's normal Development Review process, to reimburse projects for CSO volume reductions at a specified rate. The program is set to reimburse at a rate of $180,000 per acre of total drainage area to the GI. This rate is the median cost that was incurred during the Pilot (EAP and MSD owned projects). The 405 acres managed that MSD committed to is based on spending $250,000 per acre managed by GI, assuming that costs will be higher later in the program. This current reimbursement rate allows for some flexibility and adaptive management of the program.

Rainscaping Small Grants

MSD successfully completed the second full round of the Rainscaping Small Grants Program. This program continues to be administered by the Missouri Botanical Garden through an agreement with MSD. During this round MSD awarded 66 grants totaling over $165,000. MSD has continued to see increasing interest in the Rainscaping Small Grants Program. This round brought the total commitment to the small grants to $321,100 with 125 grants awarded.

Direct Projects

From continued analysis of existing LRA vacant lots, MSD has identified additional neighborhood-scale bioretention project sites. These projects are planned to begin design in 2017 and to be constructed in 2019. MSD is continuing to analyze existing vacant LRA owned lots for future projects. MSD projects to spend $2-$2.5 Million annually on direct projects in the first 5 years of the GI Program.

One concern for the Neighborhood-scale facilities in the GI Pilot Program is that redevelopment activities in the drainage, or tributary area might lead to construction sediment flowing into the facility. Therefore, knowledge of planned construction in the area is essential to ensuring the facilities are protected through erosion and sediment control and possibly blocking the inlets to the facilities during construction. Development agreements will continue to be used on tributary lots for this reason. For these parcels, MSD will review planned projects to ensure the planned imperviousness is within the amount assumed in design, and can be prepared for possible construction sediment. All development and redevelopment plans should be submitted to MSD.

Based on an analysis completed by MDC, the twelve neighborhood-scale bioretention projects completed during the Pilot have not spurred development in the areas to date. Five building permits have been issued within the twelve drainage areas as of February 2017. None of these were for new construction. Two were for exterior alteration, one for fence construction, one for porch repair, and one for converting a two-family unit to a single family unit. The first neighborhood-scale bioretention project was completed in Fall 2012 and the last in Fall 2015.

With respect to the concern of construction sediment and the findings from the analysis done by MDC, MSD is planning to look at more established residential tributary areas for locating neighborhood-scale bioretention projects. These projects would be considered an amenity, transforming a vacant lot or neighborhood eyesore to a green space also attracting birds and butterflies to the area. This would be in lieu of the initial focus during the GI Pilot Program of these projects spurring or incentivizing development.

MSD also completed a conceptual design and cost estimate for incorporating GI at the MSD headquarters building as a direct project. The conceptual design analyzed the cost of bioretention, a green roof, and pervious pavement as GI elements.

Education and Outreach

MSD has continued updating its websites with GI Grant Program information. In addition, the MSD BMP Toolkit website has also been updated with program information and documents. MSD has meet with numerous interested grant applicants and provided information to the engineering community on the potential for public private partnerships on development projects.

CONCLUSIONS

MSD successfully implemented the GI Pilot Program spending over the committed amount of $3 million and accomplishing a 4 MG CSO volume reductions, and an additional 3 MG from Early Action Program projects. Several opportunities and methods of delivery for GI were demonstrated and proven effective during the GI Pilot Program. The technology, logistics, social and political implications were successfully navigated to meet the GI Program goals. In addition, monitoring efforts confirmed better than expected results. MSD is excited about the Green Infrastructure Plan for the future, and is optimistic of its continued success.

With consideration to lessons learned during the GI Pilot Program, the major components of the full GI Program are indirect projects, direct projects, demolitions, and small grants. With over $20 million committed to public-private partnerships for indirect projects, MSD will rely heavily on these projects to meet CSO volume reduction goal through the GI Program. MSD will continue to employ an adaptive management approach to modify the program as necessary.

MSD has set out an optimistic projection for the full GI Program. While the planned expenditures and time schedule for implementation was based on knowledge from the GI Pilot Program, MSD will continue to review annual progress toward the long-term performance goals. Investments through the GI Program are believed to provide socio-economic and environmental benefits. However, long term maintenance will be important to ensuring the sustainability of all projects completed as part of the GI Program. Projecting life cycle of GI for long term planning continues to be a concern, and will be noted over the duration of the GI Program.

As MSD moves forward with the full GI Program implementation, education will be vital to convey the benefits of the GI Program to the public. The full GI Program will promote the social and economic benefits that GI practices have and will continue to add value to the City of St. Louis and its residents. It is understood that changes in City government structure could also potentially impact some components of the GI Program. The changing fabric of the City government will require continuous education of these benefits as well.

The development and implementation of the GI Program is a result of successfully bringing organizations together to meet collective goals. Many of these organizations had not worked together prior to the GI Pilot Program, and were able to come together to navigate the logistics of creating a partnership with the City of St. Louis, LCRA, LRA, MDC and MSD. The City of St. Louis has shown a great commitment to planning efforts and moving forward. Collaboration has been and will continue to be a critical element of the Program's continued success. Collaborative mindsets and positive energy related to redevelopment in the City of St. Louis has created an ideal landscape for the GI Program and MSD is excited to be a part of this momentum.

ACKNOWLEDGEMENTS

The authors gratefully acknowledge the contributions of those who were involved in the development of the final GI Program Plan and implementation of the GI Pilot Program:

MSD Team included personnel from - Planning, Project Design, Development Review, Division of Environmental Compliance, Operations, Flow Monitoring, GIS, Right-of-way, Public Information, Information Systems, Legal

City of St. Louis included personnel from – multiple City of St. Louis agencies, especially the Department of Planning and Urban Design, Land Reutilization Agency, Land Clearance for Redevelopment Authority, St. Louis Development Corporation

Missouri Department of Conservation

Consulting Team included personnel from - Jacobs, LimnoTech, M3 Engineering Group, SWT Design, Tetra Tech, Arcturis, Southern Illinois University Edwardsville, Reitz & Jens, ADS Environmental Services

REFERENCES

Brown, Jeanette. *Green infrastructure implementation : a special publication*. Alexandria, VA: Water Environment Federation, 2014. Print.

Davis, Allen P. "Field performance of bioretention: Hydrology impacts."*Journal of Hydrologic Engineering* 13.2 (2008): 90-95.

Fernandez, R. *Design of urban stormwater controls*. Alexandria, Va. New York: WEF Press/Water Environment Federation McGraw-Hill, 2012. Print.

The Metropolitan St. Louis Sewer District. Combined Sewer Overflow Long-Term Control Plan Update Report, Metropolitan St. Louis Sewer District (February, 2011) available at http://www.stlmsd.com/sites/default/files/education/448861.PDF

The Metropolitan St. Louis Sewer District. Combined Sewer Overflow Long-Term Control Plan Update Report, Metropolitan St. Louis Sewer District (February, 2011) available at http://www.stlmsd.com/sites/default/files/education/448861.PDF

The Metropolitan St. Louis Sewer District. CSO Volume Reduction Green Infrastructure program: Pilot Final Report and Full Implementation Plan. (April, 2016) available at http://www.stlmsd.com/sites/default/files/engineering/GI%20Pilot%20Final%20Section%201.pdf

U.S. EPA (2014). "Greening CSO Plans: Planning and Modeling Green Infrastructure for Combined Sewer Overflow (CSO) Control." Publication No. 832-R-14-001

Johnson County Wastewater Collection System Asset Management Program – A Road Map for Continuous Improvement

Ryan Eisele[1*], Joe Barnes[2], Patrick Beane[2], Dave Spencer[3], Peter Moody[1]

[1] HDR, Lee's Summit, Missouri
[2] Johnson County Wastewater, Johnson County, Kansas
[3] HDR, San Diego, California

ABSTRACT

As the Johnson County Wastewater (JCW) collection system and workforce continue to age, investments in maintenance, repair, rehabilitation, and knowledge transfer to sustain desired service levels and risk tolerances will continue to grow. Data from JCW's information systems has been used to determine the financial and risk management implications of various investment scenarios, develop and calibrate decision models, and develop a consistent, transparent and defensible decision making and prioritization process for sustaining collection system levels of service. This effort has enabled JCW to forecast maintenance, condition assessment, and renewal investment needs with a high level of confidence, justify appropriate investment levels, focus limited resources, and facilitate knowledge transfer.

Increasing investment needs, coupled with limited resources, has driven JCW to continuously improve the efficiency and effectiveness of service delivery while continuing to execute day-to-day work functions. To meet this challenge, JCW and HDR staff collaborated in the development and execution of JCW's Collection System Asset Management Program (CSAMP) Implementation Plan which establishes a clear, practical, and strategic path forward. This plan identifies, prioritizes, coordinates, and schedules continuous improvement initiatives at a manageable pace that strives to balance staff availability and continuous improvement objectives.

KEYWORDS: Asset Management, Collection Systems, Gravity Sewer Infrastructure, Rehabilitation Decision Making, Data-Driven Investment, Cleaning, Condition Assessment, CCTV, Confidence Level, Manhole Inflow, I/I Reduction, Deterioration Rate

INTRODUCTION

Johnson County Wastewater (JCW) owns and operates approximately 2,200 miles of gravity pipe and 60,000 manholes. As JCW's collection system and workforce continue to age, investments in maintenance, repair, rehabilitation, and knowledge transfer needed to sustain the current high level of service will continue to grow. Increasing investment needs, coupled with limited resources, has driven JCW to continuously improve the efficiency and effectiveness of service delivery while continuing to execute day-to-day work functions

JCW needed to develop a sustainable, cost effective strategy for managing aging sewer infrastructure. To meet this challenge, JCW and HDR staff collaborated in the development and implementation of what was originally called the Gravity Sewer Asset Management Program (GSAMP) which establishes a clear, practical, and strategic path forward. In 2016, forcemains and low pressure sewers (LPS) were added to the program, and the program's title was changed to the Collection System Asset Management Program (CSAMP). Hereinafter the program will be referred to as the CSAMP. The initial goals of the program were as follows:

- Development and implementation of an assessment, maintenance, and rehab/repair program based on a Business Risk Exposure (BRE) Model.
- Use data effectively to drive decisions and increase confidence in CIP and O&M investment projections in the JCW collection system.
- Refinement of decision making criteria focused on sustaining desired service levels and risk tolerances while minimizing long term costs.
- Development of a data driven 10 year budget projection for collection system maintenance and renewal with high confidence.
- Development of a realistic program Implementation Plan to strategically plan, coordinate, and implement continuous improvement and knowledge transfer over the next several years.

The CSAMP represents JCW's road map to continuous improvement in collection system asset management. This paper describes how JCW is using the CSAMP to meet these goals of developing and implementing a plan that will sustain desired long term service levels while addressing the challenges of an aging workforce, aging infrastructure, and limited resources.

COLLECTION SYSTEM ASSET MANAGEMENT PROGRAM DEVELOPMENT

Initial Assessment and Program Development

The development of the CSAMP began in the spring of 2013 with a thorough assessment and documentation of JCW's current collection system strategies, practices, and procedures. JCW had a strong foundation of strategies, information systems, and institutional knowledge from which to build. Existing informal decision making processes were documented and refined using lessons learned, industry best practices, and JCW's institutional knowledge.

The assessment findings were organized by the following core areas:

- Cleaning activities
- Pipe condition assessment and renewal activities
- Manhole inspection and renewal activities
- Information and data management systems integration

Detailed flowcharts and summary tables were developed for each of the items above to document the current workflow and decision making processes used by JCW. These were used as a starting point for the documentation of future workflow refinements and to facilitate the transfer of knowledge when employee succession occurs.

During interviews with JCW staff, over one hundred individual opportunities for improvement were identified based on staff knowledge and industry best practice. These opportunities formed a vision for how JCW would provide sustainable and cost effective service as the workforce and system continue to age.

Implementation Plan Development

The initial list of opportunities for improvement was refined and expanded over the course of the first year of the program as other key development tasks were completed. Opportunities for improvement were organized by the following core areas:

- Cleaning
- Backups/Overflows
- CCTV (Condition Assessment)
- Renewal Decision Making
- In-house Repair
- Contractor Rehab
- Manhole Inspection
- Manhole Renewal
- Data Management and Systems
- Overall Collection System Maintenance Program
- Lateral Program

The Assessment Team and JCW then further evaluated and prioritized these opportunities during a series of workshops in the fall of 2013. During this process, opportunities were either accepted or rejected. The accepted alternatives were prioritized based on order of importance and the desired time frame for implementation. Some of the opportunities were determined to be relatively low effort items that could provide immediate benefit to JCW's business practices. These low effort items, referred to as "low hanging fruit", were implemented by JCW during the first year of the program.

After developing a better understanding of the type and quantity of work that would be needed in the future to sustain the desired level of service, the next step was to develop a clear, practical, and strategic implementation plan. Based on resource availability, it was estimated that these initiatives would take multiple years to implement. Therefore, JCW prioritized each of these opportunities while considering the level of effort, timeliness, and importance of each opportunity. The project team then identified related and dependent opportunities and grouped these opportunities into initiatives. An example of related opportunities comprising an initiative is shown below in Figure A.

Initiative 7 Cleaning and CCTV Standard Operating Procedures (Phase 1)					
Opp. ID	Opportunity Title	Opportunity Description	Start Date	End Date	Champion
8	Cleaning SOP (Office)	Develop SOPs for cleaning supervisor and/or non-crew cleaning staff that may include ad-hoc Work Order (WO) intake, WO screening process, WO creation process, WO planning process, backlog management process, WO prioritization and assignment, WO reassignment, pre-job prep, field visits, performance feedback, data quality checks, post-job follow-on activities, and dependent work order management (i.e. CCTV/Cleaning/Locate).	1/1/14	4/30/14	Collection Systems Superintendent
9	Cleaning SOP (Crew)	Document/Compile Standard Operating Procedures (SOPs) for cleaning crews that may include safety procedures, quality standards, productivity standards, tools & equipment needed, nozzle selection guidelines, sequential steps to perform work, data collection standards and procedures, nozzle and equipment testing standards and procedures, vehicle inspection and maintenance procedures, and other critical information. The document should be written to the level of detail to be used for training purposes.	1/1/14	4/30/14	Collection Systems Superintendent
12	Cleaning Findings Definition	Define a consistent and documented methodology for categorizing cleaning findings.	1/1/14	4/30/14	Collection Systems Superintendent
211	CCTV SOP (Office)	Develop SOPs for CCTV supervisor and/or non-crew CCTV staff that may include ad-hoc WO intake, WO screening process, WO creation process, WO planning process, backlog management process, WO prioritization and assignment, WO reassignment, pre-job prep, field visits, performance feedback, data quality checks, post-job follow-on activities, dependent work order management (i.e. CCTV/Cleaning/Locate) and other critical information. Include prioritization of CCTV work versus other activities.	1/1/14	4/30/14	Collection Systems Superintendent
212	CCTV SOPs (Crew)	Document Standard Operating Procedures (SOPs) for CCTV crews that may include safety procedures, productivity standards, tools & equipment needed, sequential steps to perform work, data collection standards and procedures, vehicle inspection and maintenance procedures, and other critical information. The document should be written to the level of detail to be used for training purposes.	1/1/14	4/30/14	Collection Systems Superintendent

Figure A JCW Gravity Sewer Asset Management Program Initiative

The following items were defined for each initiative:

- Initiative ID
- Purpose
- Outcomes/Deliverables
- Initiative Champion
- Other Team Members
- Level of Effort (Internal and External)
- Schedule
- Schedule Flexibility
- Notes and Assumptions
- Initiative Scope/Opportunities for Improvement

Finally, based on the level of effort, resource availability, and priority of each initiative, an implementation schedule was developed. This figure details the following:

- The related JCW Strategic Business Plan goals that are supported by the initiatives
- Implementation plan schedule and dependencies between initiatives
- Initiative champions
- Key program accomplishments

The CSAMP Implementation Plan defines a realistic road map for the refinement of JCW's strategies, practices, and procedures to support continuous improvement related to aging gravity sewer infrastructure management. The CSAMP Implementation Plan is a living document that is updated over time as initiatives are completed, lessons are learned, changes or adjustments to previously identified initiatives are needed, and new challenges and opportunities present themselves.

The Implementation Plan has continuously evolved over time to meet the challenges JCW faces. For example, prior to 2016, forcemains were not a focus of the program. However, a recent increase in the rate of forcemain breaks on high consequence of failure lines led JCW to place a greater emphasis on developing a long term plan to manage these assets. Forcemains were added to the program in 2016 and a series of continuous improvement initiatives were developed.

Figure B presents the status of the implementation plan as of April 2015. Figure C presents the current status of the implementation plan, two years later in March of 2017. These figures illustrate how the plan evolves over time to meet JCW's challenges and address new opportunities.

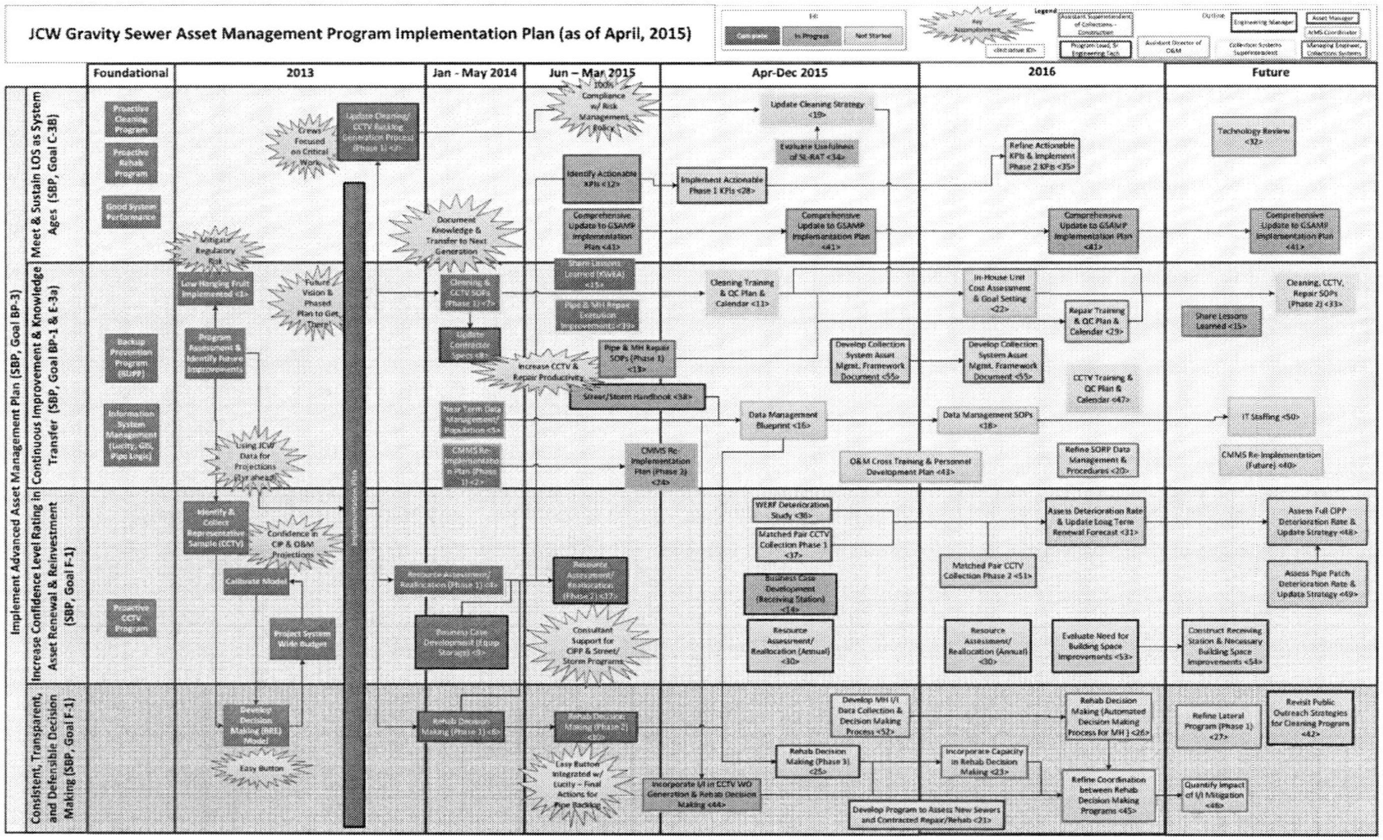

Figure B JCW Gravity Sewer Asset Management Program Implementation Plan (as of April 2015)

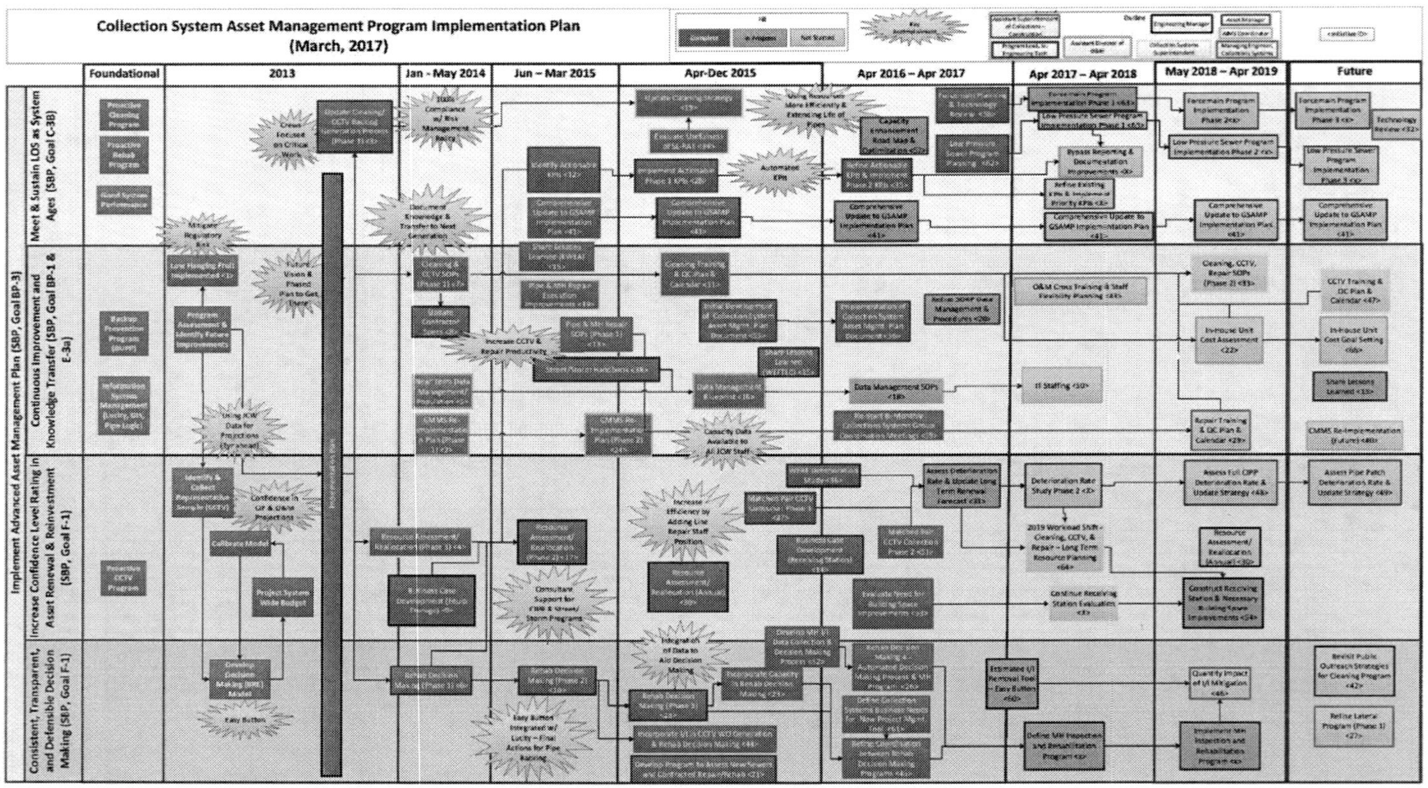

Figure C Current JCW Collection System Asset Management Program Implementation Plan (as of March 2017)

PROGRAM RESULTS – REHABILITATION AND REPAIR PROGRAM

Definition and Collection of Representative Data Sample

At the inception of the CSAMP, JCW needed to increase their confidence level in the investments needed to be made towards aging gravity sewer infrastructure to maintain the current high level of service. The available condition assessment data serves as the foundation for projecting future condition assessment, maintenance, and rehab/repair needs. A sufficient amount of data is necessary in order to establish an appropriate confidence level in the data used as the basis for system wide projections. Prior to the beginning of the CSAMP effort, JCW had developed and implemented an inspection and rehabilitation program for small diameter pipe, based on a Business Risk Exposure (BRE Model) that was developed as part of JCW's initial Asset Management Plan in 2008 and used to help make condition assessment and rehab/repair decisions. In 2013, JCW had condition assessment data on approximately 20% of the system. However, this data was concentrated in high risk areas within the small diameter system (defined by JCW as pipes smaller than 18-inches in diameter), primarily comprised of older VCP pipe that requires a higher level of investment than the overall collection system which is comprised of a mix of pipe materials and pipe age. Extrapolation of this subset of the system would have led to exaggerated investment level projections. Therefore, JCW and HDR leveraged ASTM E122 to determine how much additional inspection data was needed to reach a specified confidence level. Assets were selected and inspected to form a more "representative sample" of JCW's collection system.

The inspection crews completed the collection of the representative sample in the spring of 2013. JCW crews inspected 11.4 miles of pipe that included large diameter (defined by JCW as pipes 18-inches in diameter and larger) and lower risk pipe (PVC material). The data collected in 2013 combined with prior inspection data allowed the project team to estimate future yield and budget projections at a higher and more desirable confidence level for JCW staff. The method for estimating the confidence level in these projections leverages the following references:

- *Sampling Techniques, Third Edition, William G. Cochran, 1977*
- *ASTM E122: Calculating Sample Size to Estimate, With Specified Precision, the Average for a Characteristic of a Lot or Process, 2011*

This method calculated the level of confidence based on the size of the sample, desired confidence level, and actual yield rate. The yield rate is the percentage of pipes identified as requiring rehabilitation. Table 1 presents the calculated confidence levels and deviations based on the current sample size and yield rate for both large and small diameter pipe.

Table 1 ASTM E122 Calculation of Confidence Level for Large and Small Diameter Pipes

Pipe Size	Sample Size and Model Results			Deviation (+/-)	Confidence Level	Projected Range of Yield Rate	
	Length Inspected (miles)	Number of Pipes Inspected	Yield rate			Low	High
Large Diameter (18-Inches and Greater)	41 (22%)	877 (23%)	10.0%	14.7%	90%	8.6%	11.5%
Small Diameter (Less than 18-Inches)	596 (30%)	14,459 (28%)	16.1%	2.7%	90%	15.7%	16.6%

From a statistical perspective, this means that JCW can be 90% confident that the actual yield rate is within the Projected Range of Yield Rates.

Pipe Rehabilitation/Repair Decision Model Development and Calibration
With the achievement of an increased confidence level for projecting investments, the project team collaborated to document and refine JCW's decision making criteria for gravity sewer renewal. This effort resulted in the development of Structural Risk Score (SRS) criteria. The SRS is a numerical value between zero and one hundred representing the relative structural risk for each pipe that has been inspected based on the condition assessment findings coded by CCTV crews using standard PACP scoring, consequence of failure, and projected pipe deterioration, with 100 representing the highest possible structural risk rating. The SRS calculation was developed specifically for JCW and this program based on a combination of existing JCW decision making processes, JCW staff input, and experience with other industry leading utilities.

The SRS is determined by the pipe's probability of failure (determined by the condition assessment findings and projected deterioration over time) and consequence of failure. Historically, JCW has used consequence of failure criteria to prioritize inspections but had not directly used them to determine the priority of structural renewal actions. In JCW's new model, consequence of failure is considered during decision making and prioritization and represents 20% of the maximum possible SRS. This ratio was determined based on lessons learned from other industry leading utilities with mature decision making processes as well as JCW's engineering and field experience regarding the relative importance of the probability and consequence of failure factors. Weighting the condition assessment findings in this manner ensures that the pipes with the greatest risk of structural failure will be scored higher and prioritized for repair/rehab, while appropriately factoring consequence of failure into the decision making process.

The SRS is comprised of the individual probability and consequence of failure components listed below:

Probability of Failure factors (80% of SRS calculation):

- Defect type (e.g. hole, break, fracture) – categorized as Type 1 – 7 based on severity and type
- Defect size (e.g. clock position, length, percent blockage)
- Count of defects
- O&M issues (e.g. roots, grease, debris)
- Anticipated deterioration rate (based on pipe material and cleaning frequency)
- Presence of groundwater

Consequence of Failure factors (20% of SRS calculation):

- Land use in proximity of pipe (e.g. proximity to bodies of water and wetlands, streets or major roadways, railroads)
- Diameter
- Depth

Along with the SRS, a Structural Action decision algorithm was developed. This algorithm determines the recommended next action for each individual pipe based on JCW's preferred decision making processes and the pipe's SRS. To develop the algorithm, defects were classified based on the preferred repair action to address the defect:

- Type A – Typically addressed by trenchless rehabilitation methods using CIPP lining or Pipe Patch (e.g. fracture)
- Type B – Typically addressed by replacement (e.g. lining failure)
- Type C – Typically addressed by point repair or replacement if severe (e.g. collapse, deformation)
- Type D – Typically addressed by cutting intruding tap (JCW refers to this as "Bueler Clean")
- Type E – Abandoned inspections

Additional factors influencing the recommended Structural Action include the diameter of the pipe, the severity and distance between defects, location of defects, and specific combinations of defects that would lead JCW to complete a specific repair action.

The model was then calibrated to determine a sustainable and cost effective reinvestment level that maintains JCW's current level of service as the system continues to age and deteriorate. For pipes that have significant structural defect(s), two risk mitigation strategies were considered. The first was eliminating the risk through repair, rehab, or replacement of the pipe. The second strategy considered was to monitor the pipe at predefined intervals to ensure it remains within an acceptable risk level. Since JCW's unit costs for pipe repair, rehab, or replacement are typically

10 to 1,000 times the cost of inspection, monitoring the moderate risk pipes is seen as a cost effective strategy. High risk pipes will be addressed through repair, rehab, or replacement.

The SRS thresholds for repair, rehab, and replacement represent the risk threshold below which the pipe would be monitored in lieu of repair, rehab, or replacement. These thresholds were the primary mechanism for calibrating the model used to make renewal decisions and determine appropriate system renewal investment levels. The SRS thresholds were aligned with JCW's level of service and cost targets to maximize ratepayer investments. Recommended Structural Action outputs by the model include the following:

- CIPP Lining
- Replacement
- Point Repair
- Pipe Patch (trenchless point repair using fiberglass and epoxy liner patch)
- Install Manhole (at cleanouts or bends that cannot be lined through)
- Bueler Clean (cut intruding tap)
- 5, 10, or 20 Year Monitor (CCTV again at specified interval)

Figure D on the following page presents the decision making logic for pipe rehabilitation and repair developed and used in JCW's CSAMP.

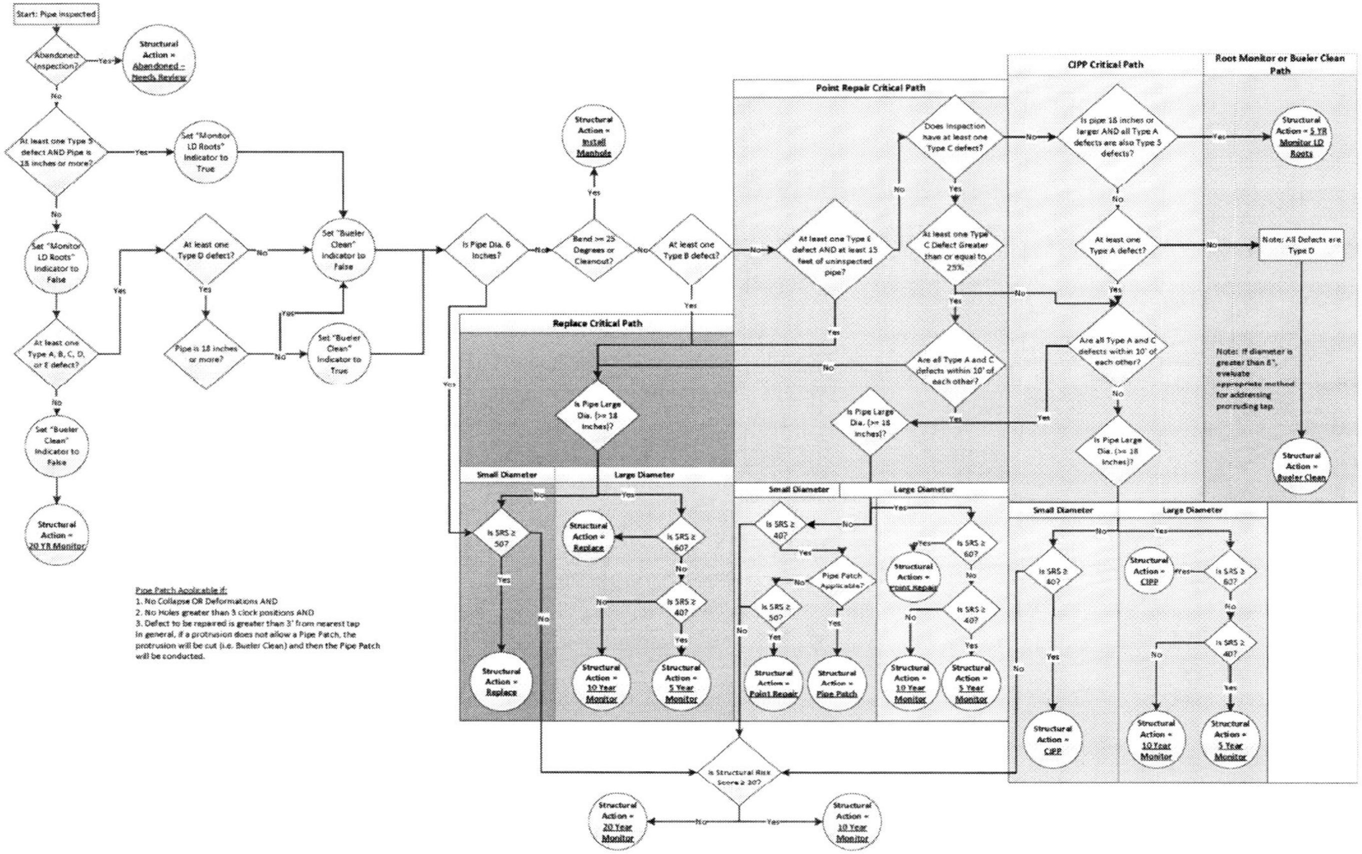

Figure D JCW Collection System Asset Management Program Decision Logic for Pipe Rehabilitation and Repair

The model algorithms were then programmed to work with the live inspection data in JCW's Computer Maintenance Management System (CMMS) to give a priority, action, and estimated repair cost to all pipes with an inspection record (note that there are currently over 19,000 pipes with an inspection record). The automated model can be run as often as needed, and is currently programmed to automatically run once a week over the weekend, after JCW's CCTV crews upload their data each Friday.

The SRS and Action algorithms used by the model were developed specifically for JCW based on their decision making processes. These algorithms can be changed as needed based on input from JCW staff, and can also be adjusted if needed based on changes to JCW's cost or level of service targets. The rehab decision making processes also align with JCW's cost effective I/I removal strategy for main line pipe.

The automated model was programmed to work directly with JCW's CMMS and the county's Automated Information Mapping System (AIMS). JCW staff can view the SRS and recommended Action for each pipe in AIMS along with available utility information, aerial images, parcel data, etc. Integration of these applications provides JCW another tool to ensure the correct rehabilitation decision is made in an efficient manner. Final actions and work orders can be assigned to pipes all on the same screen while viewing the AIMS mapping and rehabilitation model output. An example of this mapping application and CMMS integration is shown below in Figure E.

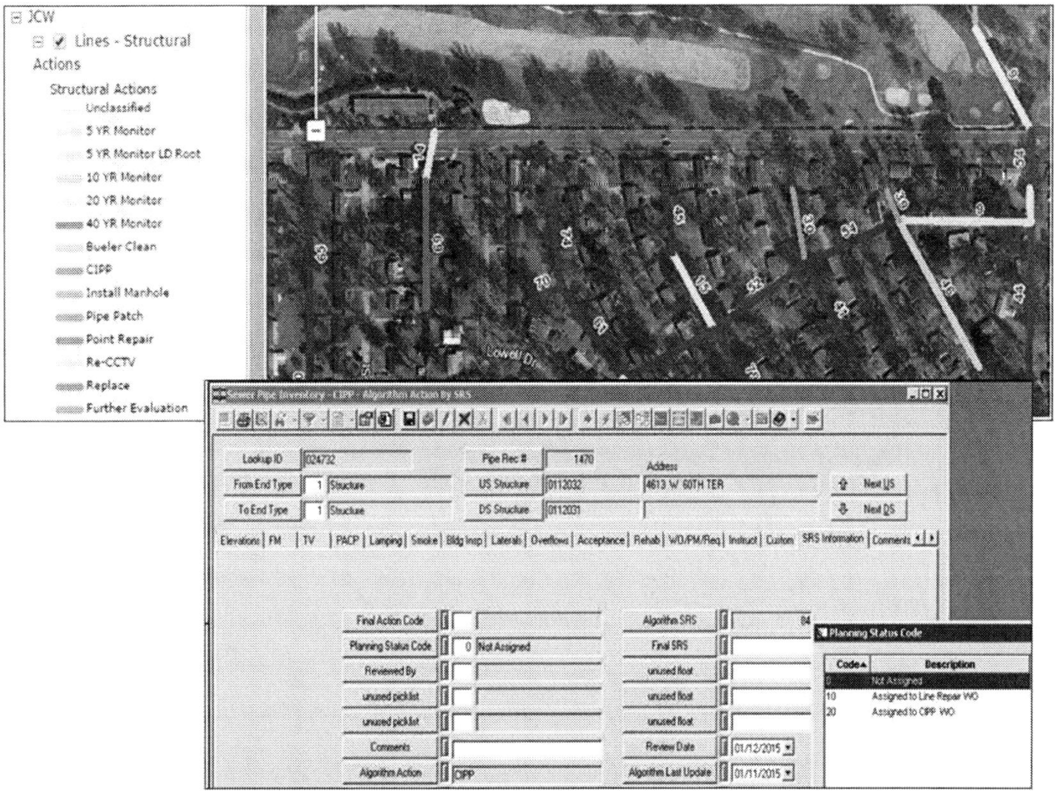

Figure E Integration of Online Mapping Application and CMMS with Rehabilitation/Repair Decision Model

Data Driven Investment Forecast

These new processes were then applied to the representative data set to forecast future rehabilitation work needed. Previously, JCW had developed age based asset replacement costs based solely on the age and physical effective life of pipes within the system. The rough estimate based on these factors assumed approximately $400 Million in asset replacement costs over the next 10 years.

The new data driven projections were developed by identifying the following:

- Current system rehabilitation needs
- The projected system rehabilitation needs that will be identified over the next 10 years

The current system rehabilitation needs were identified by applying the rehabilitation decision making logic to the pipes that have been inspected. Applying the decision making logic to the representative sample resulted in the rehabilitation action yields shown below in Figure F.

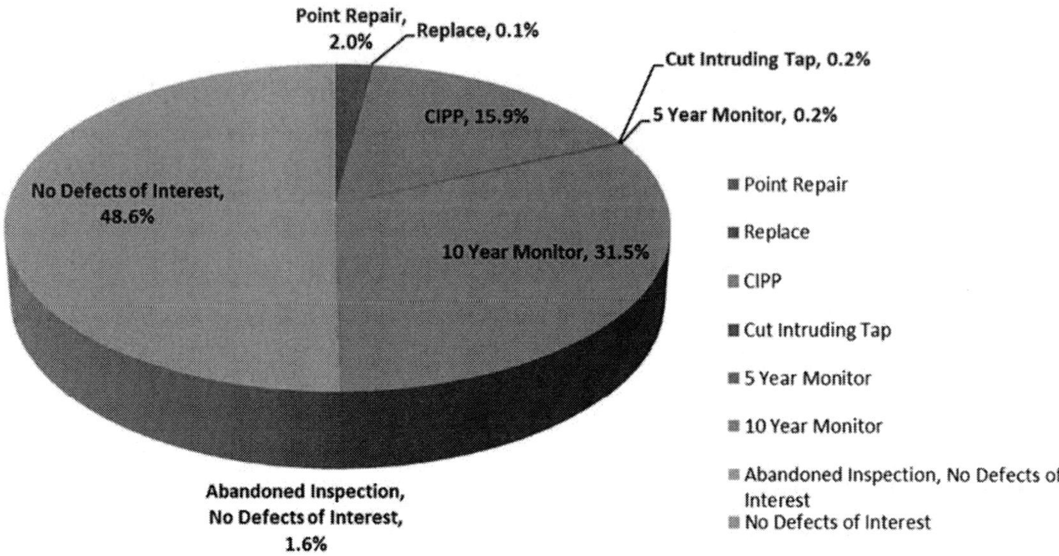

Figure F Rehabilitation/Repair Action Yields from Representative Sample

The projected system rehabilitation needs represented the anticipated amount of rehabilitation work that will be identified over the next 10 years. These projections are a function of both the rehabilitation action yield rates and the amount of CCTV inspections projected to be completed over that timeframe. The projected needs were identified by applying the rehabilitation action yield rates from the representative sample to the anticipated amount of CCTV inspections to be completed over the next 10 years.

This forecast was used to better understand the risk mitigation method and quantity of work that would be needed in the future to sustain the desired level of service and justify future investment levels. This resulted in a data driven projection of approximately $50 Million of rehabilitation/repair investments over the next ten years needed to sustain the current high level of service, at a 90% confidence level in the projections. JCW's level of confidence in the projections increased significantly while at the same time reducing the estimates from $400 Million to $50 Million dollars over the next 10 years. Over 3 years later in 2017, the original $50 Million projection has remained an accurate estimate for JCW's investment needs.

Deterioration Rate Study

JCW owns a large quantity of gravity pipe infrastructure (estimated replacement cost of approximately $1.5 billion in 2008). As the system continues to age and deteriorate, JCW will need to continue to make infrastructure renewal investments to sustain desired levels of service. JCW has developed a robust process to assess the current condition of the system and make prudent near term investment decisions. However, deterioration rates for gravity pipes are not well understood.

While other industries (e.g. pressure pipe) have made significant advances in understanding deterioration, the gravity pipe industry has struggled to develop meaningful long term deterioration curves for reasons including:

- Relationship between age and condition – Relative to pressure pipe, gravity pipe deterioration is generally believed to be less dependent on age and more event driven (e.g. installation quality, root growth, aggressive cleaning, surcharge, ground movement, infiltration, etc.). While age by itself is a poor indicator of pressure pipe condition, pressure pipe has a much stronger relationship between age and condition than gravity pipe. This is because many of the common deterioration drivers such as corrosion and cyclical loading (e.g. internal pressure, metallic pipe expansion and contraction, soil shrinkage and swelling, etc.) are age dependent.
- Definition of Failure – For the purposes of this document, the definition of failure is the criteria at which a pipe is renewed. The functional requirement of a pipe is to convey fluids under a particular operating context (e.g. pressure, flow rate). In general, gravity pipes rarely fail to provide their functional requirement due to a structural issue. The definition of failure for gravity pipes is often based on the risk of a pipe failing to provide it's functional requirements in the future. Conversely, pressure pipe (particularly lower consequence assets) are allowed to fail their functional requirement multiple times before they are considered for renewal.
- Determination of "failure" timing – In a pressure pipe system, failures are typically observed and reported within minutes to hours of the event occurrence. Conversely, gravity pipes can often operate in a state which a utility would describe as "failed" for years or even decades so long as the flow can pass and the pipe is not inspected.
- Documentation of failure history – Many distribution utilities have thirty to forty years of relatively accurate break data (date, type, and location of break) that can be used for failure modeling. Meanwhile, many gravity pipe utilities use CCTV as the basis to determine failure and only have relatively accurate data dating back five to twenty years.
- Accessibility - Gravity pipes are typically more accessible than pressure pipes for direct condition assessment.
- Cost Effective and Industry Accepted Condition Assessment Method – Gravity pipes have a cost effective and industry accepted condition assessment method (CCTV). While several direct condition assessment technologies are available for pressure pipes, the industry has not adopted a standard direct condition assessment technique. Most utilities still rely on desktop analyses (e.g. breaks, soil studies, water quality complaints, etc.) to assess the condition of low consequence pressure pipes.

To address these challenges, JCW's deterioration study relies on a "matched pair analysis" which was define as:

- A pipe has an inspection with a structural defect of interest AND
- The same pipe has a previous inspection separated by at least 5 years AND
- Both inspections have video available AND
- There is no indication the pipe was fixed between the inspections (e.g. repair work order, CIPP work, or pipe condition improved due to increase in number of material change codes)

Prior to this study, only a handful of pipes met this criteria. Therefore, JCW strategically inspected pipes that would fit this criteria. A total of 501 events were collected and analyzed with an average duration between inspections of 9.4 years. Most pipes reviewed showed little to no deterioration, indicating that deterioration is not occurring linearly in these pipes. Significant deterioration was more commonly observed for more severe defects, such as pipes where the initial video showed signs of pipe wall displacement or soil or voiding visible.

PROGRAM RESULTS – CLEANING PROGRAM

Initial Preventive Maintenance Program Scheduling Improvements

When the CSAMP began in 2013, JCW's risk management policy defined a 3, 5, or 7 year cleaning schedule based on pipe material for typical pipes (note that some grease and low slope "hot spot" areas are cleaned on an accelerated schedule). Prior to enacting this policy, JCW's annual dry weather overflow/backup rate was greater than 8 per 100 miles of pipe in the 1980s. Effective collection system cleaning based on this policy played a key role in drastically reducing this overflow rate and maintaining JCW's high level of service. Recently, this rate has been consistently below 1 overflow per 100 miles of pipe.

Prior to CSAMP development, JCW had increased cleaning output and had the capacity to clean more pipe than required to meet the cleaning schedule. Work orders were generated in bulk at a basin level and typically required many months to complete. While basins generally were prioritized and assigned based on the last time the basin was cleaned and the next cleaning schedule, individual pipes within the basin may not have been cleaned within the schedule defined by JCW's risk management policy.

In order to focus resources on risk management policy compliance, JCW moved towards smaller work order packages and prioritized them based on when pipes in that area were coming due for their next cleaning based on the risk management policy. An automated model and mapping application for cleaning work order generation was developed and programmed to work with the live data in JCW's CMMS. This application allowed JCW to identify the cleaning due date for each pipe within the system and use this information to focus cleaning efforts on the highest priority pipes that were past due or nearing their due date. This enabled JCW to ensure compliance with the preventative maintenance cleaning strategy policy, avoid cleaning basins well before their due date, and reduce the overall quantity of cleaning performed.

Concurrent with these cleaning program improvements, JCW identified a need to increase CCTV resources to mitigate the risk presented by the backlog of older pipes within the system that had not yet been inspected. This was accomplished by purchasing an additional CCTV truck in lieu of a cleaning truck (which was due to be replaced), effectively shifting one cleaning truck and crew to CCTV in 2014. Based on future cleaning workload projections and the improvements made to the cleaning program, it was determined by the project team that JCW would still have sufficient cleaning capacity to meet the risk management policy goals, as well as account for ad-hoc cleaning and the grease hot spot program.

In 2014 JCW completed the purchase of the new CCTV truck and shifted one full time cleaning crew/truck to CCTV work. Utilizing the optimized cleaning work order generation process developed through the CSAMP, JCW achieved 100% compliance with the preventative maintenance cleaning strategy policy for the first time while cleaning over 100 miles less pipe than in recent years. A comparison of the total miles cleaned by JCW versus the compliance rate with the policy is presented below in Figure G.

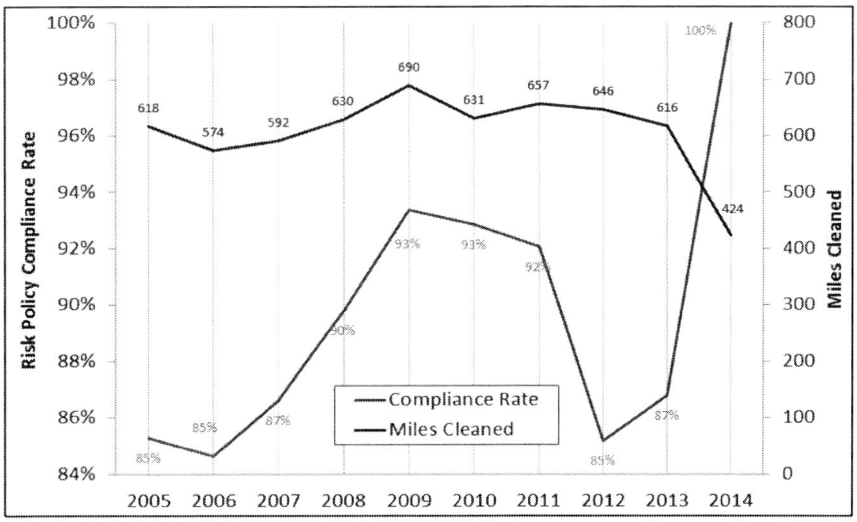

Figure G Results of Initial CSAMP Cleaning Scheduling Improvements

In order to better monitor progress towards JCW's goals, Key Performance Indicators (KPIs) for the cleaning program were refined, documented and automated to generate monthly reports using the live data in JCW's CMMS. These KPIs included the following:

- Cleaning Risk Policy Compliance (Near and Short Term)
- Cleaning Workload Forecast (Short Term and Ten Year)
- Annual Cleaning Goal Status
- Monthly Cleaning Output (Programmatic and at Crew Level)

CSAMP efforts also focused on the support of continuous improvement and knowledge transfer within the cleaning program to the next generation of leaders at JCW. As part of this effort, Standard Operating Procedures (SOPs) for the cleaning program were refined and documented.

These SOPs were completed for both field crews and office personnel involved in managing the cleaning program and collection systems.

Cleaning Strategy Update

A comprehensive review and update to JCW's preventive maintenance cleaning strategy was completed in 2015. This initiative was undertaken in order to accomplish the following goals:

- Understand why gravity pipe failures due to maintenance issues occur so resources can be focused on risk mitigation
- Maximize maintenance issue risk mitigation given existing resource levels
- Determine whether implementation of emerging technologies would support JCW in cost effectively managing risk
- Standardize existing frequencies to support long term schedule level loading and maintenance optimization
- Extend the useful life of pipes and reduce the long term cost of ownership by only cleaning pipes when needed
- Leveraging existing data (CCTV, cleaning, backup, rainfall, repair/CIPP history, etc.) to identify the appropriate cleaning frequency for every active gravity pipe owned by JCW
- Level loading the maintenance workload to eliminate severe peaks and valleys in workload
- Position cleaning crews for success in terms of cleaning productivity by minimizing drive time, minimize repeat visits to a particular neighborhood, and where practical, clean hydraulically (from the top of the system to the bottom of the system) to capture and remove maintenance issues that elude the cleaning crew.

Identification of the appropriate cleaning frequency is a critical component of a maintenance strategy. Figure H shows this concept:

- If a pipe is cleaned "too little", a utility is exposed to excessive overflow risk.
- If a pipe is cleaned "too much", it is an inefficient use of resources and will increase wear on the pipe.
- If a pipe is cleaned at the "just right" frequency, the utility is efficiently using resources, limiting the risk of overflows, extending the life of the pipe, and ultimately saving money.

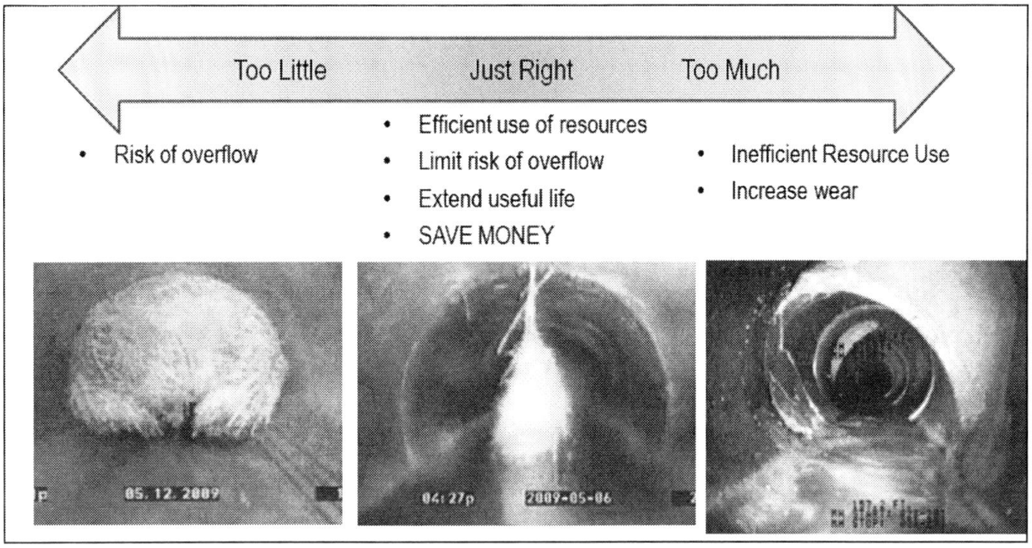

Figure H Cleaning Schedule Optimization

JCW then leveraged readily available data, industry experience, and institutional knowledge to optimize frequencies and minimize risk with the current level of cleaning resources.

Over the past 20 years, 98% of JCW's collection system has not experienced a maintenance related failure. JCW staff believed that certain portions of the system were being cleaned more often than needed. To validate this assumption, an analysis was conducted on the cleaning findings collected by crews since more robust data collection standards were put in place when the Cleaning Program SOP's were implemented in Spring of 2014. As shown below in Figure I, The cleaning findings results indicated that as much as 80% of the system could be cleaned less frequently and only about 2% of the system needs to be cleaned more frequently.

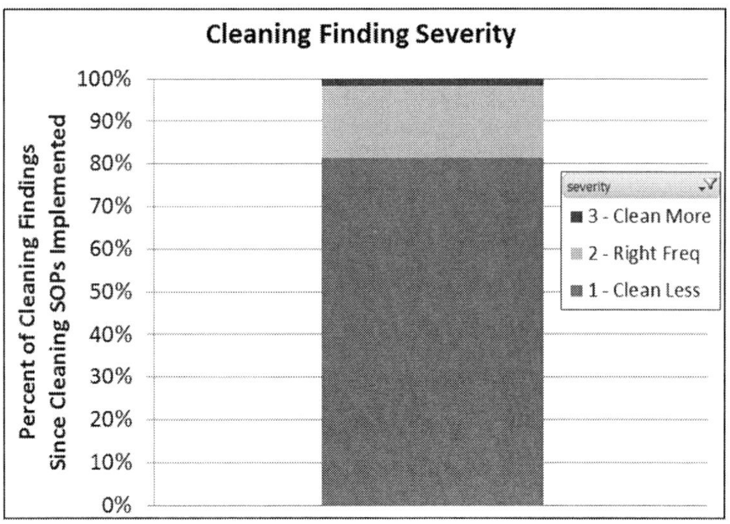

Figure I Summary of Cleaning Findings

Since pipe cleaning crews do not know the exact extent of maintenance issues before and/or after cleaning, a second analysis was conducted using CCTV data which provides visual proof of the exact extent of maintenance issues at the time of the inspection. For the purposes of this analysis, CCTV inspection data collected shortly after cleaning does not provide valuable information because it does not indicate how quickly maintenance issues are accumulating since the last cleaning on a particular pipe. With this in mind, this analysis focuses only on CCTV inspections that occurred at least 75% of the way through the cleaning frequency since the last cleaning event.

The results of this analysis is summarized below in Figure J. These findings indicated that approximately 80% of the pipes had no blockage or a very minor blockage (<5%) at the time they were due for cleaning. Based on both the cleaning and CCTV data analysis, a significant portion of the system could be cleaned less often, freeing up resources to focus on higher risk pipes and basins.

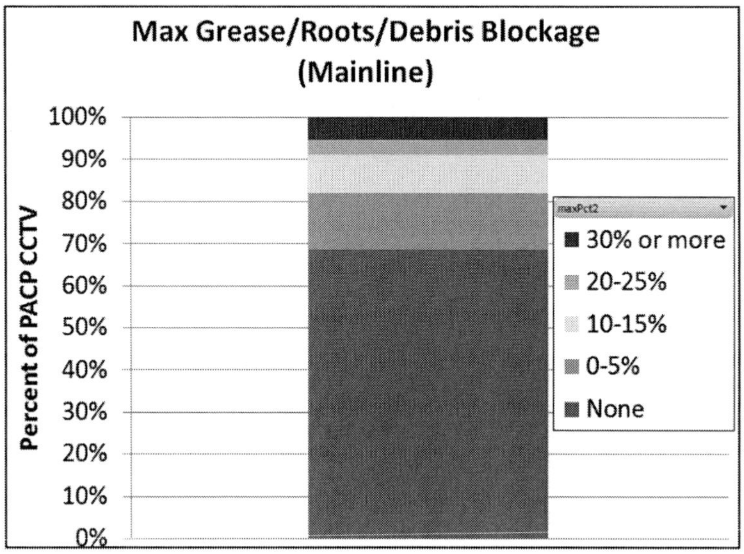

Figure J Summary of CCTV Findings

A pipe risk score was then created and used to assess cleaning risk at a pipe and sub-basins level. The risk score was developed based on backup/maintenance call history, cleaning and CCTV findings, and JCW's institutional knowledge. A summary of the entire collection system by risk score category is presented below in Figure K.

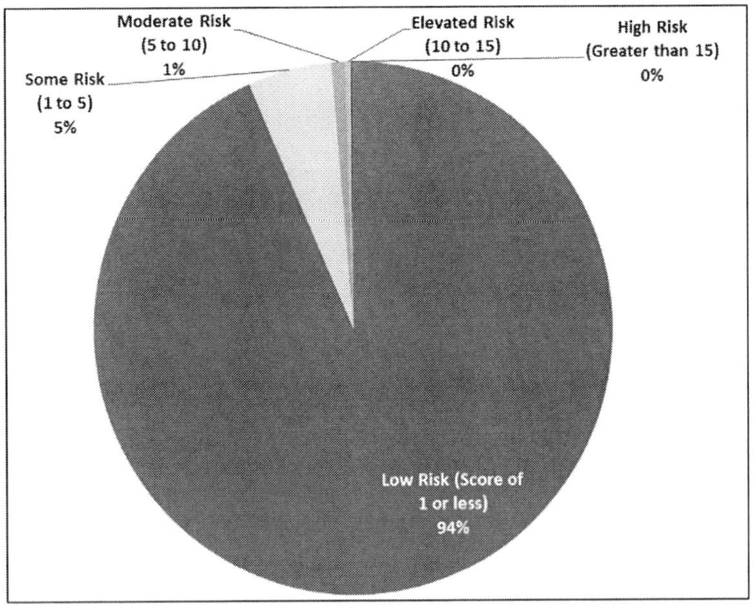

Figure K System Wide Pipe Risk Assessment Score Summary

Leveraging this data from JCW's system led to updating the preventive maintenance cleaning frequency from a 3/5/7 year schedule to a 2/4/8 year schedule, allowing JCW to accomplish the following:

- Mitigate risk by cleaning "dirty" pipes more frequently
- Reduce wear and extend life of pipe by cleaning "clean" pipes less frequently
- Clean 40 miles less per year
- Consistent workload each year aligned with existing resources (level loaded schedule)
- Maintained geographic centricity

PROGRAM RESULTS – CONDITION ASSESSMENT (CCTV PROGRAM)

Prior to CSAMP development, JCW's management strategy prioritized pipe for condition assessment using a priority basin approach. Basins were ranked based on the average Business Risk Exposure Score (BRE) score of all line segments within the basin based on BRE criteria previously developed by JCW as part of the initial Asset Management Plan created in 2008. The basins were then prioritized for condition assessment and rehab/repair work, focusing on the worst basins first. The priority basin strategy allows JCW to focus inspection efforts (as well as subsequent rehabilitation efforts) on specific geographic areas, thus decreasing travel and setup times and increasing the efficiency of inspection efforts. All pipes within the selected basins were televised, regardless of the data of the last inspection and condition of the pipe.

The decision making processes used to prioritize pipe for condition assessment and to project CCTV backlog were reviewed and refined through the CSAMP. While JCW's priority basin was determined to be an effective tool to focus condition assessment efforts on high risk basins, additional factors were desired to further focus inspection efforts on the most critical work. To improve risk management within the collection system, JCW decided to incorporate a goal of completing inspections of all pipes before they reach 40 years of age. All pipes over 40 years of age were considered due for inspection and thus part of the first inspection backlog of pipes due for inspection. The BRE basin selection approach was enhanced to include the percentage of pipes in the backlog and the relative level of infiltration/inflow (I/I) within the basin.

Forecasts of near and long term condition assessment needs were completed in order to ensure collection systems resources were aligned with JCW's future needs. The condition assessment projections accounted for all CCTV activities anticipated to be completed by JCW crews. CCTV inspections were anticipated to be comprised of the following programs:

- First Inspection Program – scheduled first inspections at or before a pipe reaches 40 years of age.
- Monitoring Program – Scheduled monitoring of pipes that were previously televised at a specific time interval based on the inspection findings, to cost effectively mitigate risk in pipes that contain defects of interest but do not yet require repair/rehab.
- Request Program - all unplanned CCTV activities initiated by a customer or JCW staff or as a reaction to findings in the field.
- Repair Acceptance Program – Quality Control (QC) inspection of the rehab/repair work to ensure JCW ratepayers maximize the value of their investment.
- New Sewer Acceptance Program - anticipated future JCW program to CCTV a sample of new privately constructed sewers, prior to the expiration of the 3-year maintenance bond to verify the pipe remains in good condition.

The CCTV work order generation process was updated to focus efforts on basins that had both a high BRE score, high percentage of backlog, and high level of I/I. An automated model and mapping application for CCTV work order generation was developed and programmed to work with the live data in JCW's CMMS. This application allowed JCW to identify the due date for each pipe within the system and use this information to focus inspection efforts on the highest priority pipes that were past due or nearing their due date. Only the pipes within the basin that were due for inspection or nearing their due date were selected. This allowed JCW to focus inspection efforts on the most critical work and realize cost savings by avoiding previously inspected pipes that are known to be in good condition.

JCW set an informal goal to eliminate the first inspection backlog over the next 6 - 8 years. In order to eliminate the backlog, it was determined that JCW will need to inspect approximately 88 miles of pipe per year. When this analysis was completed in 2013, JCW owned and operated three CCTV trucks staffed by three crews who typically could televise 60 to 65 miles per year. At this capacity, the inspection backlog would continue to increase over time as the assets within the collection system age, increasing the level of risk and the challenge of meeting JCW's level of service goals.

Therefore, JCW decided to add one additional CCTV truck and crew to allow for increased levels of proactive inspections, which in turn would help prevent costly emergency repairs, collection system overflows, and basement backups. It was determined that an additional CCTV truck and crew could be added without increasing the current O&M and staffing/equipment budget. This was accomplished by purchasing a CCTV truck in lieu of a cleaning truck that was planned for replacement, effectively shifting one cleaning truck and crew to CCTV in 2014. The addition of the 4th CCTV truck enabled JCW to focus efforts on the most critical work, mitigating the risk presented by older pipes within the collection system that had not yet been inspected.

In Quarter 3 of 2014, the 4th CCTV truck was placed into service. The addition of this truck, along with other efficiency gains enabled JCW to inspect 104 miles of pipe in 2014. In 2015 and 2016, productivity rose to 130 miles of pipes. This put JCW ahead of schedule for the original goal to eliminate the first inspection backlog and further helped mitigate risk within the collection system. This was accomplished without increasing O&M staffing or equipment budgets for CCTV inspection and cleaning activities.

By focusing cleaning and CCTV resources on the right pipe at the right time, it is estimated that JCW saved approximately $1 million dollars per year in contracted cleaning and CCTV expenditures that would have been required if JCW had elected to use the historic maintenance strategy and resources to achieve similar cleaning and CCTV policy compliance levels.

CCTV program SOPs and KPIs were developed concurrently with the Cleaning SOPs and KPIs. These allowed JCW to better monitor progress towards their collection system management goals and support continuous improvement and knowledge transfer within the CCTV program to the next generation of leaders at JCW.

PROGRAM RESULTS – MANHOLE I/I REDUCTION PROGRAM

Development of JCW's Manhole I/I Reduction Program was a major focus of CSAMP work in 2016. Prior to then, JCW's programmatic manhole inspections consisted primarily of a Tier 1 inspection completed by cleaning and CCTV crews during normal work activities. The Tier 1 inspection was focused on identifying significant structural defects, with crews using a 1-5 rating system to help identify and address manholes that required structural rehabilitation. Detailed manhole inspections were completely only on a limited basis as part of rehabilitation projects focused on specific areas within the system.

In 2016, a programmatic Tier 2 Manhole Inspection Program was developed through the CSAMP. Program development focused on the following activities:

- Development of data collection standards and practices.

- Testing of JCW's standard manhole covers to determine inflow rates for each cover type and corresponding I/I reduction strategies for cover inflow.
- Development of a rehab decision model and algorithm for manholes tailored to the manholes assets present within JCW's system.
- Piloting of JCW's manhole inspection process and QC of the rehab decision model using inspection findings.

Manhole data collection standards were developed specifically to meet the needs of JCW's program. Emphasis was placed on collecting actionable information only, limiting the number of component measurements taken to only those that will help drive rehab decisions. Separate structural and I/I component rating systems were developed. These rating systems were developed with a focus on enabling consistent, repeatable component ratings by field crews through measurable consistent ratings (e.g. ¼-inch crack, 1/8-inch gap between mortar and frame).

A Tier 2 Manhole Inspection Form was developed based on these data collection standards. It was determined that inspections would be completed using a contractor rather than internal JCW crews, and inspection findings would be documented using mobile devices linked directly to JCW's CMMS. The preliminary data collection standards, along with a 360 degree inspection camera technology were then piloted in a small 50 manhole sample pilot. After completion of the pilot, the results were revised and minor modifications were made to the data collection process and the standard procedures for manhole inspections.

After a review of the 360 camera videos taken during the pilot, it was decided to utilize this technology as part of the standard documentation and condition rating process. An example of a 360 degree manhole photo is shown below in Figure K.

Figure K Example Photo Using 360 Degree Manhole Camera

Manhole covers commonly found in JCW's system were tested to determine the inflow rates during periods of submergence. A testing apparatus was constructed that allowed JCW to simulate field conditions. This testing apparatus is shown below in Figure L.

Figure L Manhole Cover Inflow Testing

A description of the three manhole covers commonly found in JCW's system and results of the inflow testing is provided below:

- "New JCW Cover (Sunflower Style)" – This manhole cover has been JCW's standard cover for most construction since the late 1980s. This cover has one concealed pick hole. Testing determined that this cover leaked at a rate of 0.7 to 0.9 gpm at depths of 3 to 12-inches of submergence. This is relatively watertight and compares well to other commonly used manhole covers.
- "Old JCW Cover (Waffle Style)" – This manhole cover with two pick holes was JCW's standard until the late 1980's. Testing determined that this cover leaked at a rate of 7 to 10 gpm at depths of 3 to 12-inches of submergence. This is over 10 times the inflow rate of the new JCW cover.
- "Boltdown Gasketed Cover (Floodplain Cover)" – This manhole cover is JCW's standard for construction within floodplain areas. This frame and cover comes standard with an integral rubber gasket. This cover was tested with and without the gasket:
  - With a sound gasket in place, testing showed that the cover was watertight and did not allow any significant inflow into the manhole.
  - With the gasket removed or damaged, testing indicated an inflow rate of 13 to 21 gpm at depths of 3 to 12-inches of submergence.

Two primary findings from the cover inflow testing resulted in major impacts to JCW's program strategies:

- Old JCW Covers – Anecdotally, approximately 30 – 40% of the manholes in the system are believed to have this cover. As this cover leaks at over 10 times the rate of the New JCW Cover, replacing these old covers and frames found in sump areas or sheet flow areas with substantial tributary areas was determined to be a cost effective method of I/I reduction.
- Boltdown Gasketed Covers – These covers are typically installed in floodplain areas. In some locations, these manholes may be submerged for long periods of time during wet periods when creeks, river, and lake levels are above normal pool. Covers with damaged or missing gaskets were determined to be a major inflow source. Therefore, it was decided to focus some of the inspection efforts on areas where these covers are likely to be found, particularly in locations that may be underwater for longer periods of time. Replacing the integral gasket is a very cost effective way to mitigate these inflow sources. Manholes at high risk of submergence will be identified during inspections, and a plan will be developed to manage these manholes and ensure the gaskets remain sound.

All manholes within floodplain areas were identified in JCW's GIS system. An example of one such area is shown below in Figure M.

Figure M Identification of Potential Manhole Inflow Sources in Floodplain

A rehab decision process was developed for each of the three cover types commonly found in JCW's system. For each of these, a separate rehab process was developed for paved and unpaved areas. The findings described above were used as the basis to develop JCW's strategy for manhole cover renewal. Basing this strategy on the actual characteristics of JCW's specific manhole covers allowed JCW to develop an optimized I/I reduction strategy specific to the assets owned by JCW.

The rehab decision model was then developed for other manhole components. The model defines a recommended next Structural Action for each manhole component and the manhole asset itself based on the condition assessment findings. Rehabilitation strategies vary considerably based on manhole material; in addition, rehabilitation costs and strategies can be impacted based on the location of a manhole. A repair that is cost effective in an unpaved areas may not be cost effective if it requires excavation in a paved roadway. Therefore, four separate rehabilitation decision models were developed for the following manhole types and locations:

- Brick Manholes in Unpaved Areas
- Brick Manholes in Paved Areas
- Concrete Manholes in Unpaved Areas
- Concrete Manholes in Paved Areas

An example of the rehab decision making logic is shown below in Figure N.

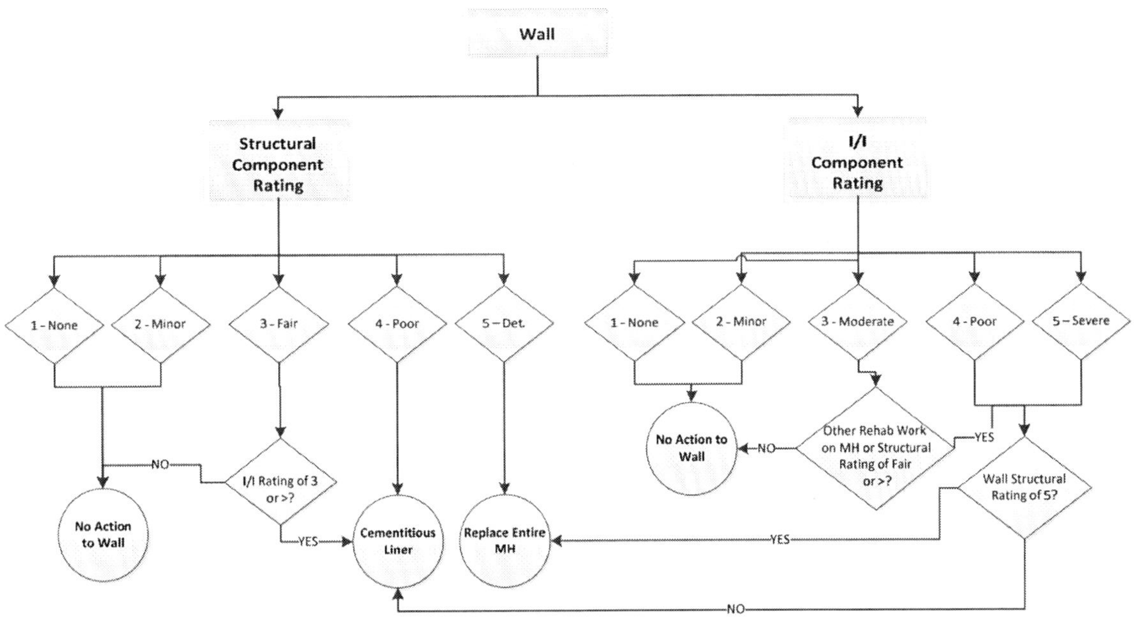

Figure N Manhole Rehab Logic Example (Brick Manholes)

After strategy development, JCW began executing manhole inspections on a larger scale. Similar to pipes, when a representative sample of data has been gathered it will be used to develop a data drive system renewal forecast for manholes.

CONCLUSIONS

Sustaining a high level of service while managing aging infrastructure, an aging workforce, and limited resources is one of the greatest challenges utilities face. JCW has developed a plan and implemented the first 4 years of this plan. The Collection System Asset Management Program represents JCW's road map to continuous improvement in collection system asset management.

The program Implementation Plan has been used to strategically plan and implement continuous improvement in collection system asset management. Data from JCW's information systems has been leveraged to understand the financial and risk management implications of various investment scenarios, develop and calibrate decision models to balance risk mitigation and costs, and develop a consistent, transparent and defensible decision making process. This enabled JCW to forecast maintenance, condition assessment, and renewal investment needs with a high level of confidence and facilitate knowledge transfer within the utility.

Approximately twice per year, the CSAMP Implementation Plan is updated to reflect past accomplishments, current areas of focus for implementation, and to add and revise future initiatives to account for changing priorities, drivers, and regulatory conditions. Below is a list of key accomplishments during the initial 4 years of this program:

- Development of a realistic program Implementation Plan to strategically plan, coordinate, and implement continuous improvement over the next several years.
- Establishment of a 90% confidence level in the 10-year cleaning, inspection, and renewal investment forecast.
- Reallocation of existing in-house resources and equipment to cost effectively meet level of service goals.
- In 2014, increased compliance with cleaning risk management policy from 88% to 100%.
- Reach full compliance with condition assessment goals by 2019. Previously, compliance would not have been reached until 2030.
- Estimated cost savings of approximately $1 million dollars per year in contracted cleaning and CCTV expenditures that would have been required if JCW had elected to use the historic maintenance strategy and resources to achieve similar cleaning and CCTV policy compliance levels.
- Increased renewal investment to sustain current service levels.
- Completed a data driven update to JCW's preventive maintenance cleaning strategy focusing cleaning resources on cleaning pipes at the right time, allowing JCW to clean "dirty" pipes more often, extend the life of pipes by cleaning "clean" pipe less often. This allows JCW to clean less mileage each year, while focusing resources on areas where cleaning is needed most frequently. Automated work order generation processes were developed to ensure cleaning resource remain focused on cleaning each sub-basin at the optimum time.
- Integration of JCW's public sector I/I removal program with the asset management program, for both pipes and manholes.
- Development of Manhole I/I Reduction program, and a rehabilitation decision process tailored specifically for the specific manhole asset types within JCW's system.
- Development of a Forcemain Asset Management Program that addresses the challenges of managing forcemains that vary widely in age, material, and size.
- Completion of a deterioration rate study of over 600 older VCP and concrete pipes within JCW's system. This study allowed JCW to better understand how quickly pipes within the system are deteriorating and refine long term investments projections in inspection and renewal of aging sewers.
- Mitigation of regulatory risk through revised data collection standards that better reflect the work performed to mitigate the risk of future overflows.
- Increased decision making transparency through the documentation, validation, and refinement of maintenance and capital reinvestment business decision making.

- Increased consistency and reduced the level of effort to make good business decisions through the semi-automation of decision making processes.
- Supported continuous improvement and knowledge transfer through the documentation and refinement of Standard Operating Procedures for office and field personnel involved in cleaning, repair, and inspection activities.
- Better monitor progress towards JCW's goals through the refinement of Key Performance Indicators (KPIs) for cleaning, inspection, and pipe and manhole repair.
- Support better decision making by integrating GIS mapping to allow easy review of project locations and easy grouping to develop and create work orders for rehab/repair, cleaning, or inspection.

ACKNOWLEDGMENTS

The authors gratefully acknowledge the contributions of the many Johnson County Wastewater employees whose efforts have made this program a success.

REFERENCES

Cochran, G. William (1977) *Sampling Techniques,* 3rd ed.; John Wiley & Sons: Hoboken, New Jersey.

ASTM E122, 2011, "Calculating Sample Size to Estimate, With Specified Precision, the Average for a Characteristic of a Lot or Process."

Balancing Existing Conditions, Clean Water Act Compliance, and Financial Realities: Development of the Unified Government's Integrated Overflow Control Plan

Trenton Foglesong, P.E., Unified Government of Wyandotte County and Kansas City, Kansas
Waldo Margheim, P.E., Burns & McDonnell

ABSTRACT

Pursuant to federal mandate, the Unified Government of Wyandotte County and Kansas City, Kansas (UG) has prepared a sewer system Integrated Overflow Control Plan (IOCP). The IOCP is designed to comply with federal requirements in a manner that meets goals supported by UG's stakeholders: protect human health, public safety, and customer property and make continued progress towards improving water quality. As required by a 2013 Partial Federal Consent Decree, in addition to developing and evaluating sewer overflow control options, the UG assessed the condition and operation of their sewer system infrastructure and identified significant repair, rehabilitation, and upgrade needs. Due to financial limitations of the economically-disadvantaged UG ratepaying community, initial investments will target continued repair and renewal of the existing sewer system and the construction of critically important early action projects. Such projects will preserve existing assets and are expected to deliver the greatest possible benefits from these critical early program dollars.

Over the course of the next decade, the UG is committed to an aggressive $200 million plan that reinvests available revenue in the higher priority existing sewer system improvements, makes continued overflow reduction progress, and grows the capability of the utility to implement and revise the overall IOCP. Once the existing system is renewed to a more sustainable condition and the early action projects are completed, the UG will reevaluate the community's financial capability, the benefits that have been achieved, and the identified goals and priorities to achieve additional sewer overflow reduction and water quality improvements. The additional controls will be submitted to the agencies in a final measures plan for approval. The initial 10-year period under the recommended plan addresses the goals and priorities of local stakeholders including significant progress on system renewal, is an aggressive financial commitment and burden on the community, and reduces combined and separate sewer overflow volumes by an estimated 20% and 85%, respectively.

KEYWORDS

Combined sewer overflow, sanitary sewer overflow, long-term control plan, financial capability assessment, CSO Control Policy

INTRODUCTION

The Unified Government of Wyandotte County and Kansas City, Kansas (UG) Water Pollution Control Division (WPCD) currently provides sewer service and local stormwater management to approximately 44,000 customers (approximately 150,000 residents) within Wyandotte County, Kansas. This relatively small rate base supports the operation and maintenance of a complex sewer system consisting of approximately 1,100 miles of combined and separate sanitary and storm sewer pipelines, five wastewater treatment plants (WWTPs) with a combined design flow of over 130,000 m^3/d (35 mgd), 71 wastewater pump stations, and nine flood pump stations. The WPCD also supports local drainage districts that maintain almost 20 miles of flood control levees along the Missouri and Kansas Rivers.

About 16% of the approximately 160 square mile UG service area is served by a combined sewer system (CSS). The remainder of the service area is served by a separate sewer system (SSS). During some wet weather events, both combined sewer overflows (CSOs) and sanitary sewer overflows (SSOs) may result in sewage diluted with stormwater being discharged into the environment and potentially into one of four CSO receiving water bodies, including the Missouri River.

Residents in the UG service area are substantially disadvantaged compared to the state and national populations for a number of key socioeconomic metrics, including median household income, unemployment, and the portion of the population living below the poverty level. The financial reality of the service area creates substantial challenges to fund the utility adequately while maintaining tolerable rate burdens. Magnifying these challenges, there has been significant loss of management experience within the utility over the last several years.

Despite significant financial and staffing challenges, since 2000 the UG spent tens of millions of dollars on sewer separation projects and WWTP and pump station capacity improvement projects to reduce sewer overflows. Considerably more funds were spent to inspect, upgrade, and renew the sewer system infrastructure. These improvements reduced the number of CSO diversion structures and reduced annual CSO overflow volume. These improvements required aggressive annual wastewater user rate increases resulting in the average annual sewer and stormwater bill for served households being 1.23% of the community-wide median household income (MHI) in 2014. Recently, the UG prepared a sewer system Integrated Overflow Control Plan (IOCP) to identify additional sewer system improvements necessary to further reduce sewer overflows.

PARTIAL CONSENT DECREE

After several years of negotiations, the UG entered into a federal consent decree with the U.S. Environmental Protection Agency (EPA) and the U.S. Department of Justice (DOJ) in March 2013, to reduce sewer overflows and comply with the Clean Water Act (CWA). The major requirement of the Partial Consent Decree (PCD) was the development of an IOCP. This comprehensive plan to upgrade the UG's sewer system integrates a CSO long-term control plan and SSO remediation plan with municipal separate storm sewer system (MS4) compliance needs.

In order to inform and prepare the plan, the PCD required the UG to perform several underlying tasks and studies including:

- Perform sewer system infrastructure condition assessments;
- characterize the existing physical characteristics and capacity of the sewer system through field inspections, monitoring, and hydraulic modeling;
- characterize receiving water quality and CSO impacts;
- perform a financial capability assessment; and

- implement a public participation program.

SEWER SYSTEM INFRASTRUCTURE CONDITION ASSESSMENT

The PCD acknowledged that the condition of the sewer system infrastructure was a fundamental concern and exacerbated the volume and frequency of system overflows. Reliability of the existing infrastructure not only impacts human health, public safety, and property, but in the SSS it also has a real impact on water quality. In addition, a majority of the collection system is over 50 years old largely composed of vitrified clay pipe. Accordingly, the UG focused on the investigation and condition and operation assessment of a large portion of the sewer system infrastructure including WWTPs, pump stations, and combined and separated collection system piping (including those considered high risk due to streambank erosion).

Initial condition assessment efforts identified substantial infrastructure upgrade, renewal, operation, and maintenance needs necessary to maintain and improve system reliability. It was determined that sustainable renewal of the system will require directing more funds to the effort than has historically been allocated and applying a logical, long-term approach that moves the utility from a reactive to a more proactive position.

SEWER SYSTEM AND RECEIVING WATER QUALITY CHARACTERIZATION

Completed in 2015, the CSS characterization identified a detailed plan of system improvements and corresponding costs to address CSOs at various levels of control (≤12, ≤7, ≤3, and zero overflow events per design year). Capital improvements in the CSS to achieve ≤12 and zero overflows events per design year levels of control were estimated to cost approximately $200 million and $980 million, respectively. These improvements included combinations of conveyance and pump station capacity improvements, storage facilities, and high rate treatment.

Similarly in 2015, the SSS characterization identified improvements necessary to address SSOs. Capital improvements in the SSS to achieve two-year and five-year levels of service were estimated to cost $85 million and $116 million, respectively. These improvements included expansion of the Wolcott WWTP; gravity sewer, force main, and pump station capacity improvements; and storage facilities.

In conjunction with the sewer system characterization effort, receiving water quality was monitored and modeled to establish baseline conditions and evaluate the impacts of overflows on the attainment of water quality standards and uses. This effort concluded that water quality standards and uses cannot consistently be met due to pollution sources other than CSOs (i.e., upstream sources).

FINANCIAL CAPABILITY ASSESSMENT

The UG completed a financial capability assessment that included the following efforts:

- Completion of a Financial Capability Assessment (FCA), following the guidelines established by the *Combined Sewer Overflows – Guidance for Financial Capability Assessment and Schedule Development*, and

- development and analysis of additional socioeconomic information consistent with the *Financial Capability Assessment Framework for Municipal Clean Water Act Requirements* to provide a more complete and accurate picture of the UG's financial capability.

Utilizing the EPA's FCA 1997 guidance, an initial financial capability assessment was performed to evaluate the financial resources the UG has available to implement additional CSO and SSO controls. To achieve the levels of control identified in the characterization effort within a 25-year timeframe, a "high" financial impact (i.e., the cost per household as a percentage of the community-wide MHI would be greater than 2.0%) to the already heavily burdened population would result. In addition to heavily burdening the ratepayers, the utility would experience an unacceptable and unsustainable debt level and risk losing significant revenue bases if higher levels of control/service were implemented over a 25-year or less period.

Funding capital projects even at the lower end of the CSO and SSO control scenarios considered will create a high burden for many in the UG service area. Thus, as encouraged by the EPA, supplemental socioeconomic indicators were presented to further demonstrate the poor financial capability of the UG's customers. Supplemental socioeconomic information shows the UG service area is disadvantaged across a wide variety of socioeconomic indicators, including the number of people with income below the poverty level and without health insurance. This information indicates funding capital projects to meet the *CSO Control Policy*, even at the lower end of overflow control scenarios considered in the IOCP, would create a very high burden for a significant number of households in the service area.

PUBLIC PARTICIPATION PROGRAM

The UG recognized the importance of a public participation program to educate and solicit comments from the public regarding the IOCP. The program included numerous public education materials, a five member community task force representing various stakeholder interests and geographic parts of the community, numerous road show presentations, an extensive community survey, and multiple public meetings.

Through this public participation effort, there was general public consensus about the following statements:

- Reducing overflows and improving water quality is important; however, other community issues are more important.
- Resources need to be spent wisely because Wyandotte County is an economically disadvantaged community already facing many other critical infrastructure needs and socioeconomic challenges.
- To avoid creating additional financial hardships on ratepayers, initial efforts should be focused on the community's highest priority of renewing the existing infrastructure.
- Keep this large-scale infrastructure investment within our community by positioning young Wyandotte County citizens for future job opportunities related to the IOCP and within the WPCD.

PROPOSED PLAN

Through development of the IOCP, the UG identified numerous uncertainties that could cloud the anticipated effectiveness, schedule, and costs of planned improvements, such as:

- The timing and magnitude of future regulatory compliance requirements, such as nutrient removal at the WWTPs and MS4 Permit requirements to address total maximum daily loads, are unclear.
- Local and regional engineer and construction contractor capacity and availability due to similar sewer system improvement efforts throughout the region are anticipated to result in unquantified (at this time) but likely construction cost escalation.
- Accuracy of current financial assumptions, such as changes in household MHI over time, impacts and degree of rate tolerance, and population growth trends (which have been negative), will affect residential affordability and the UG's ability to generate additional revenue.
- Kansas City's location at the confluence of two great rivers requires significant funding to maintain the flood control system. Federal flood control levee improvement needs that are defined in a recent U.S. Army Corps of Engineers study may exceed $100 million (local share), but the timing and level of commitment are unknown. The magnitude of this obligation will affect residential and utility affordability considerably.

Due to these uncertainties and the results of the tasks and studies completed to inform and prepare the IOCP, the UG realized the plan needed to be flexible and adaptive to ensure that the plan benefits are being achieved, priorities can be reevaluated, and community affordability is not exceeded. Reducing the occurrence of SSOs and controlling CSOs to a level meeting the *CSO Control Policy* presumption approach, both within a typical 20- to 25-year timeframe is not financially feasible for the UG. For these reasons, the UG proposed an interim, shorter-term (less than 20 years) plan that makes progress towards meeting the levels of overflow control identified in the characterization effort, while still representing an aggressive, high burden financial commitment.

The proposed plan focuses resources on infrastructure renewal and priority capital improvement projects. Through these system reliability improvements, additional overflow reduction will be achieved while maintaining sustainable, yet aggressive, annual rate increases. After the existing assets are in a more sustainable condition, future efforts, which include an update to the IOCP in year 10 of the proposed plan, are anticipated to create an approach to address the remaining overflows and achieve compliance with CWA requirements. As part of the update, the UG proposes to reevaluate the community's financial capability, the benefits that have been achieved, and the identified goals and priorities to achieve additional sewer overflow reduction and water quality improvements.

The proposed plan presented in the IOCP addresses the community's highest priorities of renewing the existing infrastructure and making progress towards meeting the CWA goals, balances near term public investments and benefits with the UG's financial capability, and represents the best level of overflow control achievable with the available public investment. The UG developed an IOCP that meets the goals supported by their community while facing the harsh realities of an economically-disadvantaged ratepaying community and a wastewater infrastructure system that is poor condition.

CONCLUSION

The IOCP was submitted to the DOJ and EPA in September, 2016. The UG is currently implementing the proposed plan and responding to follow-up information requests from the DOJ and EPA.

REFERENCES

United States Environmental Protection Agency, Combined Sewer Overflows - Guidance for Financial Capability Assessment and Schedule Development (Final), February 1997.

United States Environmental Protection Agency to Regional Administrators, Water Division Directors, and Enforcement Division Directors, November 24, 2014, Financial Capability Assessment Framework for Municipal Clean Water Acts Requirements.

Managing the Public Outreach, Regulatory, Funding, and Operations Challenges of Framingham's Infrastructure Improvements Program

James R. Barsanti, P.E., Framingham Department of Public Works

ABSTRACT

The Town of Framingham Massachusetts is a predominately residential community of 69,000 with a large business, commercial, industrial, and institutional component. The wastewater system dates back to the late 1,800s, and is now comprised of over 400 kilometers (250 miles) of sewers and 42 wastewater pump stations. Like many communities, as the population increased and demands on the system grew, the Town's Department of Public Works struggled to operate and maintain the existing system to serve the demands. The Town took an aggressive approach to the reconfiguring of its existing infrastructure to not only resolve operational problems, but also meet future needs, improve energy efficiency, incorporate sustainable design, and increase the system's overall capacity.

The Town made a significant fiscal commitment to staffing and has created a Capital Improvement Program with dedicated staff to work with DPW Operations and Municipal Engineering staff to manage and oversee the regulatory, planning, design, construction, and public outreach elements of the program. From a funding perspective, the Town has used a combination of traditional sources, such as municipal bonds and State Revolving Fund, but also secured funding through the American Recovery and Reinvestment Act (ARRA) and other grants.

KEYWORDS

Infrastructure, utility management, public outreach, capital improvements, funding.

INTRODUCTION

The Town of Framingham Massachusetts is located approximately 32 kilometers (20 miles) west of Boston and is a predominately residential community of 69,000 with a large business, commercial, industrial, recreational, and institutional component. As shown in Figure 1, the Town is comprised of several villages with Framingham Centre and South Framingham being the oldest and most urbanized parts of the community. The northern portion of the community includes the villages of Saxonville, Nobscot and North Framingham which are more rural. The Town is home to many restaurants and commercial shopping establishments, especially in its eastern section. Businesses and industries are located throughout the community and include the corporate headquarters of Genzyme, Bose Corporation, and Staples. The Town also serves as a regional hub for transportation and is served by freight and commuter rail facilities and several bus systems providing transportation to destinations in New England and the northeast.

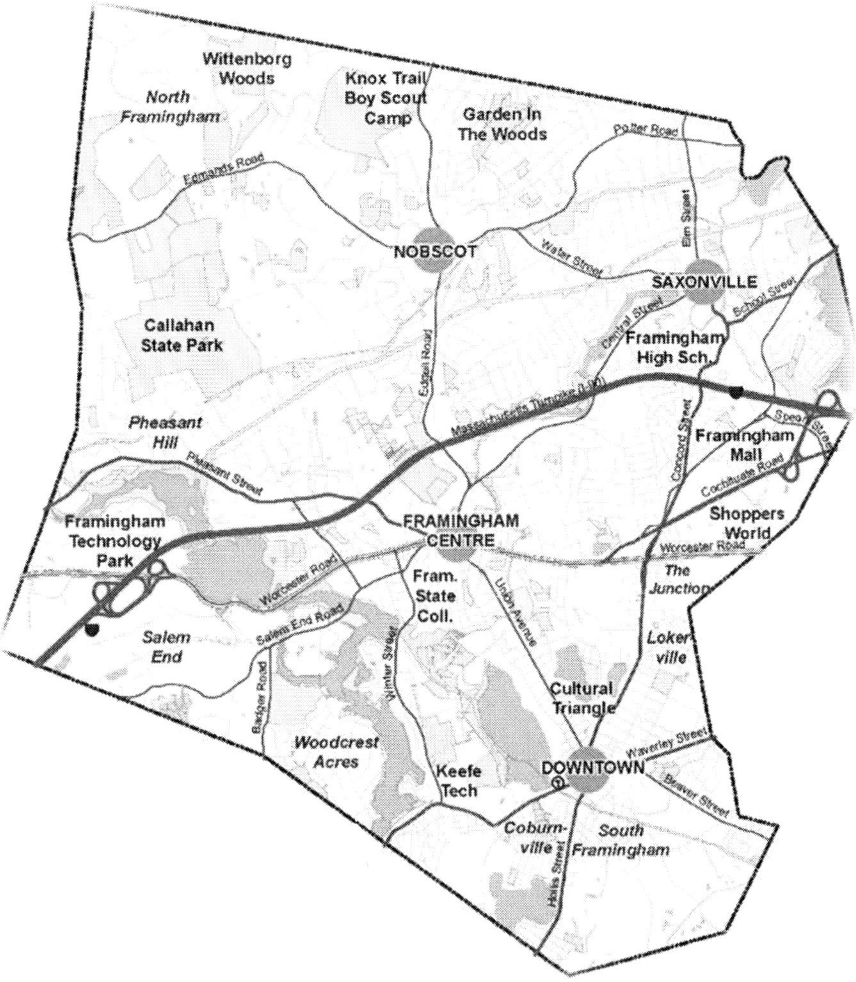

Figure 1: Town of Framingham, Massachusetts

Major features of the Town have influenced its growth and development including the Sudbury River, several manmade reservoirs intended for emergency water supply, three large aqueducts, the Massachusetts Turnpike and Worcester Road (State Route 9), and multiple at-grade rail crossings. Access to the Massachusetts Turnpike has provided for significant growth in Town as access to Boston and many other parts of the state is relatively convenient.

Framingham has a separate sanitary sewer system with an average daily flow of 27,000 cubic meters per day (7 MGD). Wastewater is collected within the Town's collection system that begins in the western and northern reaches of the system. Where necessary, the wastewater collection system employs the use of wastewater pumping stations and inverted siphons to convey flow across physical obstructions or to make up for elevation losses due to topographical changes. All flows are conveyed to the southeastern edge of Town bordered by Natick. At this point wastewater discharges to the Massachusetts Water Resources Authority's (MWRA) Framingham Extension sewer and is conveyed downstream through their system of interceptors where it is fully treated at the Nut Island and Deer Island Treatment Plants and eventually discharged into the Boston Harbor. Framingham serves as the westernmost collection area for the MWRA. Several neighboring towns utilize portions of Framingham's wastewater infrastructure to access the MWRA interceptors. Ashland disposes of all wastewater flow through MWRA system, and so all of the Town's flow passes through Framingham. However, only a portion of flow from Natick, Sherborn, and Southborough passes through Framingham, as they do not depend entire on the MWRA for their wastewater needs. Figure 2 identifies schematically how wastewater collected from Framingham and its neighbors is conveyed by the MWRA to the Deer Island Treatment Plant and the Boston Harbor.

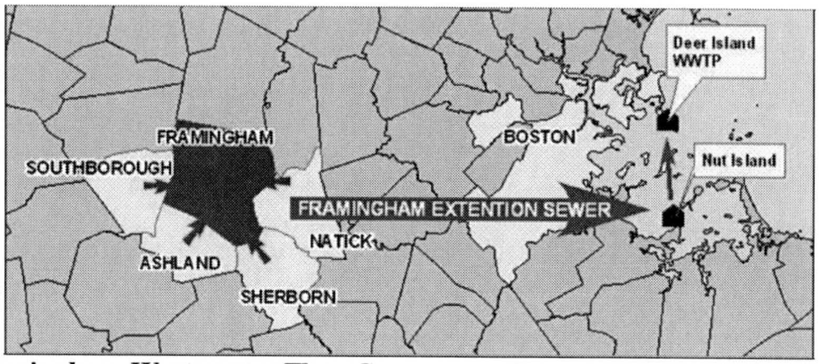

Figure 2: Framingham Wastewater Flow Conveyance Schematic

HISTORICAL BACKGROUND

Industrialization Period: 1880 – 1945

From the time the Town was incorporated in 1700 to the end of World War II in 1945, the population increased at a steady, manageable pace to 25,000 in 1945 (Figure 3).

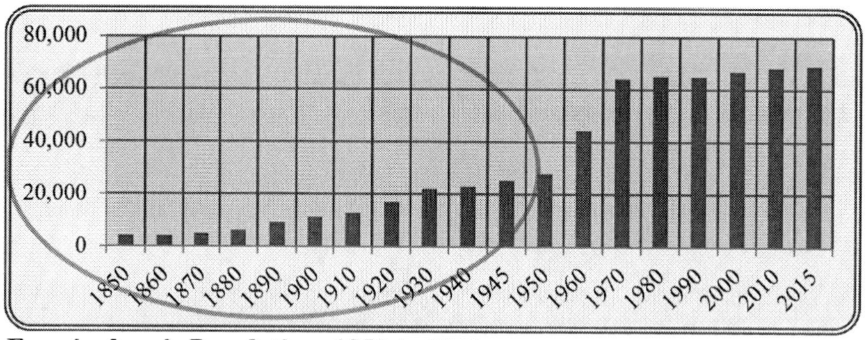

Figure 3: Framingham's Population: 1850 to 1945

During this time, the infrastructure also grew at a steady pace to meet the demands of the Town (Figure 4). For each of the periods described in the next sections, four figures show the progress of the construction of the Town's wastewater infrastructure. The infrastructure is shown on a cumulative basis, depicting where and how the Town's growth has had an impact on the infrastructure.

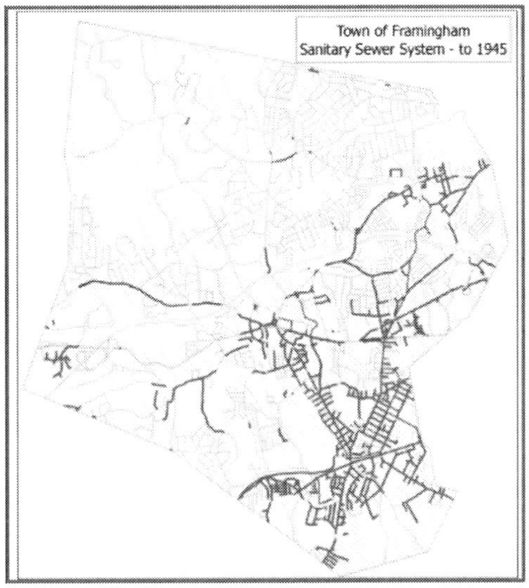

Figure 4: Framingham Wastewater Infrastructure Prior to 1946

The first wastewater sewer lines were built in 1890s, and until about 1945 consisted mostly of vitrified clay pipe except for a few locations where cast iron pipe was needed for its greater strength. About 90 kilometers (57 miles), or one-quarter of the Town's wastewater system, was installed during this time.

Post War Boom: 1945 - 1970

Shortly after World War II, manufacturing, industrial and commercial enterprise increased throughout the Town. The population almost tripled in the 25 years between 1945 and 1970 (Figure 5).

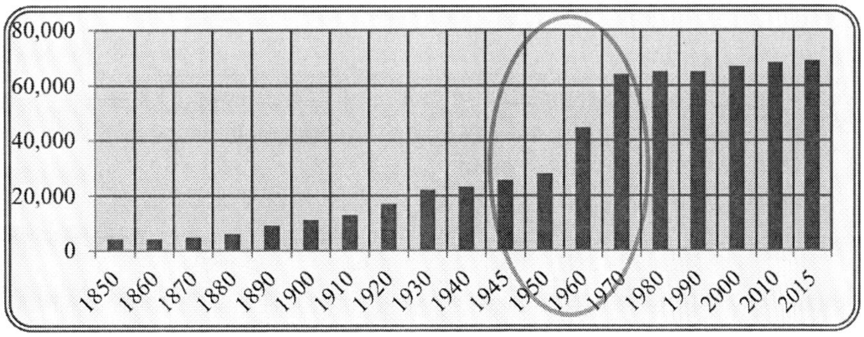

Figure 5: Framingham's Population: 1945 to 1970

It was a challenge to construct homes, roads, water, wastewater, and stormwater infrastructure to keep up with the pace of the population growth. Almost all the infrastructure during this time was installed by private developers. Each developer typically addressed the infrastructure needs of their individual development without assessing the impact on downstream systems. As a result, infrastructure was constructed in a haphazard fashion, with little overall planning or oversight.

Figure 6: Expansion of Wastewater System 1945 to 1970

After World War II the sanitary sewer system also had to expand significantly to meet the growing population needs. Figure 6 compares the wastewater infrastructure installed starting in 1946 (in orange) with those installed before 1946 (in red). Over 50%, or 215 kilometers (135 miles) of sanitary sewer lines were installed during this period. As noted above, each developer addressed the infrastructure needs of their individual development without any apparent obligation of a holistic downstream assessment. This was especially problematic for the wastewater system, and applies both to pumping stations and piping. For example, the Water Street sewer installed in 1958 began experiencing sewer overflows only a few years later as the surrounding farmlands underwent accelerated development. By 1969, the Town's consultant Metcalf and Eddy identified this sewer line as significantly undersized. Of note during this time:

- It was not until after 1945 that the proliferation of sewer pumping stations started. Developers installed pumping stations in the small developments and subdivisions as it was easier and cheaper to build a pumping station than to reconstruct the piping system to allow the use of gravity sewers to convey the wastewater, especially considering obstacles such as railroads, MWRA water transmission lines, and numerous water bodies. Over 20 pumping stations were built to support the 215 kilometers of new pipe.
- As the system expanded with the new development, there was little concern placed on the downstream sewers. Many of the existing pipes were under-sized for the new flows leading to severe sewer overflows, especially in the northeast section of Town.

- In addition, the material of asbestos was introduced in the 1920s and began being used in sanitary sewer pipes. Nearly one-third (31%) of the Town's sanitary sewer pipes were constructed using asbestos cement. Almost all this pipe was installed between 1945 and 1970. Although asbestos is considered a hazardous building material, it is safe in use as a sanitary sewer pipe while the pipe is in good condition. However, sewer pipes made of asbestos cement deteriorate relatively quickly in certain subsurface environments. When the pipe needs to be replaced the removal process involves the same costly safety requirements as used for removing asbestos building materials.

Population Stagnation, No Infrastructure Investment: 1971 – 2003

After 1970, population growth slowed as manufacturing dropped off (Figure 7) and the Town began to reach capacity under its zoning plan. In the 30-year period from 1971 to 2003, the population rose only about four percent (4%), compared with 275% during the 25-year post-World War II boom.

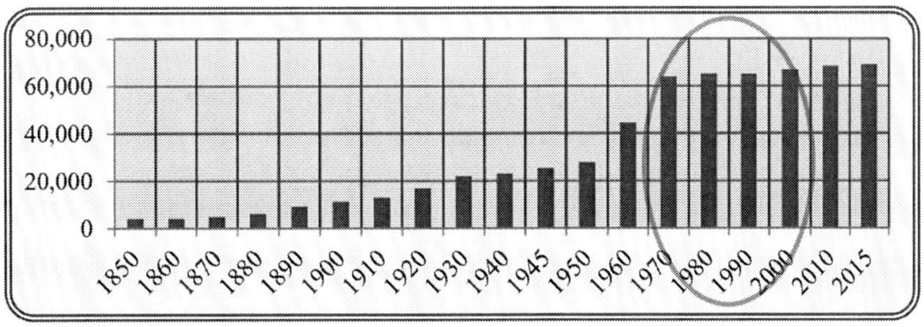

Figure 7: Framingham's Population 1971 to 2003

This pace mirrored the limited infrastructure installed during this time, shown in Figure 8.

Figure 8: Expansion of Wastewater System 1971 to 2003

After 2003 the Town began supporting the infrastructure from a financial and structural aspect. After the population boom, the wastewater system continued to expand but at a slower rate. Figure 8 compares the sewer lines installed from 1971 to 2003 (green) with those installed before 1971 (red and orange). About 25 miles of sanitary sewer lines were installed during this period. During this time:

- Compared with the 21 pumping stations installed between 1946 and 1970 to support 135 miles of new pipe, 27 pumping stations were built between 1971 and 2003 to support just 25 miles of new sewer pipe. At its height, Framingham had 48 sewer pumping stations.
- The pumping stations were small as they supported small sewersheds, and were installed in areas that had already been partially built out, where the standard gravity system would not be feasible unless existing wastewater pipes were reinstalled at a deeper level. Unfortunately the deeper sewers were not constructed.
- The proliferation of pumping stations resulted in premature failure of the system at some locations due to the production of corrosive hydrogen sulfide gasses associated with long force mains discharging from the pumping stations.

Rebuilding Framingham: 2003 to Present

As indicated in Figure 9, since 2003, Framingham's population growth has slowed even further to about 3% over the last 13 years. There are signs that it may increase at a faster rate in the future, as shown in recent increases in the school-age population. The rezoning of certain districts for transit oriented development may also spur population growth.

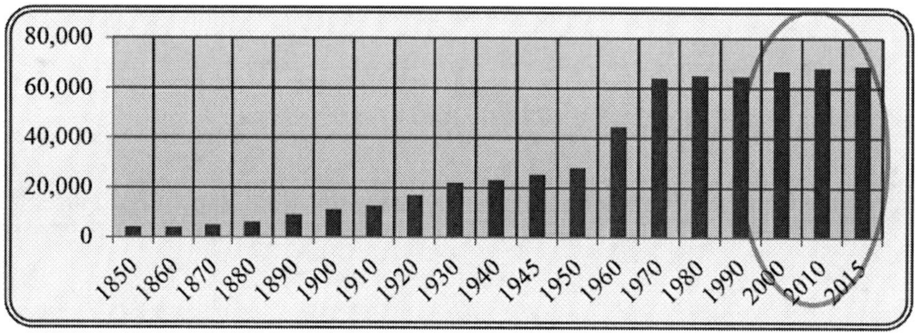

Figure 9: Framingham's Population: 2000 to 2016

EVOLUTION OF THE PUBLIC WORKS DEPARTMENT'S UTILITY MANAGEMENT PHILOSOPHY

Prior to 2003, a Board of Public Works managed the Department. The Board was composed of unpaid Town officials all of whom were part time overseers. The Chair of the Board signed all invoices and made all decisions. The Board hired a director, but the position had little authority. As a result, no director stayed in the position for long, and the Department had seven directors in 11 years. The goal of the department was to keep rates down, and spend as little as possible on operations and maintenance. Few, if any, improvements were made so as to keep tax and enterprise fund rates the lowest in the state. As a result, there was no documentation or mapping of the Town's assets, only anecdotal information regarding the systems. Operators struggled to determine a pipe's location just to fix breaks and backups. From the Town's 1973 Annual Report, the Board of Public Works stated, "Unless we provide a methodical program of maintenance and repair for our facilities we may one day be faced with a need for total replacement of all facilities. The cost would be too burdensome for the taxpayers to afford."

In 2003, authority for the Department transitioned from the Board to an Executive Director. The Department was transformed to a more professional operation in the following ways.

- The Department structure changed from hierarchical to programmatic. Each staff person was given responsibility for specific areas such as water, wastewater, facilities, etc., along with authority and accountability for their equipment and their area of responsibility. Under this structure, instead of using a department truck to fix an unspecified problem not knowing what kind of tools might be needed, a staff person has the proper equipment in their truck for their specific asset.
- The Department began recruiting professional management staff. Each position was no longer a "job" but instead became a "profession".
- To understand the needs of the Town, the Department created an inventory of all infrastructure assets. This included major equipment, vehicles, water mains, hydrants, valves, water pumping stations, water tanks, sewer manholes, sewer mains, sewer pumping stations, drainage manholes, catch basins, drainage pipes, drainage outfalls, roadways, sidewalks and traffic signs.
- The Department took the fledgling Geographic Information System (GIS) and expanded it to catalogue all the Town's infrastructure assets. Today the GIS is used in daily operations to identify problems, document work orders and document subsequent repairs.
- The Department commissioned master plans for each of the major assets of roads, water, sewer and stormwater. These plans have identified problem areas and have made recommendations for repairs, rehabilitation, or replacement.
- The Department began meeting frequently with Town officials, residents and business owners to communicate the Town's infrastructure needs, findings of the master plans and progress towards the goal of improving the quality and life expectancy of the infrastructure.

REGULATORY CHALLENGES PROVIDE IMPETUS FOR ACTION

Prior to 2003, due to the age of the system, the proliferation of pump stations and force mains, system hydraulic deficiencies, and lack of historical asset management as noted above to address infrastructure rehabilitation and renewal, Framingham's wastewater collection system was experiencing frequent sanitary sewer overflows and high hydrogen sulfide levels in portions of its system. In 2002, the MWRA required that Framingham significantly improve discharge standards due to high hydrogen sulfide levels being discharged from two long force mains to the MWRA's wastewater conveyance system to prevent damage to the pipes and not create problems downstream. Framingham entered into a Settlement Agreement with the MWRA in November 2002 to meet these standards.

In the event of an overflow or bypass of a sanitary sewer, the Massachusetts Department of Environmental Protection (MassDEP) requires a notification if the wastewater is discharged from the Town's infrastructure to private property or the environment. In 2006, MassDEP contacted the Department to discuss instances of frequent sanitary sewer system overflows from the wastewater system. At the same time, the Department was in the process of reviewing the history of the overflows and other problems with the wastewater system. The Department determined that significant system improvements needed to be performed for the wastewater system and proactively met with MassDEP to plan and schedule a long range capital improvement plan to repair and improve the system. MassDEP agreed with the plan, resulting in an Administrative Consent Order issued in 2007. While the consent order was a mandate by the state agency, there were no fines levied to the Town because of the collaborative effort shown by the Department.

The Department then developed a Comprehensive Wastewater Master Plan, completed in 2008, containing 34 critical capital improvement projects to address the issues raised in the consent order. As of 2016, more than half the projects have been completed. While this work has resulted in major improvements to the operation of the wastewater collection system, the Department still has many projects to design and construct to make the necessary improvements to the system. In addition, repairs and replacement are needed for the pipes that are part of the wastewater system, especially the 21% of the Town's wastewater pipes and manholes that are more than 75 years old.

PUBLIC WORKS DEPARTMENT ORGANIZATIONAL STRUCTURE

Due to the need to reorganize the day-to-day operation of the Department and address the capital improvements dictated by the DEP Administrative Consent Order and MWRA Settlement Agreement, the Department reconfigured its Wastewater Operations staff and augmented it with a Capital Engineering and Construction Group. The overall organizational structure as it is currently configured is summarized below:

> ***Executive Director and Deputy Director*** - Work daily with Town Manager, Selectmen, other Town Department Heads, DPW management and senior staff, and community stakeholders;
>
> ***Division Director/Assistant Director*** - Daily management and coordination of all Division activities;
>
> ***Operations Managers*** - Daily planning and quality assurance of system operation, reliability, and efficiency;
>
> ***Operations Supervisors, Technicians and Heavy and Medium Equipment Operators*** - Coordinate and oversee daily work activities of crews, operate, maintain, inspect, and repair system assets;
>
> ***Administrative*** - Logistics, billing, purchasing and procurement, budgeting and financial planning; and,
>
> ***Capital and Municipal Engineering and Communications*** - Execute planning, design, and construction of capital improvements, operations support, permitting, utility coordination, and public notifications and outreach.

Capital Program Organizational Structure

Our Municipal Engineering Department is responsible for the review and approval of development plans, street opening and trench permits, utility coordination, and public way and easement acceptances. The Capital Engineering staff is charged with the management of all aspects of capital improvements projects. We have found it is essential to building a successful program as assigning these tasks to operations staff would consume all their time and detract from their ability to operate and maintain the system and rapidly respond to the daily needs of our customers. The structure of our capital projects team is summarized below:

> ***Chief Engineer*** - Program Manager, Standards Adherence and Coordination of Information Transfer between Capital and Municipal Engineering;
>
> ***Capital Project Manager and Directors*** - Daily project activities, consultant, contractor Police/Fire /School Transportation coordination, resident and business issues/impacts, traffic management, 24/7 availability;
>
> ***Communications Project Manager*** - Information management, notifications, outreach;
>
> ***Operations Capital Projects Coordinators*** - Liaison between Operations (Water/Wastewater and Highway/Sanitation Divisions) and Capital Engineering and Contractor; and,
>
> ***Capital Administrator*** - Respond to calls from residents or business regarding construction impacts and assist with coordinating resources.

Starting in 2006, with the new management structure and staff in place, the Department embarked on a long term Capital Improvement Program to address the requirements of the Administrative Consent Order and Settlement agreement and improve is wastewater infrastructure. Since that time, the Department has replaced or improved its collection and pumping systems in several challenging problem areas. Figure 10 summarizes the improvements made to systems in the past 10 years.

System	Current System Inventory	Improvements 2006 - 2016
Wastewater	365 km of Gravity Main 6,900 Manholes 21 km of Force Main 42 Pumping Stations	Replaced or rehabilitated 43 km of pipe and associated manholes Eliminated 10 km of force mains Eliminated 7 pump stations Replaced or rehabilitated 9 stations

Figure 10: Wastewater Improvements Since 2006

CUSTOMER FOCUS

Our operators, engineers and administrators work closely to address the day to day challenges of operating and maintaining our system while also planning, designing and construction our capital improvements. While executing these objectives, we maintain a quality and customer focus that emphasizes our management and stewardship of the Town's investments in its infrastructure.

- Quality assurance – Strive to build and maintain technical and fiscal credibility by assuring projects are constructed that meet contract specifications and standards and budgets;

- Professional pride at all levels in all aspects of our work;

- Documentation - Leveraging our GIS mapping and work order system to schedule and provide consistent and timely operation and maintenance activities;

- Operations and customer focus - Acknowledge, respond, and follow up to resident and business impacts due to construction activities; and

- Communications and public and governmental outreach at all project levels: planning, design and construction.

BUILDING COMMUNITY UNDERSTANDING, TRUST, AND SUPPORT

The Department has worked diligently over the last 10 years to systematically build support from the various stakeholders in the community including its residents, businesses and institutions, developers, other public utilities, and local, state and federal regulatory agencies as shown in Figure 11. This has led to the successful approval of the projects at the planning level through the execution of the projects through construction and implementation.

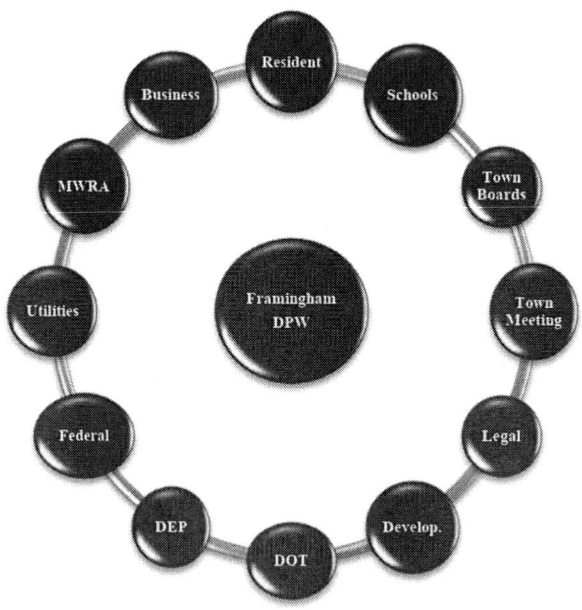

Figure 11: Coordination of Stakeholders is Critical to Program Success

This has been accomplished with a dedicated public outreach program utilizing an array of media platforms to keep the community informed about elements of the projects including:

- Notifications
    - ✓ Letters
    - ✓ Door hangers
    - ✓ Project maps
- Community/Town Meetings
- Regulatory coordination meetings
- Business liaison
- Website/Social Media
    - ✓ Traffic management plans and alerts
    - ✓ Weekly construction schedules
    - ✓ DPW capital improvements web page
    - ✓ Facebook/Twitter
    - ✓ Video
- School presentations

Our efforts include preconstruction community meetings to describe the projects to residents and businesses to listen to their concerns and to be prepared to address issues before activities begin. This allows the residents to take a level of ownership of the project and begin to feel the pride in their community that infrastructure improvement projects are being done for the benefit of their local neighborhood and the greater community. Our Communications Project Manager serves as a dedicated business liaison who meets ahead of time with the business owners to advise them of the project schedules to help with minimizing the impact of the construction to their day to day operation. We utilize all social media platforms to rapidly notify residents of project issues, either scheduled or unanticipated, so that construction activities can be described, coordinated, and impacts mitigated. Our project managers provide 24/7 on-call support during off hours and weekends in the event of a need for the contractor to return to the job site to address an issue.

FUNDING

The Town has used an array of funding sources for the capital improvements projects. Annually, we meet with the Town Manager and Board of Selectmen, Chief Financial Officer, Standing Committee of Ways and Means and Public Works to describe the proposed projects to be funded by Town Meeting for the fiscal year. The Town uses a Water and Sewer Enterprise Fund to provide funding for these projects and approval is by 2/3 vote of the Town Meeting members each spring. The Town has utilized both the State Revolving Fund and its own bonding capacity utilizing tax exempt municipal bonds for funding some projects. The Town has also benefitted from grants from the Massachusetts Water Resources Authority, Massachusetts Life Sciences, and Mass Works programs. In addition, the Town received approximately five million dollars in funding through the American Recovery and Reinvestment Act in 2010. The ability to utilize multiple funding sources has had a positive impact of the ongoing success of the program and we continue to pursue multiple options to fund the projects.

REFERENCES

2016 Infrastructure Report, Framingham Department of Public Works, May 2016

Joint Depth, Duration, Frequency Analysis of Rainfall Leading to Sanitary Sewer Overflows

Felix Belanger, PE, Vice President Water Group Lead, Central Region, Tetra Tech, Tulsa, OK
Baxter Vieux, PE, PhD, Principal Vieux & Associates, Inc., Norman, OK
Matt Vaughan, P.E., Manager of Water Pollution Control, City of Tulsa, Oklahoma

ABSTRACT

The purpose of this approach is to better characterize the probability of a rainfall event that is Tin terms of the sanitary sewer system response? The effects of rainfall on sanitary sewer overflows can have markedly different effects on the sewer system response, and the volume of an SSO, depending on antecedent rainfall and soil saturation. During periods of extended rainfall, soil moisture and groundwater levels can rise, and when relatively more intense rainfall subsequently occurs, overflows can happen which would not be expected under drier soil conditions. Understanding the probability of a given rainfall depth coupled with antecedent rainfall is termed the joint probability depth duration frequency (JDDF). An information tool was developed and continues to support ongoing JDDF analysis on-the-fly. With more complete and accurate quantification of risk, the regulatory agency did not feel compelled to fully prosecute the violation. At any time in the future, it will be possible to have ready analysis of the JDDF, and thereby gain a clearer understanding of how rainfall interacts with the collection system, and more effectively target I/I in critical areas.

KEYWORDS

Rainfall Derived Inflow Infiltration, Sanitary Sewer Overflows, Risk Management, Rainfall Depth Duration Frequency

INTRODUCTION

Classifying rainfall events according to antecedent and the more intense main rainfall period requires query and analysis of historical rainfall records. Gage-adjusted Radar Rainfall (GARR) is very beneficial to understanding rainfall depths over sewer catchments at high resolution in time and space, as described in Vieux and Vieux (2010). Finding the number of occurrences when intense rainfall is preceded by wet periods, allows us to classify the probability of joint occurrence. For example, when a 2-yr rainfall or less is preceded by a wet period of a given depth, its effect may be more consistent with the sewer system response with a 5-yr return period. Matching these events with flow information for the last 60 years would be valuable information, however, such records rarely exist, or if they do, only a dozen or fewer years may be available. Rainfall data is much more available over a longer term than is flow monitoring. To determine the probability of occurrence over a long period (more than 30 yrs), both annual recurrence interval (ARI) depths, and continuous rainfall data are needed. Table 1 presents the ARI storm depths for 1-1000 yr from NOAA Atlas 14 Vol. 8 (Perica et al., 2013) for the Tulsa International Airport (1948-2015). An important threshold is the 5-yr ARI, which for this rain gauge location is 2.59 inches in 2-hr.

The primary purpose of this approach is to understand how often the joint occurrence of antecedent rainfall occurs historically. The risk can now be quantified of having a wet antecedent period of 3-7 days followed by intense rainfall such as the 2-yr storm. The joint probability information proved useful in responding to the regulatory agency for an unusually large overflow. The City now has access to historical joint occurrences that is useful in planning, and for responding to a notice of violation for future SSO occurrences. Continuous rainfall data from gauge-adjusted-radar-rainfall (GARR) is used to augment the database of historical rainfall. This allows the client to perform ongoing analysis of joint depth duration frequency (JDDF) and respond to regulatory actions with more accurate information.

Table 1 At-station precipitation frequencies for Tulsa International Airport

Duration	Annual recurrence interval (years)									
	1	2	5	10	25	50	100	200	500	1000
30-min	1.05	1.21	1.5	1.76	2.16	2.5	2.86	3.25	3.81	4.27
60-min	1.39	1.63	2.05	2.42	2.98	3.44	3.94	4.46	5.21	5.8
2-hr	1.72	2.04	2.59	3.08	3.8	4.39	5.01	5.67	6.6	7.34

RESULTS

Understanding and anticipating the collection system response to the joint occurrence of a given return period rainfall coupled with antecedent rainfall, JDDF, provides a useful tool for planning and responding to regulatory notices. In May 2015, a storm with less than a 2-yr return period produced a large sewer overflow though none was expected for a storm of that duration and intensity. The Spring of 2015 was unusually wet, and was one of the wettest springs on record. Elevated groundwater levels and generally wet soils were more prevalent than usual. When intense rainfall occurred on saturated ground, the sewer system response was greater and the SSOs are more severe. By way of example, the Tulsa Metropolitan Utility Authority received a Notice of Violation for an SSO at Junction Box 782 (JB782). There were two events that caused significant overflows in May 2015. The first occurred on May 8th, when 1.13 inches of rain fell in 60 minutes. This, by itself is less than the 1-yr event, and expected to occur each year. Figure 1 shows the distribution of rainfall depth on the 8th. For this event and its 3-day antecedent, the hyetograph averaged for the TL-04 basin is shown in Figure 2. However, when the 3-day antecedent rainfall of 1.6 inches is added, the joint probability is greater than a 6.18-yr return period. Table 2 summarizes the number of occurrences for an event depth and antecedent of this magnitude. From the number of occurrences, the probability can be calculated by the Weibull formula (Bedient et al., 2013). The second event occurred on May 16-23 where the combination of 2.85 inches with an antecedent rainfall of 0.8 inches exceeded the 11.33-yr return period. In both cases, the sewer system response was greater than expected and resulted in SSOs from JB782. However, the City was able to show that the joint probability of these events was more than the current design standard consisting of the 5-yr return period.

Table 2 Number of occurrences each year for Event 1 (May 8, 2015) from 1948-2015

Year	Number	Year	Number
1948	2	1986	1
1949	2	1987	1
1959	3	1990	1
1961	1	1991	1
1962	1	2003	2
1971	2	2004	1
1973	2	2006	1
1974	5	2008	1
1975	1	2014	1
1978	1	2015	7
1981	1	Total	38

In conjunction with the analysis of the May 2015 SSOs, a JDDF tool was developed which provides the City with ongoing ability to classify joint events. The tool, which has been running for over a year, performs ongoing JDDF analysis on-the-fly. Table 3 presents a summary of JDDF events with antecedent periods up to 6 days, for catchment (TL-04). When an event's antecedent rainfall is considered the Joint ARI (last column) is greater than the DDF ARI (third column). In some cases, the Joint DDF ARI is more than a 5-yr event, compared to the DDF ARI of only 0.5 or 1.5-yr. This analysis provides the ability to more completely and accurately quantify the recurrence interval of the *joint* event, and helps the regulatory agency recognize the complexity of sewer system planning and provided them with justification to not take action regarding these violations.

Table 3 JDDF analysis for 3day antecedent periods in catchment TL-04

Start Date	**Depth (in)**	**DDF ARI (yr)**	**DDF Depth (in)**	**DDF Duration (hr)**	**Joint ARI (yr)**
6/12/2007	0.5	1.011	1	0.855	1.9
6/17/2015	0.5	1.859	12	2.426	5.2
3/13/2016	1.5	1.409	1	1.628	5.6
10/6/2016	0.5	1.014	3	0.255	1.2

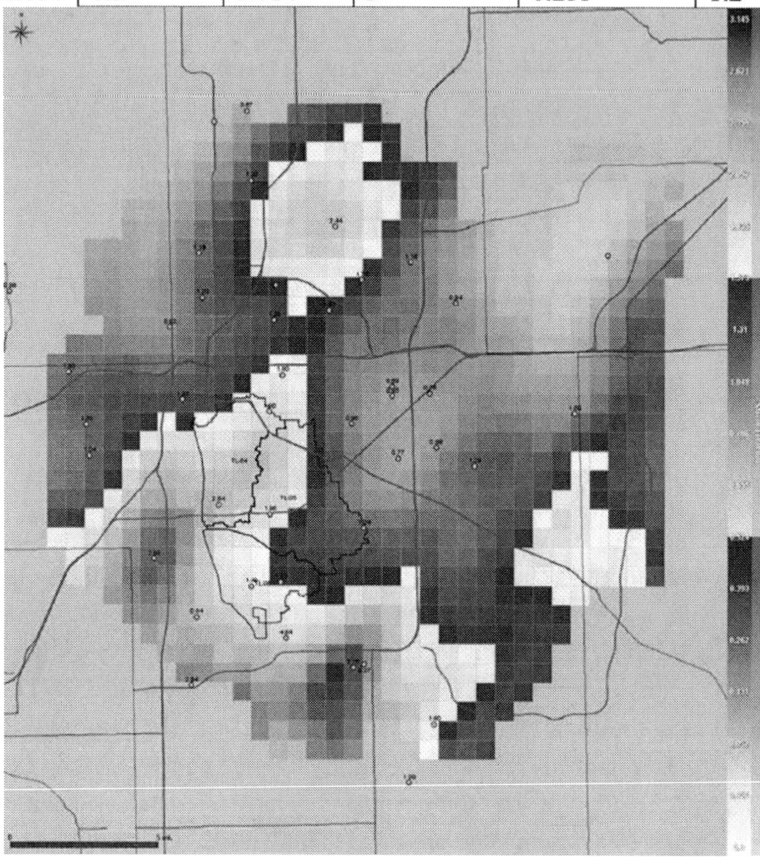

Figure 1. GARR data showing rainfall totals for the 2015-05-08 00:00 to 2015-05-09 00:00 CDT

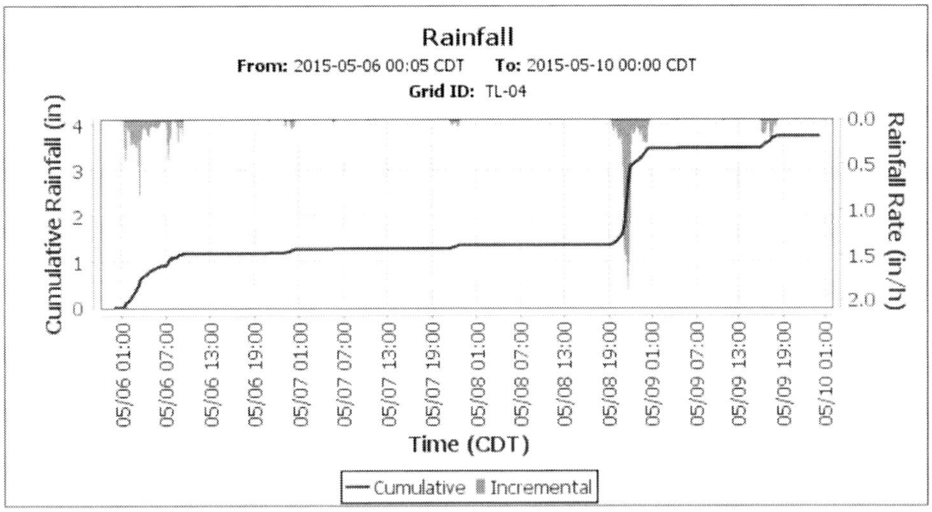

Figure 2. GARR hyetograph averaged over TL-04 showing both the antecedent rainfall on May 6, and the main event on May 8, 2015

SUMMARY

Taking account of the joint occurrence of a rainfall event and its antecedent helps understand why a sanitary sewer system responds with larger than expected overflows. The risk can now be quantified of having a wet antecedent period of 3-7 days, followed by intense rainfall such as the 1-2 yr storm. JDDF analysis is used to properly quantify risk, providing crucial information for planning and regulatory purposes. Going forward, the City and its consultant will be able to analyze past and future events, and thereby gain clearer understanding of how rainfall interacts with the collection system, and more effectively target I/I in critical areas.

REFERENCES

Bedient, P.B., W.C. Huber, B.E. Vieux, 2013. *Hydrology and Floodplain Analysis.* Fifth Edition, Prentice-Hall, Inc., One Lake St., Upper Saddle River, NJ 07458. ISBN 0-13-256796-2. p. 801.

Perica, S., Deborah Martin, Sandra Pavlovic, Ishani Roy, Michael St. Laurent, Carl Trypaluk, Dale Unruh, Michael Yekta, Geoffrey Bonnin (2013). NOAA Atlas 14 Volume 8 Version 2, Precipitation-Frequency Atlas of the United States, Midwestern States. NOAA, National Weather Service, Silver Spring, MD.

Vieux, B.E. and J.E. Vieux, 2010. Development of Regional Design Storms for Sewer System Modeling. WEFTEC, New Orleans LA, pp. 6248-6263.

SSO ALERT
Sewer Spill Prevention Using an Automated Alert System

Christopher J. Parrish[1]
[1]City of Atlanta, GA Department of Watershed Management
*Email: CParrish@atlantaga.gov

ABSTRACT

The SSO Alert software was developed to monitor over two hundred and fifty sewer level and flow meters for anomalous conditions that could indicate a sanitary sewer overflow (SSO). The application compares recent sewer level readings with historical averages and generates alerts for any anomalies. Texts and graphical email alerts are sent to on-call staff. The history, development, challenges and results are covered.

KEYWORDS: SSO, alert, Python, Eclipse, IDE, sewer, manhole, overflow, text, email, hydrograph, diurnal average, computer, smart phone, sewer spill, consent order, Isco Flowlink®, Oracle®, open-source software, statistical process control, cloud dashboard

INTRODUCTION

With its system of remote sewer level and flow meters, the city of Atlanta's Department of Watershed Management (DWM) had the infrastructure in place to monitor most of its sewer system. To maintain a high level of data quality, every morning a watershed employee reviewed the most recent data from all of the meters looking for equipment and sewer flow problems. This process took over two hours while repair crews waited for work orders to be generated. What they needed was a way to rapidly assess meter conditions and get workers into the field as quickly as possible.

An application called the Error Report was developed to compare recent meter readings with historical values, in much the same way as an operator would perform. The Error Report worked much faster, taking only a few minutes to highlight meter problems such as erratic readings, silt covered area-velocity (AV) sensors, low battery voltages and communication failures. Also, the Error Report counted the number of erroneous readings to give the operator a sense of how poorly a meter was functioning. The success of the Error Report led to an additional application called the SSO Alert.

The SSO Alert focuses on sewer levels, and how they deviate from the norm. Any deviation of sewer levels above or below the dry-weather norm is cause to suspect that a sewer might be backing up. The primary job of the SSO Alert is to detect these small level deviations and alert on-call staff of suspected blockages.

SiteName	CommStatus	BattCond	Last Data	Voltage
INC UP WQ__2105 Interface Module	NO PUSH	LOW VOLTS	2/17/2016 12:30	8.17 V
INC UP WQ__2150 Area Velocity	NO PUSH	LOW VOLTS	2/17/2016 12:30	8.43 V
INC-11__2150 Area Velocity	NO PUSH	LOW VOLTS	2/11/2016 22:15	8.84 V
INC-20__2150 Area Velocity	NO PUSH		2/23/2016 7:45	11.28 V
PR-SRV09A__2150 Area Velocity	NO PUSH	LOW VOLTS	2/21/2016 10:00	9.22 V
PRC-21__2150 Area Velocity	NO PUSH	LOW VOLTS	12/21/2015 20:00	8.32 V
PRC-21__2160 LaserFlow Module	NO PUSH	LOW VOLTS	12/21/2015 20:00	8.31 V
PTC-14__2150 Area Velocity	NO PUSH		12/29/2015 7:45	12.09 V
PTC-25__2150 Area Velocity		LOW VOLTS	2/24/2016 3:30	9.23 V
SRV-09__2150 Area Velocity	NO PUSH		2/18/2016 9:15	11.27 V
UTC-11__2150 Area Velocity	NO PUSH		2/23/2016 10:00	10.95 V

Diurnal Average Report Period: last 12 hours

Only sites with >12 errors

Site Name	50% Above Avg Level	30% Below Avg Level	30% Below Avg Velocity	70% Below Avg VSR
1stAVE__2110 Ultrasonic Module	43	0	0	0
2MCCLELLAND__2110 Ultrasonic Module	44	0	0	0
Airport-Hapeville__2110 Ultrasonic Module	0	40	0	0
Ashburton Ave__2110 Ultrasonic Module	0	23	0	0
Beechcrest__2110 Ultrasonic Module	38	0	0	0
Black Forrest__2110 Ultrasonic Module	32	0	0	0
Braebun Circle__2110 Ultrasonic Module	34	0	0	0
CASCADE02__2150 Area Velocity	20	0	0	1
Cator Woolford__2110 Ultrasonic Module	23	9	0	0
Charlie Brown__2150 Area Velocity	36	0	8	0
CMC-02M__2150 Area Velocity	20	0	0	0
Fransiscan__2110 Ultrasonic Module	38	0	0	0
Ft. Valley West__2110 Ultrasonic Module	23	0	0	0
Garrison Dr.__2110 Ultrasonic Module	18	0	0	0
Harvard 2110__2110 Ultrasonic Module	40	0	0	0
Hermance 2110__2110 Ultrasonic Module	33	0	0	0
Home Depot__2150 Area Velocity	6	30	0	0
INC-02__2150 Area Velocity	26	0	0	0
INC-07__2150 Area Velocity	19	0	0	0
NancyCreek48__2110 Ultrasonic Module	16	0	0	0
NCR-02M__NCR-01	2	0	0	1

Error Report

DEVELOPMENT

It was noticed that the Error Report could detect small level anomalies that were sometimes imperceptible to trained pollution control monitors. This effect led to the development of the SSO Alert. The software was modified to focus on level anomalies and the ability to alert on-call staff. This function was called rapid anomaly detection (RAD).

RAD Function

RAD archives ninety days of sewer level data in a comma-separated value (csv) file for each site. The data is averaged in fifteen minute increments over the ninety day period to produce one day's worth of 96 diurnal average readings per sensor. Upper and lower error limits are calculated as a percentage of each diurnal average. Every two hours, recent level data is queried from the Isco Flowlink Oracle database, appended to the ninety day average and compared to their respective diurnal averages. If a recent data value falls outside of the error limits, the site is flagged for alert. This is a fairly simple process and executes very quickly.

Data Management

Flowlink software by Teledyne-Isco is used to manage all of the city's sewer flow meters, rain gauges and water quality sites. Also, process data is imported from treatment plant Foxboro distributed control systems (DCS) and stream water quality data from the US Geological Survey (USGS). All of the data is stored in an Oracle database, and Flowlink manages data storage, display, and editing. Structured Query Language (SQL) queries were written in Python to retrieve meter/site information and stored data for analysis. Queried analysis data is stored in comma-spaced value (CSV) files for easy access.

Python

Python was selected as the development language using Eclipse as the integrated development environment (IDE). They are both open-source and free for non-commercial development. Python is easy to learn and flexible in that both functional and object-oriented methods are supported.

The SSO Alert was developed completely in-house using decades of institutional knowledge and quite a bit of programming time. It demonstrates what people can build on their own when they are given sufficient time and resources.

SSO Alert

Every two hours, recent data is queried and analyzed using RAD looking for sewer level changes that might indicate a potential blockage. Data is compared to customizable error limits and any outliers will cause an alert to be generated. Any alert generates a text and email that are sent to on-call staff.

The text alert lists flagged site names and their HIGH/LOW/NO_PUSH status. The email consists of a hydrograph pdf and a text file for each alerted site. The hydrograph is automatically generated and includes diurnal averages, error limits and recent data (figure 1). The site-specific text file displays address information, pipe and manhole dimensions, contact information and special instructions (figure 2). The on-call person does a quick review of the data, and then if the conditions warrant, calls the Sewer Division dispatcher to send out a crew to investigate.

It can be difficult to manage lists of active sites and site information. One of the labor-saving features of the software is that it automatically updates all site information from the Oracle database. A query searches for modules with recent data, and they are added to the "active unit list". As sites are installed, their data and info are automatically checked. All meter information associated with each unit is queried along with recent meter data and stored for RAD analysis.

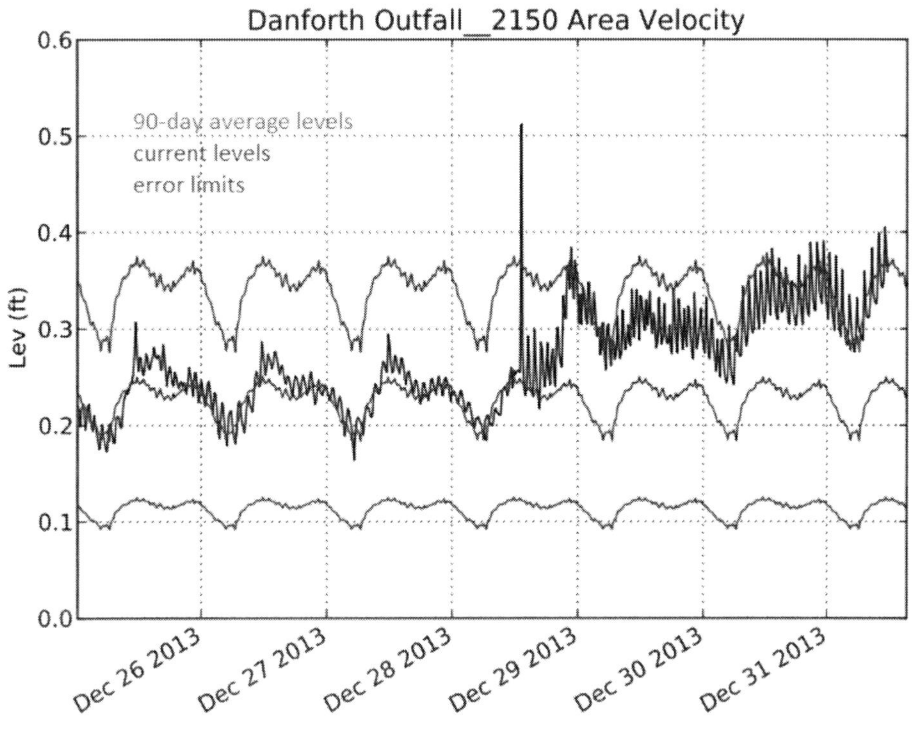

Figure 1 SSO Alert Hydrograph

```
Site name: PTC-23A
US_Node: 23280423101
DS_Node: 23280214701
Group: Modeling
Invert Elevation:
Manhole Depth: 0
Rain_Gauge: Peachtree Crk Rain Gauge
Pipe_Height: 0
Pipe_Width:
Pipe_Shape:
Primary_Device:NA
Device_Size:
Latitude:
Longitude:
Nearest Address: 36 Wakefield Drive NE
Cross St: Parkdale Way

Driving Directions:

Jurisdiction: Atlanta

Contact #: 404-658-[redacted] (Call Ctr)  or Alt. 404-546-[redacted] (Op. Ctr)

Also Notify: [redacted]@atlantaga.gov; [redacted]@atlantaga.gov; [redacted]@atlantaga.gov;
[redacted]@atlantaga.gov; [redacted]@atlantaga.gov; [redacted]@atlantaga.gov;
[redacted]@atlantaga.gov; [redacted]@atlantaga.gov; [redacted]@atlantaga.gov

Notes:
```

Figure 2 SSO Alert Site Info Text File

RESULTS

While the SSO Alert works well at showing obviously high levels or active spills, it's the software's ability to actually predict future spills that makes it truly amazing. By showing level trends and anomalous events, future potential spills can be identified days or even weeks ahead of time.

Without the complexity of modeling software which correlates high sewer levels with rainfall, the SSO Alert highlights unusually large level transitions during rain events which could indicate a restriction. Over a longer time frame, gradual increases or decreases in sewer levels may also indicate a restriction; the SSO Alert is very good at detecting them.

The SSO Alert is a relatively simple, but effective tool used to prevent sewer spills. With only a limited number of monitoring sites, it has prevented 20-30 sewer spills per year. It has reduced the volume of spills, and has helped to locate spills in remote areas that might have gone undetected.

Challenges

Typically, diurnal level patterns are very consistent, so recent data overlays the diurnal averages quite closely. However, rain events can cause sewer levels to jump significantly causing false alarms. Large level spikes due to rain also cause improper increases in the diurnal average data which can lead to missed alarms. A special function was added to remove these rain-induced spikes from the diurnal pattern generator.

FUTURE
Work is in progress to leverage the SSO Alert to its full potential. A variance algorithm is under development which will reduce false alarms caused by normal sewer level fluctuations. Work is progressing on a function to increase alarm thresholds in proportion to local rain amounts. A cloud based graphical dashboard is under development to leverage the data made available by the database export.

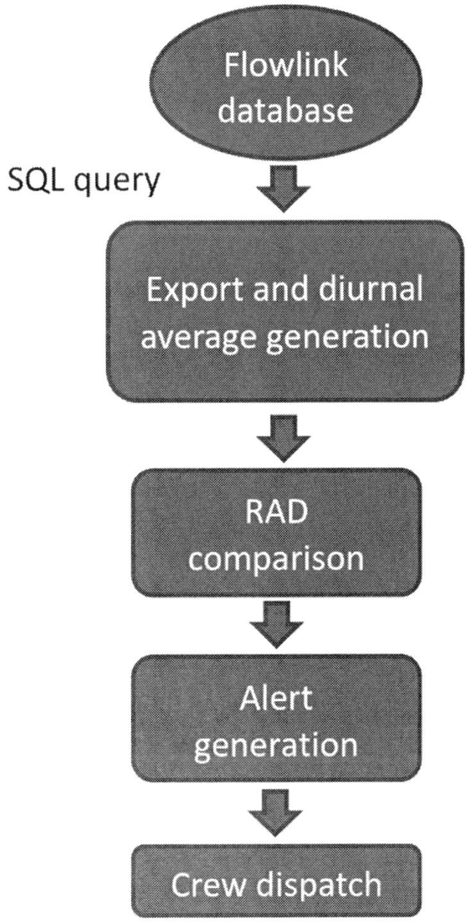

Real Time Control and Optimized Storage for Affordable Peak Flow Management

Michael Milius, PE[1*], Devin Lake[2]

[1] Olsson Associates, Kansas City, Missouri
[2] City of Sedalia, Missouri
[*] Email: mmilius@olssonsssociates.com

ABSTRACT

Optimized storage using Extraneous Flow Holding Basins (EFHBs) combined with real-time controls can provide a key peak flow management component in an affordable overflow control program to achieve agency goals for level of protection. This approach was utilized on existing basins for the City of Sedalia, Missouri. The City entered into an Administrative Order of Consent (AOC) with the Missouri Department of Natural Resources (MDNR) to address wet weather flows throughout the city's system, resulting in overflows and backups in the collection system and bypasses at the city's three Water Resource Recovery Facilities (WRRFs). A project for the City shows the outcome of the implementation of a real-time control system that allows the City to divert water during high flow periods to these basins. This approach provided a significant cost savings to the City and allows for a very flexible system for the future.

KEYWORDS: Affordable Peak Flow Management, Flexible Storage, Real-time Control of Diversion, Storm Storage??

INTRODUCTION

With the ultimate objective of compliance with the Missouri Clean Water Law, the City entered into an AOC with the MDNR, effective June 15, 2009. Citations in the AOC are due to inadequate facilities to address wet weather flows throughout the City's system, resulting in overflows and backups in the collection system and bypasses at the City's three (3) WRRFs. This led to the development and implementation of a $30 million program of improvements with an aggressive implementation schedule requirement of seven (7) years.

Sanitary Sewer Overflow (SSO) control and the minimization of Waste Water Treatment Plant (WWTP) bypasses requires a comprehensive approach of maintenance, repair, I/I source removal, and capacity improvements and Extraneous Flow Holding Basins (EFHBs) can provide more immediate and affordable relief for many common sanitary sewer overflows.

This approach looked at leveraging existing storage opportunities near the natural 'bottlenecks' and SSO locations commonly presented by pump stations and treatment facilities. Benefits include: immediate baseline of protection, improved treatment performance, benefits to normal operations, and future expansion flexibility as longer-term flow reduction and relief programs evolve.

METHODOLOGY

Due to the very quick timeline of this project, a quick and cost effective solution needed to be realized. The City had multiple locations where existing storage facilities were available, however the diversion to these structures was a manual process. In several locations, the water that was transferred into the holding basins could not be easily returned back to the treatment facilities. This allowed for evaluation of possible re-uses of these facilities.

The concept of the approach that was utilized is outlined in Figure 1 below. With the existing and proposed structures, high wastewater flows would be diverted to the storage facilities during the event. After the precipitation event was complete and while the flows are still high at the treatment facilities, the stored water would be returned to the treatment facilities to keep the treatment systems running at the increased rate and to allow the structures to be quickly lowered and be ready to receive flows from future events.

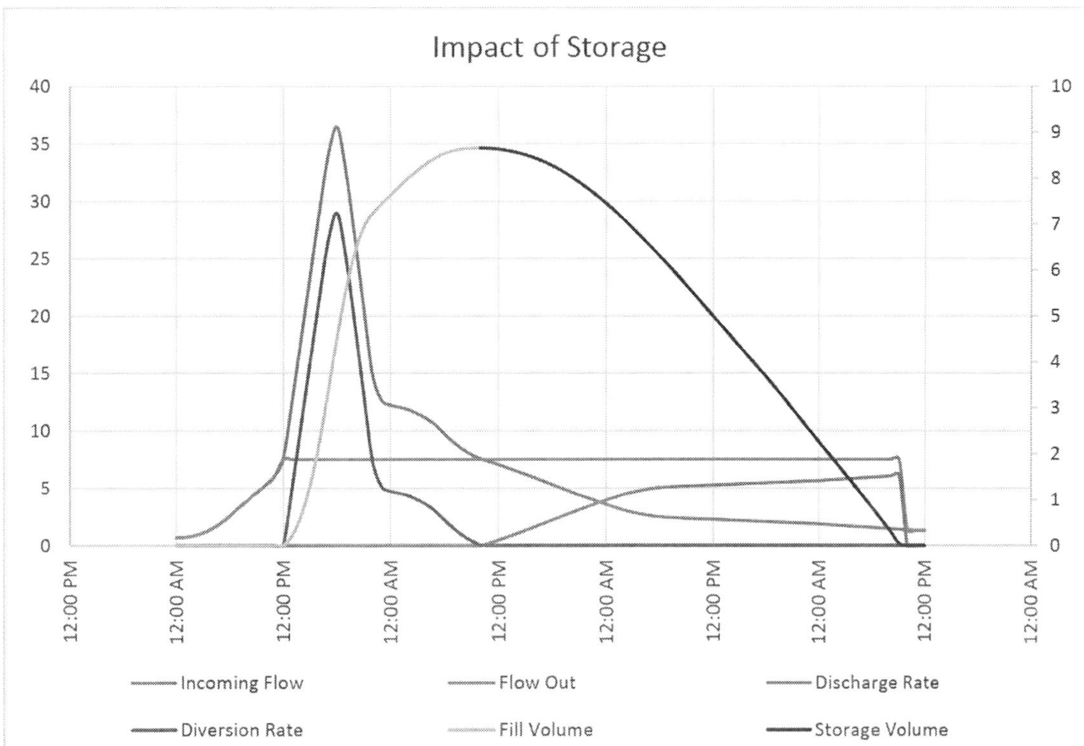

Figure 1

DESIGN APPROACH

To evaluate the possible re-use of the existing facilities, there was a number of areas that were reviewed to confirm the improvements would be cost effective and viable.

Capacity Analysis and Modeling:
Flow/rainfall monitoring, hydraulic modeling and analysis utilizing the EPA's SSOAP Toolbox program was completed. The program provided the anticipated peak flows along with the needed storage volumes. These volumes were then compared with the available storage volume in the existing facilities to determine if they were adequately sized. Additionally, the collection system upstream of the proposed facilities was evaluated for hydraulic capacity.

EPA's SSOAP Toolbox analysis was performed in conjunction with collection system rehabilitation to determine cost-effective combination of sewer capacity, storage and flow reduction to achieve the selected level of overflow protection. Elimination of overflows for up to the five (5) year return storm was selected.

Collection System Improvements:
A multiple phase program of construction contracts completed in parallel was utilized for prioritized implementation of collection system improvements in 27 of the 56 total sub-basins that comprise the collection system.

Peak Flow Improvements:
As a result of data collection and analysis completed for compliance with an administrative order for overflow/bypass elimination, the following improvement projects were developed.

Southeast WRRF - EFHB and Pump Stations:

- Re-use of existing 28 MG earthen EFHB, rehabilitation of existing 1.5 MGD duty pumps and construction of new 1 MGD return pump station at Southwest Lift Station site.
- Construction of new diversion structure along with a 1 MG Concrete EFHB with a new 25 MGD peak flow pump station that discharges to a 8 MG earthen EFHB. This is at remote site from WRRF and was coordinated with trunk sewer capacity improvements upstream of the facility. This allowed the re-use of existing lower trunk sewer to the WRRF.

North WRRF - EFHB and Pump Stations:

- Re-use of existing 18 MG earthen EFHB and construction of new 1.0 MGD return pump station adjacent to the North WWTP. Provisions at the headworks and pump station sites made to accommodate trunk sewer capacity improvements, lift station expansion and ultimate retirement of WWTP for consolidation of treatment at the Southeast WWTP.

Central WRRF - EFHB:

- Re-use of existing 7 MG earthen EFHB and construction of improvements to an existing return pump station adjacent to the storm water sedimentation basin.

All EFHB's include the addition of diversion structure controls, automated gates, variable speed pump drives, flow measurement, and supervisory control and data acquisition (SCADA) for peak flow management from the respective receiving treatment facility.

The layout of each of these sites are shown in Figures 2 and 3 below.

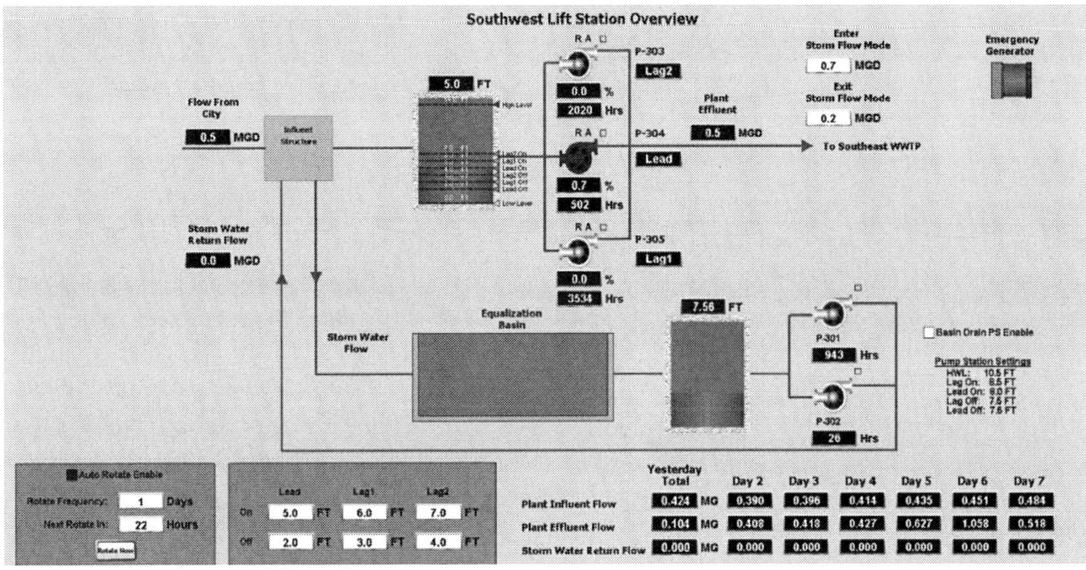

Figure 2 – Southwest Lift Station

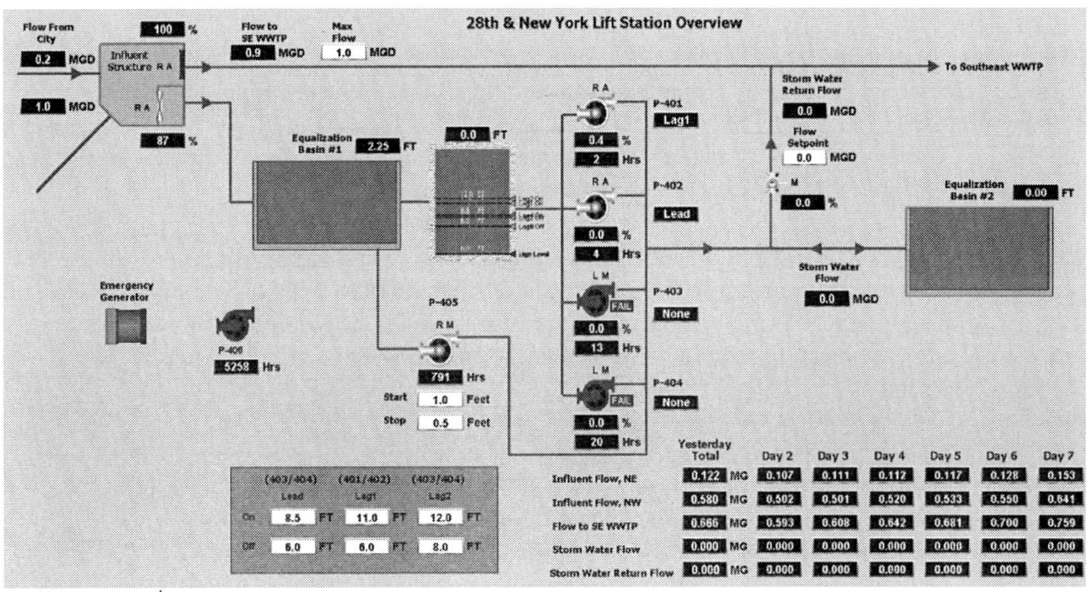

Figure 3 – 28th and New York

For all of these sites, the operations staff can input the maximum discharge rate (water that will flow directly to the WRRF). Any flows in excess of this, the system enters "storm flow mode" and flow is automatically diverted to the EFHBs. This approach allows the operations staff to continue to adjust the system depending upon the flows being received at the WRRFs.

Upon completion of the storm (when the flow goes below the maximum discharge rate) the system goes back to "normal" operation, but the water in the EFHB's is removed and discharged to the WRRF. The goal of this return to keep the flow near the maximum discharge rate so there are minimal flow swings at the WRRF and allows for the EFHBs to be drained as quickly as possible in anticipation of the next precipitation event.

RESULTS

With the re-use of facilities and construction of a new facility in the City's collection system, a significant decrease in the number of SSO and WRRF's by-passes have been realized. The upgraded system was put into service at multiple locations over the last year. This was coupled with selected improvements to the collection system up stream of the basins. The control logic continues to be slightly modified to improve the system performance.

CONCLUSIONS

While SSO control requires a comprehensive approach of maintenance, repair, I/I source removal and capacity improvements, EFHBs can provide more immediate and affordable relief for many common sanitary sewer overflows.

This approach can leverage existing storage opportunities near the natural 'bottlenecks' and SSO locations commonly presented by pump stations and treatment facilities. Benefits include: immediate baseline of protection, improved treatment performance, benefits to normal operations and future expansion flexibility as longer-term flow reduction and relief programs evolve. The key item is the location, capacity and current condition of the facilities.

Reinvestment in the conversion of decommissioned facultative lagoons or other peak flow storage facilities to EFHBs can often be an affordable alternative to relief sewer and a timelier alternative or supplement to I/I flow reduction. Analysis and incorporation of these opportunities fits readily into a typical Sanitary Sewer Evaluation Study (SSES) and the EPA's SSOAP Toolbox approach.

Also, EFHB facilities can augment normal operations or utilize existing pumping and treatment facilities, including flow monitoring equipment, SCADA, and redundant power supplies. The conversion is typically more focused on fill/return pumping systems, control/diversion structures, and piping relative to other relief/collection improvements.

Of course, biological treatment facility operations benefit by eliminating large and abrupt swings in flow as peak flows are diverted to storage and then returned for treatment as peak flows subside.

With the functionality and reliability of today's SCADA system technology, systems can be configured to provide real time control of wastewater flow. This approach needs to be evaluated throughout the design, since it can have a significant effect on the improvements needed.

For this project, the conversion of three (3) existing storage basins to EFHBs and the construction of one additional basin site eliminated the need to upsize multiple treatment facilities and install approximately 1.8 miles of interceptor relief sewer. This approach provided measurable immediate relief of SSO's and by-passes at significant cost savings while providing the City significant flexibility for operations staff in the future.

We Can Ride the Wave, But How Do We Measure It?
Flow Monitoring for Challenging Sites

Stephen T. Motley, P E, City of Virginia Beach Public Utilities
Kraig K. Moodie, FloWav, Inc.

ABSTRACT

Upon entering into a Special Order by Consent with the Virginia Department of Environmental Quality (VDEQ) the City of Virginia Beach needed to identify an effective and efficient way of conducting flow monitoring. Given the challenges that existed in Virginia Beach it was imperative to evaluate a variety of flow monitoring methodologies and technologies to determine those that would be most suitable and understand their limitations. It was also important to maximize the value and make effective use of the flow monitoring data to inform other programs.

KEYWORDS

Flow monitoring, open channel flow, pressure pipe flow, hydrologic modeling, sanitary sewer evaluation studies, SSES.

INTRODUCTION

The City of Virginia Beach is responsible for the maintenance of approximately 29,000 collection system manholes and 1,500 miles of gravity sewer infrastructure, spread among 410 sanitary sewer pump station service areas. The City's vast sewer collection system, Special Order by Consent, numerous modeling requirements, and master planning have prompted the development of a reliable and economical flow monitoring program. Given the topography in Virginia Beach there are a number of challenges encountered while flow monitoring and as a result a few unique and interesting approaches have been used to mitigate those challenges. This program

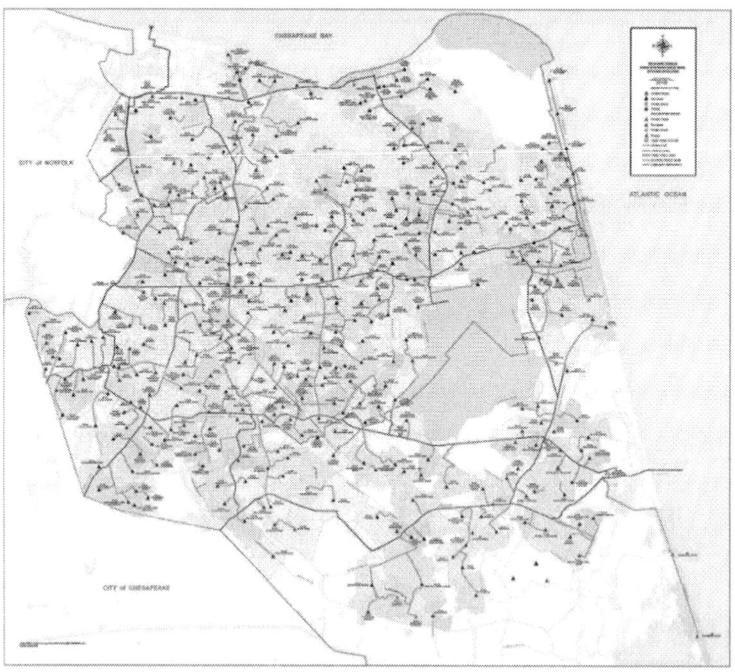

Figure 1 - City of Virginia Beach Sanitary Sewer System.

has developed strategies designed to provide accurate and reliable data in a cost efficient manner.

FLOW MONITORING CHALLENGES

With very little topography, 410 sanitary sewer pumping stations, and small diameter pipe (8") on minimum grade there are a number of challenges encountered while flow monitoring in the City of Virginia Beach. Some of those challenges are:

- Low Flows - these create limitations on many velocity and depth measurement technologies available.
- Minimum Slopes - they make the differentiation of moving and nonmoving particles with the flow difficult for velocity sensing technology.
- Poor/Complex Hydraulic Conditions – they create a challenge to determine and characterize a uniform flow profile in turbulent and rapidly changing hydraulic conditions.
- Small Diameter Pipe – due to limited space to install seasoning equipment and provide additional space to perform in-situ confirmations.
- Limited Access – this makes it difficult for field crews to access the equipment when service is required potentially causing additional delays or sensor outage between service calls.
- High Velocity – this makes sensor installation and in-situ calibrations with some technologies difficult.

Each of the challenges listed are not specifically unique to Virginia Beach, but they are common with many large coastal utilities. To address these challenges the industry has developed a variety of approaches and methodologies to ensure accurate and reliable flow monitoring data even with when dealing with these challenging conditions.

FLOW MONITORING METHODS AND TECHNOLOGIES

Volumetric Method

One of the most economical and effective methods of flow monitoring that the City of Virginia Beach uses. It simply uses a volumetric algorithm to calculate the sanitary sewer inflow:

$$(Inflow = ((Wet\ Well\ Constant\ gal/ft) \times (Change\ in\ Wet\ Well\ Level\ ft)) \div (Amount\ of\ Time\ for\ the\ change\ in\ Wet\ Well\ Level\ min) \quad \text{(Equation 1)}$$

The wet well constant is different for each pump station it is simply a 1 foot cross sectional volume of the wet well.

$$Wet\ Well\ Constant\ \frac{gal}{ft} = (Length\ ft\ \times Width\ ft\ \times 7.48\ gal/ft^3) \quad \text{(Equation 2)}$$

The change in wet well level measured in feet is simply the change in wet well level compared a specific amount of time for the change to occur. In Virginia Beach we measure the amount of time it takes for the wet well level to rise 1/10th of a foot and use that for the inflow calculation.

We use the PLC at our pump stations to perform and report this calculation. To monitor the wet well levels we use a bubbler tube or pressure transducer. We only perform this calculation in the PLC when the sewage pumps are not running. Since all of our pump stations have a PLC and a wet well level measuring device this is a quick, efficient, and cost effective manner for measuring the inflow. However there are several factors that you need to be aware of when using this method:

- If the operating zone of the wet well and pumps is within a fillet area of the wet well this method will result in an over-estimation of the inflow. This is because the actual area you are filling is smaller than the area you calculated for the wet well constant.
- If the operating zone of the wet well and pumps results in the surcharging of the influent line and gravity collection system the method will result in an under-estimation of the inflow. This is because the actual area you are filling is larger than the area you calculated for the wet well constant.
- You don't want to perform this calculation while the pumps are running as the above equation does not take into account the amount of sewer being pumped out. Thus if you have pumps that always run (i.e. a variable frequency drive for example) or run for very long times this is not a good way to measure the inflow.
- For Virginia Beach this wet well constant is also stored in the PLC at the pump station, so it is critical that each time that the PLC program is changed or updated to ensure that correct wet well constant for that particular pump station is entered and saved. An incorrect wet well constant will result in a data shift and an incorrect wet well inflow calculation (as shown in Figure 2 below).

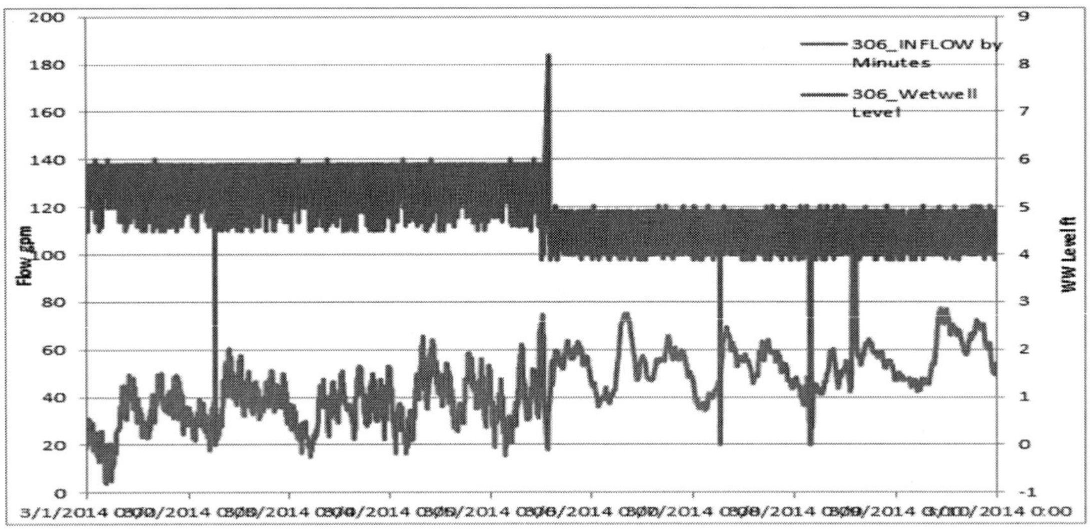

Figure 2 – Example of a fill constant being changed.

Open Channel

$$Q = a * v$$

- *where Q is the flow rate, v is the velocity of the fluid, and a is the area of the cross section of the space the fluid is moving through.*

There are two different categories of open channel flow monitoring that were evaluated for the Virginia Beach efforts, one with primary elements and one without. Each has benefits and limitations. Primary Elements require the measurement of depth only, since there is a constructed channel of know dimension that forces a predictable velocity. An example of this type of application would be a flume or weir devices with pressure or ultrasonic depth measurement sensors. A Non-Primary Device site would require sensors to measure the flow depth and velocity to calculate the flow at that site. Examples would be pressure and/or ultrasonic depth measuring sensors with doppler, radar or laser technology sensors used to measure velocity.

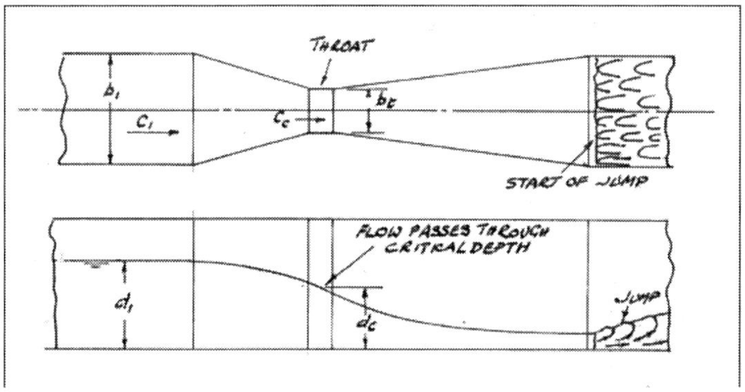

Figure 3 – Example Illustration of Venturi Flume

A common application of the primary element measurement approach includes the use of a specific type of broad crested weir known as a Venturi flume. The broad-crested weir is a device that combines the functions of measuring a flow and dissipating the potential energy where a reduction in level of a channel is required. There are many applications where the flow must be measured with a minimum of loss of energy and the Venturi flume has evolved for this purpose. The broad-crested weir's success as a measuring device is dependent upon a contraction in the flow produced by raising the bed of a channel and producing critical flow conditions on the crest.

The ease of measuring only one parameter (depth) was outweighed by the limitations. The primary concerns were the need for a constructed solution and associated costs for a temporary project and the concern that with the high amount of sand and silt that is present within the collection system of any beach community the assumed calculations for flow would be artificially high when the sand builds up in the throat.
The second category of open channel flow monitoring or non-primary element involves the measurement of the flow velocity and the flow depth. The average flow velocity can be input directly into the Q=A*V equation. The flow depth is utilized in conjunction with the host pipe geometry so the wetted area can be calculated and input as A in the above equation for flow.

The benefits to this approach allowed the City to install in almost any pipe dimension, account for the presence of silt and did not require the installation of a primary element in the channel.

Ideal Site Conditions for this type of installation included the following:

1. Consistent flow depth above 0.75" to ensure proper sensor coverage and depth to perform in-situ calibrations.
2. Flow velocity greater than 0.5 fps to allow for reliable sensor performance.
3. A "straight through" pipe or channel, this reduces the amount of turbulence with the cross-sectional area of flow that is being measured. Accuracy of your flow calculation is related to the uniform distribution of velocity across the entire cross section.
4. Access for maintenance is important so that when something needs to be changed to improve the performance at a specific site it can be completed in a timely and safe manner.
5. Manhole depth of less than 20 ft are ideal for two reasons, it supports the safe and timely access to the sensors for maintenance and, most of the major manufactures have standard cable lengths in the 25 – 35 foot range. Specialty cable lengths are available but come at a premium cost and typically require additional time to manufacture.

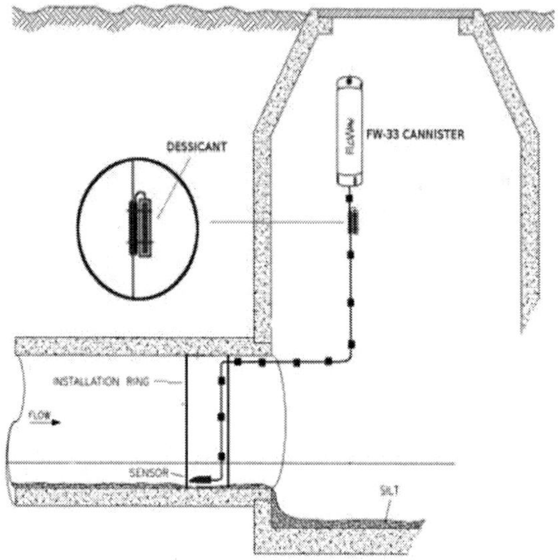

Figure 4 – Open Channel Flow Monitor Installation

6. Strong cellular voverage is important since the vast majority of flow monitoring equipment manufactured and sold today is equipped with a wireless modem to ease data collection and to facilitate timely O&M tasks. A strong signal makes the process of installing the antenna much easier.

Data Accessibility Options

As the age of cell phones, smart phones and the internet/wifi access continue to grow and touch all aspects of our lives it is also true in the world of wastewater flow monitoring. Due to these advances in technology fewer and fewer flow monitoring projects have direct connect capabilities only, where on a weekly basis a crew is deployed to download the data to a laptop wired to each flow monitor or RTU. Typically, there is a cellular modem associated with each deployed unit and the data is collected on a user defined schedule (i.e. hourly or daily) determined by data criticality, cellular signal strength and power availability.

Figure 5 – Composite Manhole Lid

Site visits to retrieve data from installed monitors are eliminated with the modem installation, with the resulting savings in download time the field technicians spend more time on repairs and inspections and less time accessing devices in harsh underground or outdoor conditions that are operating properly.

Many tools have been developed to facilitate this communication platform and many different data loggers exist to support this effort. Many of them are proprietary or technology specific. It is important to look for a platform that is not linked to one specific technology or sensor manufacturer since no one technology is ideal for all wastewater applications. In addition to the platform flexibility there are a variety of antenna options available and even traffic rated composite manhole lids that will allow for improved cellular communications (See Figure 5).

Pressure Pipe

Pressurized Pipe for the purposes of flow monitoring has an assumed depth ("full pipe") equal to the inner diameter and one variable to measure with a sensor that is the fluid velocity. Typically this type of measurement has an engineered or constructed solution.

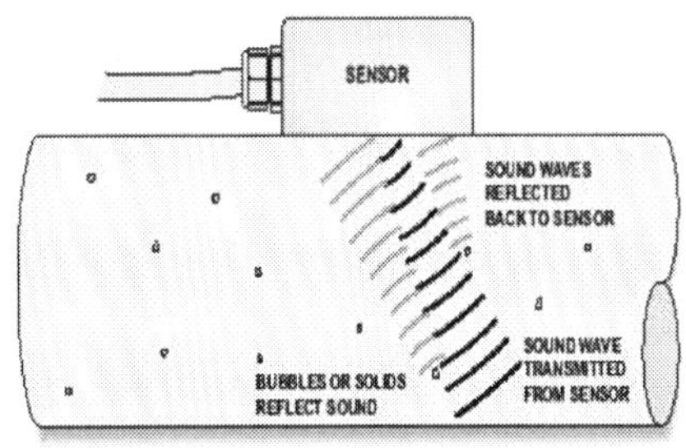

Figure 6 – Example Clamp-on Ultrasonic Velocity Sensor

To calculate the flow in a pressure pipe as show in Figure 5 only the velocity needs to be measured by a sensor, since the pipe is full the inner diameter can be determined through a couponing process or by measuring the outer diameter and determining the pipe material and manufacture the inner diameter can be derived and entered in the Q=A*V calculation for flow.

This methodology relies heavily on the known inner diameter and does not account for tuberculation or other factors that could affect the factory provided inner diameter. You can determine the inner diameter using pipe manufacturer reference material if coupons are not collected at the point of installation. This approach is typically not selected if a pipe is not always under a pressurized condition.

Monitoring Force Main Pressure

This is another cost effective and efficient method for determining the flow into a pump station. This method only works for pump stations that pump into a gravity collection system. For this method a pump drawdown test is performed for each pump at the station, this is necessary to determine the output capabilities of each pump. To perform a pump drawdown test you will need to follow these steps:

1. Be sure all the pumps are turned off.
2. Block the flow coming into the pump station by either closing a valve or installing a plug in the influent line.
3. Once the influent flow is stopped then measure the water level depth in the wet well from the top slab once the measurement is secured immediately turn a pump on.
4. Allow the pump to run for a predetermined amount of time 30 seconds, 1 minute, etc.

5. Once the predetermined amount of time expires then turn the pump off.
6. As soon as the pump is turned off then measure the water level depth in the wet well from the top slab.
7. Once you have this measurement you can open the valve or remove the plug to allow the inflow into the pump station to continue. Note while you have the valve closed or the influent line plugged you will want to ensure you are not causing a sanitary sewer overflow (SSO) upstream or any other operational problem.
8. Now you can calculate the output of that pump. You simply use the following equation
$$Pump\ Output\ gpm = \Big((First\ measurement\ taken\ before\ pump\ turned\ on\ feet - Second\ measurement\ taken\ after\ pump\ is\ turned\ off\ feet) \times Width\ of\ wet\ well\ feet\ \times Length\ of\ wet\ well\ feet\ \times 7.48\ gal/ft^3\Big) \div Amount\ of\ time\ the\ pump\ ran\ minutes$$ (Equation 3)
9. Repeat Steps 1 – 8 for each individual pump and then you will want to repeat steps 1-8 with all the pumps operating at one time.

Now that you have the pump output for each pump you are ready to move on with installing a temporary pressure recording device. In Virginia Beach we install this device on the Discharge Emergency Bypass Connection. The device is used to measure the pressure in the pump station discharge force main. We typically set the device to take a reading every 30 seconds to ensure we capture each time one of the pumps is operating. Once you download the data from this device you can tell each time a pump kicks on and for how long the pump ran based on the pressure readings. You can then simply take the pump output you calculated in the above and multiply it by the amount of time each pump ran and you will have the gallons pumped during that pump cycle. It should be evident if multiple pumps ran at one time as the pressure will be much higher than if just one pump ran. Once you have calculated the gallons pumped for each pump cycle in gallons you can then calculate the inflow by taking the gallons pump and dividing it by the preceeding cycle time (i.e. time between pump off and on cycle). You do this for each pump cycle. You now can apply a 60 minute moving average to smooth the data out and you have the inflow for this pump station. See Figure 7 for an example of data obtained using this method.

It is important to note that the methodology described above only works for pump stations that pump into a gravity system since you can be reasonably assured the pump rate is the same each time the pump is operating.

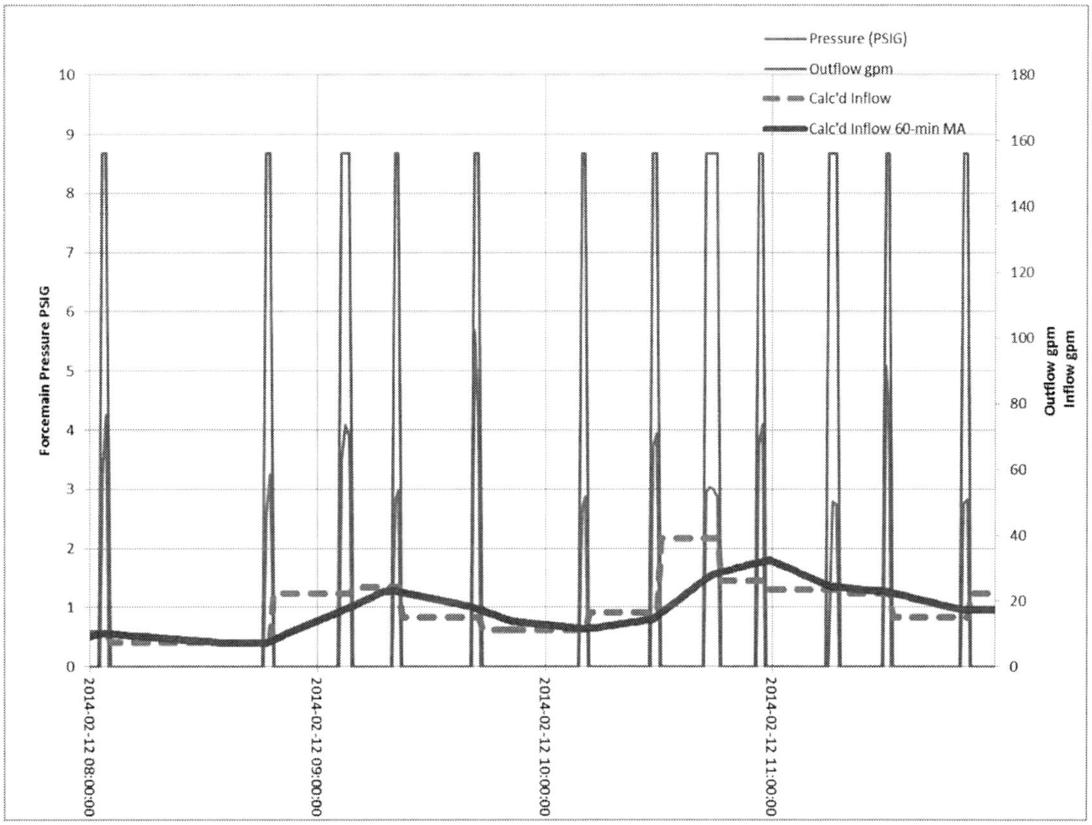

Figure 7 – Flow Data using Force Main Pressure Monitoring.

USES OF THE FLOW MONITORING DATA

Hydrologic Modeling

The city captures flow monitoring and rainfall data on a service area by service area basis. An open channel flow meter and an associated rain gauge are installed in each service area and they remain in place until the following criteria are met:.

- Captured enough data to characterize any seasonal variations and peak seasonal sanitary sewer flows.
- Period shall capture typical diurnal variations in dry-weather flow, including weekends and weekdays.
- Record 3 individual wet-weather flow events of greater than 1 inch of accumulation, including at least 1 event with at least a 1 year recurrence interval. These events shall capture system response under a variety of antecedent rainfall and groundwater conditions.
- Flow data shall continue for sufficient time between rain events for the flow to return to dry weather conditions.

An individual service area is typically monitored for between 6 months and 1 year. The biggest determining factor is capturing the necessary wet-weather events.

Flow monitoring and rainfall data is analyzed each month on a weekly basis to ensure we reach a minimum 90% data reliability per week. The City utilizes an excel spreadsheet to assist in flagging suspicious data (i.e. flat lining, drop outs, etc.). The City prepares scatter plots to compare the depth vs velocity using Manning's equation to help identify any potential issues/problems. And finally the City performs a visual inspection of all the flow, depth, and velocity time series data to check for any anomalies. Once the City has completed our QA/QC of the flow monitoring data and met all the flow monitoring criteria the City will do some preliminary model calibrations to confirm that the City can calibrate to the minimum required 3 events before the City will pull the flow meters. By doing this the City ensures that flow meters and rain gauges are not removed prematurely.

Once we have the necessary flow monitoring and rainfall data we begin our hydrologic modeling. The City utilizes a software package known as CAPE III, which is an open source Hydrologic Simulation Program – Fortran (HSPF) based hydrologic modeling software, which is used to decompose the sanitary sewer dry weather flow (DWF) into base sanitary flow (BSF) and active groundwater infiltration (GWI). Then the sanitary sewer wet weather flow inflow/infiltration is decomposed into the following four components:

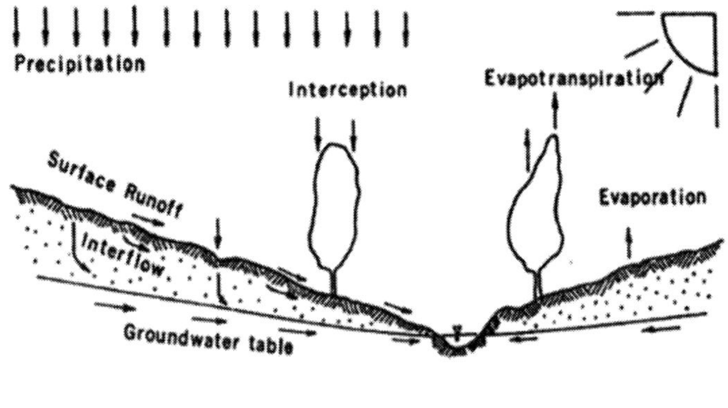

Figure 8 - Hydrologic Cycle represented in HSPF Model.

- PSURO – Pervious Land Surface Runoff
- PIFWO – Pervious Land Interflow
- PAGWO – Pervious Land Active Groundwater
- ISURO – Impervious Land Surface Runoff

Hydrologic calibration of dry and wet weather flow parameters are used to perform long term simulation of modeled sewer flows using 60-years of rainfall data, and statistical analysis of the simulated sewer flows to determine the 2, 5, and 10-year peak sanitary sewer recurring flow.

Sanitary Sewer Evaluation Studies (SSES)

The data from the hydrologic modeling is used to prioritize individual collection system service areas for SSES activities, the 10-year peak recurring sewer flow for each sewer service area is compared to the Peak Flow Threshold (PFT), or the peak design flow rate projected for the existing customers in the service area. The PFT is calculated by multiplying equivalent residential units x 775 plus commercial water consumption peaked at 4 and adding any additional major industrial flows. If the 10 year peak recurring flow was greater than the PFT the sewer service area was scheduled for SSES investigation. In addition to establishing service areas for SSES investigation and prioritizing them, the hydrologic modeling parameters are used to assist with selection of the investigative techniques used during the SSES. Table 1 shows how Virginia Beach tailors SSES activities based on the different hydrologic component parameters.

Table 1: SSES Activities Related to Modeled Hydrologic Parameters

Hydrologic Component	Potential I/I Sources	SSES Activities
Impervious Surface Runoff (ISURO) greater than 0.1%	Area drains, residential downspouts, surface cracks in manhole rims, holes in manhole covers, open cleanouts, etc.	Smoke Testing
Pervious Land Interflow (PIFWO) greater than 1.0%	Broken service connections, pipe and manhole joints and defects above the active groundwater table	Smoke Testing, CCTV, additional Flow Monitoring
Active Groundwater (AGWO) greater than 10.0%	Pipe and manhole joints and defects below the active groundwater table	Night Flow Isolations and CCTV
Constant Infiltration greater than 30.0% of the wastewater flow	Pipe and manhole joints and defects below the active groundwater table	Night Flow Isolations and CCTV

Master Planning

The flow monitoring data and hydrologic modeling information is used for sewer master planning efforts to ensure we maintain adequate capacity in our collection system and pump stations. Some examples could include upsizing pipes or changing the pump impeller sizes to allow for additional flow. It also allows us to plan for future projected redevelopment in terms of sewer flow to ensure that we do not hinder economic growth and development. We recently planned for the addition of a 19,000 seat arena in our Central Beach District, by conducting flow monitoring during the 2016 concert season at our 20,000 seat outdoor amphitheater. We were able to accurately capture peak flows at the amphitheater which assisted us in predicting peak flows from the new arena and ensuring our infrastructure was size appropriately to handle the additional flow from the new arena.

CONCLUSION

In summary when putting together a flow monitoring program it is important that you do the following:

- Determine and prioritize your flow monitoring needs.
- Consider the challenges in the sites you have selected.
- Determine the data uses, needs and accessibility.

- Choose the correct method/technology for the site to best address the challenges while meeting the data needs.
- Maximize the value of the flow monitoring data by incorporating into other programs as appropriate.

These methodologies and processes have proven to be effective and efficient way of implementing and conducting flow monitoring in the City of Virginia Beach.

HOLISTIC EVALUATION OF INTERCEPTOR ALONG CREEK – CAPACITY EVALUATION, SSES AND STREAM RESTORATION

Richard Underhill[1*], Eric Schrader[2], Keith Thompson[3]
[1] Greeley and Hansen LLC, Chicago, Illinois
[2] City of Lynchburg, Lynchburg, Virginia
[3] Wiley|Wilson, Lynchburg, Virginia
* Email: runderhill@greeley-hansen.com

ABSTRACT

This paper describes a holistic approach for evaluating an 18" combined sewer interceptor, located along a creek and installed in 1970s. The level of service (LOS) evaluation was performed to identify bottlenecks in the existing system and potential improvements were recommended for a desired LOS. Sanitary Sewer Evaluation Study (SSES) conducted in the area includes existing system evaluation, phased field investigations analyzed using NASSCOs PACP and MACP methodology, and selection of rehabilitation methods. The implementation of selected alternative for LOS improvement and sewer rehabilitation will enable the municipality to renew the asset to handle growth; prevent SSOs in the system; and restore the stream for TMDL credits.

KEYWORDS

Level of Service Evaluation, Hydraulic Model, SSES

INTRODUCTION

Burton Creek Interceptor, which runs from Wards Ferry Road near US 29 to Rhonda Road, was installed in early 60s through late 70s. The total length of the Burton Creek Interceptor, including Rock Castle Creek Interceptor, is approximately 5 miles (Figure 1a). The main line of the reinforced/un-reinforced concrete interceptor is mostly 15"-18" in diameter. With the development in the contributing sewershed, some segments of the interceptor have exhibited capacity issues. The sewershed was identified as the highest priority area for Sanitary Sewer Evaluation Study (SSES). The results of assessment of creeks within the municipality identified this creek as a candidate for restoration for TMDL credits. The City of Lynchburg (the City) is exploring interceptor improvement options to convey the increasing dry and wet weather flows associated with growth.

The City of Lynchburg developed a criticality model in 2011 to proactively identify and prioritize areas of the sanitary system that will receive asset renewal. The criticality model identified fifteen (15) high priority areas that contain between four (4) and eight (8) miles of sewer that have a high likelihood and consequence of failure. Three of the identified areas, including Areas 2, 4 & 6 (Figure 1b), are draining to the Burton Creek Interceptor. These prioritized areas received a sanitary sewer evaluation study (SSES) to develop rehabilitation

recommendations.
This paper describes a holistic approach for evaluating the sewer interceptor located along the Burton Creek.

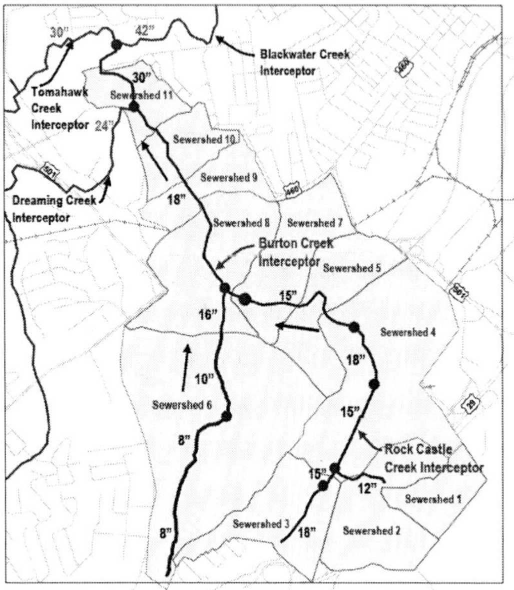

Figure 1 (a) Burton Creek Interceptor

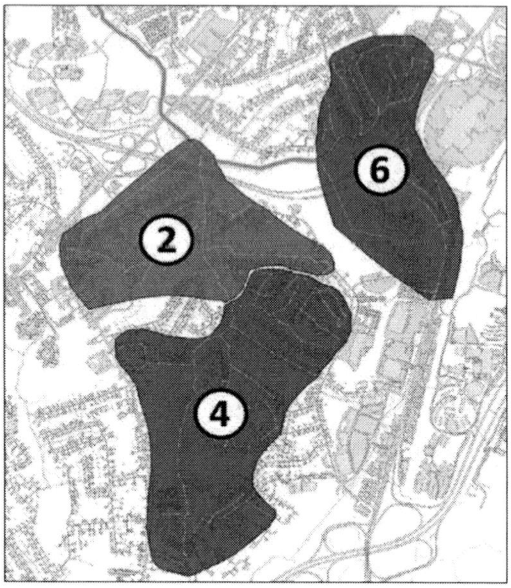

Figure 1 (b) Priority Areas 2, 4 & 6

METHODOLOGY/RESULTS

Burton Creek Interceptor Level of Service

LOS of the Existing System

The City's calibrated wastewater collection system hydrologic and hydraulic (H&H) model was used to assist LOS evaluation. The portion of the H&H model pertaining to the Burton Creek Interceptor was refined with field survey data. The capacity evaluation included determination of interceptor's existing level of service (LOS) and the potential capacity of the interceptor to convey dry and wet weather flows due to growth. The wet weather analysis included system capacity to handle 1, 2, 5, and 10-year storms. The long term historical rainfall data (1948-2013) specific to the City of Lynchburg was used to select the 1, 2, 5, and 10-year storm events.

The LOS results of the existing system is shown in Figure 2. Pipe Segments 4, 5, & 6 including the 15" and 18" pipes between Lynchburg Expressway Ramp and Igloe Dr have the lowest LOS (less than 2-yr). The remaining pipe sections have greater than 2-year LOS. Both the upstream (Segments 1, 2, 3) and downstream (Segment 9) sections have an LOS of greater than 10-yr.

Pipe Segment	Dia. (in)	(ft)	LOS
1	18	2,900	year
2	15	3,600	year
3	18	1,550	year
4	15	3,800	< 2 year
5	18	700	< 2 year
6	18	3,800	< 2 year
7	18	1,500	> 5 year
8	18	1,700	< 5 year
9	30	3,000	year

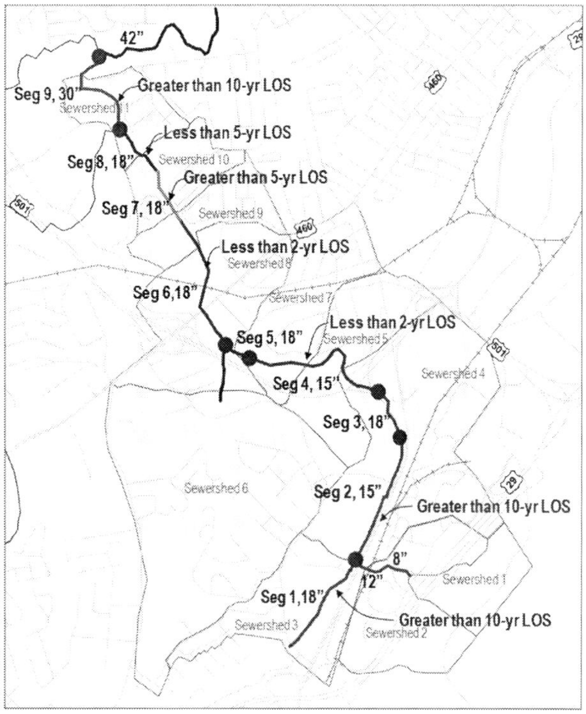

Figure 2 Existing Burton Creek Interceptor LOS Results

Alternative Evaluation

Figure 2 showed bottlenecks in the existing system that need improvements for achieving a desired LOS. Four alternatives were identified to improve the LOS of the Burton Creek Interceptor:

- Alternative 1: Increasing the interceptor pipe size for greater conveyance capacity.
- Alternative 2: Keeping sections of the existing pipes in service and installing new parallel pipes to provide additional conveyance capacity.
- Alternative 3: Installing an off-line storage facility to attenuate peak flow.
- Alternative 4: Reducing sewershed inflow/infiltration (I/I); two different I/I reduction levels were evaluated (4A and 4B). I/I reduction could reduce both wet weather peak flow and volume.

Alternative 1: Increasing Pipe Size

This alternative investigates the needs of upsizing the existing pipe size to achieve 10-yr LOS. Increasing Seg 4 (3,800 LF of 15" pipe) to 18" pipe and Seg 5-8 (7,700 LF of 18" pipe) to 30" is sufficient for the system to provide 10-yr LOS (Figure 3a).

Alternative 2: Parallel Pipe

This alternative is to explore the potential benefits of using the existing pipe as parallel conveyance to improve the LOS to 10-yr. For the existing 18" pipe segment between Sheffield Dr. and Pawnee Dr. (Segments 6 & 7), the new pipe would be located on the other side of the creek. To avoid multiple creek crossings for service laterals, the existing pipe could be kept in service to continue collecting the sanitary flows. For the section between the railway and the Highway 460, a new 'single pipe' bore will be installed under the railroad and the highway so that the existing pipe can be taken out of service. The single pipe would extend through both crossings to eliminate the potential I/I issues as this section is in a fairly swampy area. Parallel pipes will be installed on either side of each crossing. The other location that parallel pipes are not feasible is downstream from Pawnee Drive. The existing line is in the creek and should be taken out of service.

The parallel pipe alternative is shown in Figure 3b. For 10-yr LOS, most of the existing 18" pipes will be required to only increase to 24" instead of 30", which would resulted in lower new pipeline costs than Alternative 1. However, the condition of the existing 18" pipe to remain will need to be investigated to evaluate if lining or other improvements would be required. Additional connecting pipelines between the new and remaining lines would also be needed.

Alternative 3: Storage

This alternative is to explore the potential benefits of installing an off-line storage basin to improve the LOS to 10-yr. There is only one location suitable for a gravity-in/gravity-out off-line storage facility. Pump-in or pump-out storage was not considered due to maintenance concerns. The potential off-line storage would be located in Subsewershed 10, slightly north of Pawnee Drive and west of Shawnee Lane (Figure 3c, Figure 4). As shown in Figure 4, this site has a very steep segment of pipe which makes a gravity-in/gravity-out storage basin possible. The site is also owned by the City.

The storage facility, when sized appropriately, would attenuate the peak flow in the downstream pipe, thus reducing the downstream pipe size. However, storage facility would not improve the upstream LOS. Therefore, the same upsizing as in Alternative 2 would be required for segments upstream of the storage facility. For 10-yr LOS, installing a storage facility would reduce the size of the new pipelines downstream of the storage facility compared to Alternative 2.

Alternative 4: Sewershed RDII Reduction

Manhole and sewer rehabilitation will reduce rainfall derived inflow/infiltration (RDII) resulting in lower wet weather volume and peak flow. The effectiveness and quantity of RDII reduction is highly site specific and would require pre- and post-rehabilitation flow monitoring. This alternative only investigates sewer rehabilitation for the existing sewer system. 50% RDII reduction were evaluated for the existing sewer system.

With 50% RDII volume reduction from the existing sewer system, the overall system RDII volume reduction would be approximately 34% and the peak flow reduction would be about 21% based on model prediction. The evaluation showed that 50% reduction of the existing system RDII alone would not improve the system LOS; pipe upsizing would be required. However, it would only require upsizing the existing 15" to 18" and the existing 18" to 24" to achieve 10-yr LOS (Figure 4d).

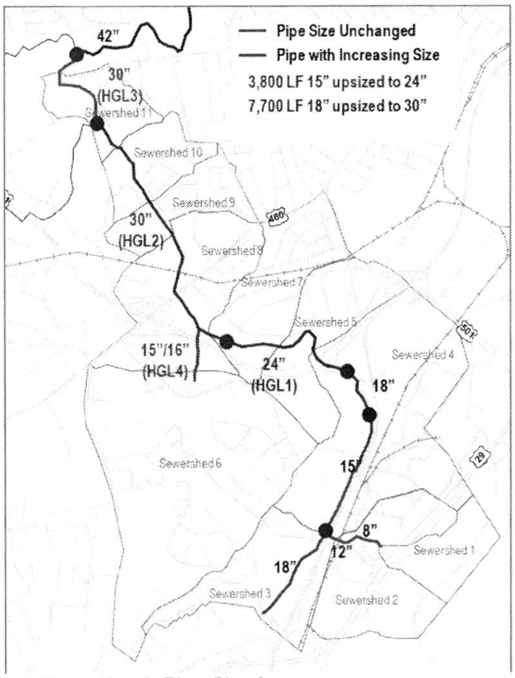

(a) Alternative 1, Pipe Size Increase

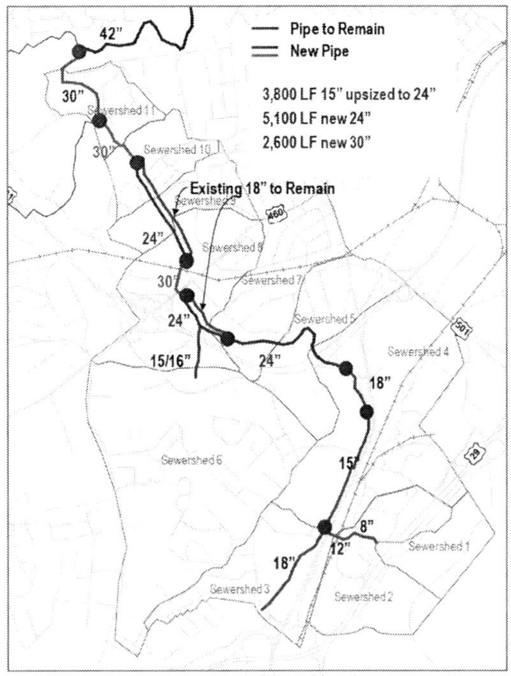

(b) Alternative 2, Parallel & Pipe Size Increase

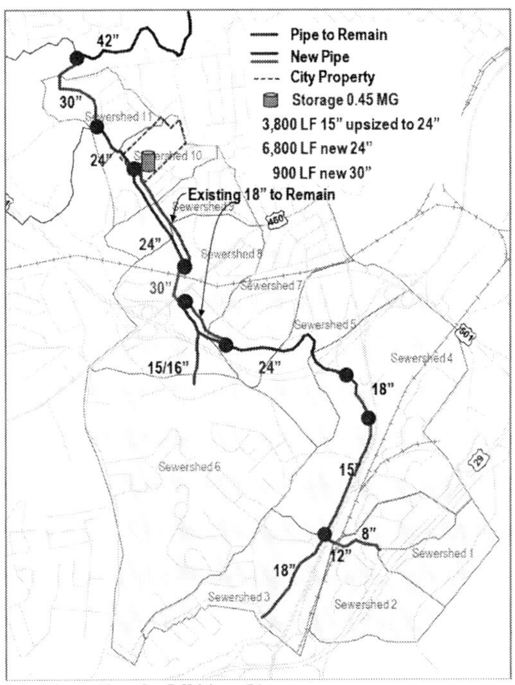

(c) Alternative 3, Off-Line Storage

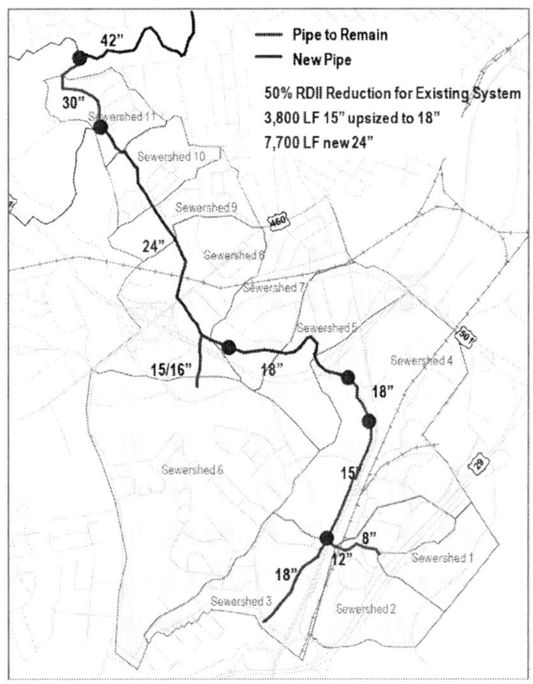

(d) Alternative 4, 50% I/I Reduction

Figure 3 Alternatives for Achieving 10-yr LOS of Burton Creek Interceptor

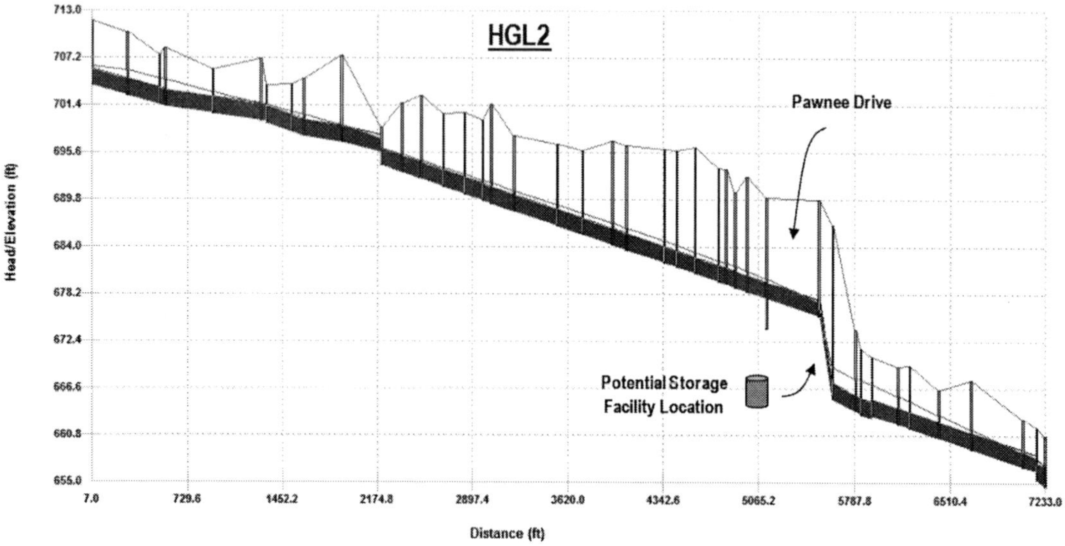

Figure 4 Profile for Pipes between Glenfield Dr. and Sunderland Pl.

Alternative Cost Comparison

For the purpose of evaluating the LOS improvement alternatives, construction costs were estimated for each alternative. The bases for cost estimates for various alternatives are presented below. All cost estimates are in September 2014 dollars.

- Construction Cost Index - The Engineering News Record (ENR) Construction Cost Index (CCI) was used to escalate the costs from other similar projects to the September 2014 dollars (ENR 9870).
- Approach for Estimating Construction Costs - Costs have been prepared using the following resources:
  - Unit costs in dollars per gallon or cost per linear foot obtained from other projects. Costs have been adjusted for relative characteristics such as complexity or location using engineering judgment.
  - Cost data from similar facilities:
  - Potential stream restoration/stabilization costs associated with the pipe installation is included

Table 1 summarizes construction cost estimates for the alternatives for 10-yr LOS. The cost comparison indicates that installing a storage facility or performing sewer rehabilitation to reduce RDII is not as cost effective as simply increasing the pipe size to improve the LOS in the Burton Creek Interceptor. System rehabilitation will still need to be conducted but the schedule will be driven by the asset renewal needs. Keeping certain segments of the existing interceptor in service could be a good option to reduce the lateral reconnection cost and reduce the upsizing requirement to achieve higher LOS, provided that the sewer line is in good condition and would only require minor rehabilitation.

Table 1 Preliminary Construction Cost Comparison for Alternatives

LOS	Alt	Description	Pipe Cost (M$)	Storage Cost (M$)	Rehab Cost (M$)	Total Cost (M$)
10-yr	1	Pipe Upsizing	$3.40			$3.4
	2	Parallel Pipe	$3.88			$3.9
	3	Storage	$3.72	$1.80		$5.5
	4	RDII Reduction 50%	$2.82		$7.92	$10.7

Burton Creek SSES

SSES areas 2, 4, and 6 (Figure 1b) include approximately 82,000 linear feet of 6-inch to 18-inch gravity sewer main. To simplify field operations and the development of remediation strategies the three (3) SSES sewer sheds were merged to create two (2) distinct project areas for this study. Area 1 includes approximately 51,000 linear feet of the gravity main associated with prioritization areas 2 and 4. Area 2 corresponds to the priority 6 sewer shed (Approximately 31,000 linear feet). Area 1 and Area 2 can be seen on Figure 5.

Figure 5 SSES Evaluation Area

The sanitary sewer evaluation was performed on each area utilizing a five phase approach. The phases included:

- Preliminary Condition Assessment
- Phase I Field Investigations
- Phase II Field Investigations
- Phase III Field Investigations
- Recommendation Development

Preliminary Condition Assessment

The preliminary condition assessment of the SSES areas consisted of a comprehensive review of the existing system records, including service call history, work order history, geographic information system (GIS) data and sanitary sewer overflow reports to DEQ. The preliminary assessment investigated sewer age, material, cleaning, repair, root control efforts, and location. The data was used to develop the scope for the Phase I field investigations.

Field Investigations

The SSES field investigations were developed utilizing an iterative approach. This phased approach was developed to efficiently identify areas of potential inflow and infiltration (I/I) without employing unnecessary investigations. Each phase of the investigation was developed based on information generated from the previous. Immediate needs identified within the basins were reported as each phase progressed. This has allowed the City to address immediate needs while developing long term asset renewal plans.

The Phase I field investigation plan employed manhole inspections, manhole checks, and smoke testing. The findings from the preliminary condition assessment were utilized to determine where each investigation technique was implemented. Table 2 provides a summary of identified deficiencies. The manholes in the SSES areas exhibit significant infiltration (Active infiltration and staining) inflow potential. The evaluation identified that over forty percent (40%) of the structures in the area had signs of infiltration and twelve percent (12%) had direct inflow potential.

Table 2 Manhole Deficiency Summary

Defect	Area 1	Area 2	Total	%
Active Infiltration	76	17	93	31%
Staining	24	12	36	12%
Inflow	36	0	36	12%
Surcharge	7	5	12	4%
Root Infiltration	25	3	28	9%
Blockage	5	2	7	2%

Phase II Field Investigations included the development and implementation of a root control plan. The results of the Phase I investigation were used to determine the probability of root infiltration for each sewer run. Proximity to trees, identified blockages, evidence of surcharge and I/I evidence were all used as markers that impacted the probability of root issues for each reach. Each reach was then classified as:

- High Probability of Root Issues
- Medium Probability of Root Issues
- Low/Minimum Probability of Root Issues

Table 3 SSES Area Root Issues Probability Statistics

SSES Area	Total Length (feet)	Root Issue Probability					
		High		Medium		Low	
		Length (ft)	Percentage (%)	Length (ft)	Percentage (%)	Length (ft)	Percentage (%)
Area 1	51,276	13,921	27%	19,347	38%	18,008	35%
Area 2	31,257	1,301	4%	12,210	39%	17,746	57%
Total	82,533	15,222	18%	31,557	38%	35,754	43%

The City determined to perform root control for sewers under a high probability characterization only. Future root control for medium and low probability areas would be determine based on the results of Phase III field investigations.

Phase III Field Investigations included the CCTV of targeted sewer reaches. During the evaluation of the CCTV scoping options it was determined that televised inspection would occur on the high priority root control areas as well as some targeted medium and low priority root control areas. A total of 13,267 feet inspected linear feet of sewers mains were inspected.

Sewer Rehabilitation

The sewer and manhole rehabilitation based on the manhole checks, manhole inspections, and CCTV inspections are summarized for Area 1 and Area 2 in Figure 6 and Figure 7, respectively.

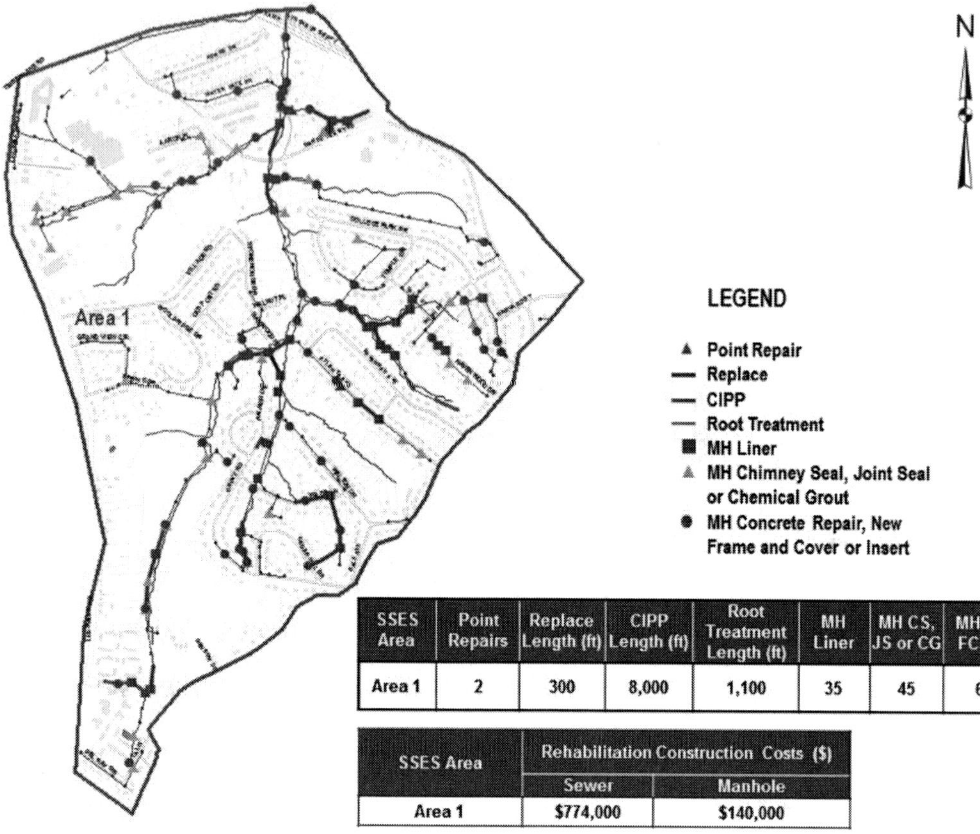

SSES Area	Point Repairs	Replace Length (ft)	CIPP Length (ft)	Root Treatment Length (ft)	MH Liner	MH CS, JS or CG	MH CR, FC or I
Area 1	2	300	8,000	1,100	35	45	67

SSES Area	Rehabilitation Construction Costs ($)	
	Sewer	Manhole
Area 1	$774,000	$140,000

Figure 6 Area 1 - Rehabilitation Work Summary

SSES Area	Point Repairs	Replace Length (ft)	CIPP Length (ft)	Root Treatment Length (ft)	MH Liner	MH CS, JS or CG	MH CR, FC or I
Area 2	2	-	700	-	15	1	21

SSES Area	Rehabilitation Construction Costs ($)	
	Sewer	Manhole
Area 2	$74,000	$47,000

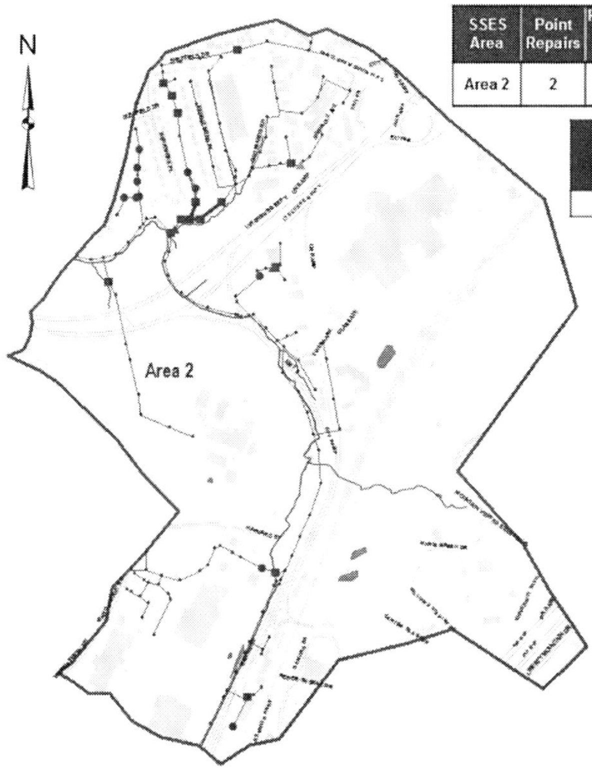

LEGEND

- ▲ Point Repair
- — CIPP
- ■ MH Liner
- ▲ MH Chimney Seal, Joint Seal or Chemical Grout
- ● MH Concrete Repair, New Frame and Cover or Insert

Figure 7 Area 2 - Rehabilitation Work Summary

The following manhole rehabilitation options were considered for manhole repairs:

- Manhole Lining
- New Manhole Frame and Cover
- Insert
- Chimney Seal
- Joint Seal
- Chemical Grout
- Concrete Repairs (bench & channel)

Open trench and trenchless repair methods were considered for sewer rehabilitation and included the following techniques:

- Open Trench Repair Methods:
- Pipe Replacement (MH to MH)
- Open-Cut Point Repairs
- Trenchless Repair Methods:
- Cured-In-Place Pipe (CIPP) Lining (MH to MH)
- Root Treatment

Rehabilitation Cost Estimates
Sanitary improvements will include both replacement and rehabilitation, as indicated. Cost data has been obtained through past construction bids. The estimates provided below include the construction cost, only. Additional engineering, inspection, and administrative cost would be incurred in execution of the sanitary sewer improvement program. The sanitary sewer and manhole improvements are listed separately. Where replacement is identified, the manholes have been included in the overall cost for sewer improvements.

The total of all sanitary sewer improvements is $945,000. They can be completed under one construction contract or broken in to multiple, depending upon funding availability.

Manhole improvement recommendations have also been provided. The improvements will provide inflow and infiltration abatement within the SSES study area. To reiterate, manholes located along segments of replacement have not been identified separately. The cost for their replacement has been included in the sanitary improvements.

The total for all manhole improvements is $167,000. They may be completed under one construction contract or divided based upon funding availability.

CONCLUSIONS

Wastewater collection system hydrologic and hydraulic (H&H) model is a useful tool to evaluate the LOS of sewer system and evaluating improvement alternatives. The LOS results identifies bottlenecks in the existing system and potential improvements for achieving a desired LOS. Selection of alternative was based on cost performance analysis; permitting; easement; synergy with stream restoration; funding opportunities; environmental impact, etc. For Burton Creek Interceptor LOS improvement, increasing sewer pipe size is the most cost-effective alternative. Selection of either Alternative 1 or 2 will depend on an assessment of existing interceptor pipe conditions.

SSES included evaluation of the existing system, phased field investigations; analysis of the field investigation results; selection of rehabilitation method and development of rehabilitation plan for interceptor reaches. SSES also included rehabilitation priority ranking for reaches based on rating criteria.

The implementation of the selected alternative for LOS improvement; reduction of I/I in the contributing sewershed by implementing SSES in the collection system; and restoration of stream will enable the municipality to renew the asset to handle growth; prevent SSOs in the system; and restore the stream for TMDL credits.

REFERENCES

None.

End of Manuscript.

OPTIMIZING KANSAS CITY'S INVESTMENT IN PUBLIC AND PRIVATE I/I REDUCTION TO REDUCE OVERFLOWS

By: John Pruss, P.E.[1], Andy Shively, P.E.[2]

[1]Burns & McDonnell, Director, Program Management, 9450 Ward Parkway, Kansas City, Missouri, 64114; PH (816) 627-4772; email: jjpruss@burnsmcd.com

[2] City of Kansas City, Missouri, Water Services Department Engineering Officer, 4800 E 63rd St., Kansas City, MO, 64130 PH (816) 513-0304; email: andy.shively@kcmo.org

Objective: This paper will present the programmatic approaches used by the City of Kansas City, MO to implement public and private inflow and Infiltration reduction throughout its sanitary sewer systems to achieve targeted reductions defined in the City's Federal Consent Decree related to the City's Overflow Control Program.

Background

Kansas City's Overflow Control Plan was developed to meet regulatory requirements related to reducing wet-weather overflows from the City's combined sewer system and preventing overflows from its separate sanitary sewer systems. The City and its regulatory partners have agreed to meet those objectives over a 25-year time period by completing a planned list of improvements targeted at capturing for treatment 88 percent of combined sewer flows and eliminating sanitary sewer overflows during a 5-year, 24 hour rainfall event. Individual elements of the City's Overflow Control Program (OCP) became part of a Federal Consent Decree on September 27, 2010, with the entry of a Consent Decree in United States District Court for the Western District of Missouri.

The City's Overflow Control Program currently contains some 121 projects at an estimated total capital cost of approximately $4.5 billion with estimated inflation at the end of the 25-year timeframe. To-date, the program is on schedule and on budget with Water Services having completed a total of 16 projects and currently implementing a total of 44 projects. The City will be investing approximately $250M in public sewer and private I/I reduction projects to reduce SSOs and the total I/I flow volume for a 5-year 24 hour storm event from 405 MG to 210 MG.

Benefits of Project

Alternative approaches for achieving Consent Decree targeted I/I reduction were defined and analyzed on a comparative basis. The results of the analysis show that the most cost-effective plan for I/I removal includes removal of I/I in both the private and public sectors of the collection system. Different combinations of private I/I reduction and public I/I reduction were evaluated for cost and I/I reduction. Unit rates for cost and flow reduction were established using flow monitoring data and evaluation, and experience with sanitary sewer system rehabilitation from numerous systems. The City's efforts to optimize the return on its investment in public and private I/I reduction will help other communities faced with having to reduce SSOs gain knowledge from the City's unique programmatic approach, lessons learned, and program findings to-date as it works to invest $178M in public sewer I/I reduction and $72M in private I/I reduction.

Status of Completion

The City began the implementation of public sewer I/I reduction projects in 2013 and private I/I reduction projects in 2016. A total of 19 public sewer I/I reduction projects, containing about 830 miles of sewers to be evaluated, are scheduled to be completed by December 31, 2021. Similarly, approximately 56,000 property parcels will be evaluated for illicit private I/I connections and the cost-effective disconnections identified will be completed by December 31, 2021. Basin-wide targeted I/I reduction ranges from 29% - 45% depending on the watershed. To achieve these targeted levels, a 50% reduction within each defined project area is being sought through a combination of public and private I/I reduction. Eight public sewer I/I reduction projects containing about 380 miles of sewers are currently under design and one project containing about 57 miles of sewers is currently under construction. One public sewer I/I reduction project has been completed containing about 57 miles of sewers resulting in a 31% I/I reduction or 12 MG. Three pilot-scale private I/I reduction projects consisting of approximately 900 property parcels are currently underway. Flow monitoring will be completed in the spring of 2017 to measure the effectives of private I/I reduction in these pilot projects. In addition, approximately 7,200 property parcels will be evaluated in 2016, with completion of identified cost-effective disconnections completed by spring 2017.

Conclusions

Balancing the investment in public and private I/I reduction is highly dependent on the condition of existing sewers and the number of cost-effective private I/I disconnections identified. Correlating measured wet-weather flows to the defects found has caused the City to adapt it programmatic approach to achieve more consistent and effective results between project teams based on lessons learned. Standard I/I flow rates have been established for various types of defects; however, assignment of standard flow rates to public and private defects identified through smoke testing, CCTV, and Manhole inspections has resulted in 20 to 40% of measured I/I flows to be unassigned. Unassigned I/I flow may be attributable to unknown private I/I sources, to defects in manholes and pipes not inspected, or to the assigned defect flow rates being too low.

For private I/I reduction, the total percentage of I/I removed could increase dramatically with an increase in the percentage of buildings having cost-effective private I/I sources disconnected. In the event a higher number of cost-effective sources are found than the 15% assumed, the City could choose between exceeding the targeted I/I reduction and reducing either public sewer I/I rehabilitation work or private I/I disconnection work. Exceeding the targeted I/I reduction may have a substantial impact on reducing downstream conveyance, storage, pumping and treatment program costs. As the City's private I/I reduction program efforts continue, Water Services will monitor and evaluate the actual number of property parcels inspected for illicit connections, the actual number of cost-effective I/I sources found and disconnected, the projected reduction in I/I flows due to these disconnections, and the overall impact of the I/I reduction program in order to adjust the program.

A focused Sanitary Sewer Survey provides actionable results for a successful sewer rehabilitation project

Joseph Kamalesh, M.S., Stantec Consulting Services Inc.
Gary Silcott Jr., P.E., Stantec Consulting Services Inc.
Nick Maniskas, City of Logan, OH

ABSTRACT

The pressures of economic stress and aging sewers require that any sewer system infrastructure analysis use a proactive and intelligent approach. To best meet their goals of accuracy and cost-effectiveness, the City of Logan's SSES program used regional flow monitoring followed by micromonitoring. Micromonitoring is a proven technology that can be easily and rapidly deployed and redeployed within the same system. Its primary advantage over traditional methods is the effective quantification of flows and detection of I&I in upstream (i.e., smaller diameter) reaches of the system where regional flow monitoring equipment doesn't always provide reliable results.

This paper describes the focused SSES approach used by the City and cost savings it achieved. It also highlights a case study where pre- and post-monitoring shows the effectiveness of the approach and completed sewer rehabilitation project. Finally, the paper provides an overview of the capital improvement recommendations for future phases of the program.

KEYWORDS
Sewer Rehabilitation, SSES, I&I, Micromonitoring, Flow monitoring.

INTRODUCTION

The key task of a Sanitary Sewer Evaluation Survey (SSES) is to maintain the structural integrity and proper conveyance of sewage to wastewater treatment facilities. A well planned SSES is a critical step for any entity responsible for managing a public sewer system. In a traditional approach, the flow monitoring results set a priority on large regional basins that are ranked for additional SSES investigations. A significant amount of costly follow-up work, including CCTV, smoke and dye testing, manhole inspections, sewer rehabilitation and private source removal, relies on these results.

Inflow/Infiltration

According to the EPA Handbook on Sewer System Infrastructure Analysis, inflow/infiltration (I&I) is the major limitation to the successful operating performance of a wastewater system. Infiltration is the water that enters the sewer pipe from the surrounding soil through foundation drains, joint cracks, defective connections, etc. Inflow is the excess water that is conveyed into the sewer lines from various sources, such as roof leaders, downspout connections, house sump-pump, yard drains, cross-connections, etc (EPA 1991).

Unwarranted I&I in a sanitary sewer system can cause hydraulic overload in sewer lines and treatment plants. This results in sewer surcharge, basement backups, and under-performing treatment systems. Controlling the I&I by removal or minimization is an important part of a sewer system management plan (Lukas et.al, 2001)

Sanitary Sewer Survey
A sewer system survey should have a systematic approach beginning from data gathering to final recommendations (EPA 1991). A typical I&I removal cycle begins with flow monitoring and data gathering to identify areas with excessive wet weather flow. The SSES field investigations are then deployed in the problem areas to assess the structures that need repair or rehabilitation. An engineering analysis on the collected field data is performed to assess sewer condition, identify defects that require repair, and develop ranking system for prioritization. The defects and rankings are then prioritized and combined into actual projects with recommendations for sewer rehabilitation.

However, the missing component of most I&I removal programs is the drive for post-rehabilitation flow monitoring data. This data is vital to measure and quantify the results achieved and effectiveness of the overall program. The need for post-rehabilitation study should be discussed and developed at the very beginning of an I&I program.

Micromonitoring
Micromonitoring is the latest addition to the SSES toolbox. It significantly reduces costs, and the follow-up fieldwork can be assigned to targeted areas instead of basin- or system-wide (McGillis et.al 2013a) If a segment of the sewer system does not show an I&I response, it can be immediately removed the from the SSES program. Other segments that show a response indicating high I&I are targeted for wet weather CCTV, smoke testing, etc., to further locate the ultimate source.

The general approach is as follows:

1. A basin is identified for I&I reduction. This is usually because of sanitary sewer overflows or from flow monitoring results showing a significant increase in flow rate during wet weather events.
2. Depending on basin size, 8-10 micromonitors are deployed. In general, the sewers on the upper reaches are selected, and the micromonitors are placed before the sewer enters a trunk line. Other branches are also selected.
3. Data from one-two storms are collected. Based on the results, the micromonitors are moved to focus on one or more pipes with high I&I or to other pipes in the basin.
4. After working through the basin, the results can be mapped with an accurate presentation of the origin of the I&I.

A typical layout for a multi-phase Micromonitoring program is shown in Figure 1. Figure 2 shows a micromonitor installed in an 8-inch pipe.

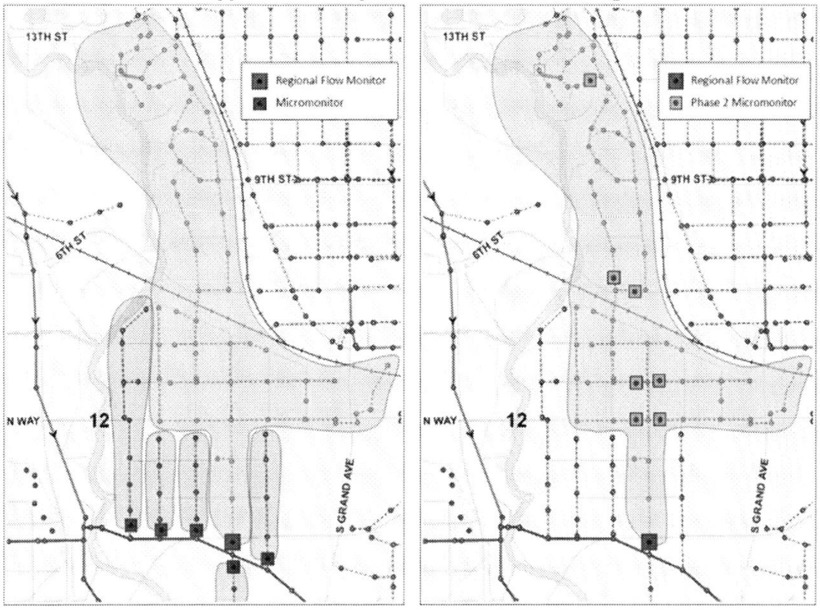

PHASE 1: Initially regional flow monitors are installed to monitor large basins. Micromonitors are installed to pick-up the small sections with no flow differencing if possible.

PHASE 2: All monitors showing no I/I response are removed. Any regional flow monitor showing significant response is subdivided by moved micromonitors

Figure 1: A typical layout of multiphase micromonitoring

Figure 2: Micromonitor installed in an 8-inch pipe

CITY OF LOGAN- A CASE STUDY

The City of Logan operates and maintains over 174,600 linear feet of gravity sewer lines, the majority of which is vitrified clay pipe (VCP). The system also contains hundreds of sanitary sewer manholes that are primarily brick construction as some of the sewers in the system are over 100 years old. In 2013, the City undertook a major improvement project for its wastewater treatment facility, which placed a large burden on the City's capital improvement program. In addition, it was decided to identify and implement capital projects designed to address ongoing issues with excessive amounts of I&I within the sanitary collection system.

The last detailed evaluation of the sewer system was done almost three decades ago. The age of the system has resulted in deterioration of the pipe, and private source clean water connections have likely occurred overtime. I&I in the City's sanitary sewer system contributes to basement flooding, overflows, and inflated operating costs. The system does not have the capacity to convey all wet weather flow to the new WWTP.

These incidents have inevitably led to compliance issues and higher operating expenses. To address the varying issues throughout the system properly and economically, a better understanding of the severity and extent of the problem was needed.

As in any traditional SSES program, the City planned to collect flow monitoring data to identify the problem areas. But in doing so, they wanted to tailor their approach to one with a more focused strategy that could identify and target the remediation of the primary sources of I&I to achieve flow reduction targets in a more cost-effective manner. The primary purpose of the project was to identify sources of I&I in the collection system so the City can properly plan capital improvements projects and other I&I remediation efforts. Although not mandatory, the City undertook this project to avoid potential compliance issues in the future.

The key goal of this project is to remove as much as I&I from the system as possible without exhausting all resources. The City would like to eliminate basement flooding, sewer overflows and be able to address future compliance possibilities. The data from this project would identify bottlenecks and other system deficiencies in the collection system. The eventual goal is to transport all the sewage to the WWTP.

APPROACH

The traditional approach for an I&I project is to conduction regional flow monitoring, rank the basins based on severity, conduct detailed SSES investigations and finally provide remediation reccomendations (McGillis 2013b)However, in customized approach such as in the City of Logan, the detailed SSES investigation is replaced with a focused SSES method. The focused SSES includes micromonitoring, wet-weather CCTV, smoke and dye testing.

The collection system was divided into 12 sewer sheds covering all 33 miles of sewers. Regional flow monitors were installed such that the extent and location of the problem areas could be characterized and the worst areas identified for further investigation through micromonitoring. After a month of regional flow monitoring, it was concluded that three of the sewer sheds could be removed from additional analysis because no significant amount of I&I was detected. The remaining sewer sheds were further investigated using a total of 67 micromonitors over a 3-month period.

This allowed the SSES to pinpoint I&I sources in the sanitary sewer system. Data from just a single storm was enough to show where the more significant I&I sources were located. It also allowed the basins to be further segmented and some of the micromonitors to be redeployed so that the I&I sources could be isolated and identified more rapidly and with greater efficiency. The micromonitoring technology proved to be effective in quickly locating the I&I sources and optimizing field investigation efforts and ultimately saving money. CCTV was completed in two rounds within the sewers found to contribute moderate or excessive quantities of I&I.

The sewer system analysis included other typical I&I investigation work such as:

- System mapping and verification
- Lift Station run time analysis
- Comparison of water use records with flow monitoring and micromonitoring data
- Interviews with City staff to gain historical knowledge of problem areas

This information was utilized in conjunction with the field work to formulate remediation recommendations.

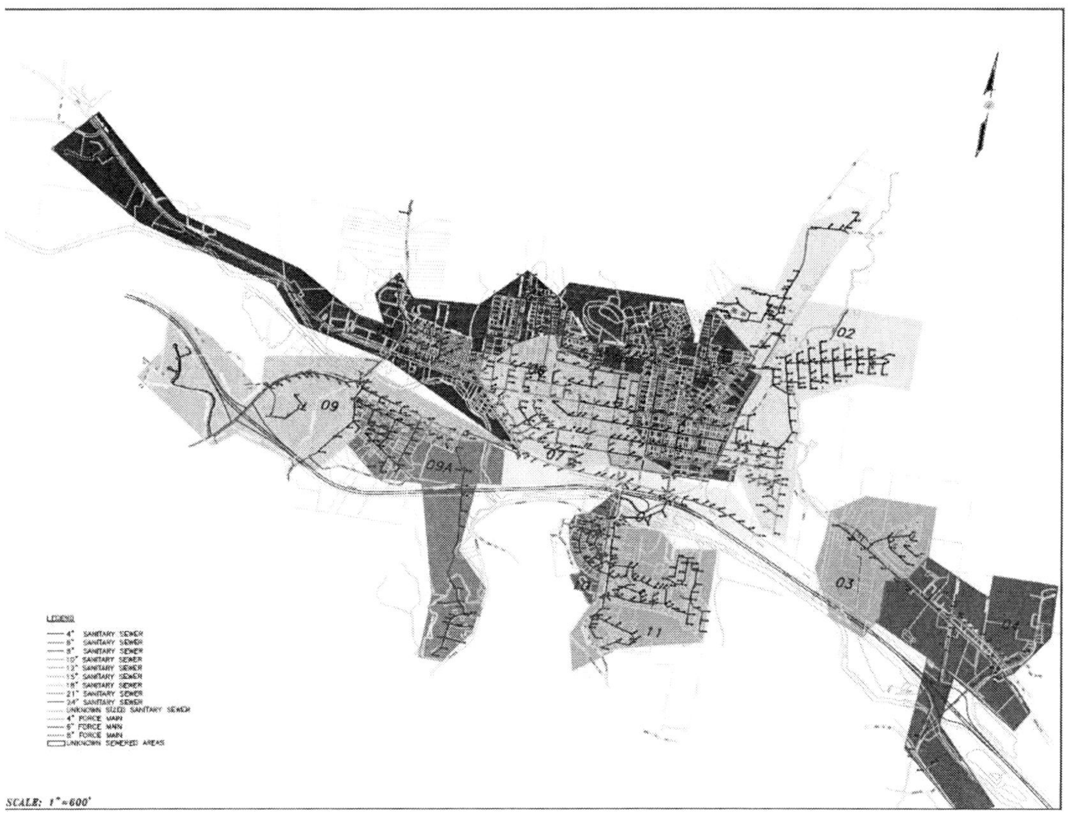

Figure 3: City of Logan flow monitoring layout

RESULTS

In addition to typical sewer system defects, the field work revealed other issues that need to be addressed through routine maintenance or capital improvements. The downstream reaches of sewer sheds 02, 05, 06, and 07 were found to contain heavy accumulation of grit and debris, which reduces system capacity and contributes to operational problems. Sewer surcharging was identified in several downstream areas of the system as well, which contributes to basement backups and the associated health concerns.

Based on the results of the micromonitoring data, CCTV inspection was performed in only those sewer sheds that ranked as having moderate to high I&I. The CCTV inspection indicated the need for sewer rehabilitation using CIPP, sewer line replacement, manhole lining, manhole replacement, and lateral replacement. Three phases of capital improvements to address the problems were developed so the City could spread out the overall costs and secure grants and low interest loans. The associated costs for each phase are: Phase 1 $550,000, Phase 2 $1,900,000, and Phase 3 $950,000.

Several sewer mains were found to contain substantial accumulations of debris, which will be thoroughly cleaned to increase capacity. Potential bottlenecks were found within the system that will be further evaluated once the lines are cleaned and the I&I remediation is complete. Some skepticism was expressed on why lines were being rehabilitated in areas where basement backups were not occurring and additional concern for the political ramifications of doing work away from the basement backups, but construction was conducted as recommended based off findings.

The following is a summary of the key results:

- Approximately 58,400 feet (33%) of the sewer system was deemed to have moderate to high I&I, contributing ~750,000 gallons of I&I.
- Approximately 116,200 feet (67%) of the sewer system was eliminated from further investigations.
- Post-monitoring in the Phase 1 sewer rehabilitation project showed a 95% reduction of I&I.

Figure 4 shows the project area for the Phase 1 sewer rehabilitation, and Figure 5 shows the flow data for a similar rain event during pre- and post-rehabilitation. The post monitoring results are clear and help reinforce the decision to perform rehabilitation upstream of the problem areas despite the political and local concerns.

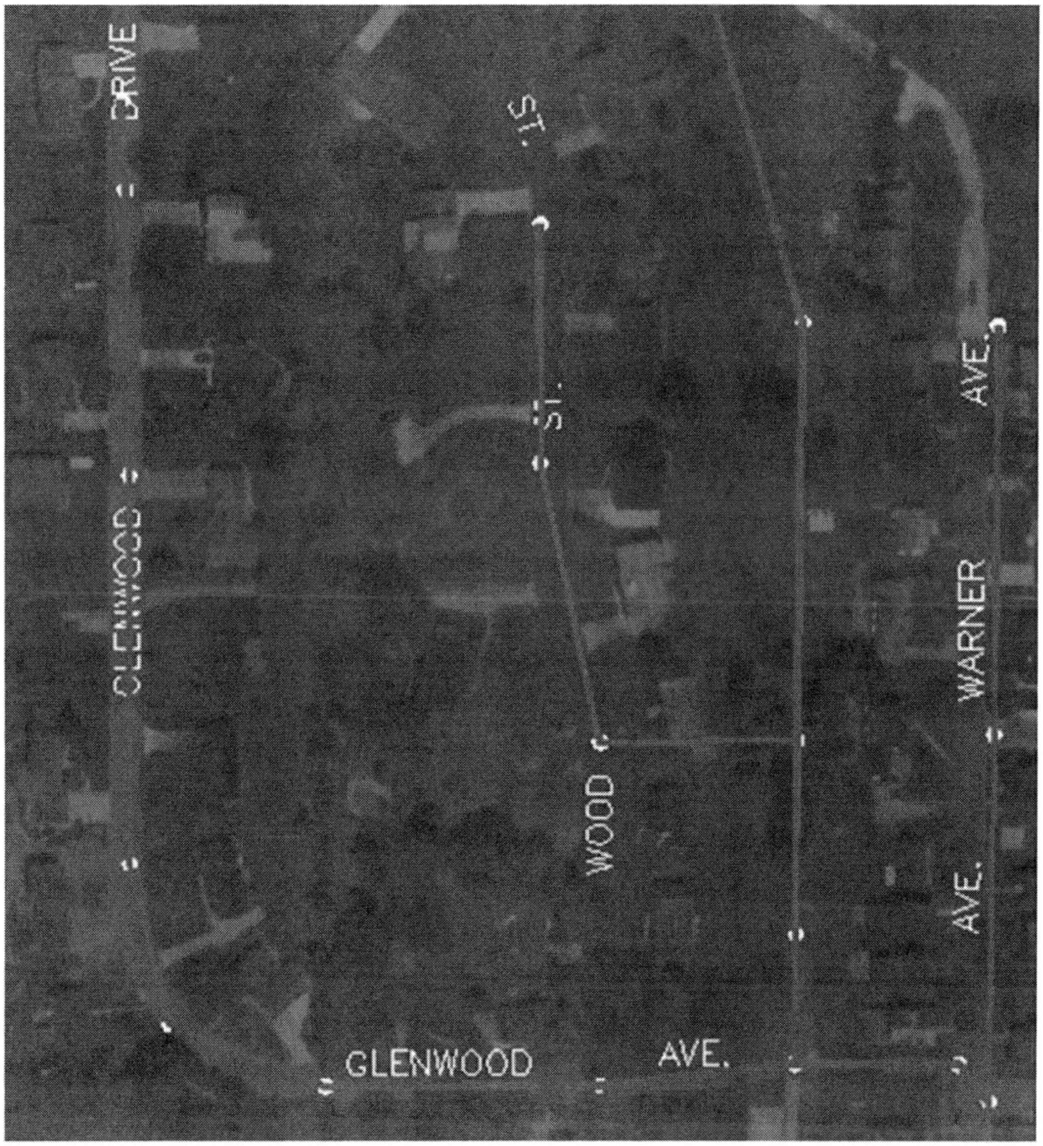

Figure 4: Project area for the Phase 1 sewer rehabilitation

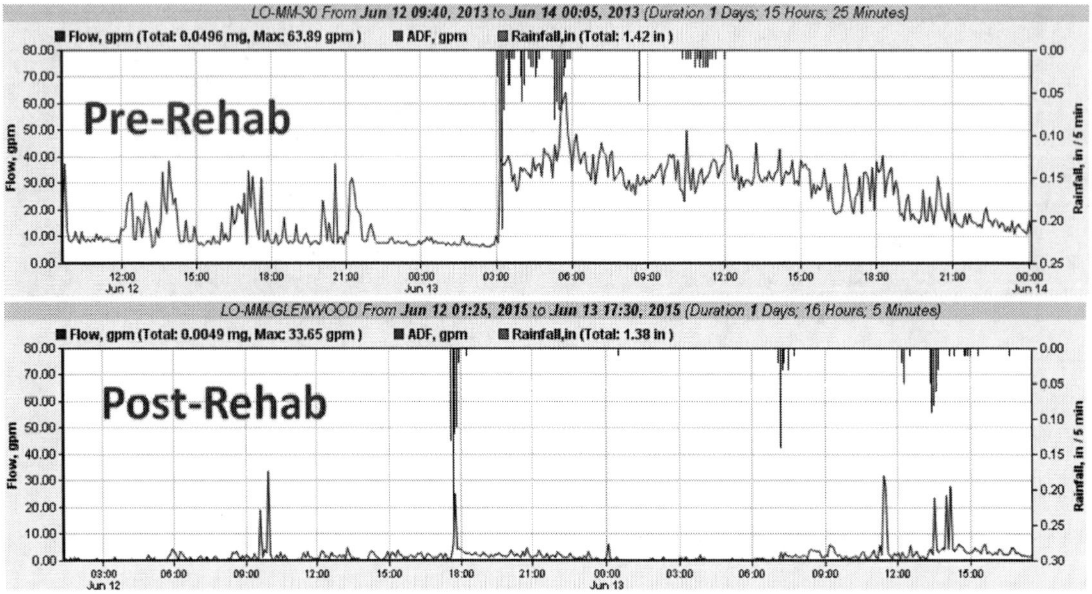

Figure 5: Flow data for a similar rain event during pre and post rehabilitation.

CONCLUSIONS

In summary, approximately 45,709 feet of the system was found to need no further investigation through regional flow monitoring work, while another 70,450 feet was considered minimal I&I contributors as result of the micromonitoring work. Therefore, approximately 116,200 feet of sewers were eliminated from consideration for further investigation, which equates to approximately 66% of the system. This results in a significant cost savings of $232,000-$350,000 to the City at a $2-$3 dollar per foot cost for television inspection work, which is the traditional approach.

Approximately 66% of the system was eliminated from consideration for further investigation using micromonitoring. The associated cost reduction for subsequent SSES investigation has allowed for the development of a manageable implementation plan designed to achieve significant I&I reduction. The post monitoring data from the Phase 1 capital improvement project clearly shows the cost savings and tangible results that can be realized with a focused SSES approach. Two phases of pipe and manhole rehabilitation were developed from the first round of CCTV work completed in 2013. The total project cost estimate was $2.15M which included manhole rehabilitation, main line CIPP, main line replacement, point repairs, spot liners, grout sealing and lateral replacement.

Stantec assisted the City with procuring grants and low interest loans to finance the rehabilitation projects. The Phase 1 capital improvement project was funded through the Ohio Public Works Commission and was designed so that construction can proceed when the funds become available. The early payoff for the City will not be in the sewers found to contribute excessive I&I, but rather in the sewers that do not contribute excessive I&I that are eliminated from consideration for additional SSES work.

ACKNOWLEDGMENT
The authors would like to thank Robert Fuller (Stantec) and the City of Logan staff for their technical contributions towards the success of this project.

REFERENCES

1. EPA Handbook: Sewer System Infrastructure Analysis and Rehabilitation; EPA/625/6-91/030 October 1991
2. Lukas, A; Merrill, S; Palmer, R; Nathan, V (2001) In Search of Valid I/I Removal Data: The Holy Grail of Sewer Rehab; WEFTEC 2001, Atlanta, GA.
3. McGillis, R; Barton, J.M.H; Kamalesh, J (2013) A New Entrant in Sewer Economics: How a Regional Sewer District Saved Scheduled Rehabilitation Budget; WEFTEC 2013, Chicago, IL
4. McGillis, R; Barton, J.M.H; Kamalesh, J; Isaacs, D (2013) Improved Proactive Asset Renewal: Targeting I&I Utilizing Micromonitoring; 2013 WEF Collections, Sacramento, CA

Ecoflow – Private I/I and Rainwater Reduction Program in Lawrence, KS

Nick Hoyt, PE Lawrence Utilities Department, Lawrence, Kansas

ABSTRACT

The City of Lawrence, Kansas has been aggressive in its efforts to maintain its wastewater collection system. The City has successfully reduced basement backups and sanitary sewer overflows through proactive maintenance and a public-sector rehabilitation program focused on cured-in-place-pipe lining. However, peak wet-weather flows of 8 to 10 times above the average daily flows continue at the wastewater treatment plant following significant storm events. These significant peak flows put a strain on the capacity of the sanitary sewer collection system and the existing wastewater treatment plant.

In 2012, the City of Lawrence Department of Utilities completed an update of its Wastewater Master Plan. One of the recommendations from the Master Plan was to reduce peak wet weather flows to the wastewater treatment plant by 19 MGD, in order to reduce or eliminate the need for conveyance improvements and expanding capacity at the existing treatment facility. The Master Plan recommended implementing a Rapid Inflow & Infiltration (I/I) Reduction Program that included a private I/I component to supplement the ongoing public sewer system rehabilitation in order to reduce I/I by 35% within about one quarter of the City.

The 5,260 acre area identified for the Private I/I program includes approximately 589,000 feet of sanitary sewer and 8,800 properties, some dating back to the late 1800s. This area is an older section of the City that drains by gravity to the wastewater treatment plant. The Private I/I removal program, branded Ecoflow – Rapid Rainwater Reduction Program, was approved as part of the Department's Capital Improvement Program. Phase 1 of 8 was launched in 2014. Phase 5 will be starting in the summer of 2017. The presentation will cover private I/I program's development, implementation, results, and key success factors.

KEYWORDS: Private I/I, Ecoflow, I/I Reduction, Private Property

INTRODUCTION

The primary objective of the Ecoflow Rapid I/I Reduction Program is to reduce peak wet weather I/I rates by approximately 19 MGD, or 35%, within the oldest parts of the sanitary sewer system near the Kansas River Wastewater Treatment Plant (KRWWTP). Reducing I/I will be accomplished by inspecting and rehabilitating both public and private sector components of the sanitary sewer system. The reduction in flow will allow the City to:

- Reduce sewer capacity restriction along the Burroughs Creek Interceptor.
- Reduce the need for capacity expansion at the KRWWTP.
- Provide significant cost savings by reducing O&M efforts.

The Ecoflow Rapid I/I Reduction Program has been implemented using a phased approach. Each phase includes inspection/evaluation and rehabilitation of both private and public assets in a particular area of the overall project area. Flows are monitored each spring to gauge the success of the Program.

This paper focuses on the private I/I investigation and rehabilitation element of the Program. The paper will cover how the Private I/I Program was developed, implemented, key results and lessons learned.

PRIVATE I/I PROGRAM DEVELOPMENT

The City Utilities Department researched existing private I/I programs for ideas and objectives to help identify the essential elements of the new Ecoflow Program. The Program's main objectives are as follows:

- Repairs must be cost-effective to reduce inflow and funded by the Department
- The program would be voluntary
- The time from initial contact with the property owner to final inspection of the repairs would be no more than 6 weeks, and
- The program had to be flexible and adaptable to the needs and schedules of both the property owners and the Department.

After defining the objectives of the program, the City Utilities Department began assembling a program team. Through a Request for Qualifications (RFQ) process, the Department selected TREKK Design Group as the consultant to lead the effort. TREKK Design Group provided the experience developing similar programs and the number of staff capable of handling all the phone calls, appointments, and data entry. The Department also added a Professional Engineer and Master Plumber as full time employees focused on the Ecoflow Program. The Department employees would provide project oversight, repair approval, building permit inspections, and minor cleanout repairs.

Later, the City selected local plumbing contractors through an initial RFQ to perform the private property I/I repairs as identified through the plumbing evaluation process. Yearly on-call unit rate contracts were negotiated with selected plumbers. The City extends the plumbing contracts each year to plumbers in good standing. Two additional plumbing RFQs have been issued to replace plumbing contractors that withdrew from the program. Over the course of the 3 years of the Program, between seven and nine plumbers have been under contract. Good relationships with local licensed plumbers are essential to the process. The Program team worked closely with the selected plumbing contractors to ensure that the City received the best possible service and I/I repairs were completed to the standards established.

The project team also includes:

- Electricians and other sub-contractors to the plumbers
- George Butler Associates to perform yearly sanitary sewer flow monitoring and tracking.
- A local drain cleaning company to provide on-call camera services.

Because of the unique nature of the Ecoflow Program, the initial step in developing the program was to gain concurrence and input from other City Departments that would be impacted by the Program. Those departments included the City Attorney's Office, Finance, Purchasing, and Development Services. These Departments provided valuable information to help set-up property owner and plumbing contracts, and streamlined the processes for building permits, purchasing, and invoicing for the plumbers. Recently, a key relationship has been established with the Rental Licensing Department to combine rental inspections and private I/I evaluation into a single appointment. The Rental Licensing Department has also shared a valuable database of property owner contract information.

Public education, outreach, and public perception of the Ecoflow Program are critical to its success. To promote awareness and the identity of the program, the Department used a local advertising firm to create the brand for the program. The Ecoflow – Rapid Rainwater Reduction Program name and logo are used on all materials associated with the Program. Program literature, field inspectors, and field vehicles all feature the logo to identify themselves and promote awareness of the Ecoflow Program. The Program Team also established an I/I hotline number, website and e-mail address for the public to obtain Program information and to schedule building evaluation appointments.

Establishing guidelines for what constitutes a cost effective repair was another key step in developing the program. After reviewing design storm rainfall intensities for Lawrence, existing design flows, existing capacities, treatment costs, and the estimated cost of significant capital projects, the following simple guidelines were established:

- Defects with direct tributary area (area drains, downspouts, driveway drains) – Cost of repair must be less than $\$25/ft^2$ of tributary area.
- Defects without direct tributary area (sump pumps and foundation drains) – Cost of repair must be less than $1,000/gpm for the project 10-year design storm.

Estimating 10-year design storm flows for particular defects like foundation drains is extremely difficult. Usually this is determined by discussing the known history of the defect with the property owner. If there is little to no known history, staff will perform follow-up visits following rain events. Also, property owners or tenants will send videos or pictures of the defect when active. Based on the amount of flow contributed during smaller rain events, assumptions are made to project a design storm flow in order to determine whether or not the repair is cost effective.

Standard repair details and specifications were established for each repair type to provide the plumbing contractors with an expectation of the repair and to set general guidelines. It should be noted that the standard details were set with the understanding that no two repairs are the same and variations are to be expected. The plumbing contractors and the Program team worked with the property owners to devise a course of action for each property and evaluated various options to find the best possible solution. The City approved any major deviation from the standard detail prior to the work starting.

PRIVATE I/I SOURCE IDENTIFICATION

TREKK Design Group implemented a combination of smoke testing, plumbing evaluations, and addition investigation methods to locate and assess private I/I sources in the system. The following sections summarize the inspection activities that took place.

Smoke Testing

Sewers within each target area were smoke tested to detect I/I sources in the public and private sectors. Smoke testing is a tool that can quickly and efficiently identify I/I defects throughout a collection system. Unfortunately, not all private sector defects can be located through a smoke testing program. TREKK Design Group conducted plumbing evaluations in conjunction with smoke testing to identify additional illicit I/I sources inside buildings. In order to eliminate smoke testing scheduling issues related to weather, smoke testing activities are now scheduled 6 to 9 months in advance of targeted plumbing evaluations.

Building Evaluations

Plumbing evaluations are the only practical means to identify contributing sources of I/I, such as leaky foundations, interior foundation drains, sump pumps and pee-trapped driveway/area drains discharging to the sanitary sewer.

A thorough public relations and notification program was implemented prior to initiating the plumbing evaluations to educate the public and minimize concerns. The notification program included public outreach and education utilizing media sources and open house meetings.

Following the open house meetings, which were held in smaller sub-areas, letters were mailed to property owners and tenants further explaining the Program and requesting they schedule an appointment for the evaluation. Evaluation team representatives called any unresponsive homeowners to schedule the appointment. A minimum of three attempts were made to contact the owner via phone. If the phone call efforts were unsuccessful, an evaluation team member made three additional visits to the property. These visits occurred on different days at various times to increase the opportunity of contacting the owner or occupant. A door hanger was left if no one was home. The door hanger provided Program information and requested the homeowner call the Ecoflow hotline number. A member of the TREKK evaluation team was available on the Ecoflow hotline to schedule an appointment or provide additional information.

Special provisions were made to communicate the legitimacy of the evaluations and importance to the property owners. The evaluation team took the time to educate property owners about the Program and to point out the benefits to the City as a whole. A majority of property owners complied with the request for inspection through the initial mailing, follow up phone calls, or property visits. However, a few remaining property owners were difficult to contact and, as a result, required extended efforts from the evaluation team.

Plumbing evaluations were conducted by a two-person crew, comprised of one male and one female in most instances, to provide a more comfortable environment for the property owners. It was important for the individuals performing the plumbing evaluations to appear professional, communicate well, and show knowledge of the Program. Communicating accurate and concise Program information was vitally important to answer property owner questions. The evaluation crew's attire featured both the Ecoflow and TREKK logos, and all crew members wore a photo identification badge. Crew members carried an introduction letter from the City, and all field vehicles were outfitted with the Ecoflow logo.

Evaluations were grouped into three categories:

- *Inspected.* The exterior of the building and the interior plumbing (basement and crawl spaces only) were inspected to locate potential I/I sources.

- *Needs Evaluation/Not Inspected.* The property owner is unresponsive to Program outreach efforts.
- *Refusal.* The owner declined to participate in the program. Attempting to complete a building evaluation at that property is immediately suspended. Ownership records for all refused properties are reviewed semi-annually. If the ownership of a previously refused property has changed, the property status is updated and efforts to evaluate the property are restarted.

Table 1 provides a summary of the program plumbing evaluation status within each targeted phases of the program. Door knocking and phone call efforts are ongoing in the Phase 4 area and outreach attempts continue in all targeted areas.

Table 1. Building Evaluation Status Summary

	Phase 1	**Phase 2**	**Phase 3**	**Phase 4**	**Total**
Phase Targeting Start Date	June 2014	Feb 2015	Sept 2015	July 2016	
Total Properties	1,287	1,313	1,337	977	4,914
Evaluations Completed	1,163 (90%)	1,108 (84%)	942 (70%)	640 (66%)	3,853 (78%)
Refusal	65	79	105	74	323
Needs Evaluation/Incomplete	59	126	290	263	738

Program efforts to complete plumbing evaluations are targeted in specific areas according to the total project area and phasing. As the Program develops a reputation and word spreads around the City, property owners in future target areas and outside the total project area have volunteered to participate. If requested, evaluations and any necessary repairs are completed at these properties. Defects identified outside the total project area are assessed at a higher cost effective relationship based on downstream facilities and other factors.

Dye Water/Camera Testing

If the intimal evaluation crew was unable to determine the termination point of a potential defect, sources were considered suspect. Follow up appointments were made to either use dye water testing or a push camera to confirm connectivity. Each suspect defect is evaluated for which technic is most likely to be successful. Whichever method to determine connectivity is utilized, it is imperative to thoroughly document the results.

Building Evaluation Documentation

Building plumbing evaluations were conducted and recorded on field forms and imported into the City's Lucity database. The following list describes data that was collected, recorded, and input into the Lucity database for record.

1. *Date.* Calendar date that inspection was performed.
2. *Line Segment.* Specific line segment to which building service lateral discharges wastewater and potential I/I to the City-owned sewer system.
3. *General Information.* Address of building, name of owner or occupant, and telephone number. Also present status of inspection and general comments.

4. *History.* History of basement flooding as reported by owner or occupant. Detailed information including flood dates, duration, suspected cause, and action taken by owner or occupant.
5. *Building Data.* General description of building including type, age, layout, and basement type.
6. *Defect Type.* Description of type of defect, including downspout, uncapped cleanout, driveway drain, stairwell drain, foundation drain, area drain, service lateral, window well w/drain, and sump pump.
7. *Tributary Area.* Estimate of square footage of runoff tributary to defect used to compute theoretical inflow rate to the sewer system based on rainfall intensity.

In addition to the data stored in the Lucity database, the evaluation crew takes digital pictures of all defects, potential defects, or related issues identified at each property. Pictures are vital for management and plumbing contractors to understand the situation without revisiting the property. The evaluation crew will also document additional information essential to understanding the specific situation at each property:

- *Access.* Defects located in crawl spaces or basements may have limited access that restrict personnel or materials from entering the area to complete the repair.
- *Electrical panel or service.* City Code requires all sump pumps to be on a dedicated electrical circuit. A complete electrical panel upgrade may be required to install a sump pump if the existing electrical service is insufficient or outdated. Extensive electrical upgrades may require the property owner to contribute to the project in order to make the repair cost effective.
- *Drainage issues.* Poor grading or a lack of green space surrounding a structure may can limit discharge options and add cost. This issue is more prevalent in commercial areas.
- *Property owner scheduling or occupancy issues.* Specific issues related to pending real estate transactions, tenant turnover, unrelated remodels or other issues can affect the desired schedule of the property owner. These issues must be communicated to the plumbing contractor in order to satisfy the property owner.
- *Existing plumbing issues related to the defect.* Repairing defects that collect both sanitary waste and rain and groundwater can be particular expensive. If these connections are not understood before a repair, sanitary waste can be discharged to the groundwater following a repair.
- *Existing Finishes.* Finished basement, elaborate tile or plantings can limit options for repairing a defect and add cost. Understanding these issues can be imperative when determining if a repair is cost effective and communicating expectations to a property owner.

Summary of Private I/I Defects

A total of 1,910 private sector defects have been identified by program efforts. All private I/I defects are summarized in Table 2.

Table 2. Summary of Identified Private I/I Sources

	Phase 1	Phase 2	Phase 3	Phase 4	Future Target Area	Outside Program Area	Total
Area Drain	0	6	2	5	2	1	**16**
Broken Cleanout	166	125	72	66	47	29	**505**
Downspout	1	5	16	11	9	0	**42**
Driveway Drain	1	1	4	2	2	0	**10**
Foundation Drain/Wall Seepage	90	120	266	182	103	54	**814**
Stairwell Drain	34	6	53	39	17	22	**171**
Sump Pump	35	71	69	73	44	40	**332**
Other	0	7	0	3	2	7	**19**
Total I/I Sources	**327**	**341**	**482**	**381**	**226**	**153**	**1,910**
Total Properties Evaluated	**1163**	**1108**	**942**	**640**	**316**	**435**	**4,604**
Properties with Defects	**263**	**289**	**394**	**302**	**179**	**125**	**1,552**
Percent of Properties with Defects	**23%**	**26%**	**42%**	**47%**	**57%**	**29%**	**34%**

Status of Private I/I Defects

After a specific defect or potential defect has been identified through a plumbing evaluation, a status is assigned. The status may be updated several times throughout the process. A summary of all identified defects is summarized in Table 3.

- *Refused Repair.* Owner does not desire to proceed with the repair.
- *Need Agreement from Owner or Owner wishes to delay repair.* A potential defect has been identified and is cost effective to repair. However, owner delays repair by not signing the agreement with the City. Also, the owner may desire to delay the repair due to other scheduling issues.
- *Does Not Qualify.* An identified defect may not be cost effective to repair.
- *Pending Repair.* The defect has been approved for repair and assigned to a plumber.
- *Suspect.* Discharge location of defect is unknown. Additional dye water testing, push camera work or other investigation is required to determine discharge location. If substantial or costly challenges exist to repairing a defect, a 2nd evaluation is completed with more experience staff in order to determine if the repair is cost effective. After the additional investigations are completed the status of each defect is updated depending on the outcome of the investigation.
- *Repair Complete.* The defect has been repaired, invoiced, and inspected

Table 3. Summary of Private I/I Defect Status

Defect Type	Total Defects Identified	Refused Repair	Need Agreement or On Hold	Does Not Qualify	Pending Repair	Suspect	Repair Complete
Area Drain	16	0 (0%)	4 (25%)	4 (25%)	1 (6%)	0 (0%)	7 (44%)
Broken Cleanout	505	4 (1%)	7 (1%)	9 (2%)	38 (8%)	0 (0%)	447 (89%)
Downspout	42	1 (2%)	1 (2%)	5 (12%)	0 (0%)	19 (45%)	16 (38%)
Driveway Drain	10	0 (0%)	1 (10%)	0 (0%)	0 (0%)	2 (20%)	7 (64%)
Foundation Drain/Wall Seepage	814	24 (3%)	59 (7%)	187 (23%)	49 (6%)	14 (2%)	481 (59%)
Stairwell Drain	171	8 (5%)	7 (4%)	109 (64%)	5 (3%)	6 (4%)	36 (21%)
Sump Pump	332	8 (2%)	21 (6%)	26 (8%)	23 (7%)	13 (4%)	241 (73%)
Other	19	0 (0%)	0 (0%)	1 (5%)	1 (5%)	3 (16%)	14 (74%)
Total I/I Sources	**1,910**	**45 (2%)**	**100 (5%)**	**341 (18%)**	**117 (6%)**	**57 (3%)**	**1,249 (65%)**

SUMMARY OF PRIVATE I/I IMPROVMENTS

Building Source Remediation

Efforts were made to transition from the plumbing evaluation and source identification to remediation as quickly and efficiently as possible. In the event that a cost-effective I/I source was identified as a part of the plumbing evaluation, the evaluation crew explained the general remediation procedures with the property owner. Because this is a voluntary program, the property owners were given the option to proceed with the repair work and execute the agreement form. They were then given the opportunity to select a City-approved plumbing contractor to complete the noted repairs.

Following authorization by the City, the selected plumbing contractor was notified to proceed with the pending work. The plumbing contractor scheduled a consultation with the property owner and a member of the Program team to discuss the pending work and schedule a date and time for completion. All repair work was coordinated between the selected plumbing contractor and the property owner.

Final Inspection and Payment

As a requirement set by the City, plumbing contractors were required to obtain a construction permit from Building Development Services and schedule a rough-in inspection upon. The construction permit was provided at no cost to the contractors and was required for repairs that involved mechanical or electrical components. Simple repairs, such as downspouts or service lateral cleanouts, did not require a City permit or City inspection. After the plumbing contractor submits an invoice for a repair, a Program team representative completes a final inspection of work to ensure that modifications met the established specifications and the property owner was completely satisfied with all aspects of the project. The Program team representative also provides the property owner with warranty documentation and other paperwork related to the repair. Each property owner was given a follow-up survey to obtain feedback on the Program and all individuals involved. Invoicing for each repair was submitted directly to the City and included a 12-month written warranty. Final payment to the plumbing contractor was processed by the City after approval and authorization by the Program Manager.

All repair work completed as part of the Program was based on unit pricing established through the RFQ process. Modifications were made to various standard repair details as the Program progressed to better correlate the repairs with local plumbing practices and readily available materials. Unit costs were revised accordingly. Table 4 summarizes the costs for all private I/I repairs. It should be noted that not all private I/I defects found during field investigations were repaired. Some property owners refused the repair, or the defects found could be not be addressed cost-effectively. Repairs not considered cost-effective included those with significant electrical upgrades/panel replacements, repairs with no practical location to install pumps/discharge pipes, dual purpose sump pumps currently servicing multiple plumbing lines in the building, and defects taking on minimal I/I.

Table 4. Summary of Phase 1 Private I/I Repair Costs

Defect Type	Total Defects Identified	Defects Repaired		Repair Cost	
		Number	(%)	Total	Average Unit
Area Drain	16	7	44%	$20,465	**$2,924**
Broken Cleanout	505	447	89%	$156,247	**$350**
Downspout	42	16	38%	$27,704	**$1,731**
Driveway Drain	11	7	64%	$27,445	**$3,921**
Foundation Drain/Wall Seepage	814	481	59%	$1,370,535	**$2,873**
Stairwell Drain	171	36	21%	$99,902	**$2,775**
Sump Pump	332	241	73%	$619,333	**$2,613**
Other	19	14	74%	$4,065	**$290**
Total Plumbing Contractors	**1,910**	**1,249**	**65%**	$2,325,697	**$1,862**

PROGRAM COST EVALUATION

The total costs for the private portion of the completed phases of Rapid I/I Reduction Program are summarized in Table 5. Cost vary for each phase depending on building type, the number of defects repaired, defect type and other factors. It should be noted that the cost ($)/address metric is based on total properties with a completed evaluation.

Table 5. Private Sector I/I Program Cost Summary

	Total Linear Ft	Evaluated Properties	% Addresses Repaired	Total Cost ($)	Average Cost ($) / Address	Average Cost ($) / Linear Ft
Phase 1	87,000	1,163	18%	$844,000	$726	$9.70
Phase 2	128,000	1,108	18%	$771,000	$696	$6.02
Phase 3	79,000	942	27%	$1,026,000	$1,089	$12.99
Total	294,000	3,213	21%	$2,641,000	$822	$8.98
* Includes Costs for Program Management, Inspections, Plumbing Evaluations, Plumber Consultations, Plumbing Repairs						

LESSONS LEARNED AND KEY SUCCESS FACTORS

Multiple procedural modifications have been made to the Program using feedback from property owners and plumbers, on-the-job experiences, and team member observations. Lessons learned and key success features of the Program include:

- Communication and input from the City Attorney, Codes, and Purchasing Departments are vital in development and implementation.
- Communicating with other City Departments such as planning and rental inspection can provide the Private I/I program with valuable contact information.
- Ideally, smoke testing is completed before the private plumbing evaluations. This can be difficult due to the soil and weather condition required for ideal smoke testing. In order to avoid this issue, smoke testing activities have been completed 6-9 months before an area is targeted for plumbing evaluations.
- Branding of the Ecoflow Program improved public perception and participation.
- Methods and successes of property owners participation:
    - Neighborhood association meetings were not effective for large scale participation.
    - Ecoflow open house meetings resulted in the initial 15% to 20% response rate.
    - Mailing campaign resulted in additional 15% to 20% response rate.
    - Phone calls/door knocking resulted in an additional response rate of 20% to 40%.
    - These standard efforts yield a total response rate between 50% and 80% within 6 to 9 months of targeting an area.
    - Depending on the number of rental properties, number of properties requiring internal entry, and overall demographics, a response rate over 50% to 80% can be difficult to reach quickly. Additional outreach efforts, neighbor to neighbor discussion and some ownership turnover is required to reach a 90% response rate. Achieving the desired 90% response rate may take several years to accomplish.

- With a streamlined find and fix approach to expedite the repair process, 7-9 pre-qualified plumbing contractors are needed to adequately manage the workload generated from two plumbing evaluation crews.
- Educating plumbers on specific I/I repairs and the overall Program is necessary to maintain consistency and to promote the Program.
- The entire Program Team must be highly responsive and deliver a consistent message to the public.

SUMMARY AND CONCLUSIONS

Implementation of the Ecoflow Program has been a success for the City of Lawrence. Much of this success can be attributed to a collaborative team effort from City officials, engineers, plumbers, and key individuals from several City departments. In addition, well-defined Program goals, work flow processes, and lines of communication were essential in planning and implementing the Ecoflow Program.

The Ecoflow Program has been well received by the public and local residents. Education, outreach, and open houses have all played a role in the public's acceptance of the Program. The emphasis placed on training each Program team member and their ability to communicate professionally, clearly and knowledgably about the Program has also helped with the public's perception of the Program. Responses and comment cards from the public are positive and have provided helpful feedback to gauge public perception.

Maintaining momentum and continuity will be key to successfully implementing the 8-year Program. The Ecoflow Program is highly visible among the public and it is important to keep everyone engaged and eager to participate in the Program as it moves forward. As Phase 4 is nearing completion, Phase 5 implementation will commence in Summer of 2017.

Noble Gulch Trunkline Relocation Project: Implementation of a Multifaceted Trenchless Technology Program

Keith Corcoran, PE, Carollo Engineers
Anne Prudhel, PE, Carollo Engineers

ABSTRACT

The Santa Cruz County Sanitation District's Noble Gulch Trunkline Relocation project was implemented to relocate and upsize 8,540 linear feet of deteriorating 12 inch asbestos cement sewer trunkline. The District needed to relocate the trunkline for several reasons: (1) it was in an environmentally sensitive gulch that required a riparian corridor permit for any maintenance activities; (2) access to most of the trunkline was available only by foot, making maintenance impossible; and (3) flow monitoring showed that the trunkline was likely leaking and contaminating downstream water bodies.

To relocate the sewer, the District obtained grant funding from California's Clean Beach Initiative. After evaluating numerous routing alternatives and considering the District's desire to eliminate an existing pump station, the only viable corridor was just outside the existing alignment through nearby higher elevation mobile home parks and underneath the scenic Pacific Coast Highway.

The selected route included many challenges including, narrow construction corridors, deep construction (up to 45 ft deep), challenging geological formations, steep terrain, and close proximity to residential neighborhoods. To address the project challenges and minimize the impact on local residents that is associated with tradition open cut construction methods, micro-tunneling (a 42-inch tunneled casing with a 15-inch carrier pipe), horizontal directional drilling (HDD) (8-inch), and cured-in-place-pipe (CIPP)(12-inch) were selected for select portions of work.

This paper evaluates and describes the implementation of multiple trenchless technologies and how they were used to overcome accessibility, environmental, groundwater, and geological formation constraints along the California Pacific Coast.

KEYWORDS

Microtunneling, horizontal directional drilling, cured-in-place pipe, manhole lining.

INTRODUCTION

The Santa Cruz County Sanitation District (SCCSD) is comprised of four sewersheds and five service areas which provide sewage collection services to 26,000 connections within the City of Capitola and unincorporated areas of Santa Cruz County which is located approximately 75 miles south of San Francisco, California. Pipelines range from 6 to 12 inches in diameter make up the Noble Gulch sewershed. The basin collects flow from primarily residential properties along with a limited number of commercial properties. Sewage collected is sent to the City of

Santa Cruz (City) wastewater treatment plant through an agreement between agencies that allows a discharge of up to 8 million gallons per day (mgd) of wastewater to the City's treatment plant.

The Noble Gulch Sewer Relocation project was developed after a meeting with the Santa Cruz County Water Quality Program Manager had revealed that that human E.coli was present in Soquel Creek downstream of Noble Gulch. With the sewer trunkline location within the gulch, it was suspected that the 50+ year old asbestos cement pipeline and/or manholes could be leaking. Because Soquel Creek discharges into the Monterey Bay Sanctuary, it was a priority to remove this potential source of contamination from the creek. Additionally, the District was able to obtain grant funding from California's Clean Beach Initiative to help fund the relocation of the sewer.

Due to the environmental sensitivity and location of the project, the following permits and easements were required:

- Riparian Permit from Santa Cruz County
- Encroachment Permit from California Department of Transportation (Caltrans)
- Coastal Development Permit from the City of Capitola
- Lake and Streambed Alteration Permit from California Department of Fish and Game
- General Construction Activity Storm Water Permit from Central Coast Regional Water Quality Control Board
- Encroachment Permits from both City of Capitola and Santa Cruz County
- Eight (8) private property easements
- Encroachment permit from the State Architect for work within mobile home community

The key challenge in replacing the trunkline was to improve accessibility while continuing to maintain gravity flow to the downstream pump station while protecting the sensitive riparian environment of the gulch and downstream beaches. In addition to the known access constraints and environmental constraints identified during the preliminary design, topography and additional geological constraints were identified which required the design team to put together and evaluate an alternative approach to the traditional dig and replace construction method.

DESIGN

The goal of the SCCSD Noble Gulch Trunkline Replacement Project design was to relocate the trunkline out of Noble Gulch to improve access for maintenance and eliminate the need for a riparian corridor permit. The first step was to identify an available corridor for the pipeline outside the gulch that would allow the District to maintain gravity flow.

Geological Challenges

A geotechnical engineering report was completed, which included site reconnaissance, borings, and desktop research for the area. The geotechnical investigation encountered a wide variety of native and suburban development conditions along the proposed alternative pipeline routes. Based on the collected information, the subsurface soil and bedrock conditions were anticipated to vary significantly along the sewer alignment. None of the subsurface conditions were considered "fatal flaws"; however, the geotechnical conditions encountered presented challenges and limitations that needed to be taken into consideration in the design to develop the most feasible and cost effective project including:

- Groundwater (including perched groundwater and seasonal drainage way flow)
- Porous and permeable soils (e.g., native sand and gravel, fill, weathered bedrock, cobbles, and granular utility bedding and trench backfill)
- Bedrock of the Purisima Formation consisting primarily of siltstone and sandstone
- Vertically and laterally variable native soil including cohesive clays and non-cohesive sands (with varying amounts of gravel)
- Vertically and laterally variable native soil density and consistency including loose to very dense sand (with varying amounts of gravel) and soft to very stiff clay
- Vertically and laterally variable native soil behavior in excavations including flowing and running non-cohesive sands that will have no stand-up time when exposed in vertical excavations, fast raveling cohesive fine-grained silts and clays that will have little to no stand-up time when exposed in vertical excavations, and soft squeezing fine-grained silts and clays
- Vertically and laterally variable bedrock character and composition including: degree of weathering, thin to thick bedding and very close to wide fracture spacing, with potentially adversely oriented shallow to steeply dipping bedding and fracture planes
- Historic fills that contain uncompacted materials and possibly deleterious debris
- Existing pipelines and utilities that contain variable trench backfill and bedding materials and variable trench-cut geometries
- Unconsolidated alluvial deposits located in the low lying gulch area
- Semi-consolidated terrace deposits located within the higher ground outside the gulch area
- Unconsolidated older flood plain deposits

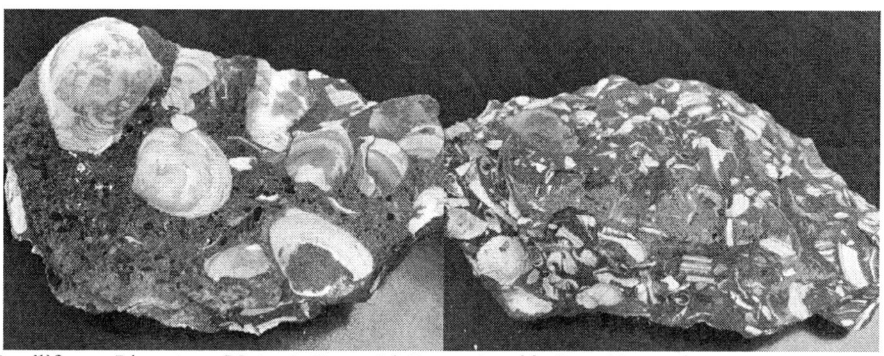

Figure 1: Fossiliferous Limestone (Nature's Concrete) Encountered in Deep Excavations

New Alignment

After evaluating a few alternatives, it was ultimately determined that an alignment at a higher elevation just outside of the gulch was the only feasible option. The alignment had numerous challenges as it traveled through two separate private mobile home parks and public land, including the scenic Pacific Coast Highway and the City of Capitola's corporation yard. This alignment presented many challenges:

- Work near/within environmentally sensitive areas
- Narrow construction corridors
- Deep construction (up to 45 ft deep)
- Challenging geological formations
- Steep terrain
- Coordination with multiple jurisdictions and private property owners
- Close proximity to residences

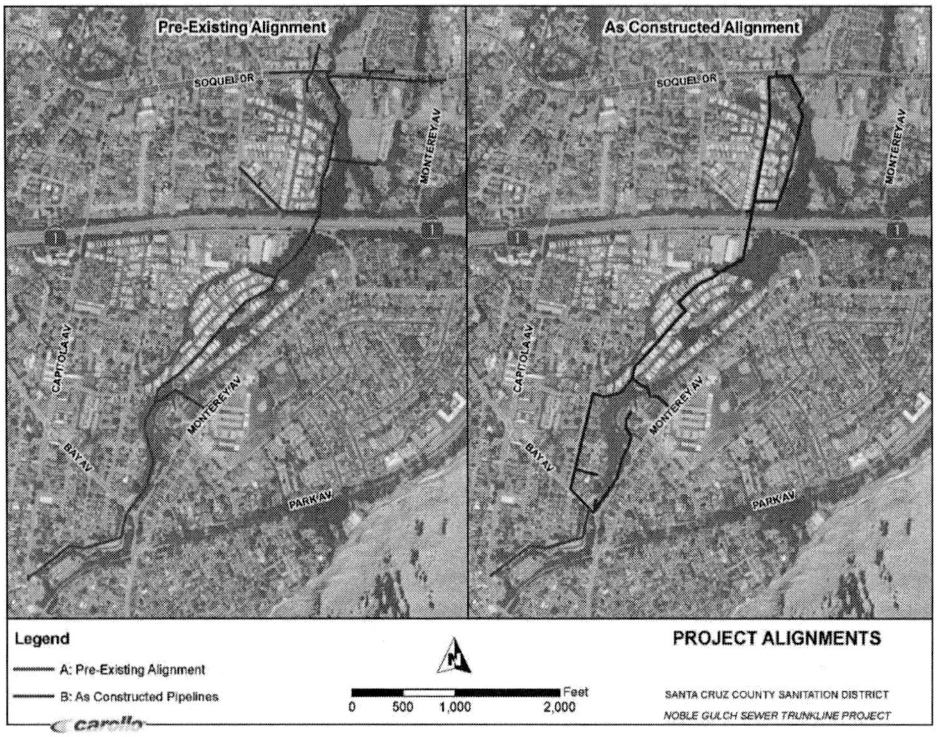

Figure 2: Final Project Alignment

Figure 3: Existing Alignment behind Residential Homes in Environmentally Sensitive Areas, Underneath Decks, Misc. Yard Structures, and Crossing Seasonal Waterways

Construction Methods

Once the alignment was selected, it was evident that in addition to open cut construction that trenchless technologies would be required in order to mitigate impact on local residence and the environment.

The replacement/rehabilitation alternatives outlined in Table 1 were identified during preliminary design.

Table 1: Construction Method Evaluation

Construction Method	Feasibility	Geotechnical Challenges/Risks
Open-Cut Trenching	-Feasible for all alignments with the exception of the HWY 1 crossing, areas of environmental concern (e.g, along creek and heavily vegetated areas), and potentially within improved easements	-Shoring -Dewatering (especially in low lying areas adjacent to the creek) -Excavation through bedrock -Will likely encounter existing, saturated utility trench backfill with little to no stand-up time and copious groundwater inflows in low lying areas adjacent to the creek and within the gulch
Tunneling: Auger bore and jack (AB), Guided Boring Method (GBM) and Microtunneling (MT)	-AB and GBM are feasible for HWY 1 crossing and Corp Yard slope, if groundwater is at or maintained below the tunnel and tunnel soils are stable -MT is feasible under high groundwater conditions and unstable tunnel soil, but may not be feasible for the alternative HWY 1 crossing if obstructions in fill and old trees roots are present	-Presence of flowing and/or unstable ground and resulting loss of ground through the open tunnel face will produce voids and sinkholes above the tunnel for AB and GBM -Obstructions in fill can stop AB and GBM requiring removal of the obstruction from the tunnel face and slowing production obstructions in fill and tree roots can stop MT requiring 911 shafts -shafts for tunnel launching and receiving will require deep shoring and dewatering (see challenges/risks for open-cut trenching)
Horizontal Directional Drilling (HDD)	-Potentially feasible for HWY 1 crossing and Corp Yard slope for pipes designed with slopes greater than 2%	-Installation accuracy limitation to +2% of depth -Risk of drilling fluids hydrofracture into adjacent Noble Gulch Creek -Risk of encountering obstructions in fill areas will slow production rates
Sliplining and Cured-in-Place Pipe Lining	-Feasible for existing alignment where decrease in diameter is not an issue	-Generally insensitive to geotechnical conditions -Will not correct sags -Pits for ingress/egress for sliplining will require shoring and dewatering (see challenges/risks for open-cut trenching)
Pipe Bursting and Pipe Reaming	-Most feasible for rehabilitation of pipelines composed of brittle materials (e.g. VCP, DI) when small upsize or same size is required -For asbestos cement (AC) pipe, environmental agencies (e.g., EPA will consider the existing pipe to be a hazardous material and the pipeline alignment (with remnant AC pipe) to be a hazardous site	-amount of feasible upsize is dependent on existing pipe zone backfill and native trench wall deposits -For 12-inch diameter pipe burst to a new 15-inch diameter pipe at depths greater than 12 feet, pipe bursting is considered challenging to moderately difficult -Will not correct sags -Ground heave can occur depending on pipe depth and/or amount of pipe upsize -Pits for ingress/egress will require shoring and dewatering (see challenges/risks for open-cut trenching) -Risk of drilling fluids hydrofracture for Pipe Reaming

Because the final alignment would be constructed in a new alignment pipe bursting and pipe reaming were eliminated from further consideration.

Trenchless Technology Evaluation

To address the project challenges and minimize the impact on the local community, numerous trenchless technologies were incorporated into the design at key locations. The design included the following trenchless technologies:

- Micro-tunneling: Installation of 1,700 feet of 42-inch micro-tunneled casing to install a 15-inch carrier pipe (high density polyethylene (HDPE) or fusible PVC allowed) was used to cross Highway 1, avoid deep (up to 45') open cut construction, and minimize impact to environmentally sensitive areas that were not conducive to other trenchless methods. The larger casing was required due geological formation and length of drive. Shoring up to 45-feet deep was required for the launching and receiving pits due to the varying California coastal topography.
- Horizontal directional drilling (HDD): HDD was utilized to install a new 8-inch diameter pipe (HDPE or fusible PVC allowed) through residents' backyards and through a park adjacent to the environmentally sensitive, Noble Gulch where access with traditional excavation equipment was not feasible without significant environmental impacts. This allowed the local connections to be connected at a higher elevation and outside the bottom of the gulch.
- Sliplining: HDPE or fusible PVC was allowed to be installed as the new carrier pipe for sections of the existing trunkline that had to remain in service to pick up a few local residential homes. The existing manholes were to be used as the insertion/receiving pits and were to be replaced with new rings and MH covers. (As part of a value engineering proposal during construction, cured-in-placed pipe was ultimately installed. Additional information concerning this is addressed at the end of the paper)

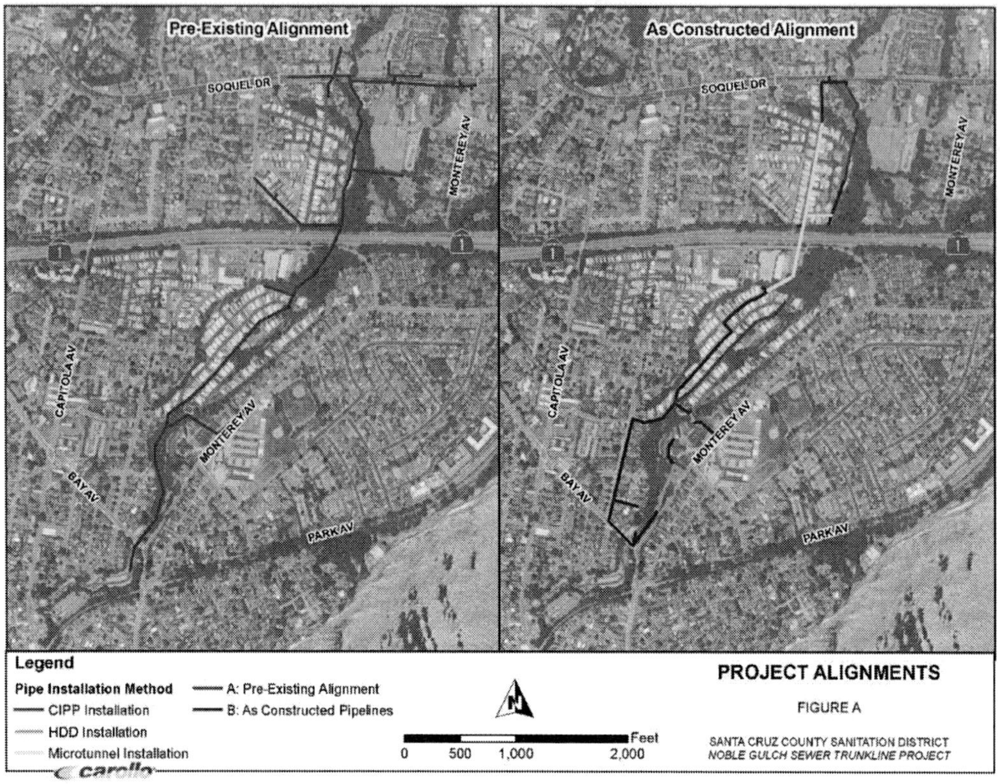

Figure 4: Project Alignment with Methods of Construction

CONSTRUCTION

The awarded general contractor, McGuire and Hester, started construction in January 2015 and was responsible for the open cut pipeline work as well as the excavation and shoring for their trenchless subcontractors. The open cut portions of the project were fairly straightforward and the original concerns of high groundwater turned out not to be an issue due to ongoing drought conditions in California at time of construction--going into the fourth year of drought at beginning of construction. This eliminated dewatering for the majority of the project except for the deep launching and receiving pits for the microtunneling portions of work.

The most challenging aspects of construction were the locations that were to be installed utilizing trenchless technology. Site access was extremely limited and both HDD and Microtunneling required a significant amount of space for the support equipment. A substantial portion of the civil work included the construction of five (5) microtunneling pits within the limits of the mobile home community and City of Capitola's Corp Yard. The two largest launching pits were 18' wide by 32' long by 45' deep. Excavation of the pits and the set-up for the microtunneling equipment was a slow process due to site constraints. Access was restricted to one way in and one way out in most locations but the contractor was able to meticulously install the necessary shoring and support equipment at each location required.

Figure 5: Micro-tunneling support equipment in tight corridors around 45' deep jacking pit

Figure 6: View of Equipment and Material Up to Neighbors Front Doors

Figure 7: Installation of 42" PermLok® Casing & Fusible PVC Carrier Pipe

During construction the contractor presented multiple value engineering (VE) opportunities to the District and Carollo. Many of the items presented were to change planned open cut sections of work to trenchless construction methods.

The key VE opportunities included:

- Additional HDD underneath the gulch within an existing right of way in order to reroute some of the deep open cut construction originally required within a road which blocked traffic to a nearby middle school. The additional HDD portion ultimately limited the traffic control requirement that was required in the original alignment from multiple months to just one (1) day.
- Cured-in-place pipe (CIPP): a 12-inch CIPP was used to line the portion of the existing trunkline that needed to be left in service along Noble Gulch to maintain a few of the residential connections that could not be rerouted. The original design called for sliplining to reduce the inside diameter due to the decrease flow; however, during construction the Contractor suggested the use of CIPP to eliminate the need to excavate existing manholes behind the residential homes within Noble Gulch. This was ultimately selected in lieu of sliplining as it was determined the size of the new carrier pipe had minimal effects on the velocity based on the low flow and slope of the existing pipe. CIPP also limited disruption of the residence to just one (1) 8-hour day where sliplining would have required multiple days.
- Elimination of a launching pit by modifying the microtunneling equipment and ordering shorter lengths of casing to fit within the narrower width of the launching pit which was originally designed as both a launching and receiving pit. This enabled the contractor to eliminate one of the launching pits and install a smaller receiving pit.

Figure 8: HDD Installation with One Day Lane Closure/Pullback

Figure 9: Installation of Cured-in-Place Pipe behind Residential Homes within the Gulch.

Figure 10: Daylighting of Microtunneling Cutting Head

The VE opportunities ultimately reduced the original open cut contract lengths from 7,050 lineal feet (LF) down to 5,900 LF by the installation of 1,150 LF feet of new sewer pipe using HDD--an approximately 15% reduction in open cut. In addition to reduction in overall open cut work, only 43 of the original planned 50 manholes were required.

CHALLENGES AND LESSON LEARNED

Numerous adverse conditions made the project quite challenging from design through construction. It took 9 years to complete the overall project due to numerous chain of events that put the project on hold. In this time, a local water agency completed another project, occupying a portion of the corridor identified for the new Noble Gulch pipeline alignment. Additionally, numerous unmarked utilities, an abandoned buried bridge, and poor as-built documentation of utilities contributed to numerous changes. This ultimately resulted in coordination with the contractor and local utility providers to identify and redesign a significant portion of the trunkline during construction.

During the installation of the shoring for one of the microtunneling launching pits, a lens/pocket of the Purisima Formation was hit. The design geotechnical engineer took samples which identified the material to have a compressive strength in excess of 10,000 PSI--a Schmidt hammer was used due to the varying sample sizes. Specialty rock coring equipment had to be brought onsite in order to penetrate the rock and continue installation of the shoring. The location of the Purisima formation was at the elevation of part of new 42-inch casing that was to be installed. There was a concern that the microtunneling machine may not be able to maintain grade and would chase the softer material. Fortunately, once the shoring was installed, it was determined that it was an isolated piece and was able to be removed as part of the planned excavation of the pit.

Site access lead to many challenges especially where construction activities cut off complete access to the mobile home community since there was only one way in and out. The mobile home community had many retired and elderly occupants, some of which required daily care, and many of the residents were home during active construction hours. Sound and vibration was a concern, especially when the microtunneling boring machine was being driven and the slurry recycle/dewatering equipment was operational as it had significant sound and vibration impacts to the community. To help mitigate, sound attenuation blankets were added as a change order.

Figure 11: Sound Attenuation Blanket around the Microtunneling Slurry/Dewatering Equipment

In addition to residential access in the mobile home community, access for USPS, Fed-Ex & UPS deliveries, garbage, and emergency services had to be maintained on a daily and weekly basis. While public outreach was not within the contractor's scope, they made significant effort to help residents from carrying groceries to rolling out garbage cans where the truck could access them.

The "Lesson Learned" was that the success of this project was attributed to the trenchless technologies implemented and the collaborative partnership, where open communication between the contractor, the designer, the construction manager, and the District were established early on. This partnership allowed the numerous challenges of the job to be openly discussed and efficiently overcome to ensure there were no impacts to the overall schedule and minimal cost impacts. In the end, the additional construction costs due to unforeseen conditions were ultimately covered by the value engineering opportunities implemented.

CONCLUSION

The Noble Gulch Trunkline Relocation project was ultimately completed on budget and 2 months ahead of schedule while overcoming many construction challenges including construction within a highly sensitive community, a congested utility corridor, within close proximity to residences and senior citizen mobile home community, and in and around environmentally sensitive waterways. The strategic use of trenchless technologies allowed us to minimize impacts to residents, environmentally sensitive areas and reduce the risk associated with deep open cut construction.

Accomplishing Sewer Separation through a Landfill and Botanical Garden with Trenchless Technology

Chris Koenig, PE, HDR, Inc.

ABSTRACT

Objectives

The City of Omaha is in the process of addressing combined sewer overflows (CSOs), as required by the Clean Water Act, U.S. Environmental Protection Agency (EPA) CSO Control Policy of 1994, and the City's Administrative Consent Order with the Nebraska Department of Environmental Quality (NDEQ), to improve water quality in the Missouri River and Papillion Creek. One of the projects outlined in the City of Omaha's CSO Long-Term Control Plan (LTCP) was the Martha Street Sewer Separation Project, located in the South Interceptor Basin.

A portion of the Martha Street Sewer Separation Project (Project) included Omaha's Lauritzen Botanical Gardens (Gardens). The Gardens were constructed over a closed balefill. Over time, the weight of the balefill had crushed the existing sewers that carried both storm water and sanitary sewage. In addition, extensive settlement of the balefill resulted in several unsightly areas in the Gardens that required constant maintenance.

The Project focused on separating the combined sewer system that ran through the Gardens by providing new sanitary and storm sewers through this area. In addition, the existing combined sewers, sanitary sewers, and leachate collection piping within the Gardens were properly abandoned. The Project also included design of a new outfall storm sewer and design of a new gravity sanitary sewer from the east end of the Gardens to the Spring Street Lift Station.

The goals of the City of Omaha were met, and a cost-effective project was delivered that improved water quality in the Missouri River and surrounding streams through the separation of the sanitary and storm sewer systems. In addition, the Gardens were enhanced by a reshaped landscape that better serves its members and guests in the future.

KEYWORDS

Balefill, microtunnel, horizontal directional drilling, combined sewer.

INTRODUCTION

Background and Location

The City of Omaha is in the process of addressing CSOs, as required by the Clean Water Act, U.S. EPA CSO Control Policy of 1994, and the City's Administrative Consent Order with the NDEQ, to improve water quality in the Missouri River and Papillion Creek. One of the projects outlined in the City of Omaha's CSO LTCP was the Martha Street Sewer Separation Project, located in the South Interceptor Basin.

The Martha Street Basin, illustrated in Figure 1, generally consists of approximately 97.1 hectares (240 acres) of the larger South Interceptor Basin. The Martha Street Sewer Separation Project was divided into two distinct areas for evaluation: the residential area and the Lauritzen Botanical Gardens (Gardens). These areas are shown in Figure 1. The Project was separated due to the type of work and complexities involved. The residential portion of the Project was similar to other sewer separation projects within an established residential neighborhood. The work within the Gardens was a more complicated type of construction due to the site conditions and the proposed use of trenchless construction methods for a substantial portion of the work.

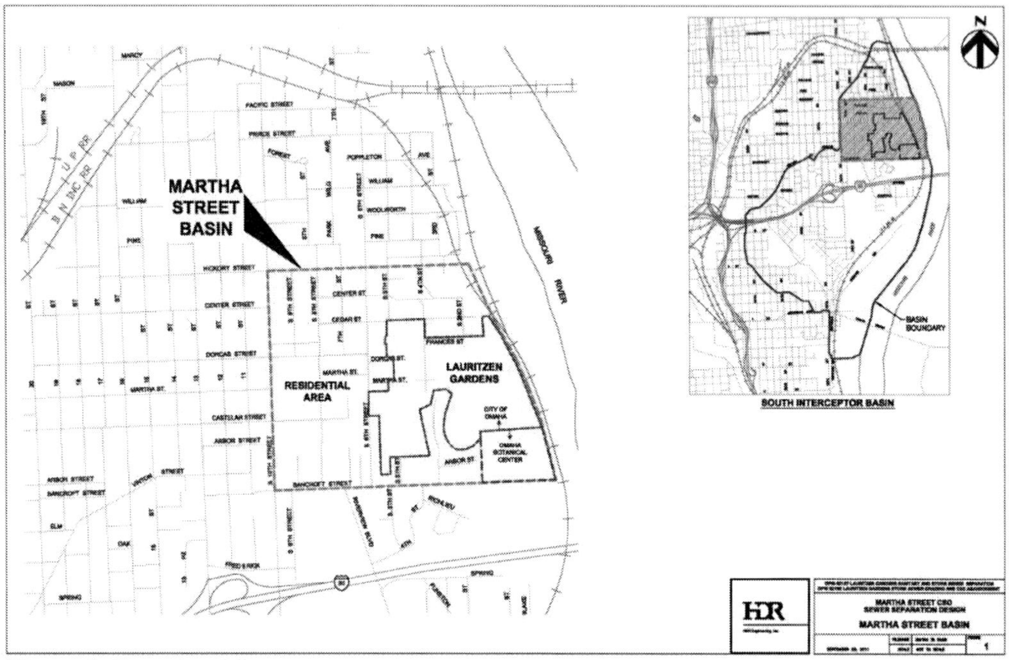

Figure 1. Location Map

Existing Infrastructure

The Gardens site consists of approximately 40.47 hectares (100 acres), of which approximately 28.33 hectares (70 acres) are owned by the City and the remaining 12.14 hectares (30 acres) owned by the Omaha Botanical Center, Inc. The Gardens were constructed over a closed solid waste balefill. A balefill is a landfill in which the domestic solid waste is compacted into bales and placed in a landfill for disposal. The balefill was developed in stages, from east to west within existing valley areas. The design drawings indicated the bottom of the balefill was graded to slope to the center of the valleys.

Original Balefill Operation – Setting Bales

The existing sewer and piping system within the Gardens, presented in Figure 2, consisted of sanitary sewers, storm sewers, combined sewers, leachate collection system, gas collection system and subdrain piping system, as described below:

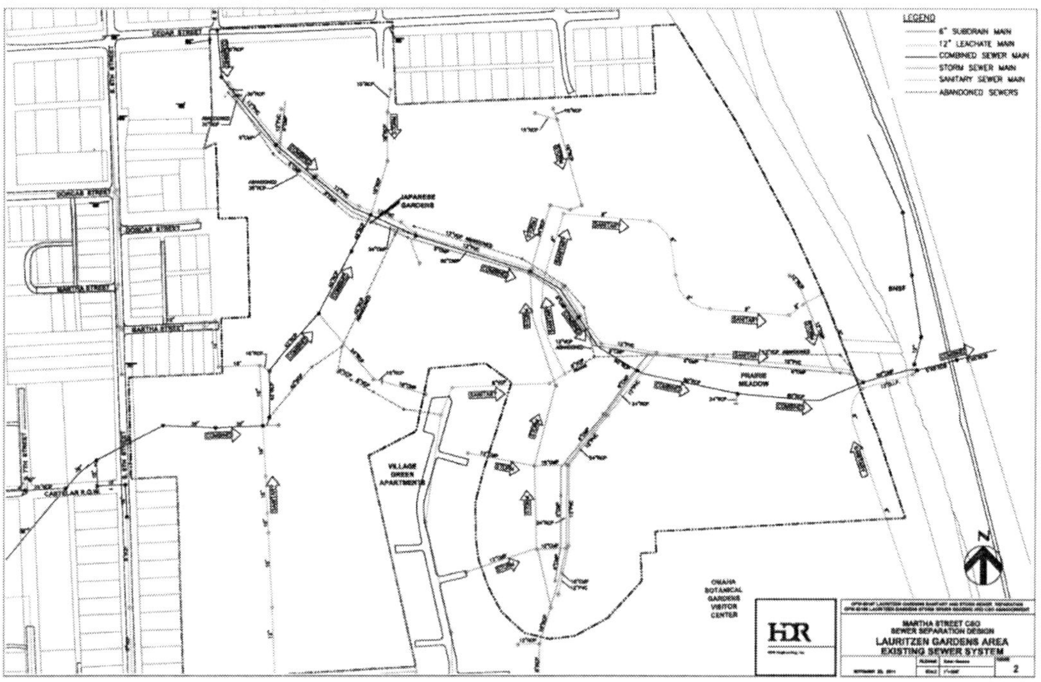

Figure 2. Existing Piping Systems

- Combined Sewer System. Storm sewer pipes were constructed at the bottom of the valleys. The pipes were reinforced concrete pipe (RCP) varying in size from 91.44 to 152.4 cm (36 to 60 IN) diameter. Subsequent projects diverted the sanitary sewer flow from the residential areas to the storm sewer pipe and portions of the original sanitary sewer pipes within the Gardens were abandoned in place. The storm sewer thus became a combined sewer.

- Sanitary Sewer System. Drawings provided by the City indicate a 30.48 cm (12 N) vitrified clay pipe (VCP) sanitary sewer extended up the unimproved valley prior to the construction of the balefill. The sewer served the Village Green Apartments located on the south side of the Gardens and the residential area to the west of 5th Street and north of Cedar Street. The drawings for the balefill closure include the construction of a new 20.32 cm (8 IN) PVC sanitary sewer across the center and northeast side of the balefill area. The sewer was intended to provide service to the Village Green Apartments but settlement of the balefill resulted in the failure of a portion of this sewer. The remainder of the 30.48 cm (12 IN) VCP sanitary sewer within the limits of the Gardens was shown as abandoned in place.
- Leachate System. The leachate collector pipes (approximately 15.24 cm (6 IN) in diameter) are located in a sand layer under the baled waste and extend laterally from the low points of the balefill. The leachate laterals connect to manholes and then enter the main leachate pipe below the landfill liner. The main leachate collection pipes extend along the bottom of the valley throughout the balefill at the lowest elevation. The historic design drawings indicate the main leachate collection pipe is 30.48 cm (12 IN) PVC. The leachate discharge collector piping connects to the combined sewer system at the east side of the Gardens.
- Sub-drain Piping System. The balefill construction included a sub-drain located adjacent to the new combined sewer pipe and a leachate collection pipe. The original design drawings indicate that the subdrain is 6-inch corrugated metal pipe (CMP) in sand bedding. It is believed the sub-drain system was originally installed to control groundwater in the valley.
- Gas Collection System. Gas collection piping was installed in the sand layer on top of the baled waste. The gas collection pipes extend to the limits of the balefill and are connected to the leachate collection manholes.
- Structures. Standard concrete manholes were constructed on all of the pipe systems. The manholes for each of these collection systems were extended to the surface as the bales were placed at the site.
- Clay Cap and Liner. As part of the balefill closure, a clay cover with a minimum thickness of 1.37 m (4.5 FT) was constructed over the baled waste and gas collection system. A 1.07 m (3.5 FT) thick clay liner was constructed under the balefill prior to the filling operations.
- Subsequent Improvements. Additional storm sewer pipes were constructed after the closure of the balefill operations during the on going development of the site by the Omaha Botanical Center.

Project Goals

The improvements envisioned within the LTCP plan involved the construction of a new sanitary sewer within the Gardens including a sanitary sewer lift station and the conversion of the existing combined sewer to a storm sewer system. However, the condition assessment of the existing sanitary and storm sewer systems found numerous defects that could not be easily repaired.

Therefore, the project focus turned to separating the combined sewer system that ran through the Gardens by providing new sanitary and storm sewers through this area. In addition, the existing combined sewer system needed to be properly abandoned.

A goal of the project was to avoid disruption to established portions of the Gardens, which are a cultural and environmental resource for the Omaha area consisting of unique four-season plant displays, meticulously maintained, and considered a hidden sanctuary in the heart of the City of Omaha.

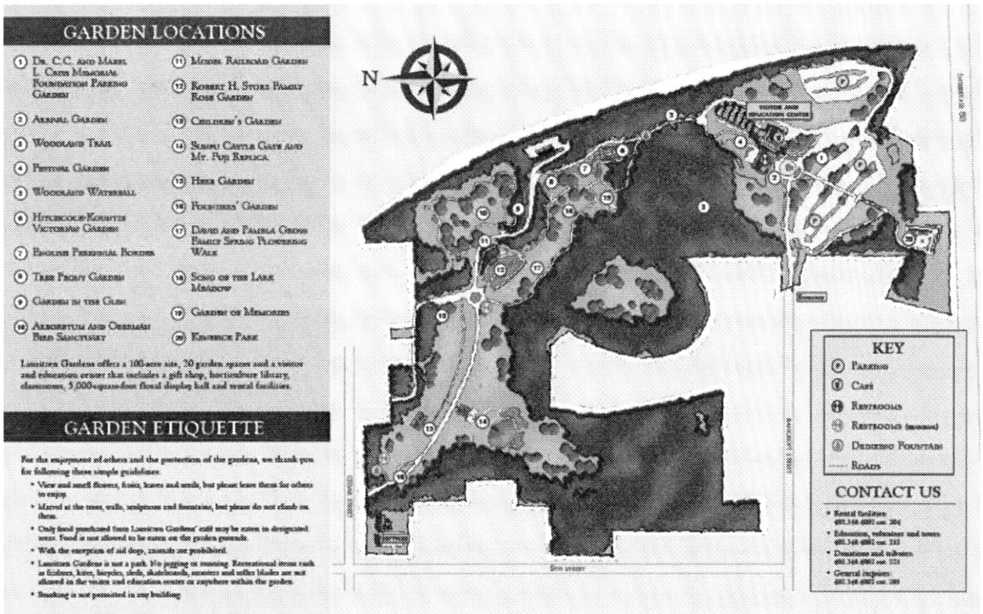

Garden Exhibit Layout

Victorian Garden

Scale Model Train Set

Rose Garden

METHODOLOGY

Field Investigations

The initial activities for this project were focused on gaining a better understanding of the existing sewer systems and piping through the Gardens so that a feasible solution could be defined. Field investigations were conducted to evaluate the condition of the various piping systems within the Gardens, obtain subsurface information to allow for completion of the design, and visual observations to evaluate construction routing and site conditions. Field investigations included:

- CCTV Inspection and Manhole/Inlet Inspection. CCTV inspection and manhole/inlet inspection was completed to the extent possible. The CCTV camera was unable to be lowered into several of the manholes due to offset manhole riser sections and/or tilting manholes. Standing water in 13 manholes did not allow for the CCTV inspection. Several of the pipe segments were in such severely deteriorated condition that they did not allow the CCTV camera to pass through them. A total of 107 manholes were inspected and 1,859 m (6,100 FT) of sewers were televised.
- Utility Coordination. A request for preliminary location information was transmitted to 15 utilities.
- Detailed Topographic Survey. A detailed topographic survey was completed along the routes of the proposed improvements and in accordance with the Omaha CSO Control Program survey standards.
- Geotechnical Investigation. The geotechnical investigation was broken down into two components: standard geotechnical investigation to obtain data for the design of the

proposed sanitary and storm sewer piping systems and the investigation related to the Balefill. Geoprobe technology was used to define the limits of the balefill, which consists of a 5.08 cm (2 IN) diameter rod to advance a plastic tube for core samples. This process allowed several holes to be drilled around the site in a short period of time to quickly identify the limits of the balefill. Traditional borings were also completed for the design of the sewer system and final sewer alignments, as well as to determine the thickness of the balefill.

Results of the investigation showed that the balefill was approximately 18.29 – 24.38 m (60-80 FT) thick and the existing combined sewer was located under the balefill, more than 24.38 m (80 FT) below grade. The weight from the balefill had crushed the existing combined sewers overtime allowing leachate to enter the sewer system. In addition, the drainage system throughout the closed balefill area had failed and was allowing storm water to flow into the balefill, resulting in substantial settlement of the balefill with adverse impacts to the Garden exhibits and roads.

The Village Green area prior to construction. The balefill had settled over 2.13 m (7 FT) in this area. The unsightly manholes extended up to 2.13 m (7 FT) above grade due to settlement of the balefill.

Based on findings from field investigations, the decision was made to abandon all of the existing piping systems in-place and construct new sewers around the perimeter of the balefill. Figure 3 presents the layout of the Gardens and the final sewer alignments.

Figure 3. Layout of the Gardens and Final Sewer Alignments

Design Criteria

The criteria for the design of the proposed sanitary and storm sewer improvements within the Gardens area of the Martha Street Basin are summarized below:

<u>Sanitary Sewer</u>

The sanitary sewer design was completed in accordance with the City's Sanitary Sewer Design Basis and Methodology Guidance, dated June 27, 2010, which outlines the use of the tributary area method: $y=0.665x^{0.7986}$ where

y = total peak flow (cfs)

x = tributary area (acres)

Individual subbasins were identified and peak flow calculated for each pipe segment based on the contributing area. Impact points were identified where pipe segments combined and the downstream pipes were reviewed for capacity requirements. The new sanitary sewers were designed for the following:

1. Capacity – sized to convey peak wet weather flows
2. Minimum flushing velocity in accordance with the guidance document

Pipe Diameter, cm (IN)	Min. Velocity, M/Sec (fps)
20.32 (8)	0.732 (2.4)
25.40 (10)	0.732 (2.4)

30.48 (12)	0.762 (2.5)
38.10 (15)	0.792 (2.6)

3. Maximum velocity – 4.57 M/Sec (15 fps)
4. Pipe material – vitrified clay pipe and HDPE
5. Roughness coefficient: n=0.013 for clay pipe and n=0.012 for HDPE

Storm Sewer
The storm sewer design was completed in accordance with the City's Storm Sewer Design Basis and Methodology Guidance, dated June 27, 2010. The guidance document requires a hydraulic analysis of the existing sewer to verify the storm sewer capacity. The hydraulic analysis was undertaken in accordance with the Omaha Regional Stormwater Design Manual. The Rational Method described in Chapter 2, Hydrology, was selected to determine storm runoff peak flows and volumes since the Gardens area of the Martha Street Basin is approximately 40.47 hectares (100 acres).

The following criteria was used for the design of the new storm sewer and the conversion of the existing combined system to a storm sewer-only system:

1. Time of concentration: 10 minutes, minimum
2. Runoff coefficient: Composite coefficients calculated per Design Manual.
3. Pipe roughness coefficient new reinforced concrete pipe: n=0.013
4. 10-year storm capacity
5. Inlet capacity designed to accept all flow during design storm event for inlets located in a sump condition.

The hydraulic analysis of the proposed storm sewer system was completed using XPSWMM. Some of the calculated pipe velocities do exceed 6.096 mps (20 fps) due to the pipe grades. The storm sewer structures were designed to address the potential movement and the pipes were designed with restrained joints.

Sewer Improvements
Balefill Crossing
The most challenging aspect of the project was the crossing of the balefill with the new storm and sanitary sewers. The initial project design consisted of constructing a tunnel under the balefill at the narrowest, shallowest location. This concept had several challenging design and construction issues including 24.38 m (80 foot) deep manholes, the potential for leachate entering the bore pits, explosive gases, massive excavations for the borrow pits, and the necessity of pipe-in-pipe construction.

Throughout the concept and design phase, the project team worked to refine the final solution to alleviate these issues. An innovative alternative was developed which consisted of constructing a pipe bridge over the balefill. The pipe bridge design was predicated on the use of an auger cast displacement pile that would maintain the integrity of the clay balefill liner and cap.

An auger cast displacement pile acts like a screw, pushing material to the side as the drill bit advances. A hollow steel shaft maintains the integrity of the hole as the drill advances. Once the drill reaches the desired depth, concrete is pumped down the hollow shaft through a hole in the drill bit. The force of the concrete entering the hole essentially backs the drill bit out of the hole and immediately fills the hole with concrete, reducing the potential of any collapse of the auger hole. The balefill liner and cap are maintained during the installation process. Reinforcement is then lowered into the concrete filled hole before the concrete achieves its initial set.

The storm sewer was designed to act as the structural component of the pipe bridge. The pipe bridge was a cast-in-place box designed to accommodate the storm sewer flow. The walls of the box provided the structural support for the completed system between the pile caps. A ledge was constructed on the side of the box to provide support for the sanitary sewer. The pipe bridge was constructed on a sand base to act as a point of collection for the gas from the balefill. Gas collection piping was then routed to the sanitary sewer manhole. The pipe bridge was buried with a minimum of 45.72 cm (18 IN) of material to allow for future plantings in the Gardens. The estimated cost savings of constructing the pipe bridge in lieu of a traditional tunnel was $4 million.

Excavation of the solid waste on the north end of the pipe bridge. The grades of the pipe bridge necessitated a portion of the bridge to be constructed in the solid waste. The clay cap was reconstructed over the balefill disturbed by construction.

Connection to Residential Area at 7th and Castelar

The Residential portion of the Project terminates east of the 7th and Castelar Street intersection. The project consists of the construction of a new storm sewer and the conversion of the existing combined sewer to a sanitary sewer. The sanitary and storm sewers connect to an existing 91.44 cm (36 IN) combined sewer at this location. The 91.44 cm (36 IN) combined sewer extends under 6th Street into the Gardens.

A new 20.32 cm (8 IN) sanitary sewer was installed under 6th Street by horizontal directional drill (HDD) construction method to the point of connection of the 91.44 cm (36 IN) combined sewer. The sanitary sewer connection to the 36-inch sewer was removed and the existing 91.44 cm (36 IN) combined sewer under South 6th Street was converted to a storm only sewer.

Connection to Existing Sewers in the Proposed Japanese Gardens

An existing 38.10 cm (15 IN) sanitary sewer extends north on South 5th Street north of Bancroft Street and connects to the existing 42-inch combined sewer. The new 20.32 cm (8 IN) sanitary sewer from 7th and Castelar Street connected to this sewer and the connection to the 106.68 cm (42 IN) sewer removed.

A new 30.48 cm (12 IN) sanitary sewer was constructed from the point of connection to the existing 38.10 cm (15 IN) sanitary sewer on South 5th Street north to the south edge of the Balefill and the start of the pipe bridge.

An existing 38.10 cm (15 IN) storm sewer serves Martha Street east of 6th Street. The storm sewer enters the Garden in the area of the proposed Japanese Gardens and connects to a 106.68 cm (42 IN) combined sewer. A new 45.72 cm (18 IN) storm sewer was constructed to carry the flow from this storm sewer and the connection to the 106.68 cm (42 IN) sewer removed. A new 152.40 cm (60 IN) diameter storm sewer collects the flow from the existing 91.44 cm (36 IN) sewer under South 6th Street and the new 45.72 cm (18 IN) sewer and conveys the flow north to south edge of the Balefill and the start of the pipe bridge.

The Gardens are proposing a detention pond to be located on the west side of the Japanese Gate. An area inlet structure was constructed on the east edge of the proposed pond. A 76.20 cm (30 IN) storm sewer was constructed from the area inlet east to a point of connection with the 15.24 cm (6 IN) storm sewer at the start of the pipe bridge. An area inlet constructed immediately west of the start of the pipe bridge to collects the flow from the top of the Balefill.

Connection to Village Green Apartments

A manhole located near the north end of the Village Green Apartments collects the sanitary sewer from the complex and a sanitary sewer extends east into the Gardens. The lateral for the north apartment complex connects to the sanitary sewer downstream (east) of the manhole. A 20.32 cm (8 IN) sanitary sewer was installed by HDD method from the northeast end of the apartment complex west into the Gardens and connects with the proposed sanitary sewer in the proposed Japanese Gardens area at approximately 5th and Castelar Street and connects to the proposed 30.48 cm (12 IN) sanitary sewer discussed previously. The existing sanitary sewer downstream from the point of connection was plugged. A new manhole was constructed over the existing north apartment complex lateral and connected to the new sanitary sewer.

The majority of the storm sewers from the Village Green Apartments drains to the east side of the complex and extend into the Village Green area of the Gardens. A 60.96 cm (24 IN) storm sewer was constructed within the Gardens, along the west edge of the Balefill in the Village Green area and connects to the existing storm sewer leads from the apartment complex. A 76.20 cm (30 IN) storm sewer was constructed west along the north end of the apartments and connects to the storm sewer located on the east side of the Japanese Gardens area. The 60.96 cm (24 IN) segment of this storm sewer was constructed by open cut methods and the 76.20 cm (30 IN) segment of this sewer was constructed by pipe jacking. The existing storm sewers downstream from the points of connection were cut and plugged.

An existing 45.72 cm (18 IN) storm sewer is located on the west side of the Village Green Apartments. The 45.72 cm (18 IN) storm sewer connects to the 152.40 cm (60 IN) storm sewer located on the east side of the proposed Japanese Gardens and the remainder of the storm sewers was abandoned in place.

<u>Connection to Residential Area at 5th and Cedar</u>
The sanitary and storm sewers in the residential area northwest of the Gardens were separated under a previous City of Omaha project. The separated sewers were recombined at the 5th and Cedar Street intersection. A new 91.44 cm (36 IN) storm sewer was connected to the existing 91.44 cm (36 IN) storm sewer in the Cedar Street right of way west of 5th Street and continues east to the intersection. A new 121.92 cm (48 IN) storm sewer was installed by pipe jacking from the 5th and Cedar Street intersection to the northeast corner of the Founder's Garden. The new storm sewer continues south and east to a point of connection to the storm sewer from the pipe bridge near the turnaround.

A new 30.48 cm (12 IN) sanitary sewer was constructed in parallel to the storm sewer from the intersection of the 5th and Cedar Streets to the point of connection to the sanitary sewers from the north end of the pipe bridge near the turnaround. The sanitary sewers were installed by microtunnel and open cut construction methods. The connections between the sanitary and storm sewers in the 5th and Cedar Street intersection were removed.

<u>Conveyance through the Garden</u>
The 152.40 cm (60 IN) storm sewer from the turnaround to the east side of the Gardens was installed by pipe jacking to minimize damage to the existing exhibits. The pipe installation was completed in one pipe jacking operation.

A storm sewer pipe receives the storm water runoff from the Prairie Meadows Gardens. The storm sewer connects to the 152.40 cm (60 IN) storm sewer at the entrance to the Garden of the Glen.

The sanitary sewer parallels the storm sewer through this portion of the Gardens. The pipe was installed by pilot tube microtunnel construction method. The pipe installation was completed in one pilot tube microtunnel operation.

Both the sanitary sewer and the storm sewer were installed along the east edge of the Gardens by open cut construction method. The storm sewer connects to the existing 152.40 cm (60 IN) combined sewer at the junction box. A new storm sewer was constructed parallel to the existing combined sewer under the BNSFR by pipe jacking. The new storm sewer connects to the existing 1.52 m x 1.52 m (5' x 5') reinforced concrete box (RCB) downstream from the existing diversion structure in the BNSFR yard. The existing RCB has the capacity to convey the storm water to the river.

The sanitary sewer terminates at the new diversion structure. The diversion structure is the point of beginning for the Martha Street to Riverview Lift Station Phase I project.

<u>Diversion Structure Operation</u>
The existing Martha Street diversion structure consists of a low wall at the beginning of the 1.52 m x 1.52 m (5' x 5') RCB. The dry weather flow is diverted to a manhole located to the north of the RCB. Storm water overtops the low wall and flows to the river. A pump station was installed in the manhole located north of the RCB. The dry weather flow is pumped through a force main to the Hickory Street Lift Station. The pump station was installed after the failure of the gravity sewer from Martha Street to the Hickory Street Lift Station.

The diversion structure diverts sanitary sewer flow to the junction box located over the existing 152.40 cm (60 IN) combined sewer. The dry weather flow continues to utilize the existing Martha Street diversion structure.

The diversion structure will allow for sanitary sewer flow to be diverted into the Martha Street to Riverview Lift Station Phase I sewer. The Martha Street to Riverview Lift Station Phase I project terminates at an existing manhole west of the Spring Street Lift Station. Sanitary sewage diverted into the Martha Street to Riverview Lift Station Phase I sewer flows by gravity to the Spring Street Lift Station. A slide gate was installed in the diversion structure to allow for the sanitary sewer flow to be diverted to the Martha Street to Riverview Lift Station Phase I sewer.

The Riverview Lift Station currently does not have the capacity to accept the additional flow from the Martha Street basin. The lift station is scheduled to be reconstructed with the capacity to accept the additional flow from the Martha Street basin. The Martha Street to Riverview Lift Station Phase II project will be constructed to coincide with the completion of the new Riverview Lift Station.

The diversion structure contains two pipe connections to the junction box over the existing 152.40 cm (60 IN) combined sewer. The lower pipe will be plugged upon the completion of the Martha Street to Riverview Lift Station Phase II project and the slide gate will be removed. All of the sanitary sewer flow will be directed to the Martha Street to Riverview Lift Station sewer. The upper pipe will remain in place to allow the system surcharged with storm water to overflow into the existing 152.40 cm (60 IN) combined sewer in the event the flow exceeds the capacity of the Martha Street to Riverview Lift Station sewer. The number of overflows will be monitored to ensure compliance with the LTCP.

The dry weather flow diversion wall located at the beginning of the 1.52 m x 1.52 m (5' x 5') RCB will be removed upon the completion of the Martha Street to Riverview Lift Station Phase II. The sewer to the manhole located to the north of the RCB will be plugged, and the temporary lift station removed.

Grading and Piping System Abandonment

The project did include the abandonment of all manholes and piping within the limits of the Balefill. Portions of the Balefill area were graded to direct the storm water runoff to new area inlets. The grading work entailed filling the area to the east of the Village Green Apartments to provide positive grade to a new area inlet located outside the limit of the Balefill. The area to the west of the Japanese Gate was lowered to allow the storm water runoff from the area northwest of the Japanese Gate to drain to the new detention pond on the west side of the Japanese Gate. The area on the north edge of the Prairie Meadow Garden was lowered to provide positive drainage to a new inlet. The grading maintained a minimum Balefill cap thickness of 1.37 m (4.5 FT).

The abandoned portions of the existing sanitary, storm and combined sewers were abandoned in-place including manholes and inlets. The upper portion of the manholes and inlets were removed to approximately the top of the baled waste. The pipe, manholes and inlets were filled with flowable fill. Any cavities or voids adjacent to the manholes or inlets were grouted to prevent long-term maintenance issues and infiltration. A minimum 4.5 foot thick compacted clay cap was placed over the abandoned manhole or inlet. Additional fill was placed to bring the site to final grades.

Subdrain and Leachate Piping and Gas Collection Systems

The existing subdrain and leachate piping systems will be abandoned in-place including manholes. The upper portions of the manholes were removed to the top of the baled waste. The pipes and manholes were filled with flowable fill. Any cavities or voids adjacent to the manholes or inlets were grouted to prevent long-term maintenance issues and infiltration. A minimum 1.37 m (4.5 FT) thick clay cap was placed over the abandoned manhole. Additional fill was placed to bring the site to final grades.

The gas collection system was not altered except where the gas collection piping connected to the leachate piping manholes. At these locations, the pipe was reconnected on both sides of the manhole. The construction of the pipe bridge did result in the removal of the existing gas collection system. The construction of the sand base under the pipe bridge and the associated vent piping replaced the gas collection system in this area.

Construction Methods

Traditional open cut construction methods were used for a limited portion of the project. Open cut construction was used in undeveloped locations, areas where the ground cover was grass, or where the new sewers were under pavement. The pipe was installed in accordance with standard bedding and backfill.

A variety of trenchless construction methods were used to minimize the impacts on the Garden exhibits. In total, four types of trenchless construction were used for installation of the new sewer systems:

- 192 m (630 LF) of micro tunneling
- 689 m (2,262 LF) of pipe jacking
- 1,141 m (3,745 LF) of horizontally directionally drilled sanitary sewer
- 349 m (1,144 LF) of pilot tube micro tunneled sanitary sewer

Micro tunneling utilizes a boring machine located at the head of the tunneling operation. The operator controls the machine from a control room on the surface of the ground. The Microtunneling machine and jacking frame are set up in a shaft at the required depth. The operator is given constant feedback about the machine's location, orientation and hydraulic devices via a computer console or CCTV camera. Augers located in the carrier pipe remove the material displaced by the tunneling machine and convey it to the jacking pit. Micro tunneling was utilized to install the 76.20 cm (30 IN) storm sewer from the Village Green to the upstream junction box of the pipe bridge located in the Japanese Gardens.

Pipe jacking utilizes a pneumatic ram to drive the pipe through the ground. A cutting head is installed as the first segment of pipe. The cutting head is open which allows the soil to enter the pipe. Personnel are located in the front of the drilling head that load the loosened material that is conveyed back to the jacking pit. Pipe jacking was used to install the steel casing under the railroad tracks and was used to install the larger storm sewer pipe through the Gardens. The longest pipe jacking operation was 255 m (835 FT) of 137.16 cm (54 IN) reinforced concrete pipe.

Pipe jacking operation

Horizontal directional drilling (HDD) is the process where a pilot hole is drilled on the proposed pipe alignment. A reamer is then pulled back through the hole which increases the diameter of the hole to allow for the carrier pipe. Depending on the pipe size, a second reaming operation may be necessary. The carrier pipe then pulled into position. HDD was used to install the various sanitary sewers under the Garden exhibits. HDPE was used for the pipe material. HDD is typically used for sanitary sewers with significant slopes due to the potential of the installed pipe varying from the design grade. However, the longest HDD operation for this project was 465 m (1,525 LF) for a 45.72 cm (18 IN) gravity sanitary sewer. The pipe was installed at a 1 percent grade.

Horizontal directional drilling machine for the 465 m (1,525 LF) 45.72 cm (18 IN) installation. The machine was set up under Interstate 80.

Pilot tube micro tunneling utilizes elements of HDD and micro tunneling techniques. A pilot tube is drilled on the proposed pipe alignment and grade. The tunneling machine is connected to the pilot tube and pulled back through hole removing the material. The material is conveyed to the jacking pit with augers. Pilot tube micro tunneling was used for the installation of the sanitary sewers that required a close grade tolerance. The longest pilot tube micro tunnel installation was 139 m (457 LF) of 30.48 cm (12 IN) sanitary sewer. Vitrified clay pipe was used for the pilot tube micro tunneling operations.

RESULTS

The City of Omaha is in the process of addressing CSOs to improve water quality in the Missouri River and Papillion Creek. The stated goal of the project was to separate the sanitary and storm sewers within the Martha Street basin to eliminate the potential of combined sewer overflows into the Missouri River while minimizing adverse impacts to the Gardens.

The Gardens Sewer Separation Project, which is one portion of the overall Martha Street Sewer Separation Project, focused on separating a combined sewer system that ran through the Gardens by providing a new sanitary sewer and storm sewer through this area. In addition, the existing combined sewer system was required to be properly abandoned. A large concern for this project was the potential for disruption to established portions of the Gardens which are a cultural and environmental resource for the Omaha area.

The project was complicated by the fact that the Gardens had been constructed over a closed solid waste balefill and the existing combined sewer was located under the balefill, more than 24.38 m (80 FT) below land surface. The project presented significant challenges to the design team not the least of which was to minimize the impact to the Gardens. The innovative solution of using a land bridge for construction of the sanitary and storm sewers across the balefill resulted in savings of approximately $4 million of construction costs while improving water quality in nearby rivers and streams.

The final installed solution achieved several other benefits that were not envisioned when the project began. The installed project reduced the potential for leachate to enter the groundwater and storm water and also further stabilized the closed balefill area. Several sources of water that were causing accelerated decomposition and settlement within the closed balefill were eliminated. The installed project eliminated the potential for leachate to enter the storm sewer system and eventually the Missouri River by constructing a new storm sewer system and abandoning the leachate collection system. The abandoned manholes and leachate collection piping which were previously leaking were filled with flowable fill.

The project also included massive regrading within the Gardens to remedy the impacts of the settlement that had occurred over time in the closed balefill area. Approximately 83,613 cubic meters (100,000 CY) of new soil was added to reshape the landscape and establish new, more effective drainage patterns to direct water away from the balefill. This regrading reduced the potential for further settlement by directing the storm water runoff to inlets located outside of the balefill, therefore protecting the aesthetic quality of the Gardens.

The project construction is complete and placed in service. The impacted areas of the Gardens are open to the public and the disturbed areas reestablished.

The completed north end of the pipe bridge. Replacement sidewalks and roads were constructed as necessary.

The completed Prairie Meadows area was replanted with native grasses.

The completed Village Green area. The Gardens is planning an amphitheater on the completed site.

CONCLUSIONS

The Gardens Sewer Separation Project provided a cost-effective, permanent solution to several critical issues:

- Separation of storm sewers and sanitary sewers, resulting in water quality benefits to the Missouri River;
- Replacement of failed sewers located under a closed balefill located below the Lauritzen Gardens with systems that are more easily inspected and maintained;
- Stabilizing the closed balefill and eliminating potential sources of environmental contamination; and
- Enhancement to the Gardens which allows planning for future exhibits in areas that were previously unusable.

This infrastructure project provided environmental benefits, improved the sustainability of the City's sewer system and provided long-term gains for the Lauritzen Gardens by allowing the Gardens to reshape its landscape to better serve its members and guests in the future.

REFERENCES

Martha Street Sewer Separation – Lauritzen Gardens Sanitary and Storm Sewer Separation (OPW 52187) and Lauritzen Gardens Storm Sewer Grading and CSO Abandonment (OPW 52188) Basis of Design dated June 28, 2012

Suburban Lancaster Sewer Authority – Transite Pipe Lining Program

Adam N. Smith, PE, CDM Smith, smithan@cdmsmith.com

ABSTRACT

Suburban Lancaster Sewer Authority (SLSA or Authority) is a municipal sewer authority located in Lancaster County, Pennsylvania. In an 18-month period from 2007 to 2009, SLSA experienced three pipe collapses that brought about much concern on the state of their collection system. After further investigation, it was discovered that deteriorated asbestos cement pipe (ACP) was the common thread among the three instances. Further investigation confirmed that certain locations with increased hydrogen sulfide (H_2S), such as downstream of pumping station forcemains, were extremely corroded and needed immediate action. From that point forward, SLSA sought a cost-effective means of repair that would prevent costly emergency action and ensure the structural integrity of the pipes.

SLSA turned to cured-in-place pipe (CIPP) lining as their repair method of choice, and began a seven-year, phased program to rehabilitate all ACP in the collection system (eight distinct projects). While not all ACP was as deteriorated as others, SLSA wanted to be as proactive as possible to eliminate the threat of collapse in all the ACP within the collection system. Utilizing SLSA's geographic information system (GIS) CDM Smith Inc. (CDM Smith) located all the ACP in the system and separated the pipes into three categories by potential H2S exposure: Downstream of a pumping station forcemain, downstream of drop manholes, and all other ACP. Further break down was done by geographic area to delineate projects for public bidding.

While the pipe collapses were unexpected and necessitated immediate action, the Authority was able to regain control by implementing their proactive Transite Pipe Lining Program. There were bumps along the way, but the result was more than 31 miles ($1/3^{rd}$) of the collection system being lined. SLSA now has the peace of mind that they now have a structurally sound and more resilient collection system for the future.

KEYWORDS

Asbestos cement pipe, ACP, CIPP, collection system, corrosion, cost-effective, Cured-In-Place Pipe, emergency repair, forcemain discharge, H_2S, Hydrogen Sulfide, Lancaster County, Pennsylvania, pipeline rehabilitation, Suburban Lancaster Sewer Authority, SLSA, transite.

INTRODUCTION

Suburban Lancaster Sewer Authority (SLSA or Authority) is a municipal sewer authority located in Lancaster County, Pennsylvania. SLSA was formed in 1970 and has a service area that covers West Lampeter Township, Pequea Township, and Lancaster Township. Figure 1 below shows the extend of the collection system. The Authority serves over 5,600 customers, has over 92 miles of gravity sewer pipeline and is made up of five representatives (two from West Lampeter

Township, one from Lancaster Township, and two from the City of Lancaster). SLSA utilizes the City of Lancaster as its provider for day-to-day system operations and maintenance, and CDM Smith Inc. (CDM Smith) for engineering services.

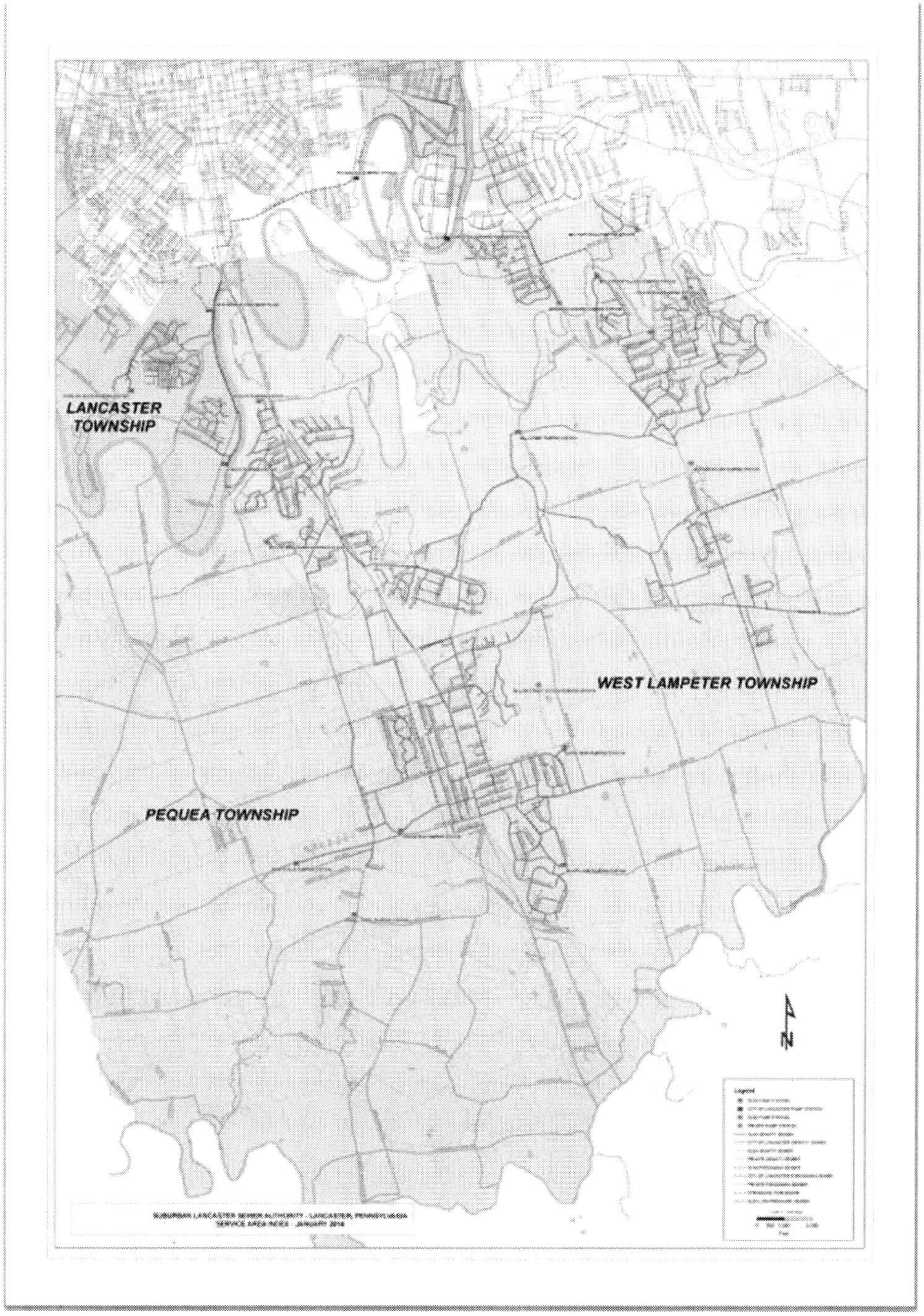

Figure 1: SLSA Service Area Index

In September 2007 and February 2008, SLSA experienced two pipe collapses along the same interceptor that were thought to be localized defects in the Hollinger Road Area. Emergency repairs were required at both locations. One of the repairs included a boring beneath Willow Street Pike (SR 0272) to install a new line and fill in the existing line under this major road as a precautionary measure to avoid a future collapse. A plan was developed to resolve this area of concern through a pipe replacement project. This retroactive approach was thought to sufficiently resolve the Authority's collection system deficiencies until on March 24, 2009, a third collapse occurred on the other side of the collection system in Penn Grant Road (SR 3028). This time, the collapse resulted in customers having flooded homes. Emergency repairs were made on March 25th and when the Authority met on March 26th, it was quickly realized that a larger problem may exist, the cause of which needed to be determined immediately. CDM Smith then analyzed the system and began development of a more proactive approach to reduce risk. In this short amount of time the Authority's approach to the collection system went from reactive to proactive.

Upon further investigation, it was found that these three pipe collapses were downstream of pumping station force main discharges and resulted from the presence of extremely deteriorated asbestos cement pipe (ACP), also known as transite pipe, the material of choice in the early to mid-1970's when the Authority's collection system was first constructed. The turbulence caused by force main discharges and drop manholes in the system had caused the release of corrosive H_2S gas from the wastewater. The ACP is particularly susceptible to this type of corrosion. When compromised by H_2S exposure, the ACP looks intact but can be extremely soft and have little structural integrity remaining as the cement holding the asbestos fibers together has deteriorated away. This quality makes it difficult to identify which pipes are worse than others during traditional CCTV inspections. The cost to dig and replace all the ACP would represent a huge unexpected financial burden to the Authority and its customers. Recognizing the even greater potential cost of inaction, SLSA requested that CDM Smith develop an overall plan for addressing risk posed by all the ACP in the collection system. Although this task had already begun to some extent by prioritizing the larger diameter pipe for televising evaluations, the Authority now recognized that an even more proactive approach must be developed to avoid future sewer line collapses, costly repairs, and potential damages resulting from sewage backing up into homes.

METHODOLOGY

As mentioned earlier, asbestos cement pipe was the pipe type of choice in the early to mid 1970's. Because this coincided with the formation of SLSA and a housing influx in the service area the full extent of the ACP was not initially known. CDM Smith used the SLSA geographic information system (GIS) database to quantify the amount of ACP in the collection system, the size of the pipe, and location. Not including the emergency replacement section in Penn Grant Road or those replaced under the Hollinger Road Sewer Interceptor Replacement contract it was determined that ACP accounted for more than 31 miles of the collection system (1/3 of all pipes), ranging from 8 to 18 inches in diameter, and was used throughout the system. The breakdown of the pipe by size can be seen in Table 1. All estimated lengths in the following tables are based on record drawing lengths from the SLSA GIS database.

Table 1: Summary of SLSA ACP Lengths by Pipe Size

Line Size (in)	Line Length (LF)	Percentage of Total Length
8	148,100	89.5%
10	8,400	5.1%
12	3,900	2.4%
15	3,500	2.1%
18	1,600	1.0%
TOTAL	**165,500**	**100%**

Since the pipe failures resulted from hydrogen sulfide (H_2S) corrosion, the pipes were further categorized based on their location relative to where the most corrosive environments are expected to exist. Corrosion damage to sanitary sewer infrastructure is most likely to occur in areas that receive "old sewage" and/or areas that experience turbulence. "Old sewage" refers to wastewater that is not fresh but has instead been sitting stagnant long enough to develop septic, or anaerobic, conditions that cause H_2S to form. These conditions arise most often in small pumping stations that do not receive much flow or in pumping stations that receive flow from other stations (sewage that is pumped more than once has an even greater opportunity to develop septic conditions, since it lays stagnant several times before entering the wastewater treatment plant). Subsequent turbulence in the system can cause the release of H_2S gas from the wastewater, and occurs most frequently in force main discharge manholes and drop manholes.

The pipes were also sorted into the following priority categories: gravity line downstream of force main discharge manholes, gravity line downstream of drop manholes and the remainder of the lines. A summary of the lengths by priority for rehabilitation or replacement is presented in Table 2.

Table 2: Summary of SLSA Transite Pipe Lengths by Priority for Rehabilitation/Replacement

Category by Priority	Line Length (LF)	Percentage of Total Length
Line downstream from force main discharges	25,900	15.6%
Line downstream from drop manholes	35,300	21.3%
Remainder of lines	104,300	63.1%
TOTAL	**165,500**	**100%**

The top two categories listed in Table 2 can be further broken down by pipe diameter, as shown in Tables 3 and 4.

Table 3: Summary of Force Main Discharge Lines by Pipe Size

Line Size (in)	Line Length (LF)	Percentage of Total Length
8	14,060	54.1%
10	5,950	22.9%
15	4,330	16.7%
18	1,650	6.3%
TOTAL	**25,990**	**100%**

Table 4: Summary of Drop Manhole Discharge Lines by Pipe Size

Line Size (in)	Line Length (LF)	Percentage of Total Length
8	31,410	88.7%
10	2,480	7.0%
12	1,530	4.3%
TOTAL	**35,420**	**100%**

Discharges from low-pressure systems or individual grinder pumps were also taken into consideration. However, these connections are all relatively recent and would not be expected to have had enough time to cause considerable corrosion damage to the ACP.

The assignment of the highest priority to the lines downstream of force main discharge manholes was verified during the Hollinger Road Interceptor Replacement. On May 11, 2009, the work approached the force main discharge point for the Mill Creek Pumping Station on Locust Lane. Excavation in this area in a heavily used intersection identified two different locations in one sewer line run where the top of the existing pipeline had completely deteriorated and the earth above the pipe location had been almost completely eroded. Upon removing the pavement from Locust Lane, it was virtually a complete chasm down to the bottom of the sewer line. These locations represented imminent sewer line collapses that were replaced just in time.

Replacement or Rehabilitation

There are several means by which the deterioration of the ACP in the collection system could have been addressed and the methods were broken down into two categories: Replacement and Rehabilitation. Replacement involves excavation of the old pipe and installation of new pipe in the ground. Rehabilitation uses the existing pipe as a conduit for repair and enables the repair to be completed with minimal or no excavation.

Replacement

Pipe replacement involves the excavation, removal and disposal of the existing ACP and replacement with new PVC pipe. One benefit of replacement is that the laterals as well as the gravity main can be replaced with PVC pipe. The drawback is the cost, which can be much higher as a result of restoration (particularly paving), asbestos pipe abatement and disposal, excavation and backfilling, and other associated project costs.

However, in some instances the only option is to replace the pipe. These instances include cases where the pipe has collapsed or is in such poor condition that it must be replaced (the condition of the pipe will be determined by the contractor and engineer during pre-lining televising inspection). Another instance when new pipe must be installed is when the sewer system hydraulic model (i.e., the SewerGEMS model for GIS) indicates that a larger line size is required. As a part of this project, the sanitary sewer system was modeled to take future development into account, so that lengths of pipe that must be increased in size could be identified. It was determined that one location needed to be increased due to capacity concerns, and this turned into the Penn Grant Road Sanitary Sewer Replacement project.

The technique of pipe bursting can sometimes be used to replace pipe with same-size or larger pipe without excavation. With this technology, equipment that breaks up pipe is pulled through the existing pipe, breaking it into pieces and pushing it back against the surrounding fill or soil. The new pipe is then pulled behind the breaking element without further excavation. This technology was not considered for this application since the pipe bursting procedure was considered by PADEP at the time to render the asbestos in the pipe friable (allowing the release of small asbestos fibers). Even though the pipe will not be excavated during pipe bursting, the procedure could create a dangerous environment for future work. Therefore, PADEP did not allow pipe bursting of large quantities of ACP as would have been required for this work.

Necessary Replacement (Open Trench) Projects

A single area was identified for replacement. This was the 8-inch gravity main along Penn Grant Road and Willow Street Pike to Nolt Avenue, which was upsized to 10-inch pipe. The need for this increase in pipe size was identified during the Silver Lane Pumping Station design to accommodate projected future sewage flows. Approximately 340 feet of the line had already been replaced with 10-inch PVC as a part of the emergency repair work performed after the line collapse in Penn Grant Road. Sewer system modeling will be required to identify any other areas that will need to be increased in size.

Rehabilitation (Pipe Lining)

Pipe lining involves the installation of a new material inside the existing pipeline. The lining material is corrosion-resistant and can also provide structural strength to prevent collapse of the pipe. The benefits of this type of technology include shorter installation times and lower costs as compared to pipe replacement. Except for areas that may require point repairs, no excavation/restoration is required, therefore overall costs can be significantly reduced. Furthermore, in the case of ACP, the pipe material itself is a hazardous waste (asbestos) and must be disposed of properly.

Pipe lining does have limitations, however. If the pipe has collapsed or if the pipe size must be increased to accommodate future flow capacity, pipe lining cannot be used. In addition, if the sewer main is lined rather than replaced, the existing laterals will not be replaced.

Lining Systems

There are a number of lining systems available for rehabilitation of sanitary sewer lines. Sliplining is one type of technology that involves the installation of new pipe inside of existing pipe. However, sliplining usually involves installation of a smaller diameter pipe inside the larger one, thereby decreasing the hydraulic capacity of the pipe. Because of the hydraulic constraints and nearly 90% of the pipe being 8 inches already this technology was not appropriate for the SLSA system.

One of the most common types of lining technology is referred to as cured-in-place pipe (CIPP). This type of technology involves the installation of the pipe liner material in a segment of pipe. The liner generally has several layers. The inside layer (the one that will come into contact with the sewage) is made of some type of corrosion-resistant material such as polyurethane or PVC. The next layer is a material, such as fiberglass or felt, which will act as a medium to contain the resin. Resin is a type of material that will undergo the "curing" process to develop hardness and strength.

Once the liner is put into place, it is then inflated up against the existing pipe using either pressurized air or water. Then the resins in the liner are activated using steam, hot water or ultraviolet light. Upon activation, the resins begin the curing process, which causes them to harden. After the resins have cured and the liner sufficiently cooled, any existing laterals reinstated and the pipe is placed back in service.

The thickness of the liners is determined based on the depth of the pipe, depth of water table and surface loadings above the pipe. The liner to be installed for SLSA should be designed to provide structural strength independent of the host ACP since the true structural integrity is unknown. Installation of a liner does result in a minor decrease in inside pipe diameter, however, since the inside surface of the liner is very smooth the overall capacity chance is through to be negligible.

There are two main types of cured-in-place pipe liners: heat cured and ultraviolet (UV) light cured. Both products are pipe liners that provide both corrosion resistance and structural strength and can be installed without the need for excavation.

Heat cured liners consists of a felt tube which is impregnated with a polyester resin at the factory. A liner on the outside of the felt/resin layer keeps the resin from rubbing off or washing out during installation. The liner is shipped from the factory to the site and is kept refrigerated to keep it from heat-curing and therefore hardening before installation. The liner is put into place in the pipe by attaching it to a manifold at one end and inverting it using either water or pressurized air. The heat-curing process is accomplished by filling the line with hot water or steam, and the resin is allowed time to cure. After the curing process is complete, a robotic crawler with a cutting head is sent into the pipe to reinstate lateral openings and place them back in service.

UV cured liners are a product that consists of a polyurethane interior layer, surrounded by a layer of resin-impregnated fiberglass. This fiberglass layer is then surrounded by a layer of material that prevents sunlight from reacting with the resin. The liner is pulled into place in the line and then inflated using air pressure. To cure the product, a UV light train is pulled at a controlled rate through the line, curing the resin as it progresses from one end to the other. After the liner has cured, the lateral openings are reinstated.

Recommendation For SLSA

Based on the fact that three collapses occurred in transite pipe since September 2007, it was known that the existing transite pipe in the SLSA sewer system was deteriorating. Due to the expected construction costs design cost, design time, and construction duration required for a traditional replacement project it was ruled out as it would not have been able to be completed fast enough and it was cost prohibitive. CIPP was chosen as the right trenchless option because the pipes were still circular with few obstructive defects, the work could begin almost immediately, the rehabilitation duration is far less than replacement, and the process is budget friendly.

As shown in Table 2, approximately 37 percent of the total pipe length was a high priority for rehabilitation. It was assumed that all pipe identified as "high priority" had experienced some degree of deterioration and should be lined quickly. Therefore, CDM Smith recommended the immediate start of a rehabilitation program on these high priority areas in lieu of expending funds on a televising-only inspection program. This was further justified knowing that during the pipe lining process it is standard practice to clean the bottom of the pipe by jetting and perform pre-lining televising to ensure that the existing pipe condition is suitable for lining, or whether it will require additional cleaning, point repairs, or replacement.

Transite Pipe Lining Program Implementation

SLSA wanted to address the risk posed by all ACP in its sewer system as a high priority situation to avoid, to the greatest extent possible, any additional pipe collapses, emergency repairs, and associated costs. For that reason, and for the purpose of developing a capital expenditures budget starting with fiscal year 2009/2010 (mid-2009 to mid-2010), SLSA set forth the following implementation plan:

- All transite pipe downstream of force main discharges will be lined during fiscal year 2009/2010
    - Phase 1A: Test Area
    - Phase 1 B: High Priority Areas Downstream of Roslyn, Goods Run, Hans Herr, Pioneer Woods, and Jefferson Square Pumping Stations
    - Phase 1C: Environmentally Sensitive Area Downstream of the Lampeter Interceptor
- All transite pipe downstream of drop manholes will be lined during fiscal year 2010/2011.
    - Phase 2: All Pipes Downstream of Drop Manholes

- All remaining transite pipe in the SLSA sewer system will be lined over a five-year period, starting in fiscal year 2011/2012 and ending in fiscal year 2015/2016.
    - Phase 3, Phase 4, Phase 5, Phase 6
- The Penn Grant Road sewer replacement project (from the termination the emergency repair to Nolt Road) will be designed, bid and constructed during fiscal year 2009/2010.

Following identification of acceptable lining technologies from several manufacturers through a request for qualifications, SLSA proceeded in 2009 to bid their first project in the Transite Pipe Lining Program, Phase 1A. Since CIPP lining was new to the Authority, this first project was limited to a relatively small length of sewer in the system for evaluating the CIPP technology and its installation.

SLSA could then determine whether to: (a) proceed with subsequent phases of the pipe lining program using the same pipe lining technology as a sole-source provider, or (b) proceed with subsequent phases of the pipe lining program allowing any of the previously approved pipe lining technologies, while reserving the right to evaluate and approve additional pipe lining technologies in the future if they meet SLSA's required qualifications.

Once Phase 1A was completed SLSA proceeded with option b above and continued to publicly bid subsequent lining projects except for the single sole sourced project, Phase 1C.

Phase 1C included the area downstream of the Lampeter Interceptor is located in farmland just south of Long Rifle Road and east of Gypsy Hill Road. During televising in 2008, it was noted that one of the manholes is now largely exposed in the creek bed, although no pipe was visible. A follow-up site visit in July 2009 has identified two additional exposed manholes and some exposed pipe. Because of the sensitive nature of this environment and the exterior exposure of the manholes and pipe to the surrounding elements, CDM Smith recommended that this section be included in the highest priority category and lined using the most encapsulated lining system with the least potential for an environmental release. This was the UV light cured system because the exterior plastic used to block sunlight provided an excellent barrier and the resin in this system did not use styrene, whose effects on natural ecosystems was not fully understood. Additionally, all manholes in this section were spray lined to rehabilitate their current state and protect from future exposure to H_2S.

Completing the Projects

Each contract presented different asset, geographic, traffic, and maintenance of sewer flow challenges and utilized the knowledge gained during the earlier projects to better manage later projects. These lessons are now being utilized on other projects for the Authority and can be used by other municipalities who may be facing similar critical decisions.

SLSA's approach to field work management was to have a full-time inspector on-site with the contractor. This approach proved to be worthwhile as there are many times where a trained inspector can assist decision making, provide invaluable knowledge of the collection system, and be the bridge contact person for the engineers in the office and contractor field staff. The portions of the work that an inspector must ensure are up to quality are all CCTV activities, pipe curing, pipe cool down, and service reinstatements.

Common Punchlist Items

Table 5 below shows some of the common punchlist items that were encountered during our lining program. These defects could take place on any lining project and are not specific to ACP. However, defects that result in ACP being unnecessarily left exposed to the sewer environment were considered to be unacceptable and against the intent of the Transite Pipe Lining Program.

Table 5: Typical Punchlist Items

Undercut Connection – The liner material was not fully removed from the opening, in this case a drop connection.	
Overcut Lateral – The liner was removed beyond the extents of the lateral opening and damage has been done to the host pipe.	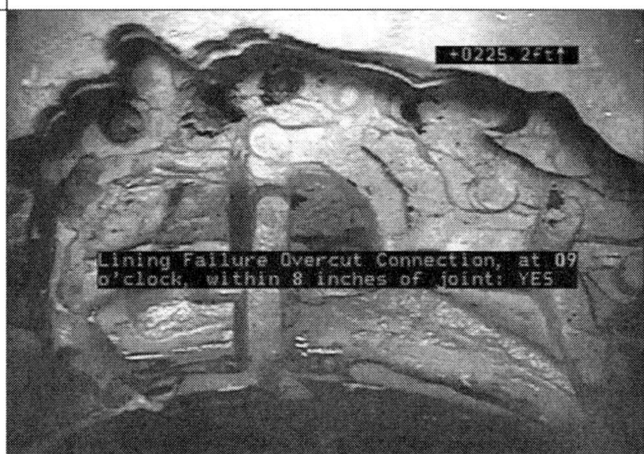

Pinhole Leaks in Liner – Either a manufacturer defect or locations damaged during shipping or installation, these pinhole leaks are weak points in the liner.	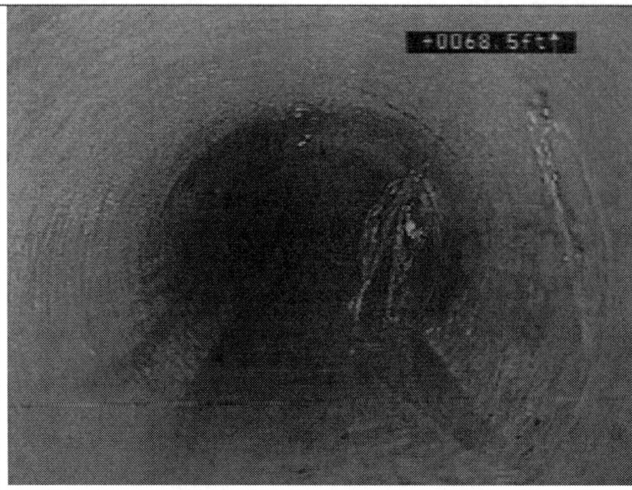
Liner Coupon Not Removed (rotated 90° for clarity) – Coupons should always be removed from the system whenever possible. Downstream facilities may not be able to process such debris.	

Liner Contraction – Many factors can cause this defect, however the contraction has now left a section of ACP unprotected and it is possible that lateral reinstatements have shifted as well.	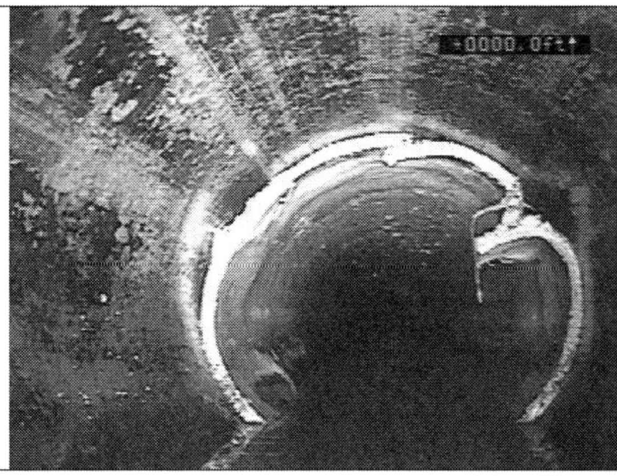

ACP Specific Items of Note

There are several specific items that were learned along the way about ACP that were valuable tools for SLSA's projects. The first and most obvious is the health concerns presented by the asbestos in the pipe. Should a point repair be necessary, it was required that the contractor follow proper protocol to minimize the exposure, especially when the pipe would need to be cut.

Another intricacy of ACP is that it has great heat retention properties, much more than other pipe materials. This means that once heat is added to the pipe it takes a long time to cool, much longer than the liner itself. This necessitated a long, gradual cool down period for the liner otherwise the thermal differential between the liner and host pipe will cause the liner to break the bond with the host pipe, contract and unexpectedly "snap-back". This "snap-back" would occasionally occur and typically cause one end of the liner to be pulled into the manhole, exposing a short section of ACP that would then need to have a CIPP point repair. If laterals were reinstated before this occurred there is also the potential that the "snap-back" would offset the lateral openings, requiring that the contractor reinstate the lateral a second time and repair the offset which now exposes a section of ACP.

END RESULTS

The program was completed in 2016 and the risk posed by this pipe type was effectively mitigated by using a trenchless technology that is now commonplace in the industry. Figure 2 below shows a section of ACP before and after lining. This pipe was extremely soft and light jetting is part of why sections of pipe are delaminating. CIPP lining gave this soft conduit many more years of service.

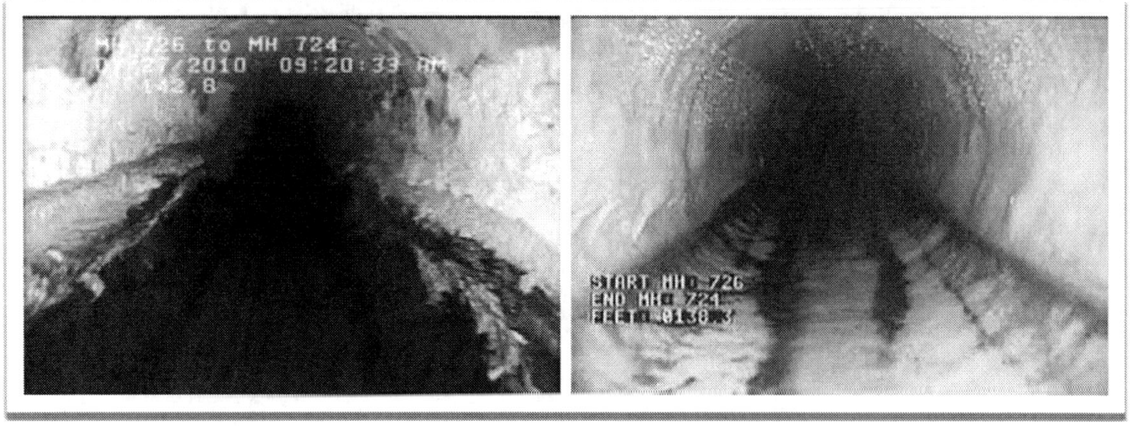

Figure 2: ACP in Wynwood Drive – Before and After (Downstream from Hans Herr Pumping Station Force Main Discharge)

The total construction cost is shown in Table 6. The whole program was just over $5,000,000, which is only a fraction of the estimated $30,000,000 (~$200/LF) it would have cost to dig and replace the pipes. For comparison purposes, the Penn Grant Road Sewer Replacement project cost $245/LF.

Table 6: Summary of Final Quantities from the Transite Pipe Lining Program

Year	Project Name	Pipe Size (in)	Lining Length (LF)	Construction Cost
DOWNSTREAM OF FORCE MAINS				
2009-2010	Phase 1A (Test Area)	10	5,950	$192,022
2010-2011	Phase 1B (General High Priority)	8	11,853	$305,404
2011	Phase 1C (Environmentally Sensitive)	8, 15, 18	8,175	$608,759
DOWNSTREAM OF DROP MANHOLES				
2011-2012	Phase 2	8, 10, 12	28,507	$808,104
REMAINING ASBESTOS CEMENT PIPE				
2011-2012	Phase 3	8	22,340	$669,050
2012-2014	Phase 4	8, 12	31,490	$888,496
2013-2014	Phase 5	8	28,010	$803,430
2014-2016	Phase 6	8	27,974	$746,758
Totals	**8 Projects - Start 2009; End 2016**	**8, 10, 12, 15, 18**	**164,299**	**$5,022,023**

CONCLUSION

Not only has this program mitigated collapses, but by implementing trenchless technology the Authority could stretch its dollars to help strengthen the worst parts of the collection system. Additionally, SLSA has received other benefits such as a reduction in infiltration, discover other defects in the system, identify manholes in need of rehabilitation, and collect modern inspection data to add to the SLSA GIS database. A map of the collection system highlighting the areas that were lined is shown in Figure 3. While the pipe collapses were unexpected and necessitated immediate action, the Authority was able to regain control by implementing their proactive Transite Pipe Lining Program. There were bumps along the way, but the result was more than 31 miles (1/3rd) of the collection system being lined. SLSA now has the peace of mind that they now have a structurally sound and more resilient collection system for the future.

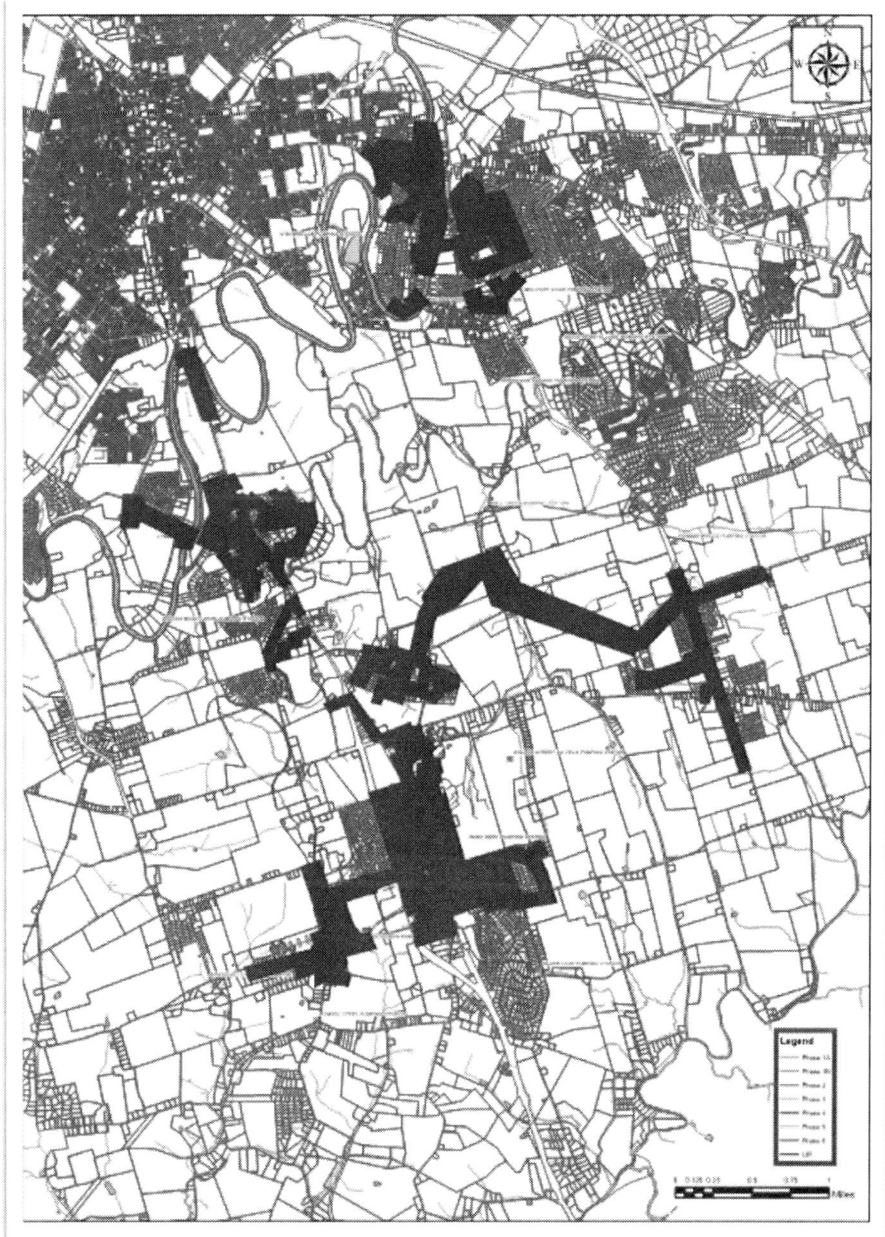

Figure 3: Transite Pipe Lining Program Area Map

REFERENCES

Robinson, Gail A (2009) *Transite Pipe Memorandum;* Presented to Suburban Lancaster Sewer Authority at the July 2009 Meeting

A Trenchless Approach to Lift Station Elimination

Scott Maughn, P.E. [1], Daniel Huffines, P.E.[2], and Lindy Higginbotham, P.E. [3]

[1]Freese and Nichols, Inc., 2711 North Haskell 33rd Floor, Dallas, TX 75204
Phone: (214) 217-2200; E-mail: asm@freese.com

[2] Freese and Nichols, Inc., 2711 North Haskell 33rd Floor, Dallas, TX 75204
Phone: (214) 217-2200; E-mail: dwh@freese.com

[3] City of Allen, Texas, 305 Century Parkway , Allen, TX 75013
Phone: (214) 509-4579; E-mail: lhigginbotham@garlandtx.govcityofallen.org

ABSTRACT

Initially in 2004, the City of Garland, Texas, approached Freese and Nichols, Inc. (FNI), about decommissioning a small lift station that was becoming a maintenance problem for the City staff. The City did not want to expend effort or funds into rehabilitating or replacing the lift station; it has preferred to eliminate lift stations, if possible, throughout the City. FNI conducted a study of the lift station service area, the gravity collection system in proximity of the lift station, and the topography of the immediate area surrounding the lift station site. The results of the study determined that it would require trench cuts in excess of 50-feet to install a gravity line by open-cut methods to the downstream system, and at the time the trenchless technology was not advanced enough to install a gravity wastewater line at grade for the lengths required for connecting to the existing collection system. FNI recommended to rehabilitate or replace the lift station due to the excessive expense and risk involved in attempting to run a new gravity line to eliminate the lift station. The City opted to wait, and attempted to maintain the existing lift station.

Fast forward to 2014. The City of Garland contacted FNI to reinvestigate the same lift station and collection system. After 10 more years of use and maintenance, the lift station was even more of a problem for the City maintenance staff. FNI reviewed the 2004 study and determined that improvements in trenchless technology capabilities made it feasible to install a new gravity line at grade. The new line would enable the City to eliminate the lift station and connect the existing collection system to the service area of the existing lift station. This project would require over 3,500 feet of pipe to be installed by trenchless methods on a grade of approximately 0.01 ft./ft. grade. The project was successfully installed and was put in service in the Spring of 2016 as designed.

Now that trenchless technologies are making it more feasible to eliminate lift stations, this paper will give utility owners a greater understanding of what trenchless installation entails and what to consider when making that decision. The paper will show the audience how the City of Garland team researched, designed and constructed the project. It will describe the installation options that the project team explored, the precautions they took during design, and the measures they implemented to minimize risks. The paper will also discuss construction issues and challenges that trenchless projects may encounter, as well as lessons learned and other guidance for utility owners.

PROJECT BACKGROUND

The City of Garland, Texas originally retained FNI in 2004 to investigate methods for decommissioning the Wynn Joyce Lift Station due to continued maintenance issues. At the conclusion of the investigations, FNI presented the City with two options. The first was to replace the existing lift station and operate the system as it had been currently. The second option was to design a gravity wastewater pipeline to convey all flow currently served by this lift station into the existing wastewater collection system, and ultimately deliver it to the downstream Windridge lift station. This option would require the construction of approximately 4,250 linear feet of pipeline to connect to an existing 15-inch wastewater pipeline. The selected corridor would require that the pipeline be installed through a hill and at depths in excess of 50-feet for a large majority of the route. Due to this, it was determined that 3,500 linear feet of the total 4,250 linear feet would need to be installed by trenchless methods. At the time of the original investigations it was decided that the trenchless technology was not advanced enough to successfully install the pipeline at the strict grade requirements for a gravity wastewater pipeline. Therefore, it was recommended by FNI to replace the Wynn Joyce Lift Station in lieu of constructing a pipeline. The City chose to leave the lift station "as-is" at that time, and continue to maintain it, as needed, with the desire to reinvestigate the trenchless technology at a future time once the capabilities and feasibility had improved.

In 2014, the City approached FNI to revisit the project and original recommendation. FNI determined that trenchless installation methods had improved in the 10 years that had passed from the original study. FNI performed an alignment study and final design services for the pipeline and decommissioning of the lift station. The overall design process will be further discussed in the following portions of the paper.

PRELIMINARY INVESTIGATIONS AND ALIGNMENT SELECTION

Alignment Study

The horizontal and vertical location of the existing utilities, existing topography, and right-of-way locations were taken into consideration when evaluating options for the gravity wastewater pipeline. Both alignments followed the Wynn Joyce Road right-of-way and experienced a hill throughout a majority of the proposed route, therefore making it very advantageous to utilize trenchless installation methods. Since both alignment options utilized the same roadway right-of-way, the main goal for the selection of a route was to find one that allows the best setup area for trenchless methods while also providing minimal impacts to the surrounding residential areas and traffic flows. The following Figure 1 shows both alignment options, along with the general project layout.

Figure 1 – Overall Project Layout

Option 1 – Northern Alignment

Option 1 begins at the existing Wynn Joyce Lift Station and travels northeasterly for approximately 4,250 linear feet along Wynn Joyce Road, ending at the downstream 15-inch gravity wastewater pipeline. The entire route is located within 2-3 feet of the northern roadway right-of-way line. The first 500 linear feet of this option was proposed to be installed by traditional open cut methods, with the remaining portion being installed by trenchless methods due to the depths the pipeline will encounter.

This option provided adequate setup area at the beginning and end of the alignment for the trenchless operations, with minimal existing utilities to contend with in these locations. The main concern with this alignment was the close proximately of the homes located on the downstream portion of the alignment for the proposed trenchless operations. The City mentioned during the kickoff meeting that a franchise entity had several issues with drill mud hydraulic fractures or inadvertent returns while installing a 12-inch high pressure gas line via directional drilling along southern side of Wynn Joyce Road. The drill mud inadvertent returns caused severe damage to one of the houses near the Wynn Joyce Lift Station, along with a large amount of drill mud escaping to the surface at another location damaging a swimming pool.

Option 2 – Southern Alignment

Option 2 begins at the Wynn Joyce Lift Station and travels along northern side of the roadway for approximately 300 linear feet, then crossing to the southern side of Wynn Joyce Road and traveling northeasterly the remaining 3,950 linear feet along the southern Wynn Joyce Road right-of-way boundary ending at the same downstream 15-inch gravity wastewater pipeline as Option 1. The entire route is located within 2-3 feet of the southern roadway right-of-way line. FNI proposed to install the first 300 linear feet of this option by open cut methods, then a steel encased bore under Wynn Joyce Road, and install the remaining portion by trenchless methods due to the pipeline depths.

The main benefit of Option 2 was the larger buffer zone between the neighboring homes along the proposed alignment in regards to the franchise utility drilling problems as mentioned in Option 1. The trenchless section of the alignment began at the large tract parcels near the Wynn Joyce Lift Station, then traveled along the front of the school property, and finally paralleled a homeowner association landscape buffer for the remainder of the route. The additional buffer would not eliminate the possibility of damages as found in the franchise utility pipeline installation, but the added spacing would possibly aid in protecting the neighboring structures.

Vertical Analysis

FNI analyzed and prepared a preliminary profile for both alignments. The controlling upstream elevation was the flowline of the existing influent 8-inch and 12-inch gravity wastewater lines at the Wynn Joyce Lift Station, at an elevation of 474.72 feet. The controlling downstream elevation was the flowline of the existing 15-inch gravity wastewater line, at an elevation of 439.11 feet. This results in an approximate grade of 0.85% for both of the alignment options, and well above the Texas Commission on Environmental Quality (TCEQ) minimum slope requirements for an 8-inch, 10-inch, or 12-inch gravity wastewater line that were investigated in the following section.

Hydraulic Loading and Pipeline Sizing

FNI received Wynn Joyce Lift Station flow data records from the City, along with a copy of their Wastewater Management Plan Update. Listed below in Table 1, is a summary of the average flow calculations, along with peak flow conditions based on the City's flow data provided.

Study Period	11/1/2013 to 12/2/2013	2/10/2014 to 3/7/2014
Total Period Flow (Gal)	1,356,600	1,341,900
Total Period Days	31	25
Avg Flow Rate	30.4 GPM / 0.044 MGD	37.3 GPM / 0.054 MGD
Peak Flow Rate*	121.6 GPM / 0.175 MGD	149.1 GPM / 0.215 MGD

**A peaking factor of 4 was utilized to calculate the existing Peak Flow rate due to the lack of any major rain events during either of the data sets.*

Table 1 - Wynn Joyce Lift Station Flow Data

As a part of the City's Wastewater Management Plan, it was determined that the Wynn Joyce Lift Station was planned for upsizing from 0.474 MGD to 0.70 MGD firm capacity in order to meet build out conditions. Due to the City provided flow data being conducted during drought conditions, the proposed pipeline was analyzed and sized to meet the future 0.70 MGD loading requirement as called for in the City's Wastewater Management Plan.

Based on the above hydraulic loading data, FNI analyzed the loading conditions on an 8-inch, 10-inch, and 12-inch pipeline. A minimum flow condition of 30.40 GPM/0.044 MGD was utilized based on the analysis under drought conditions, and a maximum flow condition of 486.11 GPM/0.70 MGD was utilized based on the City's Wastewater Management Plan since it applied to the build out conditions for the system. Based on the use of these two flow rates, an ultimate minimum and maximum flow could be analyzed for the proposed pipeline.

Table 2 below summarizes the pipeline characteristics based on the minimum/maximum loading in comparison to the TCEQ minimum allowable slope and the anticipated design slope of the proposed wastewater line.

Diameter	Roughness Coef*	Slope	Flow Rate	Flow Rate	Velocity	Normal Depth	Full Pipe Capacity
(inches)	(Manning's n)	(%)	(GPM)	(MGD)	(ft/s)	(inches)	(MGD)
8	0.013	0.33	30.40	0.044	1.26	1.69	0.45
8	0.013	0.33	486.11	0.700	N/A	N/A	
8	0.013	0.80**	30.40	0.044	1.73	1.36	0.70
8	0.013	0.80**	486.11	0.700	3.53	6.57	
8	0.013	0.85	30.40	0.044	1.76	1.34	0.72
8	0.013	0.85	486.11	0.700	3.64	6.37	
10	0.013	0.25	30.40	0.044	1.12	1.69	0.71
10	0.013	0.25	486.11	0.700	2.29	8.1	
10	0.013	0.85	30.40	0.044	1.72	1.25	1.31
10	0.013	0.85	486.11	0.700	3.77	5.21	
12	0.013	0.20	30.40	0.044	1.01	1.69	1.03
12	0.013	0.20	486.11	0.700	2.18	7.26	
12	0.013	0.85	30.40	0.044	1.67	1.19	2.12
12	0.013	0.85	486.11	0.700	3.75	4.74	

** A Roughness coefficient of 0.013 was used based on the design requirements set forth in TCEQ 217.53.l.2.A.i.*
*** The 8-inch pipeline was not analyzed at the TCEQ minimum slope since a 0.80% slope was required to convey the max design flow.*

Table 2 – Pipe Loading Characteristics

As seen in the above table, all pipeline diameters can support the max loading condition of 0.70 MGD except an 8-inch pipeline at TCEQ minimum grade. Since the 8-inch pipeline requires a minimum of a 0.80% slope to convey the max loading condition, it was chosen to eliminate this pipe size as an option since it leaves very little room for error in the trenchless installation section. Both the 10-inch and 12-inch pipelines can support the max flow condition at a grade ranging from the TCEQ minimum grade up to the proposed design grade.

Based on discussions with the City, there is very little room for development and growth in this wastewater service area, therefore it was recommended to install a 10-inch gravity wastewater line to convey flow for the proposed system.

Trenchless Installation

Due to the extensive amount of trenchless installation needed for this project, the following two primary installation methods were investigated.

Horizontal Directional Drilling

The first installation method considered to install the pipeline was horizontal directional drilling (HDD). With this being a gravity installation, additional care was needed to control the drilling accuracy since typical accuracy for this method is +/- 1.0%. Additional accuracy can be gained

by the careful selection of the guidance system, limiting the back-reaming process to prevent excess floatation, along with the use of periodic sight holes (approximately 12-inch to 18-inch in diameter) to verify the drill is on grade.
Another area of consideration was setup area requirements at the beginning and end of the drill sections in order to provide sufficient room for the contractors' equipment. For a HDD of this length, contractors typically need a 200'x200' work area at the beginning of the drill and a 100'x100' area at the end for their receiving pit and equipment. In addition to the equipment staging areas, additional room is needed to fuse and string out the proposed pipe for pull back. Generally, it is best practice to limit the number of stops during the pullback process, therefore increasing the size of area required to stage the piping.

With this installation method, FNI recommended using Fusible Polyvinyl Chloride Pipe (PVC) or High Density Polyethylene pipe (HDPE) since either pipe material can withstand the pull forces from the trenchless installation, along with its great resistance to hydrogen sulfide gas.

Pilot Tube Microtunneling
The second installation method is pilot tube microtunneling. This process is similar to standard large diameter microtunneling except modified for small diameter pipelines. The tunneling rig is placed in 8-foot to 12-foot diameter vertical shafts and can travel a max distance of 500-feet between setup pits. Once the pilot hole is drilled, the carrier pipe is pushed/jacked into place behind the jacking frame and auger. This will require multiple intermediate setup points to complete the installation of the proposed pipeline. Just as in standard microtunneling, this method can achieve accuracy within a fraction of a percent grade without the need of the periodic sight holes as mentioned in the directional drilling method.

Due to the pushing/jacking nature of this installation method, Vitrified Clay Pipe (VCP) is most commonly selected because of its strong compressive strength characteristics. Fusible PVC or HDPE can also be used in this process, but adds an extra step to the installation since a casing pipe is pushed into place first, then utilized to pull back either the PVC or HDPE pipe.

Selected Route and Installation Method
As a result of the preliminary design and discussions with the City, FNI proceeded with the final design of Alignment Option 2 – Southern Alignment. The main reason for recommending and selecting this alignment was the fact this alignment provided the greatest separation from the development in the area and an equal amount of staging room compared to the alternative alignment. In addition, both alignment options shared similar construction cost estimates. The following Figure 2 shows the selected route, along with general final design layout.

In comparison of the two installation methods, the City and FNI chose to proceed with the use of HDD for the installation of the proposed pipeline. Both mentioned methods were capable of the pipeline installation, but Pilot Tube Microtunneling was ruled out due to the extreme depths that would be required for the intermediate setup pits in relation to the proposed pipeline elevations.

Figure 2 – Selected Route and Final Design Layout

FINAL DESIGN APPROACH

Geotechnical Investigation

One of the most crucial steps of the final design was getting adequate knowledge of the geotechnical conditions. For this project, we chose to take six bores evenly spaced along the alignment. Based on the findings of the bores would determine what range of accuracy we could expect for the HDD. The results of the analysis showed that the pipeline would be installed in a shale formation for the entire installation which would aid in the accuracy of the HDD. The geotechnical information also benefited the contractor by allowing them to select the correct drill head, proper drill mud parameters, and overall equipment setup. The following Figure 3 shows the results of the geotechnical bores in relation to the proposed pipeline alignment.

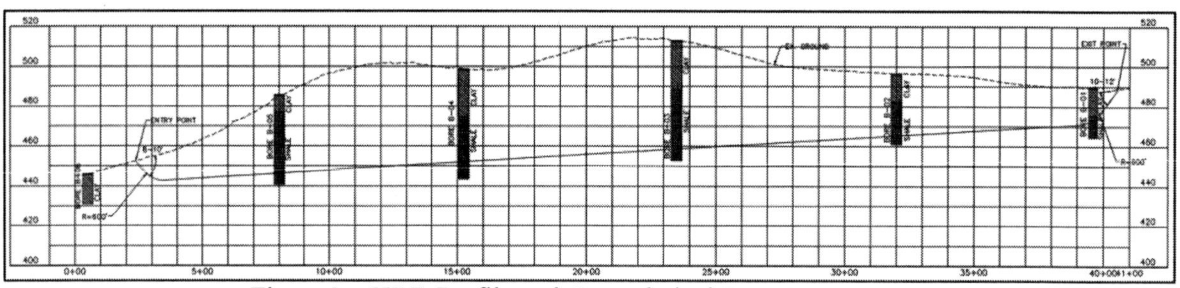
Figure 3 – HDD Profile and Geotechnical Bore Data

Client Coordination

Due to the scale and nature of this project, several items were coordinated with the City to ensure the project was designed to their preferences.

Pipe Material Preferences

The two most common materials for HDD are HDPE and Fusible PVC. Both materials were analyzed and deemed capable to meeting the design conditions for the project. Based on the City's familiarity with the material, HDPE was selected for use on the project.

Staging Areas

Early in the design phase the City informed FNI they would like to avoid acquiring any easements for the project and complete all staging within the Wynn Joyce right-of-way or existing easements along the corridor. FNI accommodated this request by utilizing existing pipeline easements located along the bank of the lake, see Figure 2. Through the use of these easements the selected contractor was able to fuse and stage all of the piping for the project. See Figure 4 and 5 below for pictures showing the pipe staging along the bank of the lake.

Figure 4 – Pipe Staging Area Along Bank of the Lake

Figure 5 – Pipe Staging Along Wynn Joyce Road

Traffic Control

With the project staging needs and City's desire to avoid additional easements, the traffic control plan was carefully coordinated with the City. Typically for HDD projects, a 200' x 200' staging area is preferred by the contractor to setup their equipment at the entry pit. This size staging area was not possible without acquiring additional temporary construction easements. In lieu of additional easements, the traffic plan design was such that the right lane of traffic was shut down so the contractor could stage their equipment end to end and allowed for moving their equipment.

Subcontractor Coordination

Based on the complexity of the HDD, FNI reached out to six HDD contractors during design for review and recommendations based on their experience in the industry. The feedback received was analyzed and incorporated accordingly into the design and ongoing internal quality control process. This was beneficial to the overall project performance and allowed for increased publicity and bidder interest.

Pipe Calculations

On any HDD design, there are multiple calculations to be made. The most critical calculations are described below.

Dead and Live Loading
This calculation is necessary for any pipeline design, but was more crucial on this project due to the 50 foot depths the pipeline would be installed at. Live loading was negligible at these depths, therefore the dead loading governed the calculations. The results were then utilized to select the proper pipe class, resulting in the use of AWWA C906, DR7 HDPE pipe. Because of the heavy walled HDPE, the pipeline was upsized to a 14-inch diameter pipeline for the HDD section to maintain the required 10-inch inner diameter for capacity needs.

Maximum Pulling Forces
The maximum pulling force is a value that is provided by the Plastic Pipes Institute for HDPE in which the max force that can be applied to the pipe during the pullback process is determined. The maximum allowable pulling force is a factor of the diameter, pressure class, HDPE material properties, and the amount of time the force will be applied. This value is then provided to the contractor in the contract documents. The HDD contractor is required to monitor the pull forces the drill equipment is inducing during the pullback process and required to stay within this limit to prevent permanent damage to the pipe. This value should also be compared against the expected pull forces that will be required to pullback the pipe which is a factor mainly based on the pipe sizing and length of the HDD. If the calculated pull forces exceed that of the selected pipe size and class, the pressure class should be increased or even possibly the diameter.

Bending Radii
A maximum bending radius should be provided in the design drawings or specifications to prevent the contractor from invoking too severe of a curve or bend on the pipe during the HDD process resulting in permanent damage to the pipe. This radius will be based on either the limits of the carrier pipe material or the steel drill stem rod utilized. The pipe manufacturers should be consulted during design to determine the bending limits of the carrier pipe, especially for plastic pipe. As a general rule of thumb, the steel drill stem rod can withstand a bending radius of 100 times the outside diameter. The HDD contractors provided guidance and confirmation of the sizing of the drill stem rod. The lesser of these radii should be provided as the bending radii limit for the HDD contractor.

Entry/Exit Pit Angles
The entry and exit pit angles are an industry standard range. For entry pit angles the allowable range is 8 to 10 degrees, while the exit pit angle range is 10 to 12 degrees. By adhering to these angles, it helps reduce the stress on the pipe during the installation process.

Additional Specification Language
For any HDD project, a thorough technical specification is required to insure a successful and accurate installation. At the time of the project, FNI's HDD technical specification was primarily written for pressure pipe installations. The following items were added to the standard specification to account for the gravity aspects and the added length of the HDD.

Vertical Control
The original specification allowed for a drill deviation up to 4 percent horizontally or vertically over a length of 100 feet. This vertical accuracy requirement was not sufficient for a gravity wastewater application. Therefore, the following statement was added:

"Vertical deflections for wastewater lines may be permitted as long as the slope is continuous (no sags), meets Texas Commission on Environmental Quality minimum slope requirements, and will flow by gravity with no surcharge and into existing or proposed downstream improvements."

Periodic Pressure Relief and Sight Holes
In conjunction with the increased vertical accuracy requirements, vertical sight holes were required every 500 linear feet along the HDD and approximately 12-inches to 18-inches in diameter. The purpose of the holes was to allow the contractor a way to perform a vertical measure down as the HDD progressed, and provide a point for pressure relief along the HDD. The periodic holes were very helpful for pressure relief, but did not meet the design expectations in the field. This is further discussed in the Lessons Learned portion of the paper.

Mud Engineer Requirement
Due to the length of the HDD, the requirement for the contractor to provide a mud engineer was added to the specifications. It was required that the mud engineer be present onsite during all drilling and reaming operations. The engineer was required to take periodic samples of the drill mud to confirm the quality of mud was within spec and would not cause any increased friction issues during the pullback phase, ultimately leading to possibly overstressing the pipe.

TCEQ Variance
Due to the long and deep trenchless sections, intermediate manholes were not provided for the trenchless section, therefore a variance was requested from TCEQ for violating the requirement that manholes are to be spaced every 500-feet. This variance was approved by TCEQ due to the 50 foot depths of the pipeline along with the fact fused HDPE was utilized therefore leaving no standard pipe joints or potential infiltration issues.

LESSONS LEARNED

Allow for Grade Flexibility
On each end of the HDD section there was a small amount of open cut installation in the shallower areas and connection points. At both open cut locations, the grade was set at TCEQ minimums to allow for the steepest slope possible for the middle HDD installation section. The thought process with this approach was to keep the HDD section as steep as possible in the event there were some slight vertical deviations. This would help prevent any sags or capacity issues.

The selected contractor chose to drill from the upstream end to the downstream end to take advantage of the large staging area near the lake. At the conclusion of the drilling operation, it was discovered that the drill was close to 1-vertical foot low on the downstream end. Prior to proceeding with the pullback phase the contractor was asked to suspend operations so that a resolution could be determined for the grade bust. The remedy was fortunately simple. The contractor was required to open cut upstream into the HDD section approximately 100-linear feet, totaling 450-linear feet of open cut installation at the downstream connection point. In this section, they were required to upsize the pipeline one pipe diameter to take advantage of the flatter grade allowances for this new pipe size, thus correcting the grade bust.

The lesson learned from this situation was to allow for some grade flexibility on the downstream connection in the event there is a deviation as was experienced on this project. If a small amount of vertical buffer (i.e. design the open cut section steeper) had been accounted for in the design, the above situation may have been a non-issue.

Daily Drill Log

The second major lesson learned was not enforcing the contractor to provide daily drill logs as the specifications required. The drill logs were requested, but not provided until completion of the drill and discovery of the above-mentioned grade bust. After review of the drill log, it was found that the vertical deviation occurred within the first 700 linear feet of the drill. If this had been discovered during the drilling operation it could have been adjusted prior to proceeding with the remainder of the drilling operations.

Vertical Sight Hole Shortcomings

As mentioned above, the contractor was required to provide periodic vertical sight holes along the HDD. One of the main goals of the holes was to provide a way to physically measure down to confirm the drill was on grade per the drawings. This proved to be an unsuccessful theory during construction due to spoil material and drill mud constantly filling the holes and making it impossible to get an accurate measured depth. Although the holes were unsuccessful from a measuring aspect, they did prove to be very beneficial for relieving pressures during the drilling operations and would still be recommended on larger and longer bores of this nature.

HDD Inadvertent Return Provisions

Another area that we will provide more emphasis on future HDD projects is requiring the contractor to provide a containment plan for inadvertent returns ahead of commencing drilling operations. Fortunately, the selected HDD contractor was properly prepared to handle any issues that arose. There were a couple of inadvertent returns during the HDD, along with some losses at the sight holes, but the contractor effectively barricaded and diverted any flows to prevent damage to neighboring development and homes.

Testing Provisions

The last lesson learned was in relation to the testing procedures to be used after the carrier pipe had been pulled into place. Standard testing procedures for gravity wastewater applications were specified in the contract documents (i.e. mandrel, low pressure hydrostatic, air testing, etc.). Due to the length of the HDD and no access points, these standard testing methods could not be utilized. As a result, the contractor was required to water jet the line and provide TV documentation that there was not ponding or pipe deformation along the pipeline. No issues were found during the revised testing procedure. On future projects, high pressure hydrostatic testing will be required in addition to the above procedures.

CONCLUSION

The decommissioning of the Wynn Joyce Lift Station and installation of the correlating pipeline infrastructure proved to be a challenging project in both design and construction, but has been considered a successful project by both the City and FNI. The pipeline was placed into service in the Spring of 2016 and the troublesome lift station was removed from service. To date, the pipeline has been functioning as designed. The valuable lessons learned from this project will be applied to future projects of this nature.

REFERENCES

ASCE HDD design guideline task committee, "Pipeline Design for Installation by Horizontal Directional Drilling," American Society of Civil Engineers, Reston, Virginia

Plastic Pipe Institute, "Handbook of Polyethylene Pipe," The Plastic Pipe Institute, Washington, DC

Najafi, Mohammed; Gokhale, Sanjiv, "Trenchless Technology," McGraw Hill, New York, New York

Sliplining Rehabilitation Approaches for Large Force Mains

Daniel I. Garcia[1], Tom Brzezinski[2], Daniel LaCross[3], Miguel Pichardo[4]

[1] Project Manager, Wade Trim, Coral Gables, FL

[2] Executive Vice President, Wade Trim, Tampa, FL

[3] Project Manager, Ric-Man Construction Florida, Deerfield Beach, FL

[4] Program Manager, Miami-Dade Water and Sewer Department, Miami, FL

ABSTRACT

As the largest water and sewer utility in the southeastern United States, the Miami-Dade Water and Sewer Department (MDWASD) serves nearly 2.3 million residents and thousands of visitors daily. The poor condition of MDWASD's system led to several United States Environmental Protection Agency (USEPA) violations and a moratorium on new development. During the next two decades, Miami-Dade County plans to invest approximately $12.6 billion to enhance and upgrade MDWASD's infrastructure under three major programs: Pump Station Plan, Ocean Outfall Legislation, Consent Decree, as well as a comprehensive Capital Improvement Plan (CIP). As part of the Consent Decree Program, an emergency repair design-build project was performed for MDWASD in the aftermath of a catastrophic failure of a section of the interceptor. The project included performing sliplining repairs on a 3.5-mile span of large 72-inch diameter force main pipe to stabilize the interceptor and restore its design life without massive surface disruptions. Much of the project transects highly-trafficked areas and residential neighborhoods. The alignment also bisects several major highways and a major railroad. To minimize disruption to traffic and neighboring residents, an HDPE liner system was employed for the repairs. Design was completed in six months and construction was completed in eight months, for a total of $14.85 million. This paper explores the benefits of the design-build delivery method, how the sliplining method was key to the success of the project, and why the sliplining method is well-suited for the scheduling rigors of typical design-build projects.

KEYWORDS

Design-build, Sliplining, Miami-Dade Water and Sewer Department (MDWASD), HDPE, Force Main, Host pipe, Bypass, Permitting

INTRODUCTION

As the largest water and sewer utility in the southeastern United States, Miami-Dade Water and Sewer Department (MDWASD) serves nearly 2.3 million residents and thousands of visitors daily. As a result of years of underfunding from Miami-Dade County tax revenues, diversion of funds allocated to MDWASD to cover budgetary shortfalls in other County departments, and a water use fee rate substantially lower than the national average, water and wastewater infrastructure has become dilapidated and accident-prone. MDWASD infrastructure woes are underscored by Southeast Florida's vulnerability to climate change and sea level rise.

The poor condition of MDWASD's system led to several United States Environmental Protection Agency (USEPA) violations as listed below:

- Between December 1, 2007, and October 6, 2012, Miami-Dade County reported to USEPA numerous sanitary sewer overflows (SSOs), totaling more than 50 million gallons. Of these SSOs, more than 28 million gallons of untreated wastewater reached navigable waters of the United States. Miami-Dade County's SSOs are not authorized by any NPDES permit and those that reached waters constitute violations of Section 301 of the CWA, 33 U.S.C. § 1311. Additionally, all of these SSOs are indicative of violations of the operation and maintenance requirements found in the NPDES Permit issued to Miami-Dade County by USEPA for the Central District WWTP (FL0024805) and in the permit for the North District WWTP (FL0032182-001).
- Miami-Dade County has experienced numerous violations of the effluent limits contained in the NPDES Permit for the Central District WWTP.
- Numerous operation and maintenance violations at Miami-Dade County's Central District WWTP, which included treatment equipment being out of service (e.g. influent grit chambers, oxygenation tanks, and/or final settling tanks) and failure to meet the pretreatment sampling and analysis requirements.
- Imminent and substantial endangerment claims pursuant to Section 504 of the CWA, 33 U.S.C. § 1364, to health and welfare of persons, as well as irreparable injury to human health, waters, and property, including animal, plant, and aquatic life of the state, due to the numerous SSOs and due to the continued threat of failure of Miami-Dade's aged and deteriorated force mains, including the deteriorated 54-inch force main underneath the Government Cut portion of Biscayne Bay (the Government Cut line) that conveys wastewater from the City of Miami Beach to the Central District WWTP. [1]

As a result of the above-mentioned violations, federal and state government officials lodged a lawsuit against Miami-Dade County. As part of the lawsuit settlement, Miami-Dade County committed to spending $1.6 billion repairing its aged sewer system. Funding for the repairs was appropriated via user fee rate increases that started in Fiscal Year 2016/2017. Rate increases will total 33% and will be rolled out gradually over five years. At MDWASD's Industry Forum held in 2014, the following statistics were presented on MDWASD's water and wastewater infrastructure. **Tables 1 and 2** summarize water and wastewater infrastructure assets managed by MDWASD.

Table 1: Existing MDWASD Water Infrastructure Assets[2]

Description	Amount
Large Regional WTPs	3
Small WWTPs	5
Treated Water Storage Tanks	11
Water Supply Wells	100
Piping (miles)	7,490
Fire Hydrants	39,000
Valves	127,000
Water Meters	455,000

Table 2: Existing MDWASD Wastewater Infrastructure Assets[2]

Description	Amount
Wastewater Treatment Plants	3
Ocean Outfalls	2
Deep Injection Wells	21
Piping (miles)	6,277
Pump Stations	1,043

MDWASD estimates that hundreds of millions of gallons of water are lost every year due to leaks in its water transmission infrastructure. Breaks in transmission infrastructure are rarely detected until they blow out. These catastrophic failures of various water and sewer transmission lines result in emergency repairs that are both expensive and cause great inconvenience to the public. Due to capacity constraints of MDWASD's current system, a moratorium for any future developments was enforced by Miami-Dade County. MDWASD estimates will require $12.6 billion to overhaul the County's existing infrastructure of pipes, treatment plants and pumping stations, as shown in **Table 3**.

Table 3: Age of MDWASD Infrastructure Assets[2]

Age of Infrastructure	Percent	Miles of Pipeline	No. of Pump Stations	No. of Valves, Meters, & Hydrants
Less than 25 years old	26%	3,579	271	161,460
25 to 50 years old	42%	5,782	438	260,820
Older than 50 years	10%	1,377	104	62,100
Unknown age	22%	3,029	229	136,620

In addition to the above the following necessary repairs for large MDWASD infrastructure assets have been identified, as shown in **Table 4**.

Table 4: Proposed Repairs to MDWASD Large Infrastructure Assets[3]

Description	Planned Improvement
Large Regional Water Treatment Plants	**Preston and Hialeah Water Treatment Plant (WTP):** Retrofitting of plants to operate under more stringent standards to address reclassification of the Northwest Wellfield (NWWF), due to potential contamination from nearby mining activities
Wastewater Treatment Plants	Proposed retrofits to existing WWTPs: • 60% reuse target for existing WWTPs with ocean outfalls (NDWWTP, CDWWTP) • Flows diverted from outfalls required to meet High Level Disinfection of treatment • Improvements to address sea level rise (3 feet by 2075) and storm surge • SDWWTP will replace three of the five small treatment plants currently operated in South Miami-Dade New WWTP proposed in West District (West Kendall)
Deep Injection Wells	Remaining flows from WWTP with ocean outfalls (NDWWTP, CDWWTP) to be disposed of via deep injections wells or outfall backup

During the next two decades, the County plans to invest approximately $12.6 billion to enhance and upgrade MDWASD's infrastructure using state-of-the-art technology. MDWASD's Multi-Year Capital Improvement Plan is comprised of projects under three major programs:

1. Pump Station Improvement Plan
2. Ocean Outfall Legislation
3. Consent Decree

In addition to the above programs, MDWASD is currently implementing a comprehensive Capital Improvement Plan (CIP) for numerous water and wastewater infrastructure projects. These improvement projects will meet the service needs of system customers, accommodate future growth, and comply with Federal and State regulations. Major projects under this program include:

- **Biosolids Processing Facility (BPF):** This facility will be designed to receive and process biosolids from the North District Wastewater Treatment Plant (NDWWTP), Central District WWTP (CDWWTP), and the South District WWTP (SDWWTP) and will allow MDWASD to produce a higher quality biosolids product, thereby lowering disposal costs.
- **South District Wastewater Treatment Plant (SDWWTP):** A design, build, finance, operate, maintain contract with a private entity is being considered for upgrades to the SDWWTP (including the BPF), as well as managing the facility after construction is complete
- **Brickell Basin Sewer Improvements:** Construction of Pump Station No. 3 and additional sewer lines to increase capacity for downtown Miami.

1. Pump Station Improvement Plan

The Pump Station Improvement Program consists of managing the upgrades to the MDWASD wastewater collection and transmission system that includes upgrades to the sanitary sewer collection system, pump stations, and force mains. Per USEPA, all pump stations are required to meet an established nominal operating time of less or equal to 10 hours per day, as well as other established criteria. Presently there are 112 pump stations that do not meet the operating criteria. The total budget for the Pump Station Improvement Program is $215 million and spans from 2014 to 2020, consisting of improvements to 111 pump stations, 27 force mains, and 78 gravity line repairs.

2. Ocean Outfall Legislation

In 2008, the Florida Legislature approved and the Governor signed a law requiring all wastewater utilities in Southeast Florida using ocean outfalls for disposal of treated wastewater to reduce nutrient discharges by 2018, cease using the outfalls by 2025, and reuse 60% of the wastewater flows by 2025. The total cost of the Ocean Outfall Legislation program is estimated to exceed $3.3 billion and consists of improvements to three existing wastewater treatment plants, construction of a new wastewater treatment plant in western Miami-Dade County, upgrades to several existing pump stations, and improvements to pipeline interconnectivity and capacity. The locations of new and existing wastewater treatment plants are identified in **Figure 1**.

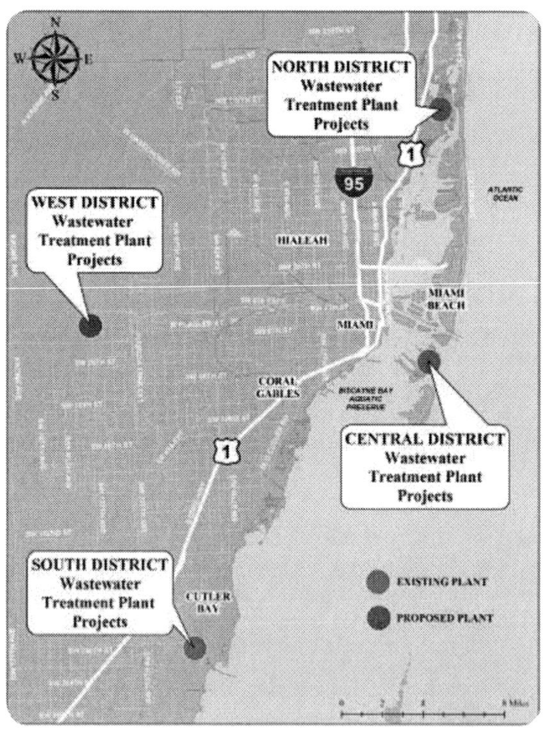

Figure 1: Wastewater treatment plant locations

Table 5 presents a summary of existing operations of MDWASD's wastewater treatment plants (WWTP) and planned improvements.

Table 5: Planned Improvements for MDWASD WWTPs[3]				
Wastewater Treatment Plant (WWTP)	**Location**	**Capacity (MGD)**	**Existing Operations**	**Planned Improvements**
North District (NDWWTP)	North Miami Beach	120	• Ocean outfall • On December 2008, NDWWTP started using injection wells to remove flow from outfall	• Reduction of nutrient discharges through ocean outfall by diverting flows away from NDWWTP • Infrastructure needed to reverse flows in collection system from east to west
Central District (CDWWTP)	Virginia Key	143	• Ocean outfall • Processing through gravity thickeners and anaerobic digesters • Disposal as Class B biosolids when weather and site availability permit or landfill • NDWWTP solids also treated; pumped to the CDWWTP headworks or gravity thickeners	• Deep injection wells • Remove nutrients from dewatering, instead of sending back to head of plant • Floridan aquifer recharge • Infrastructure needed to reverse flows in collection system from east to west
South District (SDWWTP)	South Miami Heights	131	• Processing through gravity thickeners and anaerobic digesters • Disposal as Class B biosolids • Weather permitting, biosolids are composted on site using the aerated static pile process, qualifying for Class AA marketing and distribution	• Reuse to FPL • Deep injection wells • Floridan aquifer recharge • Construct a regional Biosolids Processing Facility (BPF) to receive and process biosolids from NDWWTP, CDWWTP, and SDWWTP
West District (WDWWTP) *	Kendall West	102	Not applicable	• Reduce flows to the NDWWTP and CDWWTP • Reduce potential storm surge risks with existing WWTPs via construction of new West District WWTP • Land acquisition required for WWTP and transmission system

* Proposed plant

3. Consent Decree (CD)

On May 21, 2013, the Miami-Dade Board of County Commissioners authorized the execution of a Consent Decree between Miami-Dade County, the United States of America, the State of Florida, and the Florida Department of Environmental Protection (FDEP), in which MDWASD committed to making improvements to its wastewater collection and treatment system totaling $1.6 billion over the next 15 years. Included in the program are 82 capital projects at three wastewater treatment plants, 33 pump stations, and numerous pipelines. The proposed settlement resolves unpermitted discharges of sewage, NPDES permit effluent violations, and imminent and substantial endangerment claims due to sewer spills and the threat of a 54-inch sewer line under Biscayne Bay breaking. **Figure 2** shows a cluster of projects under the Consent Decree Program in the northern part of Miami-Dade County. CD Force Main Project 4.4 is the focus of this paper.

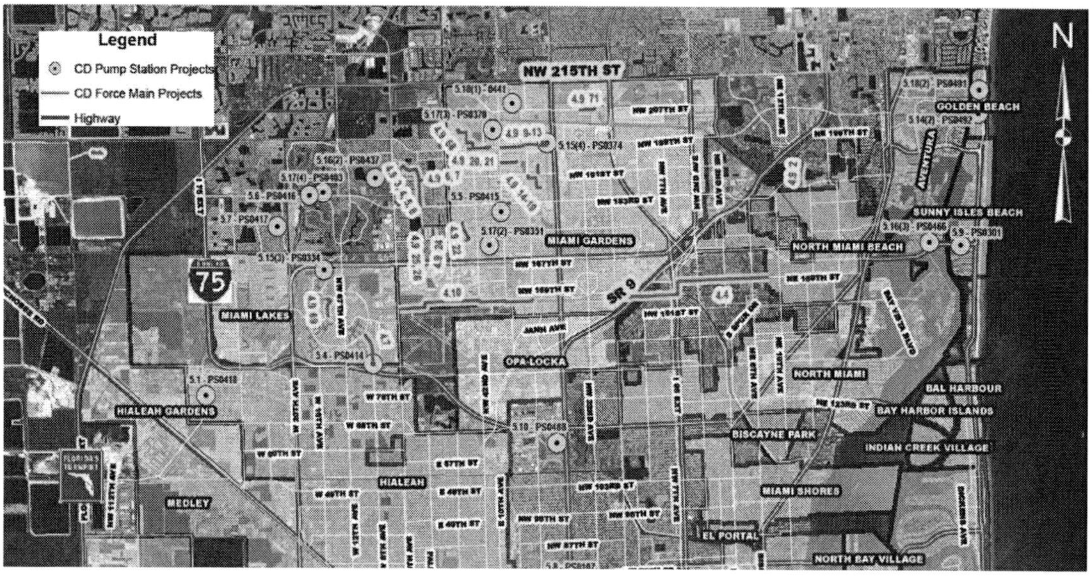

Figure 2: Consent decree projects in northern Miami-Dade County[4]

The Project

The rehabilitation of 3.5 miles of 72-inch sanitary sewer force main along NW/NE 159th Street is under the Consent Decree Program. After a catastrophic failure of a portion of the interceptor in 2010, MDWASD completed emergency repairs to a 1.5-mile section of the force main located between NW 32nd Avenue and NW 17th Avenue, just west of the location of the project segment, which is the focus of this paper. Due to concerns over the remaining sections of the force main, MDWASD commissioned the design-build team of Ric-Man Construction Florida/Wade Trim (DB) to continue repairs between NW 17th Avenue and NE 10th Avenue. Notice to Proceed (NTP) was given on March 10, 2015. On November 9, 2015, bypass force mains, diverting flows away from the project section of 72-in force main, were put into operation for repair work to begin. The design-build team is depicted in **Figure 3** (on the following page).

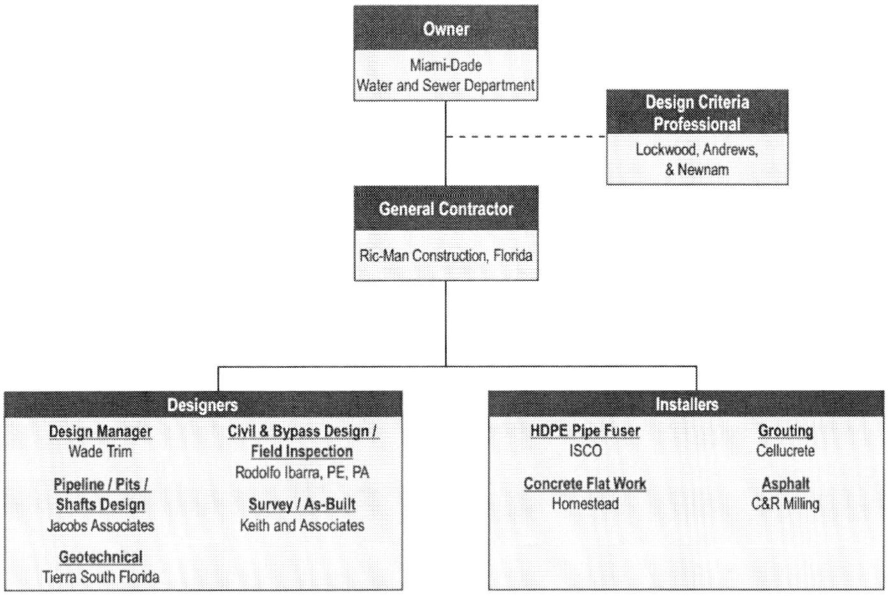

Figure 3: Ric-Man Construction / Wade Trim Design-Build Team

Project tasks included overall design, hydraulic modeling, permitting, construction management, and project certification effort, as well as civil/site engineering, maintenance of traffic, and development of insertion/pull pit design construction plans. Coordination with a variety of municipalities and permitting agencies was required to obtain a total of 18 permits from 11 regulatory agencies. These permitting agencies are described in **Table 6**.

A high-density polyethylene (HDPE) pipe liner system was selected, whereby the original 72-inch pipe remains in place, and is lined by pulling fused HDPE pipe through the existing pipe. This method was used to minimize disruption to traffic and residential neighborhoods along the alignment that crosses the cities of North Miami Beach and Miami Gardens, unincorporated

Table 6: Permitting Agencies Reviewing Project Submittals

Permitting Agency	Review Role
SFWMD	Canal right-of-way and dewatering permits
MDRER	Certification for pipeline commissioning and pipeline testing
FDOT	State Road right-of-way permits
SFRTA	Railroad right-of-way permits
City of North Miami Beach	Coordination with City on relocation of water main conflicts
City of Miami Gardens	City right-of-way permits
Miami-Dade County	Maintenance of traffic permits
MDWASD	Regular progress meetings, answers to RFIs, emergency meetings, permitting assistance when needed, and public information
USACE	Use of navigable canal waterway permits
FDEP	• NPDES Stormwater Construction Letter of Intent • Concurrence on Wastewater Collection System Commissioning (MDRER operating authority for commissioning)

areas of Miami-Dade County, several major highways, canals, and a South Florida Regional Transit Authority (SFRTA) railroad track.

The project map shown in **Figure 5** illustrates all major crossings along the project corridor and municipalities affected by the project.

Fused into a complete section for each pull, the HDPE was pulled through insertion pits at various lengths, some exceeding 3,000 feet. In addition, 14 manholes that were deteriorated, leaking, or structurally unsound were repaired and sealed.

Figure 4: HDPE liner being pulled through existing host pipe

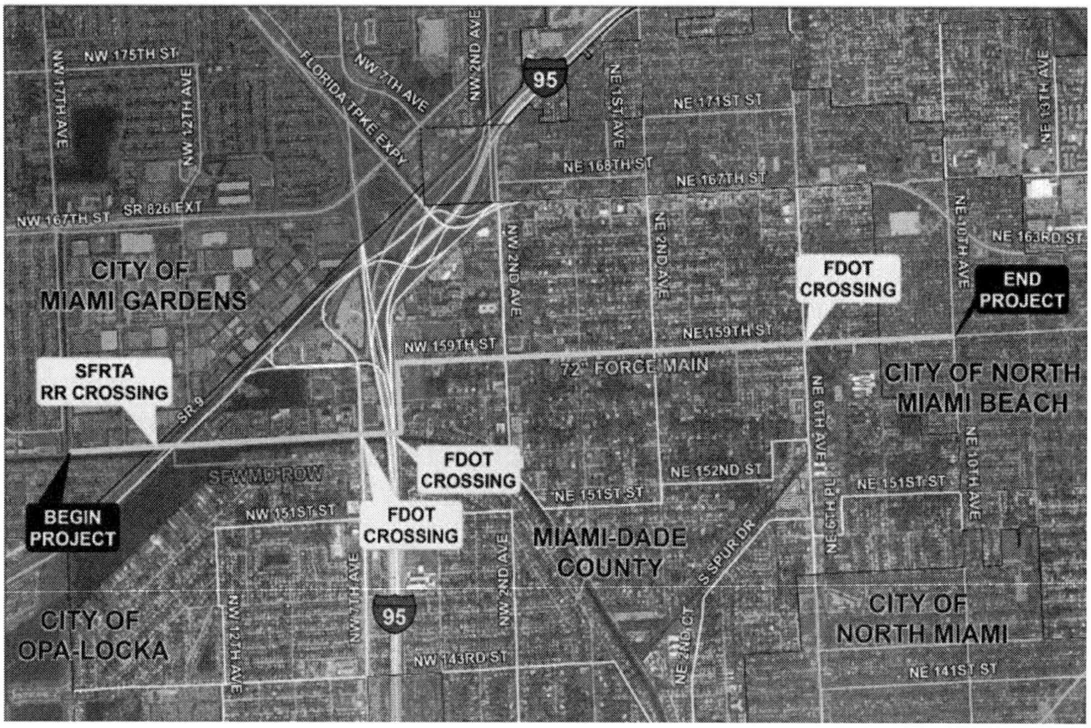

Figure 5: Major crossings of project

Objective(s) of the Work

This paper demonstrates how an alternative project delivery method, specifically design-build, can be a viable project delivery alternative to rehabilitation projects for water and sewer projects. Additionally, this paper demonstrates how the sliplining construction method can be used for rehabilitating large-diameter pipeline projects for densely-populated areas like those found in Miami-Dade County. Also explained are the strategies used to address the project's regulatory requirements in a timely manner to ensure the project remained on schedule.

METHODOLOGY

The Design-Build Delivery Method

The design-build delivery method has many benefits when compared to the traditional design-bid-build project procurement process. The most important advantages are those related to schedule, risk mitigation, communication, and added value.

Schedule

An overarching challenge for the design-build team was operating under a compressed schedule. The Design-Builder was given 14 months to complete design and construction of the project to meet the Consent Decree deadline. Three design submittal phases were used for the project as identified in the summary below. The design was completed in six months and construction was completed in eight months. Key dates of design submittals and construction are shown in **Table 7**.

Table 7: Project Milestones

Submittal Stage	Date
Notice to Proceed	March 10, 2015
60% Design	June 4, 2015
Final Design	July 20, 2015
Issued for Construction (IFC)	Sept. 4, 2015
Start Construction	Nov. 9, 2015
Substantial Completion	May 5, 2016
Final Completion	June 15, 2016

The Design-Builder was tasked with completing the design and obtaining all permits within six months from receiving a project NTP. Construction was completed during the dry season, between November and May, so wet weather flows would not overload the bypass system, which was installed to isolate the project section of the 72-inch force main. Coordination between permitting agencies and affected municipalities added an additional layer of scheduling coordination under the tight time frame. Maintenance of traffic (MOT) plans were developed for neighborhood streets, various state transportation agencies and a railroad company, via the Florida Department of Transportation (FDOT), SFRTA, and Miami-Dade County Public Works Department.

Risk Mitigation

Another important benefit of a design-build delivery method is the allocation of risk to parties that are most apt to address the risk. A risk register was a tool created by the design-build team for the purpose of communicating potential risks from the design and construction of the project. Among the potential risks identified were the following:

1. Unexpected geotechnical/soil conditions
2. Receipt of faulty materials
3. Permitting issues
4. Weather conditions, e.g. hurricane late or early in the season

For each risk item the risk register provided information that included:

1. Potential cause
2. Potential consequences
3. Likelihood of occurrence (unlikely to possible)
4. Potential impact of occurrence (low to high)
5. Project repercussion type (cost or schedule)
6. Design phase measures to minimize/eliminate risk of occurrence
7. Construction measures to mitigate or reduce consequences of occurrence

8. Mitigate likelihood of occurrence (unlikely to possible)
9. Mitigate potential impact of occurrence (very low to very high)

Risk scores were assigned to each risk item that showed a value for unmitigated and mitigated risk. This scoring method provided the Design-Builder with a measure of how critical the item was and how much control the design-builder had in managing the risk. The higher the score for unmitigated risk, the riskier the identified risk item. The bigger the difference between unmitigated and mitigated risk, the more benefit the project will have by allocating resources to address the risk.

Communication

During the design phase of the project (March 2015 to November 2015), regular bi-weekly progress meetings were held between the Owner and the Design-Builder to address questions regarding project requirements and Owner preferences on design, as well as answer any pending questions sent to the Owner via Requests for Information (RFIs) for clarification of Project Technical Specifications.

SharePoint was used as the software platform for distribution of RFIs and managing responses from the various parties reviewing RFI questions. Several emergency meetings were held with "all hands on deck" during construction, where solutions to issues with municipalities were discussed between MDWASD, the contractor, and municipal staff.

One such meeting was held on December 5, 2015, regarding higher than normal operating pressures experienced between lift stations serving the City of North Miami Beach. Unseasonably high rainfall events late in the wet season adversely affected operability of municipal lift stations feeding flows into one of the bypass force mains. Key project staff from MDWASD, Ric-Man Construction, Florida, and Wade Trim met with City staff to discuss solutions to issues faced by the City's wastewater system. Communication between all stakeholders was critical to arriving at a quick resolution to the matter at hand.

Responses by the Owner and Design Criteria Professional via bi-weekly meeting discussions or RFIs, as well as concerns raised by municipalities, were incorporated into the design and construction quickly in order to meet the aggressive design schedule, while at the same time addressing needs of all affected parties, e.g. municipalities and permitting agencies.

Added Value

The design-build delivery method allowed the Owner to incorporate several changes and additions to the project after construction began, adding significant value to the final project. Several of the requests came during the construction phase, after final construction documents were issued. Value-adding elements incorporated into the design during later design submittals and construction included:

1. Addition of two 12-inch force main connections to accommodate future connections for areas north and south of the project
2. Addition of a 63-inch by 48-inch connection to accommodate a future 48-inch plug valve connection
3. Decreasing bypass force main length at North Miami Avenue, rather than bypassing the line to the east, approximately half of the length of the entire project

After drawings were issued for construction, additional changes were incorporated to address construction safety and Owner comments. These changes included:

1. Incorporating a 20-inch connection for anticipated future connections
2. Relocating water mains at several locations due to field conflicts or design changes
3. Installing a thrust restraint block to secure an existing 60-inch valve during work at the eastern project limit

PROJECT CONSTRUCTION

Sliplining Method to Rehabilitate Deteriorated Wastewater Pipeline Infrastructure

At the core of the success of the project was the speed with which the rehabilitation was completed, through the use of sliplining method. One disadvantage of using the sliplining method was the loss of capacity, because of the reduced internal diameter of the liner within the existing host pipe. This reduction in the piping internal diameter, however, was outweighed by the speed with which sections of pipe could be lined via a combination of insertion and receiving pull pits and the ability to butt-fuse the HDPE pipe liner used for the sliplining.

Rehabilitation Method Alternatives

During the bidding phase, a combination of HDPE sliplining and cured-in-place-pipe (CIPP) was contemplated for the methods of force main rehabilitation. The HDPE sliplining method was chosen for the initial 2 miles of the project due to the accessibility of the 72-inch force main and the availability of vacant land and right-of-way to lay out the HDPE pipeline required for installation.

Figure 6: Rehabilitation methods proposed at project bid

Since almost one-third of the project's eastern limits were located within dense residential area with limited available vacant land and/or right-of-way, a CIPP system was selected for the eastern portion of the project.

As a result of negotiations with local property owners and coordination with local community leaders to find safe and adequate storage areas for the HDPE pipe section, the Owner and the Design-Builder opted to employ the HDPE slipline method for the entire length of project.

Figure 7: Congested project corridor

The flexibility offered by the design-build delivery method allowed for significant changes like this to be made, even after the project was awarded.

The HDPE liner system required for the project was fused into a complete section for each pull. The HDPE was pulled through insertion pits at various lengths, some pull lengths exceeding 3,000 feet. **Table 8** shows the lengths of the pulls with their relevant insertion and pull work area locations.

Eleven pits were proposed for the HDPE sliplining work to optimize scheduling and reduce impacts to neighborhoods, roads, and railways. Locations of the insertion and pull pits can be referenced in the map in **Figure 8**.

Table 8: Slipline Pull Information

Pull No.	Direction	Work Areas	Distance (feet)
1	East	1 – 2	1,234
2	East	3 – 4	527
3	West	5 – 6	1,559
4	East	6 – 7	1,410
5	West	8 – 9	123
6	East	10 – 11	515
7	South	11 – 12	1,142
8	East	14 – 16	3,170
9	West	18 – 19	2,189
10	East	19 – 20	1,485
11	East	22 – 23	2,382

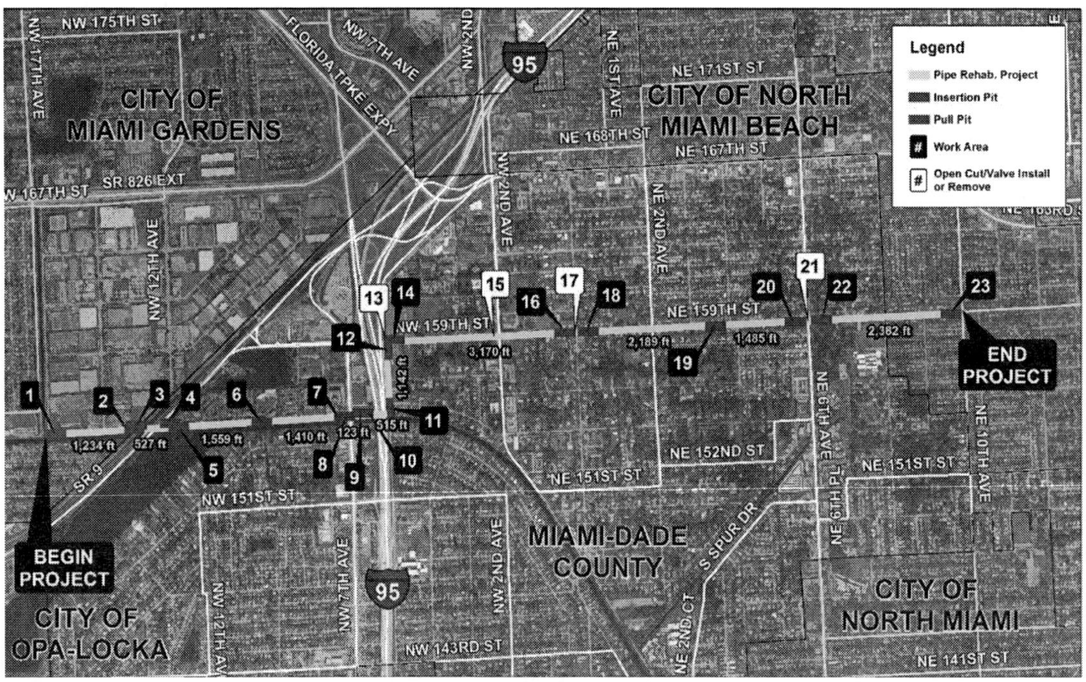

Figure 8: Location of insertion pits

Pit excavations were designed with support systems and working slabs in conjunction with dewatering wells to withstand earth and hydrostatic pressures, bottom heave, equipment loads, applicable traffic, and construction loads to allow safe insertion of the pipe without movement or settlement of the ground and prevent damage to adjacent structures, streets, and utilities.

The work also included:

1. Development of construction staging areas
2. Design and construction of pit excavations and excavation support systems
3. Ground stabilization or treatment
4. Material disposal
5. Control and disposal of groundwater and surface water
6. Placement of structural backfill and site restoration
7. Hydraulic analysis to develop a complex flow bypassing plans

The benefits of using HDPE/sliplining vs. CIPP/open-cut method for pipe rehabilitation include those related to complexity/ease of construction, schedule, safety, cost, and resilience. The HDPE/sliplining method reduces the number of insertion and receiving pits while allowing for longer pull lengths, which in turn reduces the MOT areas. The maximum length of a CIPP method rehabilitation section is 1,500 feet compared to 4,000 feet for HDPE. This results in higher safety, because it reduces the number of conflict/friction points between motorists, bicyclists, and pedestrians. HDPE also handles higher pressures: 60 psi for HDPE vs. 40 psi for CIPP. Savings in schedule and reductions in construction complexity ultimately translates to significant cost savings, as shown in **Table 9**.

Table 9: Benefits of HDPE / Sliplining Method

Advantage	Result	Benefit
Constructability: Maximum length of CIPP method section is 1,500 feet compared to 4,000 feet for HDPE	Reduces the number of insertion and receiving pits, allowing for longer pull lengths, thereby reducing footprint of MOT areas	**Cost:** Savings in schedule and reductions in construction complexity translate into significant cost savings **Safety:** Reduction of MOT areas reduces the number of conflict/friction points between motorists, bicyclists, and pedestrians
Resilience: HDPE operates under higher pressures (60 psi vs. 40 psi for CIPP)	Reduction in operation and maintenance costs	**Life-Cycle Cost Savings:** Cost savings not only realized in near term during construction, but also in long term from reduced O&M

Permitting

As with most projects of this scope, permitting requirements and the time lengths associated with the approval and certification of permits is one of the biggest unknowns in terms of scheduling. In this project, the design-build team had to obtain 18 separate permits managed by 11 separate agencies. See **Table 10** (on the following page) for a list of relevant permits with their estimated and actual approval achievement dates. The table also provides insight into the level of difficulty in obtaining issuance approval and certification of permits. Permits that were the most difficult to challenge are highlighted in red. Permits that were moderately difficult to obtain are highlighted in blue, while those that were least difficult are highlighted in green. A distinction is made between the process required to obtain permit approvals, and the process required to obtain project certification or to close-out the permit.

Table 10: Permitting Schedule & Application Level of Difficulty				
Permitting Agency	**Permit Name**	**Permit Approval Duration**		**Permit Approval Notes**
		Est.	Actual	
SFWMD	Water Use	30	17	–
	Right-of-Way	30	87	–
	Environmental Resource Permit	120	N/A	Exemption Granted
MDRER	Class V Dewater	21	131	–
	Right-of-Way	120	173	Miami-Dade County NW 17th Avenue canal crossing
	WW Collection / Transmission	21	178	72-inch force main: Submittal of monthly pressure reports at booster pump station
		21	133	8-inch force main bypass
		21	142	12-inch force main bypass
FDOT	Utility Permit: SR 7, SR 9, I-95	60	10	Approval of sliplining construction method
	Utility Permit: NE 6th Avenue	60	50	Open-cut permit across 6th Avenue along NW 159th Street
SFRTA	Right-of-Way	30	127	Transfer of railroad ownership from CSX to SFRTA
City of Miami Gardens	Right-of-Way	28	203	City ROW is on the eastern end of the project, which was scheduled for work months after NTP issued
City of North Miami	Right-of-Way	28	179	Hydraulic analysis required compensation for use of City's force main
Miami-Dade County	Right-of-Way	60	207	–
USACE	Nationwide Permit 12	60	N/A	Exemption granted
FDEP	NPDES Stormwater Construction	7	28	Letter of Intent

Level of Difficulty Green: Easy Blue: Moderate Red: Difficult

Permits that were more difficult to obtain were right-of-way permits, due to the need to provide all other permit approvals (Dewatering, Miami-Dade Regulatory and Economic (MDRER) Wastewater Collection/Transmission Construction, MDWASD certification) with applications. In regards to Project Certifications for project close-out, more difficult permits included those agencies with no formal certification process. Significant challenges arose with the final certification for MDWASD, due to as-built drawing requirements.

The best approach to ensuring timely issuance of permits involved the initiation of pre-application meetings with permitting agencies as soon as possible, and submitting permit applications as soon as minimum submission requirements were met. Another strategy involved engaging the Owner in discussions with permitting agencies, so they could support the timely review of permit applications or answer any questions from permitting agencies expeditiously. **Table 11** lays out the permitting schedule that was employed, as it relates to phase submittals.

Table 11: Permitting Application Submittal Schedule

Permit	Agency Reviewer	Phase			
		60% Design	90% Design	100% Design	Issued for Construction
Right-of-Way	SFWMD	✔			
Environmental Resource Permit [a]	SFWMD	✔			
Nationwide Permit (NWP) [12]	USACE	✔			
Domestic wastewater collection / transmission system: 72-inch	MDRER		✔		
Domestic wastewater collection / transmission system: 8-inch bypass	MDRER		✔		
Domestic wastewater collection / transmission system: 16-inch bypass	MDRER		✔		
Class III	MDRER		✔		
Right-of-Way	SFRTA		✔		
Utility Permit: SR 7, SR 9, I-95	FDOT		✔		
Utility Permit: NE 6th Avenue	FDOT		✔		
Right-of-Way [b]	City of North Miami		✔		
Water Use	SFWMD			✔	
Class V Dewater	MDRER			✔	
Right-of-Way [a]	City of North Miami Beach			✔	
Right-of-Way	City of Miami Gardens			✔	
Right-of-Way	Miami-Dade County			✔	
NPDES Stormwater	FDEP				✔
Project Certification	MDWASD				✔

Notes: [a] Exempted, [b] For use of City of North Miami force main for bypasses

RESULTS

The project met the scheduling deadlines imposed by the Consent Decree due to the use of the design-build method, early engagement of permitting stakeholders, the use of sliplining method to reduce MOT and constructability constraints, and immediate response to issues raised by affected municipalities. The structure of the design-build contract brings the Owner, Designer, and Builder into one unit, combining all key decision makers together into a single point of responsibility to address the various challenges that may arise during the design and construction phase of the project. Communication was focused between the Owner and the design-build team and other stakeholders, such as municipal and permitting agencies, who were brought in as needed.

The benefits of the design-build method not only generated cost savings (due to earlier knowledge of project costs, reduction of claims, and reduction of project charges), but also added value by allowing the introduction of design changes throughout construction. The flexible nature of the design-build contract also allowed for the allocation of risk to contractual parties best suited to handle the risk.

The reasons mentioned above led to the issuance of construction drawings within six months of NTP and accomplishment of Substantial Completion eight months after construction commencement, or one year after NTP.

SUMMARY & CONCLUSIONS

The sliplining construction method used to rehabilitate the force main resulted in rapid construction, reducing potential delays due to MOT challenges, and improving overall safety for all stakeholders and beneficiaries.

Key attributes of the design-build team that led to the successful completion of the project included:

- Focus on client needs
- Penchant for work quality
- Commitment to schedule
- Amenability in introducing design changes provided by the various project stakeholders
- Communication skills necessary to efficiently coordinate with all project stakeholders, including affected municipalities and permitting agencies
- Fully committed to providing support for the project whenever and wherever

The functionality of the repair design contributed to the rapid rehabilitation of a major sewer system arterial that directs flow from several municipalities and neighborhoods in northern Miami-Dade County to the North District WWTP. The additional connections added to the project also accommodated future connections, as MDWASD's system continues to expand.

ACKNOWLEDGMENTS
The authors wish to acknowledge the assistance of Daniel Mancini (Ric-Man Construction, Florida), Hector Meneses (Ric-Man Construction, Florida), Holly Kremers, PE (Wade Trim), and Dan Burden, PhD, PE (Wade Trim) for their valuable input on this technical paper.

REFERENCES

1. United States Environmental Protection Agency, Miami-Dade County Clean Water Act Settlement, https://www.epa.gov/enforcement/miami-dade-county-clean-water-act-settlement.
2. Miami-Dade Water and Sewer Department, Capital Improvement Plan – Overview, MDWASD Industry Forum 2014, Miami, Florida.
3. Miami-Water and Sewer Department, Ocean Outfall Legislation Draft Plan, June 5, 2013, Miami Florida.
4. Miami-Dade Water and Sewer Department. Consent Decree Projects, http://www.miamidade.gov/water/library/consent-decree-projects.pdf.

Risk Management for the Mishawaka 2,134 m (7,000-ft) Long Term Control Plan Tunnel

Gregg Sherry, P.E., M.ASCE[1] and Sarah Myers, EIT, M.ASCE[2]

[1]Brierley Associates
[2]Brierley Associates

ABSTRACT

Risk management is the primary objective in preparing underground construction documents. This paper will present the impacts and the mitigation strategies for the risk associated with the Mishawaka Long Term Control Plan Tunnel project. Successful risk management includes: The risks associated with design and construction; developing options to treat the goal of meeting cost and schedule targets; and assigning responsibilities to parties best suited to handle them. Multiple horizontal and vertical alignments were evaluated for the tunnel. The optimal horizontal alignment location was developed based on the fewest impacts to properties and structures.

KEYWORDS: Risk management, trenchless, long-term control plan

INTRODUCTION

A formal approach to risk management provides a framework for the identification, analysis, evaluation, and treatment of risks on construction projects. During the life of a construction project, many uncertainties may arise that can benefit from a procedure that identifies, assesses and treats risks. There are five main steps in the risk management approach developed for the tunnel/trenchless project elements: Establishing context of risk planning; identifying risks; analyzing risks; treating risks; and monitoring, reviewing and reporting on risks. The most notable risk management strategies considered for this project included a risk register, Geotechnical Baseline Report, project specifications, special payment provisions, contract delivery methods, and contractor prequalification. Risk management is an ongoing process during the life of a project. Risk management is the primary objective in preparing underground construction documents. Successful risk management includes:

- Identify the risks associated with design and construction;
- Develop options to treat the risks with the goal of meeting cost and schedule targets; and,
- Assign risks to parties best suited to handle them.

The tunnel is required as part of a larger, overall Long Term Control Plan (LTCP) developed by the City of Mishawaka, Indiana and its consultants. In this plan, a tunnel for storage and conveyance of combined sewage is identified as a key element of the program. The City of Mishawaka (City) entered into a Consent Decree with the U.S Environmental Protection Agency, U.S. Department of Justice and the Indiana Department of Environmental Management to implement a combined sewer overflow (CSO) LTCP to reduce the frequency of CSOs that are discharged to the waterways within the City. The tunnel project included two phases in the Consent Decree. Phase I of the project is from the Wastewater Treatment Plant (WWTP) to Main Street, and is required to be operational by December 2020. Phase II of the project is from Main Street to Merrifield Avenue, and is required to be operational by December 2022. The City's plan is to complete the project as a single phase, and meet the Phase I deadline of December 2020.

The proposed tunnel is generally described as running west from a new shaft at Merrifield Avenue, along 3rd Street, then curving to the northwest and terminating at a new shaft at the City's WWTP. The required inside diameter of the tunnel is 3,048 mm (120–inches) and the length of the tunnel is approximately 2,111 m (6,925 LF). It was recommended that an earth pressure balance machine and segments for the tunnel lining be utilized on this project. The figure below shows the tunnel alignment in red.

Figure 1. Tunnel Layout

SUBSURFACE CONDITIONS

The project site is located within the Kankakee Lacustrine Section in the Northern Lake and Moraine Region Physiographic Province. The region is comprised of glacial deposits from the Wisconsin glacial epoch that is characterized by massive rugged moraines occupied by lakes and broad moraines occupied by lakes and lacustrine plains. The region contains many small lakes. The Kankakee Lacustrine Section is a great system of sandy lacustrine plains, outwash plains, valley trains, and local enclosed till plains associated with a line of glacial drainage and ponding along the St. Joseph and other rivers. The thickness of soils is anticipated to be approximately 46 m (150 ft) and underlain by Ellsworth Shale bedrock.

The subsurface generally consisted of interbedded fine- and coarse-grained soil, and highly to slightly weathered shale with the occasional seam of siltstone. The density of the granular soil in the upper approximate 12 m (38 ft) was very loose to very dense. The consistency of the clay was generally hard with seams being stiff to very stiff. Groundwater was observed at depths ranging from approximately 1 m to 8 m (4.5 to 25 ft) below the existing ground surface.

HORIZONTAL TUNNEL ALIGNMENT

The original alignment originated at the WWTP and continued primarily under 4th Street. The original alignment considered a tunnel extending from the launch shaft to the southwest approximately 122 m (400 feet), continuing under Baker Street from Lincolnway to 3rd Street, prior to connecting the northeast-southwest oriented segment of 3rd Street to the primary tunnel drive under 3rd Street extending to Merrifield Avenue. Including the original alignment, four primary alignments were considered:

- Segmental alignment primarily under 4th Street
- Segmental alignment primarily under 3rd Street
- Curved alignment (west end) primarily under 4th Street
- Curved alignment (west end) primarily under 3rd Street

The design team considered two primary alignments between Merrifield Avenue and the WWTP. Both horizontal alignments considered a curved portion of tunnel near the westerly end at the launch shaft used to connect the straight segment along 3rd Street or 4th Street. The figure below shows the 3rd Street curve alternatives.

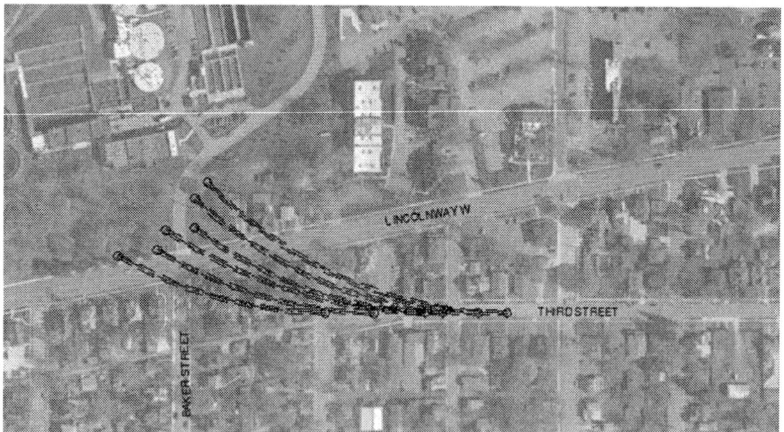

Figure 2. 3rd Street Curve Alternatives

The following outlines why the segmental alternatives were not selected.

- A curved alignment will eliminate the construction of two to three additional shafts, a significant cost savings to the project.
- A curve will eliminate the need for turntables used to align the tunnel boring machine and muck cars at shaft locations.
- A curved alignment provides a shorter tunnel, saving on overall construction costs.
- Design and installation of a large radius (> 305 m) curved tunnel is considered common practice and does not result in significantly higher construction costs per lineal foot.
- The curve eliminates the need for a 90 degree turn, a hydraulic benefit to the system.

Curve alternatives were developed to assist in determining the optimal location (fewest impacts to properties and structures). The recommended alignment is based upon an alternative of the 3rd Street option, which minimizes the impacts to private property (5 properties and 2 structures). The radius of this curve is approximately 305 m (1,000 feet). The 3rd Street alignment was selected in lieu of the 4th Street alignment as noted below:

- The total tunnel length of the recommended alignment at 3rd Street is 2,111 m (6,925 feet) whereas the 4th Street option is slightly longer at 2,171 m (7,122 feet). The shorter, 3rd Street alignment still meets project hydraulic goals.
- The 4th Street option requires a longer microtunnel section in Merrifield Avenue.
- The 4th Street alignment results in a greater number of impacted properties.

In order to avoid or reduce potential claims from affected property owners, properties on which structures are located directly over the tunnel will be purchased by the City in advance of construction operations. The City secures permanent subsurface easements for additional properties that are above the tunnel alignment. The straight segments of the two tunnel alignments considered are located below the city-owned Right of Way for 3rd and 4th Streets and should have little to no impact to private property or structures. The transition curve segments at the westerly end of the tunnel, however, require that the tunnel pass below private properties and some structures (residences).

VERTICAL TUNNEL ALIGNMENT

In addition to considering different horizontal alignments, the team also considered two different profile depths: 1) a shallower primarily gravity driven system; and 2) a deeper pumped option. Both designs will require some type of control structure at the WWTP; however, the deeper design will require a lift station, which will also control flows at the WWTP. Ultimately, the deeper solution was selected. This was primarily due to the significant decrease in risk to overlying infrastructure and more favorable subsurface conditions. The following includes additional details of the relative benefits of the pumped alternative in comparison to the gravity alternative.

- The pumped alternative contains less potential for adverse effects of settlement of existing below grade utilities and at grade infrastructure. In addition, due to more favorable soil conditions, a significant risk for major ground loss is likely avoided. The additional depth will allow for increased soil bulking prior to significant negative consequences.
- The deeper alternative places the tunnel horizon in more cohesive soils, which are more favorable to tunnel through due to increased stand-up time and less dependence on perfect machine performance. The added stand-up time will benefit the contractor by providing a greater opportunity to adapt means and methods to potential differing subsurface conditions during construction.
- An existing storm sewer system running perpendicular to the tunnel at Church Street posed a significant challenge for a gravity flow tunnel crossing due to the depth of the crossing pipe and its proximity to the tunnel under consideration. Using the gravity alternative, Church Street would require open-cutting or special considerations during tunneling which pose significant risk to the existing utility, and may impact an important transportation corridor.

The slope of the tunnel is planned at a 0.1% grade based upon hydraulic design input. This slope will provide adequate cleansing velocity. The deeper solution allows for a steeper slope than the gravity solution.

RISK MANAGEMENT

A risk is any event which is uncertain to occur, but is expected to negatively affect critical success factors if it does occur. The main steps in the risk management approach developed for the tunnel/trenchless project elements are as follows:

- Establishing context of risk planning involves forming project objectives and the critical success factors to meeting project objectives. Primary critical success factors usually include meeting cost and schedule goals for the project.
- Analyzing identified risks allows for the highest risks to be treated with greatest priority. The risk measure includes the probability of unwanted event occurrence and the severity of unwanted event consequences.
- Treatment of risks generally includes either avoiding risks, mitigating risks, transferring risks, or accepting risks.
- Throughout the project risks are monitored, reviewed and reported. The information summarized in a risk register, which is a live document that collects data risk management.

The most notable risk management strategies that the City is considering for this project include:

- Risk Register
- Geotechnical Baseline Report (GBR)
- Project Specific Specifications
- Special Payment Provisions
- Contract Delivery Methods and Contractor Prequalification

RISK REGISTER

Any summary of risks and opportunities recorded in a risk register should be considered a living document that is updated at agreed schedules to reflect current information. It was recommended that the risk register is vetted by all parties involved in the project during final design and updated on a regular basis during construction.

A risk register was assembled for construction of the Mishawaka Tunnel project. The project risk register is intended to serve as a vehicle to identify and summarize project risk and project opportunities in order to minimize and/or manage that risk.

In the effort to identify risks and prepare a risk register specific to the project, six separate sub-registers were prepared that included the following grouping of hazard types:

- Risks associated with approvals and permits: project delays and/or costs associated with obtaining permits and approvals from regulatory agencies and the City of Mishawaka.
- Stakeholder risks: general issues arising from the public, utility providers, and other Third Party interests.
- Design/Contractual/Finance risks: design delays, differing site conditions, and inadequate design.
- Logistics and construction risks: construction management, construction scheduling, differing site conditions, equipment and material delays, contractor capability and contractual liabilities.
- Safety and security: accident, injury, fire, flood, and site security.
- Operation and maintenance risks: system hydraulic performance, operational problems, structural issues, and excessive maintenance.

A quantification process is utilized to determine which risks are the most critical. Generally, each risk is assigned a probability rating from 1 to 5 as represented in Table 1.

Table 1. Risk Probability Rating

Probability Rating	Description
5	Very High (occurs every project)

4	High (occurs 1 in 10 projects)
3	Moderate (occurs 1 in 100 projects)
2	Low (occurs 1 in 1,000 projects)
1	Very Low (improbable)

Consequence ratings were developed for several criterions including financial, social environment, regulatory/legal, project schedule impacts, health/safety, operating/maintenance and natural environment. The financial risk and schedule impact consequence criterion are presented in Table 2.

Table 2. Consequence Rating

Consequence Rating	**Project Cost**	**Schedule Impact**
Very High	> $ 4,000,000	Delayed > 1 year
High	$500,000 to $4,000,000	Delayed 6 months to 1 year
Moderate	$100,000 to $500,000	Delayed 1 month to 6 months
Low	$50,000 to $100,000	Delayed < 1 month
Very Low	<$50,000	Delays not on critical path

Risk is commonly defined as failure probability times failure consequence. All potential risk hazards will be included in the register and will be prioritized based on the above ratings. The risk register contains the lists of hazards with their associated probabilities, consequences, mitigative efforts, control measures and responsibilities.

GEOTECHNICAL BASELINE REPORT (GBR)

A GBR establishes a baseline of geologic conditions to be encountered during construction. A well-defined, project-specific GBR can greatly reduce time consuming and costly litigation related to what constitutes a differing site condition. A GBR is a common and integral part of an underground project's risk management system, and is currently being completed for the Mishawaka CSO Tunnel project. Risk sharing is a common practice in tunnel construction; the Contractor accepts the risk of encountering what is baselined, and the Owner accepts the risk of potential costs associated with adverse ground conditions above what has been baselined. When a clear baseline is defined, Contractors tend to utilize lower contingencies in their bids relating to protection against unforeseen or unknown conditions. The GBR will be in general accordance with an ASCE document entitled "Geotechnical Baseline Reports for Underground Construction, Guideline and Practices," dated 2007. For the Mishawaka CSO project, a GBR would include, but not necessarily be limited to the following:

- Groundwater conditions
- Cobble and boulder quantity and strength
- Soil and bedrock strengths

- Soil conditioning
- Stickiness
- Abrasion
- Settlement
- Tunnel lining
- Ground behavior
- Shaft and tunnel construction

SPECIFICATIONS

Specifications carefully tailored to the project requirements and GBR are essential for successful project completion and risk management. Several of the specifications anticipated for the Mishawaka CSO project tunnel are listed below:

- Disputes review board
- Bid escrow documents
- Tunnel excavation by tunnel boring machine
- Shaft excavation and support
- Jet grouting
- Pipe installation and backfill grouting
- Precast concrete tunnel lining
- Contact grouting
- Geotechnical instrumentation and monitoring

In the event of conflicting information within the Contract Documents, the following order of precedence would be followed:

1. Specifications
2. Contract Drawings
3. GBR
4. Geotechnical Data Report

SPECIAL PAYMENT PROVISIONS

Risk associated with construction of the Mishawaka CSO Tunnel can be partially mitigated and managed by including project specific special payment provisions to address site or subsurface conditions with the potential of being encountered during construction. Competitive prices for these various items can be achieved by including these items in the bid schedule. When quantities are difficult to estimate, unit prices can be effective in defining costs associated with these provisions. The City is encouraged to include clear language in the General Conditions identifying special payment provisions so that these items can be deleted and not paid for if not encountered. Several examples of special payment provision items that could be included in the bid schedule related to construction of the Mishawaka CSO tunnel include:

- Ground improvement methods
- Dewatering and grouting at shaft locations
- Handling and disposal/control of contaminated ground

CONTRACT DELIVERY METHODS AND CONTRACTOR PREQUALIFICATION

Due to the size and complexity of the proposed construction, the Mishawaka CSO tunnel would benefit from value engineering and opinions regarding constructability during final design. A more detailed discussion regarding feasibility, constructability, means and methods, and risk will be critical in order to manage project risks, reallocate costs, and determine schedule contingencies during design and construction of the project. Early contractor involvement during design is recommended in order to mitigate risk and reduce differing site conditions and change order claims during construction.

Traditionally, a design-bid-build project delivery approach has been successfully utilized on projects similar to the Mishawaka CSO project. In addition, a design-bid-build approach is the City's preferred method of delivery for construction projects. Typically, this approach awards a design contract to a designer and a construction contract to a contractor after the completion of engineering and design. Design-bid-build usually requires the employment of construction management professionals to work with the contractor and represent the owner during project construction particularly on projects of this size and complexity.

The two major advantages of the design-bid-build project delivery method include: 1) being well-established, and well understood and 2) also providing the City with significant involvement during design and construction and control of the final product. The two major disadvantages of this traditional approach are: 1) a lengthy schedule due to the delivery method necessitating completing design prior to initiating construction; and 2) a lack of input from the contractor during the design phase. A traditional design-bid-build schedule will fit into the desired completion timeline for the CSO tunnel. A lack of contractor input during design has the potential to affect constructability, schedule determination, and the estimated cost for construction. Methods are recommended to overcome lack of contractor involvement during design, including scheduling design constructability and risk management workshops. These workshops should include external design and construction experts to discuss the project with the designer and owner. Site visits during design with potential contractors have proven beneficial; assuming the parties visiting the site are not precluded from bidding on construction of the project.

In addition, the City could benefit from requesting contractor qualifications (RFQ) prior to bidding for construction and selecting a contractor based on cost. The pre-qualification process would be completed prior to contract advertisement, with the list of qualified (sub)contractors listed in the contract documents. An RFQ can limit the contractor pool to firms with the proven ability to perform the proposed installation. In addition, this process separates price competition from evaluation of qualifications and reduces the potential to have bids, which are unresponsive relative to tunneling qualifications with the subsequent potential for protests.

The primary concern with the pre-qualification process is that it qualifies contractors several months in advance of the time of bidding and construction. Interest in the project from the different potential contractors is dynamic with the result that pre-qualified contractors may not be available when the project begins, and vice-versa. Based on recent project experience, we see only marginal benefit of contractor pre-qualification for this project, but recommend that it is considered.

Documents used by the contractor to prepare their bid can be escrowed to provide a baseline to back up assumptions and proposed methods and costs by the contractor in the event of a claim. They are opened only by joint consent of the contractor and owner to assist in claims resolution. For large projects, if mediation is used, it is beneficial to have escrow of bid documents.

The benefit of placing contractor bid documents in escrow is to assist with the resolution and potential avoidance of contractor claims. These documents can validate contractor claims about bidding assumptions and also prevent the contractor from claiming different assumptions later in the project. There are varying opinions as to the benefits of escrowing bid documents. In general, it is a step which is taken for larger projects and requires a moderate effort to implement and assure useful compliance by the contractor. A potential problem for public projects is that these documents, although stored in escrow, may be classified as public information with the potential loss of proprietary information by the contractor.

Dispute resolution can take several forms, including mediation, arbitration (either binding or not binding), and litigation. Mediation is a non-binding form of dispute resolution, which utilizes independent experts to render opinions. Depending on the job and the agreed upon procedures, the expert may be a single individual or a panel of several experts. Dispute resolution is not typically a component of City of Mishawaka contracts, however it could be considered further during the design phase.

Within the tunneling industry, a specific type of mediation panel termed a Dispute Review Board (DRB) has been developed and is frequently utilized. This is a board of experts (typically three) chosen jointly by the contractor and the owner to review and render opinions on claims and potential claims to avoid legal actions. The board members have periodic reviews and maintain a low-level of project involvement throughout construction. With this continual involvement, the board is in a position to rapidly respond and facilitate settlement of a potential claim. The primary difference between mediation and a DRB is that a DRB is established prior to any claims and is active throughout the duration of the project, whether or not there are any claims.

For this project, the benefits of a DRB would be quickly resolving contractor claims without litigation and in reducing the potential for spurious claims. However, the use of a DRB requires a substantial effort by the owner and the contractor. Setting up the board by choosing members and presenting the design and construction to the board is very important in having a board, which functions efficiently and fairly. If a DRB is used, the City should be prepared to be actively involved in the DRB process during the initiation phase and throughout construction.

CONCLUSIONS

The team utilized mechanisms for identifying, understanding impacts, and developing mitigation strategies for risks associated with the Mishawaka CSO tunnel project. The decision to select either the shallower gravity alternative or the deeper pumped alternative is a critical design decision and is anticipated to impact project risk significantly. The deeper pumped alternative is recommended and expected to significantly decrease overall project risk.

Two primary alternatives for tunnel alignment were considered: 3rd Street and 4th Street horizontal alignment. Both alignments considered a approximate 305 m (1000 ft) radius curve to initiate the alignment at the launch shaft. A curved segment benefits the overall alignment financially by eliminating tunneled length and intermediate shafts. The 3rd Street option recommended is shorter than the 4th street option and passes below the fewest structures and properties, all of which were primary factors in recommending this route. In determining the vertical elevation of the tunnel, consideration was given to a shallower gravity flow solution and a deeper design that would require a lift station at the WWTP. Ultimately, the deeper solution was selected as ground conditions are more favorable for tunneling and estimated settlements are less substantial for existing below grade utilities and at grade infrastructure.

A preliminary risk register has been established for the project, and should continually be reviewed and revised throughout the life of the project. In order to mitigate risks associated with the Mishawaka CSO tunnel, a geotechnical baseline report, and project specific specifications are currently being written. Incorporating special payment provisions, and a contract delivery method suited to the Owner's decision regarding risk allocation and financial responsibilities associated with risk allocation is recommended. A design-bid-build approach is recommended for this project. In addition, it is recommended that contractor prequalification is reviewed during final design to determine whether prequalifying contractors would be of benefit to the City and the successful completion of the project.

REFERENCES

Brierley Associates. (2015) *Long Term Control Plan Tunnel Preliminary Design Report,* Mishawaka, Indiana.

DLZ Indiana, LLC. (2016) *Draft LTCP Consent Decree Storage and Conveyance Tunnel and Control Structure and Pump Station Basis of Design Report,* Mishawaka, Indiana.

DLZ Indiana, LLC. (2016) *Draft Report of Geotechnical Data Report LTCP Consent Decree Storage and Conveyance Tunnel and Control Structure and Pump Station,* Mishawaka, Indiana.

United States District Court Northern District of Indiana South Bend Division (2014) *Consent Decree,* https://www.epa.gov/sites/production/files/2014-05/documents/mishawaka-cd.pdf

UTILIZING THE HAZARD MITIGATION GRANT PROGRAM TO IMPROVE SEWER PUMP STATION RESILIENCY

Timothy F. Sumner, PE, Crawford, Murphy & Tilly, Inc.

ABSTRACT

Resiliency is an increasingly common topic being discussed in the world of municipal infrastructure. The growing frequency and intensity of flood events poses a growing risk that is calling on local governments and utilities to be creative in addressing it as a part of their infrastructure planning and budgeting. The Hazard Mitigation Grant (HMG) program sponsored by the Federal Emergency Management Agency (FEMA) is a common funding mechanism utilized in flood zones primarily to mitigate flood-prone structures. However, as this case demonstrates, the HMG program is a unique source that can be creatively applied to public infrastructure that is vulnerable to extreme weather events like flooding. The challenge is in navigating a process that has limited guidance applicable to enhancing and improving the resiliency of permanent and vital infrastructure.

The Village of Rochester's primary pump station (PPS), located in Sangamon County Illinois, was designed in the early 1990s to serve the entire village populace of approximately 3,000, pumping the sewage to the Sangamon County Water Reclamation District. The structures that make up the PPS were designed and constructed to an elevation one foot higher than the regulatory base flood elevation. The PPS service was threatened twice in its history by backwater from the nearby South Fork Sangamon River. Flood waters backed up and surrounded the PPS such that access was only possible by boat. To preserve the PPS operation, Rochester Public Works was forced to sandbag the structures in 1994 and 2002. After the second flooding event and near loss of sewer service to residents, the Village began exploring avenues to fund protection of the PPS from future flooding and ensure a higher level of resiliency.

Rochester's ability to pay for improving resiliency of the PPS would hinge on obtaining grant funds. Fortunately, Sangamon County completed the Multi-Jurisdictional Natural Hazards Mitigation Plan that included addressing Rochester's PPS. FEMA's approval of the Multi-Jurisdictional Natural Hazards Mitigation Plan in 2008 and a disaster declaration in 2010 allowed Rochester to apply for funding. The Village went through the effort to apply for the grant and successfully obtained funds to cover up to 75% of the project costs. Patience paid off considering the span between the last flood event and the grant funding was more than 10 years.

Applicants should appreciate the challenge of the process that comes along with the opportunity to obtain up to 75% match for infrastructure resiliency projects through the HMG program. Learn about what is required and expected to complete a successful resiliency project with HMG funds from a community that has gone down this path before.

KEYWORDS

Hazard mitigation, resiliency, pump station, flooding, FEMA.

INTRODUCTION

Founded in 1869, the Village of Rochester, Illinois is approximately 6.4 km (4 miles) east of Springfield on State Route 29 (See Figure 1). Rochester has a population of about 3,700 people and is predominantly residential. A sanitary sewer system was constructed in the mid-1950s to serve the bedroom community, with two lagoon treatment systems located on the west and northeast sides of the village. Faced with increased regulatory pressure from the Illinois Environmental Protection Agency (IEPA) to update their lagoon treatment systems to meet current water quality standards, the Village turned to the Sangamon County Water Reclamation District (the District) for help in 1990. After much discussion and a referendum of the Rochester voters, it was decided the best course of action was to annex into the District so Village sewage could be pumped to them for treatment. While details of the annexation were finalized, the Village embarked on designs to abandon the lagoon treatment systems and build a central or primary pump station (PPS) that would send all the sewage to the District.

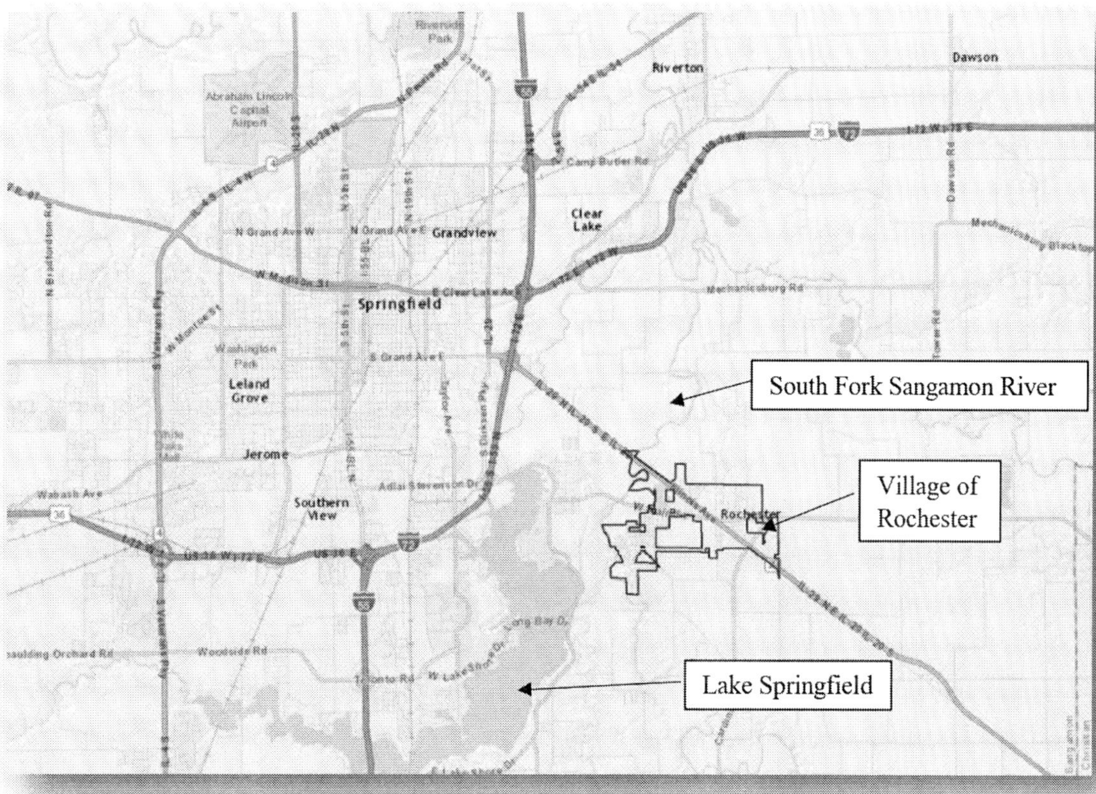

Figure 1 – Rochester Vicinity Map

While analyzing the dry and wet weather flows to be pumped to the District, it was obvious there was an inflow and infiltration (I/I) problem with the collection system. During wet weather, flows peaked at four to five times the average. The District did not want to accept the wet weather flow and required Rochester to restrict the volume of water being pumped daily. Since the Village was already spending several million dollars to abandon the lagoon systems, construct a primary pump station, force main and other sewer improvements needed to accommodate growth, they searched for a way to economically mitigate the excess flow in the system. The PPS was sited adjacent to the west lagoons and it was proposed to utilize the three lagoons as excess flow storage. The concept was to have a set of wet weather pumps at the PPS to send the excess flow to the storage lagoons temporarily until flows receded at the PPS. Approximately 11 million Liters (3 million gallons) of water can be stored in the lagoons. IEPA and the District approved the idea. As a part of construction, the lagoons were drained, solids removed and preparations made for the lagoons to receive excess flow during wet weather via the PPS. Within the PPS are two sets of submersible pumps: dry weather flow (DWF) and wet weather flow (WWF). DWF pumps have a capacity of 2,196 liters per minute (lpm) (580 gallons per minute (gpm)) and WWF pumps with a capacity of 7,098 lpm (1,875 gpm) (See Figure 2). During dry weather conditions, the pumps send a maximum of 580 gpm to the District. Any flows greater than 2,196 lpm (580 gpm) are sent to the excess flow lagoons for temporary storage and returned to the PPS when flows recede using a control valve.

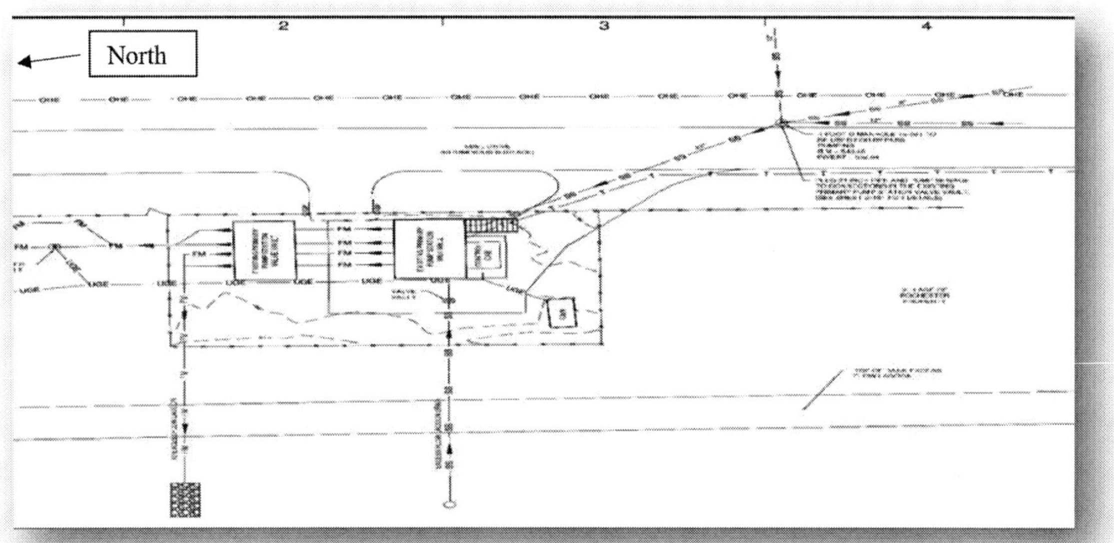

Figure 2 – PPS Schematic

SOUTH FORK SANGAMON RIVER FLOODING

To the south of the PPS is an unnamed tributary to the South Fork Sangamon River (See Figure 3). Knowing the PPS was in the floodplain, the current flood hazard boundary map was reviewed and the top of concrete (TOC) for the PPS was set at 0.3 meter (1 foot) higher than the regulatory base flood elevation (BFE), which equals elevation 165.8 meters (544.0 feet) above mean sea level (See Figure 4). Thus, the TOC for wet well, valve vault and a standby generator were constructed to elevation 165.8 m (544.0 ft.), which was a reasonable freeboard at the time in the absence of historical flood records (See Figure 5). The pump control panel sat atop the wet well in a weather proof enclosure. Shortly after the PPS was placed into service in 1994, the pump station was nearly inundated with floodwaters backing up from the South Fork Sangamon River. Flood levels reached approximately elevation 166.8 m (547.5 ft.). Rochester Public Works had to sand bag the PPS and access was only by boat. This scenario was repeated in 2002. If Public Works had not sandbagged the perimeter of the PPS, service would have been lost and all the untreated sewage from Rochester would have been overflowing into the unnamed tributary.

Figure 3 -PPS Location Map

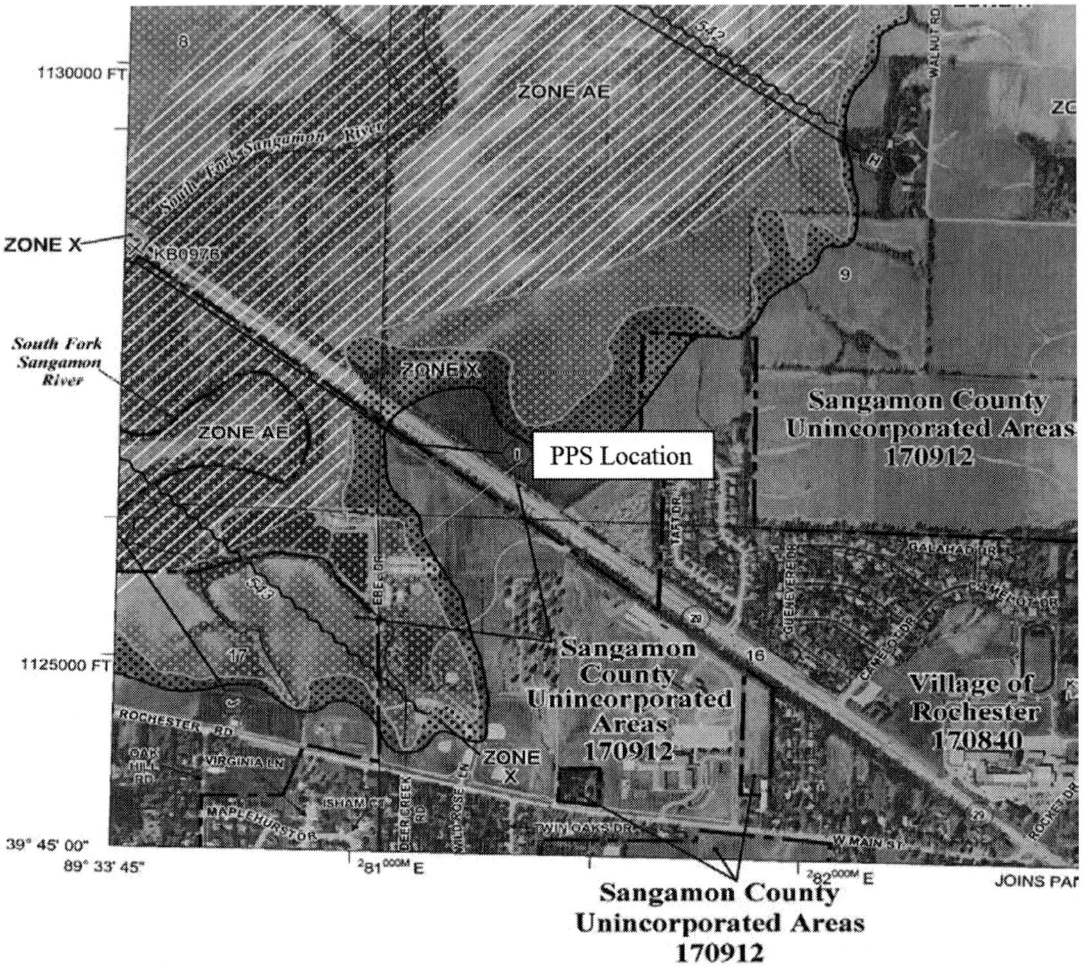

Figure 4 – Sangamon County Flood Insurance Rate Map (FIRM), Effective August 2, 2007.
Note: Flood elevations did not change between the Flood Hazard Boundary Map and the FIRM.

Figure 5 – PPS Existing Conditions, looking west.

SANGAMON COUNTY MULTI-JURISDICTIONAL NATURAL HAZARDS MITIGATION PLAN

After the second flooding event and near loss of sewer service to residents, the Village began exploring avenues to fund protection of the pump station from future flooding. High water marks were documented in the event funds became available to mitigate the impact of flooding in the future. There was a potential for the Village to obtain assistance when the Sangamon County Multi-Jurisdictional Natural Hazards Mitigation Plan (NHMP) was developed by the Springfield Sangamon County Regional Planning Commission (SSCRPC) in 2008. Through the hazard mitigation planning process, the Village could include protecting the PPS against future flooding as an eligible project. When the Sangamon County Multi-jurisdictional NHMP was approved by the Federal Emergency Management Agency (FEMA), Rochester's chances of receiving assistance was greatly improved.

The Sangamon County Multi-jurisdictional NHMP addresses the risk to communities in the county associated with dam failure, drought, earthquake, extreme heat, flood, mine subsidence, sever storm, tornado and winter storm. From the introduction, "Mitigation looks to reduce the need for response by permanently removing people and structures from harm's way when a known area of impact can be identified (such as a floodplain) or significantly reducing the impact from a known risk (such as a tornado). This Plan provides an assessment of the risks to Sangamon County from natural hazard events and a comprehensive range of mitigation projects to lessen the impact of these hazards on our communities. With the availability of mitigation grant funding from the Federal Government, communities can implement mitigation projects that would not otherwise be financially possible" (SSCRPC, 2008). This is the essence of the plan that Rochester deemed invaluable to assisting them with applying for future funding assistance.

FEMA has provided grants and encouragement to all counties within the United States to undertake the hazard mitigation planning process. A stipulation for receiving any federal grants for hazard mitigation issued to counties or communities is that the hazard mitigation plan must be current. FEMA considers hazard mitigation plans effective for five years. With the Sangamon County NHMP approved, participants were eligible to apply to FEMA for Hazard Mitigation funding. FEMA administers three programs revolving around hazard mitigation that includes:

- Flood Mitigation Assistance: The Flood Mitigation Assistance program (FMA) is a cost-share program (75% federal, 25% local match) through which communities can receive grants for the development of a comprehensive flood mitigation plan and the implementation of flood mitigation projects (FEMA – FMA, 2016).
- Hazard Mitigation Grant: The Hazard Mitigation Grant (HMG) program makes grants available to state and local governments as well as eligible private, non-profit organizations to implement cost-effective and long-term mitigation measures following a major disaster declaration (Presidential Declaration) (FEMA-HMG, 2016).
- Pre-Disaster Mitigation: The Pre-Disaster Mitigation (PDM) program makes funding available to local governments and state governments to implement cost-effective hazard mitigation activities that complement a comprehensive mitigation program (FEMA-PDM, 2016).

In Illinois, these hazard mitigation programs are implemented and overseen by the Illinois Emergency Management Agency (IEMA).

SEVERE STORMS AND FLOODING – AUGUST 2010

Presidential Disaster Declaration number 1935 was issued on August 19, 2010 due to severe storms and flooding in Illinois (FEMA, 2010). The Disaster Declaration allocated funds for HMG. As the Village Engineer, Crawford, Murphy & Tilly, Inc. (CMT), asked the Village trustees if they were interested in applying for funds through HMG to protect the PPS against future flooding. The Village was interested and CMT inquired with IEMA to see if the project would be eligible. A short project description with justification was submitted in early 2011, recommending the PPS be protected against flood waters to elevation 167.0 m (548.0 ft.), 0.15 m (0.5 ft.) higher than the high-water mark. FEMA accepted the justification and provided authorization to apply to the HMG program (HMGP), with the caveat that the PPS be protected to elevation 167.3 m (549.0 ft.) instead of 167.0 m (548.0 ft.). The Village agreed to the change.

FEMA's funding guidelines for the HMG program includes the following:

- Communities must be participating in and in good standing with the NFIP.
- Communities must have a FEMA-approved all-hazards mitigation plan (e.g., Sangamon County Multi-jurisdictional NHMP).
- Eligible projects must be environmentally sound, be cost effective, solve a problem and prevent future disaster damages (i.e., be more resilient).
- Protect public or private property.
- Approved projects may receive up to 75% funding with the remainder being a local match.

HMG PROGRM APPLICATION PROCESS

Three alternatives were investigated for possible solutions to protect the Rochester PPS from surface water flooding with varying levels of protection and sustainability. The first option would utilize 81 cm (32-inch) tall precast concrete barrier walls placed around the perimeter of the site. The second alternative involves the construction of a cast-in-place concrete flood wall around the perimeter of the site with a top of concrete elevation of 167.3 m (549.0 ft.). The third alternative raises the top of the wet well (includes electrical panel & controls), valve vault and generator to elevation 167.3 m (549.0 ft.). Details of the three alternatives presented to FEMA in late 2011 are as follows.

Alternative 1: Precast Barrier Wall:

Installation of a precast concrete barrier wall for the flood protection of the PPS had some drawbacks. To purchase, deliver and set precast concrete barrier walls at the site provides flood protection only to an approximate elevation of 166.1 m (545.0 ft.). Additional work would be required to provide a water-resistant barrier, both during construction and in a flood event. This work includes construction of a level foundation on which the precast sections would need to be tied to resist the lateral loading created by the flood water, as well as using a non-shrink grout both below and at each end of the precast sections. Also, at the corners of the precast barrier wall, areas of cast-in-place concrete are required to fill in the large voids created since these barriers are designed to be installed linearly. The standard height of the precast section is 81 cm (32 inches), which leaves over 0.9 m (3 feet) of height to make up with sandbags to reach the desired elevation of 167.3 m (549.0 ft.). Additionally, sandbags would be required to complete the flood protection at the location of the access gate. Even with this additional work, the barrier will still not be water tight. Operation of the PPS would not be interrupted by this work. An automatic sump pump and sump pit would be located at the lowest elevation on the site to allow for drainage of rainwater from the site during flooding events.

The preliminary opinion of project cost for Alternative 1 was $58,500, which included delivery of the precast barriers, construction of the foundation, grout, cast-in-place wall (filler) sections at the corners and installation of the sump pump and pit.

Alternative 2: Construct Cast-In-Place Concrete Flood Wall

In Alternative 2, a cast-in-place concrete flood wall would be constructed to elevation 167.3 m (549.0 ft.) around the entire perimeter of the site. This required the existing perimeter fence be removed. The flood wall would be topped with a security fence, where required. A gated opening would be provided in the current gate location wide enough to pass a service vehicle through. In the gated area, a manufactured flood gate will be manually installed or closed in emergency flooding situations. An automatic sump pump and sump pit will be located at the lowest elevation on the site to allow for drainage of rainwater from the site during flooding events. Constructing the flood wall would not interrupt the operation of the pump station.

The preliminary opinion of project cost for Alternative 2 was $123,000.

Alternative 3: Raise Top of Structures and Equipment

In Alternative 3, the top of concrete for the wet well, valve vault, electrical gear and standby generator would be raised to elevation 167.3 m (549.0 ft.) (See Figure 6). Alternative 3 would take the PPS out of service, thus it would have to be operated with a temporary by-pass pumping system. This work required the top slab for the wet well and valve vault be removed, the walls raised, and a new top slab constructed with replacement hatches. The electrical gear rests on a platform at the same elevation as the top of concrete for the wet well. The platform would be constructed out of steel members with a concrete-on-metal deck surface. Access to the platform would also be by ships ladder, and safety guardrail provided. The generator would be raised to elevation 167.3 m (549.0 ft.) and rest on a mass concrete slab at its current location. To accomplish the construction, the east run of the security fence would need to be removed and replaced. Bypass pumping operation would have to be maintained 24 hours per day, 7 days a week.

The preliminary opinion of project cost for Alternative 3 was $270,000.

In summary, Alternative 1 provides some flood protection to the PPS, however, it provides flood protection to an elevation less than the 2002 high water elevation and requires human intervention (active process) to prevent flood related damages. Alternative 2 is feasible for providing flood protection to the PPS up to elevation 167.3 m (549.0 ft.) and it also requires human intervention (active process) to close the gate in the event of flooding. Alternative 3 was the preferred alternative to utilize in the HMG program application because it did not require human intervention (passive) to protect it from flood waters up to elevation 167.3 m (549.0 ft.).

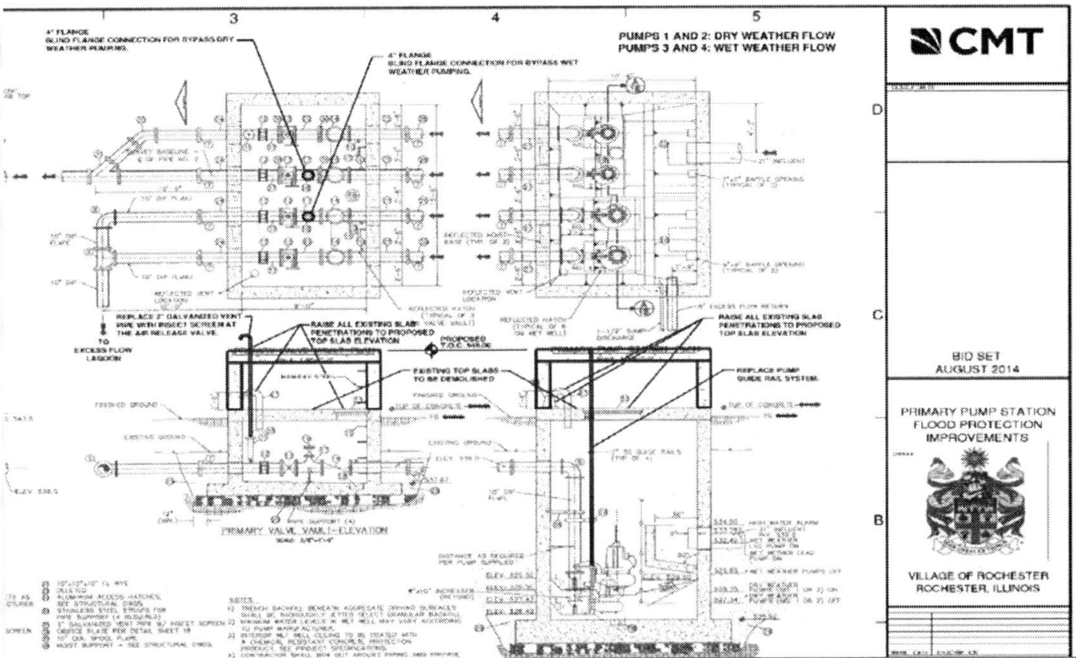

Figure 6 - PPS Flood Protection Improvements Elevation View

When preparing an application to FEMA for a HMG project involving a sewage pump station, the applicant should be prepared to be as detailed as possible with the project description and the benefits of undertaking the project. A typical HMG project for FEMA is a flood-prone property buyout. For Rochester's application, there was some explanation required with the application process to help FEMA understand the impact of pump station failure due to flooding and the approach to designing the improvements.

Rochester's application to FEMA was submitted in 2012 comparing the alternatives noted previously. The application was on-line and required a substantial amount of information about the applicant and the proposed hazard mitigation project. Due to the number of questions and details requested, the application process was started and stopped multiple times while gathering the necessary information. In total, the on-line application process required almost three days to complete.

FEMA reviewed the application and requested economic justification for raising the structures in early 2013, which meant completing a benefit-cost analysis (BCA) using FEMA software. A project is cost effective when the benefit-cost-ratio is 1.0 or greater, indicating the benefits of a prospective hazard mitigation project are sufficient to justify the costs. Using FEMA's BCA software was not a straight-forward analysis since the questions focused more on repetitive loss property buyouts, but FEMA was very helpful in completing the analysis for the Village.

During the application process, tornadoes caused damage in southern Illinois that delayed the approval process. The Village had to be patient while IEMA assisted those communities in need.

Prior to FEMA approving the HMG application, they also required sign-offs from the State Historic Preservation Officer, Illinois Department of Natural Resources for a threatened and endangered species consultation and an endangered species consultation by US Fish and Wildlife Service. Lastly, a public meeting was conducted requesting public participation and input into the project. Documentation from the public meeting was submitted to FEMA.

HMG AWARD

Once FEMA approved the grant award, IEMA began administering the grant process. A grant agreement was prepared and sent to the Village for acceptance. Important grant agreement conditions included:

- Federal language specifically prohibiting kickbacks.
- Requirement for all contractors to pay federal wages.
- Adhering to the Buy American Act which requires only steel, iron and manufactured products produced in the USA as eligible.
- Utilize a quality based selection (QBS) process to hire consultants.
- Incorporate grant agreement requirements into all contracts: consultants and contractors.
- Submit quarterly progress reports to IEMA.

CMT was charged with ensuring the grant guidelines were followed and agreement requirements were incorporated into all documents knowing the project may be audited at the conclusion.

RESILIENCY PLANNING

One definition of flood resiliency is the ability of a utility to withstand a flooding event, minimize damage and rapidly recover from the disruption to service (USEPA, 2014). When incorporating resiliency into a project, there are several aspects to consider beyond just the minimum regulatory requirements. Considerations for the PPS and pump stations in general, include:

- What should the flood protection elevation (FPE) be: the 500-year (0.5% annual chance) water surface elevation or something higher?
- Is additional freeboard needed based on historical flood elevations?

- How high can the structure(s) practically be raised?
- Does the electrical service need to be modified to protect against flooding?
- Can the controls be raised higher?
- Are upgraded stairs or steps justified?
- Should the pump type be changed, such as switching from dry pit to submersibles?
- Should a backup pump be stored on site for emergencies?
- Are control valves just as accessible as the pumps?
- Are the control valves submergence rated?
- Should the nearby stream/river water levels be connected to SCADA for monitoring?
- Should a stand-by generator be added?
- If there is an existing generator, does the pad need to be elevated?
- How will the generator be fueled during flooding?

OTHER DESIGN CONSIDERATIONS

CMT also identified pump station features in need of replacement or updating as a part of the overall improvements. Reasonable costs for features to be replaced as well as other additional improvements were included in the grant application with supporting information. Features and improvements for the PPS included the following:

- Replacing the obsolete programmable logic controller (PLC) with the latest technology
- Installing larger access hatches that included safety grates

Other considerations for improvement and/or replacement if the pump station is to be taken out of service include the following:

- Increasing the capacity of the pump station
- Performing repairs / rehabilitation / replacement to the wet well and associated piping
- Adding quick connect bypass pumping connections
- Adding emissions controls and/or noise reduction features to the standby generator
- Rewiring any of the controls to improve operation or monitoring
- Adding solar panels to power non-essential features

DESIGN AND BIDDING

IEMA is not a regulatory agency, so they did not have a means or desire to review plans, specifications, bidding and contracting documents. Preparing appropriate bidding documents is left to the grantee or its consultant to prepare. Obtaining any necessary permits from local, state and/or federal agencies is also incumbent on the grantee. In Rochester's case, no permits were required since the pump station capacity was not being increased, there was minimal ground disturbance and it is an existing structure being raised. Design was started and completed in 2014, with advertising for bids in June of that year.

Requirements for soliciting bids for the project included the following:

- Follow the bidding and contracting requirements of 44CFR13.36 (procurement)
- Sealed bidding process for contracts greater than $100,000, which was applicable to Rochester
- Advertise the project 45 for days, with at least one day in the local newspaper
- Interested contractors must advertise 15 days prior to the bid opening for disadvantaged business enterprises (DBEs) and provide proof of advertisement
- Contracts musts be awarded to the lowest responsive and responsible bidder
- Provide IEMA with copies of all documents associated with the bidding process

Bids were advertised for the PPS Flood Protection Improvements twice. The first bid opening, there was only one bidder and the bid was more than double the engineer's opinion of probable construction cost (EOPCC). Modifications were made to the bidding documents with the intent to reduce the cost of construction by implementing the following changes:

- Project was advertised in two newspapers to obtain greater exposure,
- Traffic control and bypass pumping was removed from the contract and assigned to the Rochester Public Works Department.
- Precast concrete top slabs were allowed
- Metal salvage requirements were removed
- Thickness for the wall extension were reduced from 30 to 20 cm (12 to 8 inches)
- Substantial completion time was extended by 30 days
- Liquidated damages were reduced from $2,000 to $1,000 per day

The second round of bidding was in September 2014, with five bidders and one reasonably close to the revised EOPCC. After review of the contractor's qualifications, the Village awarded a contract to the lowest, responsive and responsible bidder in December of 2014.

CONSTRUCTION

For the Village and all grantees to be reimbursed from IEMA, there are two numbers required (in order): A D-U-N-S number and SAM number. D-U-N-S® stands for Dun & Bradstreet (D&B) number and SAM is the System for Awards Management. The D&B Number is a unique nine-digit identifier for businesses and is used to establish a D&B business credit file, which is often referenced by lenders and potential business partners to help predict the reliability and/or financial stability of the company in question (D&B, 2017). A D&B number is required to obtain grants from the federal government. Following a D&B number, a SAMS number must be acquired to do business with the Federal Government (SAM, 2017). The process of applying for the D&B and SAMS numbers should be started prior to a construction contract being awarded, for it may take a while.

It was suggested the Village have the funds on hand to pay the contractor and request reimbursement from IEMA so as not to delay payments. Any change orders should be appropriately documented and applied for reimbursement. In Rochester's case, one application for payment was made to IEMA and was processed within a month.

A preconstruction meeting was held in January 2015 and the contractor was issued a notice-to-proceed shortly thereafter. Work on the pump station started in April after shop drawings and material certifications were processed. Three shop drawings were resubmitted due to inadequate documentation regarding Buy American Certifications. An extremely wet spring made for a challenging bypass pumping operation, which the Village dealt with by monitoring the operation around the clock. Substantial completion was achieved in early July of 2015 when complete functionality of the pump station was tested and restored (See Figure 7).

Implementing the proposed improvements required coordination with Public Works staff to develop contractor sequencing and mobilization requirements since the pump station was taken out of service to make the improvements. A by-pass pumping plan was developed with input from the Public Works staff to ensure smooth transitions between shut down and startup.

Figure 7 – PPS Flood Protection Improvements Completed, looking west.

LESSONS LEARNED

If the Village considers another application to FEMA for grant funds, there are some lessons learned from this project that can be applied to the next one.

- Tie high water marks as soon as possible to an elevation.
- Document the duration of flood inundation.
- Ensure the Hazard Mitigation Plan is current (plans are effective for 5 years).

- Budget several days to complete on-line application.
- Include conservative costs in the grant application for features to be replaced, with supporting information.
- The process to apply and be approved can be lengthy. If there is a natural disaster elsewhere in the state, the process may be delayed while staff assists those in need.
- Consider what other improvements or modifications should be done to the pump station while it is offline.
- Consider how the pump station will be accessed after it is raised; by ships ladder, steps or some other means.

THE HMG APPLICATION WAS WORTH THE EFFORT

When the PPS flood protection improvements were less than 6 months old, there was 10.7 cm (4.21 inches) of rain between December 26 – 28, 2015 that caused flooding and the South Fork Sangamon River backed up to the PPS and surrounding area. Once again access to the PPS was only by boat due to about 0.76 m (2.5 ft.) of water over the former top of concrete elevation. Sewer mains were inundated, but the PPS stayed operational despite the surrounding flood waters thanks to having elevated the structures (See Figure 8). The pump station would have been underwater if not for the work funded by the HMG and persistence by the Village to see the improvements through to completion.

Figure 8 – South Fork Sangamon River Flooding Around PPS, looking east.
Photo courtesy of Rochester Public Works Department

REFERENCES

Dun & Bradstreet (D&B). 2017. http://www.dnb.com/duns-number/what-is-duns.html

Federal Emergency Management Agency (FEMA). Flood Mitigation Assistance (FMA) Grant Program. Updated 08/30/2016. https://www.fema.gov/flood-mitigation-assistance-grant-program

Federal Emergency Management Agency (FEMA). Hazard Mitigation Grant (HMG) Program. Updated 09/28/2016. http://www.fema.gov/hazard-mitigation-grant-program

Federal Emergency Management Agency (FEMA). Pre-Disaster Mitigation (PDM) Grant Program. Updated 11/09/2016. https://www.fema.gov/pre-disaster-mitigation-grant-program

Federal Emergency Management Agency (FEMA). Illinois Severe Storms and Flooding. August 19, 2010. https://www.fema.gov/disaster/1935

Springfield Sangamon County Regional Planning Commission (SSCRPC). Sangamon County Multi-Jurisdictional Natural Hazards Mitigation Plan. Springfield, Illinois. August 2008.

System for Award Management (SAM). 2017. https://www.sam.gov/portal/SAM/#1

United States Environmental Protection Agency (USEPA). Flood Resilience: A Basic Guide for Water and Wastewater Utilities. EPA 817-B-14-006. Washington, D.C. September 2014.

Modal Analysis as a Tool to Resolve Pump Vibration Issues

John Koch
HDR, Bellevue, WA
jkoch@hdrinc.com

ABSTRACT

Vibration spectrum analysis and modal analysis are tools designers and plant staff can use to optimize their pumping systems. This paper specifically looks at the ways these sophisticated tools were used to diagnose and solve operational and mechanical issues on a two-stage raw sewage pumping station constructed in 1972 and upgrade twice since then to achieve its current 227 ML/d (60 mgd) design capacity.

The modal and vibration analysis resulted in solving an intermediate bearing support structure issue, but uncovered other hydraulically induced vibration issues which will require additional investigation and study to resolve.

KEYWORDS: Modal, vibration, Hydraulic Institute, pump, natural frequency, resonance

INTRODUCTION

The essential step in solving vibration problems is to understand resonance. High vibration levels can be caused by resonance in the rotating elements of the pump shaft, impeller or the motor rotor, in the structure of the pump and its volute, bearing housing or motor frame. Severe vibration, more appropriately defined as vibration above the Hydraulic Institute Standard (HI) recommended maximum limits as contained in ANSI/HI 9.6.4 can destroy bearings in pumps and motors as well as destroy mechanical seals. Resonance conditions in mechanical systems occur when the excitation forces coincide with the natural frequencies of the structure. A key to defining the source of the resonance is to determine if the source of the resonance is in the rotating elements of the pump or in the pump assembly including the pump base plate and supports and the structure the pump and pump base are attached to. To identify a resonance condition requires obtaining the operating vibration spectrums of the system and identifying the natural frequencies of the system.

The rotating elements of a pump or motor can have a critical speed close to their operating speed; however, no rotating element is ever intentionally designed to operate in its critical speed or go through its critical speed. In my more than 45 years of commissioning equipment, there has only been one instance where the rotating element was the source of excessive vibration and that was a vertical pump's motor shaft which was excited by the variable frequency drive (VFD) carrier frequency. Figure 1 illustrates the excessive vibration of this installation. The vertical axis is inches per second-root mean squared; the horizontal axis is frequency in Hz.

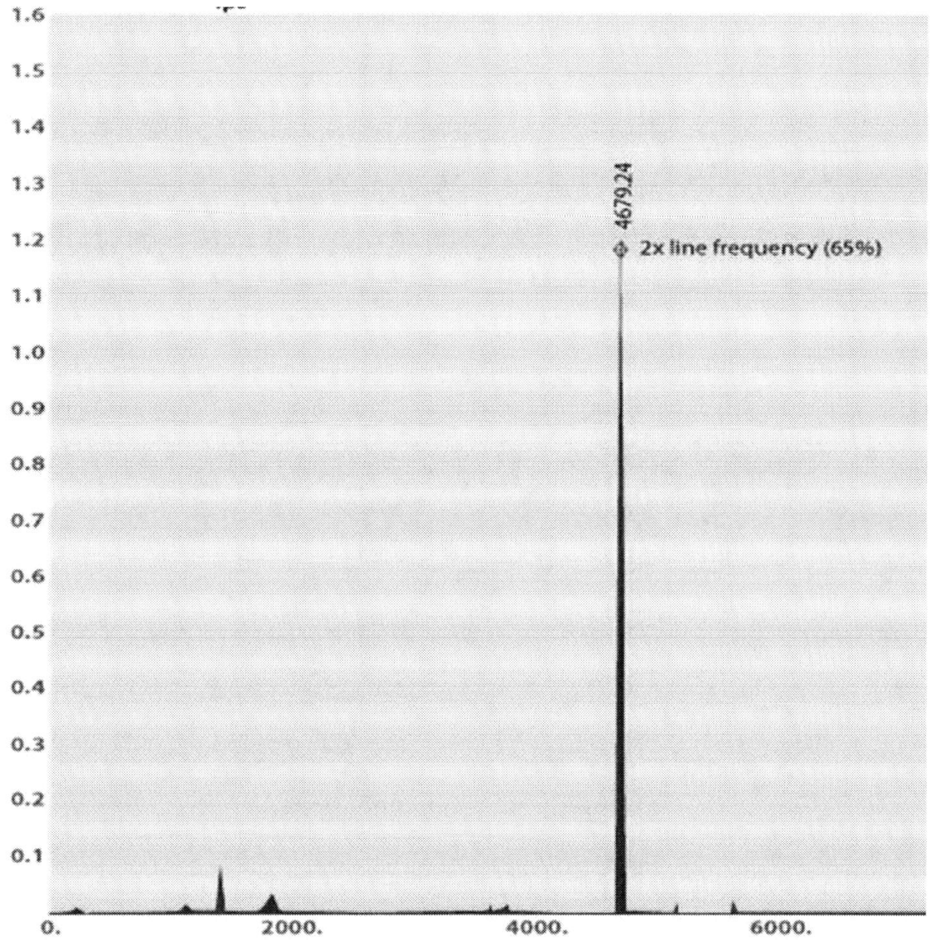

Figure 1 – Motor Shaft Vibration

All other vibration issues that I have encountered have been attributed to excitation of the pump/motor structure. Most often when a pump's vibration exceeds HI limits, the issue is with the pump and its base not with its shaft.

Almost all equipment in water and wastewater treatment plants are mounted on concrete equipment pads with grout poured in and under the equipment base as illustrated in Photograph 1.

Photograph 1- Grouted Pump Base

Grouting the pump's equipment base provides damping to the pump and motor. Damping reduces the maximum amplitude of the vibration caused by resonance and in doing so can bring the equipment into compliance with HI maximum vibration limits. This is one of the reasons to make sure the equipment base is in full contact with the concrete equipment pad and all the air bubbles are expelled from the pourable grout mixture.

The key tenet of resolving natural frequency issues is adding mass to a structure or stiffening the structure to move the natural frequencies out of the operating range of the equipment. Adding mass will typically lower the system's natural frequency; whereas, stiffening the structure of a rotating piece of equipment will raise its natural frequency. Photograph 2 illustrates where mass was added to a pump in an attempt to lower its natural frequency below the pump's operating speed.

Photograph 2 – Mass Added to Pumps

Trying to do add mass and stiffen the structure most of the time is counterproductive. The formula for calculating the reed critical frequency or natural frequency is:

$$f = \frac{1}{2\pi}\sqrt{\frac{k}{m}}$$

Where k = stiffness and m = mass

Experimental modal analysis is based on measured frequency response functions between a force applied to the structure and the response of the structure. The measured frequency response functions are post processed using modal analysis software to identify the natural frequencies of the structure. Accelerometers are used to measure the response of the structure and an instrumented hammer is used to apply the force to the structure as shown in Photograph 3.

Photograph 3 – Impact Hammer to Determine Natural Frequency

The natural frequency of a system is a property of the system that is defined by the inertial and elastic properties (mass and stiffness) of the structure. The inertial and elastic properties of a structure are a function of the material properties and the geometry of the components of the structure. Each natural frequency has a particular displacement pattern, called a mode shape, and each mode shape is independent of all other mode shapes of the system.

To identify a resonance condition in pumping systems, the measured operating vibration spectrums are used to identify high vibration levels at the natural frequencies obtained from the modal analysis. The measured vibration spectrums are reviewed to identify any vibration levels that exceed the maximum allowable levels as defined by accepted industry standards. For components that have vibration levels above the maximum allowable levels, the vibration spectrum is inspected to determine at which frequency(s) the high vibration levels reside. Once the high vibration level frequency has been identified, it is compared to the results of the modal analysis to determine if there is a resonance condition.

As important as the stiffness of the pump and structure to reduce vibration levels below a HI maximum level is the hydraulic conditions at the inlet to the pump's impeller. American National Standard (ANSI)/HI 9.6.6 recommends a minimum of 5 diameters of unobstructed straight pipe into a pump suction unless a long radius reducing elbow with more than 50% area reduction is used. For a 300 mm (12-inch) pump suction this means a long radius 300 x 450 mm (12 x 18-inch) elbow would be required.

Photograph 4 illustrates how dye in a standard radius elbow swirls into a pump inlet.

Photograph 4 – Standard Radius Elbow

Comparing the standard radius elbow to the long radius elbow in Photograph 5, the flow stream is more uniform in the long radius reducing elbow than in the standard elbow. Turbulent flow into the eye of the pump impeller will cause unstable flow conditions and potentially high vibration levels in the pump.

Photograph 5 – Long Radius Reducing Elbow

BACKGROUND

A 227 ML/d (60 mgd) two-stage high head, greater that 76m (250 feet) total dynamic head (TDH) raw sewage pumping station was evaluated for its problematic maintenance issues and perceived vibration problems. Several studies and attempts to resolve the issues were performed over the last 15 years including a hydraulic model study.

The station was configured with the discharge of the first pump routed to the intake of the second stage pump. Photograph 6 illustrates the first and second stage pumps installed in the cramped pump room.

Photograph 6 - Two Stage Pumping Units

Several of the maintenance issues in the pumping station were due to the tight quarters for performing any maintenance related activities. Many of the operational problems with the pumping units were the short suction piping and 90-degree fitting into the suction of these vertically configured pumps. The three different levels for pumps and motors presented both maintenance and operational issues with the units. Photograph 7 illustrates the motors on the next level above the pump room level.

Photograph 7 - Pump Motor Level

The installation utilized a long shaft assembly between the two floors. The shafting is supported by intermediate guide bearings approximately half way between the two floors which are illustrated in Photograph 8.

Photograph 8 - Intermediate Guide Bearing Supports

Pumps and intermediate shaft bearings were typically rebuilt and replaced every 12-18 months because of the failure of one or more of the components in each set of pumps.

METHODOLOGY

The initial vibration analysis for the two-stage raw sewage pumping station determined there were vibration issues with the intermediate bearing assemblies which could be transmitting a resonance vibration to the pumps below. The intermediate support bearing structure was a series of parallel 41 cm (16-inch) wide flange beams (WF16X26) and 30 cm (12-inch) (C12x20.7) cross support channel beams for each pump set as shown in Photograph 9.

Photograph 9 - Intermediate Bearing Supports

The multiple reference impact testing approach was used for this project where the response is measured at fixed locations and the forces are applied at all of the locations used in the modal analysis. Multiple reference impact testing is ideal for large structures like the intermediate bearing structure because it simplifies the measurement process allowing the frequency response functions to be acquired efficiently.

The intermediate bearing support frequency response function measurements were made using a Dytran Instruments Model 5802A impulse hammer and 3 PCB Piezotronics Model 604B31 tri-axial accelerometers. For each measurement point the structure was excited with the hammer and 9 frequency response functions were measured. A total of 62 and 66 points were measured on the stage 1 and stage 2 intermediate bearing support structures respectfully.

The measured frequency response functions were used to estimate the natural frequencies of the intermediate bearing support structures. The OROS Modal 5.2 modal analysis software was used to perform the modal analysis from the measured frequency response functions. For this project, the OROS BroBand modal identification module was used which is based on the Polyreference Least Squares Complex Frequency Algorithm.

The modal analysis identified four modes (natural frequencies) of interest in the 0-100 Hz frequency range of the stage 1 and 2 intermediate bearing support structures. The operating speeds of pump sets 1-4 were 1800 rpm (30 Hz) and pump sets 5-12 were 1200 rpm (20 Hz). The modes of interest identified by modal analysis are shown in Table 1 and all are in or very close to the pump operating ranges.

Table 1: Intermediate Bearing Support Natural Frequencies

Mode	Stage 1 (Hz)	Stage 2 (Hz)
1	17.08	19.38
2	19.09	21.71
3	20.97	30.60
4	29.15	33.86

The mode shape for the Stage 1 structure's first natural frequency at 17.08 Hz is shown in Figure 2. The first mode is the center section of the intermediate bearing support which supports the intermediate bearings for Pumps 5-8 deflecting in the Y (horizontal) direction.

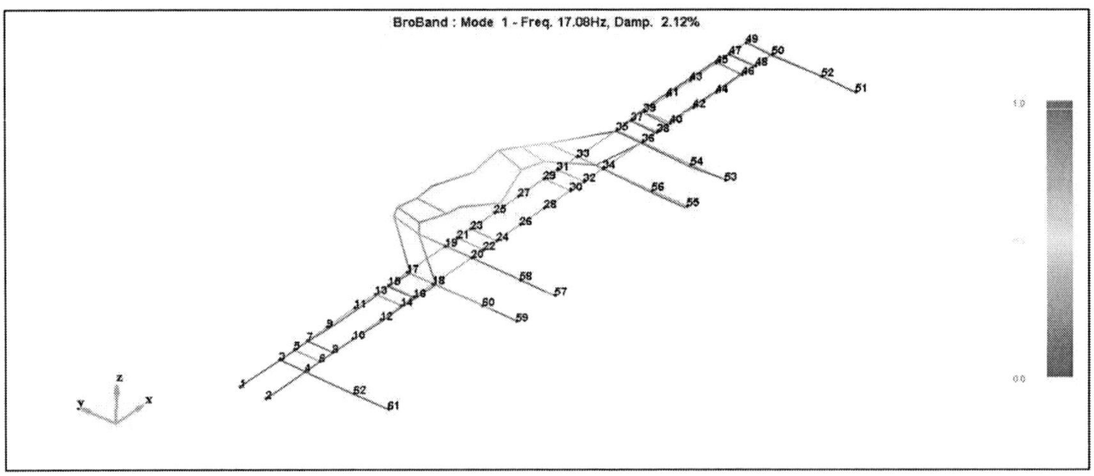

Figure 2 - Stage 1 Mode Deflecting in the Y-Axis

The mode shape for the Stage 1 structure's fourth natural frequency at 29.15 Hz is shown in Figure 3. The 29.15 Hz mode shape in the Stage 1 intermediate bearing support is the structure deflecting in the X (north-south) direction while bending in the Z direction (vertically) in a wave fashion.

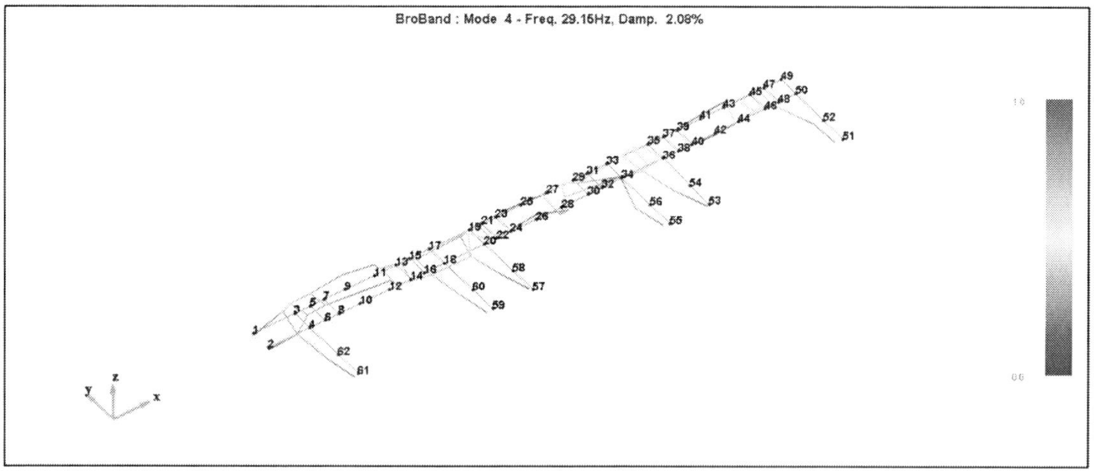

Figure 3 - Stage 1 Mode Deflecting in the X-Axis

Centrifugal pumps typically have two main sources of vibration as shown at rotating speed and at blade pass frequency. Pumps 1-4 operate at 1,800 RPM (30 Hz) and the pump impellers have 2 blades with a blade pass frequency of 60 Hz. Pumps 5-12 operate at 1,200 RPM (20 Hz) and the pump impellers have 3 blades with a blade pass frequency of 60 Hz. In Table 2, the first three rows show the operating frequency and the first two harmonics for both pumps. The last three rows in Table 2 show the blade pass frequency and its first two harmonics.

Table 2: Pump Operating and Blade Pass Frequencies (Hz)

Forcing Frequencies	Pumps 1-4	Pumps 5-15
1X	30	20
2X	60	40
3X	90	60
1X Blade Pass	60	60
2X Blade Pass	120	120
3XBlade Pass	180	180

The maximum allowable vibration levels for the pumps, drive motors and intermediate bearing support were determined using the Hydraulic Institute standard 9.6.4-2009, ISO 10816-3, Technical Associates of Charlotte, P.C. criteria for overall condition rating and experience. The maximum allowable vibration levels were set at 8.64 mm/sec (0.34 in/sec) root mean squared (RMS) for the pumps, 5.08 mm/sec (0.20 in/sec) RMS for the intermediate bearing support, 4.57 mm/sec (0.18 in/sec) RMS for 74.6 kW(100 hp) drive motors and 7.11 mm/sec (0.28 in/sec) RMS for the 298 to 373 kW (400 and 500 hp) drive motors.

All of the overall vibration levels measured on Pumps 1, 2, 5, 6, 9, and 10 were compared to the maximum allowable. The results are shown in Table 3. The vibration levels, which exceed the maximum allowable levels in Table 3, are red and the levels that are close to the maximum allowable levels and warrant a closer look are highlighted in yellow.

Table 3 - Measured Overall Vibration Levels Pumps 1, 2, 5, 6, 9, and 10 mm/sec (in/sec) RMS

Location	Pump 1	Pump 2	Pump 5	Pump 6	Pump 9	Pump 10
Motor X	2.79 (0.11)	0.25 (0.01)	1.02 (0.04)	2.29 (0.09)	2.54 (0.10)	3.56 (0.14)
Motor Y	1.52 (0.06)	1.78 (0.07)	1.02 (0.04)	3.05 (0.12)	1.78 (0.07)	3.56 (0.14)
Motor Z	1.02 (0.04)	0.76 (0.03)	6.10 (0.24)	0.51 (0.02)	0.76 (0.03)	1.78 (0.07)
Intermediate Bearing X	**7.37 (0.29)**	2.79 (0.11)	2.79 (0.11)	1.52 (0.06)	2.03 (0.08)	**5.84 (0.23)**
Intermediate Bearing Y	1.52 (0.06)	**4.06 (0.16)**	**4.83 (0.19)**	3.05 (0.12)	**21.59 (0.85)**	**16.26 (0.64)**
Intermediate Bearing Z	1.02 (0.04)	1.52 (0.06)	2.03 (0.08)	**4.32 (0.17)**	1.52 (0.06)	2.79 (0.11)
Pump X	4.32 (0.17)	**8.89 (0.35)**	**10.92 (0.43)**	**12.70 (0.50)**	4.06 (0.16)	**14.48 (0.57)**
Pump Y	5.08 (0.20)	4.83 (0.19)	7.37 (0.29)	5.84 (0.23)	**7.37 (0.29)**	3.81 (0.15)
Pump Z	0.76 (0.03)	1.27 (0.05)	2.29 (0.09)	4.83 (0.19)	2.79 (0.11)	6.35 (0.25)

Pump 1 vibration levels are below the Hydraulic Institute maximum allowable levels at all locations and directions, except the intermediate bearing X direction at 7.73 mm/sec (0.29 in/sec) which is above the 5.10 mm/sec (0.20 in/sec) (RMS) maximum vibration level for the intermediate bearing. Figure 4 shows the vibration spectrum from the Pump 1 intermediate bearing. The dominant vibration level is at operating speed (30 Hz) of the pump. The modal analysis also identified a natural frequency of the Stage 2 intermediate bearing support at 30.64 Hz. The mode shape for the Stage 2 intermediate bearing support at 30.64 Hz natural frequency is predominately in the X direction. Based on these results, the high levels of vibration of the Pump 1 intermediate bearing are due to a resonance condition.

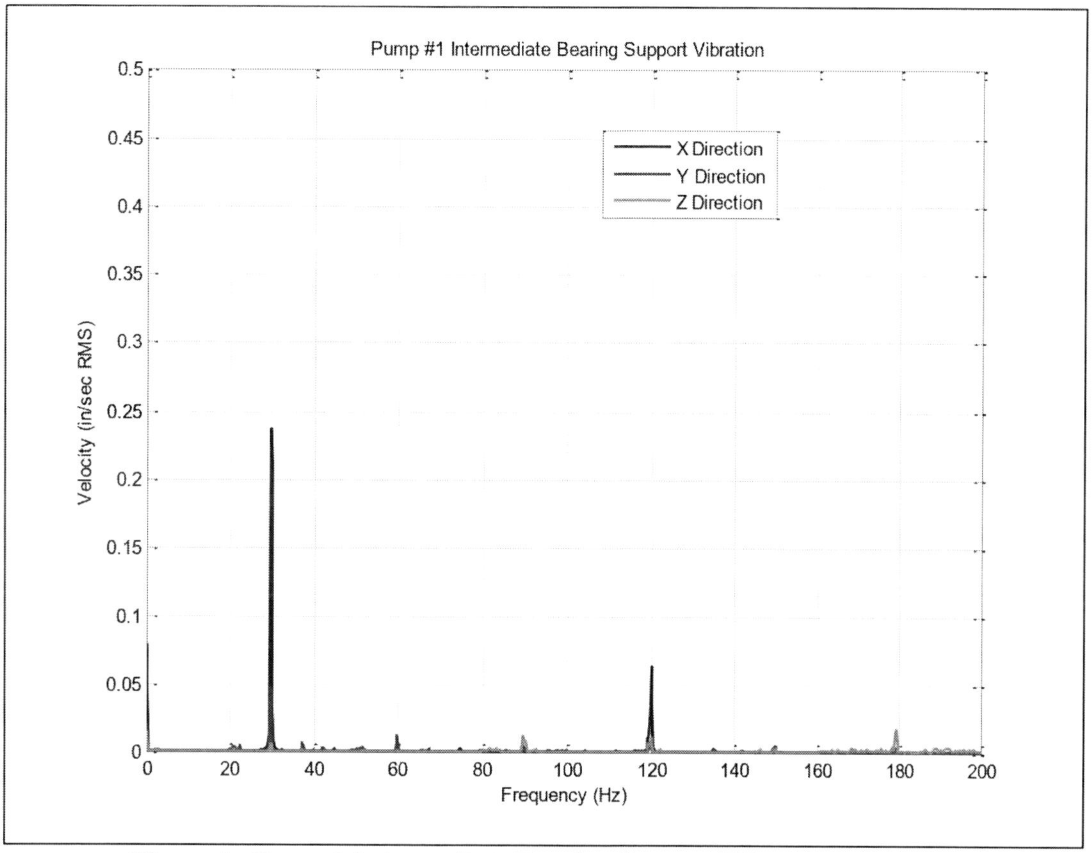

Figure 4 - Pump 1 Intermediate Bearing Support Vibration Spectrum

Pump 2 vibration levels are below the Hydraulic Institute maximum allowable levels at all locations and directions except the pump in the X direction. The vibration spectrum for Pump 2 shows that the X direction vibration is dominated by the blade pass frequency at 60 Hz and 2 times blade pass at 120 Hz as shown in Figure 5.

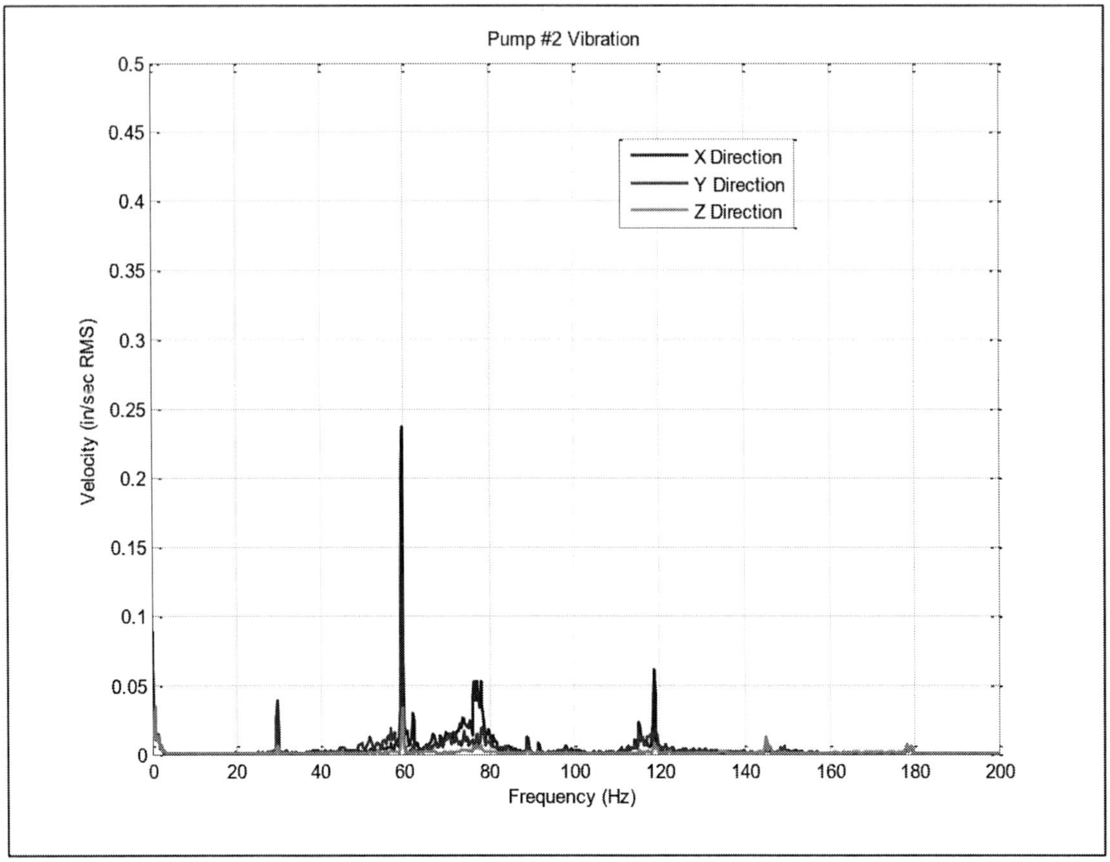

Figure 5 - Vibration Spectrum for Pump 2

The vibration at the blade pass and 2 times blade pass were below the maximum allowable vibration level of 8.64 mm/sec (0.34 in/sec) RMS. Pump 2 does not have any natural frequencies close to 60 Hz, which eliminates a resonance condition as the root cause of the high vibration levels. Pump 2 also shows signs of cavitation, which is an indicator of hydraulic issues with the pump, which will also increase the blade pass vibration levels. The cavitation induced vibration in the 60-90 Hz range increased the overall vibration level to 8.89 mm/sec RMS (0.35 in/sec RMS).

The same analysis procedure was applied to Pumps 5-10 which, like Pumps 1 and 2, found the vibration levels that exceeded the maximum allowable levels on the pumps were due to hydraulic issues with the pumps and not due to resonance conditions in the pump structure. The high vibration levels on the intermediate bearing support on Pumps 9 and 10 were due to resonance conditions where the pump operating speed of 20 Hz was exciting the 20 Hz natural frequencies in the intermediate bearing structure.

RESULTS

The modal analysis suggested that stiffening the intermediate bearing support structure would raise the reed critical speed of the supports. A finite element analysis was performed with the data from the modal analysis which determined that welding a 7mm (1/4") plate to the top and bottom of the longitudinal beam structure would provide the necessary stiffness to change the natural frequency of the support system. Photograph 10 shows the completed modifications to the beams.

Photograph 10 - Modified Intermediate Bearing Support Structure

After the suggestions to stiffen the intermediate bearing structures were performed by the City, another round of analysis of the entire pumping station was conducted to gather at one time the following information to try and pinpoint the operational and maintenance issues with this pumping station. The second round of vibration testing included:

- Vibration signature of each pump, intermediate bearing and motor.
- Flow from each two stage pump set measured with the stations flow meters that had been recently calibrated.
- Suction and discharge pressure from each pump measured with liquid filled calibrated pressure gauges at the appropriate pressure ranges and fitted with diaphragm seals.
- Power for each set of pumps measured at each motor starter.

The maximum allowable vibration levels were determined using the Hydraulic Institute standard 9.6.4-2009, ISO 10816-3, Technical Associates of Charlotte, P.C. criteria for overall condition rating. The Hydraulic Institute Standard 9.6.4-2009 addresses the solids handling pump vibration levels but does not address the motor or intermediate bearing. Technical Associates of Charlotte criteria addresses vertical pump based on the height of the motor above grade but not a solids handling pump specifically. Neither ISO 10816-3 nor Technical Associates of Charlotte specifically address the intermediate bearing support maximum allowable vibration levels. The maximum allowable vibration level for the intermediate bearing was determined from experience.

Table 4 shows the maximum allowable vibration levels defined by the two (2) standards, (HI and ISO) Technical Associates of Charlotte and experience. The last column on the right in the table shows the maximum allowable vibration levels for this project based on the two (2) standards and experience.

Table 4 - Maximum Allowable Vibration Levels (in/sec RMS)

Location	HI 9.6.4-2009	ISO 10816-3	Technical Associates of Charlotte	Criteria used
Motor – 100 HP		0.18		0.18
Motor – 400 and 500 HP		0.28		0.28
Intermediate Bearing				0.20
Pump	0.34		0.39	0.34

Vibration measurements at the intermediate support structure were recorded after the modifications were made by city personnel. The modifications had reduced the intermediate bearing vibration levels from 22.00 mm/sec (0.86 in/sec) RMS to 3.60 mm/sec (0.14 in/sec) RMS, considerably below the maximum allowable level of 5.00 mm/sec (0.2 in/sec) RMS. The City had achieved success in reducing the resonance frequency and vibration issues associated with the intermediate bearing support structure.

DISCUSSION

Vibration and modal analysis defined structural/mechanical issues with the two-stage vertical end suction pumps. Structural issues were resolved with the intermediate bearing support structure when the independent wide flanged beams were stiffened and modified to act as one large box beam as shown in Photograph 10. With this piece of the puzzle resolved, the hydraulic induced vibration issues can be further investigated.

The vibration signature of all of the pumps, once the resonance from the intermediate bearing was reduced, was at blade pass and 2x blade pass frequency. Centrifugal pumps typically have a dominant vibration at blade pass frequency.

Blade pass frequency is the number of impeller blades times the operating RPM. As an example a two bladed impeller operating at 1800 RPM (30 Hz) would have a blade pass frequency at 60 Hz (30 Hz x 2). Issues that contribute to blade pass vibration are:

1. Non-laminar fluid flow to the pump suction and impeller.
2. Net suction positive pressure required (NPSH3) exceed net positive suction pressure available (NPSHA). Or simply stated, the water level in the wet well is too low.

3. Pump operating outside of the pump's preferred operating range (POR) and/or to the far right of the acceptable operating range (AOR).

Further detailed investigation and analysis of the data collected will be required to pinpoint the root cause of the continued hydraulic vibration issues with the raw sewage two-stage vertical end suction centrifugal pumps.

CONCLUSION

Modal and vibration spectrum analysis, while not the ultimate resource for diagnosing and resolving mechanical failures, are powerful tools to assist operation and maintenance personnel in determining some of the root causes of equipment failure. As was the case in the investigation of the two-stage vertical end suction pumps, one issue was eliminated; however, several other potential problems were uncovered.

Vibration monitoring is a great evaluation tool that can be used to diagnose issues and problems. It can and should be a tool for acceptance of installed pumps to confirm they meet the minimum American National Standards Institute/Hydraulic Institute Standards.

ACKNOWLEDGEMENTS

The authors gratefully acknowledge the staff at the wastewater pumping stations for their patience while all of the issues were investigated and satisfactorily resolved. I also wish to thank the construction personnel that persevered during the issues with the suppliers and factory personnel as well as the hydraulic modeling laboratory for their continued help in evaluating pumping station hydraulics. Balancing Services Company and AVS Engineering, LLC are thanked for their continued assistance in diagnosing vibration issues with rotating equipment across the country.

REFERENCES

Hydraulic Institute (2009) *Rotodynamic Pumps for Vibration Measurements and Allowable Values*; ANSI/HI 9.6.4; Parsippany, New Jersey.

Hydraulic Institute (2009) *Rotodynamic Pumps for Pump Piping*; ANSI/HI 9.6.6; Parsippany, New Jersey.

Technical Associates of Charlotte, P.C. *Vibration Criteria*; Charlotte, North Carolina.

International Organization for Standardization, *10816-3:2009 – Mechanical Vibration*, Geneva, Switzerland

Identifying Cause of Rupture Disk Breaks at Alberta Capital Region Wastewater Commission Parkland Pump Station

Michael Georgalas, Associate, Stantec Consulting Services Inc.

Keywords: Hydraulic Transient Analysis, Water Hammer, Rupture Disk, Method of Characteristics, Discrete Vapor Cavity Model, Unsteady Vitkovsky Friction Method

ABSTRACT

This paper provides a case study to identify the cause of rupture disk breaks at the Parkland pump station and demonstrate the need to understand the limitations of hydraulic transient model software and importance of understanding the design and limitations of certain surge mitigation devices.

The Alberta Capital Region Wastewater Commission (ACRWC) commissioned a study to complete a hydraulic transient analysis of the Parkland pump station and force main to evaluate the hydraulic conditions in the force main, determine the reason for the frequent rupture disk failures, and provide recommendations to improve system performance and minimize or eliminate future rupture disk breaks.

The hydraulic transient analysis required a comprehensive data collection effort, construction and calibration of a hydraulic transient model, reviewing the failure history of the rupture disk and understanding operational conditions at the time of the failure, completing multiple scenarios to evaluate the effectiveness of each surge mitigation device, performing model runs using steady and unsteady transient friction methods due to limitations of the Method of Characteristics discrete vapor cavity model, and reviewing the design limitations of surge mitigation equipment.

Study results indicate the maximum pressures experienced in the force main do not exceed the design working pressure of the pipeline. Although column separation and surge pressures are likely contributing to the failure of the rupture disk, model results do not indicate the disk is rupturing due to pressures within the force main exceeding the disk burst pressure rating. Per discussion with the rupture disk manufacturer, the disk may be failing due to fatigue as it is desirable for pressures to remain below 70% of the burst pressure rating of the disk for the type of disk used. Replacing the existing rupture disk with a reverse bucking rupture disk, which can accommodate pressure fluctuations up to 90% of the burst pressure rating without fatigue, could minimize or eliminate future rupture disk breaks.

INTRODUCTION

The Parkland pump station is one of five regional pump stations owned and operated by the ACRWC and pumps wastewater through approximately 16 km (10 miles) of 900 mm (36-inch) diameter concrete force main. The pump station and forcemain have been in operation since 1987. The pump station, which is designed to accommodate five pumps, currently includes three pumps (two duty, one standby). There are three surge mitigation measures located on the pump station discharge header: a surge anticipation valve, a surge relief valve, and a rupture disk. Air valves are installed at 18 locations along the forcemain. A pump performance study was completed for the pump station in 2011 and indicated the station could discharge 445 L/s with one pump in operation and 642 L/s with two pumps in operation. The rupture disk at the pump station burst two times in 2014 and five times in 2015.

ACRWC commissioned a hydraulic transient analysis study of the Parkland pump station and force main to evaluate the hydraulic conditions in the force main, determine the reasons for the frequent rupture disk failures, evaluate the location and capacity of the existing air valves, identify deficiencies of the existing pumps and forcemain design or operation and operation in terms of surge control and mitigation, and provide recommendations to improve system performance and minimize or eliminate future rupture disk breaks.

This paper presents the approach used to complete the hydraulic transient analysis to determine the cause of the rupture disk breaks. Tasks completed as part of the hydraulic transient analysis included data collection and review; model construction and calibration; developing and analyzing hydraulic transient analysis scenarios; interpreting model results; including assessing limitations of the hydraulic transient analysis software; reviewing the design of surge mitigation measures; developing recommendations to reduce rupture disk breaks; and documentation of the study results.

Specifically, the paper highlights data collection efforts required to complete the analysis, development of numerous model scenarios, reviewing the hydraulic transient modeling results, and assessing the design and limitations of surge mitigation measures to determine the likely cause of the rupture disk breaks.

METHODOLOGY

Activities associated with completing the hydraulic transient analysis of the Parkland pump station and force main included Data Collection and Review; Model Construction, and Calibration; Developing and Analyzing Hydraulic Transient Analysis Scenarios; Interpreting Model Results, including assessing limitations of the hydraulic transient analysis software; reviewing the design of surge mitigation measures; developing recommendations to reduce rupture disk breaks; and documentation of the study results. Aspects associated with completing these activities are provided below.

Data Collection and Review

Data Collection and Review is a very important activity in completing a hydraulic transient analysis as it provides the information needed to construct the hydraulic transient model and evaluate operating conditions. A list of data required to complete the study was developed and sent to ACRWC. Items needed to complete the analysis included drawings of the pump station and forcemain; pump station control narrative; pressure and flow information from the supervisory control and data acquisition (SCADA) system; pump curves; rupture disk failure history and operating conditions at the time of the rupture disk breaks; the type, size, and status of air valves installed at each location on the forcemain; the type and pressure class of the forcemain pipe material; make and model of the pump station check valves, pump control valves, pressure control valve, surge relief valve, and rupture disk; pump control valve flow coefficient and flow characteristics; pump control valve opening and closing times as well as pump start-up and shut-down times; pressure relief valve operation; surge relief valve pressure setting; and rupture disk burst pressure setting. In addition, ACRWC was asked if any field testing been completed to confirm the pressure settings and operation of the pressure control and surge relief valves.

The forcemain was constructed using American Water Works Association (AWWA) C-301 (L) prestressed concrete lined cylinder pipe and the pipe class varies from Class 12 to Class 22. Per AWWA C-301 the pipe is designed to accommodate a transient pressure 40% above the pipe's working pressure rating. A review of the nine failures provided in the rupture disk failure history indicated the rupture disk failures typically occur after an uncontrolled pump shut-down, or a control valve failure, and most failures occurred with one pump in operation. Several types of air valves are installed along the forcemain, including air release valves, air/vacuum valves, and combination air valves. The operating condition of several valves is unknown.

Model Construction and Calibration

The hydraulic transient analysis was completed using Bentley Systems HAMMER V8i SELECTseries 5 software, which uses the Method of Characteristics (MOC) to solve equations for unsteady pipe flow. In addition, HAMMER uses the discrete vapor cavity model (DVCM) for areas that experience column separation. Column separation occurs when the negative pressures in the pipeline drop or fall below the vapor pressure of the fluid. This results in the formation of large pockets of vapor within the pipeline that later, during the passage of the returning upsurge, can collapse violently. The high pressure caused by the two liquid columns colliding together can cause pipeline ruptures or damage to system components. Operation of the pump control valves was modeled using a throttle control valve (TCV) model element and entering both the valve flow coefficient and flow characteristic information obtained from the manufacturer to accurately represent performance of the valve in the model.

The hydraulic transient analysis model was constructed using information obtained from the data collection task and calibrated based upon SCADA system information and the 2011 Pump Performance Study. The Hazen-Williams C factor within the steady-state hydraulic model was adjusted until model results agreed with the 2011 Pump Performance Study. The calibrated model, which used a factor of 100, indicated the Parkland pump station pumps require 61.2 meters of head to convey 456 L/s with one pump in operation and 71.4 meters of head to convey 678 L/s with two pumps in operation. **Figure 1** illustrates the hydraulic grade calculated by the model along the Parkland force main.

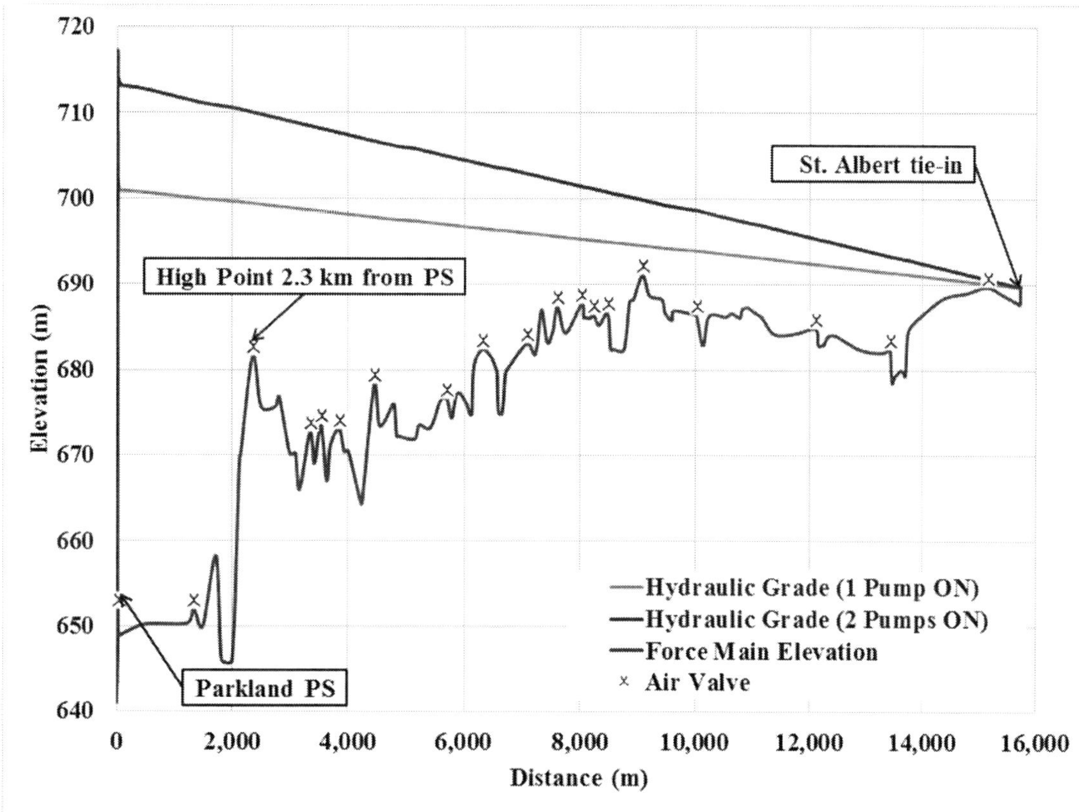

Figure 1. Parkland PS to St. Albert tie-in Hydraulic Profile Steady-state Conditions.

Scenario Development

When developing a list of hydraulic transient analysis scenarios, emphasis was placed on modeling a pump trip with one pump in operation due to rupture disk failure history. In addition, since the operating status of several of the air valves was unknown, additional model scenarios were included to complete a sensitivity analysis on the effect of the air valves on the transient head envelope (maximum and minimum head) experienced along the force main after a pump trip at the pump station. Scenarios completed to evaluate the air valves included placement of the air valves directly on the force main as well as on a lateral connected to the force main. In addition, a scenario was modeled using a 0.5 second delay to begin opening of the air valve along with a 0.5 second valve opening time. Since the air valve model element in the HAMMER software does not allow the opening of the valve to be delayed, this was accomplished by placing a throttle control valve (TCV) model element on the lateral to the air valve and creating an operating rule to open the TCV over a 0.5 second period beginning 0.5 seconds after pressure at the air valve location reached atmospheric conditions. **Figure 2** illustrates the different methods used to model the air valves.

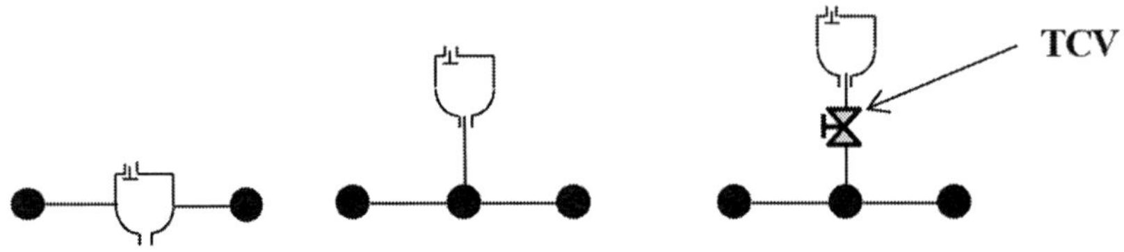

Figure 2. Methods used to model air valves.

Other modeled scenarios included a pump trip with two pumps in operation as well as a normal pump start-up and shut-down. The following is a list of modeled scenarios:

1. One Pump ON: Power failure
    a. Unprotected system (no surge mitigation devices)
    b. AVs directly on the main
    c. AVs on lateral
    d. AVs on lateral with 0.5 second delay in opening
    e. AVs and SAV
    f. AVs, SAV, and SRV
    g. AVs, SAV, SRV, and rupture disk
2. Two Pumps ON: Power failure
    a. Unprotected system
    b. Protected w/AVs, SAV, SRV, and rupture disk
3. Two Pumps ON: One pump shut-down due to pump trip (with surge mitigation devices)
4. Two Pumps ON: Normal shut-down of one pump (with surge mitigation devices)
5. One Pump ON: Normal pump shut-down
6. One Pump ON: Normal pump start-up
7. Pumps OFF: Normal pump start-up

Interpreting Model Results

Model runs of all scenarios were reviewed to confirm the results. In addition to reviewing the transient head envelope along the force main, pressure versus time was reviewed at key locations along the pump station and force main to confirm the validity of model results. For the DCVM, HAMMER assumes the size of the vapor cavity is small compared to the volume of the pipe. The default Transient Friction Method used in HAMMER is the steady-state method. Literature (Bergant et all, 2006) on the DCVM recommends the maximum vapor cavity volume should stay below 10% of the total reach volume. The literature also indicates when the DCVM model is used with the steady-state friction method one possible result is non-physical pressure spikes can occur when vapor cavities form and collapse. Interpretation of these model results requires engineering judgement. The first one or two pressure spikes after the collapse of a vapor cavity are the most reliable. Pressure spikes that appear to occur for only a single time step can typically be ignored.

Initially the steady-state transient friction method was used within the model because it yields conservative results. After reviewing model results, selected scenarios were rerun using an Unsteady - Vitkovsky friction method (Bergant, Simpson and Vitkovsky, 2001), which is the recommended unsteady friction method for use in HAMMER. This transient friction method can be used to minimize the occurrence of non-physical pressure spikes.

Reviewing Surge Mitigation Measure Design

After completing the hydraulic transient analysis and reviewing the model results the rupture disk manufacturer was contacted to review the rupture disk design and understand any limitations associated with the design. In addition, air valve manufacturer literature was reviewed to confirm the air valve size needed on the force main.

RESULTS AND DISCUSSION

Pipeline profiles with a relatively sharp increase in the pipeline elevation relative to the total elevation increase along the pipeline, or knee, a relatively short distance from a pump station are especially vulnerable to column separation. As shown in **Figure 1**, there is a knee in the Parkland force main at the high point 2.3 km from the pump station.

Model results for scenarios where the air valves were modeled directly on the force main, on a lateral connected to the force main, and with a 0.5 second delay in opening the valve indicated the method used to model the air valve did effect the minimum pressure in the immediate vicinity of the air valve and the length of force main that experienced column separation. **Figure 3** displays pressure versus time at the high point approximately 2.3 km from the pump station for each of the scenarios.

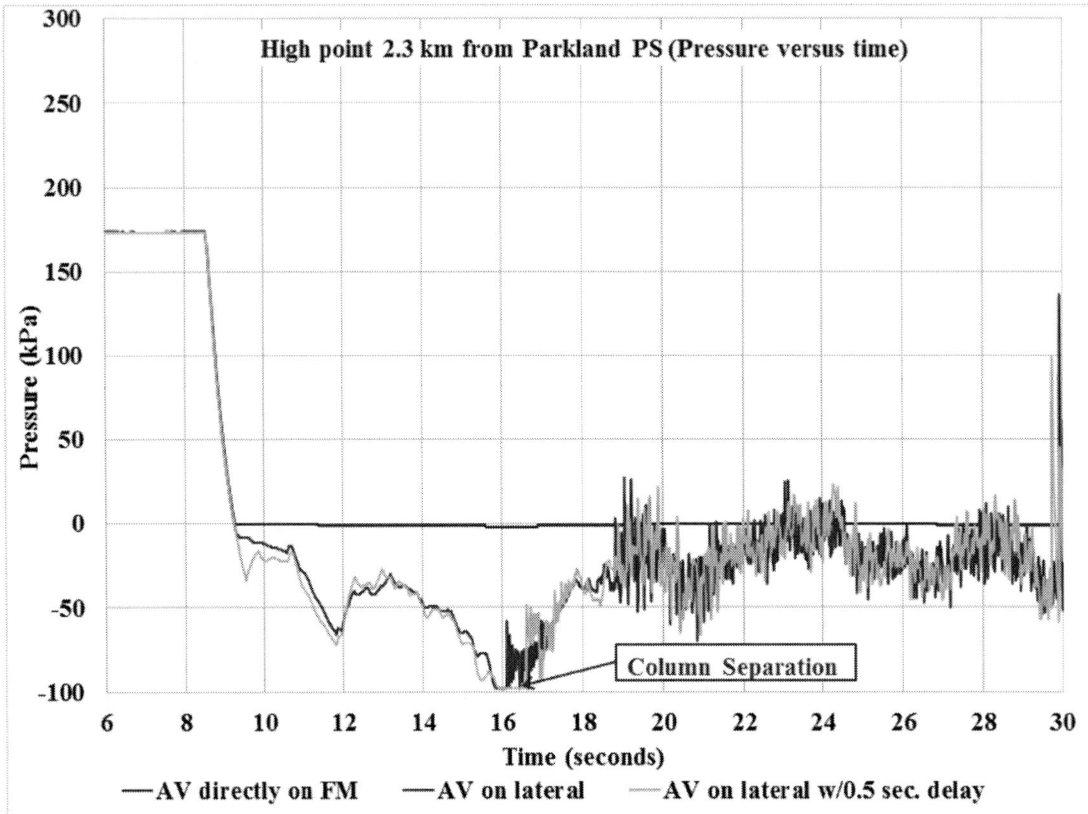

Figure 3. High point 2.3 km from Parkland PS (Pressure versus time).

Model results indicate column separation does not occur at this location when the air valve is modeled directly on the force main but does occur when the valve is located on a lateral to the forcemain. Column separation occurred along a greater portion of the force main when the air valves were modeled on a lateral connection to the force main. Therefore, the air valves were modeled on a lateral connected to the force main for all remaining scenarios as this would result in a more extreme transient head envelope. **Figure 4** displays the head envelope experienced along the Parkland force main for the scenario with the air valves placed on a lateral connected to the force main and indicates column separation occurs along a significant portion of the force main.

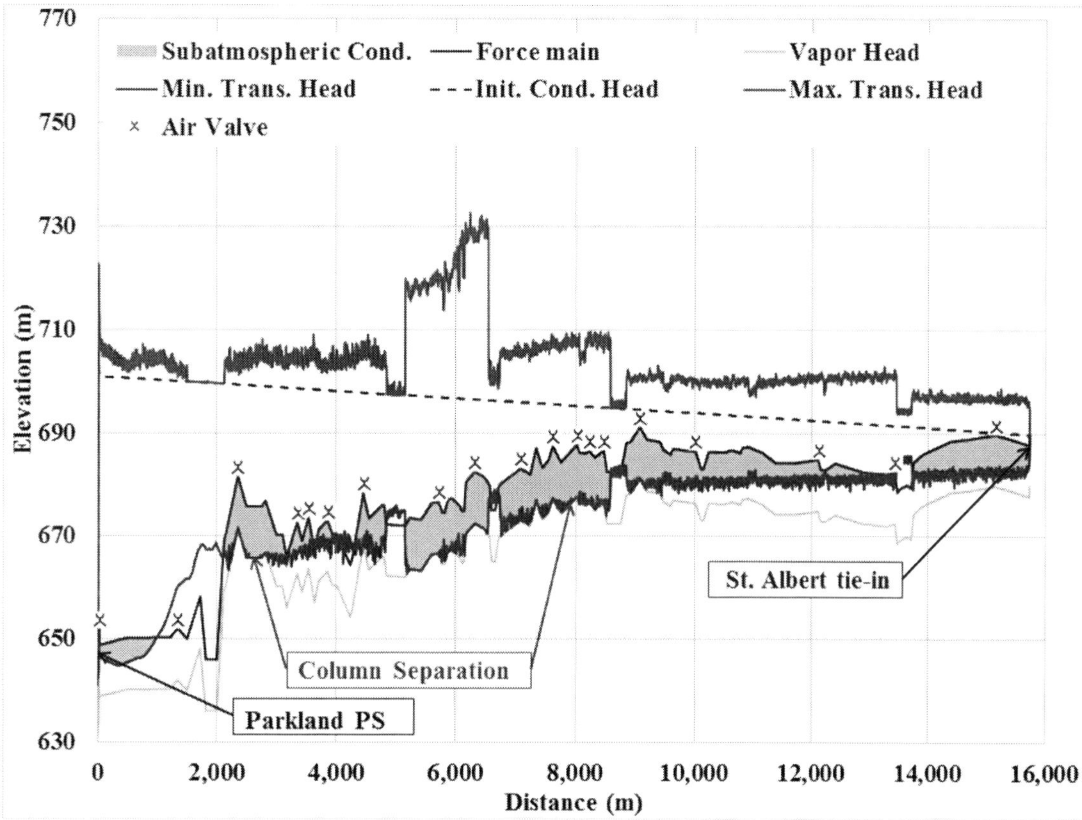

Figure 4. Transient Head Envelope w/ One Pump ON (455 L/s): Shut-down due to power failure w/air valves (modeled placed on a lateral).

The next scenario run included the air valves as well as the SAV at the Parkland pump station in the model. When activated the valve opens in 4 seconds, remains open for 20 seconds, and then closes in 30 seconds. Model results, as shown in **Figure 5**, indicate a maximum discharge of approximately 600 L/s (9,510 gpm) flows through the SAV and the increase in head that occurs upon closure of the valve is approximately 50 meters greater than the steady-state initial conditions head. The maximum pressure at the rupture disk was 1,124 kPa (163 psi), which equals the 1,127 kPa (163 psi) burst pressure.

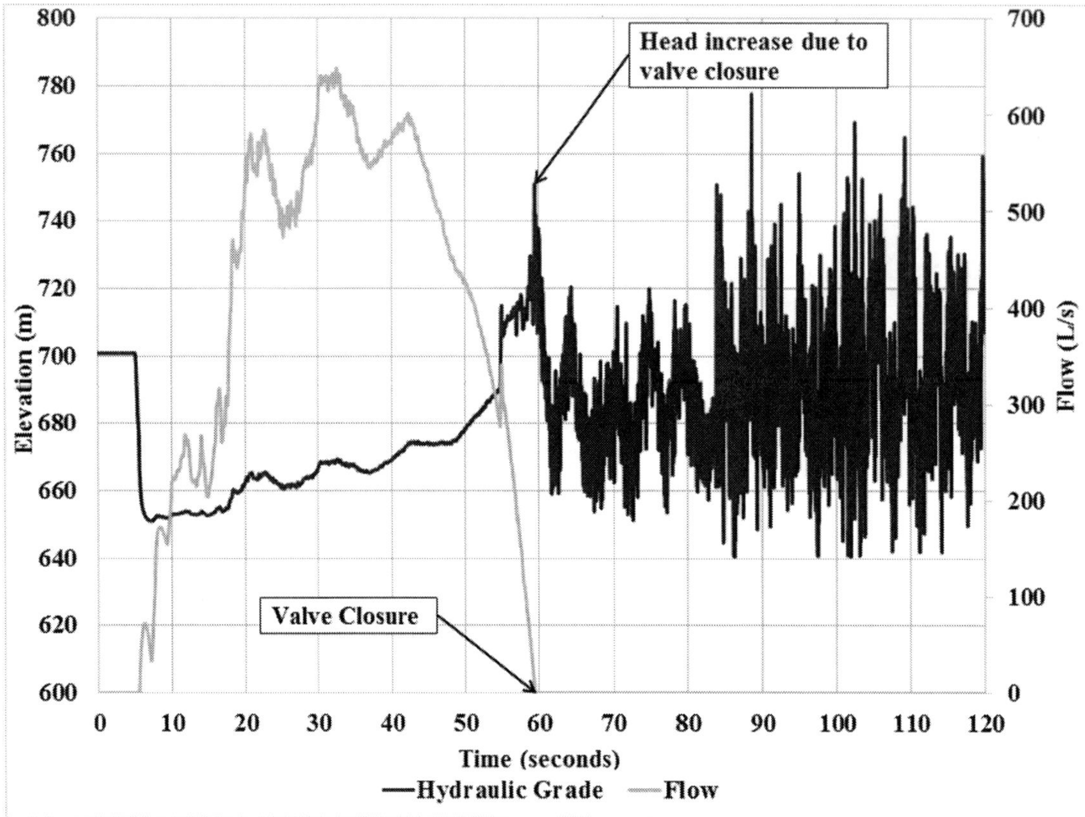

Figure 5. Flow and Head versus time at SAV w/One Pump ON (455 L/s): Shut-down due to power failure w/air valves and SAV.

A scenario was then run with both the SAV and the SRV included in the model. The pressure setting of the SRV is 760 kPa (110 psi). Model results with the SAV and SRV in the model were similar to results with only the SAV modeled in that the maximum pressure at the rupture disk was 1,037 kPa (150 psi). Although the model indicates there are a couple of occurrences of flow through the SRV, these occurrences occur after the SAV is closed and may be due to the previously referenced computational inaccuracies that can occur within the model after column separation.

Another scenario was completed with the SAV, SRV, and rupture disk. Results show the maximum pressure at the rupture disk was 1,033 kPa (150 psi) and did not exceed the burst pressure of the disk.

A pump trip event at the Parkland pump station with two pumps in operation and all surge mitigation measures functional was also modeled. Model results show column separation occurred along almost the entire length of the force main and the SRV was activated. The maximum pressure at the rupture disk was 953 kPa (138 psi).

Normal start-up and shut-down of the Parkland pump station were modeled to determine if column separation occurred within the force main under normal operation scenarios. Normal shut-down of the pump station was modeled by beginning closure of the pump control valve at t=5 seconds within the simulation and using a total closure time of 240 seconds. Once the position of the pump control valve reaches 15% OPEN, which is at t= 209 seconds within the simulation, the pump is shut-down. Since the Parkland pump station pumps are not equipped with a solid state starter, shut-down of the pump is similar to shut-down due to power failure in that shut-down of the pump itself is not a controlled shut-down. It is good design practice to open and close pump control valves over a period of four to ten times the critical time period to minimize hydraulic transients within the system. The 240 second closure time for the pump control valve is approximately 11 times the calculated 22 second critical time period of the force main. Model results, which are illustrated in **Figure 6**, indicate the maximum head experienced along the force main is less than approximately 10 meters (98 kPa), or 33 feet (14 psi), above steady-state conditions and pressures within the forcemain are only slightly less than atmospheric pressure at three locations.

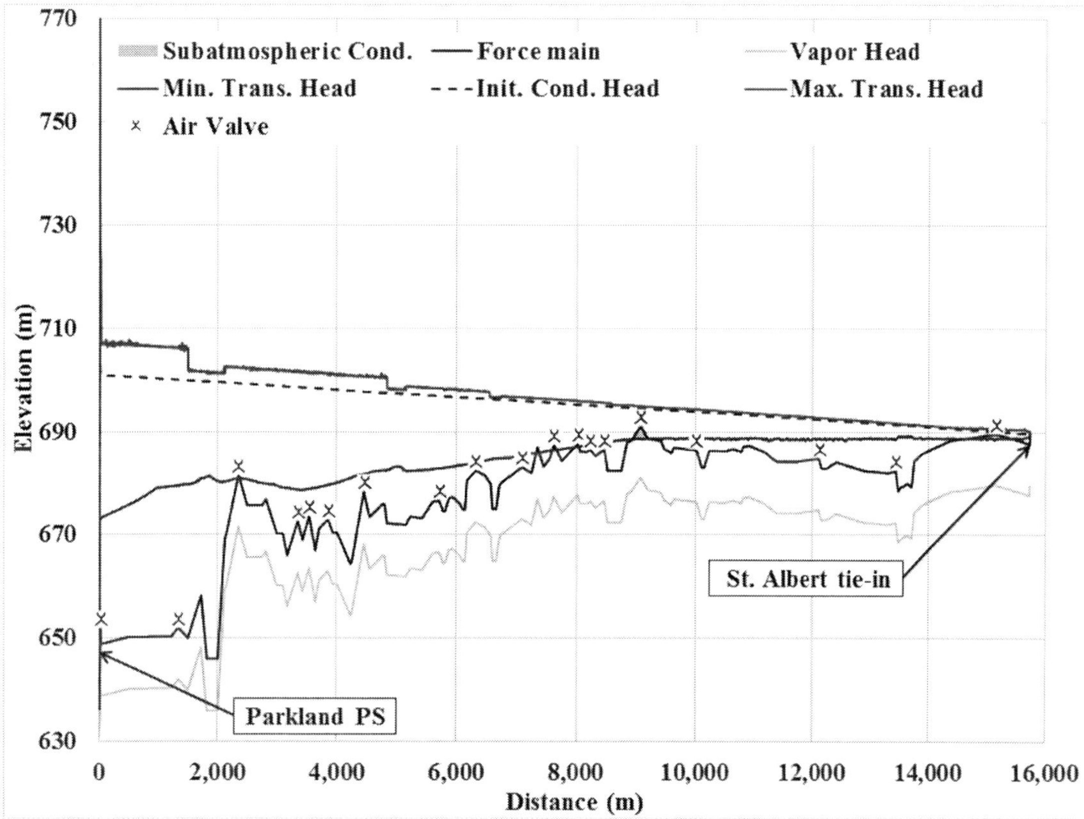

Figure 6. Transient Head Envelope w/ One Pump ON (455 L/s): Normal Pump Shut-down.

The model cannot successfully complete the simulation if initial steady-state pressures are less than vapor pressure Therefore, in order to simulate a pump startup with all pumps off, the model needed to begin the scenario with one pump on. The pump was then shut-down and started-up to complete the simulation. Model results, which are presented in **Figure 7**, indicate the maximum head along the force main does not change by more than approximately 5 meters (49 kPa), or 16 feet (7 psi), compared to initial steady-state conditions, and the minimum pressure within the force main is -11 kPa (- 2 psi). Model results for a normal pump start-up and shut-down indicated column separation did not occur during the simulation and the opening and closing time for the pump control valves was effective at limiting the pressure fluctuations within the force main.

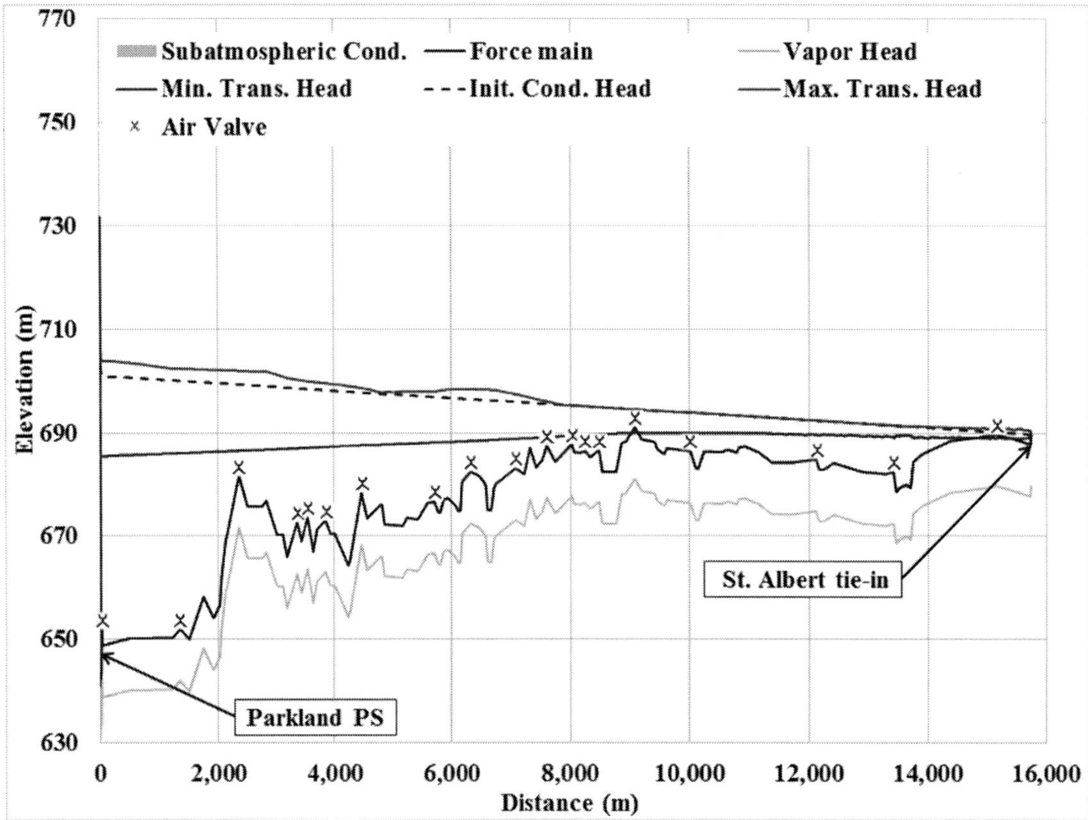

Figure 7. Transient Head Envelope w/Pumps OFF: Normal Pump Start-up.

Based upon an initial review of model results for the scenarios, it appeared the cause of the rupture disk breaks may be attributed to the upsurge that occurs in the portions of the force main that experience column separation once the vapor pockets collapse and the liquid column recombines. However, a review of the pressure trace at the rupture disk indicated the maximum pressures appear to occur for only a single time step and may be a non-physical pressure spike. **Figure 8** displays pressure versus time at the rupture disk for the modeled scenarios.

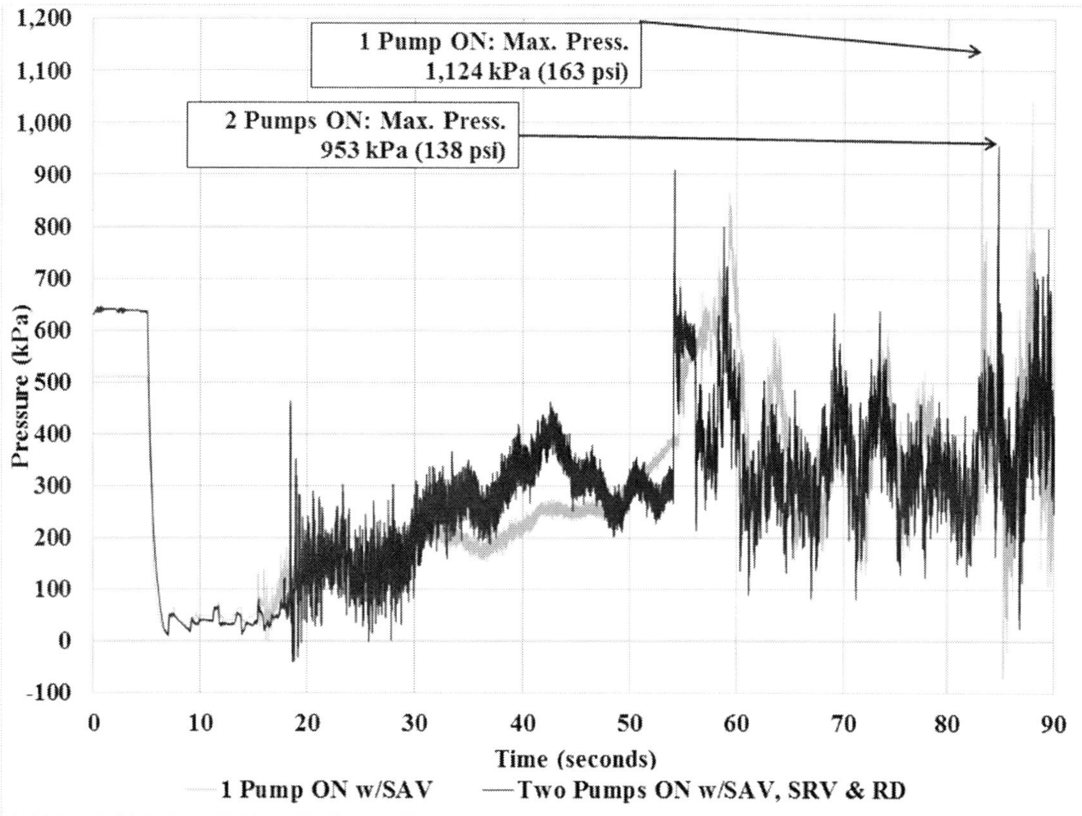

Figure 8. Parkland Pump Station Rupture Disk (Pressure versus Time).

To assess if the maximum pressures exceeding 1,100 kPa (160 psi) were attributable to non-physical pressure spikes the scenarios were rerun using the unsteady friction method from Vitkovsky. Model results using the unsteady method are shown in **Figure 9** and indicate the maximum pressure at the rupture disk is less than 1,000 kPa (145 psi), which is less than the rupture disk burst pressure. Therefore, while column separation and surge pressures are likely contributing to the failure of the rupture disk, model results do not indicate the disk is rupturing due to pressures within the force main exceeding the disk burst pressure rating.

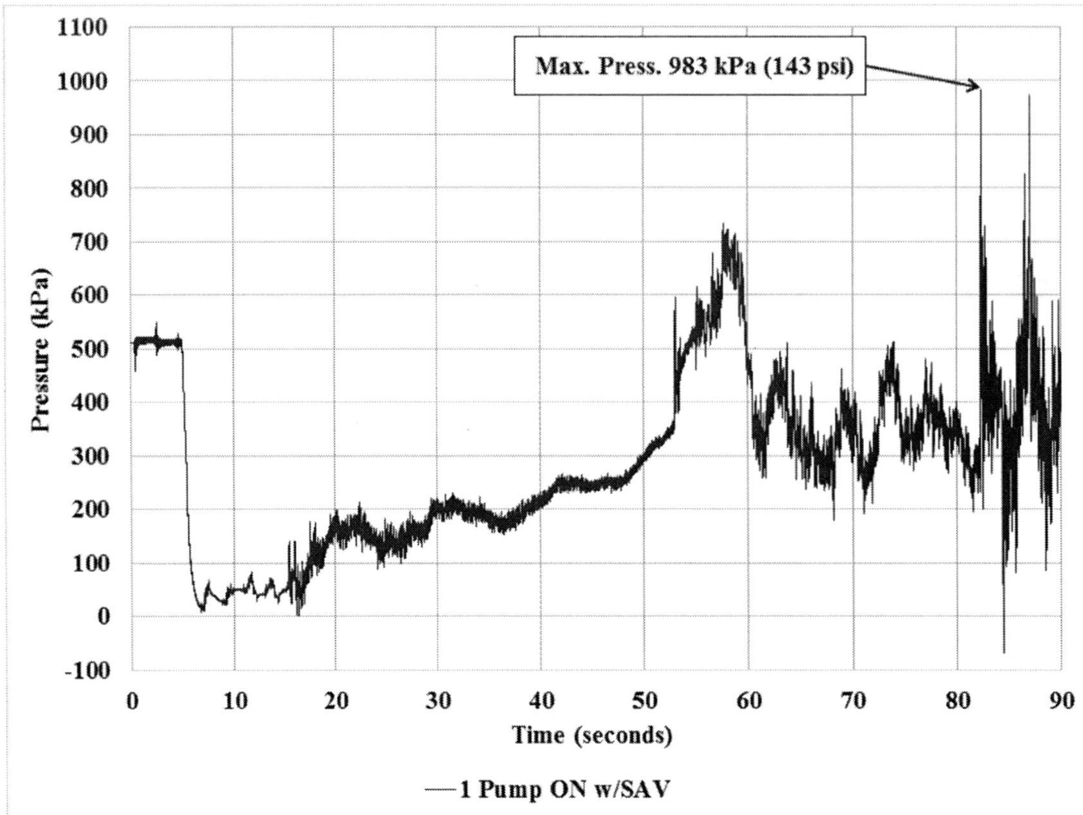

Figure 9. Parkland Pump Station Rupture Disk using Unsteady Friction Method (Pressure versus Time).

The existing surge mitigation measures currently installed at the Parkland pump station; which include the SAV, SRV, and rupture disk; do not protect against the downsurge that occurs after a pump trip at the pump station but do help mitigate the upsurge that occurs once the vapor pockets collapse and the liquid column recombines. The working pressure rating of the Parkland force main is 1,100 kPa (160 psi) from the Parkland pump station to approximately 1.75 km (5,700 feet) downstream of the pump station. The 1,127.33 kPa (163 psi) burst pressure of the rupture disk is essentially set at the working pressure rating of the Class 16 force main pipe, which is designed to withstand a total pressure with surge of 1,544 kPa (224 psi). Model results did not indicate the pressure within the force main exceeded 1,544 kPa (224 psi) during any of the modeled scenarios.

Surge Mitigation Measure Evaluation

After completing the initial hydraulic transient analysis, the surge mitigation measures were reviewed, and additional scenarios were run to evaluate the effectiveness of modifications or additions to the existing mitigation measures.

Except for the two air valves nearest the pump station, which are surge-suppression air valves, the remaining air valves are either air release valves, air and vacuum valves, or combination air valves. Special consideration must be given to the selection of air valves at locations along pipelines that may experience macro-cavitation, or column-separation. Although it is desirable to allow air to freely enter the pipeline upon a down-surge to limit subatmospheric pressures and column separation, the air should subsequently be released from the pipeline in a controlled manner to allow the air within the pipeline to act as a cushion when the line is re-pressurized by the moving liquid columns. The air and vacuum valves and combination air valves located on the forcemain do not release air from the force main in a controlled manner upon re-pressurization of the pipeline. The pump trip event with two pumps in operation scenario was rerun with surge-suppression air valves installed at all air valve locations on the force main. Model results show the maximum head along the force main was significantly reduced and the maximum pressure at the Parkland pump station discharge header did not exceed the pressure setting of the rupture disk or the SRV.

In addition, a sensitivity analysis of the air valve size was completed by running the pump trip with one pump in operation scenario using 150 mm (6-inch) air valves at all the 100 mm (4-inch) air valve locations. Model results indicated the transient head envelope along the force main did not significantly change when the size of the air valves was increased.

The pump trip with two pumps in operation scenario was rerun with a 37,854 liter (10,000 gallon) surge tank located on the discharge header of the Parkland pump station to minimize the downsurge within the force main caused by the pump trip. Model results are shown in **Figure 10** and indicate the surge tank, in combination with the surge-suppression air valves, prevent column separation from occurring within the Parkland force main.

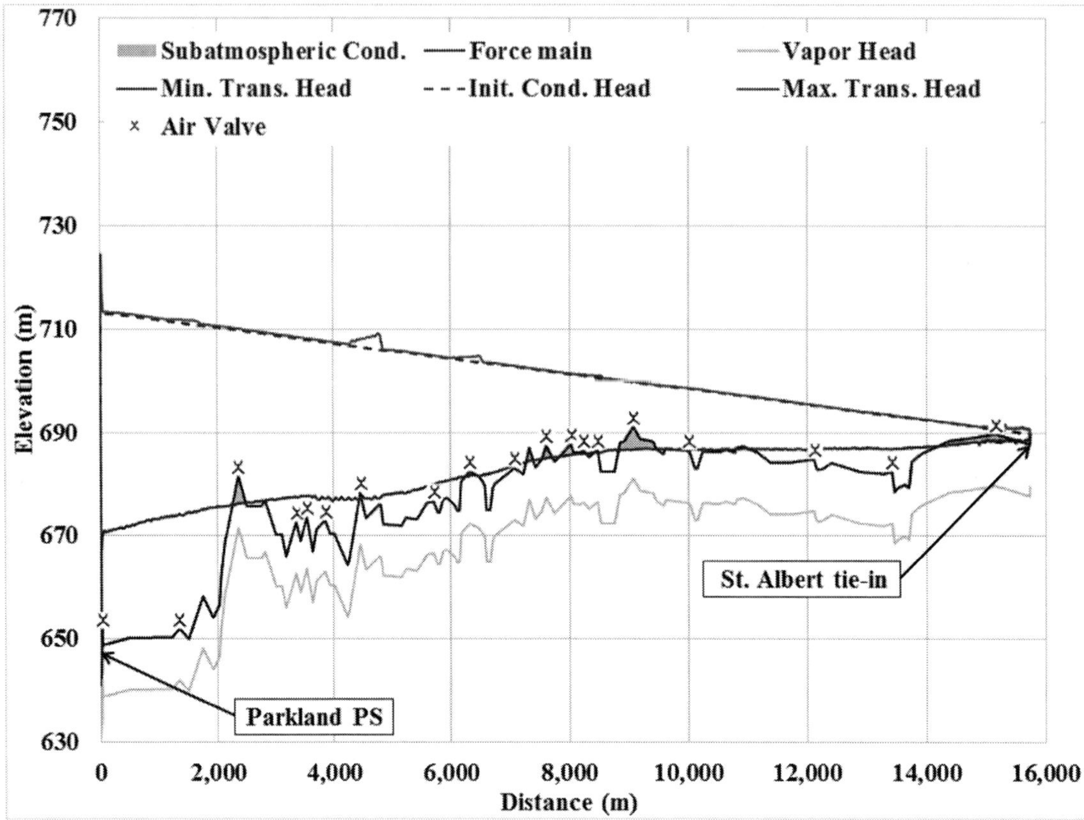

Figure 10. Transient Head Envelope w/Two Pumps ON (678 L/s): Shut-down due to power failure w/surge-suppression air valves and 37,854 liter (10,000 gallon) surge tank.

The type of rupture disk currently being used is a tension loaded rupture disk, which is an older design, and for this type of design it is desirable for pressures to remain below 70%, or 788 kPa (114 psi) of the burst pressure rating of 1,127 kPa (163 psi) of the disk. The manufacturer of the rupture disk was contacted to investigate the possibility of the existing rupture disk failing due to fatigue. The manufacturer's representative indicated it was possible the rupture disk was failing due to fatigue and suggested one available option would be to replace the tension loaded rupture disk with a reverse bucking rupture disk. The reverse buckling rupture disk can accommodate pressure fluctuations up to 90%, or 1,014 kPa (147 psi) of the burst pressure rating without fatigue. Manufacturer provided literature indicates the reverse buckling disc can withstand 1,000,000 pressure cycles from 0-90% of its marked burst pressure.

SUMMARY AND CONCLUSIONS

Results indicate the maximum pressures experienced in the force main do not exceed the design working pressure of the force main. While column separation and surge pressures are likely contributing to the failure of the rupture disk model results do not indicate the disk is rupturing because pressures within the force main are exceeding the burst pressure rating of the disk. Per discussion with the rupture disk manufacturer, the disk may be failing due to fatigue as it is desirable for pressures to remain below 70% of the burst pressure rating of the disk for the type of disk used. Replacing the existing rupture disk with a reverse bucking rupture disk, which can accommodate pressure fluctuations up to 90% of the burst pressure rating without fatigue, could minimize or eliminate future rupture disk breaks. The installation of transient pressure monitoring equipment is recommended to confirm model results.

Study results demonstrate the importance of understanding the limitations of the software, thoroughly reviewing model results, and identifying any design limitations associated with transient mitigation measures.

Based upon the results of the study the following activities were recommended:

1. Install dual orifice, triple acting, or triple-orifice air valves at all locations along the force main to release air from the pipeline in a controller manner. Install a second air valve at each of the existing air valve locations to provide redundancy in the event one of the air valves is not functioning or requires service.

2. Install either permanent or temporary transient pressure monitoring equipment at the Parkland pump station discharge header to monitor transient pressures and confirm pressures experienced at the pump station after a pump trip.

3. If the information from the transient pressure monitoring equipment indicate pressures within the force main do not exceed the burst pressure rating of the rupture disk after a pump trip the existing tension loaded rupture disk could be replaced with a reverse bucking rupture disk to minimize the number of rupture disk failures.

4. If information from the transient pressure monitoring equipment reveals pressures greater than expected (higher than the force main pressure rating with the surge allowance) then consider installing a 37,854 liter (10,000 gallon) surge tank on the discharge header of the Parkland pump station to prevent column separation within the force main.

REFERENCES

Bergant, A., Simpson, A. and Vitkovsky, J., "Developments in unsteady pipe flow friction modeling," Journal of Hydraulic Research, Vol. 39, No. 3, 2001

Bergant, A., Vítkovský, J. P., Simpson, A. R., Lambert, M. F., & Tijsseling, A. S. (2006). Discrete vapour cavity model with efficient and accurate convolution type unsteady friction term. (CASA-report; Vol. 0626). Eindhoven: Technische Universiteit Eindhoven

After the 104 Days and 104 Nights: Improvements to the N. 28th Stormwater Pumping Station Following the 2011 Missouri River Flood

Tim Schneller, P.E.
George Butler Associates, Inc.
Lenexa, Kansas

Jeff Krist, P.E.
Public Works Department
City of Council Bluffs, Iowa

ABSTRACT

Prompted by historical conveyance system flooding and the Missouri River Flood of 2011, the City of Council Bluffs, Iowa proceeded with the replacement of the N. 28th Stormwater Pumping Station. The project included permitting with the United States Army Corps of Engineers for improvements in the critical zone of the Missouri River Levee. Based on previous vandalism at the site, the design included measures to prevent vandalism in the future. During the design phase, the project evolved in two ways. Formed suction intakes were used to reduce the footprint of the pumping station. Also, the project was phased to allow for future capacity improvements to stay within funding from a sales tax initiative.

KEYWORDS

Pumping station, levee critical zone, flood, phased improvements, vandalism prevention, permitting.

INTRODUCTION

Most of the residents of the City of Council Bluffs, Iowa ("City") are protected from flooding by the Missouri River Council Bluffs Flood Protection ("Levee"). This Levee is comprised of 28 miles of earthen levees and 11 stormwater pumping stations. When the Missouri River water level rises to flood stage, the City closes gates on the gravity storm pipes through the Levee. The stormwater pumping stations provide protection against flooding inside of the Levee during these time periods.

In response to historical storm flooding upstream of the N. 28th Street Stormwater Pumping Station, the City started to make plans in 2010 to improve the pumping station and upstream sewer system. Figure 1 shows the location of the tributary area, Missouri River Levee, pumping station, and trunk storm sewer.

The original N. 28th Street Stormwater Pumping Station, shown on Figure 2, was constructed in 1957 and included a 60-inch dia. reinforced concrete pipe (RCP) gravity storm sewer through the Levee. As a bypass, the facility included a bar screen, a single 250-horsepower pump, and 30-inch dia. cast iron pipe force main over the Levee.

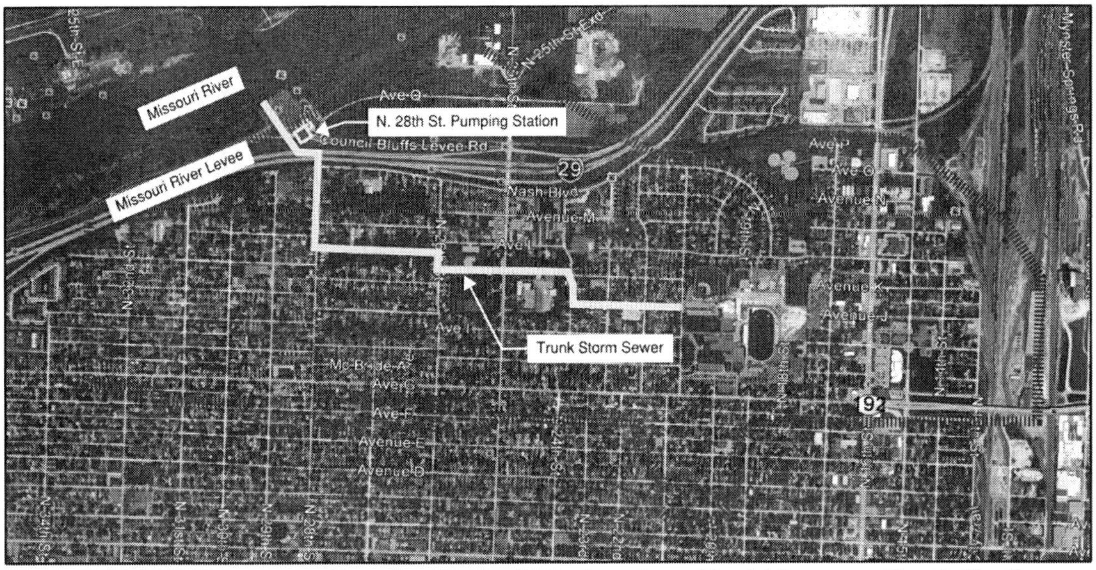

Figure 1. N. 28th Stormwater Pumping Station Tributary Area (Aerial Imagery, Google Earth, 2016)

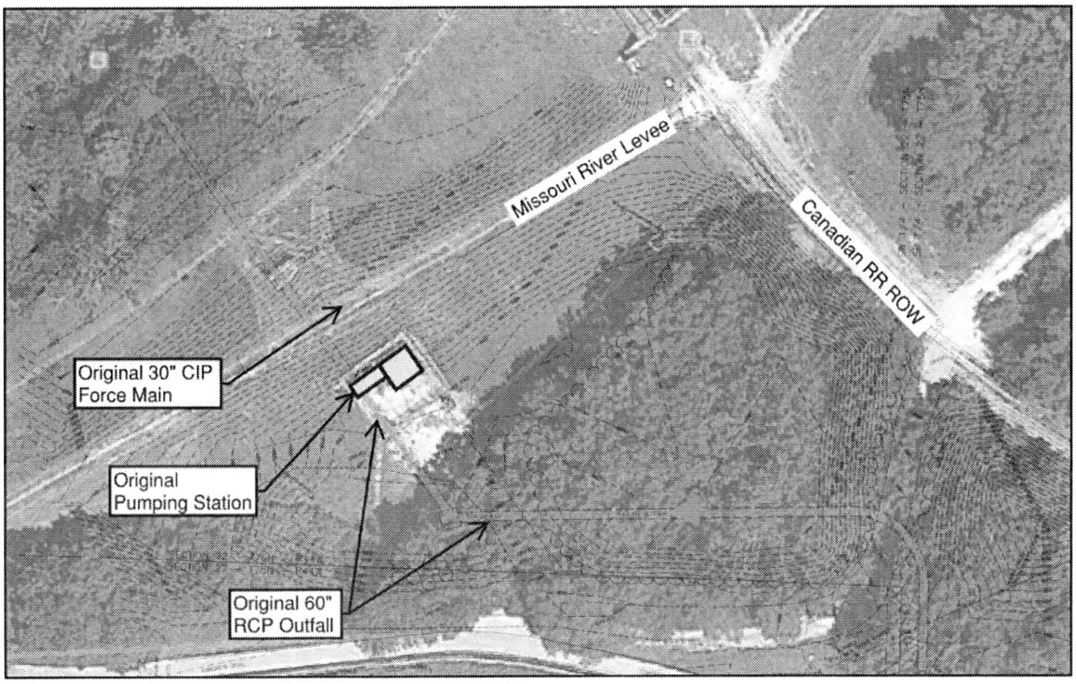

Figure 2. Original N. 28th Stormwater Pumping Station Site (Aerial Imagery Google Earth, 2009)

Previous studies had indicated that the pumping station and conveyance system were undersized for the tributary flow. In addition, the original design required the 60-inch gravity storm sewer to be at full-pipe before flow would enter the pumping station wet well. This caused a lack of capacity in the system at the beginning of storm events.

The historic 2011 Missouri River flood caused 104 days of flood levels. The extended duration of the flood besieged many communities including the City. At the N. 28th Street Stormwater Pumping Station, the City observed sinkholes over the 60-inch dia. RCP storm sewer. This prompted the City and the U.S. Army Corps of Engineers (USACE) to fill the pumping station inlet to the wet well and the 60-inch dia. RCP storm sewer under the Levee with concrete to prevent failures that would endanger the levee. To provide some relief of storm flows in the inside of the Levee, the City installed a temporary pumping station and 36-inch force main. During the height of the flood, the existing station was flooded to a depth of three feet above the ground surface as shown on Figure 3.

Figure 3. Original N. 28th Street Pumping Station After Flood of 2011

Immediately following the flood event, the USACE proceeded with a project to replace the 60-inch pipe under the Levee and repair the wet well that was full of concrete. The City also moved forward with increasing the capacity of the N. 28th Street Stormwater Pumping Station and the upstream storm sewer system. The City had the following goals for the project:

- Increase the level of protection against flooding upstream.
- Phase the project to allow completion within the available funds generated by the Local Option Sales Tax.
- Keep the new pumping station on the existing City-owned property adjacent to the original pumping station.
- Raise the top elevation of the new pumping station so it would not flood again.
- Protect the new pumping station against vandalism.

CONCEPTUAL STUDY OF CONVEYANCE IMPROVEMENTS

To begin the project, a conceptual study was performed to evaluate the scope of improvements necessary to meet the City's goals. The primary focus of the study was to determine the appropriate and affordable level of protection for the proposed improvements. A WinTR-55 model was created to evaluate storm events in the tributary area. The flows at the pumping station ranged from 150 cubic feet per second (cfs) for the 1-year storm event to 720 cfs for the 25-year storm event. As a comparison, the existing 60-inch diameter RCP storm sewer upstream of the pumping station had a capacity of 75 cfs.

To evaluate the estimated costs of the proposed levels of protection, alternate pumping station layouts were developed for the storm events. All conceptual layouts included a mechanically-cleaned bar screen, a fore bay with diverging side walls (20° maximum), individual pump bays, and a side well for a jockey pump to dewater the wet well and help with grit removal.

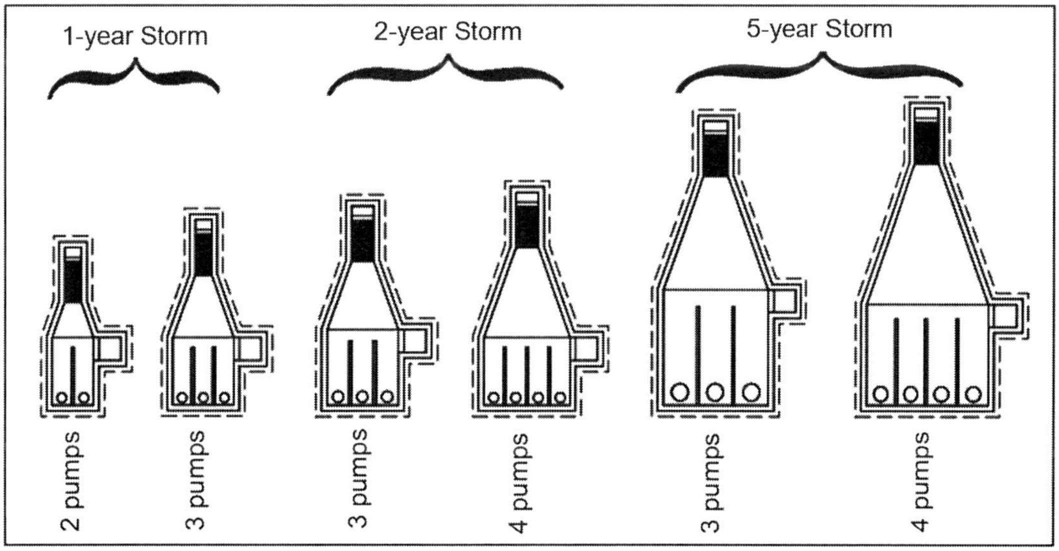

Figure 4. Alternative Pumping Station Sizes Considered in Conceptual Study

The study included development of cost estimates for the new pumping station. The study also included an analysis of incremental cost increases for pumping capacity and storm sewer conveyance that would convey one-year, two-year, and five-year storm events. The City requested that the cost evaluation look strictly at the pumping station and storm pipe costs. The cost estimates were not to include ancillary costs (e.g. streets, utility relocations, etc.) which would be common to all alternatives. Table 1 was developed to show varying storm events and the comparable costs for the necessary infrastructure to convey the flows.

Table 1. Conceptual Opinions of Probable Construction Costs

Event	Discharge, Q (cfs)	New Pumping Station Costs ($M)	Upstream Storm Pipe Costs ($M)	Total Costs ($M)
1-year	150	$6.6	$2.2	$8.8
2-year	250	$8.4	$4.7	$13.1
5-year	420	$11.9	$6.0	$17.9
10-year	560			
25-year	720			

The City selected the 2-year storm event based upon available budget from the Local Option Sales Tax fund. The City also chose to move forward with the design of a pumping station with the following elements:

- Four (4) pumps would convey a peak flow of 250 cfs (112,000 gallons per minute)
- The existing 60-inch dia. RCP through the Levee (as replaced the USACE) would remain in service
- A mechanically-cleaned bar screen would be installed to protect the pumps
- A bridge crane would be included to remove the pumps
- An electrical building would be installed to house the electrical equipment
- The station would be protected against vandalism

Initial discussions were held with the USACE staff to determine the regulations for the design of the pumping station and discharge piping. Based on the requirements for a pressurized conduit through the Levee and the cost of excavation through the Levee, the City decided to proceed with design of a pumping station that would pump flows through force mains over the levee.

DESIGNING FOR A SMALL FOOTPRINT

The conceptual pumping station wet well layout shown in Figure 4 was patterned after the Hydraulic Institute (HI) American National Standard for Pump Intake Design (1998) layout for expanding concentrated influent flows. This wet well design, as shown in Figure 5, reduces the potential for non-uniform flow at the pump intakes. Non-uniform flow can include swirling, vortex formation, unbalanced flow distribution, and air ingestion, which results in pump vibration and losses in pump efficiencies. This design was successfully used by the City on a previous stormwater pumping station project. Based on the HI design parameters, the length of a 4-pump wet well sized for a 2-year storm stretched to 80 feet.

As shown on Figure 2, the existing site was constricted by the Levee to the north, the existing 60-inch dia. RCP to the south, and the Canadian RR Right-of-Way (ROW) to the east. During the pumping station layout, it became apparent that the HI Intake Design for expanding concentrated influent flows would not fit on the current site. An alternative design was sought to reduce the pumping station footprint.

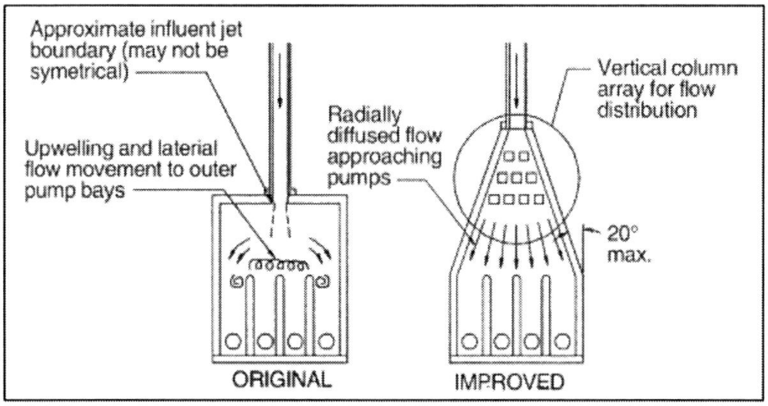

Figure 5. Pump Intakes for Expanding Concentrated Flows (Hydraulic Institute, 1998)

As an alternative to the Pump Intake Design for expanding concentrated influent flows, the HI suggests the use of formed suction intakes (FSI). Developed by the USACE, the FSI design, as shown in Figure 6, promotes better flow patterns into pump intakes. The FSI design provides two benefits. First, approach flow orientation and velocity is of less concern, thus the length of the pumping station can be reduced. Second, a FSI reduces the required minimum pump submergence. This results in a reduction in overall wet well depth.

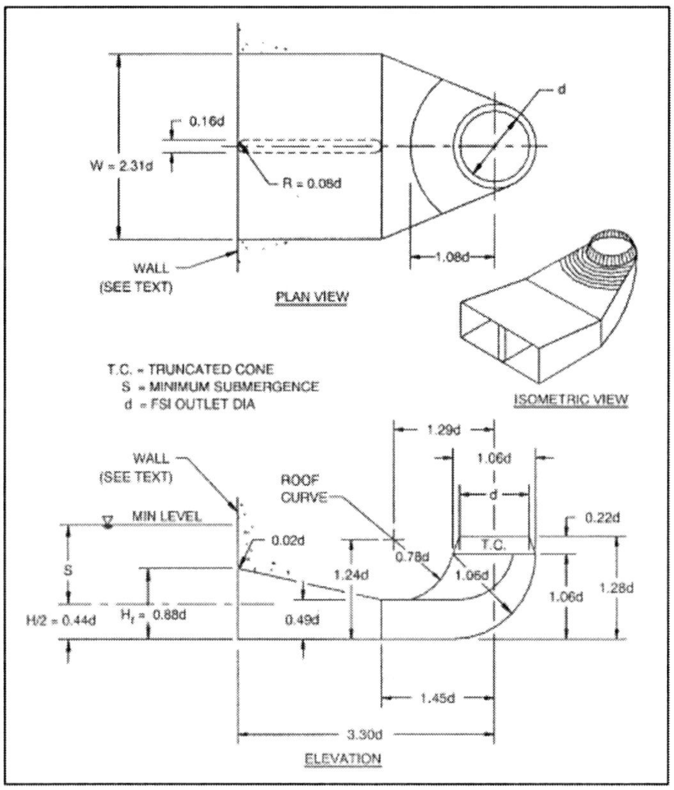

Figure 6. USACE Formed Suction Intake No. 10 (Hydraulic Institute, 1998)

The City has standardized on Xylem Flygt pumps for its stormwater pumping stations. Discussions with the local representatives for Xylem Flygt revealed that Xylem Flygt had developed a design of a compact wet well for their submersible stormwater pumps that employed FSI intakes. This design incorporated a baffle wall to dissipate the velocity of the incoming flow and spread the flow across the wet well. Based on dimensions of the compact wet well design, the station length was reduced by 20 feet.

PHASING OF CAPACITY IMPROVEMENTS

Proceeding into the design phase, the costs associated with the pumping station prompted the City to phase the capacity improvements. The pumping station would be constructed first. Upstream storm sewer improvements would be constructed in later years as funds become available.

GBA presented a preliminary design submittal to the City with an ultimate pumping capacity of 112,000 gallons per minute (gpm) or 250 cfs. This design featured four (4) submersible stormwater pumps and a submersible non-clog jockey pump to dewater the wet well and remove grit. The jockey pump was designed in a separate well on the side of the main wet well.

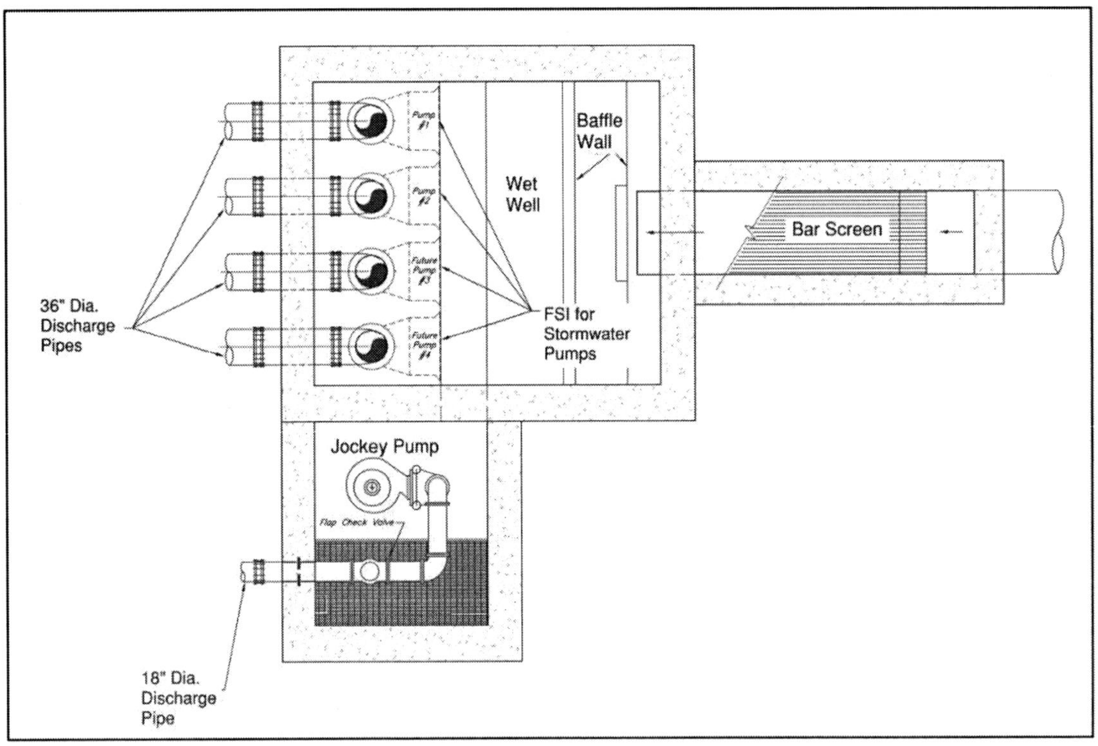

Figure 7. Preliminary Wet Well Design

Based on initial cost estimates, the City decided to phase the pumping capacity to match the capacity of the upstream storm conveyance system which would be phased in over time. This approach also complemented the City's need to spread out the funding for the project. In addition, the decision was made to install two (2) non-clog submersible pumps in the main wet well and eliminate the jockey pump in the separate well. Figure 8 shows the layout of the wet well for Phase 1.

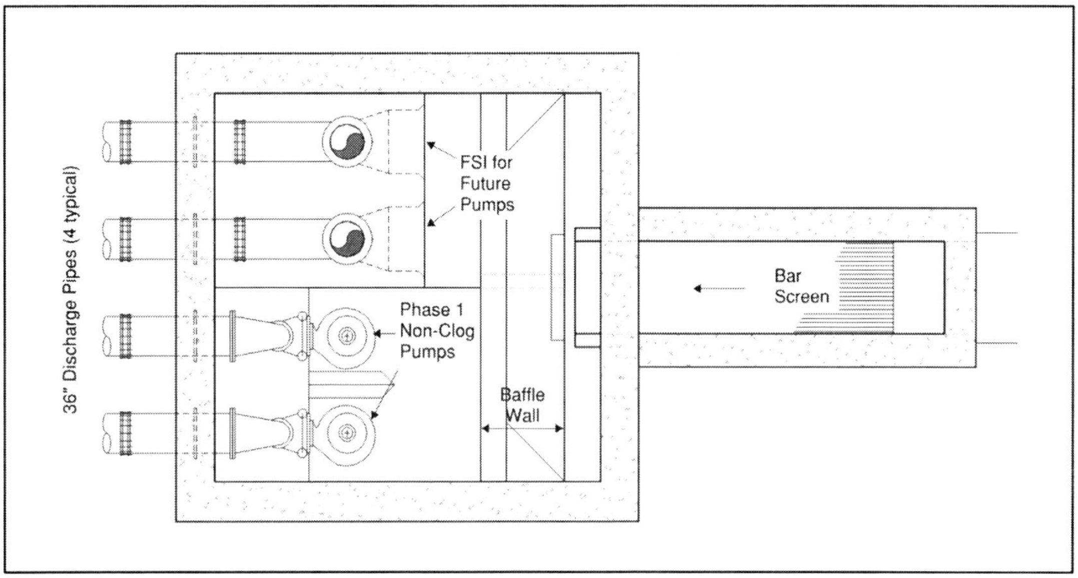

Figure 8. Final Wet Well Design

Based on the City's direction, a design was proposed that featured a combination of non-clog pumps and mixed flow pumps with variable frequency drives (VFDs) to provide the following pumping capacities:

- Phase 1
  - Install two (2) 20-inch diameter submersible non-clog pumps and VFDs
  - Total flow capacity of 6,000 gpm to 34,000 gpm
  - Matches existing upstream storm sewer capacity of 75 cfs
  - Install FSI and pump tubes for the Phase 2 submersible mixed-flow pumps

The City will implement following phased improvements as increases in the upstream conveyance capacity are completed:

- Phase 2 (Future)
  - Install two (2) 44-inch diameter submersible mixed-flow pumps and VFDs
  - Total flow capacity of 6,000 gpm to 75,000 gpm

- Phase 3 (Future)
  - Replace the two (2) non-clog pumps with 44-inch diameter submersible mixed-flow pumps
  - Total flow capacity of 5,000 gpm to 112,000 gpm (250 cfs)

The Phase 1 project also included installation of the following features that would make the future phases relatively easy:

- The electrical system was designed for the future electrical load. The exception is the emergency power engine generator which was designed for the current load, but can be replaced or paralleled with a 2nd generator.
- Phase 1 also included installation of all four (4) discharge pipes over the Levee.
- The bar screen and solid waste collection system was designed for ultimate flows.

INCORPORATING LESSONS LEARNED FROM 2011 FLOOD

During the 2011 Flood, the City experienced several problems with their stormwater pumping stations. The following items were incorporated into the design of the N. 28th Stormwater Pumping Station based on those lessons learned.

In southern Council Bluffs, The Veterans Memorial Highway Stormwater Pumping Station failed during the 2011 Flood. The cause of the failure was determined to be forces from station buoyancy and hydraulic uplift that overwhelmed the foundation and caused the entire pumping station to shift.

The City's desire to raise the N. 28th Stormwater Pumping Station site due to the flooding in 2011 increased the potential of settlement due to consolidation. For these reasons, the foundation was designed to be supported by piles.

The use of auger displacement piles, as show in Figure 9, is a new construction technique that helped reduced the number of piles required for this project. Auger displacement piles are similar to auger cast piles but have two added benefits. First, during pile construction, soil is pushed outward to make room for the concrete pile. This provides more support due to the densification of the surrounding soil, thus increased friction for pile strength. Pressurized concrete is then introduced as the drill is withdrawn. The second benefit is the elimination of the excavated soils typical of traditional auger cast piles. Test piles were installed and load tested at the beginning of the pile construction.

Figure 9. Auger Displacement Pile Installation

The flood waters during the 2011 Flood obliterated the outfall structure for the original N. 28th Stormwater Pumping Station. Because of this failure, the City wanted the outfall and discharge pipes for the new pumping station to be secure, even if a future flood washed away the soil around the outfall and pipes. For this reason, the outfall structure was constructed on steel H-piles. The four (4) discharge pipes were constructed of welded steel pipes supported by bents on H-piles, as shown in Figure 10. The H-piles were driven 80 to 100 feet below grade.

Figure 10. Bents on the 36-inch Dia. Discharge Pipes

NAVIGATING WORK IN LEVEE CRITICAL ZONES

The construction of the N. 28th Stormwater Pumping Station impacted the Missouri River Council Bluffs Flood Protection Levee. The Levee is constructed of an earthen berm with thick clay blankets on either side of the berm to control underseepage. The Levee has a critical zone that spans 300 feet from the centerline of the levee on the river side (wet side) to 500 feet from the centerline on the landward side (dry side).

In preliminary discussions with the USACE, the City explored the possibility of installing a larger pipe that could be pressurized through the Levee. The USACE indicated that a pressurized pipe crossing through the Levee must be designed to meet the following criteria:

1. Resiliency: Pipe must have shutoff valves on both sides of the levee
2. Redundancy: Must be a pipe within a pipe
3. Robustness: Must be able to function under a wide variety of conditions

Based on the anticipated cost to meet the USACE requirements, the City decided instead to install separate pressurized discharge pipes for each of the four (4) stormwater pumps over the top of the Levee. Even with this plan, the proposed pumping station and discharge pipes still involved excavation into the clay blankets on either side of the Levee as well as excavation into the projected slope of the Levee for the construction of the pumping station wet well.

To obtain USACE approval for the project, the City was required to submit a request for approval of a Minor Section 408 Modifications of Flood Risk Reduction Projects. This process involved three (3) approvals during the design and construction phases. The first was the request for approval of the design documents. In addition to construction drawings and specifications, the request addressed the following:

- The scope of the modifications
- A discussion of the construction impacts on the Levee including the level of protection of the existing Levee, excavation during construction, groundwater, underseepage, uplift,

slope stability, erosion protection, hydraulic conveyance, encroachment into the foreshore, and ponding areas
- The proposed mitigation measures that would be taken to address these impacts
- The operation and maintenance requirements associated with the improvements

The second submittal was the Emergency Action Plan (EAP) prepared by the Contractor. This submittal included the following:

- A description of the proposed construction sequence
- Procedures that would be used during normal operations as well as during flood conditions
- Emergency contacts; a layout of the construction site including excavations, stockpiles, and staging areas
- Equipment and materials maintained on site to address emergency conditions
- A professionally-engineered design of the coffer dam for construction of the wet well
- Dewatering details
- A project schedule
- Haul routes and soil fill materials.

The third and final submittal was the Construction Report. This report described the modifications made to the Levee and included the following:

- Record drawings
- A construction history
- The Contractor's EAP
- Comprehensive documentation of the test results and observations
- Design and construction changes
- Photography of the construction

In addition to the Section 408 permit, the following permits were obtained for the project:

- Section 404 Nationwide Permits No. 12 and 13—USACE
- Floodplain Permit—Iowa Department of Natural Resources
- Sovereign Lands Permit—Iowa Department of Natural Resources
- Storm Water Discharge for Construction Activities—Iowa Department of Natural Resources
- Air Quality Construction Permit—Iowa Department of Natural Resources

One of the major aspects of the Section 408 permit was the construction schedule. The allowed window for excavation on levees is from November 1st through March 31st. In addition, Section 404 permitting with the USACE determined that construction on the banks of the Missouri River could affect the Pallid Sturgeon and would be further limited to July 1st through February 28th. These dates were included in the bidding and construction documents.

ARMORING THE FACILITY TO STOP VANDALS

Due to the relative remoteness of the site, the original N. 28th Stormwater Pumping Station had a history of vandalism. Most of the vandalism at the City's pumping stations was related to the theft of precious metals in the electrical system and generator radiators. The City desired that the new pumping station provide protections against vandalism. The following elements were incorporated into the project.

Addressing visibility of the site was a major effort. The original pumping station site was set back and down from Interstate 29. To make the new pumping station more visible, the site was raised up to be roughly level with Interstate 29, the trees surrounding the site were removed, and sufficient site lighting was provided to make the site visible at night. In addition, the City elected to install four (4) video cameras and a high speed SCADA system. This allows the City to remotely monitor the site around the building from their Operations Center.

Figure 11. Video Camera on Trolley Support Frame

In addition, the City wanted the majority of the equipment housed inside of a building to prevent access to the equipment. Equipment that could not be housed in a building should be protected and highly visible. The design included an electrical building over the wet well. The building houses the electrical equipment, the access to the pumps, and the hoist and trolley that allow removal of the pumps. The building was designed to limit access points with glass block in lieu of windows, security bars at ventilation openings, and no hatches on the roof. In addition, intrusion alarm circuits were added to the doorways and to the man-entry access hatch to the wet well.

The exterior equipment included the emergency generator, sluice gate operators, mechanically-cleaned bar screen, dumpster to receive the screenings, and a winch to move the dumpster at the bar screen. The generator enclosure was protected by security bars in front of the radiator and by an intrusion alarm circuit on the enclosure doors. The bar screen was provided with an enclosure to prevent access to the motor and electrical cables.

LESSONS LEARNED

There were many lessons learned during the design and construction of the project. The following paragraphs outline the more challenging issues and how they were addressed.

The biggest hurdle for the project was obtaining USACE approval before construction and after construction. The request for the Minor Section 408 Modifications of Flood Risk Reduction Projects involves complying with 11 guidance documents that addressed design requirements, calculations, and construction requirements. The designers significantly underestimated the time and effort necessary to prepare and submit the construction permit request and the Construction Report to USACE. However, to facilitate the development of the request document and agency review, the request was outlined in the same order and numbering as the "Checklist Construction in the Critical Area of Flood Control Projects Constructed by The Corps of Engineers."

During installation of the guide rails for the submersible non-clog pumps, the contractor noted that the rails were not plumb and did not align with the pumps suspended on the overhead trolley. The City was concerned about the need to "wrangle" the 4.5-ton pumps and cables into position to mate with the guide rails. After several iterations, the guide rail connections were modified to allow pump installation by lowering and rotating the pump into place. However, the guide rails were angled from vertical to allow installation. After construction, the issue was determined to be the hatches for the non-clog pump bases were mistakenly aligned with the centers of the future mixed flow pumps. In hindsight, it would have been helpful to create a detail showing the alignment of the hatch, pumps, piping, and guide rails following submittal of the pumps, hatches, and top slab reinforcement.

The City's plan for a high speed SCADA system was based on the communications between the Allen-Bradley programmable logic controller (PLC) at the pumping station and the Ethernet-based communication system to the Operations Center. Two (2) switches were installed in the Pump Station Control Panel. The first switch facilitated communication between the PLC and pumping station components. The second switch provided a connection between the first switch, the SCADA system, and the video cameras. For several months, the City experienced communication problems with the Operations Center. This issue was finally resolved when it was determined that the second switch needed to be a Layer 3 Distribution Switch to allow for Ethernet communications with the Operations Center.

SUMMARY

Through this project, the City took the first step in addressed a needed capacity upgrade in their stormwater conveyance system. Careful planning by the City allowed for the station to address issues exposed by the 2011 Flood, vandalism at the site, and phase improvements at the station as the upstream conveyance system is improved.

Figure 12. New N. 28th Stormwater Pumping Station

The lowest construction bid for the project was $7,836,052.60. Construction of the pumping station, shown Figure 12, in was completed with four (4) change orders, resulting in increase in the contract price of $11,700.04, or 0.15% of the total contract amount. The Contractor met the intermediate deadlines required by the USACE.

The N. 28th Stormwater Pumping Station was completed 3½ years after the end of the flood. The construction of the first phase of upstream conveyance system improvements is slated to begin in 2017.

In 2016, the N. 28th Stormwater Pumping Station project won the Iowa APWA Project of the Year in the Environment Category for projects \$5 Million to \$25 Million.

Integrated intelligence – A breakthrough in wastewater pumping

By Kristel Zaman, Lisa Riles

Xylem, Inc.

ABSTRACT

Aging infrastructure, global urbanization, increasing energy costs and the need for sustainable solutions make cost reduction and reliability a priority in wastewater pumping. Small wastewater pumps can be given a new level of functionality and intelligence by integrating advanced software functions and state-of-the-art hardware into a conventional submersible design. Sensing the operating conditions and the environment, technology exist today that can adapt the pumping performance in real time, making smart decisions and providing feedback to the operator. This can result in significant OpEx and CapEx savings for both the end user and specifier. This paper will give the audience an understanding of the breakthrough technology available as well as the substantial and tangible benefits of integrated intelligent wastewater pumping through proven installation case stories.

Small wastewater pumps are commonly found in duplex wastewater pump stations in sewage collection systems at municipalities and private operators. Pump station design code ensuring adequate capacity for peak flow situations often leads to a significant oversizing of the pumps resulting in excessive energy usage and unnecessarily high equipment wear.

By employing an increased level of component integration it is possible to add intelligent electronic hardware and specialty software inside the shell of a submersible wastewater pump. This technological breakthrough delivers many valuable pump system functions and customer values resulting in OpEx and CapEx savings.

Installation case stories will be presented substantiating the significant benefits of asset management through a broad performance field as opposed to a fixed curve; trouble free pumping with built-in algorithms for sump/pipe cleaning and clog detection/auto-reverse pump cleaning; energy savings with IE4 motor efficiency, energy minimizing algorithms and hydraulic design; and reduction in total investment with a compact, integrated design eliminating the need for traditional cabinets components such as motor protection, soft starters, variable frequency drives and climate control equipment.

KEYWORDS

Wastewater pumping; Intelligent pumping system; Clog-free pumping; Energy savings;

1.0 INTRODUCTION

Global water trends like aging infrastructure, urbanization, increasing energy costs and the need for sustainable solutions are making cost reduction an urgent priority for all. The wastewater pumping sector is no stranger to these trends as they represent both a challenge and an incentive to develop new technologies.

Another major global trend is the current revolution of integrated software, which impacts every industry and demands sophisticated, user-friendly and leaner technologies. For example, most people are familiar with smart phones, including the increasing functionalities they have and the way they simplify our daily tasks. The wastewater industry is also moving towards this trend. The new smart and intelligent technology enables the collection, analysis, and sharing of huge amounts of data generated inside and outside of the products. This amount of data allows integrated products to make smart decisions ultimately making our life easier.

This white paper was written to describe developing the technology that adapts to these challenges and helps the pump users to deal with these new demands. Intelligent pumping systems can be made to meet the future requirements of all wastewater pumping applications and duty conditions as well as system requirements that change over time. Intelligent wastewater pumping system will address common pain points, such as improved system reliability, reduced energy consumption, operational flexibility, reduced footprint, connectivity and a significantly lower pump system lifecycle cost can be achieved with integrated intelligent wastewater pumping systems, the ultimate solution for trouble-free pumping.

This white paper describes the revolutionary technological advances and functionality: reduced asset management and the total reduction in operational lifecycle costs. The results will show the significant benefits that owners and operators of wastewater pumping systems will experience and appreciate.

2.0 A NEW LEVEL OF THINKING

2.1 Traditional pumping solution

Submersible wastewater pumps available on the global market in the 3-10 HP range are mature products typically consisting of a close-coupled pump and induction motor integrated in a package (enclosure) that can withstand continuous submergence. Different pump technologies are used and these products have typically been on the market for over 10-20 years with little or no design innovation.

Small wastewater pumps are commonly found in duplex wastewater pump stations in sewage collection systems at municipalities and private operators around the world. Pump station design code ensuring adequate capacity for peak flow situations might lead to a significant oversizing of the pumps resulting in excessive energy usage and unnecessarily high equipment wear.

Today's typical small pump station control panel is often fitted with simple relay logic or basic 'black-box controllers' that operate standard contactors in an on/off pumping mode. Pump station monitoring and protection is often limited to thermal overload and pump leakage. Because the inflow of wastewater is varying over time and is often only a fraction of a pump's required capacity engineers have looked for ways to operate wastewater pumps at a reduced capacity in order to optimize performance while reducing operating costs. During the last 10–15 years the industry has seen a significant increase in the adaptation and acceptance of variable frequency drives (VFD's) in wastewater transport systems. A general desire is to optimize the wastewater flow through the pump stations.

Solids handling capability is one of the most important features for any wastewater pump. Conventional channel impeller wastewater pumps, both single- and multi-vane as well as vortex impellers are prone to clogging due to soft and fibrous objects (a.k.a. modern trash) accumulating on the leading edges of the impeller or at the center of the impeller.

Recent attempts to improve pump efficiency, other than using self-cleaning hydraulics, has been realized by replacing standard induction motors with premium efficiency motors, a minor improvement that results in few percentage points better pump system efficiency.

2.2 Understanding the pain points in pumping wastewater

Pump user survey responses indicate that many pump users are constantly looking to lower their energy consumption and overall cost while they want to have higher reliability and longer lifetime of the products.

They also said that they don't want to send out people to remote locations in bad weather and hazardous environments to manage unplanned maintenace. Instead they are seeing a trend towards more remote supervision and more integrated systems.

Pump clogging is a common wastewater pumping challenge. The key design criterion for a non-clog pump is its ability to pass solids without clogging the pump. Clogging can be a full or a partial clog of the impeller or the volute. A full clog exists when the pump has ceased pumping; this condition is easy to detect and highly undesirable. A partially clogged pump however, is harder to detect and most often goes unnoticed because the pump still delivers flow, although it is reduced. This might result in reduced pump efficiency and substantially increased energy consumption. Conventional style pumps may operate for extended periods of time partially clogged until the clog becomes a full blockage and results in an un-planned maintenance call.

Environmental concerns are getting increasingly more important, such as carbon footprint size and complaints about odors at their local sewage pump station. Also data connectivity is often mentioned as a requirement to help with environmental compliance and better data reporting capability.

While the pump system's upfront cost is still a major consideration, the operational costs over the life of the pump station might help to differentiate between competing pump system designs at the time of purchase. Increases in pump system efficiency alone may not always convince a buyer to switch to a better performing system, but a single unscheduled call to clean a clogged pump will easily tip the scales in favor of a system that has proven to be consistently clog-free.

Clean and odor free pump stations are always desirable, yet many pumping systems frequently require expensive pump outs – for example vacuum cleaning call-outs, to remove bottom sediment, floatables and fat build-up that create odor problems, and can cost thousands of dollars annually.

2.3 Integrating intelligence

An answer to these challenges is the creation of integrated intelligence in wastewater pumping. This is made possible by innovation and adaptation of components in a submersible wastewater pump: integrating electronics such as processor, software, sensors, synchronous electric motor and self-cleaning hydraulics inside a submersible shell.

The word 'intelligent' in wastewater pumping is a result of factors that together contribute to a new level of operational autonomy and efficiency. Pumping systems that are designed to automatically deliver the desired performance at minimum energy consumption while reducing the total cost of ownership are considered intelligent.

The practical implication of an integrated intelligent system translates into functionality such as soft-start, pump clog detection, pump cleaning, advanced motor protection and 'always correct' impeller rotation. These are features that increase the pump and system life, reducing downtime and giving users peace of mind. The pump station controller provides pump system management functions including: energy minimizer, sump cleaning function, pipe cleaning function, etc.

Lastly, overall reliability of the entire pumping system is high on operator's agenda. Systems that can deliver true operational peace of mind will always be in demand. This leads to the desire to build 'future proof' stations that are flexible and resilient, meaning that they can easily adapt to changing conditions and can handle extreme flow conditions without failures.

3.0 INTELLIGENT PUMPING SYSTEM

Among hardware components we can highlight: integrated control electronics, new generation non-clog self-cleaning hydraulics, and IE4 rated motor efficiency. The result of this combination is a sophisticated wastewater pumping system that simplifies selection, installation and commissioning while reducing total investment costs.

3.1 Hydraulics

The innovative N-technology revolutionized wastewater pumping with sustained high efficiency and self-cleaning operation, thereby enabling lower total cost of ownership.

The key feature in the design is the N-impellers horizontal backswept leading edges together with the relief groove in the insert ring and the guide pin. The backswept leading edges work together with the guide pin to sweep solids from the center to the perimeter of the inlet. When solids arrive at the perimeter of the inlet, they get transported inside the relief groove, sliding along

Figure 1. N-technology impeller with insert ring

the edge of the impeller vane, through the volute and out of the pump.

The mechanical self-cleaning design means that the pump will have the same efficiency in pumping wastewater as when pumping clean water. Typical for all pumps the stated efficiency (pump curve) is measured in clean water, and this value is a lot higher than the actual efficiency in a real application due to soft clogging etc. but with the N-technology we can have the same high efficiency also in real applications, why we call this sustained high efficiency.

3.2 Advanced drive unit

The pump's drive unit is an assembly consisting of monitoring and control circuitry, software, power electronics and a synchronous motor in a single package. By utilizing a new concentrated winding motor design, benefits such as increased motor efficiency, controllability, and reduced motor size are achieved.

The stator consists of multiple identical stator lamination packages with individual coils that are placed side-by-side inside the stator housing. The concentrated winding stator can be produced efficiently and at a shorter height than an induction motor of corresponding rating.

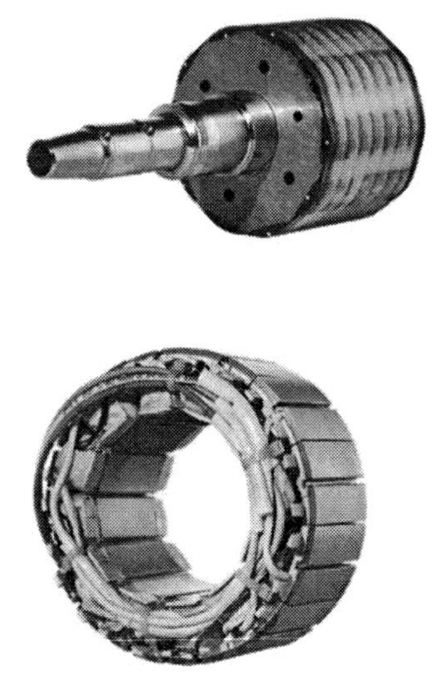

Figure 2. Concentrated synchronous motor (rotor and stator)

The rotor is equipped with permanent magnets that create and maintain the rotor's magnetic field, yielding very low rotor losses during operation at synchronous speed. The low losses result in virtually no heat being generated in the rotor and no heat transferred to the bearings via the shaft. The cooler running motor increases bearing life due to less thermal stress on the bearings.

Due to the high motor efficiency the winding temperature is lower than normal, add to this the advanced temperature protection system where motor power is reduced, and should circumstances warrant it, and the result of the design is a longer motor life.

The concentrated winding synchronous motor requires power electronics to start and run the motor. The power electronics offers full control of the pump shaft speed and torque; therefore the wastewater pumping system can operate within a large flow/head field. The embedded software serves to control the motor's speed and torque as well as to read the momentary load requirements.

A new, simpler and more compact high performance synchronous motor further improves pumping system efficiency allowing operation at reduced pump capacity with maintained high motor efficiency. The concentrated winding synchronous motor does this and meets the proposed future super-premium motor efficiency standards (IE4).

3.3 The intelligence of a wastewater pump

Intelligence refers to a pump system that can sense the environment it is working in as well as the load it is subjected to and adjust its behavior to meet the end-user optimization targets. By collecting and analyzing relevant data the pump system can make smart decisions about how it operates and what feedback it will provide the user. This results in precise process control, reduced risk of clogging, clean and odor free pump sumps, substantial reductions in energy consumption, reduced asset management, comprehensive data reporting and lots more. In short, it delivers peace of mind to the end user.

The in-pump processor controls the power electronics to achieve variable pump performance, always meeting the demand at hand. A single impeller size per volute size minimizes the need for multiple spare impellers and yields maximum hydraulic efficiency. Instead of having to remove the pump to trim or change an impeller, a different duty point can simply be met by varying the pump speed by the touch of a button.

When power electronics and a processor with control software are integrated into a submersible pump, costs are driven down and new functionality added. The challenge is the 'packaging' and integration of new components, i.e. the size, shape and performance within the submersible envelope. In order to achieve efficient component integration some components need to be re-designed to fit the environment of a submersible pump and adapted to function as a system. The power and control electronics which in a conventional pump control panel are mounted on a back plate must be adapted and fitted into the head of a wastewater pump.

Placing power electronics inside a submerged wastewater pump is beneficial since they are protected from the exposure to heat, dust, dirt and extreme temperature variations. The cooling is reliable, compact and stable and not dependent on fans or A/C units. The motor leads are as short as possible removing any issues with wave reflections and voltage spikes.

The entire system is pre-engineered and the equipment start-up and commissioning time is drastically reduced by simple set-up wizards. When power electronics are placed between the grid and the motor the system becomes frequency independent and voltage tolerant.

The pump station controller allows for data communication and measurement/analysis, providing communication to the world via standard communication protocols. The pumping station can easily be connected to remote station monitoring systems including Supervisory Control and Data Acquisition (SCADA) systems.

4.0 RESULTS

4.1 Case story – Stony Brook, Suffolk County, NY

4.1.1 Problem

Suffolk County, Long Island, New York, boasts 21 sewage treatment plants and sewer districts, over 70 pumping stations, and more than 1,250 miles of sewers.

The Suffolk County Department of Public Works Stony Brook Pump Station was facing on-going interruptions to pumping operations as a result of "ragging"— a common challenge in wastewater pumping where a build-up of fibrous materials leads to frequent pump blockages. The problem-prone 3 million gallons per day (MGD) facility had been built years ago, when the nearby campus of Stony Brook University was smaller and generated less flow. Then, the influent pump station and its original equipment simply could not keep up with the demands placed on it by the much larger university center and three outlying areas that comprised the Stony Brook of today. Although it would appear that the station's three pumps had ample capacity, the size of these units and their variable speed drives definitely contributed to the clogging problem during low flow periods, which at the time was most of the day. Unplanned downtime as a result of ragging was becoming a frequent and significant challenge. Clearing a pump is a costly, dirty, and unpleasant task that requires a maintenance team and often a crane.

4.1.2 Solution

The prototype intelligent pumping system was proposed as a potential solution and it was installed in August 2015. The pump station was a standard duplex pump station with two pumps installed. Suffolk County Department of Public Works personnel simply removed one pump from the pump station and installed the new prototype pump in its place. A factory technician then installed the controller and input the settings. The entire process only took about one hour.

Figure 3. Prototype pump ready to be installed

4.1.3 Results

The pumping system was monitored remotely by research and development team in Sweden. Local operators closely monitored the pumping system and positive results were clear within a matter of a month. Before installing the system, it was necessary to lift the pump regularly to remove fibrous materials caught in the impeller. Maintenance call-outs were as frequent as once every three months. Maintenance cost reduction was about $1,500.00 for personnel and $2,500.00 in equipment costs for the year.

With the new wastewater pumping system, the pump clogging was completely eliminated and it was no longer necessary to check the rotation of the pump, saving time and money and offering new-found peace of mind. This intelligence system included automatic pump down of the well, reducing the need to clean the well of grease accumulations and reduced level control rod failure.

The pumping system aims to deliver proven reliability at the lowest total cost of ownership and to achieve this it also, among other benefits, dramatically reduces energy consumption. Operators of the Stony Brook pumping station noted a significant reduction in energy consumption following the installation. Energy consumption was reduced about 20-30%, the exact data was not available at the time of writing this paper due to billing specifics of the municipality.

4.2 Case story – Fond Du Lac, WI

4.2.1 Problem

The Fond du Lac (WI) Regional Wastewater Treatment Facility treats the City of Fond du Lac wastewater along with that of 18 neighboring entities. The water is then discharged to nearby Lake Winnebago. The city's population with neighboring communities included is about 75,000. The average hydraulic flow capacity is 9.84 MGD with a peak flow of 50 MGD. Neighborhood wastewater collection systems are submersible. Of the 17 pump lift stations, 11 are submersible, none of which use variable frequency drives. Overseeing the maintenance of these submersible stations as well as rising energy costs are two stiff challenges for Fond du Lac wastewater treatment facility.

4.2.2 Solution

The prototype intelligent pumping system was proposed as a potential solution and it was installed in July 2015 in a duplex pump station located at Wild Life Acres, replacing one of the old pumps (5 HP pumps each rated for 140 gallons per minute). The intelligent pump is the lead pump and it is expected to handle all flows in normal operation.

4.2.3 Results

The City of Fond du Lac estimates energy savings at 20 percent at the Wild Life Acres lift station after installing the new system. The pump is running at 3.3 horsepower, as compared to the original pump, which was running at about four horsepower.

Bothersome, time-consuming, and expensive cleaning call outs have also been reduced, as confirmed by monitoring system tracking the lift station operation both before and after the

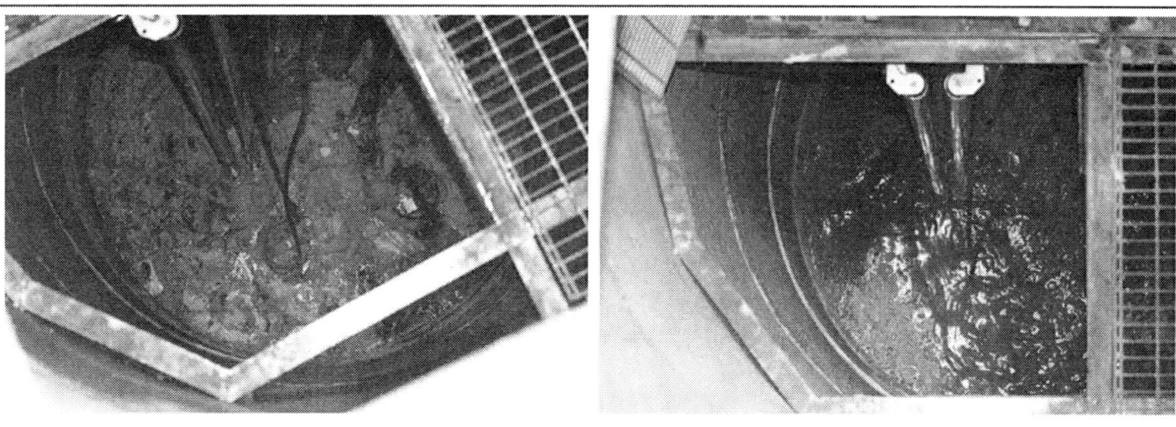

Figure 4. Wetwell view before and after the intelligent pumping system installation; results in 2 weeks.

intelligent pump installation. Over the course of one year, the intelligent pump has undergone over fifty cleaning cycles, e.g., the pump detected some kind of blockage and completed its cleaning cycle of reversing the pump rotation at a reduced speed, jogging back and forth until the blockage cleared. This resulted in a visually cleaner sump and eliminated costly call outs. In addition to cost savings, these cleaning cycles prevent build up and potential clogs that would eventually have to be cleared manually, an unpleasant and unsafe process.

Due to integrated intelligence many conventional control panel components are integral to the submersible pump motor housing, resulting in a much smaller control cabinet enclosure. The city stated that of the 17 pump lift stations, most of these are in residential areas and reduced panel size is considered visual improvement and desired by neighboring residents.

5.0 CONCLUSION

An intelligent wastewater pumping system offers unprecedented flexibility in meeting typical wastewater pumping challenges as well as changing pump system requirements. It offers a vast array of features and meets the needs of many different pump users worldwide.

Desired wastewater pumping features, such as improved system reliability, operational flexibility, better asset management, reduced footprint, connectivity and significantly lower lifecycle costs are met by combining different elements of the intelligent wastewater pump platform to meet a specific application challenge.

Further research and data collection is planned to substantiate and quantify the user benefits over the life cycle length of the intelligent wastewater pumping systems. The prototype testing provided assurance that significant life cycle cost reduction is possible and desired by the pump users.

OVERCOMING CONSTRUCTION CHALLENGES AT THE TURKEY CREEK WASTEWATER PUMP STATION

[1] Suzie Carpenter, CDM Smith, carpenters2@cdmsmith.com*
[2] Bon Marie Gardner, City of Kansas City, MO Water Services

ABSTRACT

This paper will focus on the planning, processes, and solutions the client and project team implemented to overcome challenges faced during construction of improvements and upgrades to the Turkey Creek Wastewater Pump Station in Kansas City, MO. The challenges included requirements to keep the facility in operation during construction and constructing a new structure around a double box culvert built almost 100-years ago. The pump station has a 166,558 cubic meters/day (44 mgd) capacity and operates within a combined sewer system. The construction project included improvements for grit removal, new bar screens, replacement of all electrical gear, a second power feed, installation of VFDs, new pumps, piping, and valves, and hydraulic modifications in the wetwell and pump intakes. The construction improvements utilized ground freezing technology during open pit excavation as well as a temporary pumping system that incorporated the installation of a Linestop ® on an active forcemain.

KEY WORDS

Pump station rehabilitation, construction staging, construction planning

INTRODUCTION

The Turkey Creek Wastewater Pump Station (TCWWPS) is located in the Central Industrial District of Kansas City, Missouri and pumps dry and wet weather flows received from the OK Creek Sewer to the Westside Wastewater Treatment Plant. All dry weather flow is pumped to the wastewater treatment plant, but during wet weather events, excess flow is sent through flap gates to an outfall at the Kansas River. The pump station feeds into a common forcemain before discharging into a splitter box at the wastewater treatment plant which also receives flow from other pump stations.

The TCWWPS, constructed in the 1960s, required renovation and modifications to meet projected future wastewater demands in the area and operate as a key facility as part of the City's Overflow Control Plan. The improvements project was challenging because modifications had to be completed while keeping the pump station operational or providing temporary pumping, while relying on the aging equipment for some staging, and while utilizing the existing, active forcemain to deliver the flow from the temporary pumping system while isolating the pump station.

The modifications included increasing the station's capacity to 166,558 cubic meters/day (44 million gallons per day (mgd)) of sanitary sewage using four (4) pumps by the year 2025. The current design includes the installation of three (3) pumps that provide a capacity of 83,279 cubic meters/day (22 mgd) to 105,992 cubic meters per day (28 mgd) depending on flow conditions.

The current operational plan is to allow only one (1) pump to operate at reduced flow until the Westside Wastewater Treatment Plant capacity is increased.

The OK Creek Sewer includes twin double box culverts, each of which is 5.2 meters by 5.5 meters (17 feet x 18 feet). The combined collection system flow carries a substantial amount of non-soluble debris into the TCWWPS. This debris includes rags, wood, bottles, grease balls, leaves, grit, and other items that typically enter the large combined sewer system. If this debris is transported to the pump station, it can adversely affect the downstream equipment including accumulating in the wetwell, thus blocking pump suction or clogging the pumps, valves, and piping. In some cases, this material, especially rocks and grit, can even cause equipment damage, thereby reducing the capacity of the pumping station.

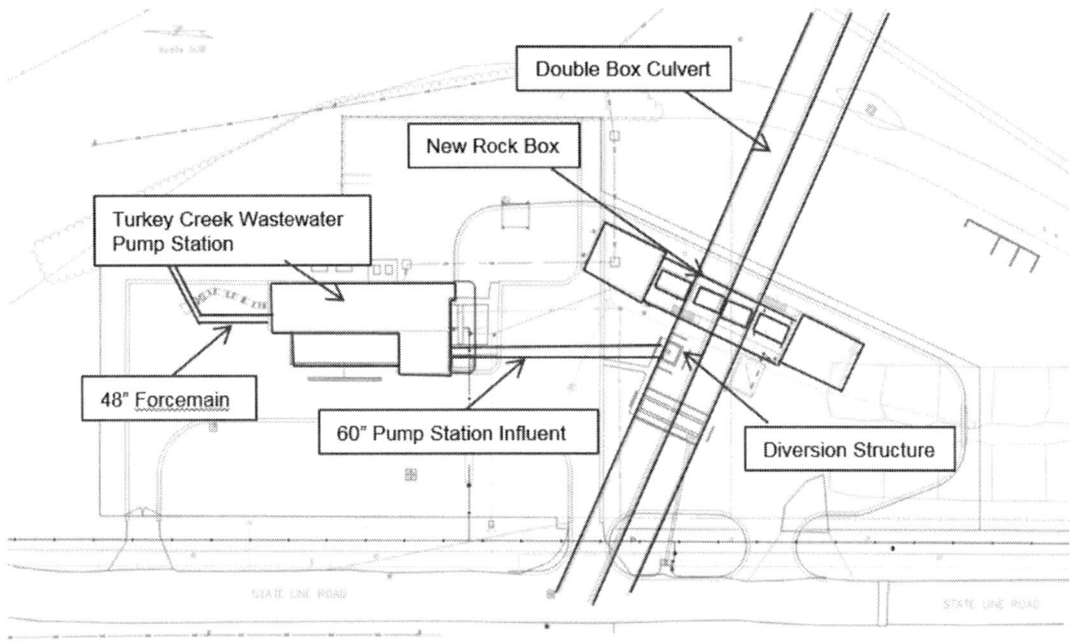

Figure 1 – Turkey Creek WWPS Site Plan

To reduce the impact of debris entering the pump station, the improvements incorporated two (2) debris collection boxes (Rock Box) upstream of the pump station. The Rock Box utilizes a monorail hoist and clam shell removal system. The clam shell system includes a double girder bucket monorail with a single drum hoist and hydraulic clam shell bucket that removes debris from the two (2) rock box storage pits and dumps the debris directly into two (2) debris removal dump pads for removal by front end loaders.

Additionally, as part of this project, two (2) new mechanical bar screens with washer/compactors were installed downstream of the debris removal system to further reduce the impact of debris entering the wetwell. The screenings are dewatered by the washer/compactors and then discarded into trash dumpsters for haul off and disposal.

The three (3) newly installed pumps are vertical non-clog, 223, 710 Watts (300 horsepower (hp)), shaft driven wastewater pumps. Each of the pumps are controlled by a variable frequency drive (VFD) that controls the speed and flow discharged from the pump station. The pumps are protected from surge by electric-actuated ball valves installed on each pump discharge in tandem with an existing surge tower at the pump station. The discharge valves are powered through an industrial type uninterruptible power supply (UPS) which will close the open ball valve in the event of a power outage.

A Programable Logic Control (PLC) is installed at the station for remote monitoring and controlling the pump station. A local Touch Screen Operator Interface Terminal (OIT) allows the operator to view the process parameters and make station control decisions. At the Westside Wastewater Plant, Blue River Wastewater Treatment Plant, and at the Briarcliff Water Treatment Plant, the supervisory control and data acquisition (SCADA) system can also be used to monitor and control the pump station and associated equipment at the TCWWPS, with the same information that the operator has at the pump station through the OIT. The pump station modifications and upgrades also included new heating, ventilation and air condition (HVAC) systems, a new electrical room, new switchboard and new motor control center (MCC), and new power feed and transformers from Kansas City Power and Light.

Due to the criticality of this pump station to the overall system, the equipment and controls needed to be able to maintain operation of the pump station in the event of a power failure. Two separate power feeds with an automatic transfer switch (ATS) provided reliable power. In the event that one power feed is lost, the pump station is automatically switched to the other power feed and all systems come back on-line. The new pump motors, drives, and controls needed to be installed with the ability to automatically start back up after power outages.

The pump station modifications were completed while keeping the existing capacity of the pump station intact either through use of existing equipment or supplying temporary pumping for short periods of time. This required planning, communication, and teamwork among all parties to keep the construction of the improvements moving.

MANAGEMENT ELEMENTS FOR TACKLING CONSTRUCTION CHALLENGES

The key component to managing construction challenges for this aging facility was planning. For the new equipment to be installed without interruption of the operating pump station, it was important to include construction constraints in the design documents in addition to clear communication between the City, contractor, and engineer during construction.

The design documents noted the impact of seasonal flow from the combined sewer discharging to the pump station. The specifications limited temporary pumping activities in addition to deep excavation work to certain times of the year. Additionally, the design documents provided the contractor with the requirements for pumping capacity to be kept available throughout the construction period. Planning required not only an understanding of rain events, but an understanding of how the level of the Kansas River could impact work in the wetwell as there were times when the river level was high enough to push back through the OK Creek Sewer and into the wetwell.

The design requirements required that prior to any activity that could impact operations, a detailed Maintenance of Plant Operations (MOPO) Plan be developed by the contractor with input from the owner and engineer. The MOPO provided details of the activity to occur, the planned duration, contact information for those involved, and identified risks and an action plan in the event something happened. For instance, when a bar screen was to be removed, the plan identified the timing, which gates needed to be closed for removal, and what would be done if the gate did not function correctly. The development of the MOPO was a combined effort of all parties involved. The contractor initiated the plan which was then discussed with the City's operations staff and engineering staff. All parties worked together to identify potential risks and ways to mitigate the risks. This teamed effort allowed for the outage periods to be as short as possible and well-planned.

Additionally, the City's Operations Group has implemented a requirement to also prepare an Operational Change Control Plan (OCCP). This City document is similar to the MOPO but goes into further detail on impacts to the overall system operation and provides contacts for those within the operations group in the event of an issue during the planned activity. The OCCP was developed by the City staff based on the MOPO and reviewed by all parties. On future City projects, only an OCCP will be required and it will be initiated by the Contractor similar to the MOPO.

Another critical component to implementation of this project was communication. At the beginning of the project, it was recognized that in addition to a monthly progress meeting, a weekly site meeting would be necessary. The monthly progress meeting was utilized to keep all team members, including management, updated on work progress, change orders, and critical upcoming work. The weekly site meetings were utilized to work through any potential issues in a timely, collaborative way. The success of the weekly meetings was dependent on having the decision makers on-site and required communication and planning prior to the meeting to allow this to happen.

As mentioned earlier, another important factor was building a responsive, collaborative team. This team included both the City's project manager, City engineering staff, the City's operations staff, the contractor and its subcontractors, and the engineering design team. When issues arose, specific identification and data was captured to provide an understanding of the factors involved in the issue. At the next weekly meeting, following occurrence of an issue, the team members, who could assist with trouble-shooting, participated either in person or via conference call to work towards a solution. This collaborative effort by all team members allowed brainstorming and implementation of practical solutions. Developing a communication plan that allowed the team to work together was critical to the success of the project. For this project, when an issue arose, everyone worked together towards a common solution rather than wait for others to respond. This collaborative effort kept the project moving even when difficult challenges occurred. The project managers for the City, contractor, and engineer recognized that working towards a common goal was an important component of the project and developed a strong, trusting working relationship.

Finally, and equally as important, having the appropriate team members present at system and equipment startup and testing was necessary to provide a fully functional system. Since this pump station was being added to the City's SCADA system, various operational scenarios needed to be tested and proven before the new system was accepted. These startup activities required participation from the City, general contractor, electrical subcontractor, system integrator, programmer (provided by the engineer), and the equipment representatives. Due to the complexity of having so many individuals on-site, testing and startup activities had to be planned well in-advance and the schedule had to be met for these activities. A detailed functional testing plan was developed to make sure the system was operating as intended under multiple potential scenarios. The functional plan was extensive and required that the SCADA system be checked at multiple locations for monitoring and control functionality, and required sign-off by multiple persons involved in the testing including City, contractor, and engineering staff.

CONSTRUCTION CHALLENGES

As mentioned earlier, there were several construction challenges associated with this project. The first challenge was that flow to the facility is constant and the rate varies by season and weather conditions. Since this pump station is fed by a combined sewer system, the flow is dependent on the amount of rain in the area. Since the flow cannot be routed elsewhere or shut-off to the pump station, the pump station dry-weather capacity had to be maintained. To meet this requirement, the design documents required for the existing pumps to remain in operation while new pumps, piping, motors, and drives were installed. This required problem solving and team coordination during design to determine how this could be accomplished. The original pump station was designed for five (5) pumps, however, at the time of construction, only two (2) pumps were in operation. This allowed for three (3) new pumps to be installed where the three (3) out-of-service pumps were located. Additionally, a new electrical room was constructed for the new VFD's, PLC, and discharge valve UPS. This room was constructed in a space which was not utilized so that it could be built and all equipment installed while the existing equipment remained in operation.

Figure 2 – Staging of the Pump Station Improvements Utilizing Existing Pumps While New Pumps Were Installed

An unexpected challenge that arose was the failure of the existing equipment planned for use during the construction. At the time the contractor started work, there were two (2) existing pumps utilized by the City. Shortly after mobilization, one of these pumps began to fail. Since two (2) pumps were needed for reliable operations, an emergency repair was performed on the failed pump by the contractor by using parts from the other unused pumps. Due to the collaborative effort of the team, this work was completed quickly, on a time-and-materials basis, to allow the City to continue meeting operational requirements.

Another challenge arose when it was discovered that the flap gates on the OK Creek Sewer were leaking and allowing the Kansas River water level to impact construction activities even when there was not high flow in the sewer. A change order was issued to repair the gate seals and minimize the impact of the back water. The work for this change order was identified by the City and contractor working together to inspect the gates and identify repairs needed.

Two (2) temporary pump-arounds were planned to allow for work in the wetwell. The first pump-around lasted eight days and was needed to install a temporary bulkhead and chute for isolation of the wetwell. This allowed for work to occur in the dry on one side of the existing wetwell while the other side had active flow. Also, during this pump-around, electrical work was completed, the pumps no longer in service were isolated from the existing discharge header, and a temporary connection was made to the surge tower. The second pump-around lasted seven days and was needed to remove the bulkhhead and chutes and other temporary piping, install new gates to isolate the bar screen channels, and to demolish the old switchgear which was no longer in service.

The pump-around equipment had to be installed to pump flow from the double-box culvert which feeds the lift station. The design documents allowed for the contractor to cut the top of the box if needed for access. The contractor cut out two sections of each side of the box culvert for access to install the pump around system and a temporary screen to protect the pumps. As a result of the construction activities, the City operations staff identified the advantage of having access into the double box culvert and requested that a new concrete box be constructed over the openings. The box was tall enough for flood protection. This box was added at the end of the project due to the open communication between the project team.

For the electrical phasing, the existing electrical gear was located in the center of a large room. The new MCC and switchgear was installed along the walls to minimize downtime. Additionally, the power phasing was carefully planned so that only a short power shut down was needed to switch the old and new gear. The electrical subcontractor was heavily involved in meetings to plan this work and identify any potential issues.

Figure 3 – Aerial View of the Site During Construction Showing Temporary Openings in the Double Box Culvert and The New Debris Collection Structure

A second challenge involved construction of the debris collection system upstream of the pump station. This system included a structure to be built around the twin 5.2 meters by 5.5 meters (17 feet x 18 feet) box culverts which were originally constructed in the 1920's. A structural and geotechnical evaluation of the box culverts and surrounding soils was performed during initial design of the new structure to determine if it could span an excavation wide enough to accommodate the proposed debris control structure width of 7.3 meters (24 feet). The structural analysis determined that some form of excavation support would be required to meet the structural limitations of the culvert and to provide groundwater control during construction. To provide the needed support, the design documents included supporting the box culvert from the top using a hanger rod system and from the bottom using ground freezing. Implementation of the ground freezing process required that a subcontractor specializing in this technology be hired to perform the work. Additionally, 22.6 meters (74 feet) deep excavations, each 3 square meters (32-square feet) were needed on each side of the double box culvert to prepare for the ground-freezing. These shafts were constructed using sheet piles and wales. Once the ground was frozen, excavation took place below the double-box culverts to construct the debris collection structure.

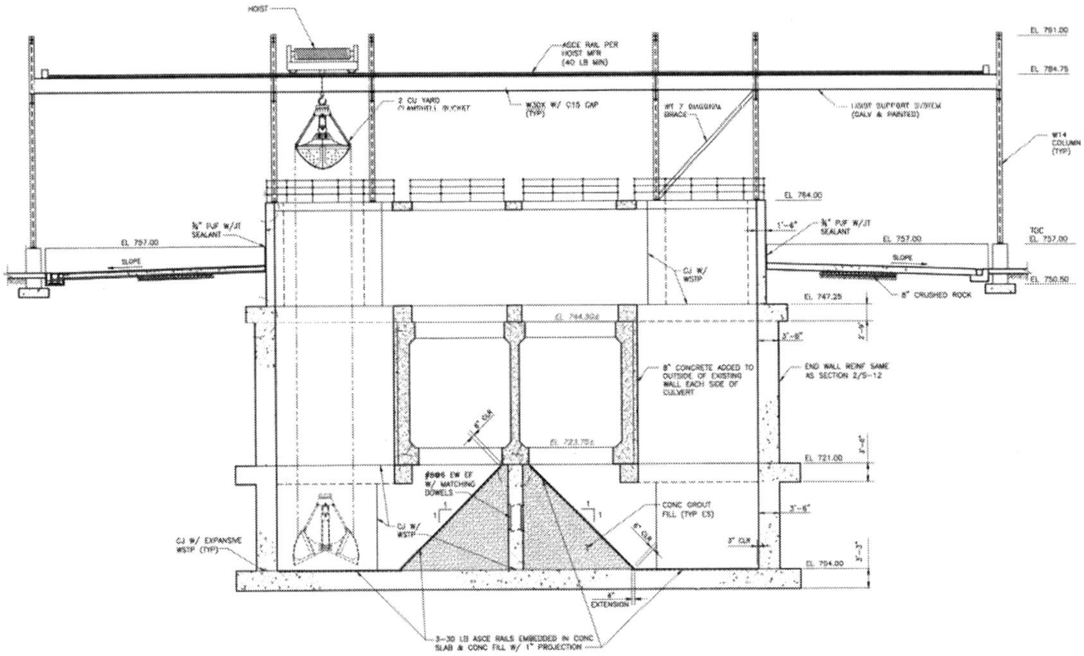

Figure 4 – Debris Collection Structure and Existing Double Box Culvert

Another known challenge for this facility included utilizing the existing forcemain from the pump station to the Westside Wastewater Treatment Plant during pump-around operations. Since the forcemain is fed by another pump station, the forcemain had to be isolated during periods of time that the pump station was taken off-line. In order to utilize temporary pumps, a hot tap was needed on the active forcemain. This hot-tap was completed while the pump station was in operation. It was planned to use an existing gate valve located on the header inside the pump station to isolate the forcemain from the existing pumps during temporary pumping, however, during construction, it was discovered that the gate valve was no longer operational. This required quick collaboration by the team to find a way to isolate the pump station. As a result, a Linestop ® was installed outside of the building. The Linestop ® is a hot tap with a valve which is installed on a live pipeline. Because the Linestop ® was installed by a specialty contractor, the team had to work quickly to evaluate options and costs impacts to keep the project moving. The Linestop ® was removed after the first pump around and reinstalled for the second pump around which required additional planning and mobilization for completion. The City decided to install a permanent valve in this location to allow for future isolation if needed. The City and contractor were able to work together to add this valve near the end of the construction period before the Linestop ® tap was removed.

This pump station is sized for future conditions and is limited on the flow which can currently be delivered to the wastewater treatment plant. During startup of the pump station, the control system was modified to set the maximum speed of the operating pump to less than 100% so that the treatment plant will not be overloaded. Additionally, only one pump can operate at a time which is also programmed into the operational scheme. Since the control system is monitored at multiple locations, it was important for operators at each facility to understand how the pump station operates, what alarms may occur, and how to appropriately respond. This required training for each location. Due to the willingness to work together, modifications were made as needed during startup to make sure the pumping capacity was limited to what the treatment plant could handle and the limitation on one pump in operation was included. Additionally, the City staff was able to receive the training while the equipment suppliers and programmer were on-site.

SUMMARY AND CONCLUSIONS

The TCWWPS project reached substantial completion in 2016 with a final cost of $17.2 million. The total change orders for the project were under 6% which included emergency repairs of existing equipment, addition of a Linestop ® to isolate flow back into the pump station from the active forcemain, and installation of an additional concrete box at the request of the City operations staff for future access into the OK Creek Sewer.

Although there were several challenges to overcome for completion of the project, planning, communication, and collaboration allowed for the pump station to stay in operation while the facility was completely rehabilitated. As issues arose, the City, contractor, and engineer worked together to reach a resolution. Due to the working relationship, all parties could trust that a favorable solution would be reached. This relationship resulted in developing quick, well planned solutions which benefited the City by minimizing schedule impacts resulting in less cost and less impact to operation of the sewer system.

Title: Putting it All Together: City of Columbia Pump Station Maintenance Program

Authors: Simon Watson, Brown and Caldwell; Jay Fulmer, Brown and Caldwell; Ashley Dove, City of Columbia

Abstract

The City of Columbia, South Carolina developed and implemented an update to its pump station maintenance program. The City owns and operates 57 pump stations ranging in size from .03 MGD to 32 MGD. The goal of the program is to standardize and enhance pump station maintenance procedures in compliance with requirements of an EPA Consent Decree. The standardization helped the City in focusing resources for a more efficient maintenance program.

Key Words

Pump Station, Maintenance Management, Standardized Procedures, Criticality, Inventory Management, Effective Resource, Consent Decree

Introduction

The City of Columbia needed to enhance the existing maintenance program to be in compliance with the requirements of an EPA Consent Decree. To enhance the program a series of workshops were completed to identify critical equipment, develop standard operating procedures, improve routine inspection/maintenance tasks and frequencies, and to identify critical spare parts and equipment. The project resulted in the development of an annual maintenance master plan, staffing justification and the purchase of critical spare parts and equipment.

Methodology

The first set of workshops focused on assessing the pump stations to develop a criticality list using the criticality assessment development process below.

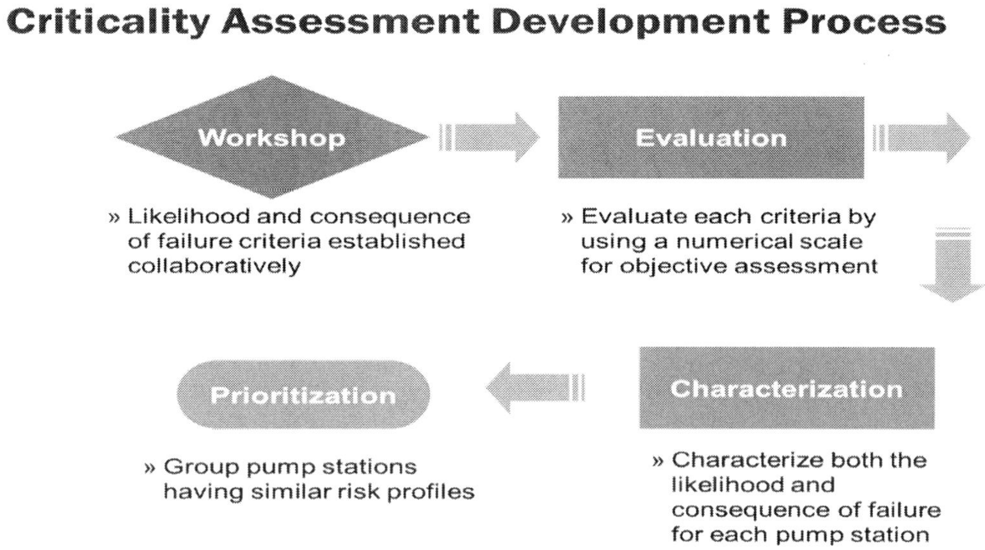

The results of the workshop provided a list of pump stations ranked by criticality and failure mode. This list was then used to see what was the most effective way to reduce or mediate the potential failures including installation of additional equipment, improved maintenance practices, SCADA enhancements, and enhancing the inventory management system for critical spare equipment and parts.

Building on the results of the criticality workshop a pump station maintenance task analysis was performed that identified existing maintenance practices and gaps that needed to be filled to better manage the potential failures of critical pumps stations. Using the table below as a foundation a list of new maintenance activities and standard operating procedures (SOPs) was developed.

Activities of a Well Managed Utility

Activity	Activity
Normal Pump Station Operations	Station Checks ~ Daily/ Weekly
Alternate Pump Station Operations	Visual Inspections
Pump PM (packing, seals, lubrication)	Facilities Maintenance
Pump Rehab/Replacement	Wet Well Cleaning
Pump PdM	Pump Station Housekeeping
Pump Motor Rehab/ Replacement	Telemetry Maintenance/ Repair
Pump Motor PM/PdM	Control Panel Maintenance/ Repair
Driveline Maintenance/ Repair	Stand-by Power Testing
Pumping Efficiency Testing	Electrical Equipment Infrared Scanning
Valve PM	SCADA / Alarm Components Maintenance/ Repair
Bypass Pumping PM	
Bypass Pump Check	
Unclog / De-rag	

In all eleven new SOPs were developed along with a schedule and frequency for each activity. In order to improve the efficiency and ensure the effective use of the City's resources annual maintenance schedules were developed to assist with load leveling of resources across three different pump station route teams. This provided the staff with a clear picture of what needed to be accomplished each month and across the entire year.

The last effort was to make sure that the staff had the adequate spare parts and equipment in the warehouse. Spare equipment included reserve pumps and generators. Spare parts included level monitors, pump seals, check valves and various electrical and instrumentation equipment. To determine what should be kept in the stores the following process was used.

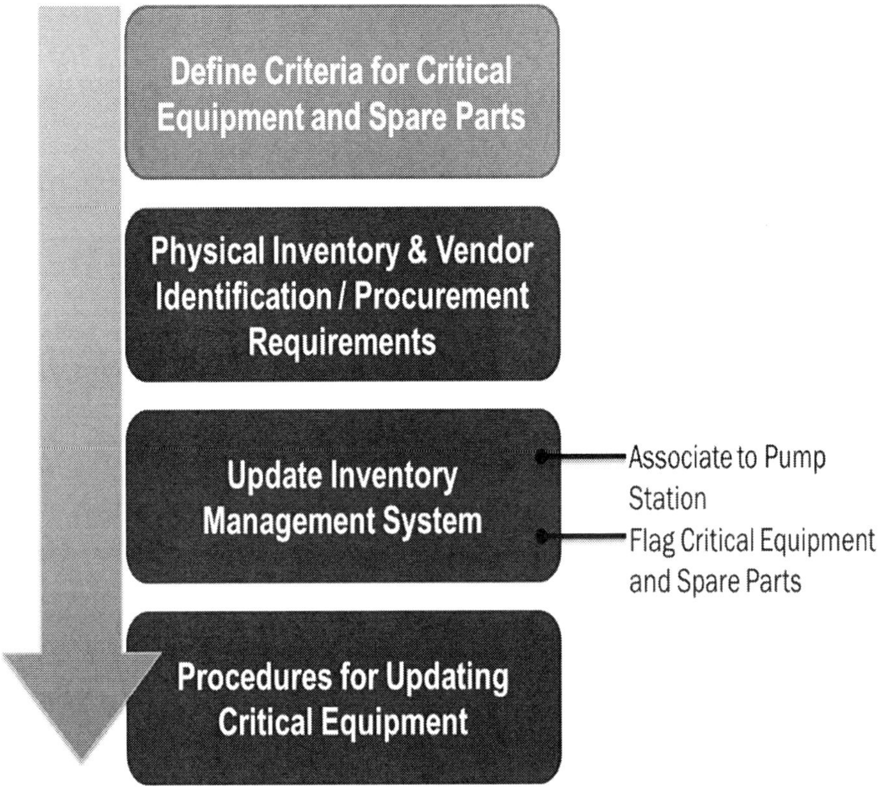

A continuous improvement process was also implemented to ensure that the items stored and managed in the warehouse were always up to date. This included an annual dormant stock review that would allow for the identification of parts that are no longer needed, resulting in only required parts and equipment being stored and maintained.

Results

The focus of this project was to meet the requirements of the Consent Decree which was accomplished. Along the journey the City gained much more including more effective use of limited resources and standardization of processes. In addition, it identified improvements to the City's extensive SCADA system to enhance capabilities of remote monitoring of pump station status and performance. One of the key reasons the project succeeded is because of the appropriate engagement of the collections staff in the workshops. This collaborative approach in the workshops provided the grounds for the easy adoption of the improved program.

Use of CFD Modeling for Designing Permanent Canal Closures and Pump Stations

Fangbiao Lin, Ph.D., PE, Stantec Consulting Services, Inc. (Email: Fangbiao.Lin@Stantec.com)
John Take, PE, Stantec Consulting Services, Inc.

ABSTRACT

This paper presents an advanced modeling approach, computational fluid dynamics (CFD) modeling, used for designing three drainage pump stations in New Orleans, Louisiana. The CFD modeling was used for designing the pump wet wells, designing pump discharge conduits, evaluating flow patterns upstream and downstream of bypass gate structure channels, predicting velocity and water surface profiles in the outfall canals, and sizing riprap both upstream and downstream of the pump stations.

The Permanent Canal Closures and Pump Stations project (PCCP) is a design-build (DB) project, which includes three large pump stations –17th Street Pump Station (12,600 cfs), London Avenue Pump Station (9,000 cfs) and Orleans Avenue Pump Station (2,700 cfs) – and associated flood gate structures. This project will replace temporary closures and pump stations in the three outfall canals and reduce hurricane and storm damage risks for New Orleans. Stormwater normally flows by gravity through bypass gate structures to Lake Pontchartrain and is pumped to the Lake with the gates closed when water levels surge high during a hurricane event.

The use of CFD modeling approach, complemented by physical modeling and other traditional 1D and 2D hydraulic modeling, enabled Stantec to develop satisfactory designs with confidence.

KEYWORDS: CFD Modeling, Pump Stations, Intake, Discharge Conduit, Flood Gate, Riprap Sizing.

INTRODUCTION

Project Description

The New Orleans Permanent Canal Closures and Pumps (PCCP) Project requires the construction of permanent gated storm surge barriers to prevent the 100-year storm surge in Lake Pontchartrain from entering the 17th Street Canal, Orleans Avenue Canal, and London Avenue Canal. The current Interim Closure Structures (ICS) on the three outfall canals will be permanently replaced with three pump stations at the discharge ends of the outfall canals to evacuate storm water, unimpeded, from the City under the Current Condition (CC) and Possible Future Condition (PFC) scenarios.

The CC scenario requires 100-year storm surge barriers near Lake Pontchartrain for each of the three outfall canals. These storm surge barriers include bypass gates and pumping stations to form a new permanent risk reduction system. Each PCCP facility will allow the respective canal to be isolated from Lake Pontchartrain and evacuate unimpeded canal flows. The new PCCP facilities will operate in series with the existing interior Drainage Pump Stations (DPS). The PFC scenario, which is currently neither authorized by Congress nor funded, involves removal of the PCCP gate structures and interior DPS facilities, along with deepening of the outfall canals from the DPS to the PCCP facilities to allow drainage to flow to the PCCP pumping stations by gravity.

These pump stations use formed suction intakes (FSI) adapted from the United States Army Corps of Engineers (USACE) Type 10 design as presented in the Hydraulic Institute (HI) Standards (ANSI/HI 9.8-2012). While not completely unique, these FSIs have an opening width approximately equal to the width of the pump bay, whereas the FSI opening, in many cases, is less than the width of the pump bay. There are two sizes of pumps manufactured by Patterson Pumps: TMF 159, with capacity of 1,800 cubic feet per second (large pump), and TMF 113, with capacity of 900 cfs for the Current Conditions (small pump). It should be noted that the small pump FSIs slightly differ from the standard Type 10 in that the length of the upstream section is longer. For each type of pump, a siphon discharge conduit is constructed, which includes two components: a steel conduit directly connecting to pump flange and a downstream concrete conduit discharging water to the canals.

The project design point pumping capacities for current conditions of the three canals are: 12,600 cfs for 17th Street; 2,700 cfs for Orleans Avenue; and 9,000 cfs for London Avenue. The maximum potential pumping capacities for possible future conditions of the three canals are: 14,560 cfs for 17th Street; 3,213 cfs for Orleans Avenue; and 9,873 cfs for London Avenue. The descriptions of each pump station project are summarized below and locations of these pump stations are shown in Figure 1.

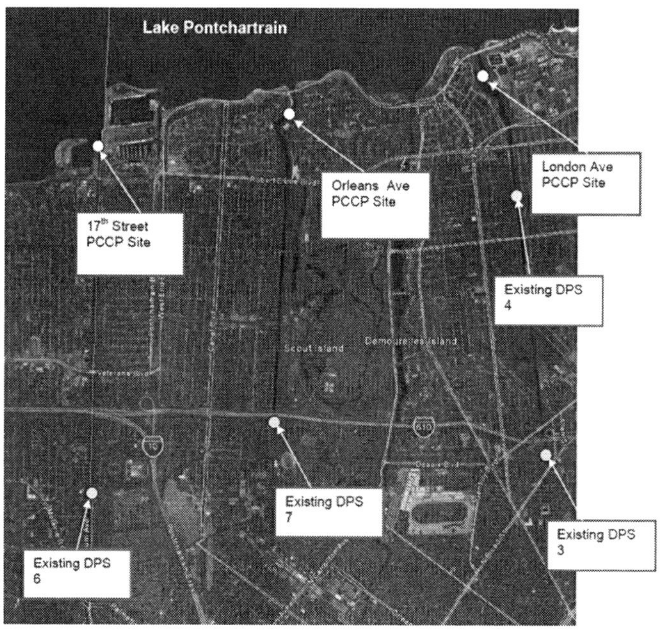

Figure 1. PCCP Pump Station Locations

The 17th Street Pump Station is located approximately 400 feet (ft) north of the temporary ICS. The new pump station will discharge stormwater past a Coast Guard navigational channel and into Lake Pontchartrain. The pump station will house eight (8) pumps: six (6) large (TMF 159) pumps, and two (2) small (TMF 113) pumps. One small pump will not be operated for Possible Future Conditions; it will be replaced by two smaller 120 cfs pumps equipped with variable frequency drives (VFD) to handle dry weather flows.

The London Avenue Pump Station is located approximately 900 ft north of the temporary ICS. The new pump station will discharge stormwater into the canal 1,100 ft upstream of the Lakeshore Drive Bridge. The pump station will house six (6) pumps: four (4) large (TMF 159) pumps, and two (2) small (TMF 113) pumps. One small pump will not be operated for Possible Future Conditions; it will be replaced by two smaller 120 cfs pumps equipped with VFD to handle dry weather flows.

The Orleans Avenue Pump Station is located approximately 1,000 ft north of the temporary ICS. The new pump station will discharge stormwater into the canal 500 ft upstream of the Lakeshore Drive Bridge. The pump station will house three small (TMF 113) pumps. An empty bay is provided for the future installation of two smaller 120 cfs pumps equipped with VFD to handle dry weather flows.

CFD Modeling Objectives

A comprehensive computational fluid dynamics (CFD) modeling study was performed to address a variety of hydraulic challenges of the project from the proposal preparation to detailed engineering design. This paper focuses on the CFD modeling work to achieve the following four objectives:

1. Evaluate initial layouts and designs of the three pump stations by predicting flow patterns in the approach canals, in the canal transitions, in front of the trash racks, in the pump forebays, and through the modified USACE Type 10 FSI before physical modeling verification and refinement. (Note: results of the CFD modeling were used to provide boundary conditions for physical modeling of the pump stations).

2. Develop initial designs of the discharge conduits for physical modeling verification and refinement, which have the potential to meet the requirements for system head losses and minimum pressures with minimal modifications.

3. Develop designs that, at each project site, meet with the requirements of average gate velocity and water surface elevation criteria at the interface between the Hydrologic Engineering Centers Rivers Analysis System (HEC-RAS) and CFD models (the model interfaces) so that Maximum Operating Water Level (MOWL) criteria are not exceeded in the upstream canals.

4. Evaluate the flow patterns in the forebays, upstream approaching canals, and downstream canals to provide velocity and water surface elevations for sizing ripraps to control erosions within the project limits.

CFD Modeling Approach

For each CFD model, a three dimensional (3D) geometrical model was first prepared based on computer aided design (CAD) files or design sketches using Gambit. This software was then used to generate computational meshes by dividing the 3D geometrical model into millions of small elements within which the Navier-Stokes equations and turbulence equations were solved with proper boundary conditions.

The state-of-the-art CFD software, ANSYS FLUENT, was used to solve the full set of the Navier-Stokes Equation coupled with a two-equation k-ε turbulence model. FLUENT can model both the pressurized and open channel flows for the PCCP project.

The CFD solution data were analyzed and plotted using post-processing software, FieldView, including flow streamlines, velocity contours and vectors, water level contours, and pressure contours. CFD results also include performance parameters such as head losses, swirl angles at pumps, minimum pressures in discharge piping, average gate velocities, and water surface elevations at model interface locations.

In the sections below, pump intake modeling will be first discussed, followed by pump discharge conduits modeling, gravity flow modeling and pumped flow modeling. For brevity, only the work for the largest pump station – 17th Street PS – will be presented. The CFD modeling approach and results for other two stations are similar to those for the 17th Street PS.

PUMP INTAKE MODELING

The pump intake CFD modeling during the proposal stage developed initial pump intake layouts and provided upstream boundary conditions for the pump station physical modeling. The physical models conducted at Northwest Hydraulic Consultants (NHC)'s laboratory in Vancouver, BC, were used to refine the initial designs and demonstrate that the final designs of these pump intakes meet with the requirements of Hydraulic Institute Standards ANSI/HI 9.8-2012. The only design modifications developed with the physical models are thirteen vortex breaker bars installed above the FSI entrances for all three pump stations.

CFD Model Description and Boundary Conditions

The CC model for the 17th Street pump intake covers the area shown in Figure 2 – from about 860 ft upstream of the FSI. The CC model includes details of the pump intake design such as the grade separation wall, sloped bottom connecting the pump bays to the canal, pump bays, and FSIs, as shown in Figure 3. Since this model was only used for predicting approach flow conditions and flow patterns around the pump intake, modeling of pump impellers was not necessary, and straight pipes were added onto the FSI exits to form the downstream outflow boundary. Figure 3 shows locations of swirl meters and velocity probes and the pump numbers: six 1,800 cfs large pumps (LP1 to LP6) and two 900 cfs small pumps (SP1 and SP2). The initial design does not include the vortex breaker bars shown in purple as these breaker bars were developed through subsequent physical modeling at NHC.

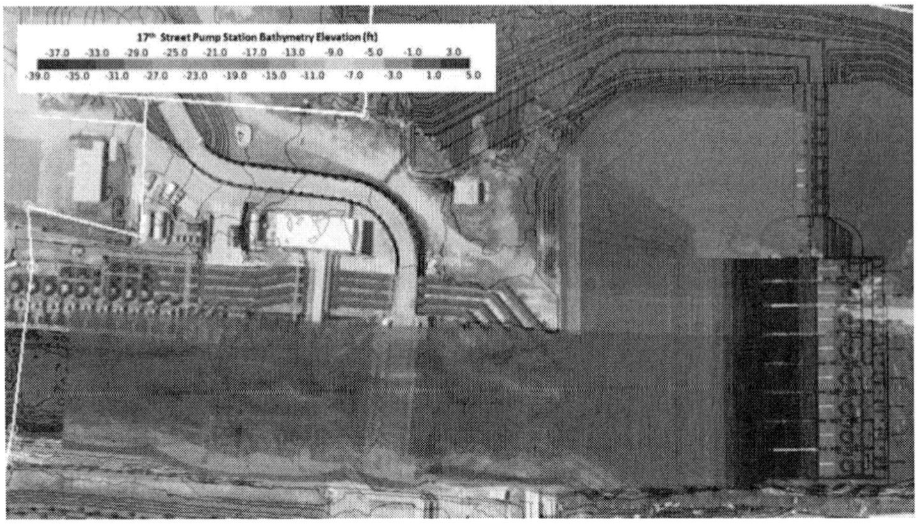

Figure 2. CFD Model Extent for the 17th Street Pump Intake

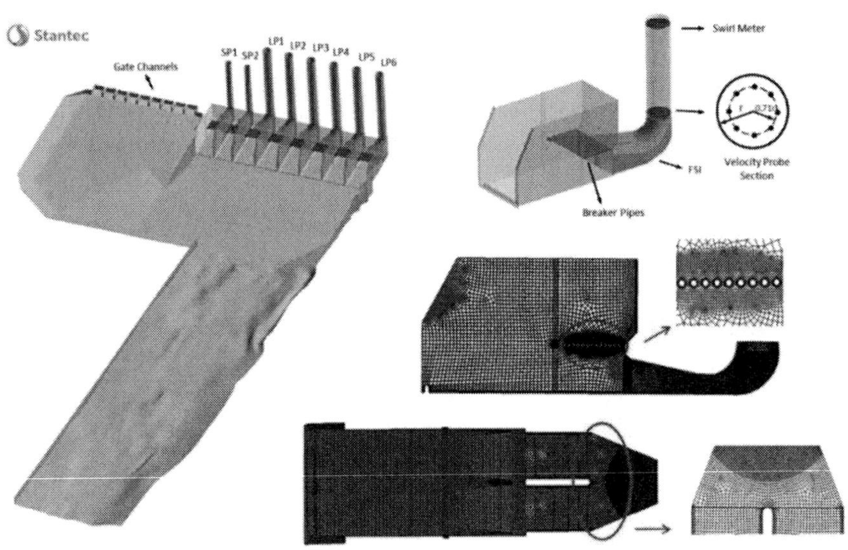

Figure 3. CC Model of the 17th Street Pump Intake

The 3D pump intake models were sub-divided, using Gambit, into meshes of about 7 million. These figures also show that finer cell meshes were clustered near walls and areas where high velocity and pressure gradients were expected.

Four types of boundary conditions were used in the model:

- Slip wall boundary condition with zero shear stress was specified on water surface, which allows water to freely move on the water surface. The water surface profile was not modeled because the objective of this study was to predict flow patterns approaching the pump entrances and slight change of water surface does not impact the study conclusions.

- No-slip wall boundary conditions (i.e., zero velocity) were specified at all physical walls, such as the intake bottom, banks, pump intake walls, FSI walls and pipe walls.
- Velocity inlet boundary conditions were specified at the upstream canal inlet, which took the form of a uniform velocity to match the flow rate into the pump intake.
- Outflow boundary conditions were used to model discharge flows at the ends of straight pipes where fractions of the total inflow were specified.

CFD Results/Discussion

Two simulations for the initial design (ID7 and ID13) are first performed to provide velocity distributions for setting up the upstream boundary of the Orleans Avenue Pump Station physical model. The operating conditions of these tests are summarized in Table 1.

Table 1. Operating Conditions of Initial Tests

Test No.	Condition	Intake Water Level (ft)	Pool-to-Pool Head (ft)	Operating Pumps and Flow Rates (cfs)							
				SP1	SP2	LP1	LP2	LP3	LP4	LP5	LP6
17th-ID7	High Head	-1.0	13.8	811	811	1,615	1,615	1,615	1,615	1,615	1,615
17th-ID13	Design Head	+2.0	8.2	909	909	1,814	1,814	1,814	1,814	1,814	1,814

The velocity distributions shown in Figures 4 and 5 are used by NHC to set up velocity distributions at the upstream boundary of the physical model which is located at about 425 ft upstream of the FSI entrances. These figures show that the velocity of flow is higher in the area close to the left bank, indicating that more water flows on the left side of the canal. Use of the velocity distribution at the upstream boundary significantly reduce the area modeled in NHC's hydraulic laboratory.

Eleven scenarios for the CC conditions were modeled to cover different pool-to-pool head and pump operating modes. For brevity, only the CFD results of one CC scenario, Case No. 17th-CC-DDOC1, are presented. This case simulates a scenario with all pumps in operation at a pool-to-pool head of 13.8 ft and intake water level of -1 ft. The pump flows are 1,615 cfs for large pumps and 811 cfs for small pump.

Figure 6 plots streamlines in the pump intake along with swirl angles at the FSI exits. From this figure, it is evident that there are large flow recirculation zones in front of the gate structures and pumps SP1 to LP2, and the swirl angles are all less than 5°. Figure 7 shows details of the velocity distributions at the FSI exits and FSI entrances. Both maximum and minimum velocities at the same eight measurement points of the FSI exit of the 17th Street PS physical model are all within 10% of the average velocity. Figure 8 shows velocity contours and vectors on the water surface and velocity distributions at Transect T1 where velocity measurements were made in the physical model - points shown in red. Finally, Figure 9 shows velocity contours and vectors at four different elevations. In these results, water approaches the pump bays at angles of attack up to 45°. These results are self-explanatory and will not be discussed in detail.

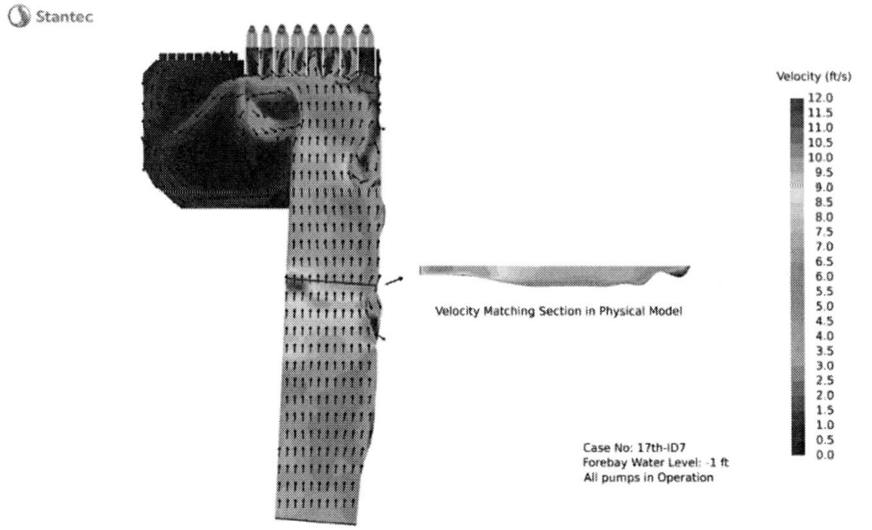

Figure 4. Velocity Contours and Vectors on Water Surface and Matching Section of 17th Street PS (Orleans- ID7)

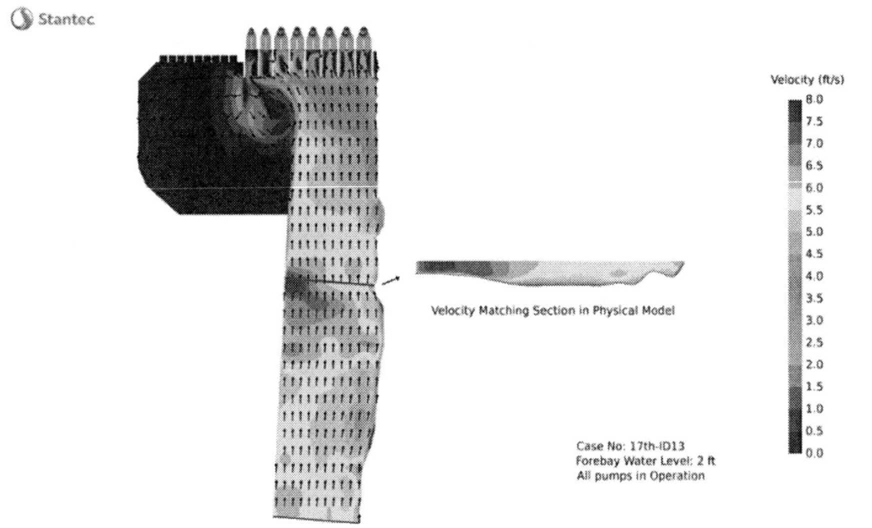

Figure 5. Velocity Contours and Vectors on Water Surface and Matching Section of 17th Street PS (Orleans-ID13)

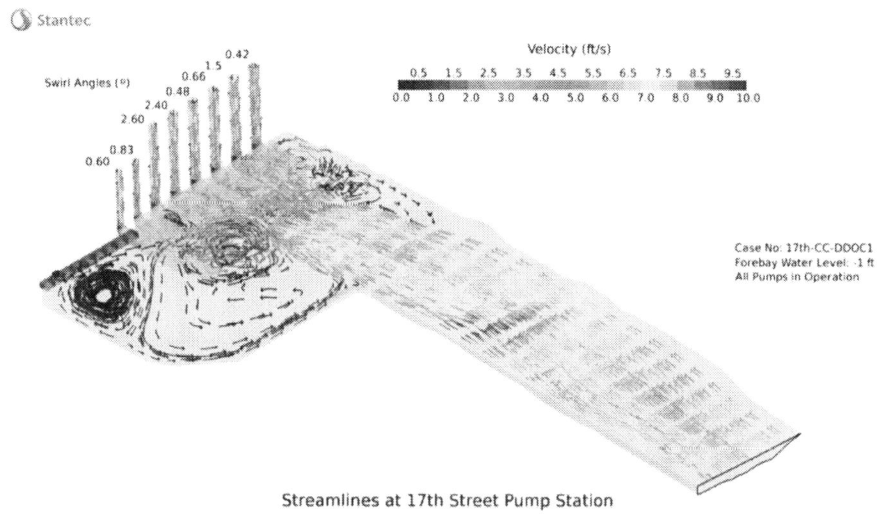

Figure 6. Streamlines at 17th Street PS

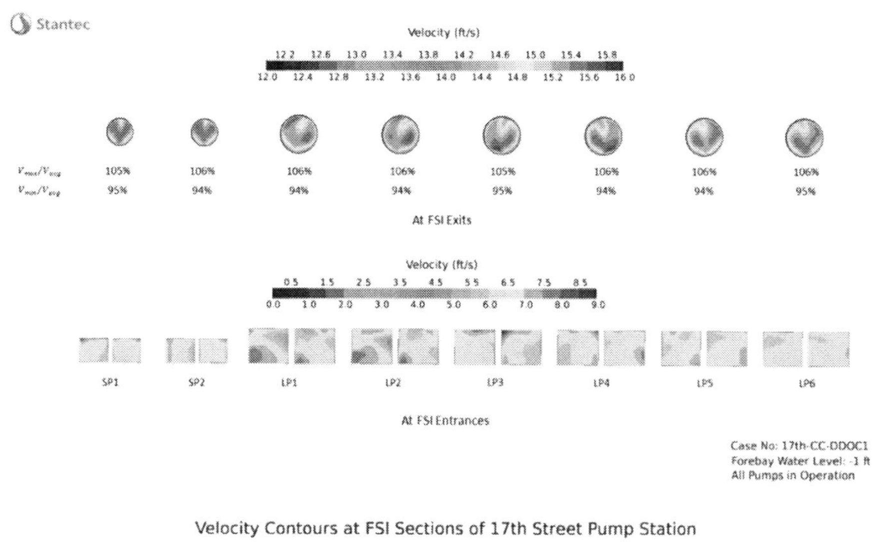

Figure 7. Velocity Contours at FSI Sections of 17th Street PS

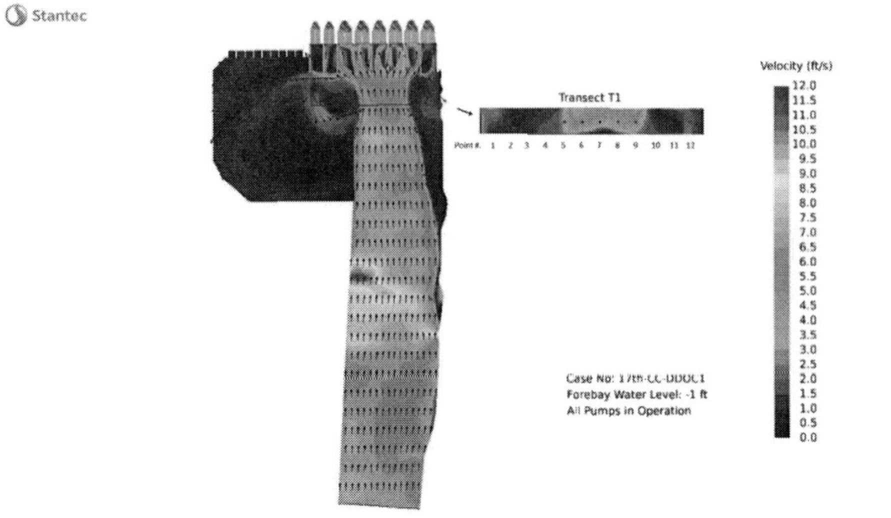

Velocity Contours and Vectors on Water Surface of 17th Street Pump Station

Figure 8. Velocity Contours on Water Surface of 17th Street PS

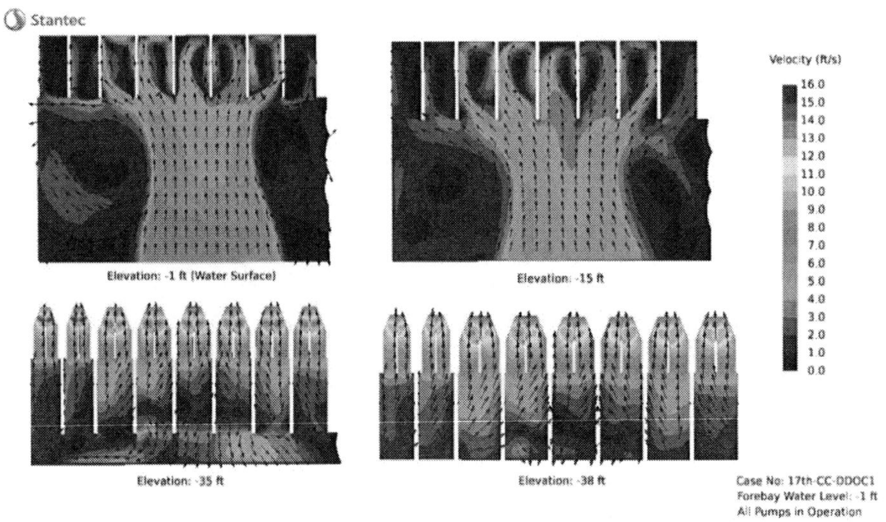

Velocity Contours and Vectors at Different Elevations of 17th Street Pump Station

Figure 9. Velocity Contours and Vectors at Four Elevations of 17th Street PS

Table 2 compares CFD predicted swirl angles to those measured in the physical model. In this case, the differences between CFD predicted swirl angles and physical model measurements are all less than 2°. Both CFD and physical models demonstrate that the swirl angles are less than the 5° required by HI Standards.

Table 3 compares CFD predicted ratios of maximum and minimum velocities over average velocities to those measured in the physical model. These results show that the difference between CFD ratios and physical model ratios are within 3%, and all results satisfy the requirements of the HI Standard (within 10% of the average velocity). These results were based on velocity measurements at eight points equally spaced around a circle of 0.71 radius of the FSI exit.

Table 2. Comparison of CFD Predicted Swirl Angles to Physical Model Measurements for 17th Street PS (CC)

Test Number	Swirl Angle (°)															
	SP1		SP2		LP1		LP2		LP3		LP4		LP5		LP6	
	Physical Model	CFD	Physical Model	CFD	Physical Model	CFD	Physical Model	CFD	Physical Model	CFD	Physical Model	CFD	Physical Model	CFD	Physical Model	CFD
DDOC1	1	0.60	<0.5	0.83	1	2.6	1	2.4	<0.5	0.48	<0.5	0.66	<0.5	1.49	<0.5	0.42

Table 3. Comparison of CFD Predicted Velocities to Physical Model Measurements for 17th Street PS (CC)

Test Number	Pump Throat Velocity Vmax/Vavg (%)						Pump Throat Velocity Vmin/Vavg (%)					
	SP2		LP1		LP6		SP2		LP1		LP6	
	Physical Model	CFD	Physical Model	CFD	Physical Model	CFD	Physical Model	CFD	Physical Model	CFD	Physical Model	CFD
DDOC1	106	105	107	106	106	106	94	94	93	94	92	95

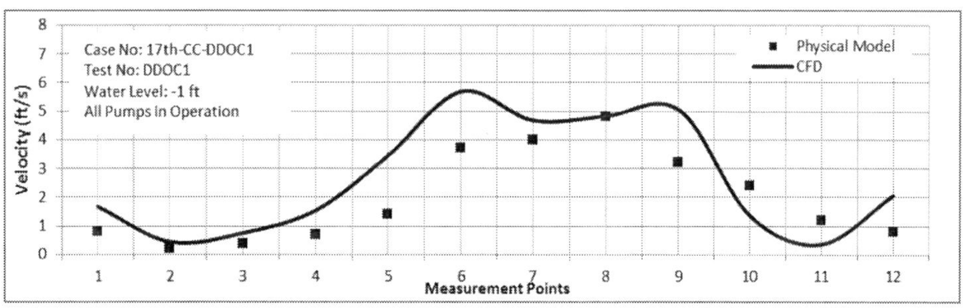

Figure 10. Velocity Comparison between CFD and Physical Model

Figure 10 compares the CFD predicted velocities at the Transect T1 (see Figure 8 for the transect location) to those measured in the physical model. Both sets of velocities match quite well. However, velocities measured in the physical model are generally lower than the CFD velocities.

Summary of Pump Intake Modeling

CFD modeling of the pump intakes produced initial designs of the 17th Street, London Avenue, and Orleans Avenue Pump Stations for physical model testing and subsequent development of design modifications (thirteen vortex breaker bars in each pump bay, see Figure 3) to meet with the requirements of ANSI/HI Standard (9.8-2012).

- CFD modeling provided velocity distributions used to setup the inflow boundaries of the three physical models used to study current conditions.

- CFD predicted swirl angles and deviations of the maximum and minimum velocities at FSI exits from average velocity compare very well to those measured in the physical models.

- CFD predicted velocities compare well with those measured in the physical models at sections where flow recirculation does not exist. The differences between CFD and physical model results become more pronounced in areas of flow recirculation where accurately measuring velocities is more challenging.

PUMP DISCHARGE CONDUIT MODELING

CFD modeling for pump siphon discharge conduits were performed to develop recommended designs before verification and refinement was conducted with physical models. These designs were intended to meet the requirements for system head losses and provide sufficiently high positive pressures to avoid cavitation within the siphon discharge conduit. CFD modeling results were also used for locating more than fifty (50) pressure taps in the physical models.

Figures 11 shows the discharge conduit for the large pump that consists of a steel portion (yellow) and a concrete portion (cyan). For brevity, only the large pump conduit will be discussed. The steel conduits connect to the pump flanges, and the concrete conduits discharge the water to the downstream canal. At the beginning of the study, Patterson Pumps provided the steel portion design with mitered bends, and Stantec used CFD modeling to develop a design for the concrete conduits (initial design). NHC tested the initial design of the discharge conduit for the 1,800 cfs pumps. During a witness test in October 2013, it was found that the pressure in the steel conduit was extremely low and could cause cavitation. Subsequently a recommended siphon discharge designs with curved bends and flow guide vanes was developed using CFD modeling to refine the layout of the two vanes, adjust the height of the rectangular section connecting to the concrete conduits, and re-design the matching concrete conduits.

The recommended discharge conduit was then tested at the physical models by NHC to confirm their hydraulic performance. No change is required for the recommended design of discharge conduit for the 1,800 cfs pumps developed with CFD models.

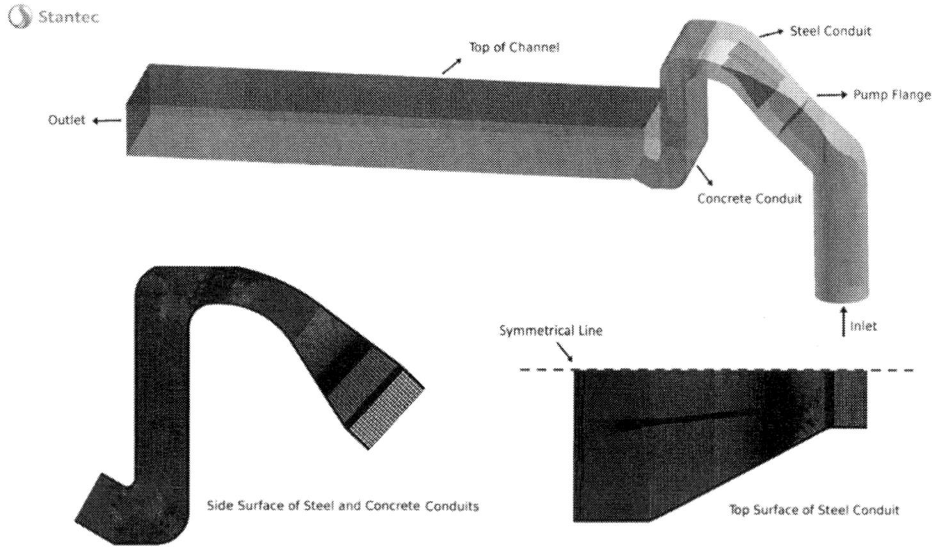

Figure 11. CFD Model and Mesh for 159 TMF Pump Discharge Conduit

CFD MODEL DESCRIPTION AND BOUNDARY CONDITIONS

The CFD model includes the pump discharges up to the pump flange, the steel conduit, the concrete conduit, and downstream canal. A straight pipe was used to represent the pump discharge, and the pump impeller was not modeled. Since the discharge conduit is symmetrical to the vertical central plane, only half the discharge conduit needed to be modeled, and the CFD results could be mirrored to the other half of the conduit.

The CFD model was developed using ANSYS FLUENT (version 14.5) and the computational cell meshes were developed using Gambit (version 3.2.26). A mesh, containing about 3 million cells, was used for each model with finer meshing around the vanes and close to the walls. To predict the free surface in the canal, a volume-of-fluid (VOF) model was used to solve two phase flows (water and air) and using a VOF fraction of water of 0.5 to define the water surface. Again the K-epsilon (k-ε) turbulence model with the standard wall-function was used in all simulations.

Five types of boundary conditions were used in these models:

- Velocity boundary conditions were specified at the inlet, which used a uniform velocity to match the flow rate into the pump discharge.
- Pressure boundary conditions were used at the outlets, which specified a lake level and allowed water to flow out of the computational domains.
- No-slip wall conditions were used at all physical walls, such as conduits walls, vanes, and the canal bottom on which velocities were set as zero.

- Slip wall boundary condition with zero shear stress was specified on top surfaces of the channels, which allows air to freely move on those surfaces.
- A symmetry boundary was used at the central plane of the conduits, as indicated by the red "symmetrical lines" in Figures 32 and 33. This boundary condition assumes a zero flux of all quantities across the boundary; i.e., zero normal velocity and zero normal gradients of all variables.

CFD Results/Discussion

The physical modeling demonstrated that no design modification is required to the recommended design for the 159 TMF (1,800 cfs) pump discharge conduit; therefore, the recommended design was adopted as the final design. The recommended design was developed by modifying an initial design provided by Patterson Pumps through CFD modeling with adding vanes and adjusting height of the conduits, etc.

Three CFD simulations were performed with the operating conditions detailed in Table 4, including a PFC low head condition for evaluating minimum pressure, CC design head condition for assessing head losses, and the CC pump exercise condition. For brevity, only the results for the first case was presented.

Table 4. Operating Conditions for 159 TMF Discharge Model

Case No.	Condition	Lake Level (ft)	Discharge (cfs)
159TMF-Discharge-1	PFC Low Head	-1.0	2,530
159TMF-Discharge-2	CC Design Head	10.2	1,800
159TMF-Discharge-3	CC Pump Exercise	-1.0	2,065

Figure 12 shows pressure contours on top, bottom and side surfaces of the discharge piping. The minimum pressure occurs near the beginning of the 90° arc transition from horizontal to vertical. The pressure distributions of this figure were also used to locate pressure taps and the pressure measurements in the physical model confirmed that this design did not produce excessive negative pressures that could result in cavitation.

Figure 13 plots velocity contours and vectors at different sections in the discharge conduits of the large pump, which shows large flow recirculation zones near the lower part of the steel conduit and inside part of the concrete conduit, and that the velocities on both sides of the leading and trailing edges of the two supporting vanes are not equal.

Figure 14 shows pressure distribution on both sides of the supporting vane. The average pressure difference between two sides of the vane is 0.48 pounds per square inch (psi). The pressure differentials for other operating scenarios are less because the flow rate is lower. These pressure distribution data could be used as an input for structural analysis of the vane.

The discharge conduit should have a hydraulic loss of 6.171 ft from the pump flange to downstream of the pump discharge at the CC design head condition. The NHC model demonstrated that the discharge conduit for the large pumps designed with CFD models provides the required head loss and no pressure is lower than -20.7 ft.

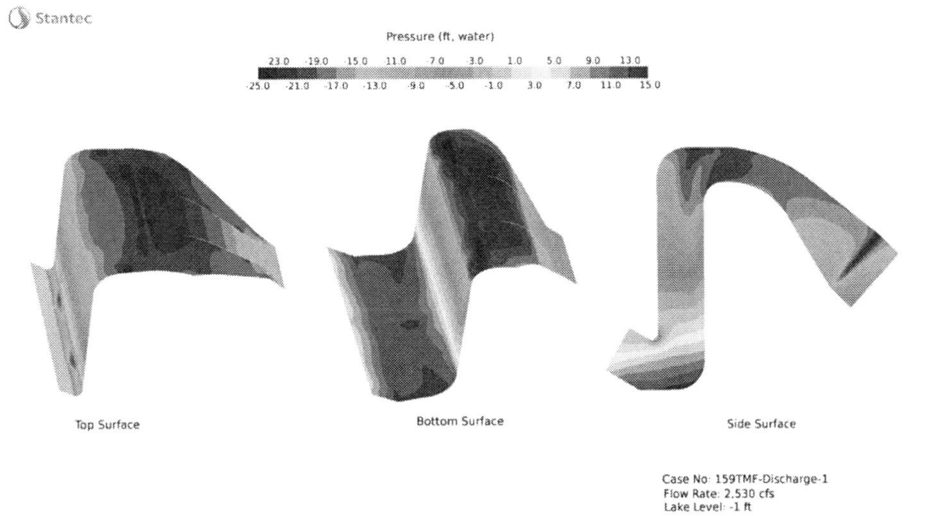

Figure 12. Gauge Pressure Distributions on Surfaces (Case No. 159TMF-Discharge-1)

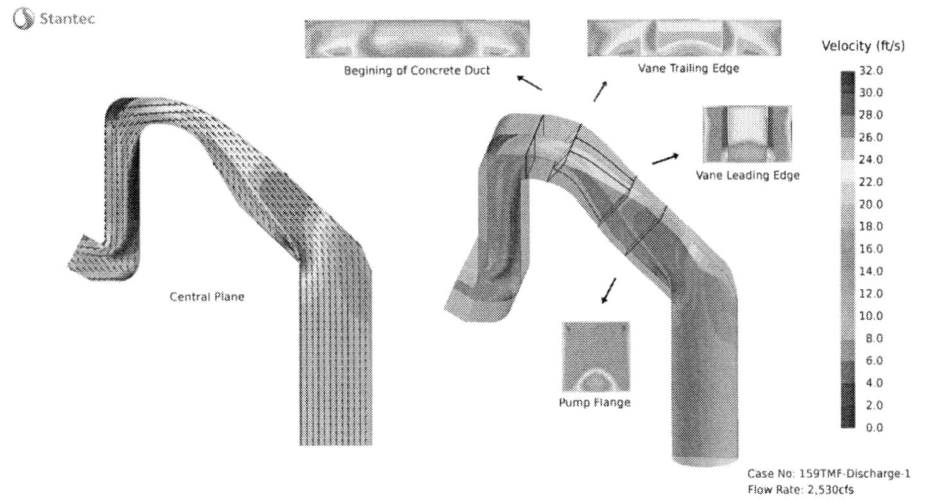

Figure 13. Velocity Distributions at Sections (Case No. 159TMF-Discharge-1)

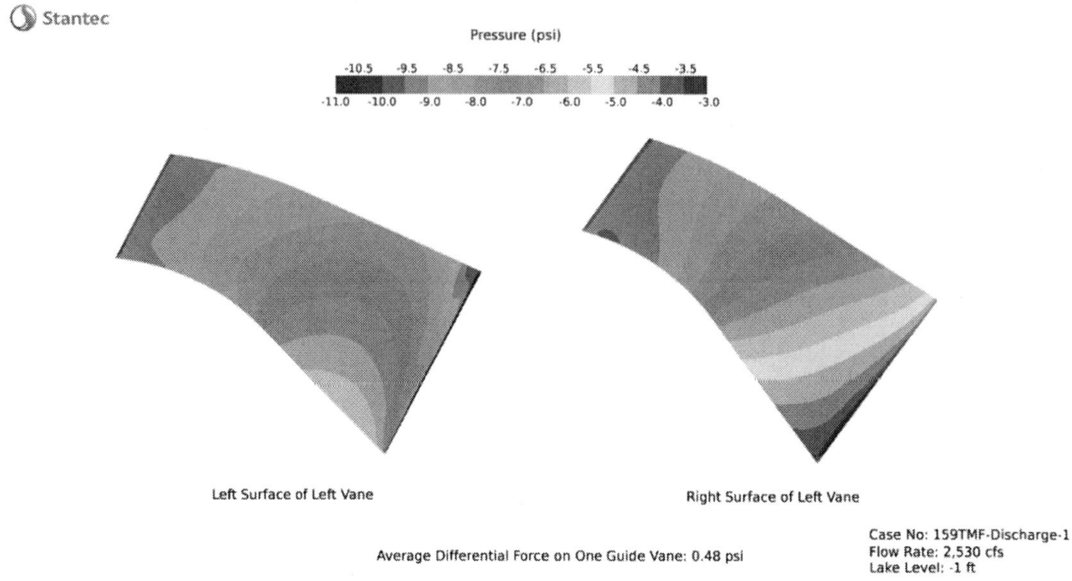

Figure 14. Gauge Pressure Distributions on Supporting Vane (Case No. 159TMF-Discharge-1)

Summary of Discharge Conduit Modeling

- The pump discharge modeling developed a siphon discharge design for each pump size to be physically modeled.
- The CFD results were used to locate pressure taps in the physical models.
- The physical model results confirm that the CFD developed designs for both the large and small pumps provide the required head losses and do not produce excessively negative pressures that could cause cavitation.

GRAVITY FLOW MODELING

This section presents CFD modeling approach and results for the gravity flow through flood gates. The objective of the gravity flow modeling is to develop a design at each pump station that has acceptable riprap protection within the project limits and meets the requirements of both MOWL and average gate velocity.

The CFD modeling provides velocity and water surface elevation for designing riprap. The details of the riprap design process will not be discussed in this paper. The proposed designs with additional gates were developed after modeling many design alternatives. The water level at an interface section was used as the downstream boundary for HEC-RAS model to predict upstream water surface profile. For brevity, only the CFD results for the proposed design of the 17th Street site will be presented.

CFD MODEL DESCRIPTION AND BOUNDARY CONDITIONS

The CFD model for each site includes the details of approach for canals, forebay, the pump bays, gate channels, and downstream canals.

Figure 15 shows the CFD model for 17th Street Pump Station, and Figure 16 shows riprap placements and excavation details upstream and downstream of the gate channels. There are eleven gate channels, which are 14 ft wide. The most western gate is numbered gate #1, and the most eastern gate is numbered gate #11. The bottom of the western nine gates (gate # 1 to 9) is at Elevation -8 ft, and the bottom of two eastern gates (gate #10 and 11) is at Elevation -11 ft. The forebay area in front of gates 10 and 11 is excavated to Elevation -13 ft, and the other forebay area invert is at Elevation -8 ft. An area immediately downstream of gates 10 and 11 is excavated to Elevation -11 ft, so water exiting gates 10 and 11 can have some room to expand.

These CFD models were developed using ANSYS FLUENT (version 16.1), and the computational meshes were developed using Gambit (version 3.2.26). About 4.3 million cells were used for the 17th Street PS. To predict the free surface, the VOF model was used to solve two phase flows (water and air) and use the VOF fraction of water of 0.5 to define the water surface. Again, the k-ε turbulence model with the standard wall-function was used in all simulations.

Four types of boundary conditions were used in the sectional model:

- Velocity boundary conditions were specified at the upstream canal inlet, which used a uniform velocity to match the flow rate into the pump intake.
- Pressure boundary conditions were used at the downstream outlets, which specified a lake level and allowed water to flow out of the computational domains. The water levels were based on the HEC-RAS model predictions.
- No-slip wall conditions were specified at all physical walls, such as piers, channel bottom, and walls. On these walls, velocities were set to zero. Roughness heights of the riprap (=D_{50}), concrete, and natural canal were specified on each wall surface to model the hydraulic resistances and their impacts on the flow.
- Slip wall boundary conditions with zero shear stress were specified on top surfaces of the channels, which allow air to freely move on those surfaces.

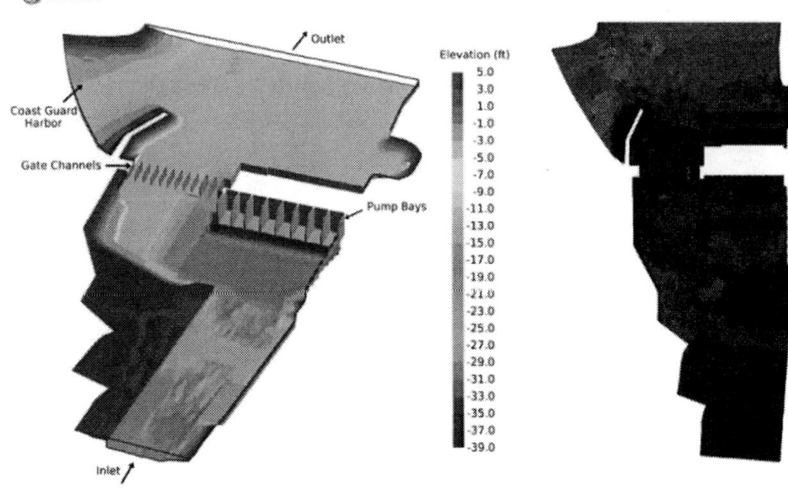

Figure 15. CFD Model of 17th Street Pump Station and Meshes

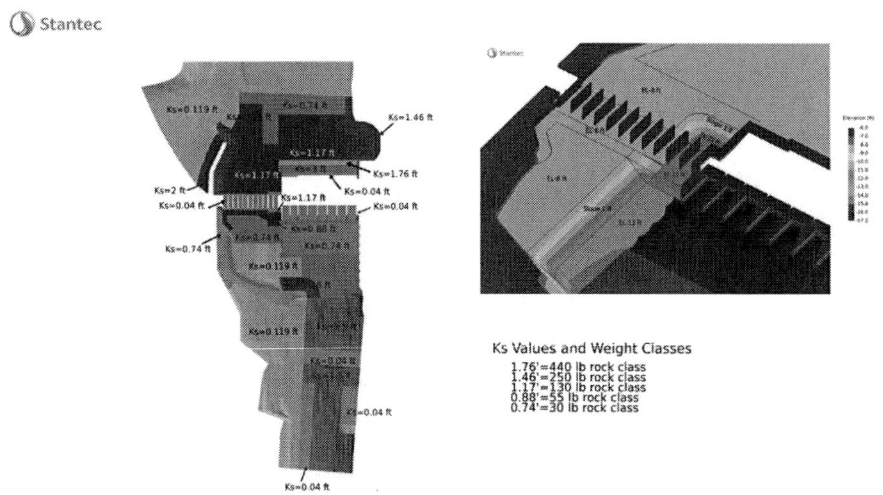

Figure 16. Riprap Map and Excavation of 17th Street Pump Station

<u>Model Verification</u>

The capability of the CFD models for predicting water surface and flow patterns was assessed using the physical modeling data taken from a sectional gate channel model completed at NHC, as shown in Figure 17. This model includes a full gate channel, two half gate channels, a downstream basin, and a block at the entrance of the central channel. The block was included to skew the flow to mimic the flow pattern with more flow diverted to the left side and generate a velocity at the channel exit close to 17.9 ft/s. The final design of the bypass gates and upstream and downstream dredging is different from the one modelled and without the downstream basin, but with ripraps. The final design also has much lower velocity at the gate channel exits because more gate channels are added.

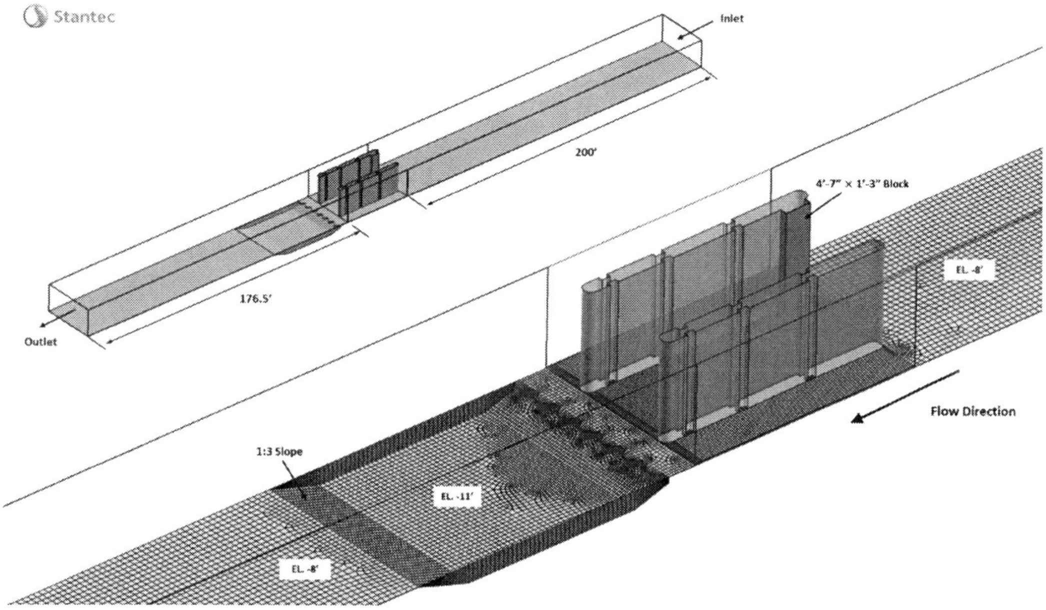

Figure 17. Sectional Model Details and Meshes

Two CFD simulations were performed for a flow rate of 2,960 cfs at -1 ft and +4 ft water level at the outlet. Figures 18 and 19 show velocity contours on the water surfaces for these two cases.

Detailed comparison of the longitudinal water surface profiles along the centerline of the gate channel is plotted in Figures 20 and 21. In these figure CFD results were compared to the water surface profiles measured with both pressure taps and staff gauges. This figure confirms, to a certain degree, that the CFD models can predict the water surface profiles with reasonable accuracy. In particular, the water surface elevation drop after entering the gate channel matches well between the CFD model and physical model. The CFD model is also able to predict the first wave height with good accuracy, but the waves damp faster as water travels downstream than in the physical model.

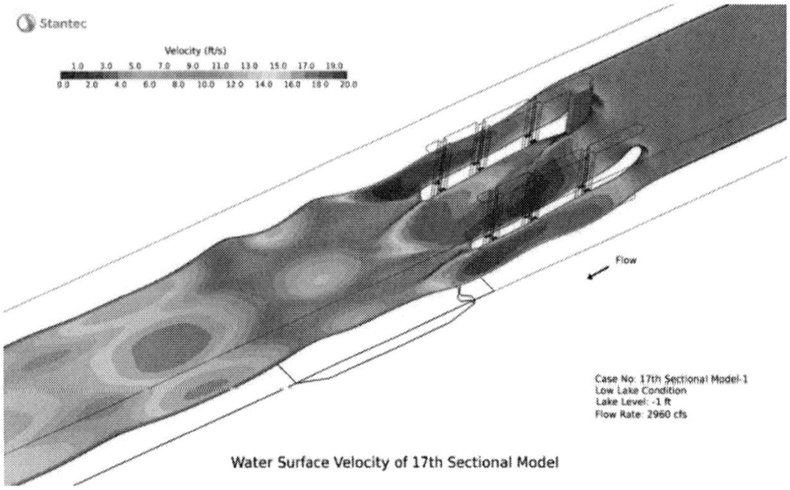

Figure 18. Velocity Contours on Water Surface for -1 ft Water Level at Outlet

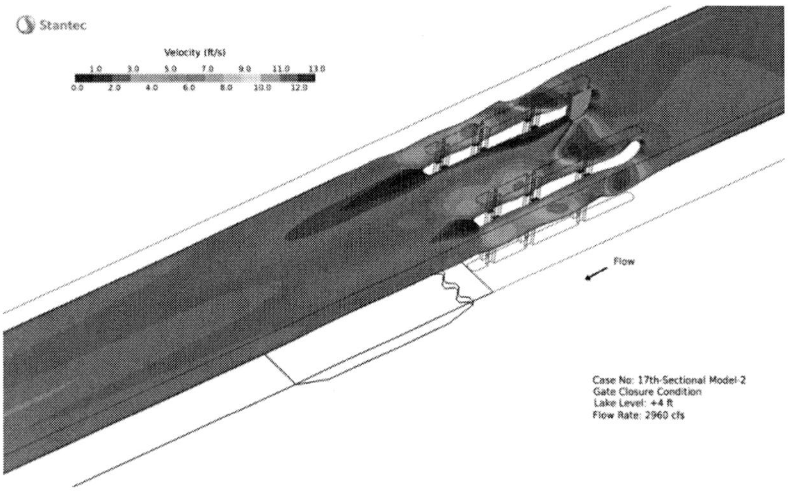

Figure 19. Velocity Contours on Water Surface for +4 ft Water Level at Outlet

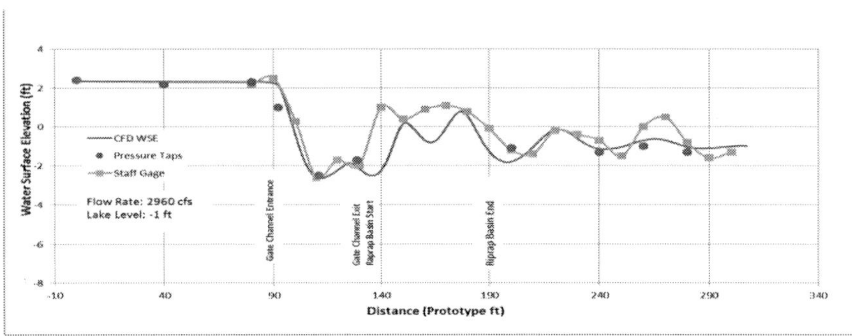

Figure 20. Comparison of Water Surface Profiles (-1 ft Water Level)

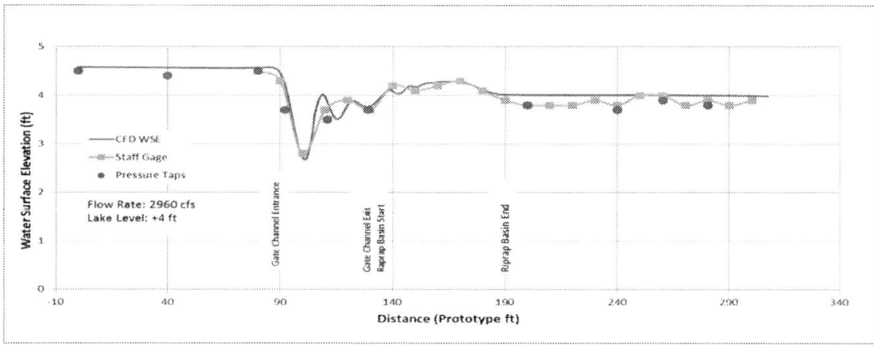

Figure 21. Comparison of Water Surface Profiles (+4 ft Water Level)

CFD Results and Discussions

For each pump station, two CFD simulations were performed – one at gate closure lake level and one at the low lake level. The results for the gate closure levels are used to assess if the design meets with the requirement of MOWL, and the results for the low lake level are used for demonstrating that the design provides an average gate velocity less than 12 ft/s. The water surface profile along the upstream canal of each site was predicted using the HEC-RAS program with the CFD-predicted water surface elevation at the model interface as its downstream boundary condition. The water surface profile should stay below the MOWL with a margin of 0.5 ft, the desired tolerance for uncertainty.

MOWL

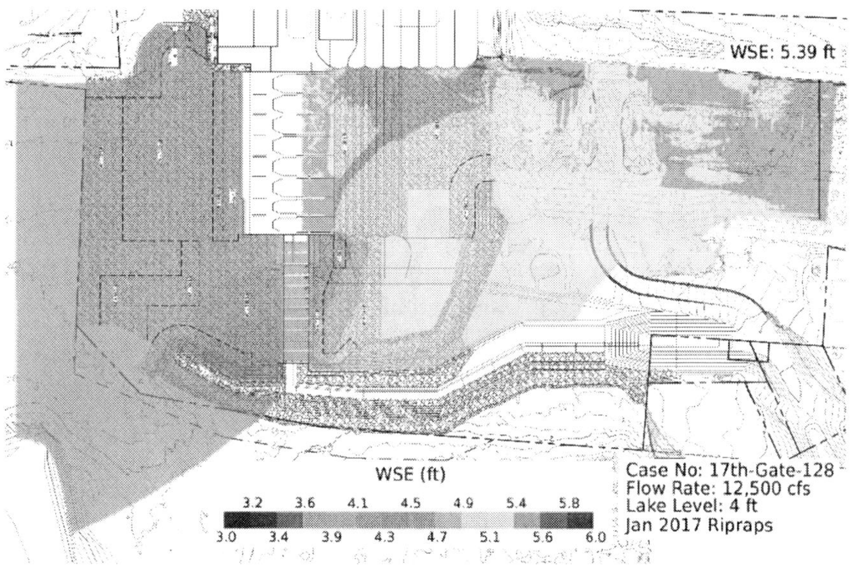

Figure 22. Water Surface Profile (Case No. 17th-Gate-Final-128)

Case No. 17th-Gate-Final-128 was performed at the gate closure lake level of +4 ft and 12,500 cfs.

Figure 22 plots the elevation contours on water surface and shows that the water surface elevation at the model interface is 5.39 ft, which was used as an input to HEC-RAS models to demonstrate compliance with the MOWL requirement.

Average Gate Velocity

Case No. 17th-Gate-Final-129 was performed at the low lake level of -1 ft and 12,500 cfs. Its results were used to demonstrate the compliance with the average gate velocity requirement, i.e., the average gate velocity should be less than 12 ft/s.

Figure 23 plots the elevation contours on water surface and shows that the water surface elevation at the model interface is 2.42 ft.

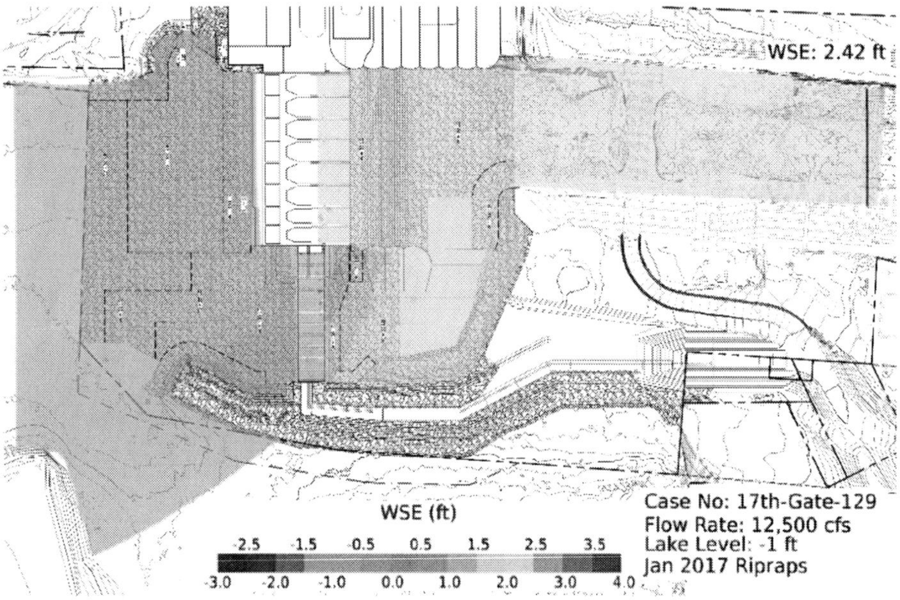

Figure 23. Water Surface Profile (Case No. 17th-Gate-Final-129)

The results shown on Figure 24 show that water must make an S-turn to get into the gate channels. In addition, water flows into the gate channels with an angle towards the west for each gate channel. This figure also shows that gate 11 is not as effective because it sits in a low velocity zone generated by the west pump bay wall of the first small pump.

Figure 25 shows the velocity contours on the upstream face of each gate slot. The average velocity over the entire gate area is equal to 9.87 ft/s, meeting the average gate velocity requirement. This figure also shows that gate 11 is not as effective because it sits in the shadow of the western wall of the pump bay for the first small pump (see Figure 24).

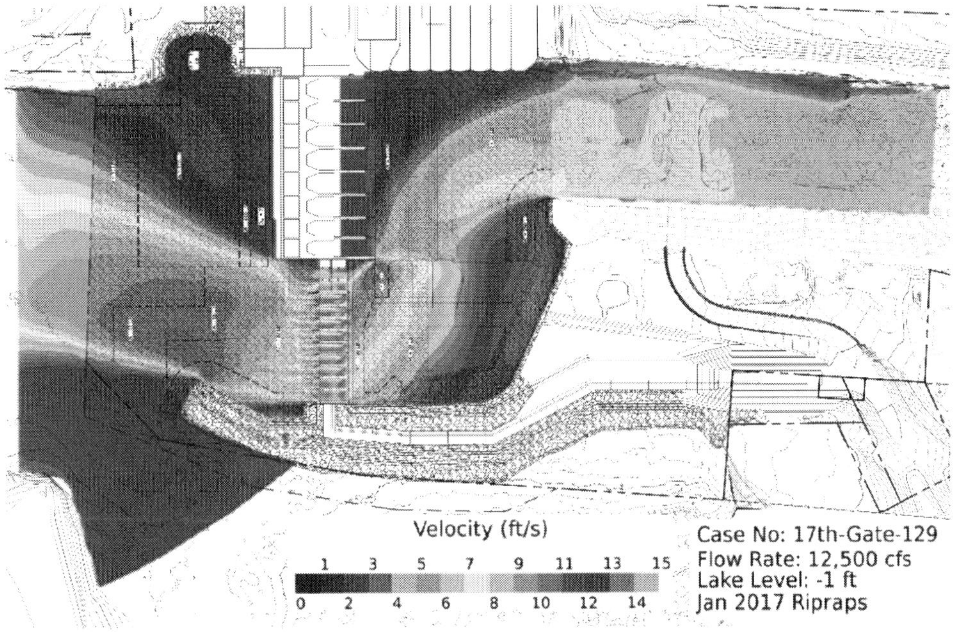

Figure 24. Velocity Contours on Water Surface (Case No. 17th-Gate-Final-129)

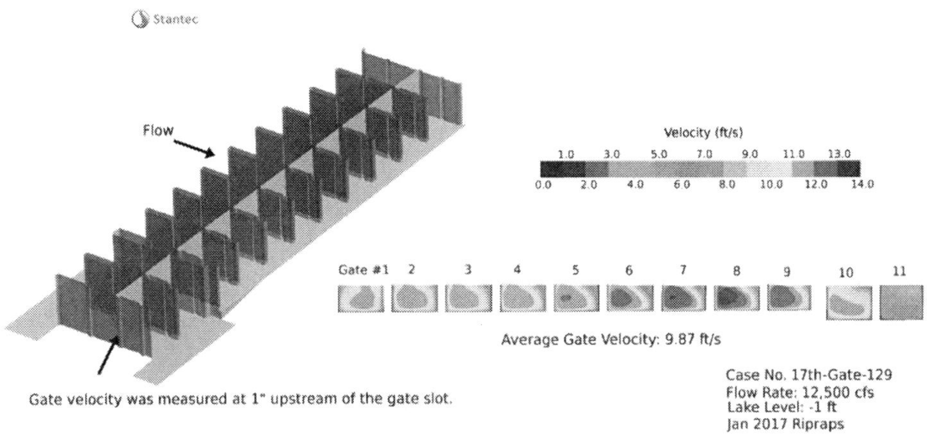

Figure 25. Gate Velocity and Flow Distribution (Case No. 17th-Gate-Final-129)

Figure 26 shows velocity contours at different sections illustrating that flow from the gate exits will diffuse as water flows downstream.

Figure 27 plots depth averaged velocity contours that were used in sizing riprap for scour protection.

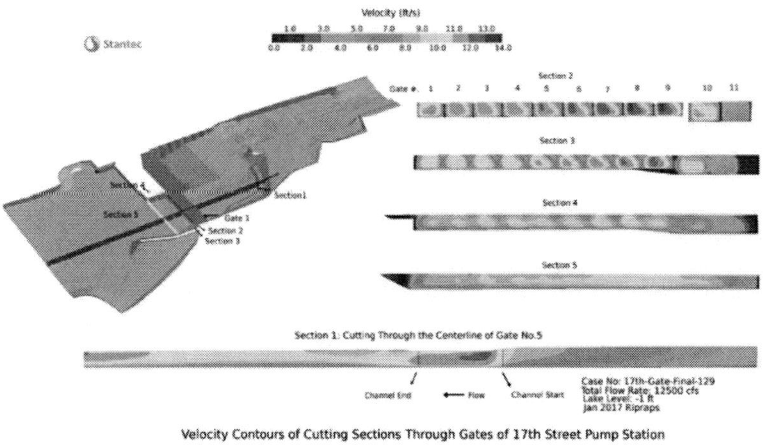

Figure 26. Velocity Contours at Sections (Case No. 17th-Gate-Final-129)

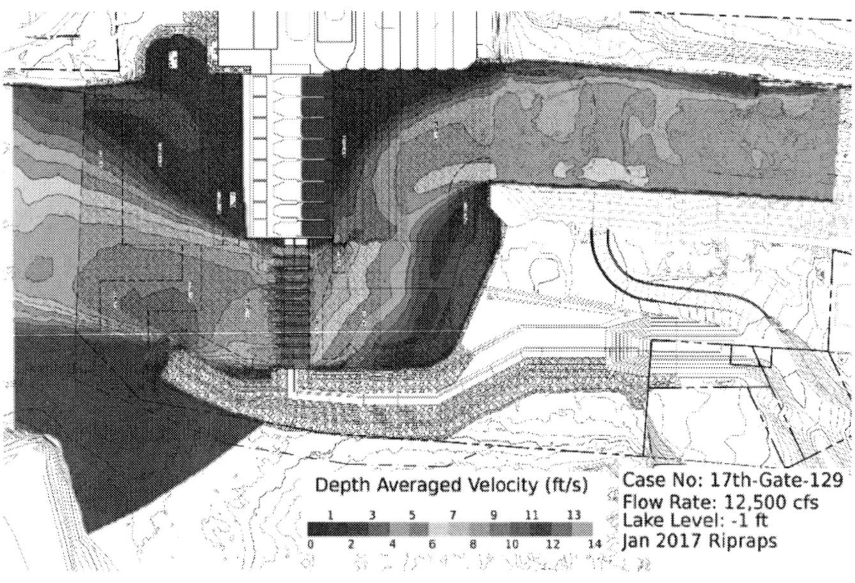

Figure 27. Depth Averaged Velocity Contours (Case No. 17th-Gate-Final-129)

<u>Summary of Gravity Flow Modeling</u>

- Comparison of CFD results to physical modeling data demonstrate that the CFD model can predict flow patterns and water surface profile with reasonable accuracy.

- The CFD modeling of the gravity flow through gate channels at each site provided water surface elevations at the model interfaces and average gate velocities.
- The modeling study demonstrates that the eleven gate design for the 17th Street Pump Station, the seven gate design for London Avenue Pump Stations, and the two and half gate design for the Orleans Avenue Pump Station are fully compliant with the MOWL and average gate velocity requirements.
- The CFD models at lower lake levels provided velocity data for sizing riprap to protect upstream canals, pump forebays, and downstream canals from scour.

PUMPED FLOW MODELING

This section describes the CFD modeling approach and results of pumped flows. The results were used to predict velocity distributions and water surface profiles downstream of the pump discharges to be used for sizing riprap. The riprap designs for the areas immediately downstream of the pump discharges were proven using the physical models. For brevity, only the 17th Street PS will be discussed.

CFD Model Description and Boundary Conditions

Figure 28 shows the CFD model for 17th Street Pump Stations. Similar to the CFD models for the gravity flows, it was developed using Gambit and simulated with ANSYS FLUENT. The free surface was modeled with the VOF model. The number of cells within the meshes for these models are about 3.4 million.

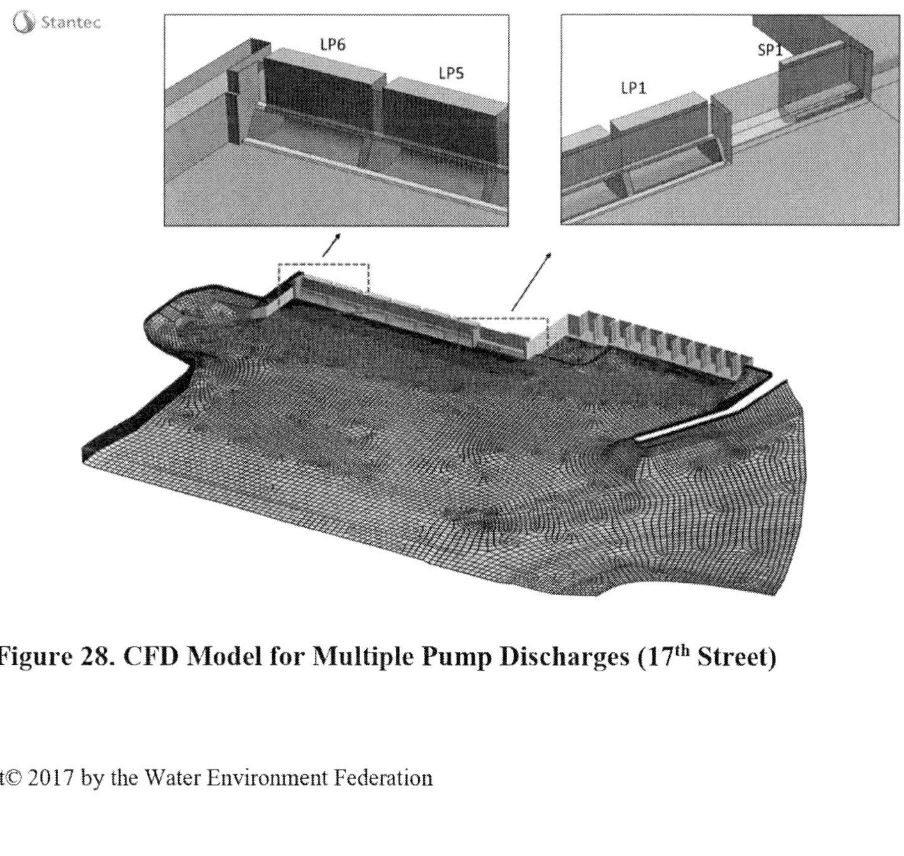

Figure 28. CFD Model for Multiple Pump Discharges (17th Street)

These models include the downstream sections of pump discharge conduits so the pump flows can be specified at the inlets (shown in gold). For the PFC condition, these pumps have higher flows than the flows for the CC; therefore, the PFC configurations were also modeled. For both 17th Street and London Avenue, there will be no pump in the pump bays for the small pump, SP2. The ripraps shown in Figures 16 were modeled.

The boundary conditions used for these models are similar to those used for the gravity flow models and will not repeated here for brevity.

<u>CFD Results and Discussions</u>

CFD simulations were performed for CC, pump exercise, and PFC conditions. For each simulation, three plots will be presented: water surface profile, velocity contours on water surface, and depth-averaged velocity contours. These plots were used as design inputs for scour protection with ripraps. For brevity, only the results of CC Pumping Condition are presented.

A CC pumping condition was modeled with SP1 pumping 216 cfs and LP1 to LP6 each pumping 2,064 cfs at lake level of +4 ft (Case No. 17th-Pumped Flow-21). The total flow is 12,600 cfs.

Figure 29 plots the water surface profile that is slightly higher in the area downstream of the pump discharges. Figure 30 shows velocity contours on the water surface, and Figure 31 shows depth averaged velocity contours that were used for designing ripraps. These plots are self-explanatory and will not be discussed in detail.

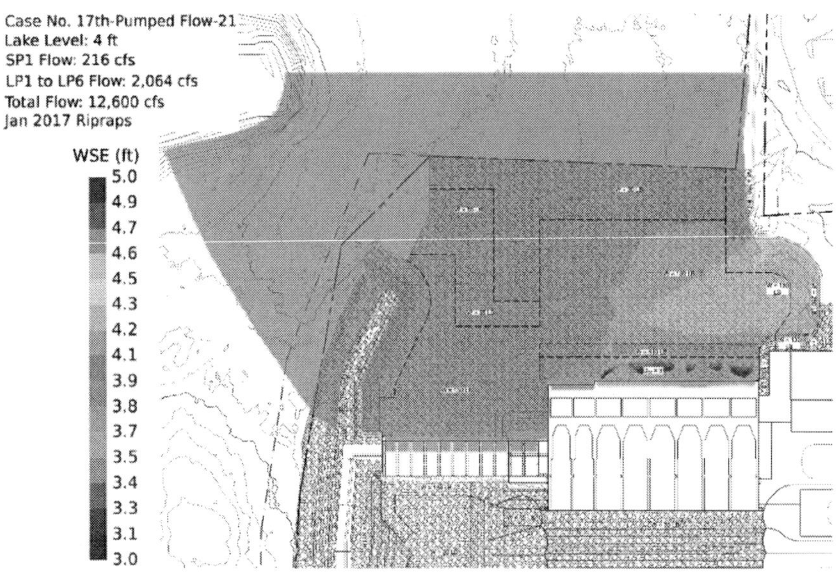

Figure 29. Water Surface Profile (Case No. 17th-Pumped Flow-21)

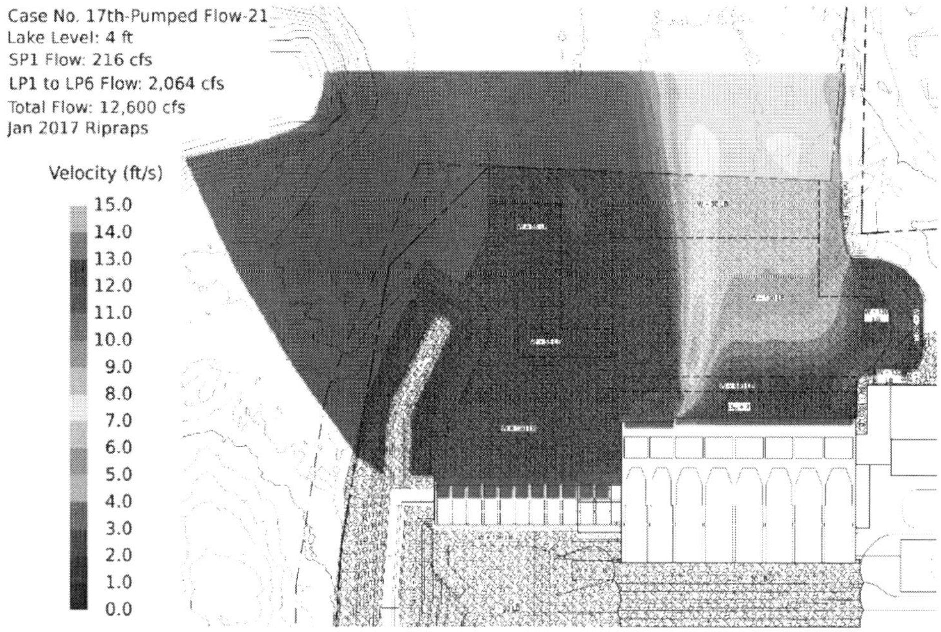

Figure 30. Velocity Contours on Water Surface (Case No. 17th-Pumped Flow-21)

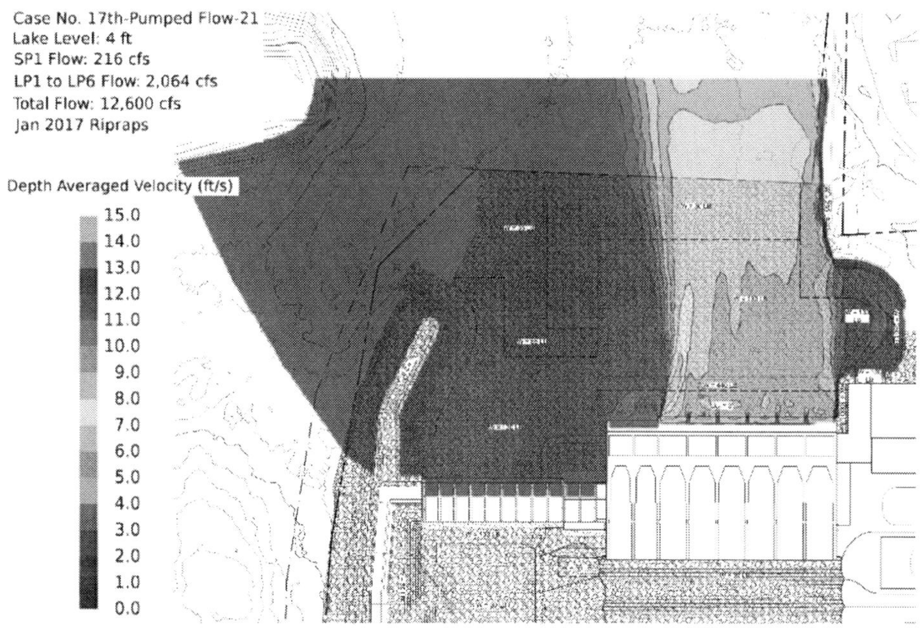

Figure 31. Depth Averaged Velocity Contours (Case No. 17th-Pumped Flow-21)

Summary of Pumped Flow Modeling

The CFD results of the multiple pump discharge models provide velocity information for designing riprap for scour protection.

CONCLUSIONS

This paper demonstrates that use of CFD modeling is a useful tool in pump station design for addressing a variety of hydraulic issues for pump intake, pump discharge piping, flood gate and canals.

The use of CFD modeling greatly improved the confidence level of the pump station designs, reduced the scopes of physical modeling, and accelerated the design schedule. CFD modeling of gravity flows downstream of the gates and pump discharges provided depth-averaged velocity and water depth for riprap designs. The CFD models also provide downstream boundary conditions for HEC-RAS models to predict water surface profiles upstream of the project limits.

REFERENCES

Hydraulic Institute (2012) *American National Standard for Rotodynamic Pumps for Pump Intake Design,* ANSI/HI 9.8, Parsippany, New Jersey.

Attacking the Big Data Challenge: Integrating Collection System Flow Monitoring and Modeling into Financial Billing

Yuanhang Mel Meng[1], Charles Schroeder[2], Mike Lunn[2], Ben Whitehead[1], Steve Cook[1]

[1] Black & Veatch, [2] Environmental Services Department, City of Grand Rapids, MI

ABSTRACT

The City of Grand Rapids, MI calculates the annual sewer rate for each customer community based on the cost of maintaining and building the pipes along the flow paths conveying the flow from the customer to the Water Resources Recovery Facility (WRRF).

A financial model was first developed by the City for the rate study in 2003. To track major capital and maintenance spending activities, the collection system is divided into more than 100 zones to track the flows from each customer community, a process the City refers to as zonegating. With only 39 short-term flow meters and limited collection system GIS data, the source and quantity of the flows at most of the zones were estimated.

With over a decade's investment in a data centric utility management approach, the City has built a data asset including a high quality GIS database, 100 permanent flow meters providing real-time flow data since 2013, and an all-pipe sanitary sewer collection system model calibrated under the 2014 flow condition.

This paper presents a case study of a recent zonegating sewer rate update that took advantage of the data asset to greatly improve the accuracy of flow estimates. This paper demonstrates the challenges, lessons and opportunities of bridging the gap between better business intelligence and the ever growing large data set generated by smart infrastructure monitoring and modeling technologies.

KEYWORDS

Flow monitoring, modeling, big data, financial billing, smart city, smart infrastructure

INTRODUCTION

Big data analytics popularized by the technology industry in recent years are getting more and more attention from the water utility owners as a promising technology that can potentially transform the business in the near future. However, a 2016 survey showed that despite the consensus that water utilities have an integral role in developing smart cities, these plans frequently do not include water's role in forging smart city blueprints (Black & Veatch, 2016). The survey showed that a little more than half of respondents are either at the very earliest stages of smart city planning or are not reporting much progress in smart city coordination with their counterparts from other industries (Black & Veatch, 2016).

The intent of this paper is to serve as a case study of implementing big data analytics, and inspire more advanced applications to speed up the adoption of big data in the water industry.

The City of Grand Rapids Wastewater Collection System sewer service area is approximately 127 square miles (329 km^2) as shown in Figure 1.The City performs an annual sewer rate study to determine proposed sewer rates, effective the following January 1st, for all sewer system users. The annual rate study is scheduled for completion by the first week in November and is typically distributed to customer communities for review and comment by mid-November. Customer communities included in the rate study consist of two user groups; wholesale and retail customers. The flows from wholesale customers are monitored by master flow meters, which include the City of East Grand Rapids, Ada Township, Caledonia Township and Gaines Charter Township. For retail customers, flow enters the Grand Rapids Wastewater Collection System through multiple pipes, most of which are not monitored. Therefore retail customer flows have been estimated, until recent technology improvements were implemented. The retail customers include the City of Grand Rapids, City of Walker, City of Kentwood, Cascade Charter Township, Grand Rapids Charter Township, Tallmadge Charter Township, and Wright Township. The rate study provides that non-integrated revenue requirements are allocated between customer communities on the basis of their share of relative flow. The City of Grand Rapids refers to this allocation process as 'zonegating', and requires adjustments from time to time to reflect changes in flows.

A zonegating financial model was first developed by the City in 2003 to set annual sewer rates. A zonegating framework that divided the collection systems into more than 100 zones was developed to separately track flow paths from individual communities to the Water Resource Recovery Facility (WRRF) through the backbone of the system. With only 39 temporary flow meters and limited GIS data at that time, the flows through many zones were estimated with imprecise assumptions. Despite the limitations, it is the consensus of all the customer communities that the flows developed in the 2003 study have been the best flow estimates based on the available information.

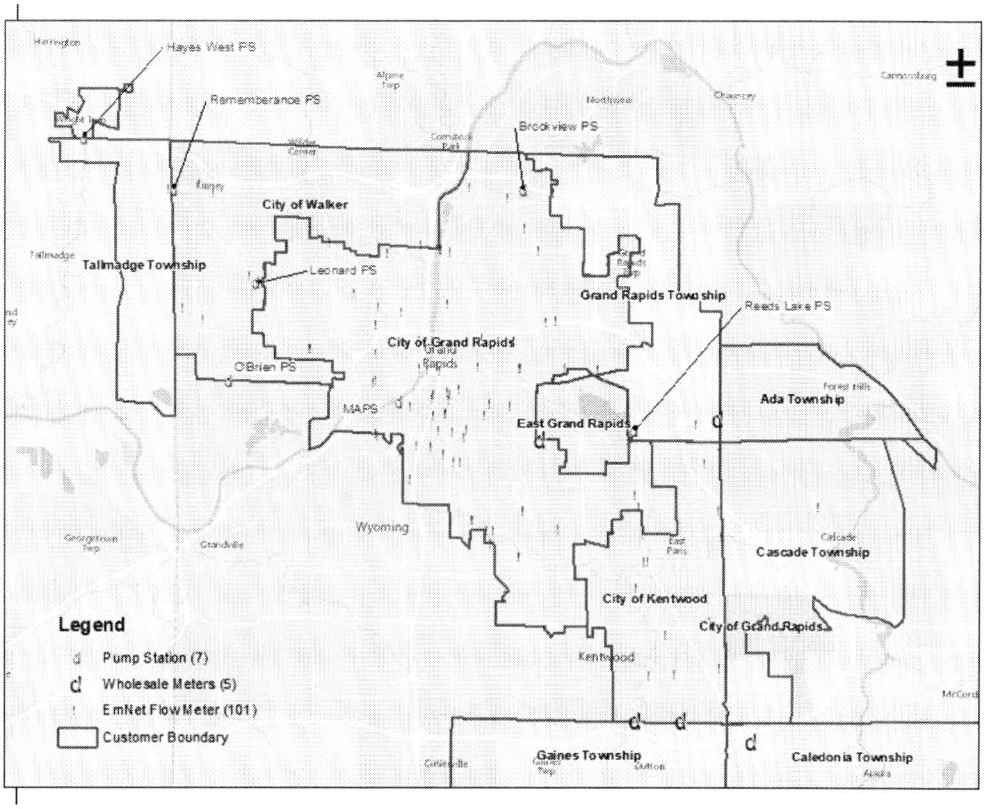

Figure 1 Customer Community Map

The situation has changed in recent years. With a vision of adopting a data centric management through technological innovations, the City carefully scoped projects to build a digital infrastructure that can fulfill the vision and meet future customer needs and regulatory requirements. Since the completion of the 2003 rate study, the City has built an accurate GIS database with all the pipes in the system; installed 100 permanent flow meters to monitor the system in real-time since 2013; and constructed an all-pipe sanitary sewer collection system model, calibrated using the 2014 flow data.

The recent improvements to the City's data asset, especially the flow monitoring data at more than 100 locations, prompted an update of the zonegating flows to improve the accuracy of the flow estimates and sewer rates.

METHODOLOGY

The sewer rate calculation is rather sophisticated. Instead of calculating a flat rate and charging customer communities by volume, the rate is set by the actual use of the infrastructure at the pipe level, which requires a very detailed flow analysis of the collection system. Back in 2003, this was manually calculated using a spreadsheet program. An example is shown in Figure 2. For a system with 2 customers and 3 pipe segments, the spending for each pipe is $100, and the cost is divided between the two customers based on the relative flows through each pipe. The result is a spreadsheet of more than 200 rows and 40 columns, with most of the cells manually calculated to route the flow segment by segment, customer by customer. This is a tedious and error prone analysis for a system with more than 24,000 pipes.

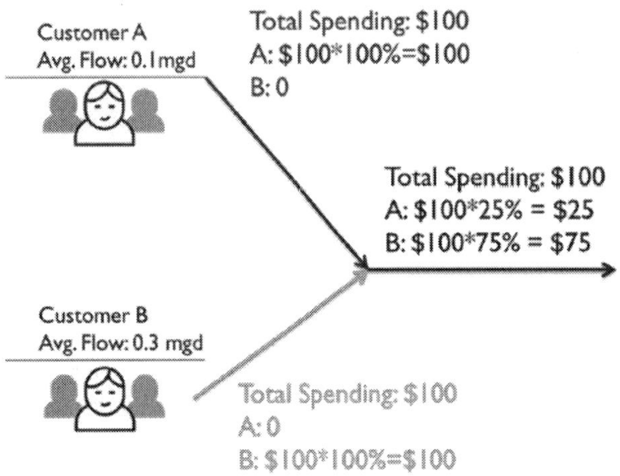

Figure 2 Zonegating Calculations of 2003 Study

As shown in the example, the sewer rate calculation is based on two pieces of information at each flow segment, the cost associated with the flow segment, and the loadings from all customers to the flow segment. The development of costing information is beyond the scope of this paper. The calculations of the loadings and the routing and tracking of the flows through the flow segments are explained in the following sections.

Flow Segments

To incorporate improvements to system completed since the 2003 study, the flow segments delineated in 2003 were reviewed; abandoned flow segments were removed, and new projects were delineated as new segments. Both the all-pipe model and the latest GIS database were referenced to confirm and update flow splits and parallel sewers. The updated flow segments are shown in Figure 3. 151 zones with corresponding costing information in the financial records were delineated.

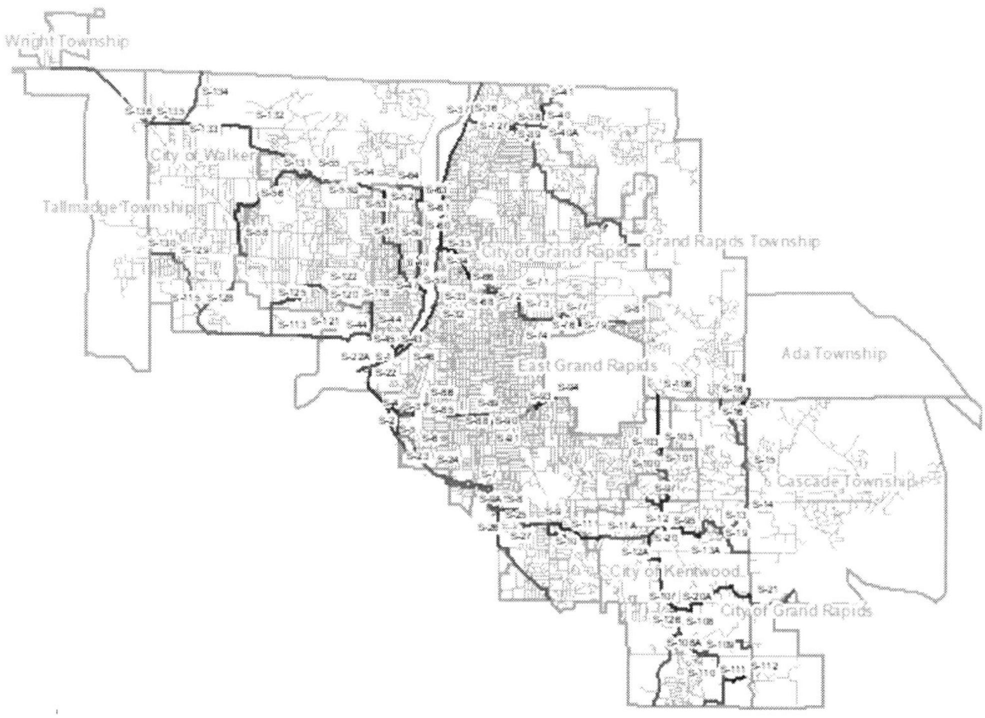

Figure 3 City Defined Flow Segments

Flow Monitoring Data

The best source of customer loadings are directly monitored flows. In 2016 the City established an advanced monitoring network with more than 100 locations as shown in Figure 1. It consists the following,

- 5 Master flow meters for wholesale customer communities
- 101 permanent flow meters within the City of Grand Rapids service area
- 7 SCADA runtime data feed from pump station and storage facilities

The flow monitoring data from April 2015 to May2016 provided by the City were reviewed and processed to remove missing and erroneous data points. When flow data are missing for short periods of time, the data were interpolated if good correlation can be established between good flow data nearby. Once the flow monitoring data were cleaned and the recoverable missing data interpolated, a simple average was calculated for each flow meter to represent the annual average flow.

Customer Loadings

Customer loadings need to be developed for each segment to route the flows through the system. Loadings from 260 submetersheds throughout the system were delineated to capture all the flow sources. An example of submetershed delineation is shown in Figure 4. Submetersheds were delineated as discussed below:

- Along boundaries of customer communities to track the flow from the customer
  - For example, there are 10 submetersheds from Kentwood labeled as KEN. The grey lines are pipes, the thin blue lines delineate submetershed boundaries, and the thick blue line represents the boundary between Grand Rapids and Kentwood. The outlet of each submetershed is labeled and shown as points and the flow path between submetersheds outlets are shown as dashed lines.
- Along boundaries of the flow segments to match the financial costing records
  - For example, submetersheds are delineated for segments S-96 through S-105 shown as thick lines with different colors

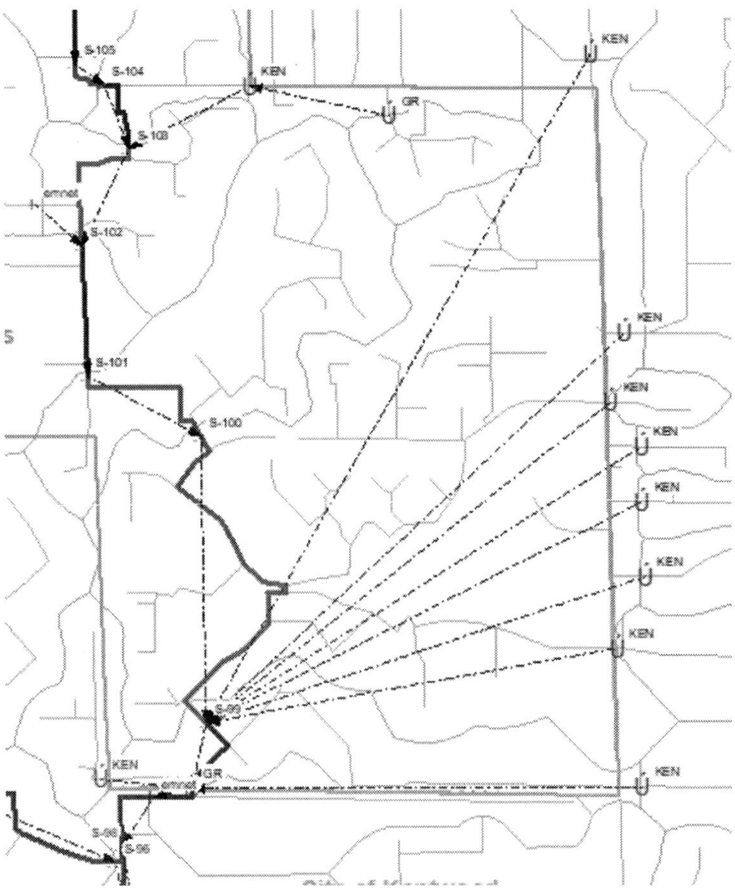

Figure 4 Submetershed Delineation

Augmenting Monitored Flows with Simulated Flows

With only 101 monitoring locations, many of which not at the zonegating locations, the ability to confidently estimate loadings at locations without direct flow data is crucial. To fully explain the method of estimating flows at 260 delineated submetersheds locations from 101 flow metering locations, it is necessary to take a step back and review the dry weather flow (DWF) modeling methodology used in the all-pipe model.

Before DWF flow loadings at modeled manholes can be calculated, metersheds are delineated, which is the tributary area between the upstream and downstream flow meters. Then, the local flow is calculated as the difference between the downstream flow and all upstream flows to represent the local loading from the delineated metershed.

As shown in Figure 5, the dry weather flow (DWF) includes sewer contributions from all customers which is called base sanitary flow (BSF) with varying magnitude throughout the day (referred to as diurnal pattern), combined with groundwater infiltration (GWI), the constant ground water inflow into the pipes through defects.

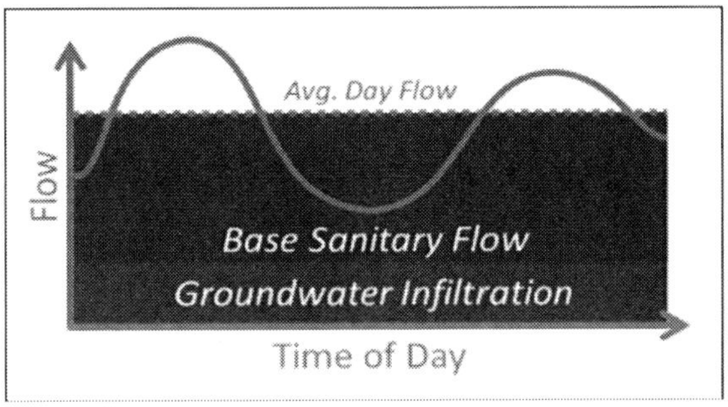

Figure 5 Dry Weather Flow Schematic

The GWI and BSF components are first calculated for each metershed from flow monitoring data. Then allocated to each modeled manhole (node) as described below,

- GWI is allocated using the cross sectional area of all the pipes in the manhole's tributary area (areas immediately upstream of the manhole, excluding areas flowing through upstream manholes). In other words, it is assumed the infiltration rate into the system per pipe surface area is the same for all the pipes in the same metershed.
- BSF is allocated to each modeled manhole using the winter month average water billing records from customers in the tributary of the manhole, with the assumption the same percentage of water used by the entire customer in the metershed is returned to the sanitary sewer system. An example is shown in Figure 6. The observed 0.32 mgd at the flow meter is allocated to the two manholes based on the water billing records.

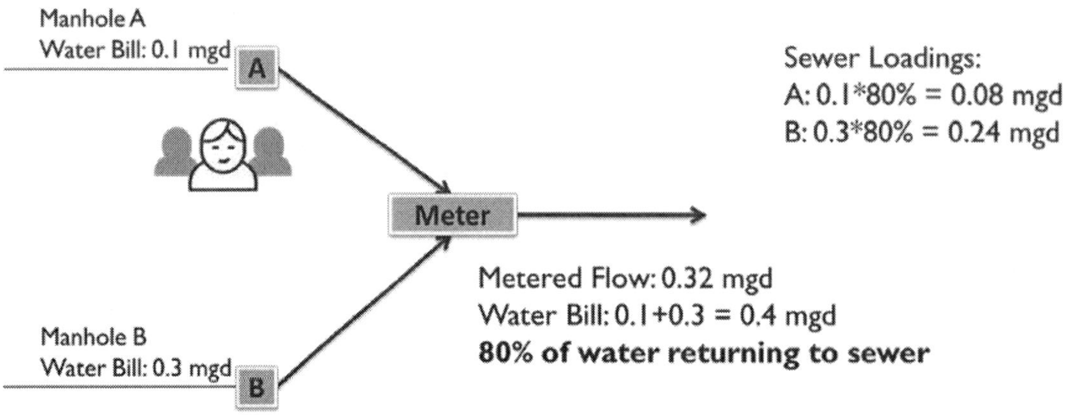

Figure 6 Dry Weather Hydrology Example

Similar to the method used for estimating flow from each manhole using water billing data, the loadings from the 2016 monitoring data can be allocated to each modeled manhole using simulated loadings from the all-pipe model, which also include the ground water infiltration components. The rigorous process of calibrating and allocating dry weather flows using observed flow monitoring data and water billing records provides high confidence in the simulated DWF at the modeled manhole level. This methodology allowed the all-pipe model to greatly increase the resolution of the flow data from a single measurement at the flow meter to manhole level.

It should be noted that the method implies the same metershed delineation is used during the calibration and the 2016 flow monitoring. Due to data quality issues, the metershed delineations differ at a few locations, but for most metersheds, the delineation did not change in 2016.

SWMM5 Zonegating Model

With the customer loadings developed, the next step is routing the loadings through the system. The flows from customers were manually routed and tracked in the 2003 spreadsheet, resulting a table with 200 rows and 40 columns. This is a time consuming manual process; moreover, it is very difficult to validate the results. To improve the accuracy and efficiency, the calculation must be automated.

The solution is a very simple EPA SWMM5 zonegating model. Tracing the source of the flows from customer communities is achieved using the water quality module of SWMM5, which is the equivalent of a dye test. By adding different dyes from each customer community, SWMM5 tracks the concentration of the dye as the flows are routed downstream. By setting the concentration of the dyes at 1.0 at all the loading nodes, a mass balance calculation below shows that the simulated dye concentration at the junction is the same as the percentage of total flows from the dye source:

- The mass of the dye can be calculated as: [mass of dye i at junction] = [flow from source i]*[dye concentration at source(set as 1)] = [concentration of dye i at jucntion]*[total flow]
- Thus, [flow from source i] = [concentration of dye i at junction]*[total flow]
- [concentration of dye i] = [flow from source i] /[total flow], which makes the concentration of dye calculated at the junction the same as the percentage of flows from the source of the dye

The zonegating flow segments are represented in the model as follows:

- Flow segments, and customer loading node are modeled as junctions.
- The loading into each junction is the observed flow if available, otherwise augmented by the simulated flows. A tracer chemical labeled the same name as the customer is added to the junction, and the concentration is set as 1.0 so that the mass of the chemical at the loading point is the same as the flow rate.
- The flow paths between the submetersheds are established based on the all-pipe model.
- Dividers are created where flow diversion occurs; flow diversion rules are developed based on simulation results and the input from the City engineers. An example of the flow split from S-70 to S-88 and S-89 is shown in Figure 7.
- The simulation uses a steady-state routing method with water quality enabled. The model routes the flow from all submetersheds through the network to the outlet. The flow through each zonegating flow segment is reported at the corresponding junction. Because the concentration of the trace chemicals is set as 1 at all sources, the concentration reported at each flow segment is the same as the percentage of the total flow from that label community, and the mass of the tracer chemical is the same as the flow from the label community.

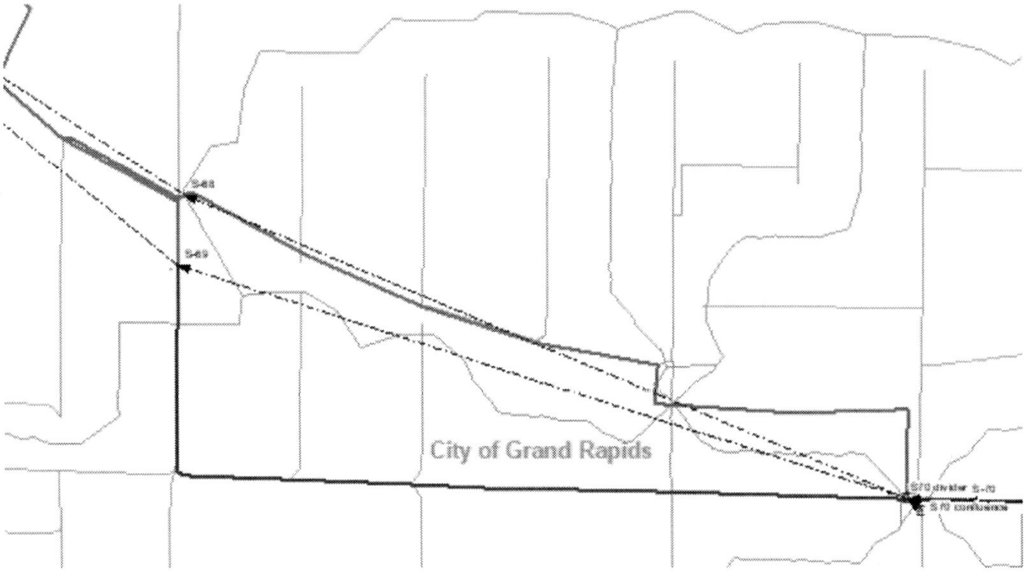

Figure 7 Flow Split at S-70

The resulting SWMM5 zonegating model has 254 junctions, 6 dividers and 264 conduits as shown in Figure 8.

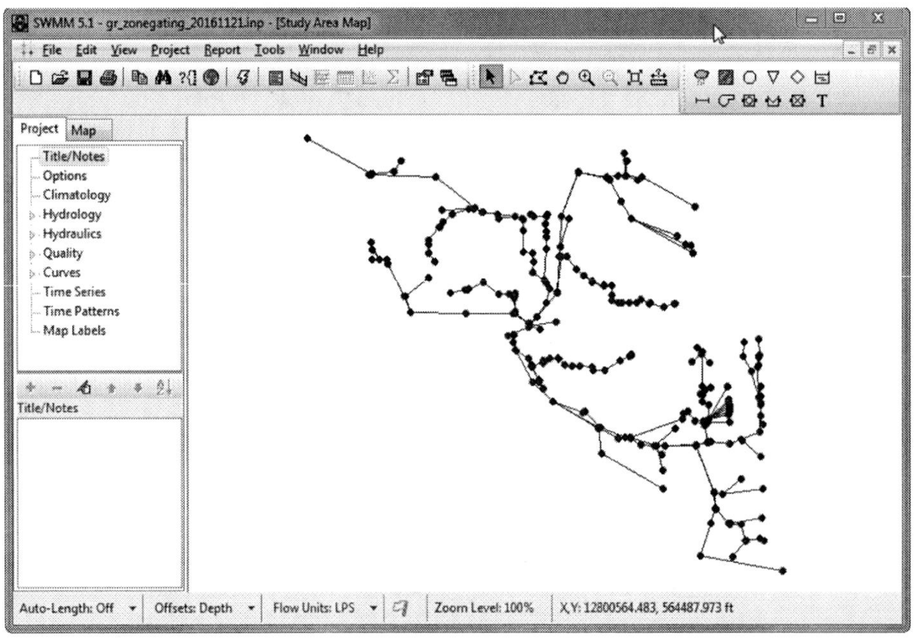

Figure 8 SWMM5 Model

Figure 9 shows the input windows of inflows. Flows from submetersheds are entered as baseline and tracer chemicals as constituent with a concentration of 1.0 (baseline).

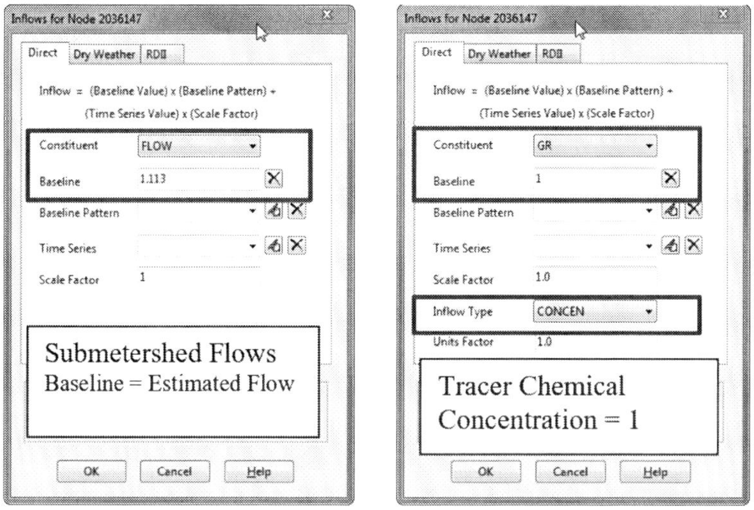

Figure 9 Junction Inflow Input Windows

RESULTS

The zonegating model simulation results were extracted from the SWMM5 output file and processed in a spreadsheet. A sample of the reported results is shown in Figure 10. The first column is the obfuscated flow segment name; the lateral flow column is the flow from the tributary of the segment; the total flow is the flow through the segment including both the lateral flow and upstream flows; remaining columns are the percentage of the total flow from each customer community.

Segment	Lateral Flow (mgd)	Total Flow (mgd)	ADA	CAL	CAS	EGR	GAI	GR	GRT	KEN	TAL	WLK	WRT
[illegible]	0.001	0.645	0%	0%	0%	0%	0%	8%	0%	0%	11%	81%	0%
[illegible]	0.094	0.643	0%	0%	0%	0%	0%	8%	0%	0%	11%	81%	0%
[illegible]	0.107	0.156	0%	0%	0%	0%	0%	31%	0%	0%	0%	69%	0%
[illegible]	0.000	1.210	0%	0%	0%	0%	0%	97%	0%	0%	0%	3%	0%
[illegible]	0.654	1.177	0%	0%	0%	0%	0%	97%	0%	0%	0%	3%	0%
[illegible]	0.079	0.523	0%	0%	0%	0%	0%	94%	0%	0%	0%	6%	0%
[illegible]	0.008	0.008	0%	0%	0%	0%	0%	100%	0%	0%	0%	0%	0%
[illegible]	0.017	7.416	12%	1%	14%	0%	5%	13%	7%	48%	0%	0%	0%
[illegible]	0.437	3.894	23%	0%	22%	0%	0%	0%	12%	43%	0%	0%	0%
[illegible]	0.007	0.444	0%	0%	0%	0%	0%	93%	0%	0%	0%	7%	0%
[illegible]	0.176	0.437	0%	0%	0%	0%	0%	92%	0%	0%	0%	8%	0%
[illegible]	0.000	0.261	0%	0%	0%	0%	0%	87%	0%	0%	0%	13%	0%
[illegible]	0.047	0.261	0%	0%	0%	0%	0%	87%	0%	0%	0%	13%	0%
[illegible]	0.153	0.186	0%	0%	0%	0%	0%	82%	0%	0%	0%	18%	0%
[illegible]	0.033	0.033	0%	0%	0%	0%	0%	0%	0%	0%	0%	100%	0%
[illegible]	0.003	0.612	0%	14%	0%	0%	55%	0%	0%	31%	0%	0%	0%
[illegible]	0.000	0.613	0%	0%	0%	0%	0%	96%	4%	0%	0%	0%	0%
[illegible]	0.243	0.393	0%	0%	0%	0%	0%	0%	0%	0%	18%	82%	0%
[illegible]	[illegible]	[illegible]	[illegible]	[illegible]	[illegible]	[illegible]	[illegible]	[illegible]	[illegible]	[illegible]	[illegible]	[illegible]	[illegible]

Figure 10 Sample Zonegating Results

DISCUSSION

Although the problem solved in this study is unique to the City of Grand Rapids, the method of augmenting the flow monitoring data with model simulation results can be applied to a much wider range of issues. The ability to seamlessly integrate flow monitoring data and modeling results to gain insights into the collection system is at the heart of big data analytics in the water industry.

Despite the relatively small size of the data used in this study, quite a few common challenges discussed in the literature were encountered, and are discussed below (SAS, 2017).

Addressing Data Quality

Big data is famously characterized by its properties of volume, velocity, variety, variability, and veracity (Wikipedia, 2017). To evaluate the quality of the data, all those dimensions should be considered. Despite the great technology advancement in the water industry, it is still common getting flow data in the collection system with less than desired accuracy due to limitation of hardware and the constrains of the environment. Thus the quality of the data in most cases is not always reliable and direct instrument readings usually need to be reviewed by trained staff. To address the data quality issues, a deep understanding of the data is required. A few examples are listed below:

- Flow meters at the WRRF and main pump station are considered accurate in general.
- Area-velocity flow meters installed in manholes are considered less accurate, especially for those installed with undesirable conditions. For example, at locations upstream of the WRRF where the pipes are surcharged most of the time, the water depth is high and the velocity is low, a condition the flow meter cannot accurately measure.
- Some pump flow is calculated from runtime data. Accuracy of this data is dependent on the testing results of the pump flow rate.
- Seasonal flow variation was observed at some sites due to factors such as ground water impacts. For locations with only a few months of flow data, the average flow might not be representative of actual conditions.

In an effort to evaluate the accuracy of the flow data used in this study, flow monitoring data were cross referenced to the results of the simulation results. Flow data with significant differences were reviewed and data deemed as of poor quality were rejected.

Understanding the Data

As discussed above, it is common to have flow data recorded with very different hardware, collected and processed by people with different expertise. It is rare to find a person possessing the required expertise and knowledge to understand all of the data. Building a team of domain experts to develop a deep understanding of where the data comes from, how it can be properly used in is needed in many cases.

Analyzing the Data

In order to gain insights into the business with big data, models need to be built that can be used to learn from the past and predict future scenarios.

Contrary to common belief that big data require more advanced algorithm, many researchers concluded collecting more data and letting the "data to speak for itself" often is a better strategy than focusing on sophisticated algorithms (Wu, 2013). From a practical perspective, building simple models is the key to ensure the success of big data project, as Veeramachaneni pointed out (Veeramachaneni, 2016).

"If companies want to get value from their data, they need to focus on accelerating human understanding of data, scaling the number of modeling questions they can ask of that data in a short amount of time, and assessing their implications."

The choice of building a skeleton zonegating model instead of using the all-pipe model to route the flows is a good example. Instead of dealing with a complicated fully developed hydraulic model with more than 24,000 loadings to update, the team built a simple steady-state zonegating model, and treated the simulated flows from the all-pipe model as input data instead. This move greatly simplified the analysis and calculation, and created the opportunity for automation, another recommended key for success of big data project (Veeramachaneni, 2016).

Automation saved the project team weeks of valuable time to quickly iterate through submetershed delineation and loading calculations as assumptions continued to change, and new data issues arose. The procedure of estimating loadings based on simulated flows, although tedious, is a well-defined and can be easily automated. A script was developed to update the loadings calculation when the submetershed delineation changes, which took only minutes to run.

Leadership

More importantly, to move the industry forward with innovation, it takes leadership. As Schrage wrote in a Harvard Business Review article:

"Organizations need a fundamental paradigm shift. Innovation will no longer be based on an R&D pipeline, but on an 'experiment and scale' model. Experiments are the future of business innovation." (Schrage, 2016)

As the City of Grand Rapids demonstrated, the City has a vision that data is a critical asset that can create values for the customers. It took the City almost a decade to build a high quality GIS database, install real-time monitoring network, and develop an all-pipe model. Without a clear vision and leadership on building a data asset with long term benefits, the outcome could have been very different. As the data asset is shaping up, the City is in a new phase of experimenting with the data to find new ways maximizing the return on those investments.

CONCLUSION

A zonegating model was built using EPA SWMM5 with steady-state routing method and water quality module enabled which can route and trace the sources of flows from customer communities through the collection system. The zonegating simulation results reported the total flows and the sources of the flow from each customer community at each delineated zonegating flow segment, which is the basis for calculating the sewer rate for each customer by allocating the cost of building and maintaining the infrastructure.

The loadings used in the model were developed by augmenting the flow monitoring data with simulated flows from an all-pipe model calibrated using 2014 flow data. The method discussed in the paper allowed the City to take advantage of the modeling work and confidently interpolate the flow monitoring data in areas without flow meters installed.

The updated zonegating results greatly improved the accuracy of the flow estimation from the previous study.

The authors hope this study can serve as a case study of how the City of Grand Rapids is implementing "big data". With over a decade's investment in a data centric utility management approach, this study is the first experimentation attempting to extract values out of the investments outside of the original scopes of those projects, and the results showed that although not easy, the accuracy of the zonegating flow estimates were greatly improved.

REFERENCES

Veeramachaneni, K;2016;Why you are not getting value from your data Science, Harvard Business Review. 2016, Dec. 07, 2016

SAS;2017; Five big data challenges and how to overcome them with visual analytics. https://www.sas.com/resources/asset/five-big-data-challenges-article.pdf (accessed March 30, 2017)

Wikipedia;2017; Big Data. https://en.wikipedia.org/wiki/Big_data (accessed March 30, 2017)

Garrett Wu.2013. Why More Data and Simple Algorithms Beat Complex Analytics Models. http://data-informed.com/why-more-data-and-simple-algorithms-beat-complex-analytics-models/

Michael Schrage. 2016; Leadership and Big Data Innovation, Harvard Business Review. 2016, Dec. 13, 2016, https://www.sas.com/content/dam/SAS/en_us/doc/whitepaper2/hbr-leadership-big-data-innovation-108667.pdf

Black & Veatch Insights Group. 2016. 2016 Strategic Directions: Water Industry Report, https://www.bv.com/docs/reports-studies/sdr-water-industry.pdf?mkt_tok=eyJpIjoiWkRnNVpqa3dZVFE1TXpBMiIsInQiOiJKZktxXC9QRG1yYW1HemlieXBNVitIK3NFRklneDdieDFWUk82d1ZNc1RiU2ozbzh5NHlqNURoWFVicUdzRHVZeGFMdmVWSjcwMTJId2I4S0VSaWZpNlhDN2ttSitSdVJIbEZKYjBEQVplXC9RPSJ9

Moving Beyond a Plan - Precision Modeling in Boston, MA

Sharon Tsay[1], Lawrence Soucie[1], Deanna Clark[1], Paul Keohan[2]

[1] AECOM, Chelmsford, MA
[2] Boston Water and Sewer Commission

ABSTRACT

The Boston Water and Sewer Commission (Commission or BWSC) is updating its collection system model to reflect recent improvements in the collections system. The hydraulic conditions simulated by these models will be used eventually by the Commission's design and operational staff. The overall goal is to make these models reflect real world conditions. When using a model to support sewer design and/or to make operational decisions, the expectations for the model's accuracy are greater than for a planning model. Model calibration will increase the accuracy of the model. However, an accurate representation of the various hydrologic features across the model is also necessary and care should be taken that parameters are adjusted within reasonable ranges. Selecting the appropriate parameters to adjust during the calibration is critical. This paper will illustrate the techniques used to update the sewer model for the Boston Water and Sewer Commission. These techniques include comparing model parameters to data such as water consumption information, reviewing seasonal trends, and regional inflow characteristics.

KEYWORDS: sewer modeling, calibration, RDII, SWMM, sewer, CSO

INTRODUCTION

The Boston Water and Sewer Commission (the Commission or BWSC) maintains a SWMM model of their sewer and drainage network. SWMM modeling has been used to evaluate the Commission's sewer system for CSO control and other planning-level measures since the early 1990's and this modeling has served as a valuable tool for the Commission. The sewer model for Boston was last updated in 2007 (BWSC, 2009) and calibrated to approximately 68 meters throughout the system. Since then the model has been utilized by the Commission for planning evaluations. Recently the Commission has undertaken an update to this model to prepare the model for use by the Commission's design and operation staff. These updates include modifications to the network and hydrology in the model and a recalibration effort to more recent meter information.

A key goal of the Commission is that the model updates and calibration reflect the actual conditions within the collection system. Too often a parameter such as percent impervious are adjusted without examining if the changes are justifiable. This paper covers some techniques used to refine the model to reflect actual conditions. This includes:

- Use of other information sources to supplement meter data during model calibration.
- Calibration to different times of the year to account for seasonal variation.
- Accounting for regional differences in the model.

METHODOLOGY

Software

The modeling conducted for this project uses PCSWMM 2017, a graphical user interface and pre/post-processing tool for EPA SWMM 5. The version of EPA SWMM utilized for this project within this interface is SWMM 5.1.011.

Model Updates

Prior to model calibration, numerous updates were made to the model network and hydrology. These included:

- Shifting manhole locations to coincide with the Commission's updated GIS, and replacing ground elevations based on information from Massachusetts GIS LiDAR (2013-2014).
- Updating the model in thirteen (13) regions where sewer separation has occurred since the last model update.
- Refining the representation of other areas with recent development in the model.
- Replacing pump station and other facilities with more detailed representations based on information provided by the Massachusetts Water Resources Authority (MWRA).
- Catchments were redrawn to more accurately depict drainage patterns.

A map showing the updated model network and delineation is presented in Figure 1.

Data Used for Calibration

The model was calibrated to depth and flow meters that were in place from November 2015 through May 2016. This effort coincided with multiple metering programs being conducted throughout the city. These programs are summarized below:

- <u>MWRA Meters</u>: Boston is the largest customer served by the MWRA, the agency responsible for providing sewer and water treatment to 43 member communities in the greater Boston area. As the largest and most downstream community to the MWRA's Deer Island Treatment Plant, the Boston sewer system is intertwined with MWRA's interceptors. The MWRA uses several meters to measure flow and depth throughout the city. Data for 42 MWRA meters were obtained for the calibration period.
- <u>CSO Notification Meters</u>: The Commission maintains a CSO Notification Program that uses meters at the influent and overflow lines of ten major regulators. These meters are used to quantify the amount of flow coming into the regulators and determine when overflows occur.

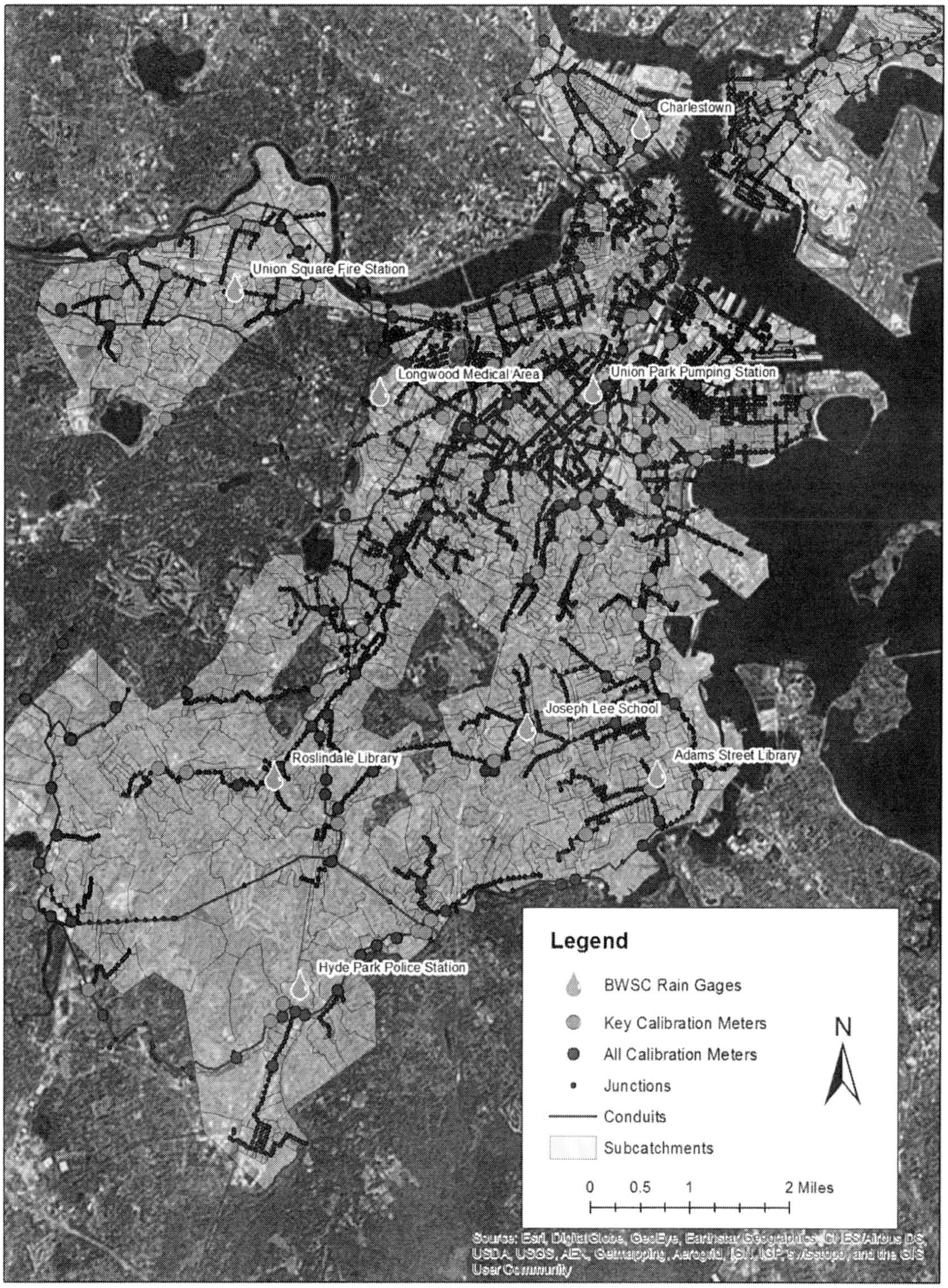

Figure 1. Boston Model Network, Hydrology, and Data Collection Points

- BWSC I/I Meters. In the summer of 2015 the Commission began an Infiltration and Inflow Analysis Program. As part of this program meters were installed throughout the Commission sewer system. Ninety-nine (99) flow meters were installed in the fall of 2015 and 115 meters were installed in the spring of 2016. This meter data was used to calibrate the model as well as for the I/I analysis.
- Project Meters. An additional 15 meters were installed as part of this project. The purpose of these meters was to 1) evaluate the effectiveness of sewer separation in updated areas and 2) add meters where needed to support future alternatives evaluations.

Not all of the meters were used for calibration. If the location was not on a modeled pipe or the data was of questionable quality the meter was not used. In total 160 meters were used for the model calibration effort. Dry- and wet-weather flow was calibrated to all 160 meters for the fall and spring. Sixty-eight (68) of these meters were identified as "key" meters throughout the system and calibration statistics were generated for those locations. Calibration plots and statistics were generated for two periods: 11/8/15 – 11/25/15 for fall and 4/1/16-4/15/16 for spring. The meters used for calibration are shown in Figure 1.

Rainfall Events

For model calibration, continuous simulations were conducted for the period of January 1, 2015 through April 15, 2016. Rainfall for this period was provided by the Commission, who maintains eight rainfall gages throughout the City; the locations of these gages are shown in Figure 1. The data for these gages were reviewed to assess their general quality. In some cases, the rainfall data was determined to be not useable. For example, the Joseph Lee School gage was missing data for 2/12/2015 21:25 - 3/12/2015 9:00. In these cases, data from the closest Commission gage was substituted. Also, the Commission's gages did not have useable data during the spring of 2015. If none of the Commission's gages were useable, then hourly data from Logan Airport converted to 15-minute data were substituted.

Care should be taken in selecting storms for model calibration. Sometimes it is not advisable to only calibrate to the single largest storm of the monitoring period. For example, if the largest storm cause widespread surcharging and flooding, then it may be difficult to calibrate the model hydrology because it is not known the degree to which the observed flow is influenced by the pipe hydraulics. Usually multiple storms and periods are used to evaluate a model calibration.

Two rainfall periods within the continuous simulation were used for model calibration. The period of 11/8 through 11/25/15 was selected for the fall period. This timeframe was selected because it contained a reasonable range of storms that did not cause flooding. The largest storm in this fall period, November 20-21, 2016, was in particular useful for calibration as it:

- Had a fairly uniform depth (0.83 to 0.92 inches) across all gages, minimizing the impact rainfall variation would have on results.

- Was preceded and followed by a reasonable dry weather period, so the impact of the storm could be assessed.
- Was not large enough to cause widespread surcharging or flooding. If a system floods during a storm, then the amount of flow that a meter sees may not reflect the actual flow coming from the sewershed.

A second period of 04/05 through 04/15/2016 was selected to evaluate the model under spring conditions. A spring evaluation period is important in the northeast because wet weather response is more influenced by antecedent rainfall conditions during this period. The 04/05 through 04/15 timeframe was selected as it contained three storms exceeding 0.25 inches in depth within a seven day period. Having a number of storms close together is useful for evaluating the impact of antecedent conditions on wet-weather response. This period was also late enough in the season to avoid snow melt influencing meter results.

The total rainfall for each calibration period is shown in Table 1.

Table 1. Total Rainfall for Fall and Spring Calibration Periods.

Storm	Gage	Total Rainfall (in)
Fall: 11/8/2015 – 11/25/2015	Allston	1.83
	Charlestown	1.77
	Adams Street Library	2.17
	Joseph Lee School	2.02
	Hyde Park	2.04
	Longwood	1.78
	Roslindale	2.0
	Union Park	1.94
Spring: 4/5/2016-4/15/2016	Allston	1.97
	Charlestown	1.91
	Adams Street Library	1.78
	Joseph Lee School	2.01
	Hyde Park	2.01
	Longwood	1.84
	Roslindale	1.97
	Union Park	1.94

Dry Weather Calibration

Dry weather flow is divided into sanitary and infiltration components in the Commission's model. Sanitary flows were estimated using the Commission's geocoded water use data, assuming a return rate of approximately 85%, meaning 85% of the used water enters the sanitary system. The sanitary flows are multiplied by a diurnal pattern developed from available metering data. Base infiltration was assumed to be the average dry weather flow minus the sanitary flow estimated from the water consumption data. The base infiltration was distributed upstream of meters by assuming the infiltration was proportional to the sanitary flow.

It is typical during calibration to adjust estimated sanitary and infiltration flows to match meter data. However, this approach could mask connectivity or other issues with the model. Large differences between dry weather flow measured in the pipe and water consumption data were used to identify connectivity issues in the Commission's model. An example is illustrated as follows:

The Dorchester Brook Sewer (DBS) system is comprised of approximately 514 acres of combined sewer area, 270 acres of separated sewer area, and 237 acres of stormwater area tributary to the CSO outfall. The flow in the DBS is conveyed to the Boston Main Interceptor (BMI) at regulator RE 070/7-2. The DBS system is shown in Figure 2.

A flow meter was present at the downstream end of the system, just upstream of regulator RE 070/7-2. The model showed significantly higher flow at this point than was recorded at the meter. Although one solution would be to simply lower the sanitary flow in the model, this would result in return rates that are less than typically expected. Examination of GIS data compared to the model showed a 12" cross-connection diverting flow away from the meter and around the regulator. Coding this cross-connection in the model led to modeled flows that matched metered estimates. Figure 2 shows the location of the cross-connection. Figure 3 shows the model calibration with and without this cross-connection present.

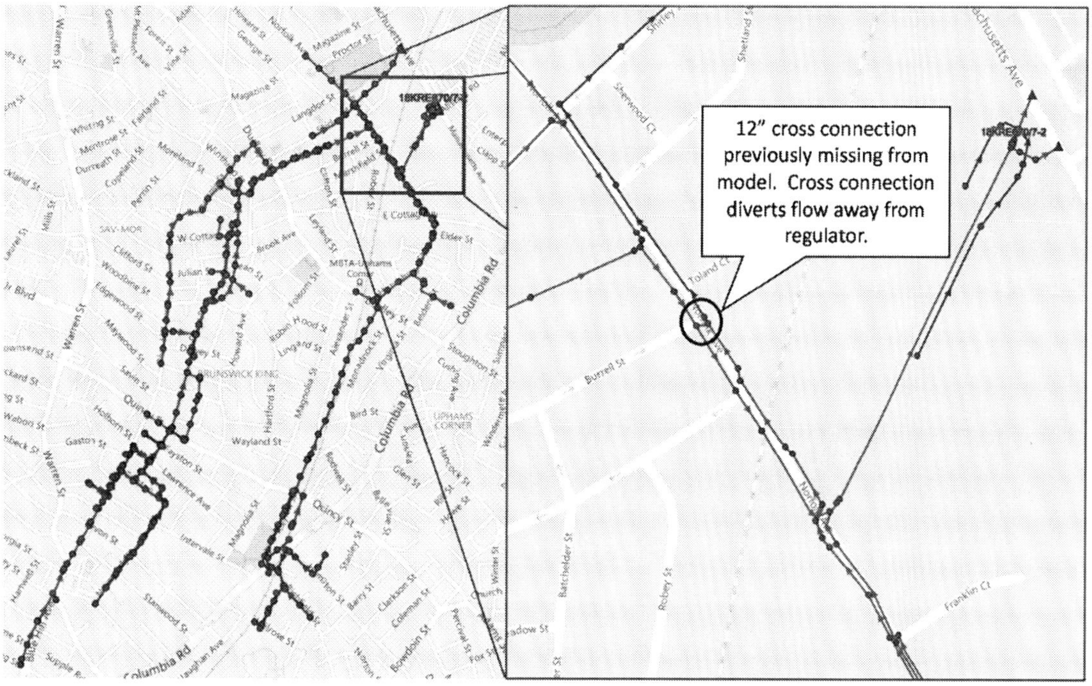

Figure 2. Map of Dorchester Brook System and Cross-connection that Impacts Dry Weather Flows

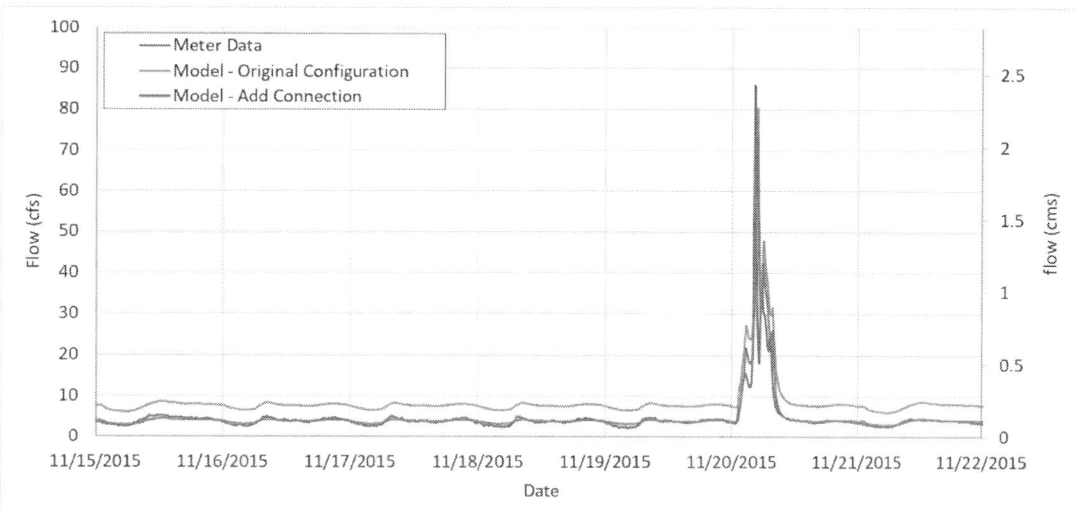

Figure 3. Meter vs. Model Inflow to RE 070/7-2 With and Without Cross-Connection

Checks like these were performed throughout the system. Comparing the water use data to the measured sewer flows provided an additional check on the flow data and helped the model better reflect actual conditions.

Wet Weather Calibration

After the dry weather calibration was complete, the model was calibrated to wet weather conditions. The wet weather response to rainfall can vary substantially by season in Eastern Massachusetts; wetter soils and high groundwater conditions can significantly increase the response of the system to rain events. In Massachusetts this often occurs in the spring after snow has melted. In Figure 4, flow versus time for a meter in the downstream most portion of Dorchester is shown. The figure shows the same location in the fall and spring. Note that there are significant differences in both the base flow and the wet weather response. A 1.28 in/hr peak intensity storm in the fall shows a smaller response compared to the 0.44 in/hr peak intensity storm in the spring. In addition the base DWF is higher in the spring as well.

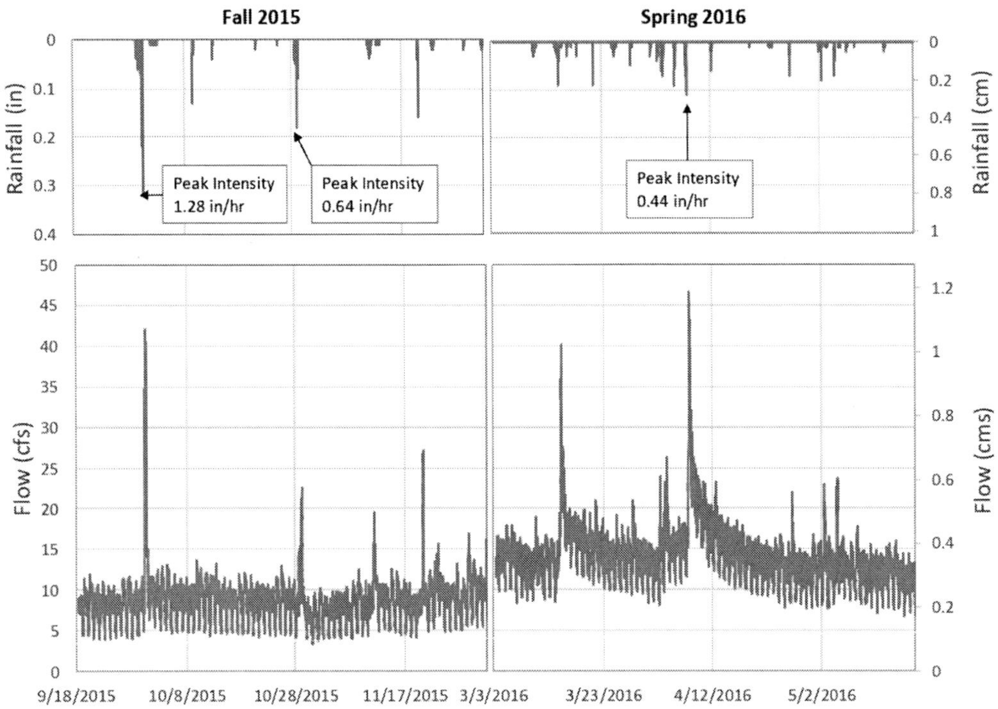

Figure 4. Fall 2015 and Spring 2016 Meter Data for Meter 50 in Dorchester

What Figure 4 shows is that seasonal variation must be taken into account when developing the model. Some model calibration efforts may calibrate to only fall or spring flows, but this may lead to a model that shows inaccurate results for other seasons. The Commission model was developed to simulate both fall and spring inflow conditions.

Fall Calibration

There are two components to wet weather response in the developed SWMM model: the fixed response and the ground infiltration/rainfall-derived infiltration/inflow (RDII) response. The fixed response is represented in the model by the SWMM runoff routine and represents the rainfall response that occurs during storm events independent of antecedent rainfall. For separated systems this response may be due to illicit stormwater connections to the sanitary system such as roof leaders or stray catch basins. For combined and stormwater areas the fixed response is generally proportional to the amount of impervious area in the watershed.

The wet weather flow for the fall monitoring period was assumed to be primarily fixed response; peak flows were generally proportional to the amount of rainfall that occurred, and the response rapidly decreased back to dry-weather conditions. Therefore the model was first calibrated to fall flows using only the SWMM Runoff routines. The main SWMM calibration parameters were percent capture, which represents the amount of inflow into the system and the basin width, which impacts the routing time and peak flow from the subbasin. For combined and stormwater areas, the total impervious area for each sub-catchment was estimated from the Commission's impervious GIS coverage. Figure 5 shows an example of the impervious area coverage for a portion of South Boston.

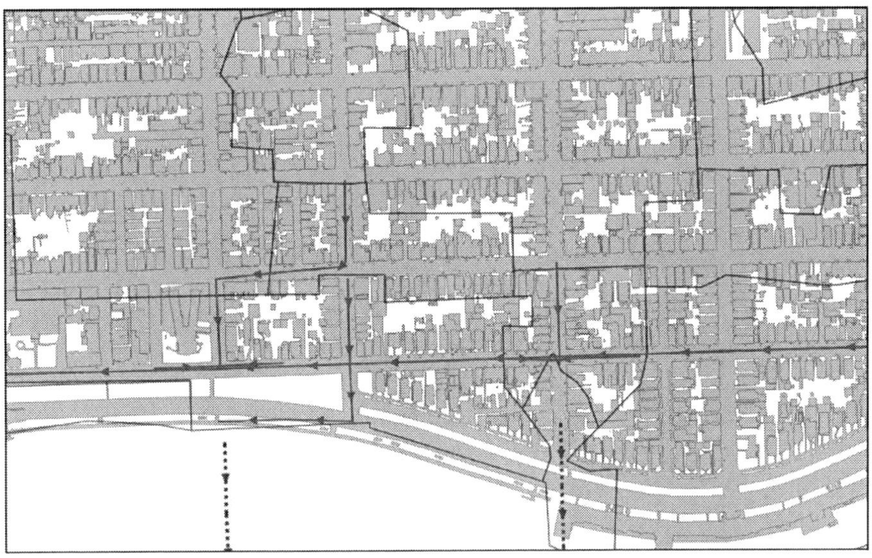

Figure 5. Example of Impervious Coverage for Portion of South Boston System

The total impervious area (TIA) calculated from the impervious coverage was specified in the model. However, only a portion of this impervious area contributes flow to the collection system. The fraction of impervious area that contributes flow to the combined or stormwater system is often referred to as directly connected impervious area (DCIA). The SWMM model allows this to be modeled using the routing feature; a percentage of the impervious area can be routed to the pervious area with the remainder flowing into the collection system. Based on previous modeling of the Commission's system, the initial routing was set at a value of 65% (Heineman, 2008). The interpretation of this is that 65% of the impervious area is routed to pervious surfaces, and 35% directly to the sewer system. Thus, the DCIA is equal to the TIA x (100 - routing percent). The routing percentage was adjusted as part of the model calibration. Significant modifications to the routing percentage were investigated for connectivity concerns.

For separated sewers, the routing feature was not used. The percent impervious was used to represent potential direct storm connections; in other words, the portion of the system that contributes a rainfall response to the system during every storm event. This percent is usually small – within the range of 1 to 10%.

Figure 6 shows an example meter where the model was calibrated to fall conditions.

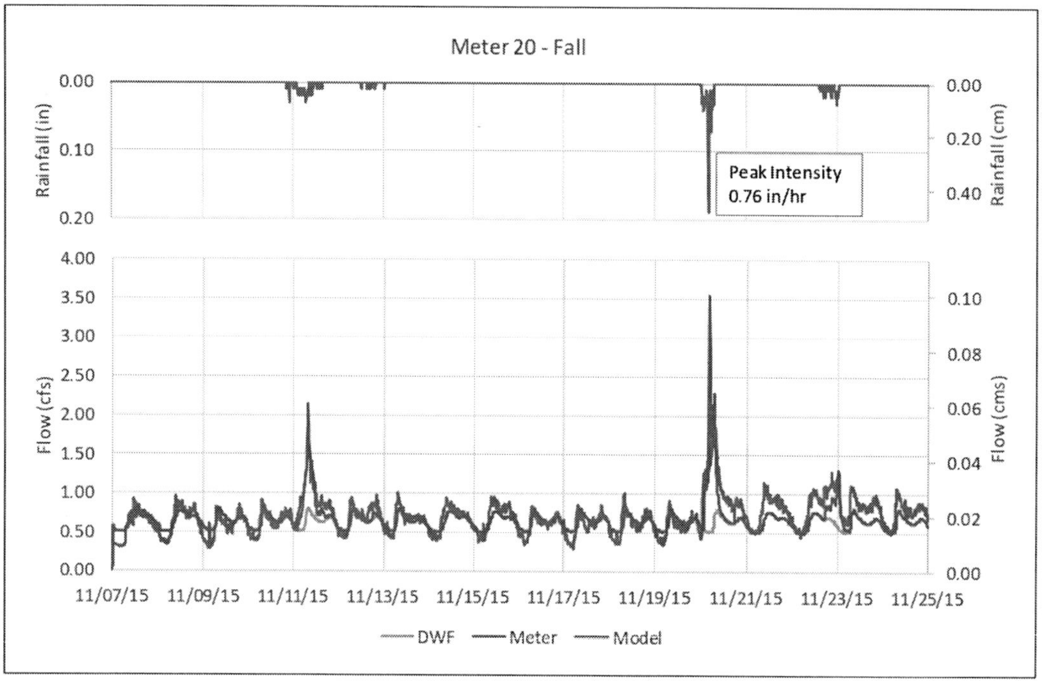

Figure 6. Fall Calibration Plot of Meter 20

<u>Spring Calibration</u>

Once the meters were considered calibrated for the fall, the meters were calibrated for the spring. As previously discussed, spring flows show considerably more wet weather response compared to fall in much of the Boston area. If the model developed during the fall calibration is used without modification, the spring flows are significantly under predicted. Figure 7 is the same meter represented in Figure 6, but shows meter versus model spring flows if only the fall calibration is used. Note that in addition to having an insufficient rainfall response, the dry-weather flow is under-predicted as well.

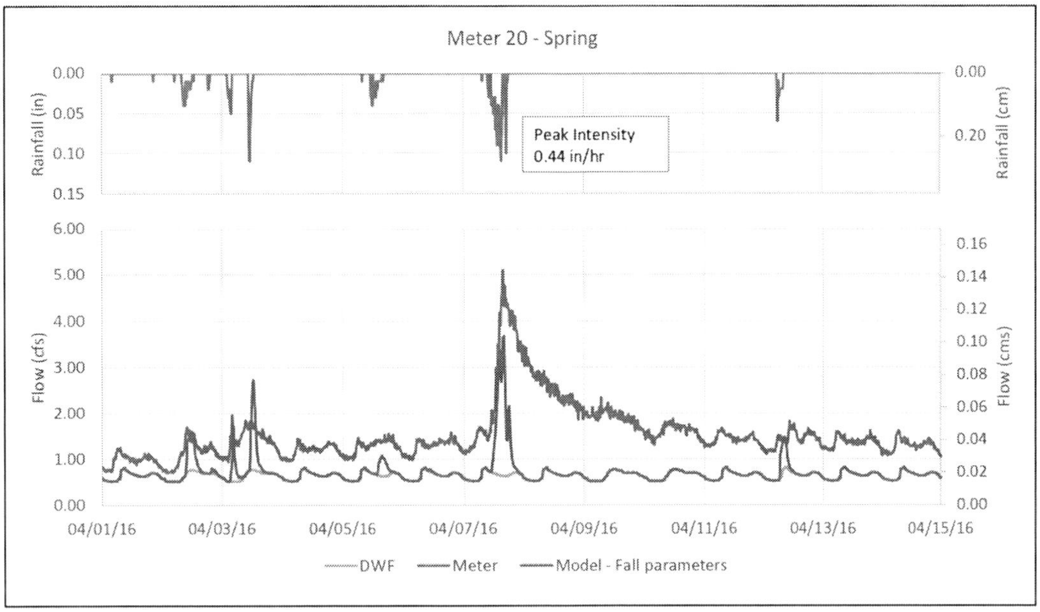

Figure 7. Spring Calibration Plot of Meter 20 Using Only Fall Calibration Parameters

SWMM ground infiltration routines were turned on and a continuous simulation was conducted from January 1, 2015 through April 15, 2016. In EPA SWMM, the groundwater flow is represented by a two zone model. An illustration of this model is presented in Figure 8, which is taken from the EPA Stormwater Management Model Reference Manual (2016). The groundwater infiltration routines in SWMM were not used in previous modeling of the Commission's sewer system, but they were added to the current model to improve model calibration.

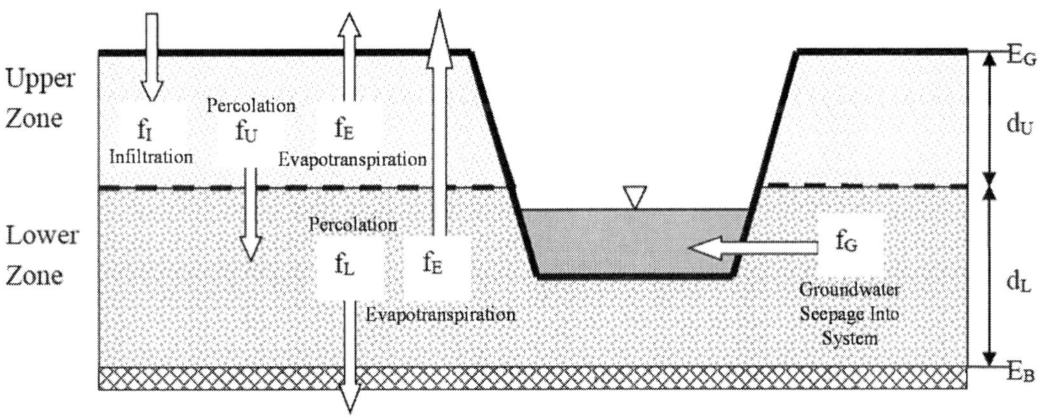

Figure 8. Illustration of EPA SWMM's Two-Zone Groundwater Inflow Model (EPA, 2016)

The groundwater inflow module in SWMM treats the soil surrounding a pipe like two boxes: one for the unsaturated zone and one for the lower zone. Water enters and leaves these boxes through a series of fluxes. When the water level in each box reaches a user-specified elevation, water can enter the collection system as groundwater infiltration. Rainfall enters this box through surface infiltration and leaves via percolation, evapotranspiration, or flow into the collection system. Evapotranspiration varies significantly throughout the year as it is dependent on both temperature and vegetation which are both seasonally dependent. In much of the US evapotranspiration is highest in the summer and early fall when temperature is the highest. During this period the ground is less saturated and the groundwater response is lower. However in the winter and early spring evapotranspiration is low and the ground is more saturated, leading to more events where the groundwater response is larger.

For locations where the fall calibration parameters underpredicted the spring flows, parameters to represent groundwater inflow were developed for the Commission's model. These parameters were adjusted so that the model would simulate the increased wet-weather response under spring conditions but would not change the fall calibration.

Figure 9 shows the model calibration at meter 20 with the groundwater RDII turned on. Three continuous simulations are shown in the graph, all of which started on January 1, 2015 and continued until April 15, 2016. The green is dry weather flow only, the blue is the meter data, the red is the model with the groundwater routine turned on, and the purple is with the fall calibration parameters only (groundwater routine turned off). With the groundwater routine turned off, the model significantly under-predicts the hydrologic response. Turning the groundwater on resulted in the model more accurately representing the observed seasonal hydraulic response variability. It is noted that turning the groundwater on did not impact the fall calibration because the simulated groundwater level was below the threshold resulting in flow entering the collection system.

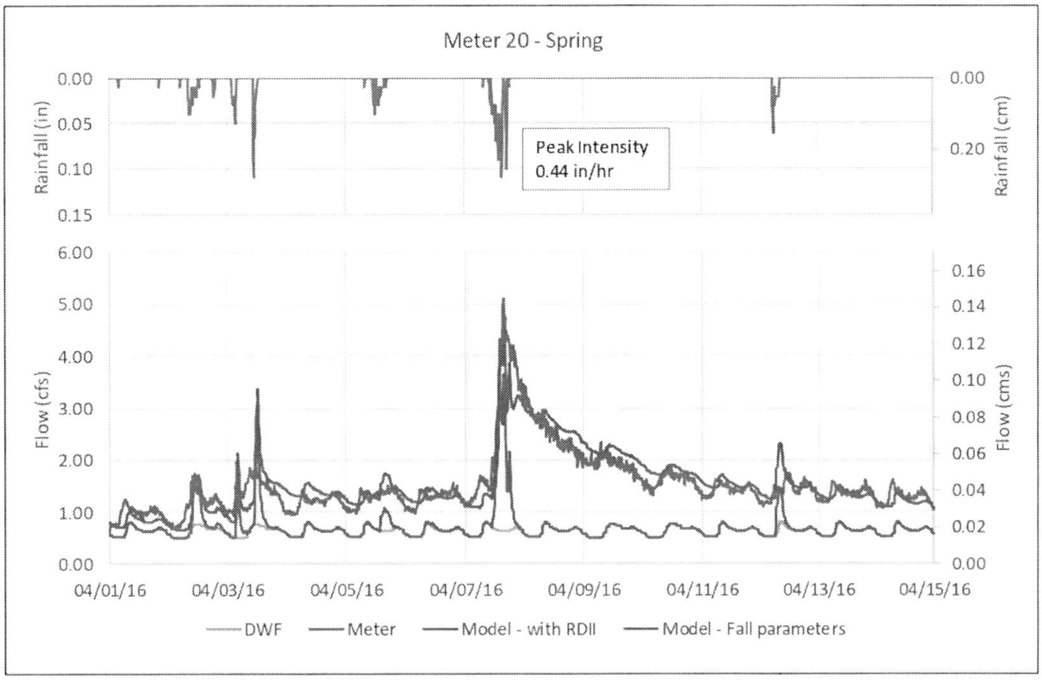

Figure 9. Spring Calibration for Meter 20 Showing Impact of Groundwater RDII Routines

Accounting for Regional Features

The calibration methodology described in this paper was used throughout the city. However, it was important to note that each region of Boston has individual characteristics that should be accounted for. Planning level models treat many aspects of flow uniformly throughout the study area; for example using a uniform diurnal pattern or starting with the same initial values for infiltration. However, individual neighborhood characteristics should be accounted for more detailed simulations. Accordingly, the model was broken into ten geographic regions that represent areas with specific characteristics.

Regions were defined in the model based on a few criteria. First, areas were selected if they were somewhat hydraulically independent and could be made into submodels. Areas were grouped together if they were expected to have somewhat uniform flow characteristics. Finally the presence of a downstream metering point was also a consideration as having such a downstream point is useful for evaluating the overall calibration for the region.

A map of the defined regions is shown in Figure 10.

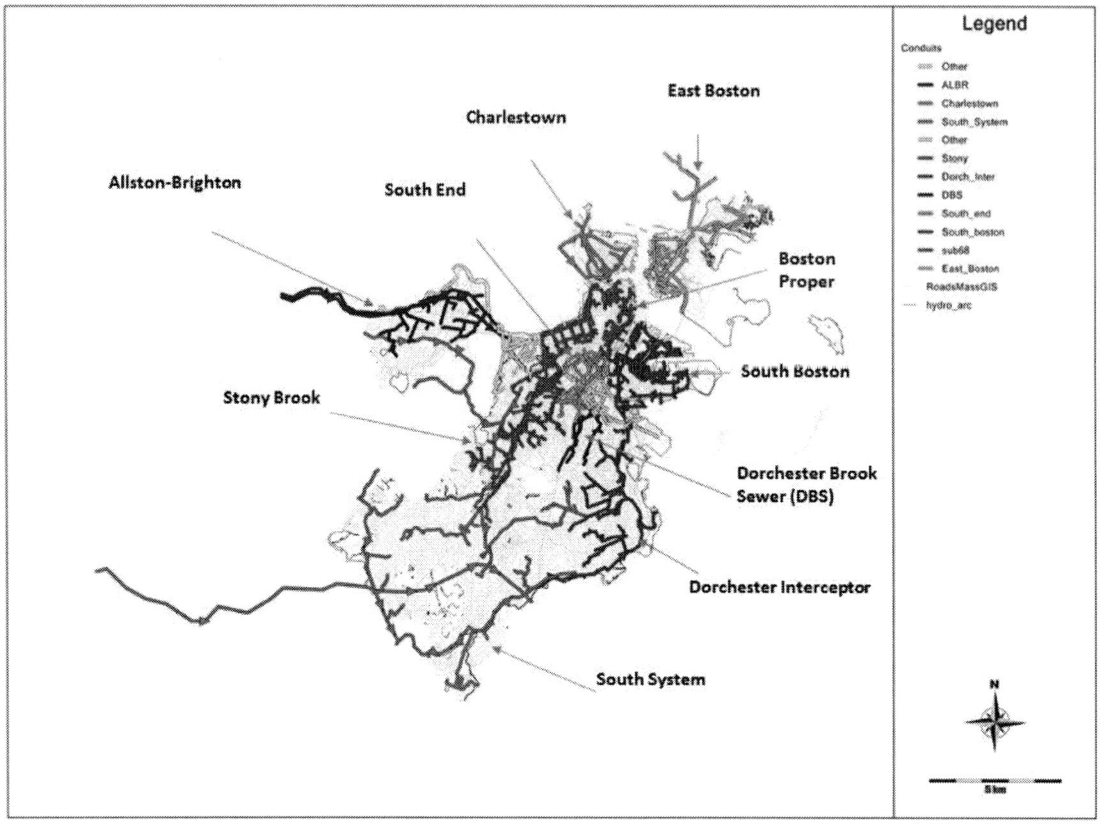

Figure 10. Boston Regional Submodels

Each region has distinctive characteristics that should be treated differently from one another. Table 2 is a summary of the major regions of Boston with descriptions of some of their unique characteristics. For example, within South Boston there has been significant recent development; sewers in areas with recent development tend to be newer and experience less groundwater infiltration. As such meters in this area shows little difference between fall and spring flows and no groundwater infiltration was used for this area. However, in areas such as the South System, the collection system is almost completely separated with few direct connections. Therefore most of the inflow is due to groundwater. Meters were placed in areas of new development (such as recent sewer separation projects) in order to separate the inflow characterization of these areas compared to older portions of Boston.

Table 2. Summary of Regional Characteristics

Submodel	Tidal	Groundwater Inflow	Pump Station	Combined Sewer	Separated Sewer	Average Fall Percent Inflow	Average Spring Percent Inflow
East Boston	Yes	Yes	Yes	Yes	Yes	16.6	25.4
South Boston	Yes	No	No	Yes	Yes	20.4	20.4
Dorchester Brook Sewer	No	Yes	No	Yes	Yes	19.3	26.5
Dorchester Interceptor	No	Yes	No	No	Yes	2.9	38.0
South System	No	Yes	No	No	Yes	1.9	23.5
Stony Brook	No	Yes	No	Yes	Yes	13.9	29.9
Allston-Brighton	No	Yes	No	No	Yes	9.4	34.3
South End	No	No	Yes	Yes	Yes	18.3	18.3

Tidal influences in inflow were also accounted for in regions close to the ocean. For example, the influent meter tributary to regulator RE-057-6 shows evidence of tidal inflow, which was accounted for in the model (Figure 11).

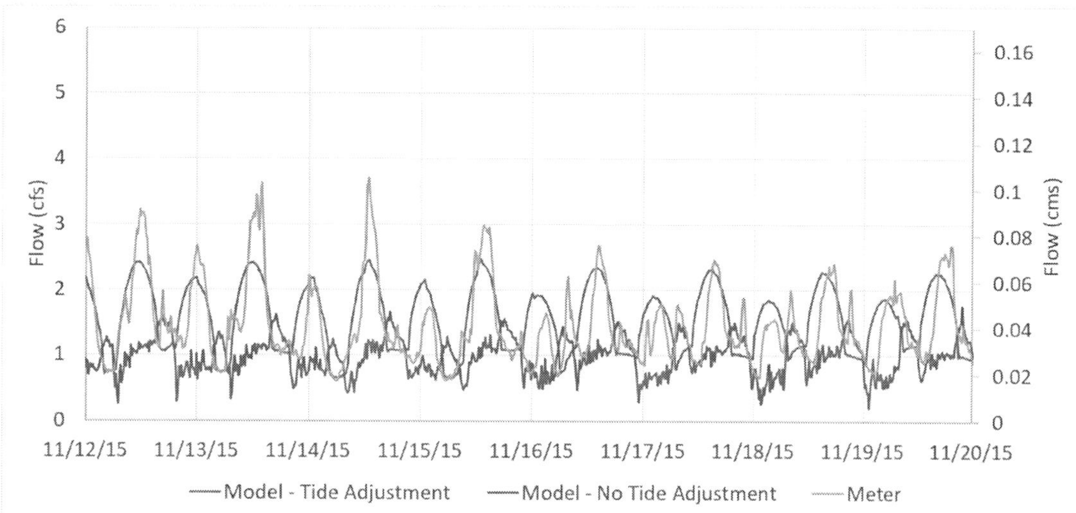

Figure 11. Influent Meter to RE-057-6 During Dry Weather With and Without Tide Adjustment

The individual characteristics of each region was taken into account during calibration.

RESULTS

This methodology was used to calibrate each meter for the project. Calibration plots were generated for each of the 160 meters used for this project. Calibration statistics were calculated for the 68 key calibration meters for both the fall and spring monitoring periods. Two statistics often compared are peak flow and total volume. Figures 12 and 13 show metered versus modeled peak flow and volume for each meter location calibrated in the model.

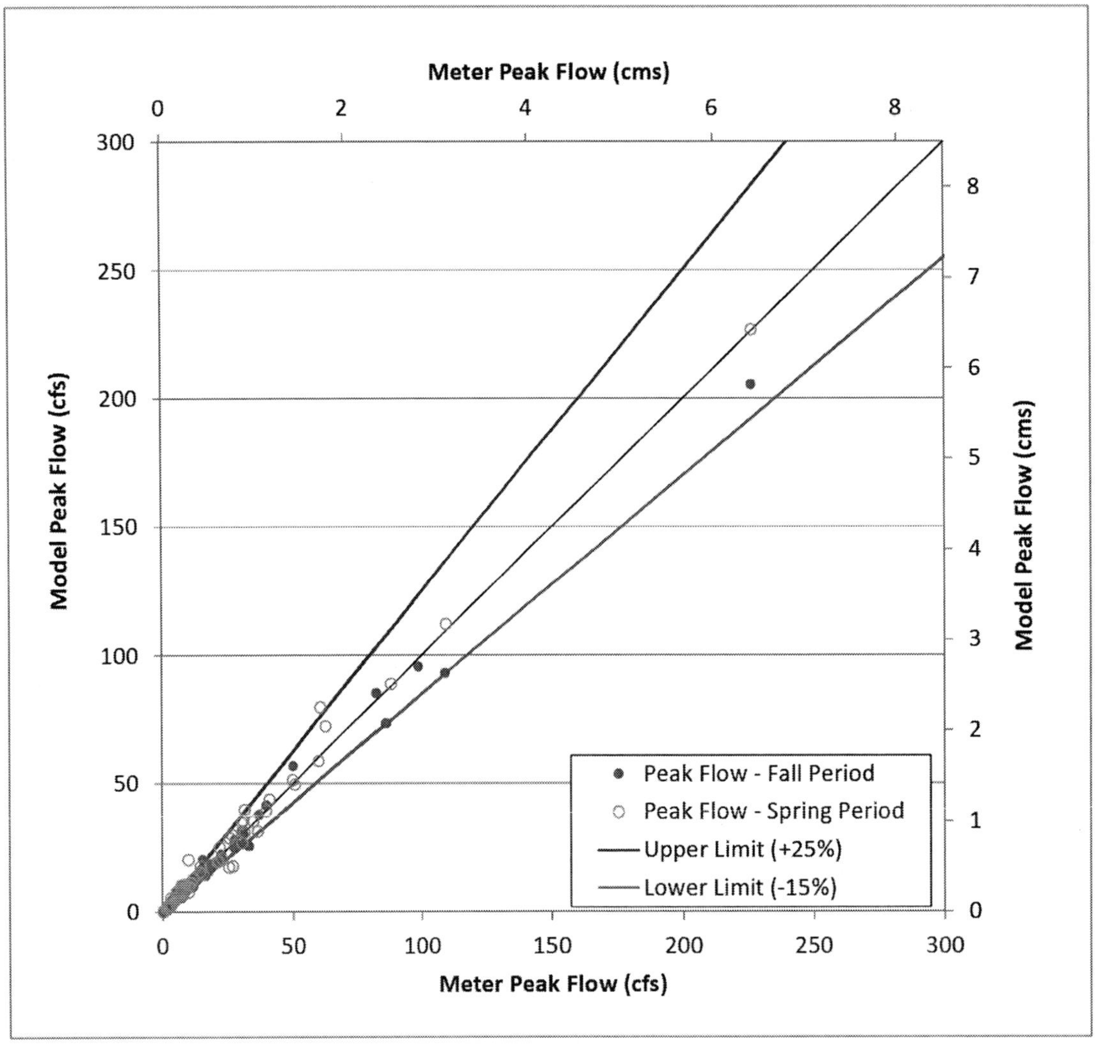

Figure 12. Metered vs Modeled Peak Flow for Key Calibration Meters

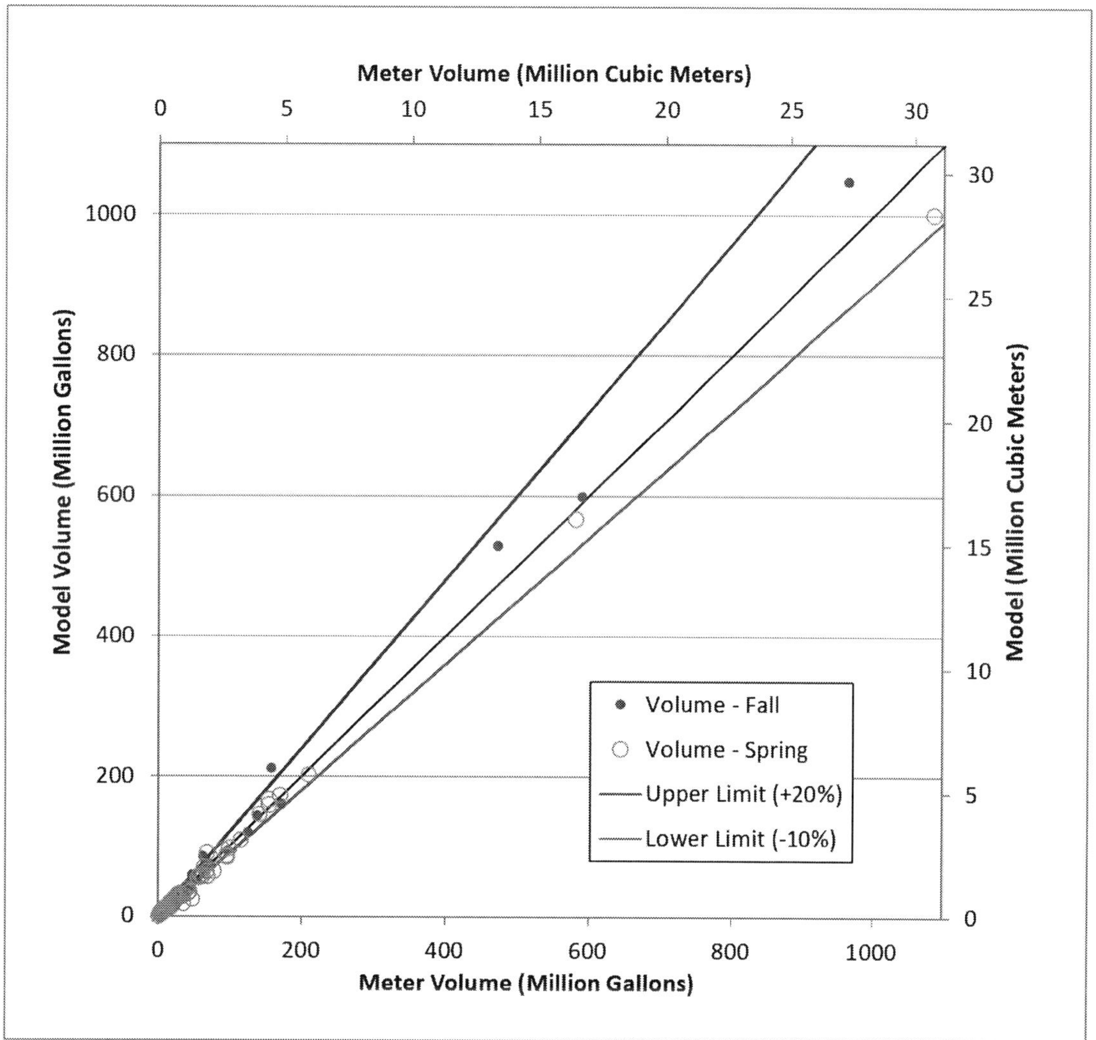

Figure 13. Metered vs Modeled Total Volume for Key Calibration Meters

Each point in Figures 12 and 13 represents one meter in the study. The upper and lower boundaries show the calibration criteria set forth by the WaPUG Code of Practice for the Hydraulic Modeling of Sewer Systems. These criteria recommend modeled flow volume to be with +20% and -10% of the meter and for peak flow to fall between +25% and -15% of the meter. As the graphs show, the model meets these criteria for most meter locations in both the fall and spring.

Based on the review of the individual calibration plots, and an analysis of the statistics presented in Figures 12 and 13, the model was considered suitably calibrated.

CONCLUSIONS

This paper summarizes the methodology used to calibrate flows in the updated sewer model for the Commission. The purpose of the update was to prepare the model for use by Commission's design and operations staff. When using a model to support sewer design and/or to make operational decisions, the expectations for the model's accuracy are greater than for a planning model. Model calibration will increase the accuracy of the model. However, an accurate representation of the various hydrologic features across the model is also necessary. Selecting the appropriate parameters to adjust during the calibration is critical. This means numbers should not be adjusted without comparing them to other reasonable data to determine the suitability of the changes. These checks can include:

- Comparing meter dry-weather flow to water consumption data, which can identify potential connectivity issues. Similar checks in other parameters, such as percent impervious, should be conducted to check the reasonableness of model results.
- Accounting for seasonal variation in inflow properly. In areas where there is significant seasonal variation, calibrating to drier periods only may lead to models that underpredict flow whereas calibrating to wet conditions only may lead to models that overpredict during drier periods.
- Examining regional differences in a larger system, to make sure different locations take into account unique characteristics properly.

Following these guidelines will result in a model that reflects real-world conditions more accurately compared to just adjusting model parameters to match meter data.

REFERENCES

Boston Water and Sewer Commission. 2009. Systemwide Modeling Project User Manual. (February 2009)

Environmental Protection Agency. 2016. Storm Water Management Model Reference Manual Volume 1 – Hydrology (Revised). (January 2016)

Heineman, Mitchell C. 2008. Collection System Model Development Using Raster Impervious Data, World Environmental and Water Resources Congress 2008 Ahupua'a, ASCE, 2008

Wastewater Planning Users Group. 2002. Code of Practice for the Hydraulic Modelling of Sewer Systems. Version 3.001. (December 2002)

Getting It Right: The Hampton Roads Approach to Hydraulic Model Data Collection and Management

Priyanka Mohandoss, PE, CDM Smith
Dr.-Ing. Matthias Wittenberg, PE, D.WRE, REM, CDM Smith
Robert J. Martz, PE, Hampton Roads Sanitary District

CDM Smith
5041 Corporate Woods Drive, Suite 250
Virginia Beach, VA 23452
(757) 318-9800

ABSTRACT

HRSD is a regional wastewater authority serving 1.7 million people in southeastern Virginia. HRSD's customer localities own and operate most of the more than 9,650 km (6,000 miles) of sanitary sewer pipe and 1,600 pumping stations. HRSD provides the backbone of the sanitary sewer conveyance and treatment system with over 690 km (430 miles) of force mains, 70 pumping stations, 20 pressure control facilities, and nine major wastewater treatment plants. HRSD maintains two dynamic hydraulic models for their regional sanitary sewer collection systems and 14 additional hydraulic models for the upstream locality-owned collection systems. These models support the development of a Regional Wet Weather Management Plan (RWWMP) required under a Consent Decree with the Unites States Environmental Protection Agency (U.S. EPA).

The Regional Hydraulic Models (RHMs) and Locality Hydraulic Models (LHMs) are relied upon for various planning and design analyses and will need to be updated periodically to reflect current asset inventory including ownership and asset conditions. This paper provides an overview of the collaborative standardized asset management approach developed by HRSD with its customer Localities. The approach allows for the development of a RWWMP that identifies the true needs of the regional system by accurately reflecting those facilities in the regional sanitary sewer system that are relevant for capacity and operations.

KEYWORDS

Hydraulic model, regional sanitary sewer system, asset management, template geodatabase, wastewater collection system, sewer, force main, SSO, capacity assessment, consent decree, wet-weather management plan

INTRODUCTION

HRSD is a regional wastewater authority that serves a population of 1.7 million people living in 17 cities and counties in southeastern Virginia. HRSD owns and operates nine major and four smaller sewage treatment plants (STPs) with a combined treatment capacity of 940 million liters per day (249 million gallons per day). HRSD and 13 of the 17 localities that comprise the HRSD service area entered into a Special Order by Consent with the Virginia Department of

Environmental Quality (VDEQ) in September 2007. One other locality that is part of the regional system that HRSD serves was already under a separate Consent Order with VDEQ. HRSD was subsequently placed under a Consent Decree with the Unites States Environmental Protection Agency (U.S. EPA). As part of this ongoing program to control sanitary sewer overflows (SSOs), HRSD and the localities were required to assess the condition and capacity of the regional sanitary sewer system and to develop a Regional Wet Weather Management Plan (RWWMP) for future flow conditions. Under the third amendment to the Consent Decree, HRSD is solely responsible for designing, funding, and implementing the RWWMP.

The primary tool for developing the RWWMP are complex hydrologic and hydraulic computer models of the regional sanitary sewer system. HRSD maintains two dynamic hydraulic models that represent the backbone of the regional sanitary sewer collection systems, the North Shore and South Shore Regional Hydraulic Models (RHMs), and 14 additional hydraulic models for the upstream locality-owned collection systems; one Locality Hydraulic Model (LHM) for each of the 14 localities involved.

The regulatory orders required development of the models using a dynamic hydraulic modeling software that models both gravity and pressurized flows simultaneously. The RHMs include all conveyance facilities owned and operated by HRSD and the major locality facilities that are directly connected and manifolded with these facilities. Since HRSD became responsible for capacity in all facilities, it took over the development of the separate LHMs that include selected portions of the upstream locality sanitary sewer facilities. The model development and calibration of the RHMs was especially complex due to the influence of approximately 900 terminal pump stations that discharge to HRSD's 692 km (430 mile) manifolded interceptor force main system. HRSD and the localities chose MIKE URBAN CS® by the Danish Hydraulic Institute for this application and have been innovatively utilizing geodatabase protocols and Web-based applications to facilitate the extensive process of collecting sewer system GIS data and operational information for integration in the hydraulic model development approach.

Collaboration between HRSD and the localities extended through all technical elements of the development and calibration of the models. Each Locality collected flow and rainfall data to develop hydrologic and hydraulic model input parameters. Within the modeling approach, localities had the responsibility of developing dry weather and wet weather flow inputs. In addition to locality flow monitoring, HRSD implemented a substantial program to install flow and pressure meters at more than 250 locations throughout the HRSD interceptor system and to provide a regional rainfall gauge network. The HRSD and locality flow, pressure, and rainfall (FPR) monitoring programs provided the necessary data to reliably calibrate the RHMs and LHMs.

MODEL EXTENT

The hydraulic modeling focuses on the highly-developed areas of Hampton Roads, including service areas in 14 of the 19 customer localities and nine STPs. The locality service areas and the extent of the sewer collection system included in the RHMs is shown in Figure 1. The RHMs were developed as two hydraulically independent models identified as the North Shore and South Shore because of the natural split of the two systems by the James River where it enters the Chesapeake Bay.

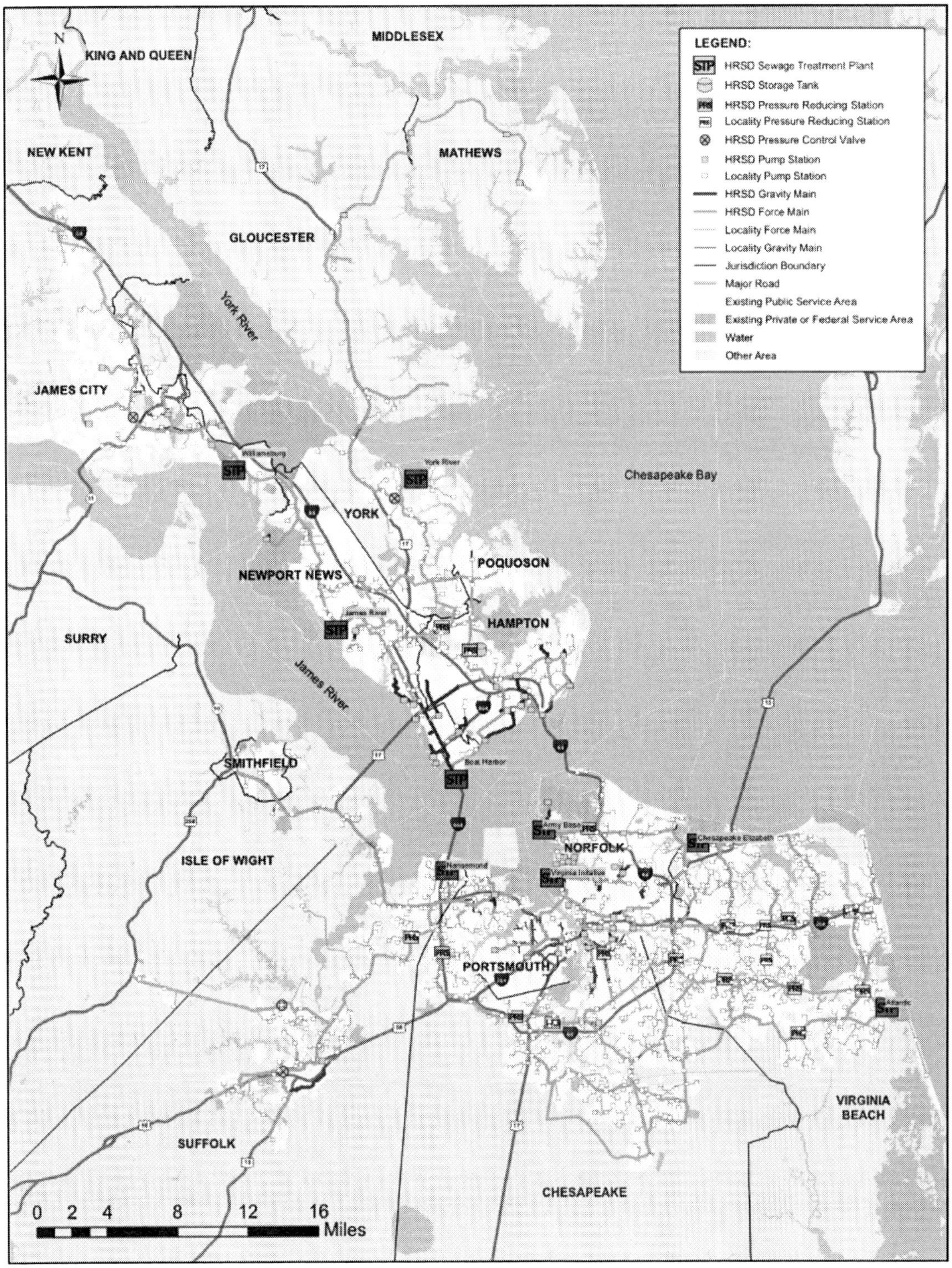

Figure 1. Regional Hydraulic Model Extent

RHM Extent

The extent of the RHM includes the following components of the Regional Sanitary Sewer (SS) System as shown on Figure 2:

- All operating HRSD force mains, gravity sewers, pressure control valves (PCVs), pressure reducing stations (PRSs), pump stations, and mainline valves.
- All operating Locality terminal pump stations and force mains that connect directly into HRSD force mains or gravity sewers. This includes Locality lift stations that connect to HRSD gravity sewers.
- Open and closed Locality valves operated by the Locality to divert flows during normal operations
- Locality gravity sewers, usually limited to the first sewer segment and manhole upstream from modeled terminal pump stations or the point of connection to an HRSD gravity sewer
- Locality facilities up to Locality Overflow Points (LOPs) identified in Appendix 1 of the Consent Decree

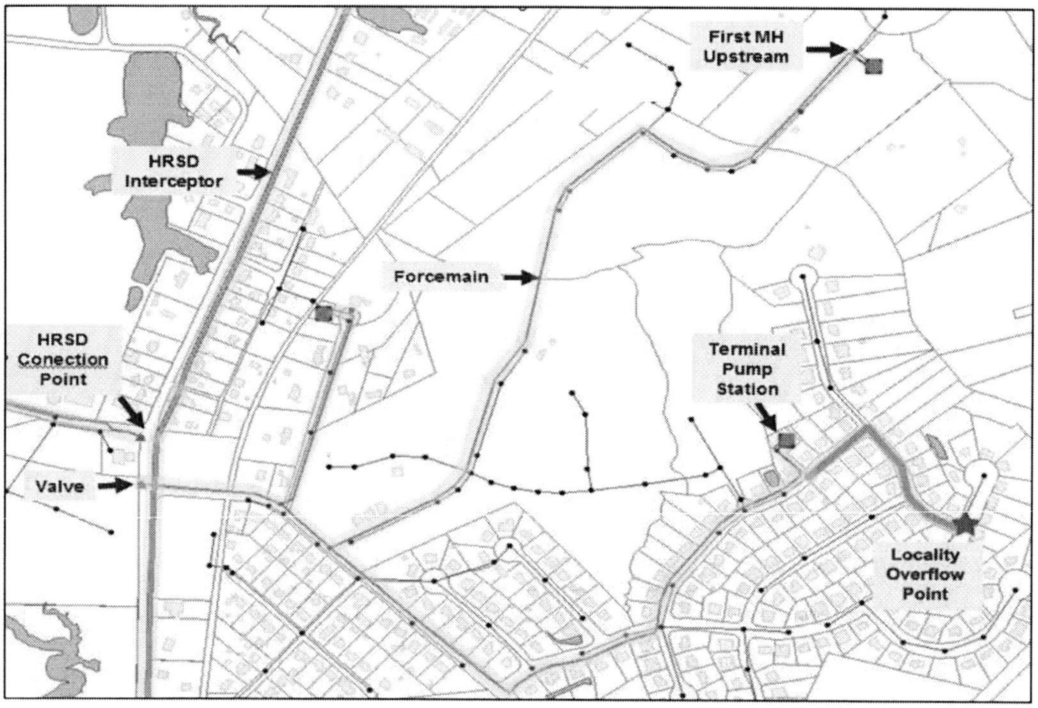

Figure 2. RHM Extent

The storage volume available in the Locality gravity sewer system upstream of these facilities is included in the RHM in most cases.

LHM Extent

The LHMs are intended to supplement the RHMs. The LHMs generally include Locality facilities upstream from the RHM with very limited overlap upstream of terminal pump stations and HRSD gravity sewers. Terminal pump stations and terminal lift stations where the force main discharges directly to an HRSD facility are modeled in the RHM. All remaining Locality lift stations are modeled within the LHMs. Where a hydraulic analysis of the Locality gravity sewer system is warranted, the LHMs start at the Locality or HRSD terminal pump station wet well. In some cases, the LHMs start where the Locality gravity sewer connects to an HRSD gravity sewer.

A screening process determined the service areas where a detailed LHM hydraulic analysis is warranted. This screening process eliminated terminal pump station service areas, lift station service areas, and gravity-to-gravity connection service areas where **all** of the following are true:

- There are no capacity related overflows or Locality Overflow Points (LOPs) documented in the Consent Decree except during extreme events (e.g. greater than 10-year peak flow recurrence). The reported incidents were tied to a manhole or other sewer system facility in a geodatabase. HRSD flagged those incidents and associated locations that would warrant further evaluation with a LHM.
- The 10-year recurrence peak flow can be conveyed by an 200 mm (8-inch) diameter pipe at minimum slope (22 l/s (350 gpm))
- There are no upstream lift stations.

A service area was not included in separate LHM, where the RHM could be extended to include the service area of interest without further complicating the allocation of flows from catchments to loading manholes.

For service areas that did not meet the above criteria, all gravity sewers and manholes were modeled where accurate and complete data are available or could be obtained with a reasonable effort. Where complete and accurate gravity system data (especially including pipe diameters and inverts) was not available for a service area of interest, an extensive effort was expended to include the following minimum locality facilities in the LHM:

- 250 mm (10-inch) and larger diameter gravity sewers and the first 200 mm (8-inch) sewer segments upstream from the 250 mm (10-inch) sewer.
- Gravity sewers downstream from a LOP or 'capacity related SSO' and the first sewer segment and manhole upstream of this location.
- Force mains and gravity sewers downstream from lift stations
- All lift stations that do not discharge to an HRSD sewer and, at a minimum, one pipe segment and manhole upstream from the lift station.
- Closed valves and force mains operated to divert flows during normal operations and the gravity sewers required to convey the flows downstream from these valves
- For service areas where Locality gravity sewers connect to HRSD sewers, the sewer segment and manhole upstream from the point of connection.
- For service areas where a cross-connection exists, the gravity sewers required to convey the flows downstream from this connection.

Figure 3 illustrates the minimum extent for service areas of interest.

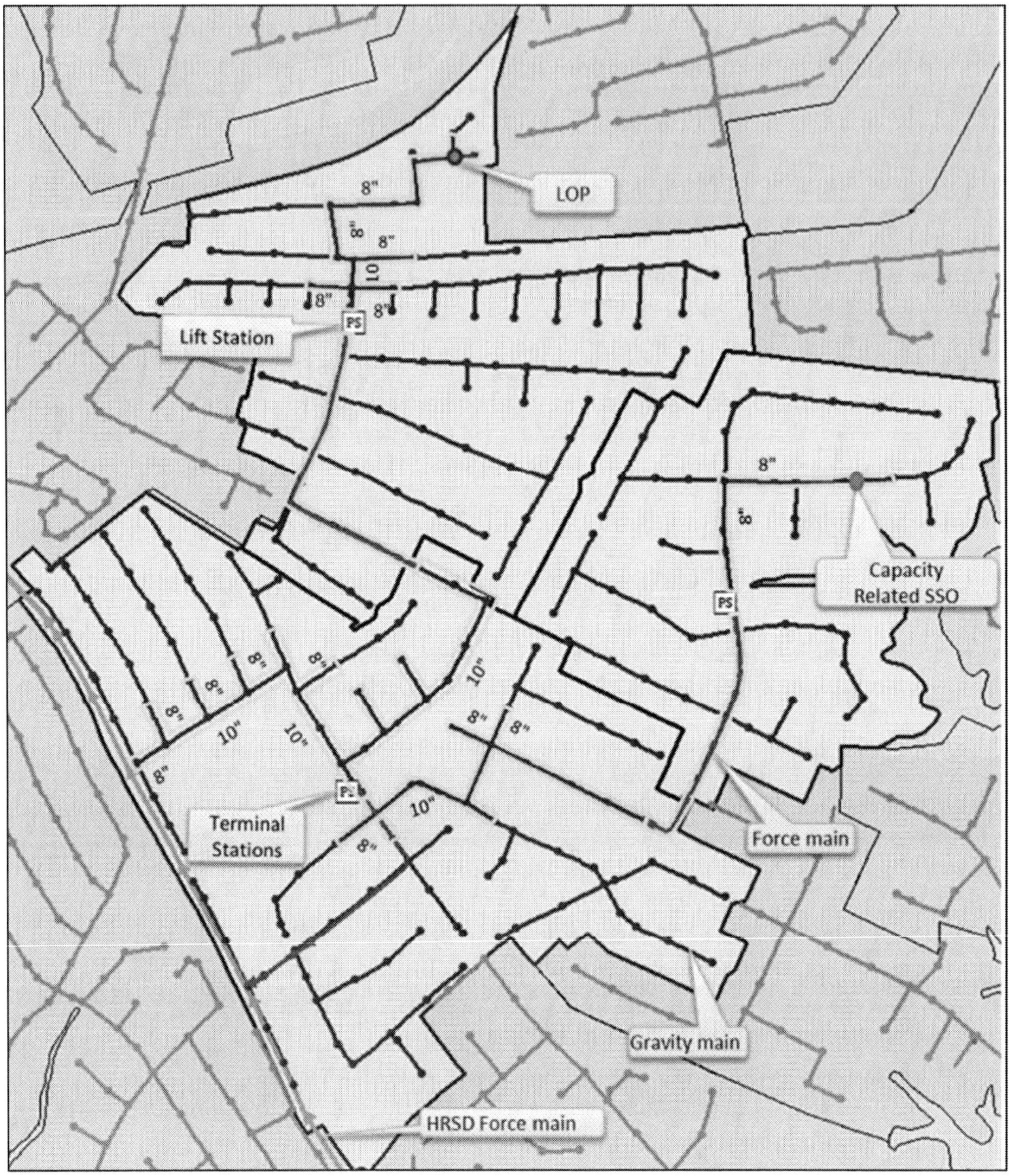

Figure 3. LHM Minimum Extent

Model Elements

For a perspective on the quantity of data included in the models, Table 1 provides a summary of elements included in the RHMs and LHMs.

Table 1

Facilities in the North Shore and South Shore RHMs and the 14 LHMs

Facility	North Shore RHM	South Shore RHM	LHMs
Number of STPs	4	5	-
Number of HRSD Pump Stations	29	37	-
Number of HRSD Pressure Reducing Stations	2	13	-
Number of HRSD Pumps	88	160	-
Number of HRSD Gravity Sewer Pipes	811	673	-
Length of HRSD Gravity Sewer, km (miles)	46.5 (28.9)	38.0 (23.6)	-
Number of HRSD Force Main Pipes	2,735	5,444	-
Length of HRSD Force Mains, km (miles)	253 (157)	442 (275)	-
Number of HRSD Pressure Control Valves (PCVs)	3	2	-
Number of HRSD Valves, Gates, Weirs and PCVs	522	1,022	-
Number of HRSD Storage Tanks	1	-	-
Number of Localities	**7**	**7**	**14**
Number of Locality Pump Stations	208	692	665
Number of Locality Pressure Reducing Stations	-	2	-
Number of Locality Pumps	431	1,413	1320
Length of Locality Gravity Sewer, km (miles)	12.7 (7.9)	27.3 (17.0)	3,130 (1,950)
Length of Locality Force Main, km (miles)	145 (90)	555 (345)	360 (220)
Number of Locality Valves and Gates	2	68	2
Number of Catchments	661	1,212	1402
Total Catchment Area, km^2 (square miles)	564 (218)	784 (303)	920 (360)
Total Sewered Area, km^2 (Square Miles)	318 (123)	515 (199)	680 (260)

FACILITY DATA COLLECTION AND COLLABORATION

HRSD supported the development of the RHMs with an ArcSDE enterprise sanitary sewer facility GIS. The HRSD GIS contains record drawing data describing the physical characteristics and location of HRSD force mains, gravity sewers, manholes, force main junctions, valves, bends, and many other features. HRSD worked diligently to obtain accurate HRSD GIS and facility operation information for input to the RHMs. All HRSD facility data are derived from record drawings, field inspections, pump station testing and field survey of the HRSD gravity sewers and manholes. HRSD successfully organized the HRSD facilities data and developed custom procedures for representing and updating that data in the RHM.

To collect the locality facility data listed in Table 1, HRSD developed a standard facility geodatabase template and associated metadata template file that must be used by the localities for all data submittals. This puts the vast amount of locality submitted data into one common format and facilitates the continuous efforts for validating, correcting, and updating of that data. The Locality Facility Geodatabase (FGDB) template was developed in ESRI's ArcGIS environment and includes MS Access queries and other automated check tools that allow the user to validate the data prior to submission to HRSD and further use.

The Locality FGDB template includes the following five feature classes and five associated data tables as shown in Figure 4:

- **Nodes** — Point feature class that includes gravity system manholes, junction chambers, and force main nodes (T-fittings, reducers, air release valves, bends, etc.). The primary data provided for these features include the invert elevation of gravity manholes, invert elevation of pressure system nodes, and the rim elevation of gravity manholes.
- **Valves** — Point feature class that includes Locality valves that can be operated to divert flows to different modeled facilities and therefore must be included in the RHM. Many Localities included all pressure system valves in this data table, including those not operated to route flows. A procedure was developed for the model administrator to move non-critical valve point features to the nodes table.
- **Pipes** — Line feature class that includes gravity sewer and force main pipes. Important data include the pipe size, upstream invert elevation, and downstream invert elevation.
- **Pump Stations** — Point feature class that represents the approximate pump station location. The pump station data table includes the pump station identifier, name, capacity, etc. Associated tables contain data for the pump station wet well, pumps, and pump controls.
- **Catchments or Service Areas** — Polygon feature class that outlines sewered parcels upstream from each model loading point. The catchments typically load to the most upstream Locality feature in the model.
- **Wet Well Table** — Table that includes data describing the wet well(s) for each modeled terminal Locality pump station. The primary data are the wet well surface area (if not provided as a depth-surface area table), and top and bottom elevation. This table is linked to the pump station feature class through the pump station identifier.
- **Wet Well Surface Area Table** — Table that can be used to define the depth versus surface area relationship for wet wells where the horizontal surface area changes significantly with depth. This table is linked to the wet well table through the wet well identifier. A procedure was developed for the model administrator to assign a sequence identifier to each surface area value based on increasing depth as required in the RHM.
- **Pumps Table** — Table that provides information on the individual pumps within each modeled pump station linked to the pump station feature class through the pump station identifier. Important data includes the type of drive (constant, dual, or variable speed) and the pump start and pump stop elevations.
- **Pump Curves Table** — Table that includes pumped flow versus total dynamic head points describing the pump operating characteristics. One curve is required for constant speed pumps and maximum and minimum curves are required for dual or variable speed pumps. The points that describe a curve for a particular pump are linked to the pump table through the pump identifier. Procedures were developed for the model administrator to create a

unique identifier for each pump curve based on the pump identifier and pump speed and to assign a sequence identifier to each flow value based on increasing head as required by the modeling software.

- **Gravity Storage Table** — Table that provides information on the storage volume available in the Locality gravity collection system upstream of a modeled HRSD or Locality gravity manhole in the RHM. Important data include the manhole identifier to which the storage volume will be assigned to and the gravity pipe and manhole storage volume determined by the procedure. The procedure uses the LHM system information available in the FGDB and additional GIS information if needed. It excludes non-modeled locality pressure systems and gravity systems upstream therefrom. A procedure allows for the model administrator to convert the potential storage volume to a depth versus surface area relationship based on the depth between the invert at the connection to the model and the overflow elevation. The model administrator will also assign a sequence identifier to each surface area value based on increasing depth as required in the RHM. The overflow elevation is the lowest manhole rim of upstream system or at the point of model connection. The procedure determines the storage volume available for wet weather conditions and therefore excludes storage volume that is being utilized during dry weather conditions.

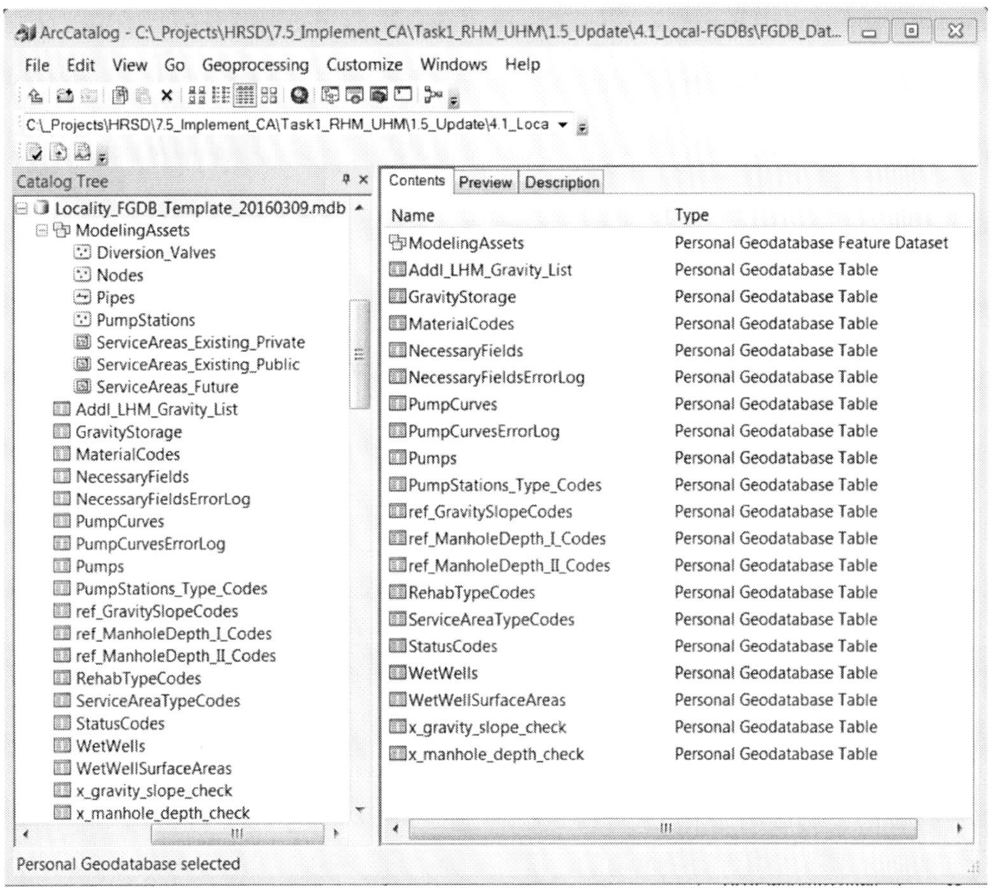

Figure 4. Locality FGDB Template

A series of technical reports and memoranda document the RHM and LHM locality facility data needs and the procedures localities used to deliver these data to HRSD. Localities are asked to provide all facility data using the facility geodatabase template and in accordance to the standards and guidance provided in these documents.

Attribute information for each feature class and data table type was classified as "necessary", "highly desirable", "desirable" or "informational" to assist HRSD and the localities in setting priorities for data collection.

These are listed below in decreasing importance or priority and shown in Figure 5.

- **Necessary Data** are absolutely required for RHM and LHM development. These data should receive highest priority in data collection within the modeled portions of the regional and locality sanitary sewer systems. These fields have red text in Figure 5.
- **Highly Desirable Data** will assist in developing an accurate RHM/LHM, but are not absolutely necessary. These fields have blue text in Figure 5.
- **Desirable Data** will assist in model development, but only need to be delivered if readily available. These fields have black text in Figure 5.
- **Informational Data** documents features and attribute values developed by HRSD for the RHM/LHM that should not be modified by the Localities.

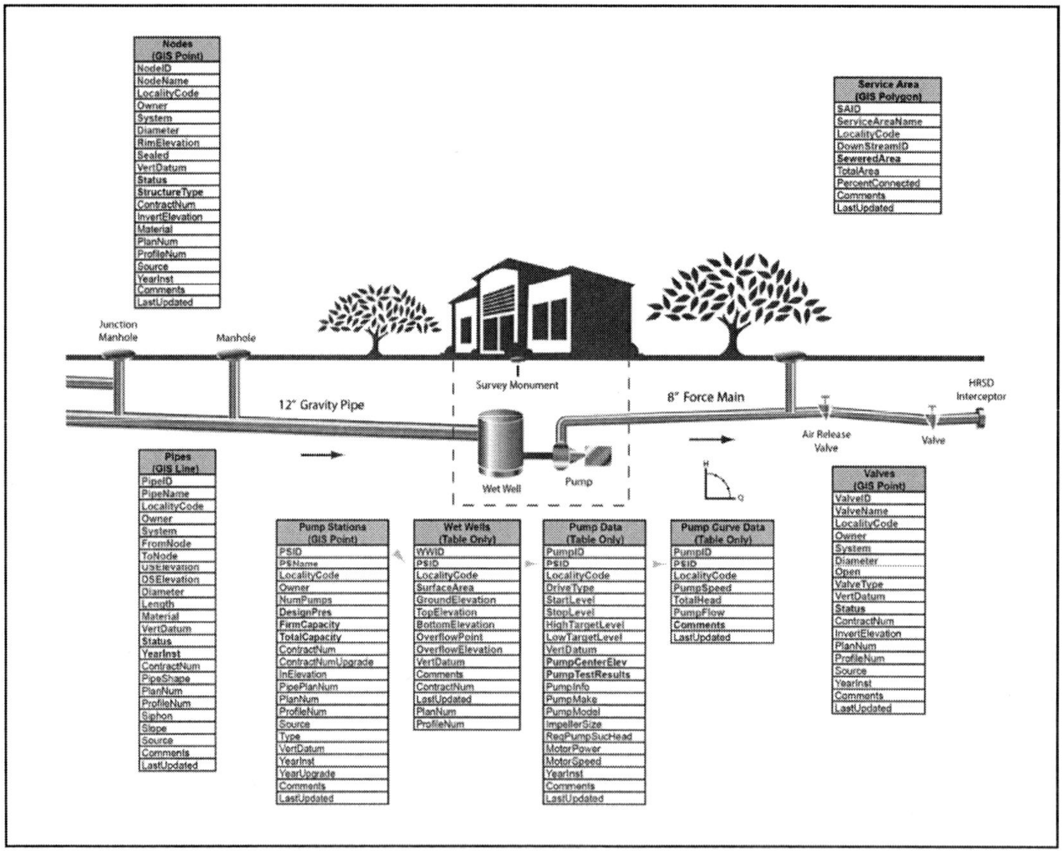

Figure 5. Classification of Attribute Information within the Locality FGDB Template

The facility geodatabase template was developed in ESRI's ArcGIS environment and includes Microsoft Access queries and other automated check tools that allow the user to validate the data prior to the submittal to HRSD. The Microsoft Access queries perform detailed review of the data to find duplicates, check that nodes associated with pipes are in the database, check the number of pumps, wet wells, and pump curve points, check pump settings, and perform a variety of other data checks for the "necessary" attribute information.

To further facilitate the data collection process, HRSD maintains a Locality Data Delivery Portal that the Localities must use at all times when providing the facility data updates for the RHM and LHMs. Figure 6 shows the Portal Web site.

A login username and password is needed to upload data. Detailed upload instructions and instructions to obtain login information are available on the Portal.

Figure 6. HRSD Locality Data Delivery Portal

An important practice to promote consistency among HRSD and the localities during development and update of the models was to formulate and publish standards and guidance documents for data content and format. Listed below are some of the documents distributed over the course of project phases:

- **Regional Wet Weather Management Plan Locality Sanitary Sewer System Facility Data Standards:** Report that documented the data needed and the format to be used to deliver the facility data and metadata to document the data sources
- **Answers to Frequently Asked Questions Regarding Locality GIS Data Delivery Requirements:** Memorandum that provided answers to common questions from the localities regarding the delivery of locality GIS and operations data

- **Regional Hydraulic Model Catchment Data and Delineation Requirements:** Report that clarified issues regarding the catchment delineations that had to be provided for all areas served by the localities. The report noted that catchments are required for terminal and lift station service areas, gravity sewer connections to an HRSD gravity sewer, and that small gravity sewer connections can be aggregated to a single catchment if desired.
- **Locality Gravity Sewer Storage:** Guidance document that described the procedures for calculating available storage in Locality gravity collection systems for inclusion in the RHM and documented the format to be used to deliver the computed data and metadata file.

In addition, HRSD also scheduled meetings with the locality engineering and GIS staff in advance of data submittal due dates. During these meetings, individual cases that qualify as exceptions to the standard procedures were discussed and questions and concerns voiced by the locality representatives regarding the data request were addressed.

Some examples of exceptions to the standard approach include:

- If all Locality owned assets under construction as of a set date are requested, how should facilities that are under design and ready for construction be treated?
- If a facility is under construction but not included in a Locality's GIS, how should the identifiers be assigned for such assets?

These meetings served as a great platform for knowledge sharing between locality and HRSD staff and helped resolve complex data issues prior to the hydraulic model development. In addition to establishing a sustainable technical basis, the collaborative process promoted ownership and confidence in the completed models, which is critical for modeling products according to Fugal et al. (2009). With these best practices in mind, the RHMs and LHMs (including data and supporting tools) were developed to empower and benefit HRSD and the localities (Carolan and Robichaud, 2008).

MODEL DEVELOPMENT

The combination of using a defined data format, transfer scripts and automated model building tools effectively streamlines the addition of the quality assured locality data to the model. Figure 7 illustrates the robust work flow process that was developed to enter the locality facility data into the MIKE URBAN CS model geodatabase and manage facility data in the RHM (Wittenberg and Carolan, 2012). These data were updated several times during model development, calibration and application to maintain the correlation between the models and the localities' GIS facilities. The RHM and LHMs were updated with GIS and data from both the localities and HRSD before each round of major simulations to maintain consistency between the models and the locality and HRSD facility data stored in their respective GIS databases.

The processes developed to load the locality GIS and non-GIS data into the models facilitated the RHM and LHM development and now supports ongoing updates.

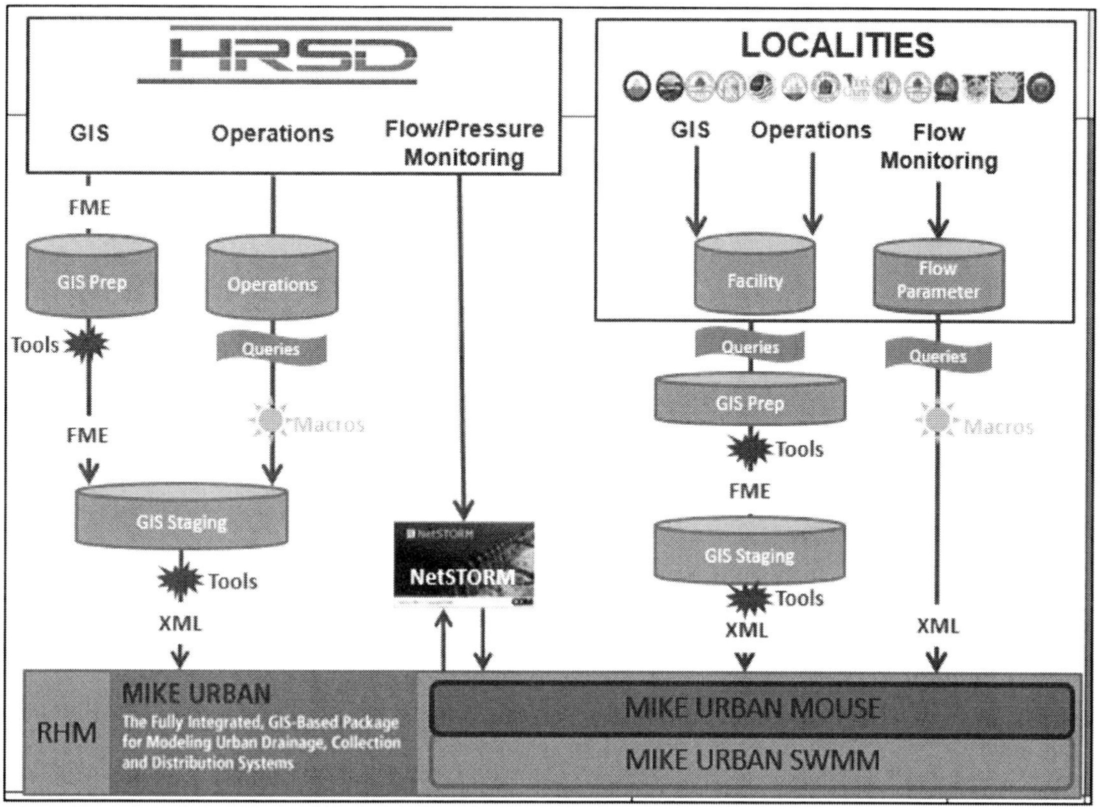

Figure 7. Model Development Work Flow

Locality Facility Data Preparation

The data in the Locality FGDBs require several transformations before they can be entered into the RHM. HRSD developed procedures to streamline these data conversions, but the transformation process is still primarily manual work. The designated model administrator must apply diligent quality control and have a complete understanding of the processing rules, transformation tools, and MIKE URBAN CS software.

The data from each Locality RHM Facility Geodatabase are first transformed to a so-called GIS Prep geodatabase that contains additional feature classes (i.e., wet well point, pump line, valve line, and merged pipe), additional fields in the existing tables, a domain table that includes a model status code later used to transfer the data as needed, and Microsoft Access queries. The Microsoft Access queries review the data to find duplicates; check that pipe nodes are in the database; check the number of pumps, wet wells, and pump curve points; check pump settings; and perform other data checks.

Tools developed to process the HRSD data are then used to transform and check the Locality data within the GIS Prep database. An example of a GIS transformation tool is the Locality Pump Station Conversion Tool. The Locality FGDB contains a single point for each pump station. The RHMs and LHMs require additional features to represent these pump stations. The automated Locality Pump Station GIS tool performs the following major conversions on selected pump station features within the GIS Prep database:

- Include a wet well point feature that connects to the gravity sewers entering the pump station and populate the feature attribute information from the Locality facility data Wet Well table, including depth versus storage data. The wet well is placed at the pump station point provided by the Locality.
- Include a new node point feature at the upstream end of the force main pipe connected to the pump station feature. The attribute data for this feature are populated using MIKE URBAN CS default values and connected force main data.
- Include pump links that connect the wet well to the first force main segment. The attribute data for the pump links are populated using the pump and pump curve data in the Locality RHM Facility Geodatabase.

Figure 8 illustrates the conversions to Locality pump stations performed on the Locality GIS Prep database.

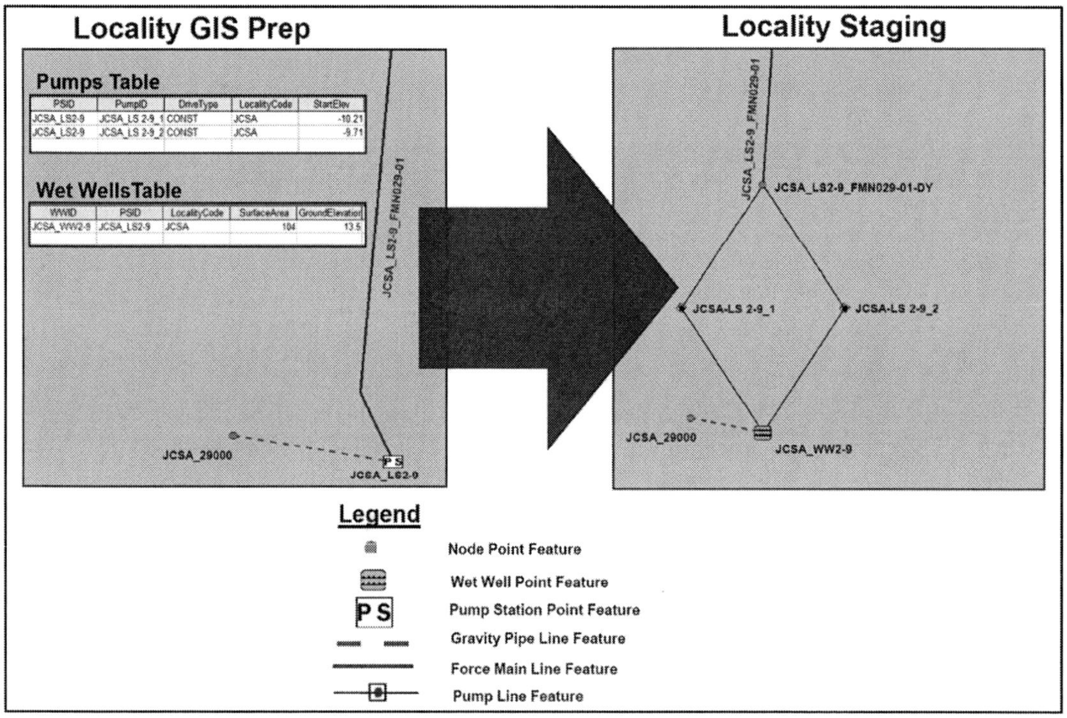

Figure 8. Locality Pump Station Conversions

The following other data adjustments may be performed on the Locality GIS Prep database to provide a valid model dataset:

- Convert the valve point features provided by the localities to line features to be compatible with the MIKE URBAN CS software.

- Linearly interpolate force main elevations where manifold force mains connect from the nearest feature with invert elevations. When the locality obtains more detailed information, the force main data are updated in the models as part of a resubmittal.
- Assign **Model Status** codes to the features in the GIS Prep geodatabase to describe the data source and indicate which data are to be transferred to Staging tables for input to the RHMs or LHMs. These codes also control whether feature attribute data in the GIS Prep database are overwritten when updated from future Locality FGDB submittals. In addition, they identify new features selected by the model administrator for model input.
- Assign a diameter to selected pressure nodes based on the largest diameter of the connected force main. A GIS tool was developed to perform this operation. In some cases, the force main junction diameter may be increased to improve model stability.
- Add a point or line feature with an assumed identifier and attribute information to maintain connectivity where not provided by the Locality.
- Merge a line feature to avoid the addition of a new point feature where the Localities did not provide nodes at force main junctions.
- Assume unique identifiers for features with duplicate identifiers in the data provided by the Localities. The assumed value is updated if the Locality provides the information in a subsequent submittal.
- Update the 'ToNode' field, 'DownstreamID' field, or other feature attribute information based on HRSD data or the model administrator's understanding of the modeled facilities and RHM requirements.

The Locality facility database was sometimes populated with duplicate HRSD features, duplicate neighboring locality features, or other features from the locality GIS that are not required for the RHM. The status code of these features was set to indicate and ensure that these are not imported into the RHM. The status code of all remaining, previously untouched features in the Locality GIS Prep geodatabase was set to indicate that they are to be included in the model database.

Locality GIS Data Staging

Once processing is completed, the facility data in the Locality GIS Prep geodatabases are loaded to ArcGIS Staging geodatabases prior to being input to the RHMs or LHMs.
This step accomplishes the following tasks:

- Places the facility data in a format compatible with the MIKE URBAN CS database structure prior to loading into the RHMs or LHMs
- Excludes GIS features not required or necessary in the models based on the model status code
- Adds temporary node point features to maintain connectivity between modeled pipes to replace temporary HRSD or locality GIS 'Network Junction' features
- Performs other data manipulations necessary to place the data in a format compatible with MIKE URBAN CS (e.g., check connectivity and populate 'To Node' and 'From Node' fields, sort pump curve and wet well values in sequence, check and create unique identifiers for GIS and other modeled features)
- Incorporates attribute data not in the GIS (e.g., pump and wet well data, pressure system node point diameter, roughness coefficients, etc.) from other databases

ArcObjects Feature Manipulation Engine (FME) scripts are used to transfer the data from GIS Prep to the staging tables perform many of these manipulations automatically. MIKE URBAN Extensible Markup Language (XML) is used to import the data from the staging tables into the modeling software geodatabase. An XML script is generated automatically by the software when the staging table data fields are manually mapped to those in MIKE URBAN CS during the import process. HRSD subsequently modified this script to perform additional functions that assign default values and filter the data when imported into the RHM or LHM databases.

Locality Data Update and Correction

HRSD performed reviews of the Locality data when delivered. Additional questions and errors were identified as the data were reviewed and processed for entry to the RHMs and LHMs. Also, once these data were in the models, engineering and other data checks identified additional concerns. These data concerns were transmitted to the Localities for review and correction in memoranda and updated geodatabases. This feedback and correction process keeps the models, locality data, and interim databases in sync and allows errors to be corrected by the localities. This work also allows the RHMs and LHMs to be updated using the GIS data.

PROJECT HISTORY AND CURRENT STATUS

The locality facility data collection approach and model update processes were successfully applied by HRSD in 2008 and 2009 to collect the data necessary to develop the initial RHMs, in 2010 and 2011 to support the calibration of the RHMs to the flow and pressure meter collected by HRSD and the localities (Morgan et al., 2012). The approach was expanded from the RHM extent to the entire regional system in 2015 to update the RHMs, develop the LHMs and perform a capacity assessment of the regional system with the RHMs and LHMs (Hubbard et al., 2015). The approach was applied again to quickly update the models prior to the next RWWMP milestone in the summer of 2016 that included the definition and testing of capacity improvements in support of an alternatives analysis and level of service selection. The tools and procedures were recently applied for the final RHM and LHM update prior to the development of the RWWMP due in October of 2017.

HRSD plans to continue using the data collection procedures through the implementation of the RWWMP and beyond to keep the model updated for all other engineering applications.

CONCLUSIONS

HRSD established a standardized approach to collect regional collection system data from its 14 customer Localities and uses ArcGIS tools and FME and XML transfer scripts for development of two RHMs and fourteen LHMs.

A pre-defined ArcGIS geodatabase template allows the application of custom GIS and Microsoft Access tools that perform data consistency and completeness checks, track data sources and history, and automate the data collection process, ensuring quality and resulting in time saving in excess of 400 hours each time the models need a major data update.

The template geodatabase was accompanied by a detailed guidance report that defined the data quality standards the localities were required to adhere to. This comprehensive document summarized allowable assumptions, required data standards, and computational procedures including how to calculate gravity system storage volume for non-modeled upstream portions of the sanitary sewer system. Supplemental memoranda provide answers to frequently asked questions to minimize ambiguity and to produce consistent data submittals to HRSD regardless of individual locality standards, protocols of data collection and asset management databases.

The data collected in the template geodatabase simplified the visual data review and coordination with each responsible party. Once all facility data and operation information are complete, ArcGIS tools are applied to transform the data into the format needed for model input and a code is assigned to each feature to inform the model development tools if a feature needs to be updated in the RHM, the LHMs, or both. These tools and transfer processes automate the model update process, ensuring data consistency and resulting in time saving in excess of 500 hours each time the models are updated.

The RHMs and LHMs are relied upon for various planning and design analyses and need to be updated periodically to reflect current asset inventory including ownership and asset conditions. Establishing a standard approach and template facility geodatabase based process continues to save hundreds of hours in data review and processing time, while allowing a consistent data format for the models even though Localities maintain custom individual formats.
The template geodatabase allows for effective communication and collaboration between HRSD and the localities, ultimately leading to hydrologic and hydraulic models that accurately represent field conditions.

REFERENCES

Carolan, J., Robichaud, J. (2008). "Integrating Regional and Local Municipality Geographic Information Systems (GIS) Data in Support of Regional Wastewater Hydraulic Modeling." *Proceedings of AWWA WaterJAM.* Virginia Beach, Virginia, September 7-11, 2008.

Fugal, A., Fyock, B., Shay, R. (2009). "Model Coaching To Create a Collection System Model." *Proceedings of WEF Collection Systems Specialty Conference.* Louisville, Kentucky, April 19-22, 2009.

Hubbard, P.L.; Martz, R.J.; Moore, C.I.; Morgan, M.C. and Wittenberg, M. "Hitting the Peak Flow Recurrence Target; A Hydrologic Approach for the Hampton Roads Regional Hydraulic Model." *Proceedings of WEF Specialty Conference.* Cincinnati, Ohio, April 19-22, 2015.

Morgan, M., Hubbard, P., Martz, R.; Moore, C., Wittenberg, M. "A Collaborative Approach to Modeling the Hampton Roads Regional Wastewater Collection System." *Proceedings of WEF Specialty Conference.* St. Louis, Missouri, June 3-6, 2012.

Wittenberg, M., Carolan, J. "Leveraging Geographic Information and Work Order Systems Data for Hydraulic Model Development and Maintenance: An Effective, Sustainable Methodology." *Proceedings of Chesapeake AWWA, CWEA and Maryland, Delaware and District of Columbia WWOA Tri-Association Conference (Tri-Con 2012).* Ocean City, Maryland, August 28-31, 2012.

Using SWMM LID Tools To Calibrate Existing Condition Wet Weather Response

Mark R. Delisio, P.E., CT Consultants, Inc.
Robert H. Greytak, P.E., CT Consultants, Inc.
Christopher P. Rybak, P.E., CT Consultants, Inc.

ABSTRACT

SWMM models are one of the tools used to simulate rainfall impacts on storm and sanitary sewer systems. The model replicates a subcatchment's stormwater runoff characteristics by applying measured, calculated or estimated values for various parameters and calibrating the model against measured sewer flow data. In urban areas the values are typically an average representation of an overall area consisting of hundreds of individual residential lots each with their own specific runoff characteristics. Stormwater that does not find its way to the storm sewers may find its way to the sanitary sewers as I/I by the way of illicit connections, pipe defects, or groundwater seepage through joints. There is not a direct method to route these conditions to sanitary sewers in SWMM and so they are estimated through the use of RTK values applied to a sewershed. This paper investigates utilizing SWMM LID tools as a method for modeling a small residential sewershed, analyzing private property interaction between sanitary and storm sewers, and evaluating alternatives for reducing I/I.

KEYWORDS: SWMM, LID Tools, Small Area Modeling, Model Calibration, I/I Reduction

INTRODUCTION

The development of SWMM models (EPA Storm Water Management Model, Ver. 5.1.009) to analyze the interaction between storm and sanitary systems on residential property is complex due to the high number of variables and uncertainties associated with each variable. Models representing small residential areas must consider the individual parcel features that collect and route rainfall to the public storm sewers with a portion of the rainfall routed to the sanitary sewer by way of I/I (Infiltration and Inflow). The model should represent features such as:

- Steep roofs with gutter, downspout, and lateral sewer routing systems.
- Impervious driveways with area drains routed through connected piping.
- Impervious driveways with surface routing.
- Pervious rear and front yards with surface routing to pervious and impervious areas.
- Impervious street flow routing.

Note that street drainage, though not technically part of a residential parcel, is a feature that needs to be included in the analysis so that the study area output represents the entire study area subcatchment.

This paper explores how to develop a residential parcel-scale model using LID (Low Impact Development) tools that allow evaluation of existing conditions, provides a valid mechanism for calibration, and incorporates an efficient routing of the parcel features. Street drainage was not represented by LID tools in the model but rather as individual subcatchments. The aggregate of the parcel-scale model and street drainage output must be consistent with larger subcatchment scale models encompassing the same sewershed.

Description Of The Study Area

The study area includes a neighborhood of 102 parcels in Lakewood, Ohio (Figure 1). A common manhole on Eldred Avenue was isolated with sandbags for the study purposes. Sandbags were also used to block a cross connection at the intersection of Delaware and Eldred Avenues.

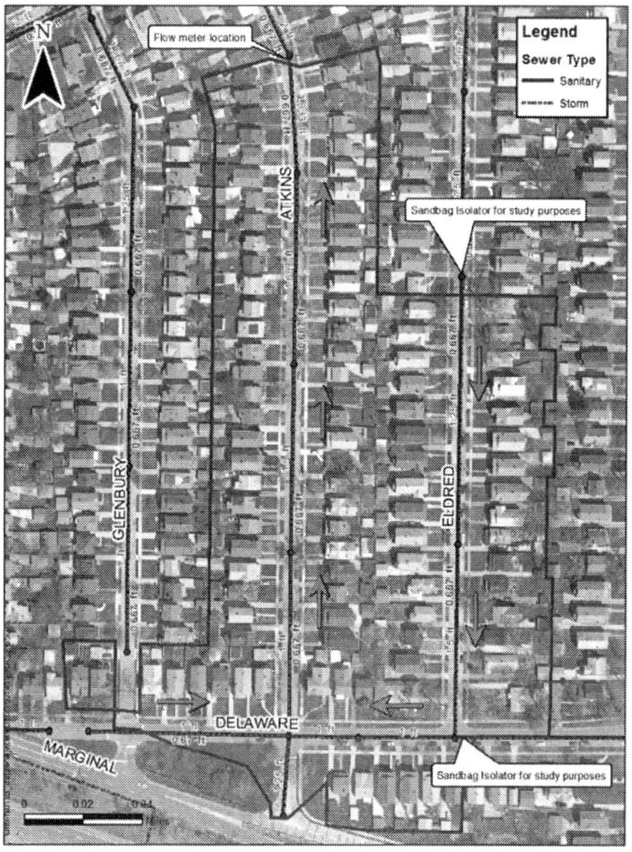

Figure 1: 102-parcel study area in Lakewood, Ohio showing sewer configuration.

The study area sewer construction consists of over/under (O/U) sewers with the storm sewer directly above the sanitary using a single manhole to access both pipes. Figure 2 illustrates the typical configuration of the O/U sewers. An invert plate in the "over" sewer allows access to the "under" sewer. O/U sewers were built predominantly during the 1920s and 30s although O/U sewers were installed until the 1950s.

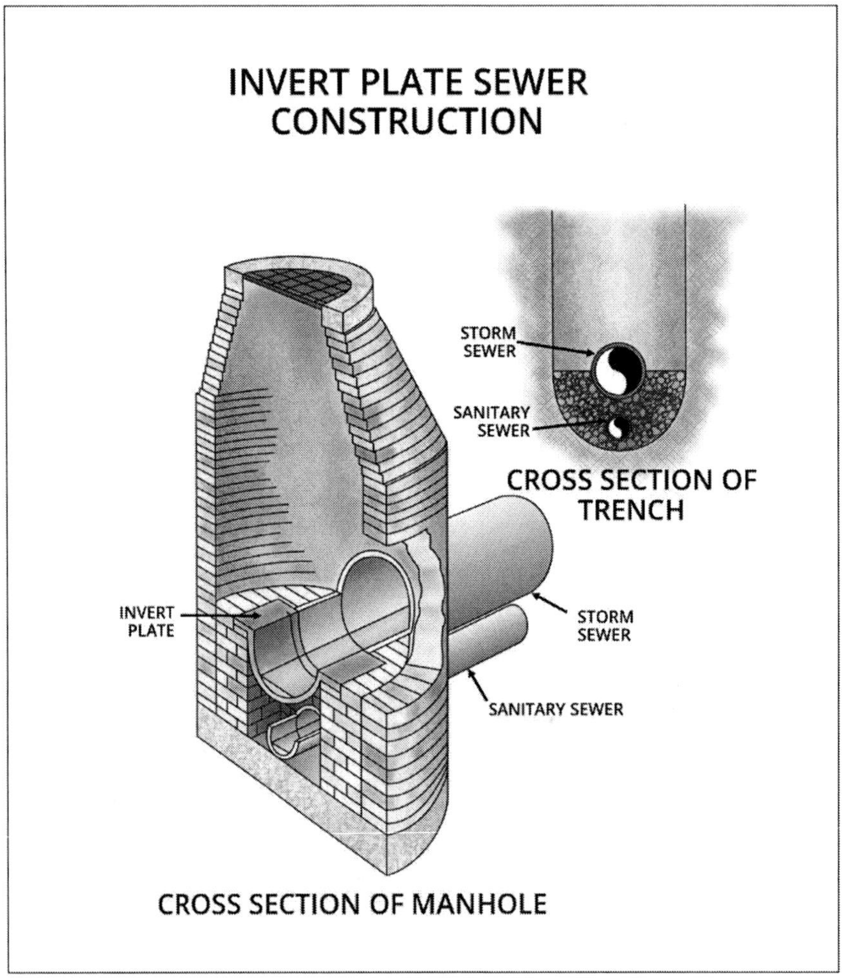

Figure 2: Sketch of Over/Under Sewer Configuration

Private Property Investigations and Development Conditions

Houses within the study area were assumed to have been built using similar construction methods. Each home was constructed with a storm lateral and a sanitary lateral (in a common trench) connected to its respective public sewer at the street. The city led field investigations of each of the individual 102 properties to evaluate if cross connections existed between the storm drainage system and the sanitary laterals. Field investigation concluded that there may be cross connections between the storm drainage system (downspouts, area drains, and foundation drains) and the sanitary laterals as shown in Figure 3.

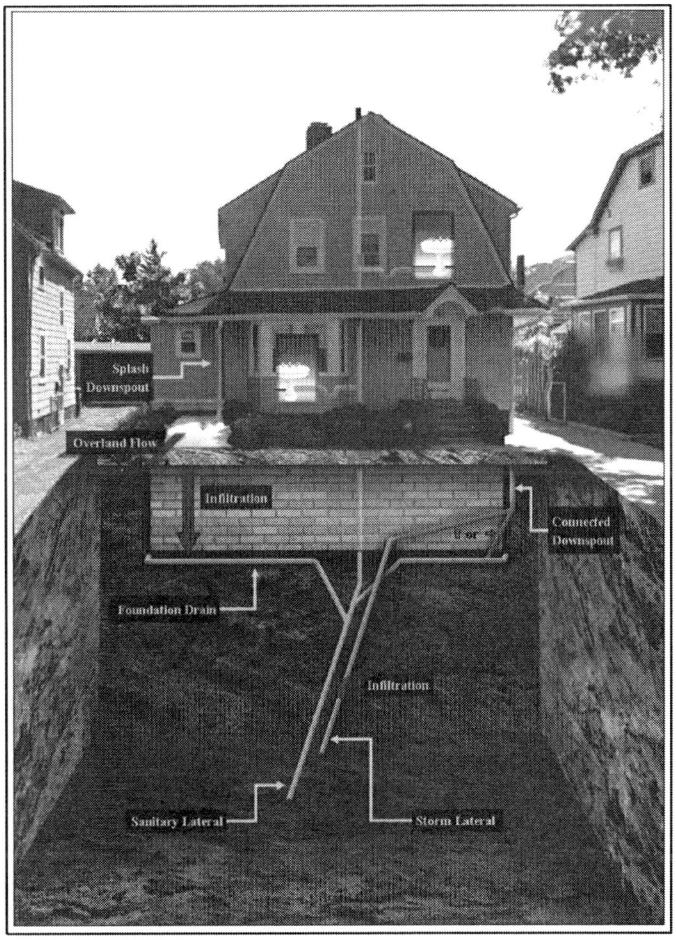

Figure 3: Private parcel connections based on field investigations.

Closed Circuit Televising (CCTV) investigations also revealed sanitary sewer laterals were in disrepair with joint offsets, cracked pipe, and root intrusion. Some of the worst infiltration points were noted to be near the foundation of the structure or at the point of connection to the sanitary sewer at the street.

Pilot Study Flow Metering

Sanitary sewer flow metering was performed to estimate and document any I/I to provide a benchmark for evaluation of the post construction flow.

Flow metering occurred from March 27, 2015 to May 2, 2015 and June 4, 2015 to June 29, 2015 and captured approximately 16 wet weather events of significance. The break in metering was due to access issues in May. Representative flow meter data was selected from the observed data set and is shown in Figure 4 and Figure 5. The June rain events were significantly larger than those in March and May.

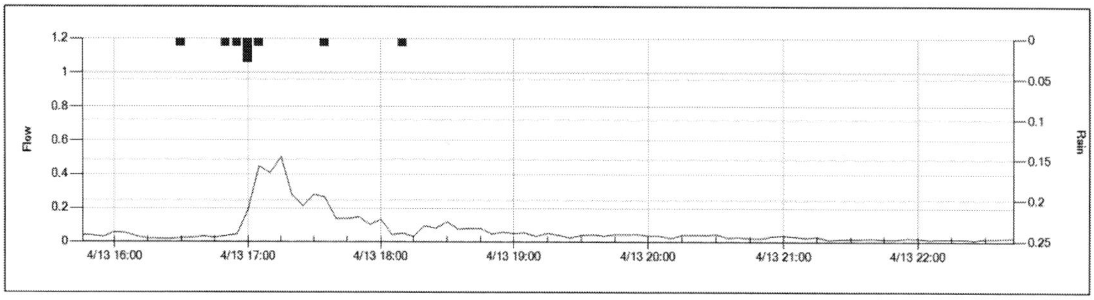

Figure 4: Representative Sample Storm (March to May)

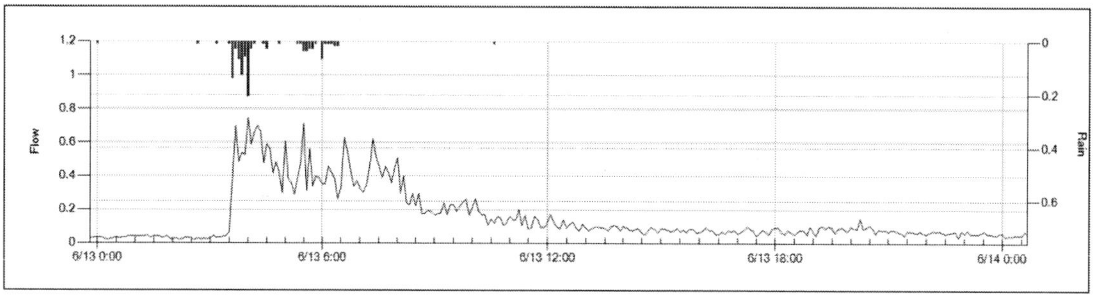

Figure 5: Representative Sample Storm (June)

The meter data suggests that the flow in the sanitary sewer is impacted by Rain Derived I/I (RDII). The question to be answered is how much of the RDII is emanating from private property; how much from the public right of way; and what sanitary sewer flow reduction could result from an aggressive I/I removal program on both private and public property.

Initial Model Development

The model was developed based upon the premise that the sanitary sewer system is isolated from the storm sewer system and only the impacts to the sanitary sewer were to be evaluated. It was assumed that any stormwater removed from the sanitary sewer would be routed to the storm sewer. The model also needed to quantify the impacts from private property since it is the intent of the city to remove all sources of I/I within the study area including those sources on private property. The data from the flow meter in the sanitary sewer at the north end of Atkins Avenue is to be used to calibrate the model.

METHODOLOGY

The general runoff characteristics typically used for modeling subcatchments were determined to not be sensitive enough to evaluate the magnitude of I/I from each feature on a parcel. From the flow meter data, I/I was shown to have a direct impact on sanitary sewer flow. It became important to develop an enhanced model that evaluated the impact of I/I from each feature within the parcel. A runoff hydrograph for each parcel was needed representing the impact of each incorrectly connected or deficient feature. The parcel hydrograph would recognize the contribution of roof drainage, foundation drains, area drains, and other features.

The study area was chosen in part because of the uniform parcel sizes, comparable size and age of the houses, and similar parcel features. Each parcel consisted of a lot of roughly 0.045 hectares (0.11 acres) with a house, detached garage (with a few exceptions), driveway, sidewalk, and lawn. (Figure 6)

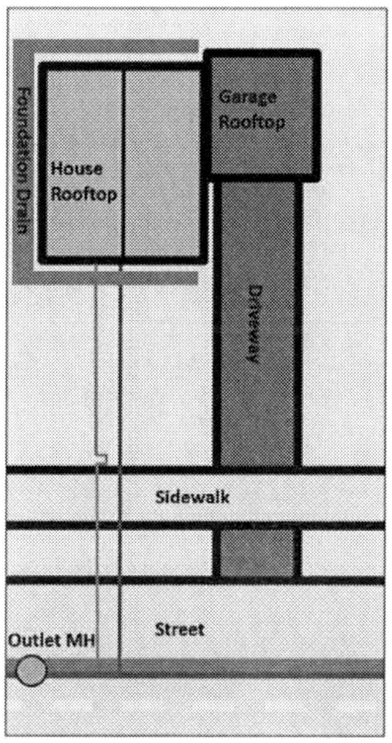

Figure 6: Typical Parcel Features

Each parcel was reviewed in the field to identify its features and probable connectivity to one of the two sewers. It was determined that there were 19 scenarios reflecting different combinations of features and connectivity overlaying an individual parcel. Modeling of these scenarios could be done in two different ways.

Modeling Method 1

In method 1, each parcel is subdivided into subcatchments representing features that replicate the runoff characteristics of the parcel. The concept was to create a feature for each impervious surface area (house roofs, garage roofs, driveways, and walkways) and a feature that represents foundation drains. Separate pervious features would be created for the rear yard and front yard. Each feature would outlet to the storm sewer unless they were found to outlet to the sanitary lateral. The exception being if a downspout from the house roof or garage roof discharged to the surface, then that feature would outlet to the pervious lawn feature. The study area consists of 102 parcels, and each parcel has between 3 and 6 features routing either to the storm sewer, sanitary sewer, or another feature (Figure 7).

For the study area, tracking hundreds of individual subcatchments and routing them to the observed outlet would be prohibitively time consuming and expensive. This method was determined to be too costly to develop and manage for the study area.

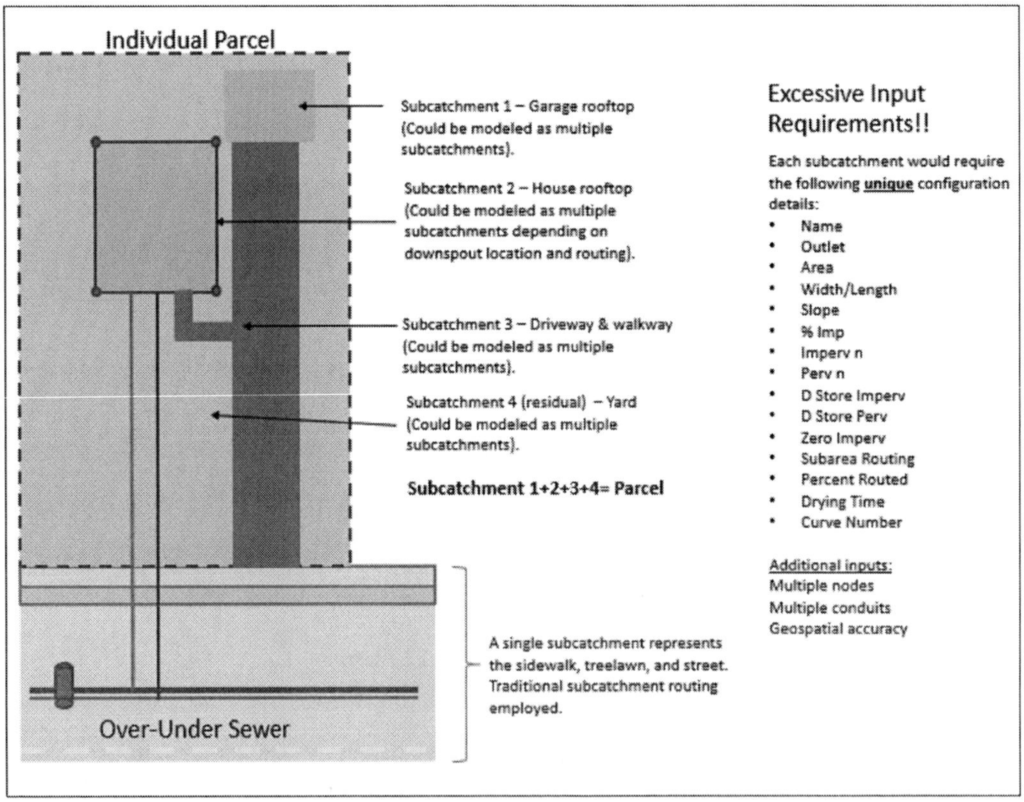

Figure 7: Modeling Method 1

Modeling Method 2

In method 2, each parcel is segregated into LIDs representing the features of the parcel. The LID tools in SWMM have controls that overlay the typical subcatchment with their own values.

This method considers every parcel a pervious lawn (*Imperv* % = 0) overlaid by various LIDs representing impervious features. Each parcel has at least 3 impervious LIDs representing: (1) the house roof; (2) the garage roof; and (3) the driveway plus any impervious walkways. This allows the LIDs to represent existing conditions and routing as well as the results from work to reduce I/I. For example, an LID can be used to represent a roof connected to a sanitary lateral via a cross connected downspout, and the same LID can represent a downspout with a surface discharge to the lawn merely by changing the values of the LID. Controlling where an LID discharges replicates what portion of a feature may directly drain to another feature such as a lawn, a driveway, a storm sewer, or a sanitary sewer (Figure 8).

An added benefit is that calibration of the model is simpler since a LID type is assigned to each feature on the parcel initially and then calibrated by adjustments to the LID values rather than redefining each feature.

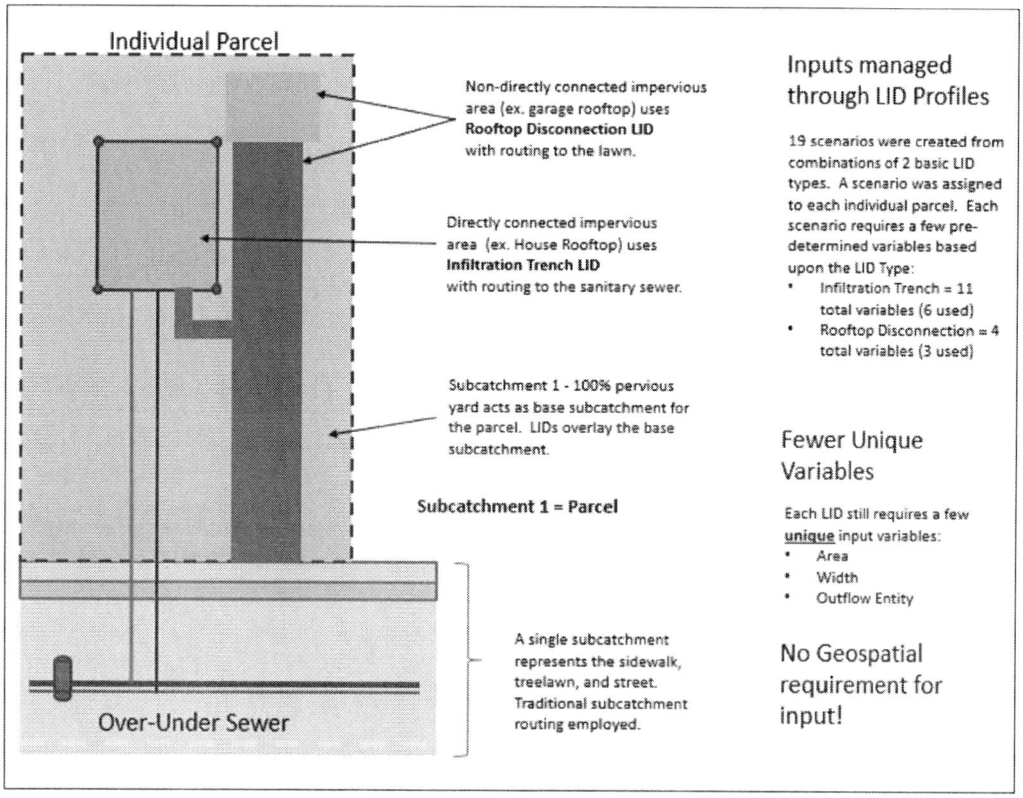

Figure 8: Modeling Method 2

Model Development

Two of the 8 LIDs available in SWMM were chosen to represent the features on each lot:
- Infiltration Trench
- Rooftop Disconnection

These LIDs were selected not for their nomenclature but for the variables associated with them and the ability to modify those variables to replicate a feature of the parcel. For example, using an "Infiltration Trench" LID does not mean that there are infiltration trenches on the parcel, but it represents the response associated with foundation drains and sanitary laterals that are in poor condition. The Infiltration Trench LID variables can be adjusted such that their effect on the system is the same as foundation drains and leaking sanitary sewer laterals.

Field investigations determined that each of the 102 parcels could be assigned a scenario that represented the features and routing of the parcel. Multiple parcels that have common features and routing could be assigned the same scenario. Other parcels could be assigned another scenario representing other combinations. Table 1 identifies the profiles which reflect the routing of rooftops.

Table 1: Rooftop LID Profiles

Name	LID Type	Routed To
1/2ConnectedGarageRoof	Infiltration Trench	Sanitary Manhole
ConnectedGarageRoof	Infiltration Trench	Sanitary Manhole
1/2ConnectedHouseRoof	Infiltration Trench	Sanitary Manhole
3/4ConnectedHouseRoof	Infiltration Trench	Sanitary Manhole
ConnectedHouseRoof	Infiltration Trench	Sanitary Manhole
1/2DisconnectedGarageRoof	Rooftop Disconnection	Pervious Yard
DisconnectedGarageRoof	Rooftop Disconnection	Pervious Yard
1/4DisconnectedHouseRoof	Rooftop Disconnection	Pervious Yard
1/2DisconnectedHouseRoof	Rooftop Disconnection	Pervious Yard
DisconnectedHouseRoof	Rooftop Disconnection	Pervious Yard

The term "Connected" identifies any feature cross connected to a sanitary lateral. The term "Disconnected" refers to any feature that discharges to a pervious surface (lawn or garden) or storm lateral. Figure 9 illustrates both types of downspout conditions.

Figure 9: Downspout Configurations

Driveways with area drains connected to the sanitary lateral assumed that only the rear portion of the driveway (between the house and the garage) was draining to an inlet and the balance of the driveway drained toward the street. Table 2 identifies profiles which reflect the routing of driveways and walkways.

Table 2: Pavement LID Profiles

Name	LID Type	Routed To
1/2ConnectedDrive	Infiltration Trench	Sanitary Manhole
1/2DisconnectedDrive	Rooftop Disconnection	Storm Manhole
DisconnectedDrive	Rooftop Disconnection	Storm Manhole

The Infiltration Trench LID has 11 adjustable parameters of which 6 were used in the model. The Rooftop Disconnection LID has 4 adjustable parameters, of which 3 were used (Table 3).

Table 3: LID Parameters Used In The Model

Name	Infiltration Trench	Rooftop Disconnection
Berm height	×	
Vegetative cover	×	
Surface roughness	✓	
Surface slope	✓	
Storage thickness (height)	✓	
Storage void ratio	✓	
Seepage rate	×	
Storage clogging factor	×	
Drain coefficient	✓	
Drain exponent	✓	
Drain offset height	×	
Storage depth		×
Surface roughness		✓
Surface slope		✓
Flow capacity		✓

Each parcel in the study area was assigned a combination of LID profiles that created a scenario representing the parcel. Overall, 19 scenarios were needed to represent the combination of features and routing on the 102 parcels. Table 4 is a summary of the scenarios created for the model. Five scenarios represented over 70% of the properties. One of the five scenarios (Scenario 7) represented about 30% of the parcels in the project area and was modeled using LID profiles for 3 features; a connected house roof, a disconnected garage, and disconnected drives.

Table 4: Scenario Development

⇩LID Profiles/Scenario ⇨	1	2	3	4	5	6	7	8	9	10	11	12	13	14	15	16	17	18	19
ConnectedHouseRoof	X	X					X		X			X			X			X	
3/4ConnectedHouseRoof			X	X		X							X						
1/2ConnectedHouseRoof					X			X						X			X		
ConnectedGarageRoof		X				X			X						X				X
1/2ConnectedGarageRoof	X		X		X								X	X		X	X	X	
1/2ConnectedDrive	X	X	X		X	X			X		X	X				X	X		X
DisconnectedHouseRoof										X	X					X			X
1/2DisconnectedHouseRoof					X			X						X			X		
1/4DisconnectedHouseRoof			X	X		X							X						
DisconnectedGarageRoof				X			X			X	X	X							
1/2DisconnectedGarageRoof	X		X		X								X	X		X	X	X	
1/2DisconnectedDrive	X		X		X	X			X		X	X				X			X
DisconnectedDrive		X		X			X	X		X			X	X	X			X	

MODEL CALIBRATION

The flow metering showed that there was significant wet weather response in the project indicating I/I in the sanitary sewers. The hydrograph of the wet weather flow was typical of systems that had inflow sources as well as infiltration sources. Dye testing performed in the project area indicated that most downspouts appeared to have some sort of connection to the sanitary sewers (direct or indirect). Most foundation drains were assumed to be connected to the sanitary sewers although dye testing was mostly inconclusive.

During the televising investigation of the sanitary laterals, it was determined that laterals were generally in disrepair with evidence of settling, root intrusion, and debris. It was assumed that the adjacent storm laterals were in similar or worse condition than the sanitary laterals. It was also assumed that debris, such as leaves and sediment, had built up in the storm lateral over time. If enough debris builds up in the storm lateral then capacity is lost. As capacity is lost, the private property storm lateral collections system begins to store wet weather volumes that can migrate to the sanitary lateral and/or the cross connected foundation drains.

Impervious surfaces within a parcel were initially modeled as a Rooftop Disconnection LIDs routed to the sanitary sewer. A limit of 500 mm/hr (20 in/hr) was set for the rooftop disconnection LID flow capacity parameter. This assured that all rainfall to the rooftop was routed to the sanitary lateral, rather than spilling onto the pervious area.

For the purpose of comparison, RTK unit hydrograph components are used as reference to describe the short, medium, and long term components of a LID's response. The aggregate response at the flow meter from all 102 parcels modeled in this way showed an RTK hydrograph with too high of a peak flow (R_1) and with a falling leg that was too steep (K_1) (Figure 10).

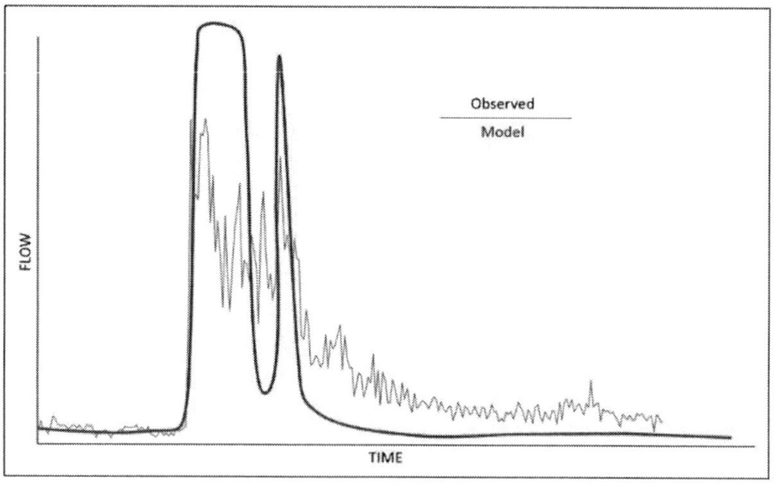

Figure 10: RTK Hydrograph

Analyzing the existing metered hydrograph, it appeared that there was some storage in the sanitary sewers within each parcel. A hypothesis was developed that many of the connected impervious surfaces had inherent storage within conveyances such as clogged gutters, clogged downspouts, or blockage in the laterals. The Rooftop Disconnection LID does not allow sufficient storage release control. Therefore, the approach was changed to modeling of the connected impervious surfaces within the parcels as an Infiltration Trench LID.

Using primarily the Infiltration Trench *Drain Coefficient*, *Drain Exponent* and the *Storage Thickness* (height) parameters, the values were adjusted to develop a model hydrograph that resembled the metered hydrograph. Adjusting the *Drain Coefficient* allowed control of the outflow such that smaller events passed through the LID while larger events were limited and began to store runoff volume in the LID. This shaped the short term hydrograph for the peak flow, time to peak, and falling leg. The *Drain Exponent* determines the shape of the hydrograph recession.

Adjusting the *Storage Thickness* controlled how long the LID would drain for after a rain event. It became apparent that during larger or back to back events, the R_2 of the storm responses became similar and did not become larger with increased rainfall amounts. Therefore it was understood that the inherent storage volume had a limit. This volume was controlled by the *Storage Thickness* parameter. By limiting the LID *Storage Thickness*, the excess volume in the LID overflows and is routed to the pervious area of the parcel. This shaped the medium term hydrograph to resemble the sanitary sewer hydrograph (Figure 11). Storing too much volume in the LID allowed more volume to drain for a longer period than in the metered hydrograph. *Storage Thickness* calibration resulted in a hydrograph that was a much closer match to the metered hydrograph.

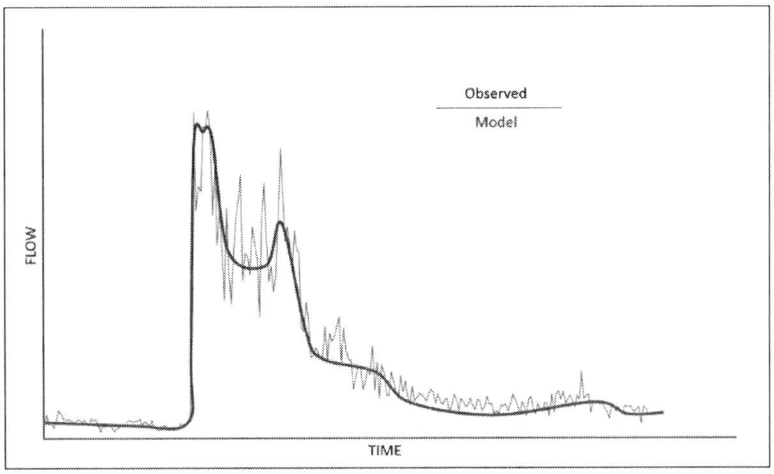

Figure 11: Hydrograph With Rooftop Disconnection & Infiltration Trench LIDs

Further adjustment to match the metered hydrograph required inclusion of an RTK long term response (R_3, T_3, and K_3) for saturated ground infiltration to the sanitary sewer and the storm sewer (Figure 12).

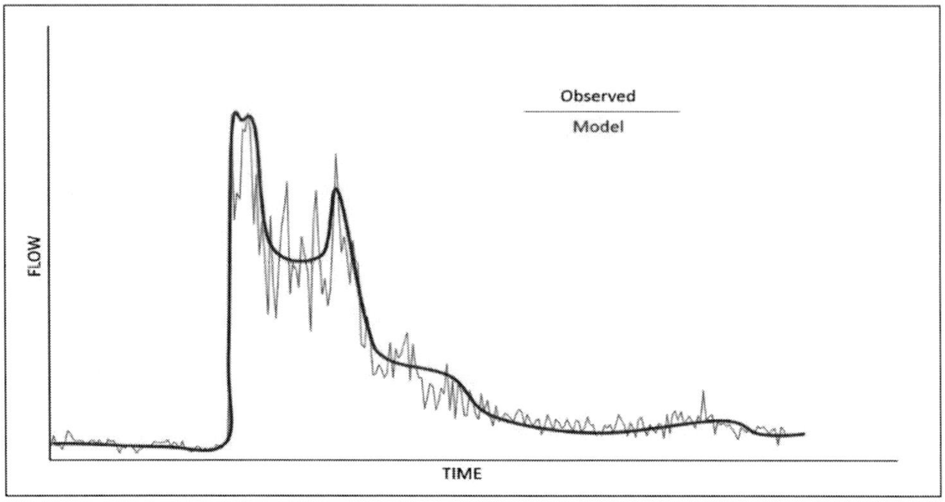

Figure 12: Hydrograph Illustrating Rooftop Disconnection & Infiltration Trench LIDS + $R_3T_3K_3$

These adjustments created a modeled hydrograph that closely matched the metered hydrograph. However, the adjustments were based on theory and not actual observations. To validate the model, evidence was needed from the field that there were conditions that created storage on the parcels. During property corrections undertaken by the City, it was found that almost all of the downspouts were connected to the storm lateral, not the sanitary lateral as was assumed. It was also found that the storm lateral was constructed with a "P" trap at the right of way line and that the majority of traps were clogged with sediment and debris (Figure 13).

The clogged traps backed up storm flow and created storage around the foundation that slowly migrated and seeped into defects in the sanitary lateral. Interviews with residents also confirmed that in major storm events, the crock receiving the downspouts would back up and overflow onto the ground. This overflow would find its way to the sanitary sewer by seepage along the foundation; overland flow across the lawn with seepage into the sanitary sewer trench; and overland flow to the storm sewer in the street where it could seep into the sanitary sewer below it. Seepage was found to be rapid due to lenses that have developed over time in the clay soils from the migration of flow from the storm lateral to the sanitary lateral.

Figure 13: Trap clogged with sediment

MODELING I/I REMOVAL

A residual benefit of calibrating with LIDs was the ease of making model adjustments to reflect the effects of construction in ideal proposed conditions and actual post construction conditions.

In ideal proposed conditions, all homes associated with the 102-home pilot project would be corrected, and removal of I/I would have been 100% successful at each property. To reflect this in the model, any rooftop or driveway LIDs which were originally connected to the sanitary system would be rerouted to the storm system, and the RTK would be reduced to small values reflective of a tight system. The modeler has the luxury of splashing the downspouts (i.e. routing the rooftop LIDs to the pervious area), or connecting downspouts to the storm lateral (i.e. routing the rooftop LIDs to the storm system). This accurately portrays the relining of the sanitary lateral and removal of illicit connections. Additionally, the flow capacity could be increased for the rooftop LIDs modeled as Infiltration Trenches to represent the removal of the clogged "P" traps.

Actual post construction conditions will not likely achieve the same results as the ideal proposed (100% correction & success) conditions. Some reasons for this could be:

- Inability to get 100% participation in program (if it is not mandatory), leaving some homes unaddressed.
- Inability or failure to correct the entire property, leaving some downspouts or drains connected to the sanitary lateral.

- Partial improvement/partial completion success, meaning that the targeted results are achieved prior to completion of the planned project and no additional construction is necessary.

The model can be easily adjusted to reflect actual post construction conditions. If a parcel was not addressed, that particular parcel's LIDs could simply remain unchanged. Likewise, if correction was found to be no longer needed, those parcels' LIDs could remain unchanged.

For parcels where private property drainage and connections were re-piped, LIDs could be completely rerouted and discharged to the storm system. For parcels where private property drainage and connections were partially re-piped, the area of the rooftop for any particular LID could be altered based upon contractor observation. For example, if only the front downspouts which collect the front half of the roof were re-piped, the rooftop LID area could be reduced by half. This conceptually leaves the back half of the roof as contributor to the existing Infiltration Trench LID while the front re-piped half could be assigned to a new LID which represents the front half of the roof being routed to the storm or to the pervious area (representing splashed downspouts). These adjustments noted can be done using a whole scale approach so as to perform a sensitivity study, or one parcel at a time to best replicate known conditions.

It is expected that some level of infiltration exists in all systems. The likelihood of completely removing infiltration is low, and therefore some R_3 component should remain in the model loaded to the sanitary sewer. While a placeholder value of the modeler's choice can be used to complete the model, it is recommended that post construction data be analyzed to refine the R_3 which remains after construction.

Once an understanding is formed regarding the different hydrograph components, the model could also be used in a predictive sense to determine where I/I is coming from by associating lessons learned regarding hydrograph components in completed project areas to newly obtained flow data. For example, a very peaked response may be indicative of downspouts which are connected to the sanitary. Alternatively, if a long drawn out tail exists, lateral relining may be necessary. Lastly, if a bench like response is witnessed on larger and back to back events, it is possible that some private property drainage system storage is present in the project area.

CONCLUSIONS

Utilization of LIDs for existing conditions modeling provides options not available when utilizing a traditional subcatchment. In order to provide similar routing, control, and storage options, a single parcel subcatchment would need to be subdivided into multiple subcatchments with connection nodes, conduits, storage nodes, and control pipes. It is estimated that a subdivided parcel would require an additional 6 to 9 of each subcatchment, each node, and each conduit along with 1 storage node to control pipe outflow. This creates a large model that becomes cumbersome and expensive to calibrate to existing conditions and to use for analysis of sewer system improvements. LIDs provide efficient options due to their ability to route flows to nodes outside the subcatchment, other subcatchments, and to the subcatchment's pervious area. The LID options exceed a subcatchment's ability to drain portions of runoff between pervious and impervious areas before draining to its outlet node.

The process of simulating sources of I/I using LIDs was very effective in matching the actual metering data for the project area. The corrections instituted during the city's sewer system rehabilitation program were easy to apply in the model and provided a quantification of the impact of the program on I/I.

The private property corrections will be expanded to other parts of the city and the development of this LID modeling technique will allow the city to project the I/I removal for future projects.

Managing the Useful Life of Your Segmented Block Combined Sewer

Dianne M Sumego, P.E. Black & Veatch

ABSTRACT

Following two road collapses, as a result of deterioration within the City's segmented block combined sewer system, the City needed to understand the condition, useful life and potential risk of their combined sewer system. Black & Veatch was retained by the City to perform a detailed investigation and condition assessment of 40,000 lineal feet of their segmented block combined sewer system. The goal of this study was to understand the condition and associated risk to the city in this identified area.

KEYWORDS

Planning, Condition Assessment, Risk Analysis

INTRODUCTION

One of the biggest challenges our industry is facing is finding resources, both financially and staffing, to maintain our underground infrastructure. Across the globe these combined sewer infrastructure systems were installed 80 plus years ago and have already surpassed their useful life. However, what defines useful life and what are the factors impacting the determination of useful life? Each sewer system varies and with that factors impacting useful like include soil conditions, loading, age and make of pipe material, composition of wastewater, and operation and maintenance to name a few. The challenge, as mentioned above, is having the available resources to assess and maintain this critical infrastructure on a regular basis.

The City of Lima's (City) segmented block sewers are 70 years or older and like most communities have had minimal inspection and cleaning over the years. To add to the challenge, over the past several years the City has experienced a couple of collapsed sewers within their segmented block sewer system with the most recent collapse occurring in 2012, requiring an emergency repair to the sewer. Both collapse ended up with cars in the hole with passengers. Fortunately no one was hurt. These occurrences elevated the need for the city to understand not only the condition but what changes in their O&M should take place to extend the useful life of their segmented block combined sewer system.

BACKGROUND

In 2014 Black & Veatch in cooperation with Terra Contracting Services undertook the task of inspecting approximately 40,000 feet of mainline segmented block sewers ranging in size from 24-inch to 102-inches and laterals. The inspection areas were identified by the City in maps provided to Black & Veatch and included only segmented block sewers. These segmented block sewers were divided into thirteen (13) inspection areas for planning of the work and are shown on Figure 1.

Figure 1: Inspection Areas

These combined sewers react quickly to storm conditions and therefore fill quickly once a storm event occurs. The impact time is minimal and during storm events portions of this combined sewer are surcharged creating internally forces on the sewer system.

Segmented block sewers were used for construction during the transition from brick to pre-cast pipe and were made of terra cotta and approximately 18" x 9" in size. Segmented block sewers are double walled with the exterior wall serving as the structural support for the sewer. These tiles are staggered and interlocked and failure typical occurs at the spring line of the pipe. When failure occurs in a segmented block sewer, typically it is not one tile that fails but many tiles that fall allowing the external wall to be exposed. These older systems have lateral connections and many were "cut in" during construction. This construction methodology weakens these points within the system and often creates points of failure. When performing a condition assessment of these sewers the NASSCo coding system was applied but some interpretation was required due to the type of materials used.

INSPECTION OF COMBINED SEWER AND LATERALS

The inspection was conducted by technicians certified by the National Association of Sewer Service Companies (NASSCO). The technicians used the NASSCO standard rating of structural defects and operation and maintenance defects using the Pipeline Assessment Certification Program (PACP). The code for PACP identifies defects in two categories, structural and operation and maintenance (O&M). The structural defects include cracks, broken pipe, surface damage and holes and are identified with an "S" followed by a rating score. The O&M defects include deposits, infiltration/inflow, roots, obstacles and vermin and are identified with an "M" followed by a rating score. The rating scores are based on a scale of 1 to 5 and are summarized in Table 1. The term failure is considered to mean the pipe is no longer able to furnish the expected level of service. These guidelines are general in nature and give the technician leeway to interpret the results based on their experience and provide a basis for consistency within the inspection.

Table 1: Overview of NASSCO Grading

Pipeline Grade	Classification	General Description	Guideline Relating to Failure
5	Immediate Attention	Defects requiring immediate attention	Has failed or will likely fail within 5 years
4	Poor	Severe defects that will become grade 5 within the foreseeable future	Pipe will probably fail in 5 to 10 years
3	Fair	Moderate defects that will continue to deteriorate	Pipe may fail in 10 to 20 years
2	Good	Defects that have not begun to deteriorate	Pipe unlikely to fail for at least 20 years
1	Excellent	Minor defects	Failure unlikely in the foreseeable future

INSPECTION METHODOLOGY AND PHOTOS

Pipelines less than 72-inches were inspected with CCTV mounted on a robotic platform, shown in Photo 1. This is an IBAK CCTV unit with an Orpheus Camera and 1,000 feet of cable. Pipe diameters 72-inch and larger were inspected with an IBak Man-Cam system.

The laterals on pipe with a diameter less than 42-inches were inspected by a side launch CCTV unit and laterals 42-inch and larger were inspected by manned entry into the pipeline and pictures taken of the connection and inside the lateral as described above. Photos 2 and 3 below are representative of condition observed in the mainline sewer and Photos 4 and 5 are those of the lateral connections.

Photo 1: IBAK Man -CAM

Photo 2: Missing Tiles at Spring Line of Main Line Sewer

Photo 3: Missing Tiles at Spring Line of Main Line Sewer

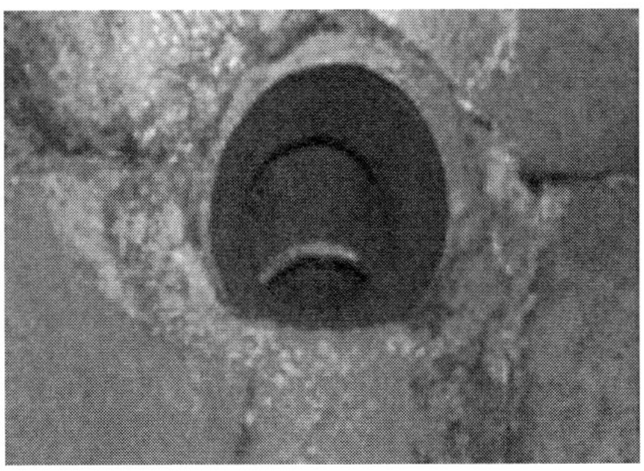

Photo 4: Collapsed Lateral Connection

Photo 5: Collapsed Lateral Connection

RECOMMENDATON FOR FUTURE PREVENTATIVE OPERATION AND MAINTENANCE

Based on our physical inspection and review of the results of the CCTV inspection, we can see a correlation between the structural deficiencies and the build-up of debris in the segments and "break in" installation of taps for the laterals. The pipe is typically failing along the spring line of the pipe and in many cases is initiated at the "break in" taps for the laterals. As the pipe wall fails large pieces of segmented block become obstacles that collect debris creating a "dam" affect in the flow.

Due to the age of the pipe and the amount of debris build up in the segmented block sewer system Black & Veatch recommended a rigorous cleaning and inspection schedule to be performed over the next three years to track any changes to the infrastructure and to lengthen the useful life through preventative maintenance.

RISK ASSESSMENT EVALUATIONS

Risk-based capital spending prioritization is one of the core components of good asset management practice. By evaluating the consequence and likelihood of failure for each pipe section, limited available capital and operational budgets can be directed to those projects and activities that will give the greatest reduction in risk for each dollar spent.

<u>Consequence of Failure (CoF):</u> The failure of large diameter sewers like those inspected as part of this project can have serious negative impacts on customers and lead to expensive emergency operations and reputational damage for Cities. The relative severity of these consequences is an important variable considered when prioritizing capital spending and can be estimated using pipe attributes, operational information, and geographic information.

<u>Likelihood of Failure (LoF):</u> This analysis examines the likelihood of two modes of failure for inspected pipe segments. Structural pipe failures may result in pipe collapses which can, in turn, result in subsidence, backups, and overflows. Structural failures are typically caused by deterioration of the pipe wall material, changing environmental stresses, and construction strikes. Operational pipe failures are defined as blockages that reduce pipe capacity cause conveyance system problems like surcharging, overflows, and basement backups. These blockages are caused by debris or fats, roots, and grease (FOG) deposits.

RESULTS

After the scores are calculated, they are plotted on the two-dimensional risk matrix below with likelihood of failure represented on one axis and consequence of failure represented on the other.

Structural: Structural failures can range from broken blocks to a collapse of a road. Based on the below assessments approximately 8.4% of the City's segmented block sewers inspected as a part of this study are in need of immediate attention. A quick overview of the results based on 35,946 LF inspected is summarized as follows:

Immediate (Red)	- 8.4% of segmented block sewers
High Risk (Orange)	- 5.1% of segmented block sewers
Medium Risk (Yellow)	- 14.2% of segmented block sewers
Low Risk/Maintenance (Green)	- 69.1% of segmented block sewers
Not Inspected (Due to Issues)	- 3.2% of segmented block sewers
	100% of segmented block sewers

Tables 2 and 3 show the breakdown of the Structural Risk Profile and Risk Matrix:

Table 2: Structural Failure Risk Profile - Segments

Likelihood of Failure					
5	0	0	0	5	2
4	0	0	0	3	0
3	0	0	0	4	0
2	0	0	0	8	2
1	0	0	0	77	8
Consequence of Failure	1	2	3	4	5

Table 3: Structural Failure Risk Matrix

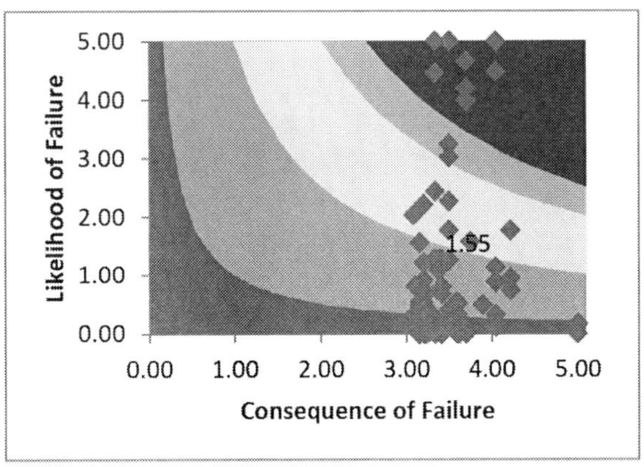

O&M: O&M failures can range from debris build up to FOG causing a reduction of flow through the pipe with the potential to cause backups and potential overflows. Based on the below assessments approximately 9.9% of the City's segmented block sewers inspected as part of this report are in need of immediate attention. Some of these sewers are also identified as immediate for structural at risk. A quick over of the results based on the 34,804 LF inspected:

Immediate (Red)	- 9.9 % of segmented block sewers
High Risk (Orange)	- 4.5 % of segmented block sewers
Medium Risk (Yellow)	- 59.3 % of segmented block sewers
Low Risk/Maintenance (Green)	- 23.1 % of segmented block sewers
Not Inspected (Due to Issues)	- 3.2 % of segmented block sewers
	100 % of segmented block sewers

Tables 4, 5 and 6 show a breakdown of the O&M Risk Profile and Risk Matrix:

Table 4: O&M Failure Risk Profile - Length of Pipe

Likelihood of Failure					
5	0	0	0	2,788	0
4	0	0	0	1,605	782
3	0	0	0	12,714	0
2	0	0	0	7,187	1,431
1	0	0	0	7,188	1,109
	1	2	3	4	5

Consequence of Failure

Table 5: O&M Failure Risk Profile - Segments

Likelihood of Failure					
5	0	0	0	5	0
4	0	0	0	5	2
3	0	0	0	29	0
2	0	0	0	24	5
1	0	0	0	34	5
-1	1	2	3	4	5

Consequence of Failure

Table 6: O&M Failure Risk Matrix

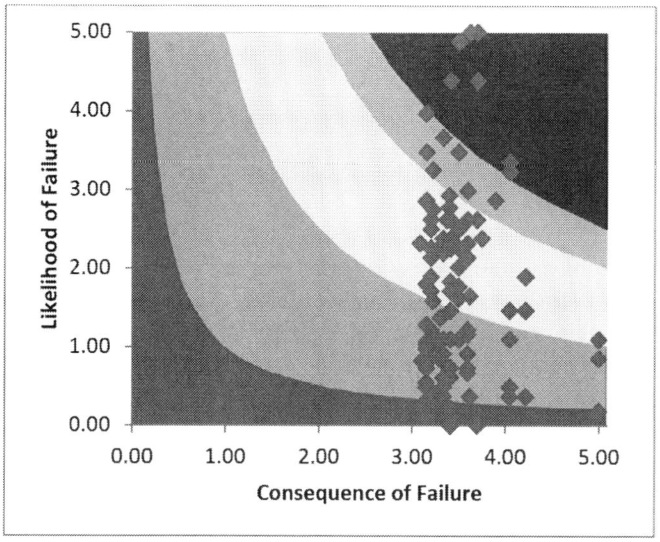

PRIORITIZATION AND RECOMMENDATIONS

Of the sewers evaluated, 3,035-feet received the highest ranking of combined structural risk and consequence of failure, which should be addressed immediately. The second highest priority sewers are considered high risk and are comprised of 1,817-feet, followed by 5,089-feet of medium risk, and nearly 25,000-feet of low risk sewers.

Table 7: Structural Failure Risk Profile – Length of Pipe

Likelihood of Failure	1	2	3	4	5
5	0	0	0	2,303	732
4	0	0	0	1,817	0
3	0	0	0	1,252	0
2	0	0	0	2,985	852
1	0	0	0	23,125	1,738

Consequence of Failure

Recommended projects were developed by referencing the structural likelihood of failure ranking, consequence of failure, and resulting Business Risk Exposure (BRE) for each segment. However project limits cannot be defined using only a numerical ranking system. Detailed inspection reports compliment the numerical rankings of each segment, along with rating characteristics of surrounding sewers. Project limits and project priority were also developed considering total length, pipe diameter, cost, and construction conditions, such as minimizing the number of mobilizations required for each project.

How to Prevent Your Wastewater CIP from Bing Flushed – The Intersection of Master Planning and Business Planning

Richard Weatherly, PE, Freese and Nichols, Inc.
Kendall Ryan, PE, Freese and Nichols, Inc.
Carol Reed, City of Huntsville, TX

ABSTRACT

The City of Huntsville, Texas (City) owns and maintains an aging wastewater system dating back to the 1930's that includes three wastewater treatment plants and 26 lift stations, many of which are nearing the end of their design life. The aging collection system also contains a large percentage of clay tile gravity sewer lines and experiences high levels of inflow and infiltration. The Freese and Nichols, Inc. (FNI) project team and City staff evaluated the condition and capacity of the wastewater system through a holistic approach to a wastewater master plan. This project incorporated frequent communication with City staff at all levels: operations, engineering, public works, and management. The healthy, frequent communication and multiple analysis methods utilized on this project helped to build a convincing business case for the wastewater system CIP and the expenditure of the City's dollars. The audience will walk away with ideas about how to intersect master planning with business planning to develop a defendable and implementable wastewater capital improvements plan.

KEYWORDS

Master planning, business case, capital improvements plan, modeling, asset management, risk based assessment, lift stations.

INTRODUCTION

The City of Huntsville (City) is a growing community located in Walker County, Texas. The City currently provides water and wastewater service to approximately 40,000 people, including five Texas Department of Criminal Justice (TDCJ) units and Sam Houston State University (SHSU). The population within the wastewater service area is projected to grow from 39,894 to 54,949 by 2041. Additionally, the planned expansion of a major interstate highway will increase connectivity to the fourth largest city in the United States: Houston, Texas. The City's aging wastewater system dates back to the 1930's, and includes three wastewater treatment plants and 26 lift stations, many of which are nearing the end of their design life. The aging collection system also contains a large percentage of clay tile gravity sewer lines and experiences high levels of inflow and infiltration. The City's existing wastewater collection system is shown on **Figure 1**.

Figure 1: Existing Wastewater System

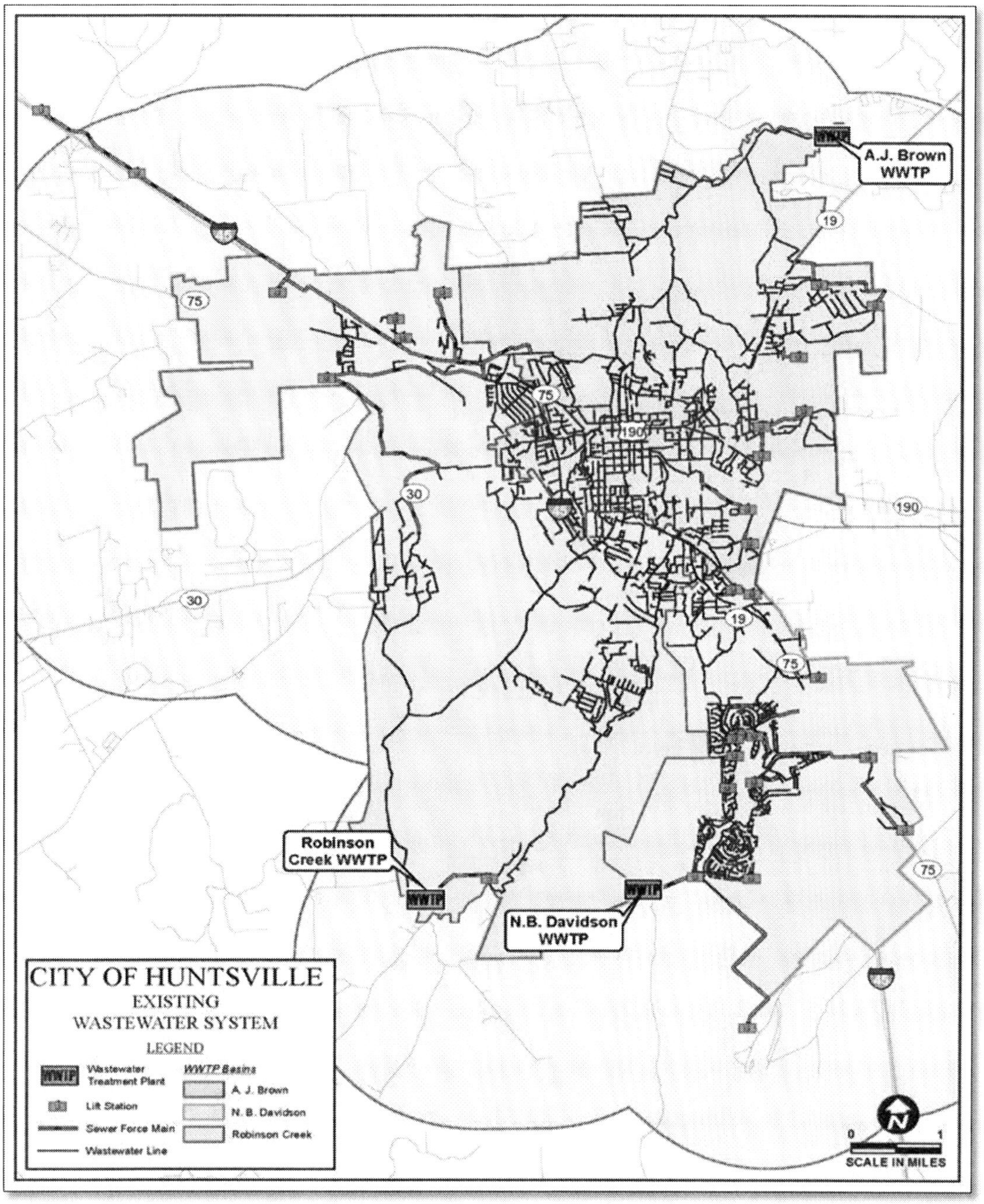

The Freese and Nichols, Inc. (FNI) project team and City staff evaluated the condition and capacity of the wastewater system through a holistic approach to a wastewater master plan. Multiple methods were utilized that built a convincing business case for capital improvements plan (CIP) recommendations to be presented to City management. These included:

- A thorough risk based assessment approach to develop facility recommendations
- Dynamic hydraulic modeling based on flow monitoring to develop capacity based collection system improvements
- A Pilot Sanitary Sewer Evaluation Survey (SSES) carried out as part of the overall master plan and the development of subsequent SSES and rehabilitation recommendations

A hallmark of this project was the frequent communication between the FNI project team and City staff at all levels: operations, engineering, public works, and management. The FNI project manager and engineering staff held weekly conference calls with City staff to discuss the project status and direction and to confirm that everyone was on the same page.

RISK BASED ASSESSMENT OF LIFT STATIONS

A risk based assessment was performed on all of the City's lift stations. The risk based assessment considered condition and criticality components to evaluate the risk of failure of each lift station. The results of this assessment were used to develop prioritized projects for the rehabilitation and capacity capital improvement plans. The condition assessment included a visual inspection of each lift station and its components. The criticality assessment, or consequence of failure, included an analysis of the proximity of each lift station to critical or environmentally sensitive areas, as well as the residential population served. Each lift station was assigned overall condition and criticality scores based on the results of the assessments. These scores were used to assign a risk category (high, moderate, or low) to each asset. Lift station rehabilitation projects were developed based on the results of the risk based assessment and included in the wastewater CIP.

Lift Station Condition Assessment

Site visits for the 30 lift stations were performed on June 9 and 10, 2015. The condition inspections were conducted by a team of design, electrical and modeling engineers from FNI, as well as wastewater operations staff from the City. FNI developed a list of electrical, structural, mechanical, and site components to be inspected at each lift station site. A condition weighting factor was assigned to each component group based on the relative importance of the component to the overall function of the lift station. Major components in each of these categories were evaluated separately. **Table 1** illustrates the condition component groups, parameters, and weighting for the lift station facilities.

Table 1: Condition Assessment Component Groups and Weightings

Component Group	Weight Factor
Electrical – MCC, Back-up Power, Cables	20%
Instrumentation – SCADA & alarms	15%
Pumps and Motors	20%
Structure - Hatches, Corrosion, Cracks, Leaking, Ventilation, Odor Control	20%
Piping and Valves	15%
Site - Drainage, Access Drive, Security, Fencing	10%
Total Weighting	**100%**

FNI assigned numerical scores ranging from 1 to 5 to each component group based on the physical condition as seen during the inspection and information provided by the City staff relating to its operational performance. **Table 2** shows the guidelines used during assignment of numerical scores for component groups.

Table 2: Guidelines for Condition Scores

Condition Rating	Scoring Guidelines
1	**Very good** condition; no improvements recommended to maintain function
2	**Good** condition; minor improvements recommended to maintain function
3	**Fair** condition; improvements recommended to improve performance or efficiency
4	**Poor** condition; improvements recommended to maintain reliability
5	**Very Poor** Condition; rehabilitation or replacement required

In conjunction with the condition assessment site visits, FNI reviewed approximately two years (March 2013 to December 2014) of lift station work order history provided by the City. This information was screened for maintenance issues including instances of de-ragging, cleaning of return lines, failures of the SCADA system, issues with transducers, and operational issues with pumps or generators. It was found that the average number of work orders per lift station was 4.9. Based on this data, FNI developed the work order impact criteria shown in **Table 3**. The inspected condition score of lift stations with greater than five work orders was increased by 0.25 points. Eleven of the 30 lift stations were penalized as such to reflect above average maintenance history. The work order impact criteria was developed to have a performance component influence on the inspected condition scores since the assessments were visual, and no equipment testing was performed.

Table 3: Work Order Impact Scores

No. of Work Orders	Impact to Condition Score
≥ 5	+ 0.25
Less than 5	0

Once the visual site inspections and work order history analysis were completed, ranges were assigned for the condition scores, and categories were designated from very good to very poor as shown in **Table 4**. Final condition scores for each lift station are shown in **Table 5**.

Table 4: Condition Score Ranges

Condition Rating	Min	Max
Very Good	0.00	1.50
Good	1.51	2.25
Fair	2.26	3.00
Poor	3.01	3.50
Very Poor	3.51	5.00

Table 5: Lift Station Condition Score Summary

Lift Station	Condition Score	Rating
Airport	2.75	Fair
Bayes	2.40	Fair
Bearkat Blvd	2.75	Fair
Brook Hollow	3.10	Poor
Elkins Lake Dam	3.55	Very Poor
Elkins Lake #1	3.50	Poor
Elkins Lake #2	3.35	Poor
Elkins Lake #3	2.65	Fair
Elkins Lake Post Office	1.55	Good
Elkins Lake Equestrian Center	3.35	Poor
Hitchin' Post	3.50	Poor
Highway 190	2.65	Fair
Mallery Lake	2.35	Fair
McCoy's	2.70	Fair
McGary Creek	3.00	Fair
Old Colony	3.75	Very Poor
Simmons Street	2.35	Fair
Southwood Drive	2.10	Good
Tanyard Creek	2.25	Good
TDCJ BOT	3.70	Very Poor
Waters Edge	1.90	Good
Park Road 40	2.25	Good
McDonald Creek	2.20	Good
Transfer Station	2.60	Fair
Sims	2.85	Fair
Sterlingbrook	1.75	Good
TXDOT #1 (Rest Area 45)	1.90	Good
TXDOT #2 (Rest Area 1696)	3.60	Very Poor
Badger Lane	1.55	Good
Huntsville State Park	2.25	Good

Lift Station Criticality Assessment
A criticality assessment was performed for each lift station based on three categories:

- Proximity to High Impact Areas
- Population Served
- Proximity to Environmentally Sensitive Areas

Table 6 shows the scoring parameters used and the weighting factors assigned to each lift station criticality category.

Table 6: Condition Assessment Component Groups and Weightings

CRITICALITY CATEGORIES & WEIGHTING SYSTEM	
Proximity to High Impact Areas (25%)	**Population Served (25%)**
≤ 1000 ft from school or university = 5 ≤ 1000 ft from golf course = 4 ≤ 1000 ft from commercial structure = 3 ≤ 2,000 ft from any structure = 2 >2,000 ft from any structure = 0	> 2,000 = 5 1,001 – 2,000 = 4 501 – 1,000 = 3 251 – 500 = 2 ≤ 250 = 1
Proximity to Environmentally Sensitive Areas (50%)	
> 75 ft from any water body, or inside Sam Houston National Forest = 5 76 – 150 ft from any water body, or < 500 ft from Sam Houston National Forest = 4 151 – 300 ft from any water body, or < 1,500 ft from Sam Houston National Forest = 3 301 – 600 ft from any water body = 2 > 600 ft from any water body, or > 1,500 from Sam Houston National Forest = 0	

GIS tools were utilized to determine the distance from each lift station to the high impact areas and environmentally sensitive areas. Geocoded water meters were used to determine the existing population within each lift station service area.

Once the proximity and population analyses were completed, ranges were assigned for the criticality scores, and categories were designated from very low impact to very high impact. **Table 7** shows the scores associated with the rating categories of the criticality assessment. Final criticality scores for each lift station are shown in **Table 8**.

Table 7: Criticality Score Ranges

Criticality Rating	Min	Max
Very Low Impact	0.00	1.00
Low Impact	1.01	3.00
Moderate Impact	3.01	3.49
High Impact	3.5	3.99
Very High Impact	4.00	5.00

Table 8: Lift Station Criticality Score Summary

Lift Station	Criticality Score	Rating
Airport	0.25	Very Low
Bayes	4.00	Very High
Bearkat Blvd	4.25	Very High
Brook Hollow	3.75	High
Elkins Lake Dam	4.00	Very High
Elkins Lake #1	3.25	Moderate
Elkins Lake #2	2.75	Low
Elkins Lake #3	3.75	High
Elkins Lake Post Office	3.75	High
Elkins Lake Equestrian Center	3.50	High
Hitchin' Post	1.00	Very Low
Highway 190	4.00	Very High
Mallory Lake	4.00	Very High
McCoy's	2.50	Low
McGary Creek	3.25	Moderate
Old Colony	4.00	Very High
Simmons Street	3.50	High
Southwood Drive	0.75	Very Low
Tanyard Creek	3.50	High
TDCJ BOT	0.75	Very Low
Waters Edge	4.00	Very High
Park Road 40	3.25	Moderate
McDonald Creek	3.25	Moderate
Transfer Station	0.25	Very Low
Sims	3.25	Moderate
Sterlingbrook	3.50	High
TXDOT #1 (Rest Area 45)	0.25	Very Low
TXDOT #2 (Rest Area 1696)	2.75	Low
Badger Lane	3.50	High
Huntsville State Park	3.25	Moderate

Lift Station Risk Assessment

FNI utilized the results of the condition and criticality assessments to develop a risk based assessment of the City's lift stations. Risk scores were calculated by the summation of the condition and criticality scores for each lift station. These risk scores were divided into four ranges (extreme, high, moderate, and low) to assign a risk of failure to each lift station. The resulting risk scores, ranges, and risk of failure ratings are shown in **Table 9**.

Table 9: Risk Score Ranges

Risk Rating	Min	Max
Low Risk	0.00	3.00
Moderate Risk	3.01	5.75
High Risk	5.76	7.00
Extreme Risk	7.01	10.00

The lift stations were arranged into a risk rating matrix, which graphically shows the condition and criticality ratings, as well as the overall risk of failure rating. The risk matrix is shown on **Figure 2**. **Table 10** and **Figure 3** show the condition, criticality, and overall risk of failure ratings in a tabular and graphical format, respectively.

Several lift stations that received a good condition score received a moderate risk rating due to the high criticality score. It was noted that these lift stations should continue to be well maintained to minimize lift station downtime given the high consequence of failure of these facilities.

Figure 2: Lift Station Risk Rating Matrix

Risk Rating = Condition + Criticality		Condition Ranking				
		Very Good	Good	Fair	Poor	Very Poor
Criticality Ranking	Very Low Impact		TXDOT #1 (Rest Area 45) Southwood Drive	Transfer Station Airport	Hitchin' Post	TDCJ BOT (1)
	Low Impact			McCoy's	Elkins Lake #2	TXDOT #2 (Rest Area 1696)
	Moderate Impact		McDonald Creek Huntsville State Park Park Road 40	Sims (1) McGary Creek	Elkins Lake #1	
	High Impact		Badger Lane (1) Sterlingbrook Elkins Lake Post Office Tanyard Creek	Simmons Street Elkins Lake #3	Equestrian Center Brook Hollow	
	Very High Impact		Waters Edge	*Mallery Lake* (WFTC) Bayes *Highway 190* (WFTC) *Bearkat* (WFTC)		Elkins Lake Dam Old Colony (1)

Table 10: Summary of Lift Station Risk Based Assessment (Table)

Lift Station	Condition Rating	Criticality Rating	Risk Rating
Old Colony	Very Poor	Very High	Extreme Risk
Elkins Lake Dam	Very Poor	Very High	Extreme Risk
Bearkat Blvd	Fair	Very High	High Risk
Brook Hollow	Poor	High	High Risk
Elkins Lake Equestrian Center	Poor	High	High Risk
Elkins Lake #1	Poor	Moderate	High Risk
Highway 190	Fair	Very High	High Risk
Bayes	Fair	Very High	High Risk
Elkins Lake #3	Fair	High	High Risk
TXDOT #2 (Rest Area 1696)	Very Poor	Low	High Risk
Mallery Lake	Fair	Very High	High Risk
McGary Creek	Fair	Moderate	High Risk
Elkins Lake #2	Poor	Low	High Risk
Sims	Fair	Moderate	High Risk
Waters Edge	Good	Very High	High Risk
Simmons Street	Fair	High	High Risk
Tanyard Creek	Good	High	Moderate Risk
Park Road 40	Good	Moderate	Moderate Risk
Huntsville State Park	Good	Moderate	Moderate Risk
McDonald Creek	Good	Moderate	Moderate Risk
Elkins Lake Post Office	Good	High	Moderate Risk
Sterlingbrook	Good	High	Moderate Risk
McCoy's	Fair	Low	Moderate Risk
Badger Lane	Good	High	Moderate Risk
Hitchin' Post	Poor	Very Low	Moderate Risk
TDCJ BOT	Very Poor	Very Low	Moderate Risk
Airport	Fair	Very Low	Low Risk
Transfer Station	Fair	Very Low	Low Risk
Southwood Drive	Good	Very Low	Low Risk
TXDOT #1 (Rest Area 45)	Good	Very Low	Low Risk

Figure 3: Summary of Lift Station Risk Based Assessment (Graph)

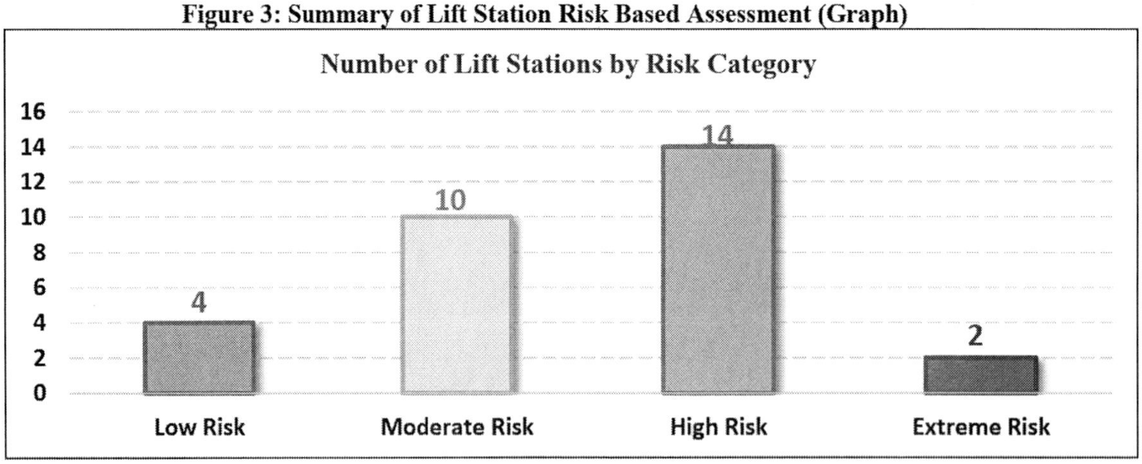

FLOW MONITORING AND HYDRAULIC MODELING

FNI conducted flow monitoring and recorded rainfall throughout the wastewater system as part of this project. The flow monitoring and rainfall data was used to characterize dry weather and wet weather flows at key points within the wastewater system, evaluate wet weather inflow and infiltration (I/I), calibrate the hydraulic model of the wastewater collection system, and select basins for Sanitary Sewer Evaluation Surveys (SSES).

Dry weather and wet weather system responses within the three WWTP service areas were evaluated by installing wastewater flow monitors to observe and document existing flow conditions. Rainfall data was simultaneously collected with rain gauges. A total of 12 flow monitors and three rain gauges were used for this study. All flow monitors and rain gauges were installed and maintained by ADS Environmental Services (ADS). Wastewater flow monitoring was performed using area-velocity flow monitors manufactured, installed and maintained by ADS. Rainfall during the study period was captured using a standard tipping bucket rain gauge. Each tip of the bucket generates an electronic pulse that is recorded by an ADS RainAlert II data logger. Flow monitoring and rain gauge data were collected in five-minute time step intervals. An example of the resulting flow and rainfall data is shown on **Figure 4**.

Figure 4: Flow Monitor and Rain Gauge Data

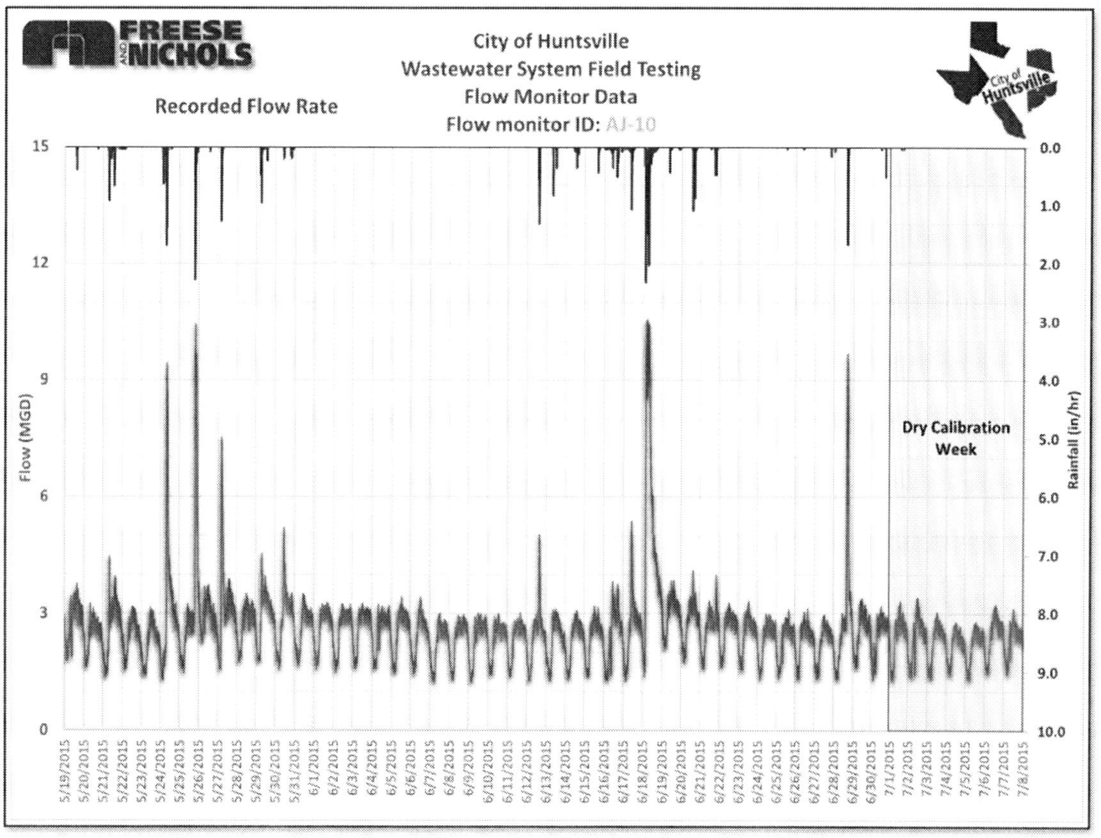

Flow Monitor and Rain Gauge Data Evaluation

Wastewater flow rates (dry weather and wet weather), peaking factors, depth to diameter ratios, and maximum depths of flow were calculated for each flow monitor site. The observed rainfall and the associated measured flow responses were used to calibrate the hydraulic model to observed wet weather conditions.

Inflow and Infiltration (I/I) Analysis

A wet weather analysis was performed by ADS and FNI to calculate the volume of I/I in each flow monitor basin. Calculating the I/I per flow monitor basin helped identify which areas of the wastewater system were prioritized for future condition assessment (SSES) work later in the project. This information was combined with hydraulic capacity modeling of the projected future wastewater system loads to prioritize system improvements in the integrated CIP. The 12 flow meter basins were ranked according to the relative volumes of discrete I/I measured during a selected storm event. The discrete volume of I/I within each basin was categorized as high, moderate, or low. The flow meter basins and their respective I/I volumes and classifications are shown on **Figure 5**. The flow meter basins, SSES priority rankings, discrete I/I volumes, and categories of I/I are summarized in **Table 11**.

Figure 5: Inflow/Infiltration by Flow Monitor Basin

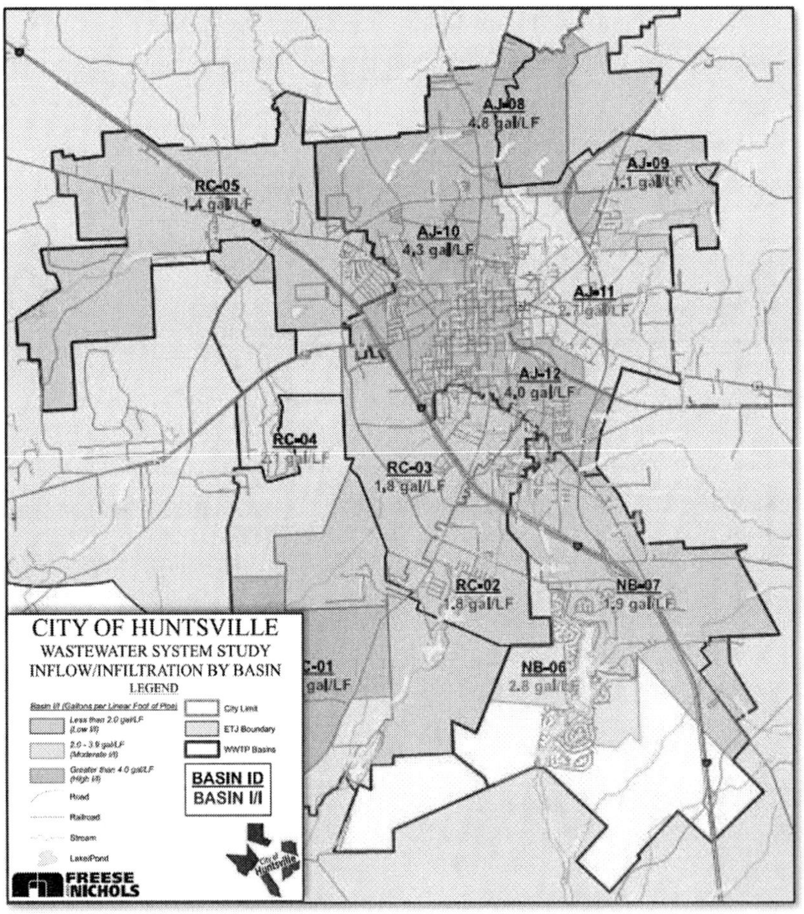

Table 11: Summary of I/I by Flow Monitor Basin and Categories of I/I

Flow Monitor ID	WWTP Basin	Basin Priority Ranking	Basin I/I (gal/LF)
RC-01	Robinson Creek	1	30.9
AJ-12(1)	A.J. Brown	2	4.0
AJ-08	A.J. Brown	3	4.8
AJ-10	A.J. Brown	4	4.3
NB-06	N.B. Davidson	5	2.8
AJ-11	A.J. Brown	6	2.7
RC-04	Robinson Creek	7	2.1
NB-07	N.B. Davidson	8	1.9
RC-02	Robinson Creek	9	1.8
RC-03	Robinson Creek	10	1.8
RC-05	Robinson Creek	11	1.4
AJ-09	A.J. Brown	12	1.1

Categories of I/I

(gal/LF)	Description
I/I Greater than 4.0	*High* amount of I/I
I/I Between 2.0 - 3.9	*Moderate* amount of I/I
I/I Less than 2.0	*Low* amount of I/I

(1) The AJ-12 Basin was moved to Priority Ranking 2 due to shallow lines and the subsequent high risk for sanitary sewer overflows.

Wastewater Hydraulic Model Development and Calibration

The City selected the Bentley SewerGEMS® hydraulic modeling software, which has dynamic modeling capabilities as well as GIS interoperability. The City completed an update of the City's wastewater GIS database at the beginning of this project, and this updated GIS data formed the basis of the hydraulic model. The completed hydraulic model consists of approximately 2,642 wastewater lines, 2,629 manholes, 3 outfalls, 5,385 catchments, 21 lift stations, and 42 pumps. FNI reviewed the modeled wastewater network for proper connectivity. Lift stations and force mains were input into the model based on GIS data, information provided by the City, and general knowledge of the lift station layouts, including wet well dimensions, obtained during the condition assessment site visits. Twenty-one of the City's lift stations were included in the hydraulic model. SewerGEMS® stores hydrologic runoff information important to wet weather calibration in catchments. FNI generated catchments around all manholes in the system. The catchments were generated as Thiessen polygons around each manhole and were bounded by the limits of the associated flow meter basins. Loads were allocated in the hydraulic model using geocoded water customer billing account information. The model was calibrated against the observed flow monitor data for dry and wet weather flows. Dry weather average flow rates were calibrated to within +/- 5%, and wet weather peak flow rates were calibrated to within +/- 10%. The dry and wet weather calibration results provided a high level of confidence that the model was closely matching real world conditions and suitable to use for hydraulic analyses and CIP development.

Design Storm and System Analyses
Hydraulic analyses were conducted to identify deficiencies in the City's existing wastewater collection system and establish a capital improvements plan to address deficiencies in the existing system and accommodate the projected wastewater flows through 2041. Various combinations of improvements and modifications were investigated to determine the most appropriate approach for conveying projected flows. Considerations in developing the CIP included increasing system reliability, simplifying system operations, consolidation of lift stations, conveying peak wet weather flows, and reducing surcharging and sanitary sewer overflows.

A 5-year, 6-hour design storm was utilized for the existing and future system analyses. This design storm is commonly used in Texas and provides a reasonable balance between level of service and wastewater infrastructure cost. The size of the 5-year, 6-hour storm was determined using historical storm event data for the region of Huntsville. The historical data was analyzed and summarized in the United States Geological Survey (USGS) Atlas of Depth-Duration Frequency of Precipitation Annual Maxima for Texas (Asquith and Roussel, 2004). The 5-year, 6-hour design storm for the City of Huntsville is a 3.95-inch rainfall event. The shape of the design storm hyetograph (rainfall vs. time) was developed using the alternating block method. The 5-year, 6-hour design storm hyetograph that was applied to the wastewater model is shown on **Figure 6**.

Figure 6: 5-Year, 6-Hour Design Storm Hyetograph

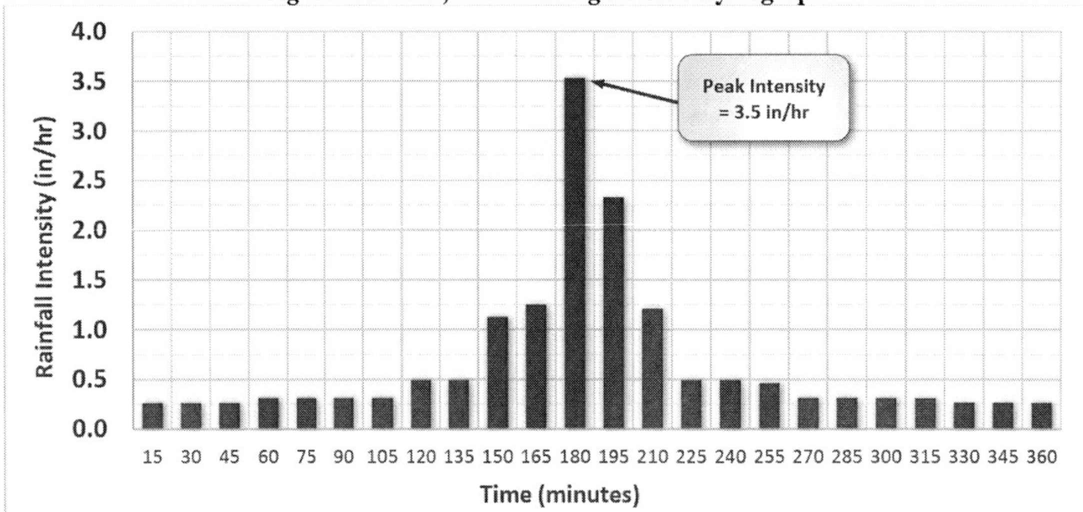

The critical flow condition for analyzing a wastewater collection system is peak wet weather. When the design storm is applied to the calibrated model, the effects of I/I in the system can be seen. As the storm intensifies (shown through time on **Figure 6**), additional flow enters the system. The model determines the point in time at which the amount of water from the design storm event reaches the peak within the system. This peak represents the most taxing load the system experiences under the design storm event.

FNI analyzed the collection system under existing conditions as well as three future planning periods as part of this project. Future hydraulic model scenarios were created to analyze needed capital improvements. These scenarios added the projected peak wastewater loads for each planning period to the existing system. This approach shows what improvements are needed in each of the three CIP planning periods (years 2021, 2026, and 2041). The results of the existing and future system analyses were presented on color-coded maps illustrating the surcharged state of modeled lines and manholes under the peak flow conditions of a 5-year, 6-hour design storm event.

SANITARY SEWER EVALUATION SURVEY (SSES)

A unique component of this project was the inclusion of a pilot SSES study within the scope of the master plan project. A traditional wastewater master plan will include recommendations to conduct SSES projects, as shown on **Figure 7**. For this project, forward-thinking City staff included a pilot SSES in the scope of the master plan, as shown on **Figure 8**.

Figure 7: Traditional Approach

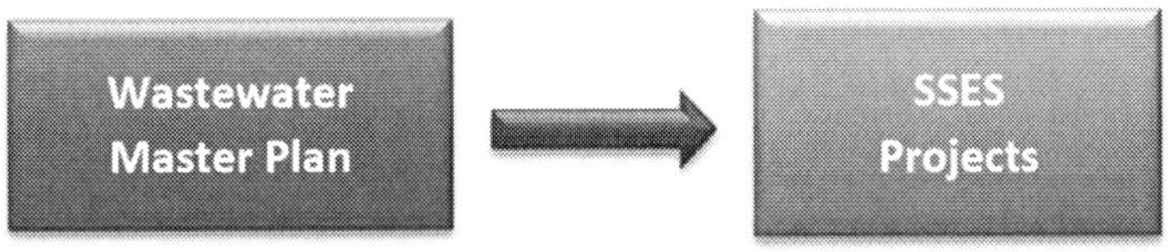

Figure 8: Huntsville Master Plan Approach

The pilot SSES study was conducted in two of the twelve flow monitor basins. The SSES efforts carried out in each basin are described in **Table 12**. The results of these SSES efforts were used to develop rehabilitation projects with the goals of reducing I/I and sanitary sewer overflows and extending the life of sewer infrastructure.

Table 12: Wastewater Sub Basin SSES Efforts

Wastewater Basin	SSES Efforts Conducted	Notes
RC-01	• Manhole Inspections	This basin was identified as having the highest level of I/I **(30.9 Gal/LF)** identified during the June – July 2015 wastewater flow monitoring period. Smoke testing was not conducted in this sub basin due to the relatively good condition of the 36-inch wastewater line.
AJ-12	• Flow Monitoring • Manhole Inspections • Smoke Testing	This basin was identified as having a high level of I/I (**4.0 Gal/LF**) and was prioritized for SSES efforts during this study due to shallow lines and the subsequent high risk for sanitary sewer overflows due to surcharging.

Focused flow monitoring, manhole inspections, and smoke testing were performed in the AJ-12 Basin. These are all typical components of an SSES. SSES flow monitoring is conducted over smaller areas than the system-wide monitoring performed as part of the flow monitoring evaluation discussed previously. The standard order of these SSES activities is 1) flow monitoring, 2) manhole inspections, and 3) smoke testing.

The focused SSES flow monitoring is used to identify where the majority of I/I is coming from within a given basin. The remaining SSES efforts can then be focused on that portion of the wastewater basin to maximize rehabilitation and renewal dollars to reduce I/I. As part of this study, SSES flow monitoring was conducted in the AJ-12 Basin to identify the portion of the network (sub basin) contributing the most I/I. The subsequent manhole inspections and smoke testing focused on that sub basin.

SSES Results

The flow monitoring and rainfall data within the AJ-12 Sub Basin were evaluated similarly to the system-wide data discussed previously. Five flow monitors are a rain gauge were used to monitor the AS-12 Sub Basin. The results of the flow monitoring I/I evaluation are presented in **Figure 9**.

Figure 9: Normalized I/I per Sub Basin by Storm Event

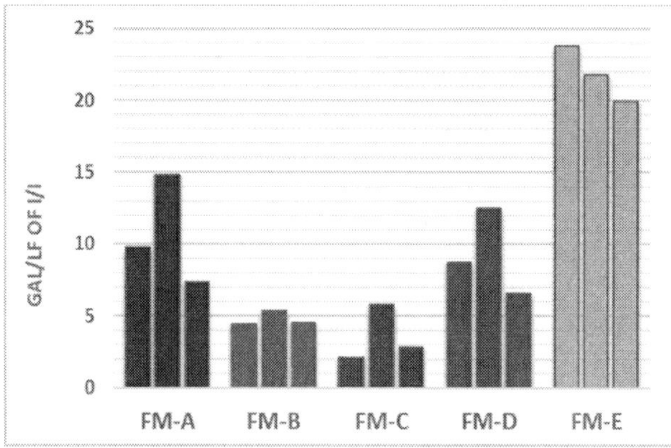

The area with the most I/I in all three storm events was Sub Basin E. Therefore, subsequent manhole inspections and smoke testing were carried out in this sub basin.

Manhole inspections were performed on 90 manholes in AJ-12 Sub Basin E. ADS performed visual manhole inspections during which the condition of each manhole was inspected from cover to invert. Structural defects, operation and maintenance (O&M) concerns, and infiltration sources were noted. The manholes were then scored on a 100-point system, with 0 being the best and 100 being the worst. The results of the manhole inspections are shown in **Table 13**.

Table 13: AJ-12 Sub Basin E Manhole Inspection Results

Manhole Condition	Number of Manholes	Score Range
Good	74	0 - 30
Fair	9	31 - 60
Poor	7	61 - 100
Total	90	-

Smoke testing is a common diagnostic method used to locate and identify potential sources of inflow and infiltration within a sanitary sewer system. During smoke testing, a special non-toxic, non-staining smoke is blown into a selected portion of the system to check for defects. These defects indicate sources of inflow and infiltration in the collection system. Approximately 23,000 linear feet (93%) of the sewer lines in Sub Basin E were checked for defects via smoke testing. A total of 20 defects were identified. Examples of the defects found in AJ-12 Sub Basin E are shown on **Figure 10**. The results of the smoke testing are shown in **Table 14**.

Table 14: AJ-12 Sub Basin E Smoke Testing Results

Defect Type	Number of Defects
Manhole	8
Sewer Line	4
Service Lateral	8
Total	20

Figure 10: Examples of AJ-12 Sub Basin E Smoke Testing Defects

Sewer Line Leak & Manhole Wall Leak — Sewer Line Leak — Service Lateral Leak — Inflow Through Pick Holes

CAPITAL IMPROVEMENT PLANS

Capital Improvement plans were developed for Capacity Projects, Lift Station Rehabilitation, and Sewer Basin SSES Rehabilitation. These plans phased the recommended improvements over the 5-Year, 10-Year, and 25-Year planning periods.

BUSINESS CASE

Through risk based assessment, flow monitoring, hydraulic modeling, SSES work, and a close partnership with multiple levels of City staff based on weekly communication, a convincing business case was developed for the wastewater system CIP and the expenditure of the City's dollars. The wastewater system was evaluated for deficiencies in condition and capacity, and the results of each evaluation were considered holistically when developing the CIPs. This multi-faceted approach helped to focus the City's dollars more efficiently, and provided defendable business cases for project implementation. Examples include:

- The City's rehabilitation dollars wouldn't be spent on a facility that was identified to be replaced with a larger capacity facility in the foreseeable future.
- Capacity wastewater line projects could be shown to be necessary based on real-world flow monitor data demonstrating existing system surcharging conditions.
- Verified wastewater line defects identified in the pilot SSES study established the usefulness of the SSES process.

Following this master plan project, the City of Huntsville held a bond election asking voters to approve spending for the infrastructure recommendations in the CIP. The results of this study were used as justification to the public regarding the need for and benefits of the recommended wastewater infrastructure. The bond election passed in November of 2016, and the City is currently planning to implement the initial phase of wastewater CIP recommendations.

RESULTS AND CONCLUSIONS

The project scope incorporated a number of system evaluation methodologies and tools that allowed for a holistic view of the condition and capacity of the City's existing collection system. These tools also allowed for a future evaluation of capacity needs and the opportunity to develop an ongoing SSES plan based on empirical data to manage the condition of the future system.

These evaluation methodologies and tools were combined with weekly communication with City staff at all levels to create buy-in and project understanding throughout the life of the master plan. This resulted in a defendable wastewater Capital Improvement Plan that the City is currently implementing.

REFERENCES

Asquith, W.H; Roussel, M.C; (2004) *Atlas of Depth-Duration Frequency of Precipitation Annual Maxima for Texas*; U.S. Department of the Interior; U.S. Geological Survey, Reston, VA.

Memorandum on Principles for Wet Weather Flow Design Decision Making

John C. Hall, Hall & Associates, Washington, D.C.
Gary B. Cohen, Hall & Associates, Washington, D.C.

ABSTRACT

Whether designing a new publicly owned treatment works ("POTW"), including the collection system and municipal wastewater treatment plant ("WWTP"), or upgrading an existing POTW, a municipality is presented with the issue of how large to size the POTW. Regardless of the decision, there is always the possibility that a larger storm event will occur that exceeds the design capabilities. At what point is the municipality's efforts deemed to be reasonable such that the municipality should not be deemed culpable or otherwise liable for overflows or other events exceeding the POTW design?

KEYWORDS

Design standard, wet weather, liability, consent decree, collection system.

INTRODUCTION

For the past four years, following the 8th Circuit's favorable ruling in *Iowa League of Cities v. Environmental Protection Agency* (711 F.3d 844 (8th Cir. 2013)), a group of Iowa communities have sought to resolve the proper selection of design flows to be utilized in the sizing of key wastewater treatment system components and for selecting remedial measures associated with elimination of sanitary sewer overflows and basement backups. These concerns were recently addressed in a comprehensive manner in the Waterloo Federal Consent Decree ("Waterloo decree") and approved by the Federal District Court of Northern District of Iowa as "consistent with the goals of the Clean Water Act and Iowa law." (Court Ratification at 12).

DISCUSSION

The various components of the Waterloo decree, approved by both the Environmental Protection Agency ("EPA") and Iowa Department of Natural Resources ("IDNR"), were structured to ensure that the collection system and WWTP were designed to comply with applicable laws by providing "adequate capacity" to cost-effectively address a range of wet weather conditions while avoiding the need to design for "extreme events" (*e.g.*, flooding conditions, extreme rainfall events, unanticipated conditions).

The decree also provided a basis for ensuring proper operation for the collection system, thereby providing an objective basis for allowing communities to have defenses to unforeseen events. Because of (1) the comprehensive nature of the Waterloo decree and (2) the explicit recognition of the design conditions to be evaluated to ensure cost-effective wet weather compliance and adequate system capacity sufficient to meet the requirements of both State and Federal law, the elements of this decree could provide a basis for statewide program implementation. If this can be accomplished, it should resolve any ongoing dispute with respect to the selection of the appropriate peak flow for collection system and wastewater treatment unit process design.

The following briefly describes the major elements and design conditions of the Waterloo decree:

Waterloo Decree Provisions

The Waterloo decree embodies several major principles that serve to (1) ensure proper system design and operation and, importantly, and (2) provide defenses to conditions (overflows/backups/WWTP effluent exceedances) that may, nonetheless, occur when the overall treatment facilities are properly designed and operated, as follows:

1. A range of design storm conditions are specified for evaluation and, upon final selection by the permittee, constitute the basis for identifying the required capacity of system components (2-10 year, 6 hour storm under average groundwater conditions). In essence, this approach uses a cost-effective/cost benefit analysis matrix to identify system improvements within the 2-10 years level of service (potential system overflow return frequency). Furthermore, different areas may have different levels of service. The approach provides an appropriate "upper limit" on the proper design of a collection system (and by implication the maximum flow condition to be processed at the WWTP).

2. Defenses to unanticipated conditions that may cause sanitary system overflows are recognized (upset, bypass, flooding, proper operation and maintenance ("O&M")). Development of local program management documents/procedures are required to provide an objective basis for evaluating defense applicability and ensure that proper collection system O&M/management is actually occurring (*e.g.*, Capacity, Management, Operation, and Maintenance ("CMOM") Program Document, Emergency Spill/Overflow Protocol, and, if necessary, a Capacity Assurance Plan). The Consent Decree only imposes stipulated penalties for an overflow or bypass that reaches waters of the United States (or waters of the State). The Consent Decree, however, recognizes that penalties should not be imposed where the SSO's (or prohibited bypasses) occur when the City is in compliance with the terms of the standard conditions of its NPDES permit. Furthermore, the overflow must result from a lack of adequate capacity (as discussed above) or deficiencies in O&M. As such, implementation of a CMOM O&M program or other proper O&M would be a defense to an overflow – even if that overflow reaches waters of the U.S. or State.

3. Blending is allowed, as authorized/recognized by a city's NPDES permit. A peak flow management protocol for WWTP operation provides the basis for using blending to minimize system overflows while also protecting biological system operations under high flow events. This combination of approaches allows for cost-effective management of wet weather flows and ensures that the regulators and regulated community both understand and objectively determine when a system is being properly designed, operated, and maintained. The design storm evaluation matrix approach allows for a more flexible selection of adequately protective system performance expectations, depending upon the collection system setting.

Under the Consent Decree approach, if, for example, topography or other physical conditions create a large cost differential between the 2 versus 10 year "level of service" the lower protective level may be selected. However, if the conditions allow for a 10 year level of service (*e.g.*, in newer collection system areas with limited infiltration/inflow ("I/I")) then that section of the system receives the higher level of protection. In this manner, the peak flow to be processed by the wastewater facility will be somewhere between the 2 and 10 year storm condition. The approved collection system and WWTP design will not include "flooding" or other extreme events, which would constitute potential "upset" or "anticipated bypass" conditions, assuming the system is being properly operated at the time.

With regard to treatment plant component design, the facility designs may incorporate "blending" of biological and some other physical and/or chemical process to accommodate the peak flow while meeting permit conditions, within the engineer's discretion. In essence, the decree recognizes that a collection system is part of the "technology" employed to properly process sanitary wastewater. As a technology, it is not "infallible" and even the best operated and maintained systems may have periodic overflows and backups. Therefore, defenses are allowed for such events, assuming proper design and operation of that system - just as is allowable at the WWTP. It should be noted that the "proper O&M defense" could apply even to overflow/backup conditions occurring during non-wet weather events. By demonstrating that the approved CMOM program was being properly implemented at the time, one obtains the defense for the unforeseen condition (*i.e.*, unanticipated line plugging, unusual mechanical failure).

CONCLUSION

We would suggest that the approach used in the Waterloo decree be applied in other consent decrees. As the approach has been acknowledged as being consistent with the Clean Water Act, municipalities with less favorable existing consent decree requirements should seek to have their consent decrees modified.

New Approach to Design Level of Service for Wichita's Collection System

The typical year analysis was originally developed for Oklahoma City, Oklahoma by Imran Khan for their master plan in 2008. The typical year analysis was applied to the City of Wichita's (City) wastewater master plan due to the size of their system. The capital improvement savings significantly offset the engineering fees associated with the analysis, and therefore made this an easy decision during project scoping. The City's wastewater system is comprised of four (4) reclamation facilities and over fifty (50) pump stations. Peak flows are largely impacted by the hydrology of the sewer system. At wastewater facilities, issues occur more frequently during periods of multiple, low intensity, back-to-back storms. The Soil Conservation Service (SCS), now known as the National Resources Conservation Service (NRCS), developed four (4) standard synthetic rainfall distributions for different regions of the United States. The City of Wichita (City) falls into the type II distribution. This distribution is the most intense, short duration rainfall. This type of storm conflicts with the rainfall events that typically cause issues in sanitary sewer systems and facilities. Therefore, for the Wichita Study Area, synoptic typical years of rainfall were developed using 65 years of rainfall data for the 75th-, 85th-, 95th-, and 100th-percentile of non-exceedance. A synoptic typical year aggregates the historical rainfall data to determine the average typical years of rainfall at the varying percentiles of non-exceedance. These typical years have a few short duration, high intensity events, but they also contain multiple low intensity, back-to-back storms.

The SCS Type II 5-year, 24-hour rainfall hyetograph for Wichita, Kansas is displayed in Figure 1. The peak rainfall depth during this event is 1.64 inches.

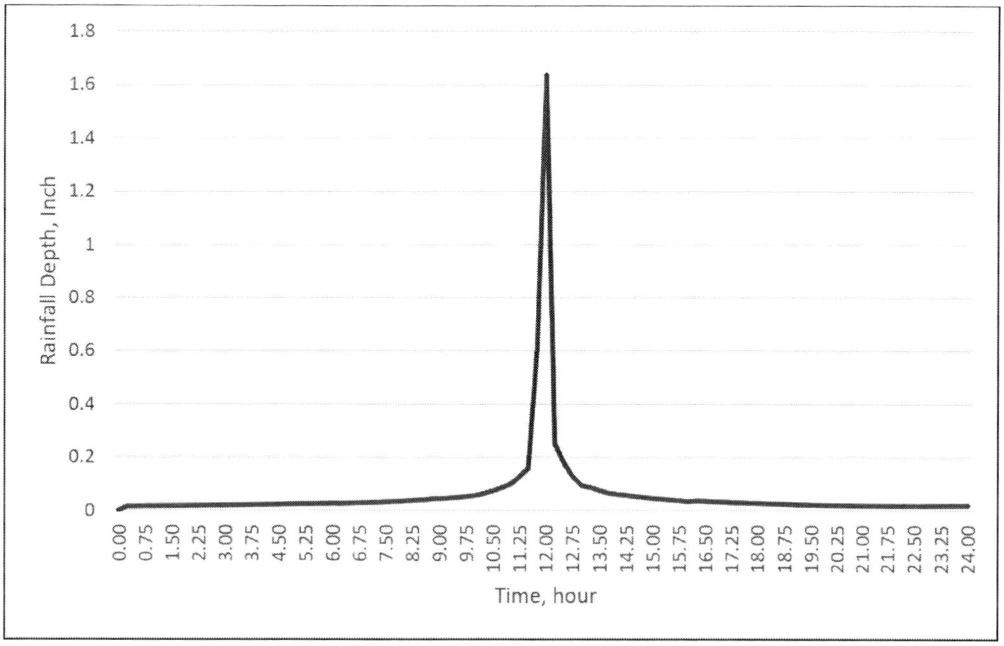

Figure 1: NRCS (SCS) Type II 5-Year, 24-Hour Rainfall Distribution for Wichita, Kansas

The 95th-percentile typical year has a peak intensity less than that of a standard 5-year, 24-hour SCS type II event. As recommended by the Environmental Protection Agency (EPA), sixty-five years of historical rainfall data (1948-2013) from the City was used to analyze the typical year. The rainfall data was in 1-hour intervals and totaled 23,343 records. An excel spreadsheet was used to determine the annual rainfall totals. Figure 2 shows the total rainfall during each year for the City.

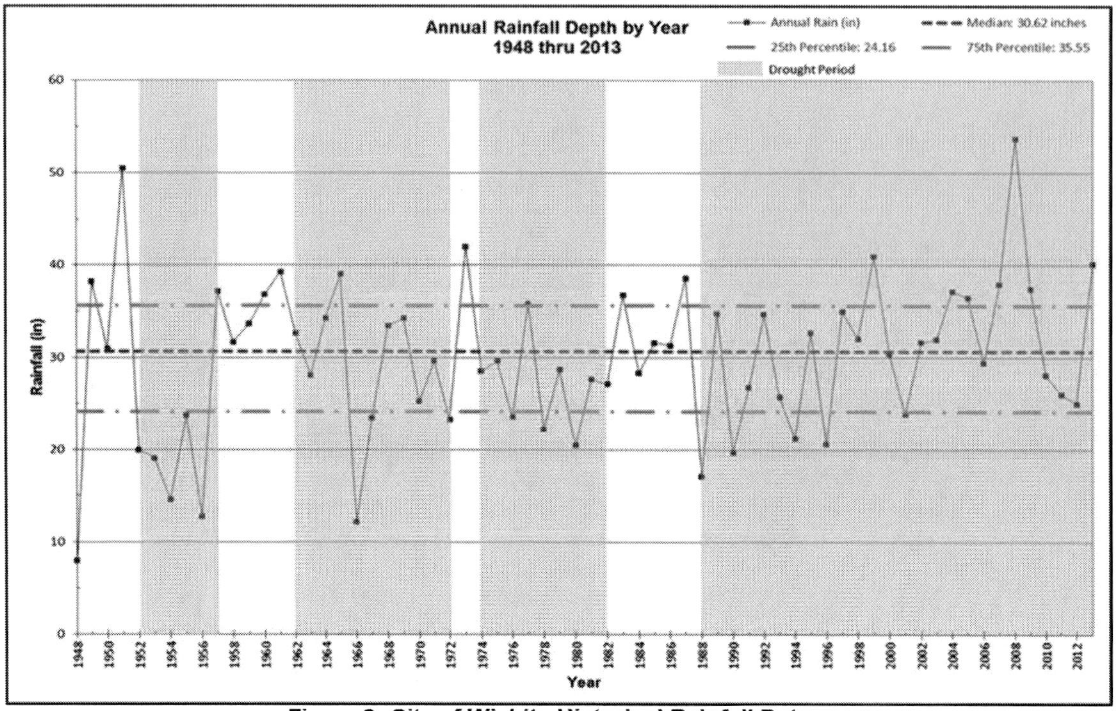

Figure 2: City of Wichita Historical Rainfall Data

Each month of each year was ranked against the same month in other years, i.e. January 2012 to the total January rainfall from the remaining 64 years. From the rankings, a percentage of non-exceedance was developed for each month of each year.

Figure 3 ranks the annual rainfall totals for each year of the 65-year period from 1948-2013. During the 65-year record period, the maximum total yearly rainfall occurred in 2008 with a total rainfall of 53.74-inches. The minimum annual rainfall occurred in 1948 with a total rainfall of 7.96-inches although the records for 1948 did not include rainfall from the months of January through July. The median annual rainfall for this 65-year period was 30.62-inches. The graph also shows that 25-percent of the annual rainfall was less than or equal to 24.16-inches with seventeen (17) years not exceeding this amount, and 75-percent of annual rainfall was less than or equal to 35.55-inches, which was exceeded in 17 years of the 65-year period.

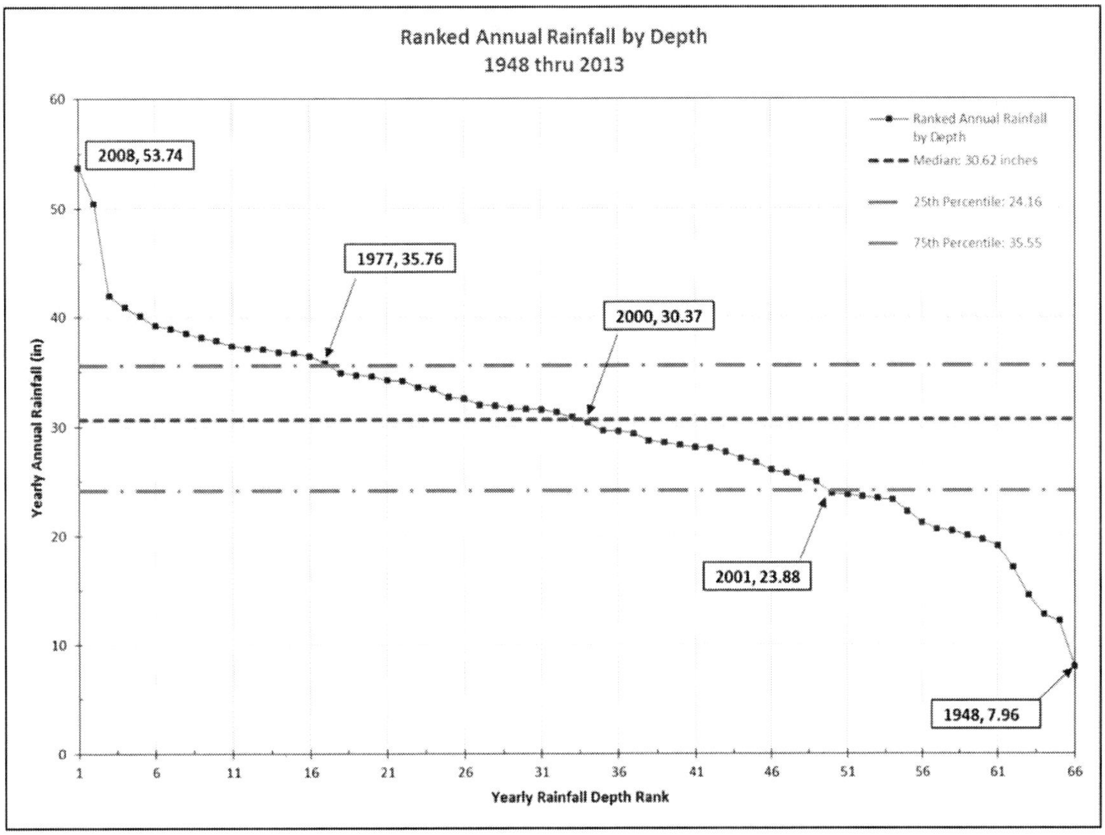

Figure 3: Ranked Annual Rainfall by Depth 1948-2013

An aggregate 75th-, 85th-, 95th- and 100th-percentile synoptic year was developed from the percentage of non-exceedance. Figure 4 shows the total rainfall depth in each month of a synoptic typical year for the 75th-, 85th-, 95th- and 100th-percentiles. They hyetograph for the 95th-percentile synoptic typical year is displayed in Figure 5. These synoptic typical years use data from the City to develop a typical year that is representative of the historic rainfall seen in the City. Figure 4 confirms that the City traditionally has the largest amount of rainfall during the months of May, June, and July. The peak intensity of the 95th-percentile synoptic typical year is less than that of even the SCS 2-year, 24-hour rainfall for the City of Wichita.

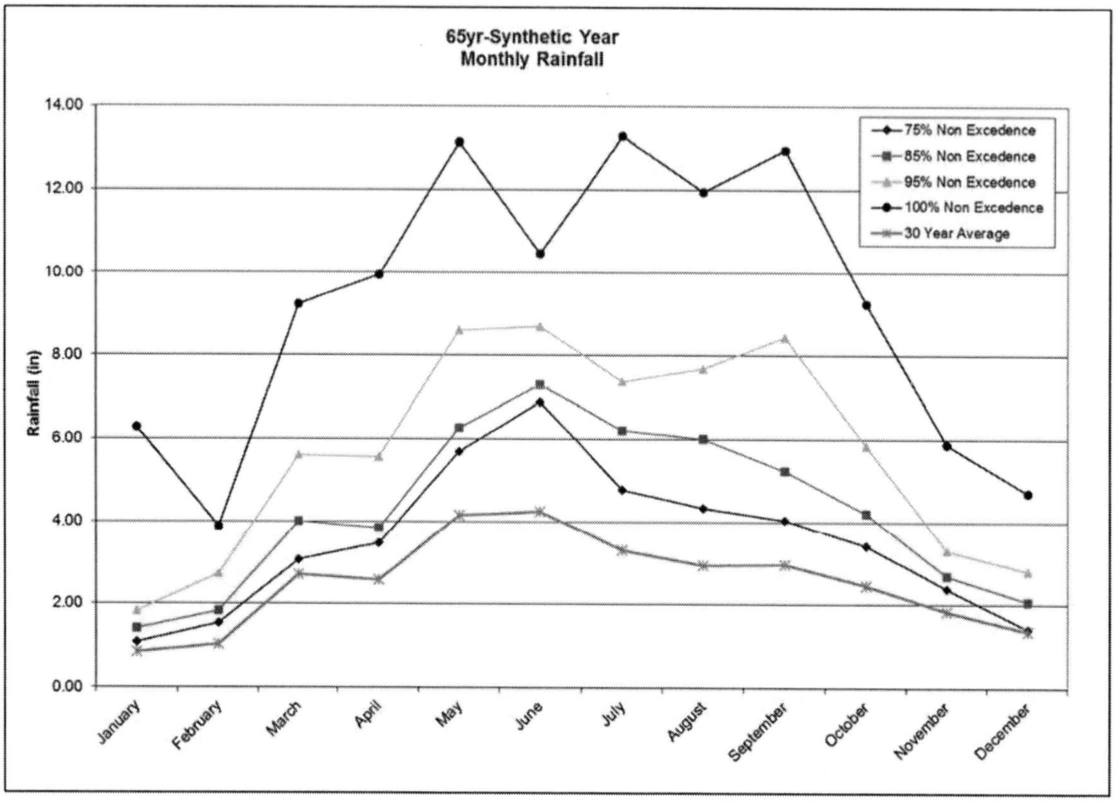

Figure 4: Synoptic Typical Years for Various Percentile Rain Events

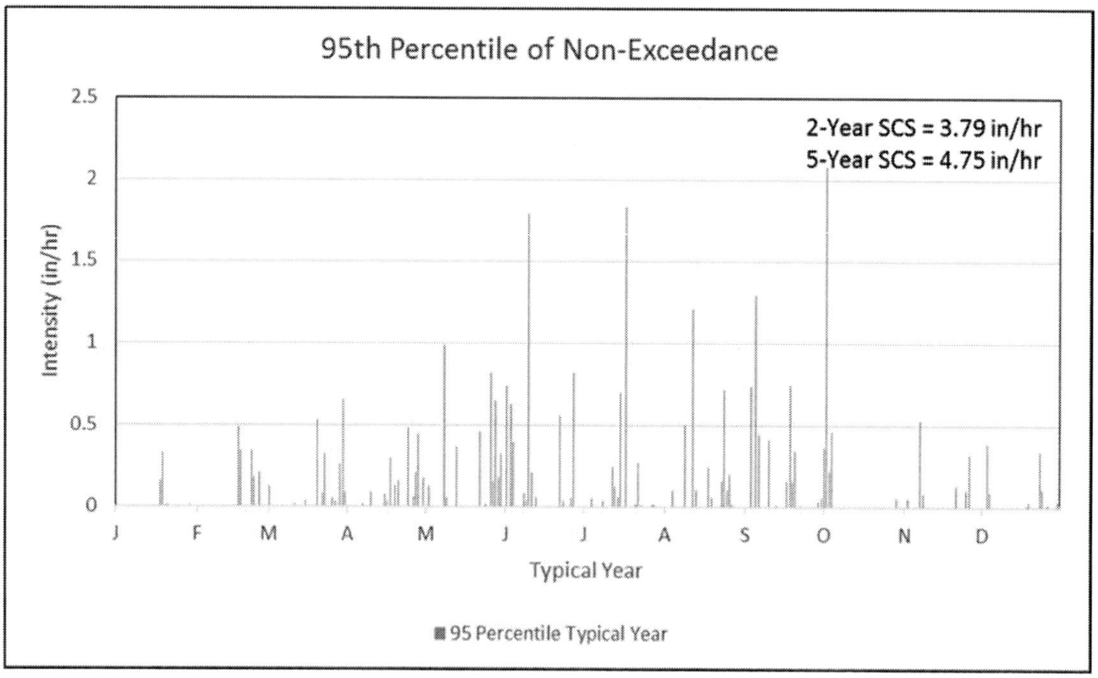

Figure 5: 95th Percentile Synoptic Typical Year

Due to the high peak intensity of the traditional SCS design storm, many capital improvement projects (CIPs) can be oversized. By using the synoptic typical year that is specific to the City, there is a level of confidence in the anticipated CIPs. The City selected the 95th-percentile synoptic typical year, which is equivalent to a return period of greater than 20 years. The anticipated costs of CIPs for this return period are less than the cost of CIPs that are developed using the NRCS (SCS) rainfall distribution with a 5-year return period as a design storm.

The Four Mile Creek basin in the Wichita Study Area was analyzed using both the 95th-percentile typical year and the traditional SCS 5-year, 24-hour storm. This basin currently serves just over 27,000 citizens and has over 33 miles of sanitary piping. The analyzed scenario during 2045 is estimated to serve 60,000 citizens and includes roughly 37 miles of sanitary piping. Figure 6 and Figure 7 show the results of the SCS Type II 5-year, 24-hour design storm and the 95th-percentile design storm, respectively. The pipes displayed in blue are under free-flow conditions, the pipes displayed in pink have a downstream restriction, and the pipes displayed in green are the cause of the restriction. It can be observed that Figure 6, demonstrating the SCS storm, has more pipes displayed in green or pink indicating more locations of backups and/or surcharge.

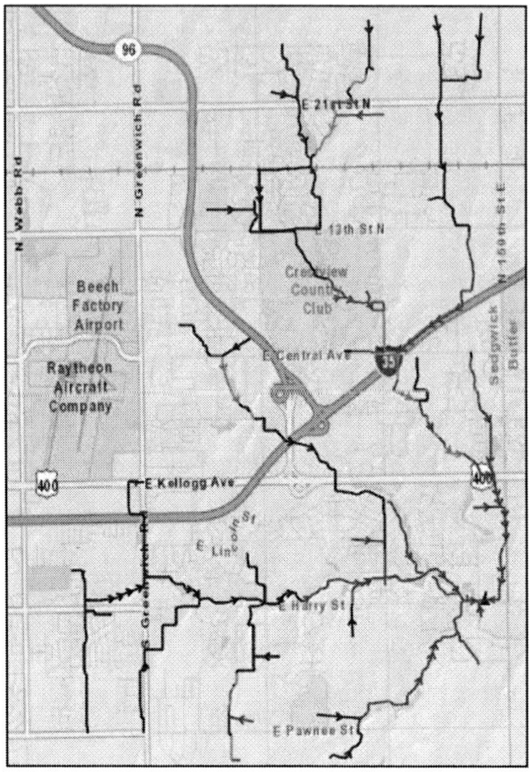

Figure 6: 2045 Four Mile Creek Basin with SCS 5-Year, 24-Hour design storm

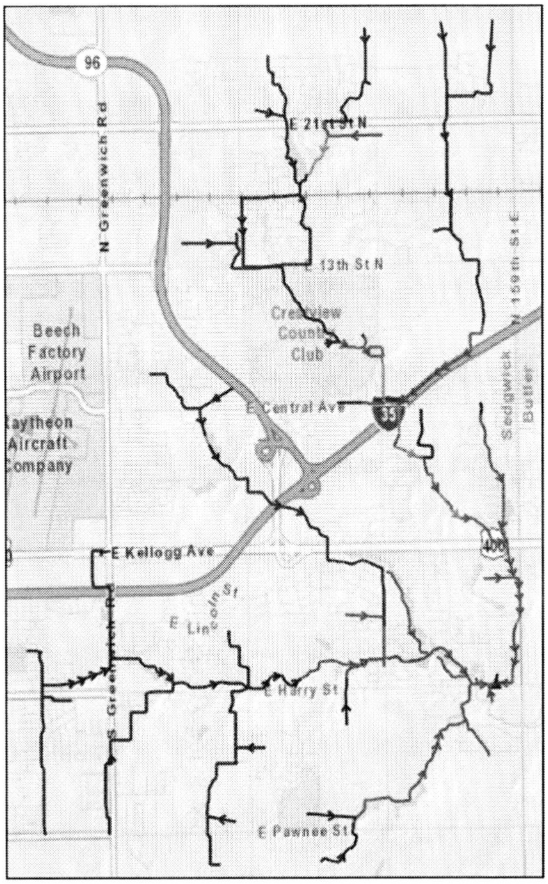

Figure 7: 2045 Four Mile Creek Basin with 95th-percentile design storm

Table 1 summarizes the mileage and count of sewers within the Four Mile Creek Basin that are under free-flow conditions, have a downstream restriction, or are the restriction for the 95th-percentile and the SCS 5-year, 24-hour rain events respectively.

Table 1: Summary of Surcharge for Four Mile Creek Basin

	95th-Percentile		SCS 5-year, 24-hour	
	Length (mi)	Count	Length (mi)	Count
Free-Flow	27.4	520	24.5	578
Downstream Restriction	10.0	228	11.6	198
Restriction	0.2	33	1.5	5

Capital improvement projects were developed for both scenarios using a trigger point approach and Figure 8 shows the cost comparison in just one basin. The trigger point approach assigns an anticipated year for improvement, but also focuses on assigning a trigger, to each capital improvement project. The triggers can vary, but a few common triggers are capacity, regulatory, or growth. Similar to a cause and effect situation, when the trigger happens, then the capital improvement should follow or occur simultaneously. The SCS Type II 5-year, 24-hour design storm is anticipated to require an additional 5 miles of pipe in the Four Mile Creek Basin for capacity increases when compared to the same basin with the 95th-percentile design storm. The change in design storm resulted in a nearly 50% reduction in capacity improvement projects within the Four Mile Creek Basin.

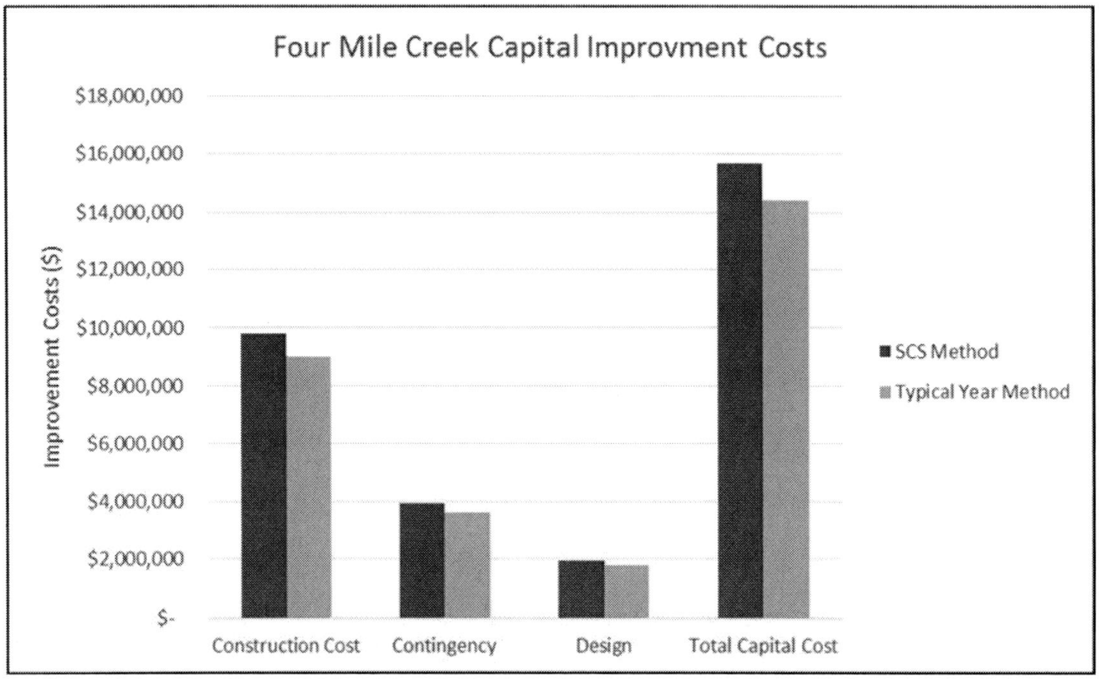

Figure 8: Four Mile Creek Capital Improvement Project Cost Comparison

The City of Wichita is projected to save over $1 million in just one basin by utilizing this innovative, synoptic typical year analysis. It is anticipated the City of Wichita is saving about $14 million city-wide on capital improvements through the next 25 years and providing customers with a higher level of service.

Into the Information Age: Application of Internet of Things for Sewer Maintenance Optimization

Tamsen McNarie[1], Gregory Quist*[2], Keith Lewinger[2], Richard Haimann[3]

1. San Antonio Water System (SAWS), TX
2. SmartCover Systems, Escondido, CA
3. Haimann Engineering, Huntington Beach, CA

*gquist@smartcoversystems.com

Abstract
The application of real-time sewer monitoring allows a much lower cost of prevention of sanitary sewer overflows. This has been shown with a pilot test using SmartCover® (Escondido, CA) technology in the San Antonio Water System (SAWS). The traditional method being employed prior to the pilot test was high frequency cleaning of sewer sections deemed to be high risk – typically due to historical overflows. Most overflows are caused by buildup of fats, oils and grease (FOG), root intrusion, pipe decay, or other capacity constraints. A strategy was employed to clean the sewers based on real-time data from SmartCover remote level monitoring units instead of relying on the calendar ..

SmartCovers were installed and water elevations in the manholes at these high risk locations were monitored in real time over a 12 month period. Trend analysis was conducted on the water level patterns seen in the manholes adjacent to high risk pipes. This analysis allowed the utility to clean when data showed that debris was building up, rather than on a fixed schedule. SAWS was able to reduce cleaning frequency by 94% without having any overflows.

The application of real time monitoring to sewer systems provides an immediate return on investment to sewer utilities. These data shows definitively that using real time monitoring allows reduction in cleaning schedules, which frees up resources for other needs and lowers the costs of sewer maintenance dramatically, while achieving the same or better SSO reduction objectives. The dollars not spent are significantly greater than the cost of instruments and data acquisition. This benefits rate payers, the environment, and utility staff dramatically.

Introduction
Unpermitted discharges from a sanitary sewer system to waters of the United States are prohibited under the Clean Water Act (33 USC §1311(a), *Id.* §1362(6), *Id.* §1362(14)). The United States Environmental Protection Agency (USEPA), responsible for enforcing the Clean Water Act, calls these discharges, when coming from a sanitary sewer system, sanitary sewer

overflows (SSOs). To date, sewer utilities have attempted to reduce SSOs through high frequency cleaning (HFC) in areas having a history of spills. These higher risk locations can include areas susceptible to fats, oils, and grease (FOG) buildup, and root intrusions, pipe deformations, decaying pipe and siphons. Many sewer utilities have been able to reduce SSOs through aggressive HFC programs. However, the costs of HFC programs can be expensive and consume a large share of sewer utilities' budgets.

In lieu of HFCs, there have been proposals for the use of real time monitoring of water elevations in Wastewater Collection and Transmission Systems (WCTS) to inform system operators to conduct cleanings in time, but not more frequently than necessary, to prevent SSOs. San Antonio, Texas has been piloting real time monitoring to identify areas where they can reduce the frequency of cleaning without increasing the frequency of SSO.

A typical sewer cleaning schedule program is based on observed pipe conditions and spill history. Such cleaning schedules can be ,for example, one, three, six, 12, 24, 60, and 120 month cleaning frequencies. Pipe conditions are assessed from a data obtained from CCTV inspections using the National Association of Sewer Service Companies (NASSCO) Pipeline Assessment & Certification Program (PACP) codes, cleaning crew field observations FOG, debris and roots, and prior SSO frequencies.

Significant success has been observed when applying frequent inspection and cleaning schedules as shown in Figure 1, which shows SSO history in San Diego from 2000 through 2014[1,2].

There are concerns associated with determining the correct cleaning frequency for a pipe.

1. The application of standard scoring systems regarding pipe condition are subjective, leaving room for underestimating or overestimating potential problems.
2. Assessment of cleaning needs from field observations during cleaning are subjective.
3. Conditions between inspections are unknown and may change without warning. In order to decrease this risk, a conservative system manager would be motivated to move to higher cleaning frequencies in order to minimize the risk of spills.

Costs associated with SSOs can include fines of thousands to millions of dollars, cleanup costs, damages from third party claims, and indirect costs of loss of public good will.

Because of these large SSO costs, agencies may be more willing to increase cleaning frequencies to minimize risks of SSOs within their systems. This can effectively lower SSO risk, as shown in Figure 1., but the HFC costs put an upward pressure on rates. HFC costs include: fuel, labor, equipment use. Additionally, HFC increases the carbon footprint by operating equipment and rolling stock, and the risk of worker injury and traffic accidents when crews are working in the right of way.

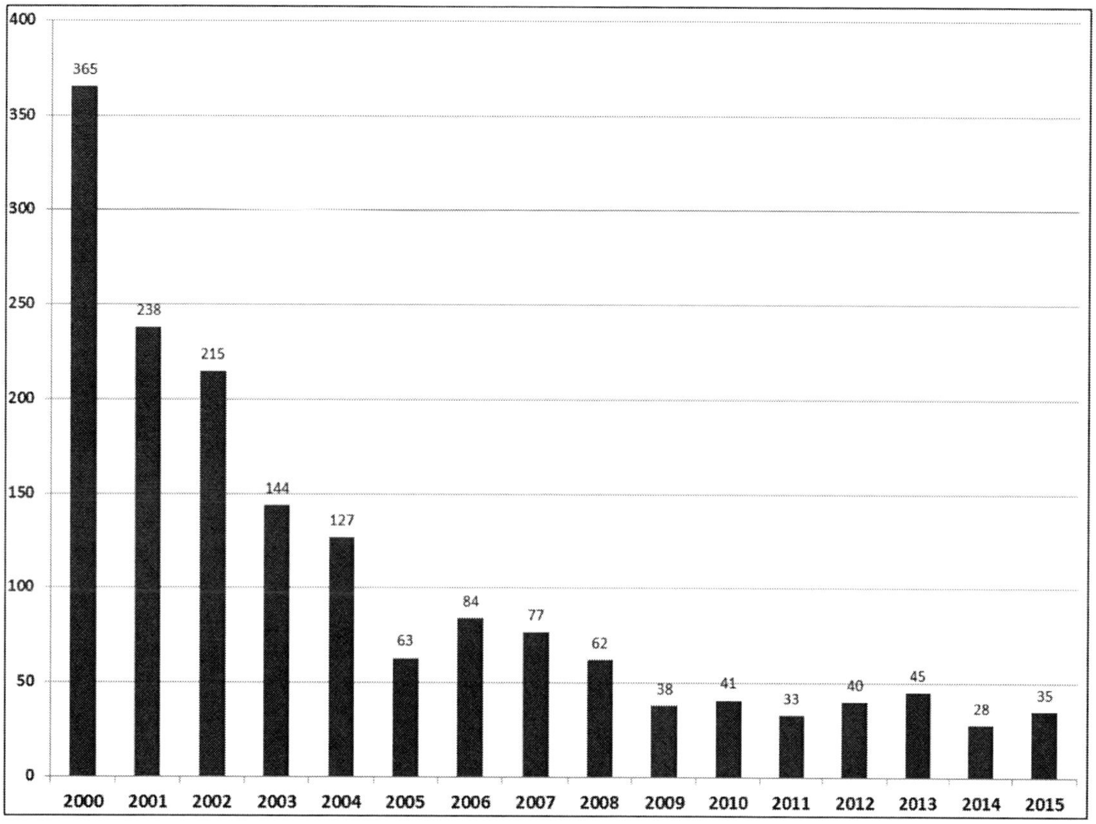

Figure 1: City of San Diego sewer spill history, 2000 – 2014.

Over the last several years, new "internet of things" technology has been commercialized that allows real time continuous monitoring of water surface elevations within WCTS. After a decade of deploying this technology in sewer systems and enhancing capabilities, it has become highly resilient, reliable and durable.

This sensing technology used in concert with data analytics that determines when inspection and/or cleaning is needed has the potential to significantly lower cleaning costs with no increased risk of SSOs.

To test this hypothesis, in 2015, small test numbers (10 units each) of real-time remote robust sensor networks were deployed at HFC locations in San Antonio, Texas .

Methodology

The San Antonio Water System (SAWS) installed SmartCover® (Escondido, CA) water level remote field units (RFU) at 10 HFC locations in June and July of 2015. The SmartCover® system consists of:

- Ultrasonic water surface level sensor with 1.59mm (0.0625inch) resolution and recording level measurements every 12 minutes
- Satellite transceiver for two-way wireless communications
- Two-year battery
- Rugged bracketry connecting the system to a standard manhole cover
- Cloud servers for data storage and real-time analysis
- Web based data viewing dashboard showing trends, current and historical water surface information, real-time alarms when elevation thresholds are exceeded, battery charge information, system transmission information, sensor location, system installation location information such as manhole bottom elevation, manhole top elevation, inlet invert elevation, inlet pipe diameter and slope, outlet invert elevation, and outlet pipe diameter and slope.

The operating rules for these sites during the test were to suspend regular HFC and use real-time level data to prompt cleaning events. The SmartCover® system provided automated real-time water level trend analysis at these locations, enabling SAWS to monitor - remotely, automatically, and in real-time - conditions at these HFC locations. The trending analysis indicates small but systematic changes in water levels representing systematic variances from "normal" diurnal fluctuations that suggest the need for inspection or cleaning. This system provides sufficient time – typically one to many weeks – to schedule a cleaning crew to clean the pipe prior to a critical event that could lead to an SSO.

The effects of FOG, debris, or roots on the water levels could be seen in real-time. See Figure 2. The SmartCover® system automatically monitored each of these sites by performing a daily analysis of the water level trends at these sites. In addition, the SmartCover® system provided a back-up water level alarm in cases where water levels may rise quickly as may happen during a significant rain event to provide time for a crew to respond and prevent an SSO.

In July 2015, SAWS installed 10 RFUs at one month HFC locations. These were locations with known FOG and or root issues. At these sites, the monthly cleaning was suspended. Figure 3 shows the location of the sensors that were used for the SAWS study.

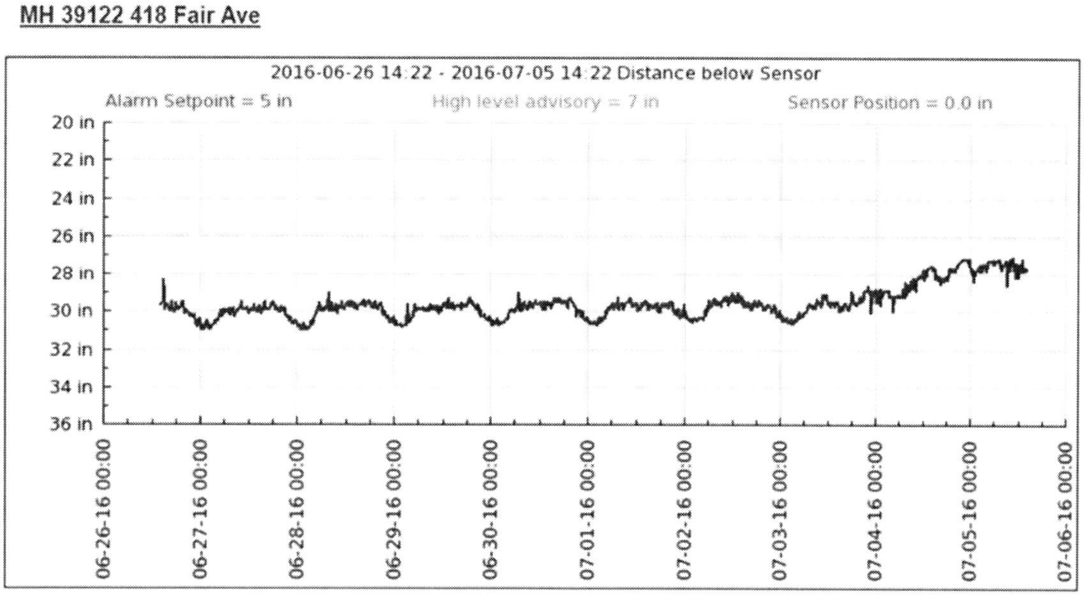

Figure 2: Sample manhole data showing rising trend. SAWS MH 39122 level data over a period of nine days. Normal diurnal flow is followed by a period of rising levels, suggesting a blockage downstream or a local grit or debris buildup potentially requiring near-t

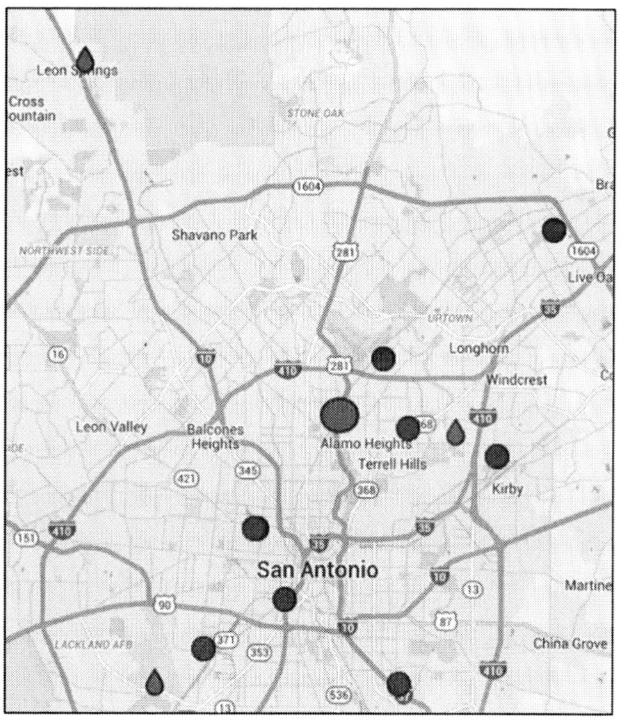

Figure 3: Locations of SAWS Pilot Program manhole sensors (green dots) and NOAA radar rain gauge sites (blue drops).

Results

Representative graphs and interpretations of the water levels at example pilot sites in San Antonio (SAWS) and are shown below. The complete10 pilot sites with level interpretations are shown in Appendix A.

In the following graphics and the graphics in the Appendix displaying the water level histories of the SmartCover® monitored locations: the solid blue line is the water level measured down from the sensor; rainfall at NOAA radar-based sites (Figure 3) is also charted in various dotted line colors; actions taken by SAWS, for example CCTV recordings and the results, are annotated on the graphs; and notable level trends or level events (e.g. brief surcharging) are separately called out and explained in the chart descriptions. (Note: All dimensions on the water level and rain level graphs are English, as this is the native system for SAWS and these are screen shots from the SmartCover web interface for these WCTS.)

At the 10 SAWS test sites, monthly cleaning was suspended and inspections and cleaning were completed when the level data provided by the SmartCover® level monitoring system showed a trend that would suggest a clog was forming. Water level data from a representative monitored location is plotted below in Figure 4 (the remainder are shown in Appendix A). During the second half of the pilot test, San Antonio experienced unusually high rainfall, including a period in late May and early June when nearly 16 inches of rain fell in one week. During these periods, there was local flooding and level trend signatures in several of the pilot locations were overwhelmed by inflow. These periods are noted on the level charts.

During the 12 months of the pilot test, only seven cleanings were performed, five at a single particularly problematic site, as opposed to the 120 that would have been called for under normal scheduling, indicating a drop of (120 - 7)/120 = 94% in cleanings. See Figure 5. No associated spills occurred during the non-flooding period of the pilot test.

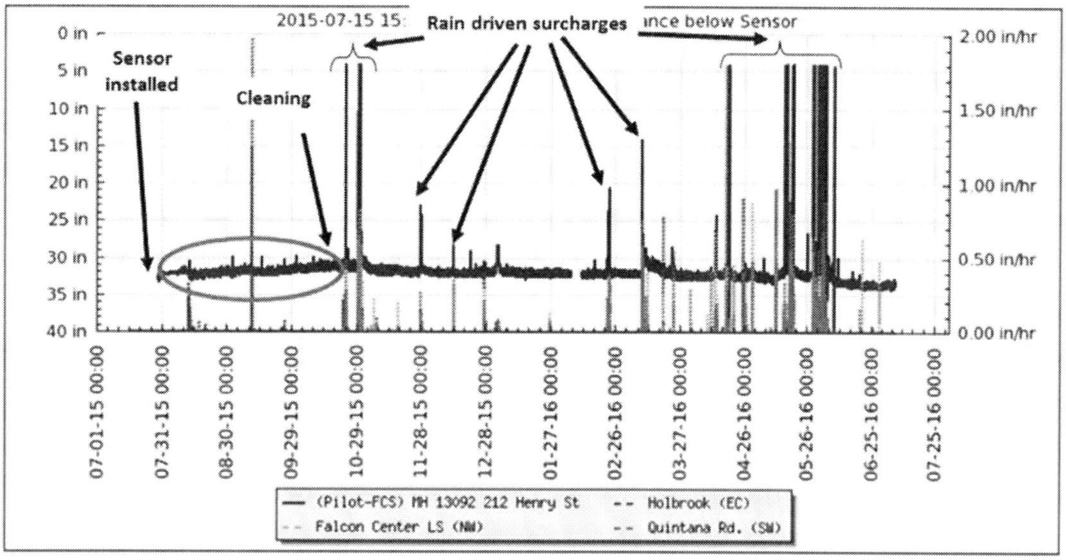

Figure 4: Water level history (blue), from SAWS Pilot Site #1, MH 13092, a monthly cleaning site. The level sensor was installed on 29 July 2015. Over the next three months, there was a very slow, but distinctive rise in average water levels, indicated in the green circle and expanded in the figure below. Based on this level data, SAWS crews cleared out the manhole and prevented an overflow. Local NOAA radar rain data is shown (dotted red, orange, green, and pink) for reference. There were several rain correlated surcharges at this location, including several during the intense rain storms in late May through early June 2016

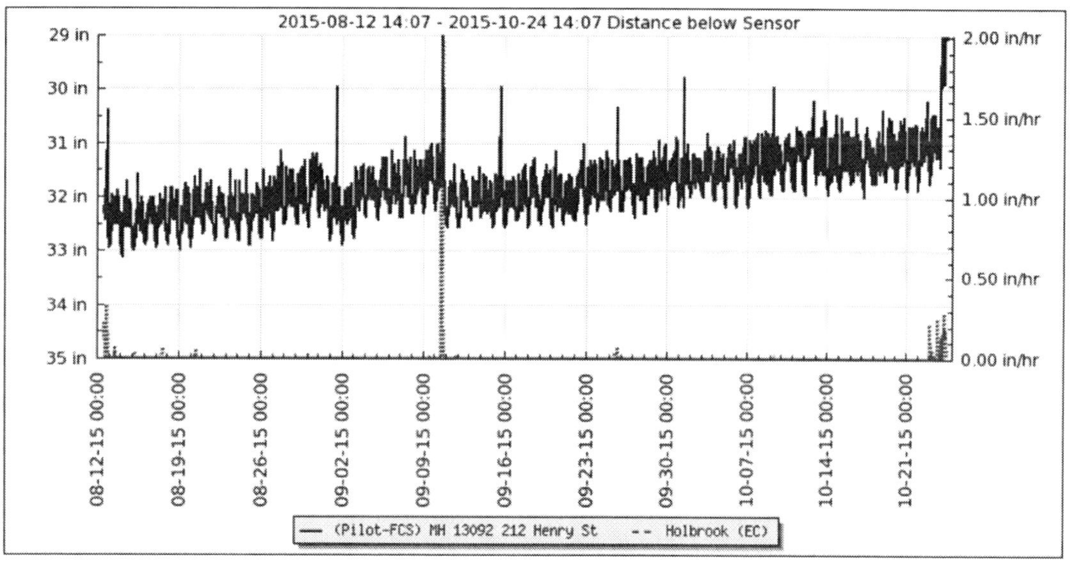

Figure 5: Expanded view of water levels (blue) at SAWS Pilot Site #1 between 8/12/15 and 10/26/15. Diurnal flows are seen, with a relatively constant rise suggesting either a blockage downstream or local grit or debris buildup. This level change would not be detected manually, it corresponds to 0.021" of rise per day.

Discussion

The results from this pilot using internet connected real-time level monitoring to support data-driven maintenance rather than calendar driven maintenance suggest cost savings can be attained for SSO management. As shown in this pilot test and displayed Figure 6, the number of times maintenance trucks are mobilized per year at HFC sites (monthly) can be dropped on the order of up to 90%, freeing up human and equipment resources for more pressing needs. The results suggest that even with the success of aggressive scheduled cleaning, WCTS maintenance programs can be dramatically improved, resulting in fewer spills, better enforcement of illicit discharges, much more efficient use of resources for cleaning, and use of funds for capacity improvements within the system, rather than calendar-driven cleaning.

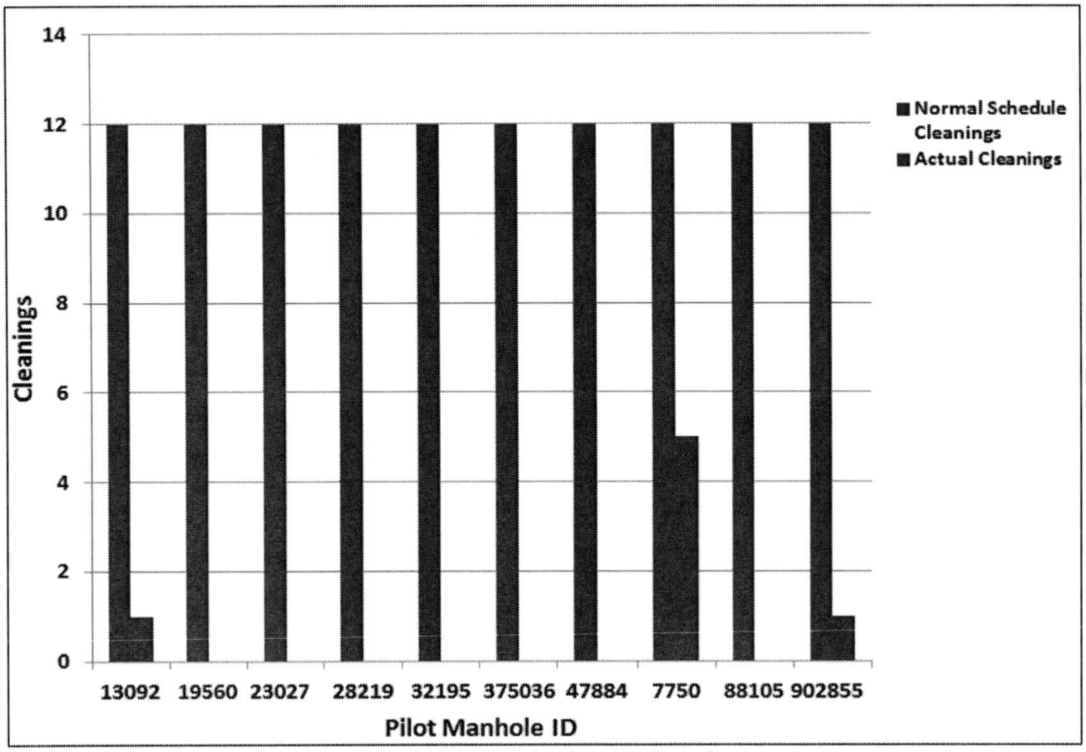

Figure 6: Expected versus actual cleanings for SAWS HFC maintenance program, over 12 months of program, starting in June 2015. All locations were monthly schedule. 120 cleanings were scheduled, 7 were performed, a decrease of 94%.

Costs for cleaning are reported to vary from $0.43 to $1.46 per lineal foot of pipe between manholes. This depends on if pipes are contiguous, diameter of pipe, and other factors that may adjust the time it takes for crews to clean pipe.

The average distance between manholes in San Antonio is 285 feet. Thus, the average cost per cleaning of one manhole to manhole section of pipe is estimated to vary from $123 to

$416/manhole per cleaning. 12 cleanings per year (monthly) would make an annual cost per manhole range from $1,471 to $4,993. With annual sensor costs on the order of $800, which assumes 10 year useful life, annual inspections, biennial maintenance (battery replacement and other component replacement) and monthly data service, reducing from 12 to 1 cleaning per year, can result in a return on investment of 66% to 465% (savings divided by cost of annual costs of sensors). One need only reduce cleanings by 2 times per year in the higher cost case to 7 times per year in the lower cost case to pay for the sensors with savings through reduced cleanings.

Most importantly, the risk of SSOs is no higher using sensor driven cleaning schedules, than with HFC – in fact likely lower because the sensors can provide early warning of a change that HFC approaches cannot.

Beyond straightforward cost savings associated with fuel costs, labor rates, and equipment wear and tear, lowering cleaning frequencies has other direct positive consequences to the utility, including:

- Lower risk of Workers Compensation claims. In Texas (category 9402) and California (category 7580) field staff performing cleaning would be exposed to traffic risks, H_2S exposure, back injuries, extremity trauma, slips and falls. Lower cleaning frequencies could translate into lower exposure to these risks. Since workers compensation rates range from 5% to 40% of payroll, reduced experience modification rates could occur from lower cleaning time. A typical workers compensation claim is $20,000 to $100,000, so even a single avoided claim provides a huge payback for lower cleaning intensity.
- Combo trucks not in traffic don't have collisions or generate liability claims from the public.
- Lower cleaning frequencies also means a lower burden on management.
- Less work in night hours reduces traffic risk and reduces overtime.
- Lower cleaning frequencies can result in a lower carbon footprint for the WCTS.

The technology is now available, is resilient and durable for field applications, and has been shown to effectively improve operational efficiency for WCTS.

Additionally, with increasing regulations, less tolerance for spills from regulators and the public, and pressures on cost control, the ability to provide data to "right clean" at the right time can help WCTS managers save money, reduce risk, and successfully navigate the difficult waters of the 21st century wastewater utility.

Conclusion

The emergence of low cost, reliable and pervasive sensing and real-time data processing and analytics has led to a world-wide expansion in the areas of the "internet of things". The results from this pilot test in San Antonio illustrate, for WCTS, monitoring and real-time remote data create valuable visibility into the collection system where data were incomplete, subjective, and sporadic, thereby enabling better and timelier decisions.

As regulators push forward seeking compliance with the Clean Water Act, utilities have responded with aggressive, high cost, often expanding cleaning programs. This study suggests there is a lower cost, real-time solution that also adds 24/7 spill protection. In a larger sense, there also exists the potential for gaining valuable insights into the behavior of the collection system under the broad spectrum of influences affecting the performance of the WCTS. Acquisition of data from a network of monitors has the potential to quantify ongoing flow and level patterns that can be associated with, for example, weather data, local water usage, and climate change. In turn, these insights may reveal vulnerabilities. With this new "sewer intelligence" planners can be armed with data that enable much improved targeting of capital. Monetarily, the savings effects could be profound.

While this study demonstrated clear advantages in terms of continuous visibility into the pipe conditions in a WCTS, more work is being done to further link these type of results with hydrological models, CCTV camera data, and level signatures. With tools and technology at hand and able to support a strong, fast return on investment, the last challenge for utilities is to assess and embrace this potential for transformational change using the emerging world of the internet of things.

REFERENCES

1. http://www.sandiego.gov/mwwd/sewerspill/index.shtml
2. City of San Diego Press Release, February 4, 2010. "**MAYOR APPLAUDS CITY WORKERS FOR TEN-FOLD DROP IN SEWER SPILLS** - *Spill rate down from one per day in 2000 to one per ten days in 2009"*

APPENDIX A - SAWS Level Data Results, All 10 Pilot Sites

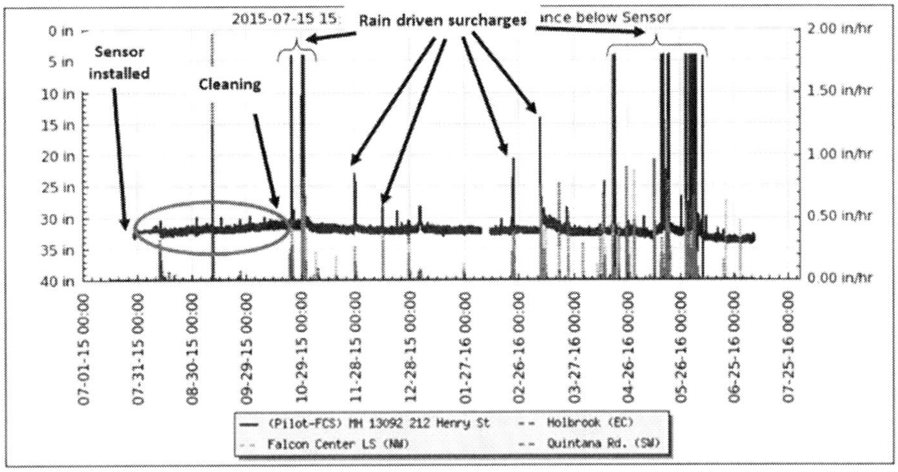

Figure A1a. Water level history (blue), from SAWS Pilot Site #1, MH 13092, a monthly cleaning site. The level sensor was installed on 29 July 2015. Over the next three months, there was a very slow, but distinctive rise in average water levels, indicated in the green circle and expanded in Figure 7b below. Based on this level data, SAWS crews cleared out the manhole and prevented an overflow. Local NOAA radar rain data is shown (dotted red, orange, green, and pink) for reference. There were several rain correlated surcharges at this location, including several during the intense rain storms in late May through early June 2016. 203mm (8inch) line in, 203mm (8inch) line out.

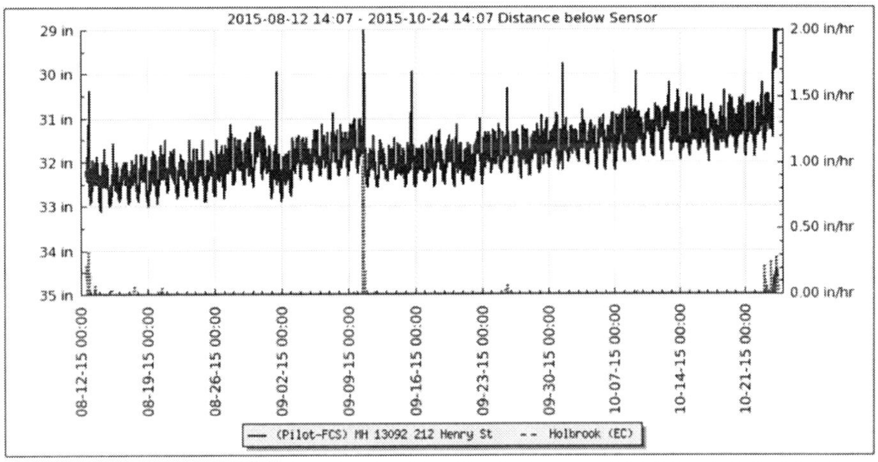

Figure A1b. Expanded view of water levels at SAWS Pilot Site #1 between 8/12/15 and 10/26/15. Diurnal flows are seen, with a relatively constant rise suggesting either a blockage downstream or local grit or debris buildup. This level change would not be detected manually, it corresponds to 0.021" of rise per day.

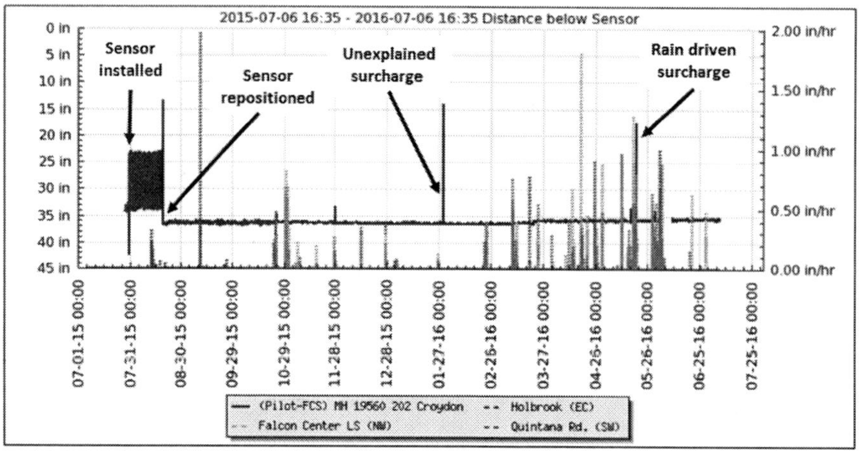

Figure A2. Water level history, from SAWS Pilot Site #2, MH 19560, a monthly cleaning site. The sensor was installed on 28 July 2015 and required repositioning on 8/19/15 to capture the water levels in the invert. There are no notable trends detected over the next seven months. After significant rain occurred starting in late February, there is a slow rise in baseline levels, suggesting either ground water infiltration or grit/debris build-up. An unexplained brief surcharge, unrelated to rain, occurred on January 29, 2016. No cleaning events and no spills occurred here during the pilot. 381mm (15inch) line in, 381mm (15inch) line out.

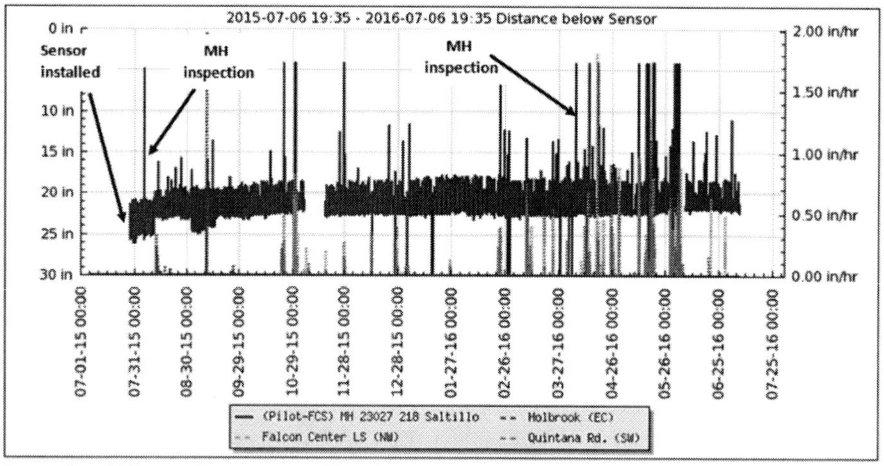

Figure A3. Water level history, from SAWS Pilot Site #3, MH 23027, a monthly cleaning site. The sensor was installed on 28 July 2015. Recorded manhole inspections took place on 8/5/15 and 4/6/16. Other visits also occurred when alarms were received as a consequence of rain events or other unknown reasons. There is a small upward trend in the minima of the data over the period of the year, especially in the late May-June 2016 time frame. No cleaning events and no spills occurred here during the pilot. 203mm (8inch) line in, 203mm (8inch) line out.

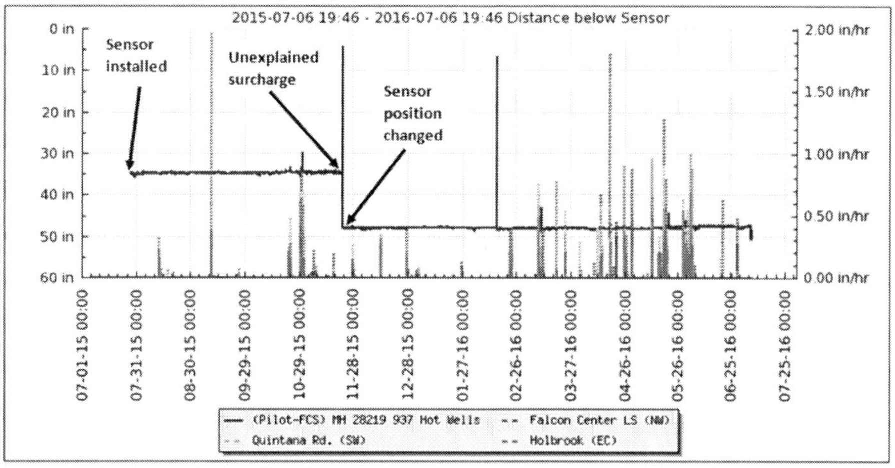

Figure A4. Water level history, from SAWS Pilot Site #4, MH 28219, a monthly cleaning site. The sensor was installed on 7/28/15. An unexplained short duration surcharge occurred on 11/22/15. Crews responding to the alarm noted no problems, but as a consequence of the visit, the sensor position was changed as is reflected in the data. There are no notable trends indicating potential blockages or reduced capacity. No cleaning took place here during the pilot and there were no spills. 203mm (8inch) line in, 203mm (8inch) line out.

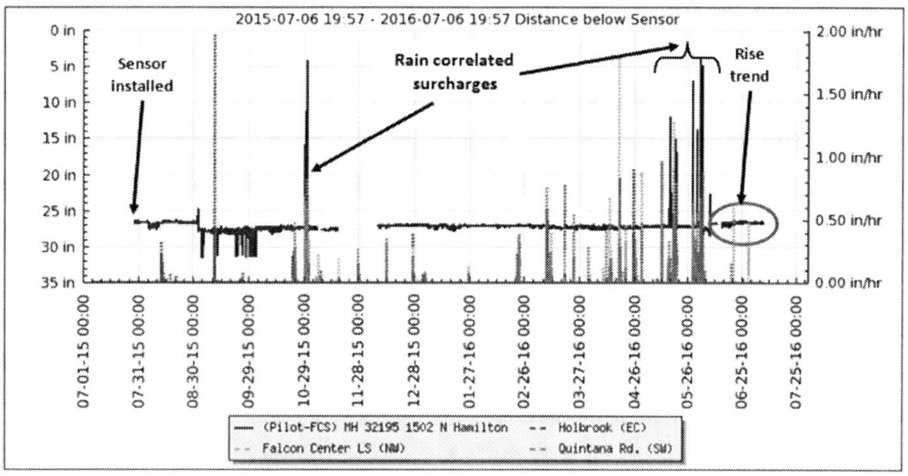

Figure A5a. Water level history, from SAWS Pilot Site #5, MH 32195, a monthly cleaning site. The sensor was installed on 7/29/15. Rain correlated surcharges occurred on 10/31/15 and during the May-June deluge. After the severe rain, there appears to be a rising trend, suggesting either CCTV investigation or cleaning. No cleaning took place during the pilot period. 203mm (8inch) line in, 203mm (8inch) line out.

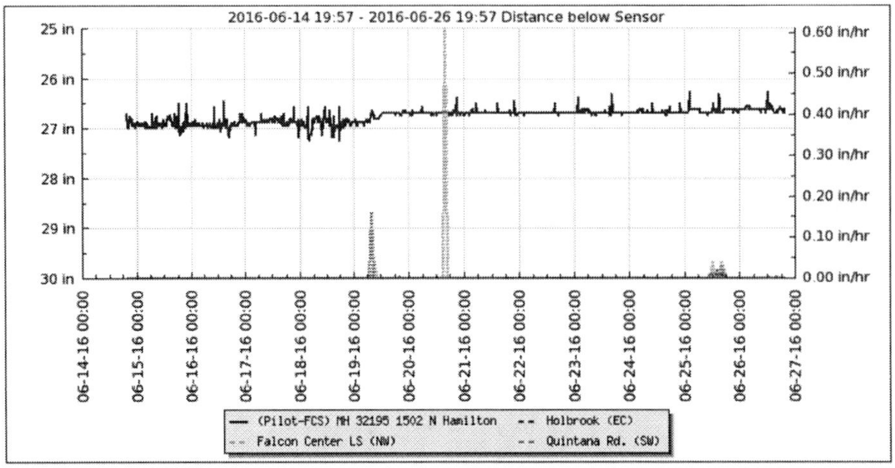

Figure A5b. Expanded view of water levels at SAWS Pilot Site #5 between 6/14/16 and 6/27/16 suggesting either a downstream buildup or local grit/debris buildup. A CCTV investigation or cleanout would be recommended.

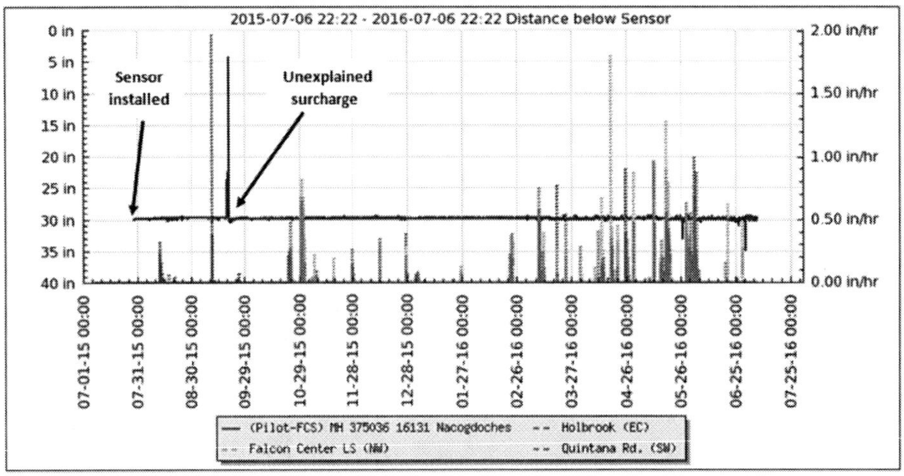

Figure A6a. Water level history, from SAWS Pilot Site #6, MH 375036, a monthly cleaning site. The sensor was installed on 7/29/15. An unexplained short duration surcharge (no overflow) was detected on 9/20/15. No other notable trends occurred during the pilot. No cleaning events took place here during the pilot, and there were no spills at this location. 203mm (8inch) line in, 203mm (8inch) line out.

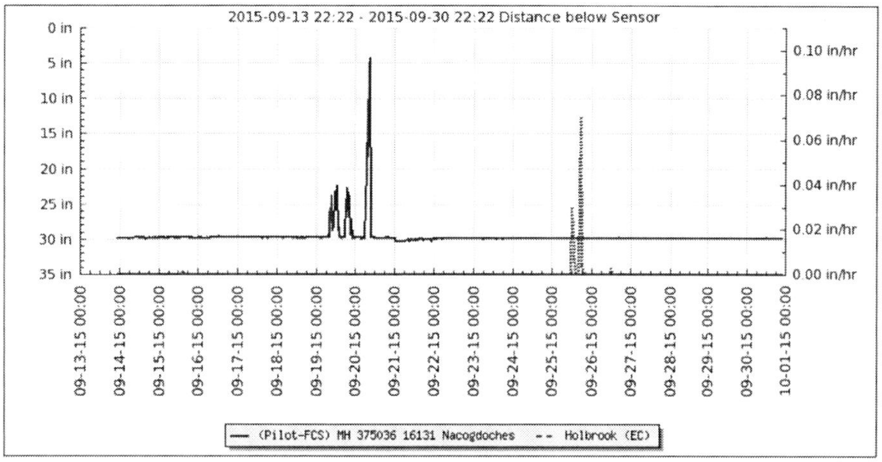

Figure A6b. Expanded view of water levels at SAWS Pilot Site #6 between 9/14/15 and 10/1/15. Two small surcharges occurred before the surcharge that led to a high level alarm. The sources of these surcharges is unknown, but baseline levels remained relatively flat and without trends for the remainder of the pilot period.

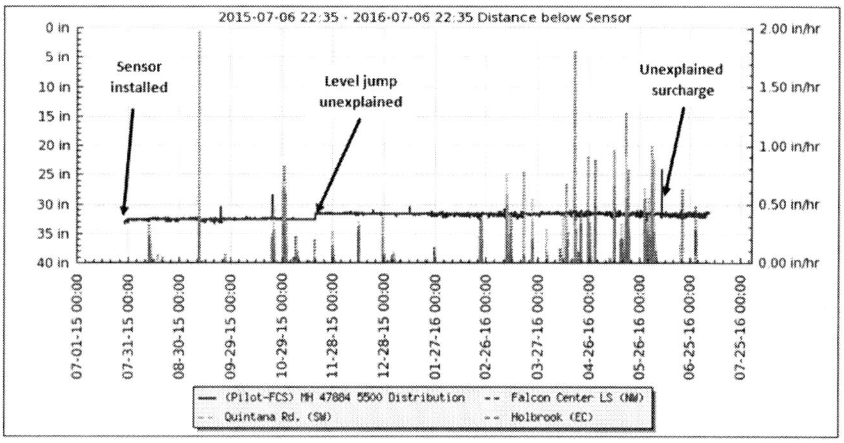

Figure A7. Water level history, from SAWS Pilot Site #7, MH 47884, a monthly cleaning site. The sensor was installed on 7/29/15. A rapid level jump – unexplained, but perhaps due to manhole inspection - took place on 11/18/15. An unexplained surcharge, sub-alarm level took place on 6/8/16. At around this time, levels appeared to become more variable, suggesting inspection or cleaning was in order. No cleaning events and no spills took place during the pilot. 203mm (8inch) line in, 203mm (8inch) line out.

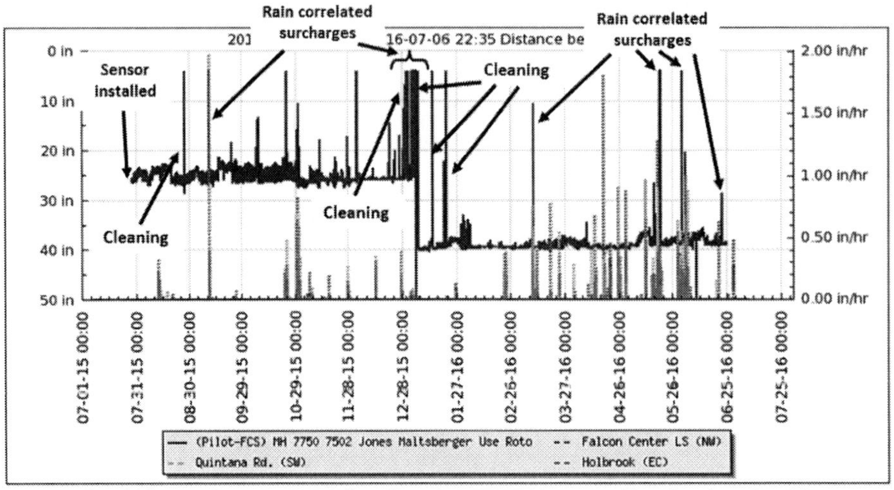

Figure A8a. Water level history, from SAWS Pilot Site #8, MH 7750, a monthly cleaning site. The sensor was installed on 7/28/15. Based on a high level alarm that occurred on 8/27/15, SAWS crew cleared out a grease buildup and prevented a spill. An expanded view of this period is shown below. In the next few months, between 8/30/15 and 12/31/15 other surcharges occur, including some rain correlated surcharges and others during dry periods. See expanded view of this period below. Causes of the dry period surcharges are unknown. Between 12/31 and 1/6/16 there were a series of surcharges and associated cleanings. Repairs to this site were performed on 1/16/16, mitigating capacity issues that were causing the surcharging. 203mm (8inch) line in, 203mm (8inch) line out.

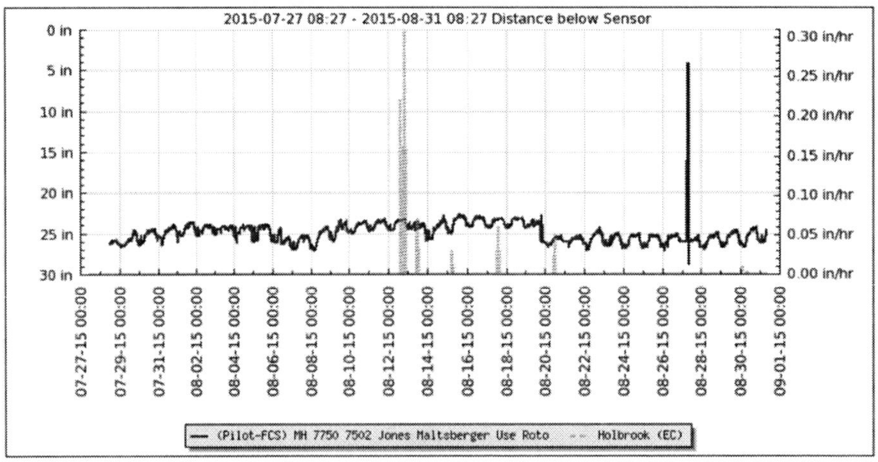

Figure A8b. Expanded view of water levels at SAWS Pilot Site #8 between 7/28/15 and 8/31/15. Diurnal flows are seen, but baseline lines are variable, the source of this variability is unknown. The surcharge on 8/27/15 prompted a visit by SAWS crews, who discovered grease buildup and cleaned out the downstream lines.

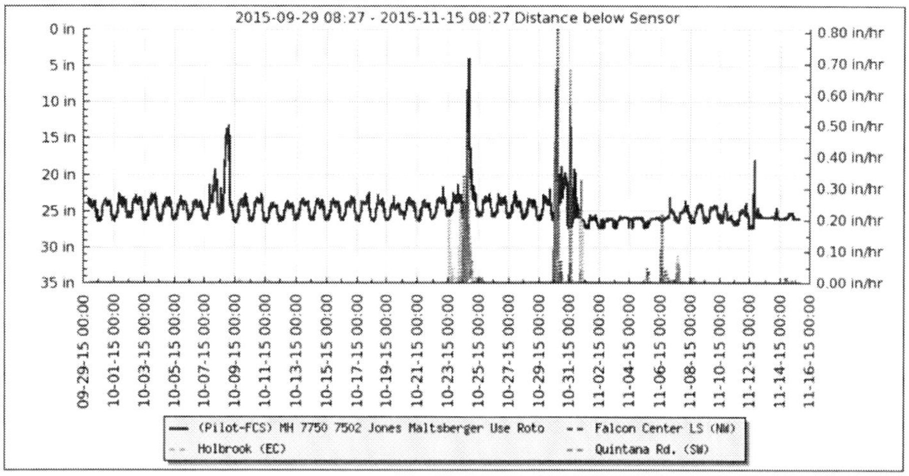

Figure A8c. Expanded view of water levels at SAWS Pilot Site #8 between 9/29/15 and 11/15/15. Two dry weather surcharges occur on October 7 and 8, respectively. The cause of this surcharge is unknown, but is should be noted that level patterns before and after these surcharges are similar, suggesting these surcharges are temporary capacity issues and not FOG or root constrictions that would have a longer term effect. Response of this location to rain events is seen on 10/24/15 and 10/30-15 – 11/1/15.

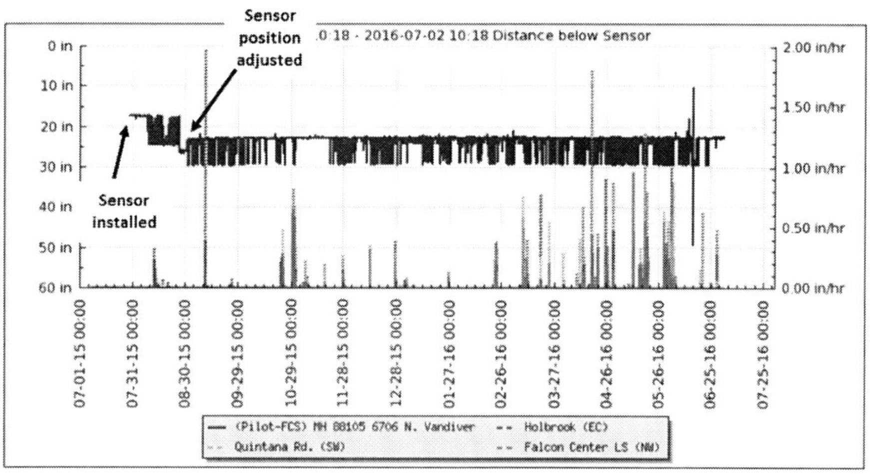

Figure A9. Water level history, from SAWS Pilot Site #9, MH 88105, a monthly cleaning site. Based on the noisy signature, the sensor was repositioned on 8/28/15. The level signature continued to be noisy throughout the term of the pilot program, making trend analysis difficult. Nevertheless, this site showed little response to inflow from rain events and had no significant surcharges during the test period. No cleanings were performed here and there were no spills. 203mm (8inch) line in, 203mm (8inch) line out.

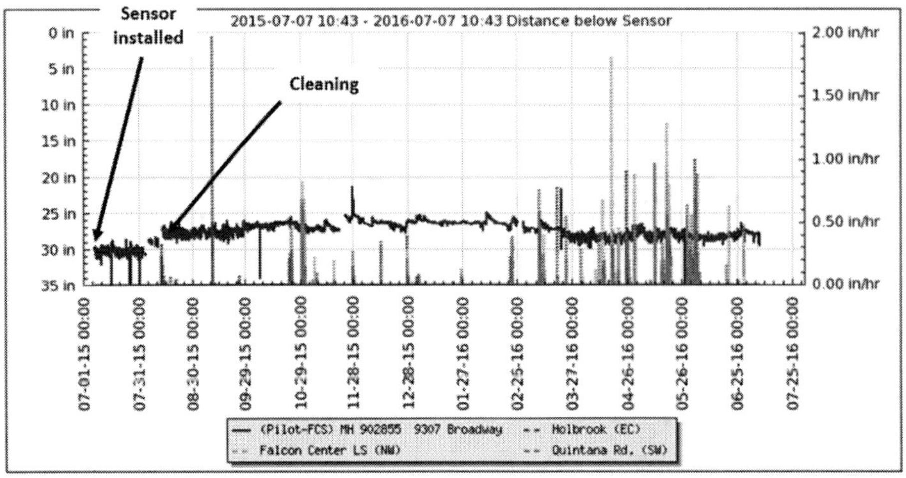

Figure A10. Water level history, from SAWS Pilot Site #10, MH 902855, a monthly cleaning site. This location was already in service prior to the commencement of the pilot program, and sited at a one month frequency location. A cleaning event took place on 8/15/15. No other cleanings took place at this site during the remainder of the pilot program, and there were no significant surcharges or spills at this site. 203mm (8inch) line in, 203mm (8inch) line out. ***

The Digital Water/Wastewater Utility of the Future: Case Studies in Leveraging Smart Utility Technology and Best Management Practices

Patrick Woodall (City of Atlanta), John Abrera, Srini Vallabinini, Katherine Y. Bell, and Allegra da Silva

ABSTRACT

Utility leaders are constantly presented with nearly impossible "do more with less" demands facing an increasing regulatory environment; and complex issues of hiring the right talent in both the short and long term. The Internet of Things (IoT), Big Data, Cloud Computing, Business Analytics, Utilities of the Future and Smart Cites are terms that increasingly becoming part of the daily vernacular for utility managers and their staff. In order to clarify these terminologies and how the key concepts of IoT, Big Data and Business Analytics are brought together into a highly useable and cost effective framework called, Intelligent Platforms, this paper will provide an overview of two relevant case studies from the UK and the USA, Thames Water (United Kingdom) and the City of Atlanta, Department of Watershed Management. In both of these case studies, an Intelligent Platform was deployed to bring together real-time and other data from a multitude of source systems to produce unprecedented results and improvements in both OPEX and CAPEX, particularly in the areas of sewer, storm water monitoring and water production.

KEYWORDS

Water Utilities, Intelligent platform, Big Data, data analytics, utility of the future, real-time data, predictive analytics, business intelligence, Sewer Systems, Stormwater, Sewage, Overflows, Spills, CAPEX/OPEX, TOTEX.

INTRODUCTION

Many leading utilities are trying to figure out how and where new technologies will make it easier to solve problems. While there is an underlying assumption that if the answer is to be found in the data, then more data will provide better answers. In most cases, the response is to provide more data, for less money.

Intelligent Platforms Provide Utility Managers with New Methods and Tools

- Meeting regulatory compliance and safety goals
- Enhancing operations and maintenance
- Improving service levels for customers
- Enabling decisions to be made in real time/near real time with more accurate information
- Enabling more effective capital planning and repair replacement programs

However, this is just part of a complete solution for solving real operational and management problems as demonstrated in **Figure 1**. In addition, these solutions can be implemented much more efficiently and with a much faster return on investment (ROI).

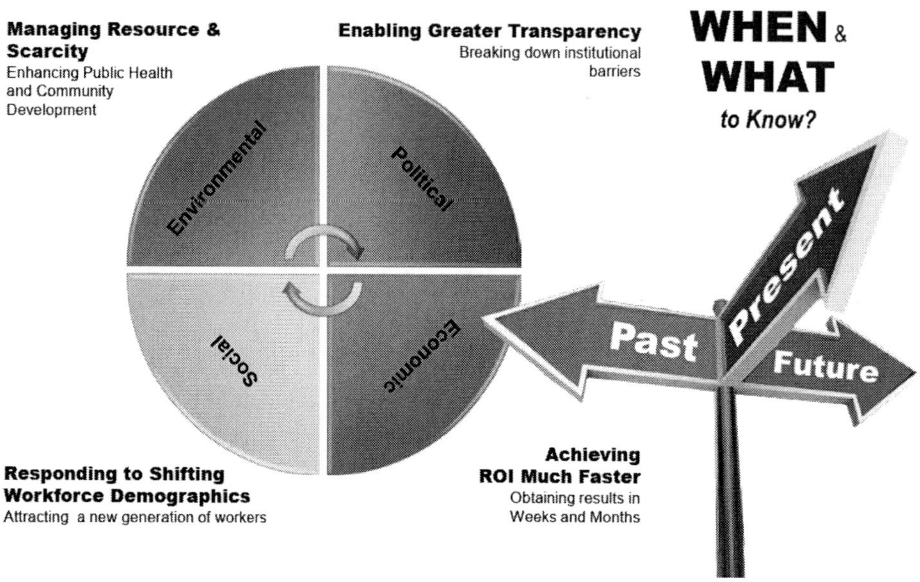

Figure 1. More effective ways are emerging in managing competing/complex utility challenges

Big data solutions must address the challenges of design and operational data, such as customer billing, geographical information systems (GIS), hydraulic modeling databases, supervisory control and data analysis (SCADA), laboratory data, computerized maintenance management software (CMMS), and other data from various sources which is often stored in disassociated platforms. If leveraging existing data for specific management and operational goals is difficult, then adding more data may prove especially challenging.

Until existing data can be used efficiently and prove value, investments in expanded or improved instrumentation is a difficult sell. It is common to find data on a USB drive, a comma-separated-value (CSV) file in a spreadsheet, separate laboratory information management systems, data historians, and other formats. Crucial operational data are in stored in the SCADA historian, but always require filtering, transformations, or tag editing to correlate asset identifications. If data are to be used for decision-support, they need to be readily accessible, manageable and users must have a platform that enables analysis of these enormous data sets.

Use of intelligent platforms allows a shift from conventional systems integration **(Figure 2)** to data integration **(Figure 3).** An integrated data approach supports analytics for improved decision-making. Predictive analytics applied to enormous data sets that are generated at water utilities will enable process performance optimization that assures reliability and compliance, and allows optimization to maximize efficiency, reducing operations and maintenance costs.

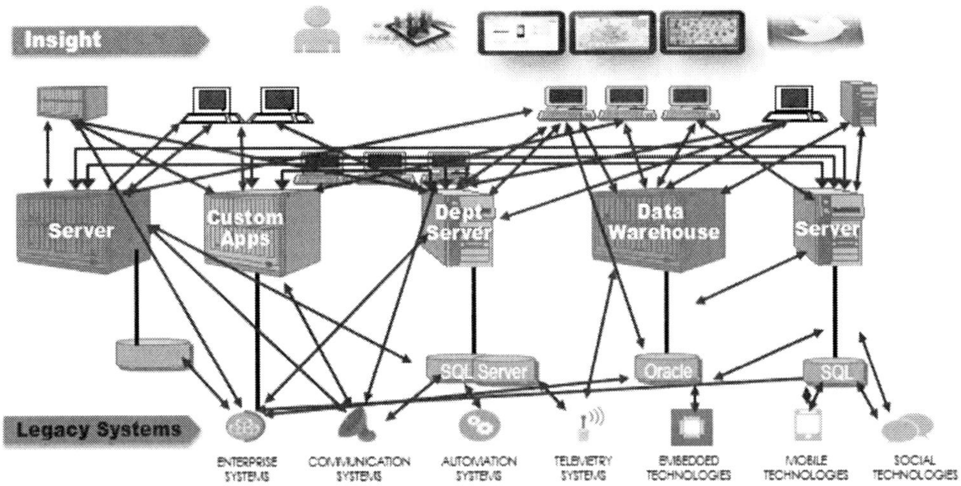

Figure 2. Traditional systems integration approach is costly to maintain and slow to implement

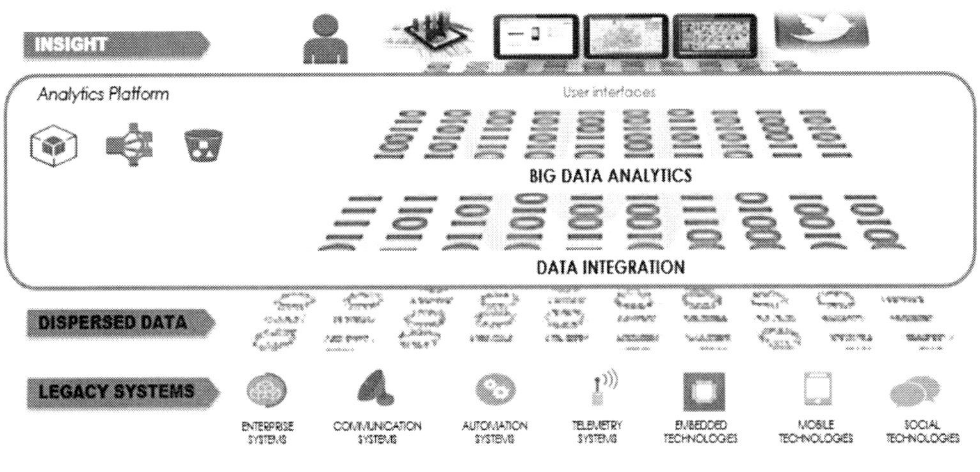

Figure 3. Data integration supports maximization of compliance and efficiency and less expensive to maintain

DATA PLATFORMS SUPPORT UTILITY TOTEX MANAGEMENT (CAPEX and OPEX)

While intelligent platforms and big data analytics have been used in other industries for decades, e.g., transportation, banking and finance, power, and retailers such as Amazon, the water industry has lagged in maximizing the value of existing data. MWH implements data management techniques that advance clients toward application of the concepts of Smart Utilities and Intelligent platforms. The details of this approach include application of an iterative process that follows six steps as shown in **Figure 4.** The first step, project visioning, is critical to shaping the direction of the project and allows the project team to develop a plan that is tested before advancing to full implementation.

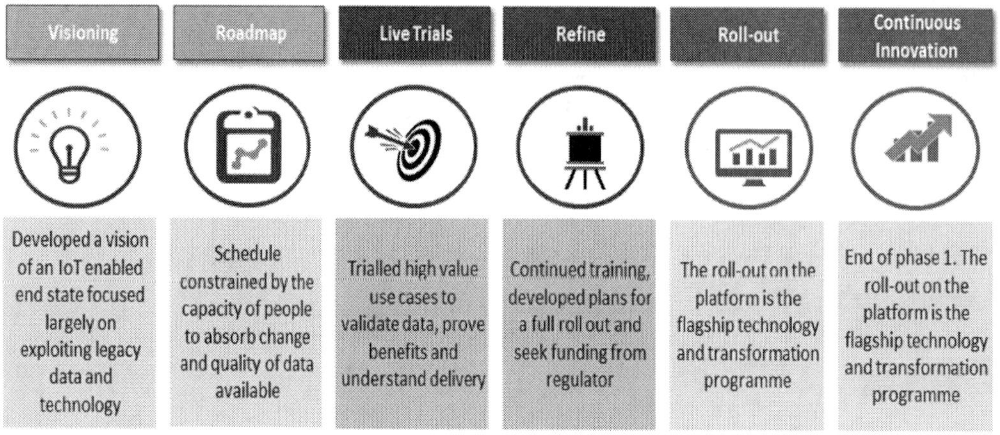

Figure 4. Approach to applying concepts of Smart Utilities and Intelligent Platforms enables both immediate and long term success in TOTEX

During the project visioning, focus is on answering important questions including:

- How prepared are you to leverage and apply new innovations in smart technologies and intelligent platforms?
- Where do you begin and align those efforts with your business priorities?
- Do you have the foundational components and systems in place to deploy new innovations and management approaches in a timely and cost effective manner?
- What training and other organizational improvements are needed to effectively apply this new approach?
- How best can you capture and leverage the knowledge about our utility's operations, processes and assets and share it with the next generation of utility staff.

Once the roadmap has been developed, it is possible to conduct live trials that are refined through an iterative approach. The roll out of use cases is then conducted in a step-wise fashion to demonstrate the value of the systematic approach allowing continuous innovation which ultimately can transform the utility.

Case Study 1: City of Atlanta, Sewer Overflow Analytics and Consent Decree Management

The City of Atlanta Department of Watershed Management is the water and wastewater utility for the City of Atlanta, Georgia and adjacent communities. The Department has two drinking water plants, four wastewater treatment plants, six combined sewer overflow treatment facilities, and 16 sewage pump stations. There are approximately 1,600 miles of water distribution pipelines lines and roughly 2,000 miles of wastewater and combined sewers. The city's wastewater system serves more than 1.5 million people who commute to the city to work or who live outside the city in DeKalb, Fulton, and Clayton counties.

MWH is currently the Program Manager for Consent Decree (CD) projects for the City of Atlanta Department of Watershed Management. The CD Projects include the combined sewer and wastewater systems. As part of this program management effort, the City engaged MWH in the implementation of an intelligent data platform to track Consent Decree data and to expand its use to other functional areas of the Department. This intelligent platform solution utilizes innovative technologies including the Big Data, cloud based computing, predictive analytics and strategic sensors and incorporates best management practices in the areas of program and storm water management.

The overall goals for this effort include:

- Leveraging the City's existing infrastructure and systems to enable more effective decision making
- Predicting both dry and wet weather overflows before they happen
- Enabling more effective capital planning

Following below in **Figure 5** is the City's Implementation Roadmap which was developed using a similar approach as described in **Figure 4.** The implementation roadmap consisted of three distinct phase with each phase establishing success as well as a building block for the next phases.

The three phases included:

1) Consent Decree Closeout
2) Asset Management Focused Utility
3) Customer Benefits Driven Utility

The City is currently in Phase 2 of the project.

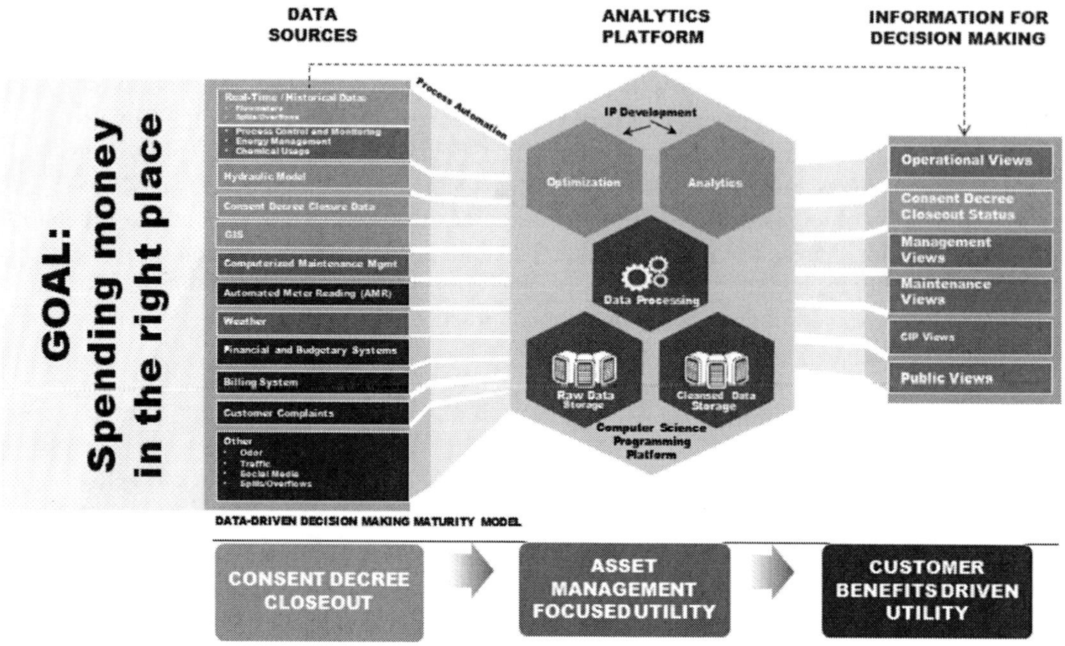

Figure 5: City of Atlanta's Intelligent Platform Implementation Roadmap based on Three Phased Approach

A state of the art analytics platform has been developed that receives near-time unstructured information from The City's SCADA, CMMS, GIS, and other business systems and disseminates useful and meaningful information through enterprise wide dashboards. Atlanta was able to see immediate benefits in just a few weeks of the platform launch.

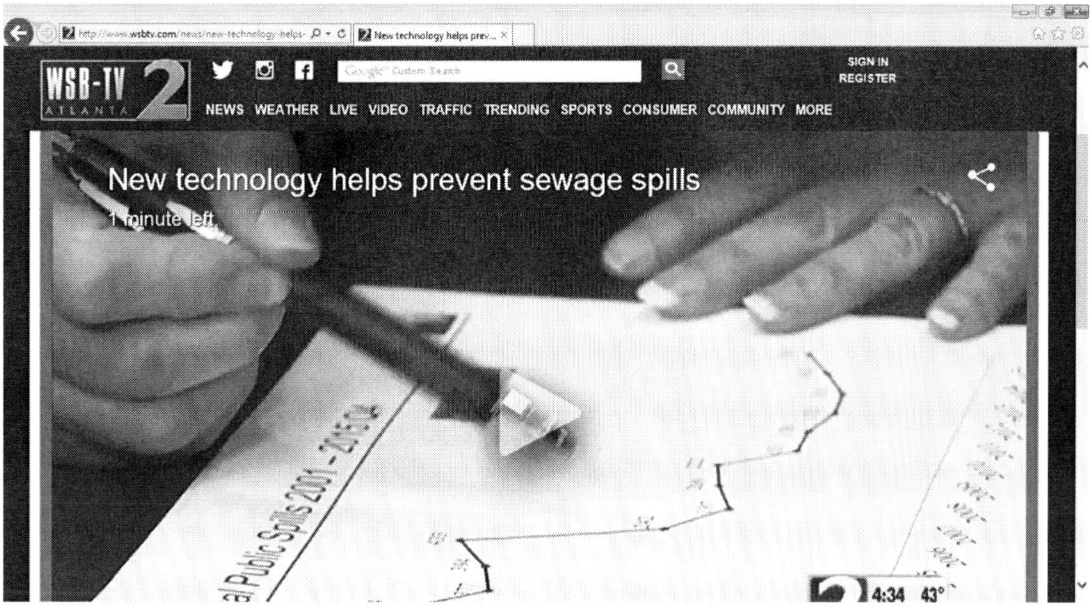

January 2016 new report on project: http://www.wsbtv.com/news/new-technology-helps-prevent-sewage-spills/54666169

Implementation the analytics platform allowed the City to view a number of data sets - some relatively static (from GIS, models), some changing less frequently (project data), and some dynamic (SCADA, flow and level monitoring) to allow for real-time, comprehensive decision-making related to the sewer monitoring.. Predictive analytics, based on large data analysis and statistical regression analysis on sewer level, allow operators to see if there are any actions required at specific points in the network, reducing the amount of wet weather and dry weather spills. Additionally, daily sewer spill information, shown overlaid with completed projects show the effectiveness of completed projects, while also helping prioritize upcoming projects. These types of analytics enable the City to be more proactive, and in some cases, predictive in their analysis of data being produced from their systems, allowing them to make better, more timely decisions.

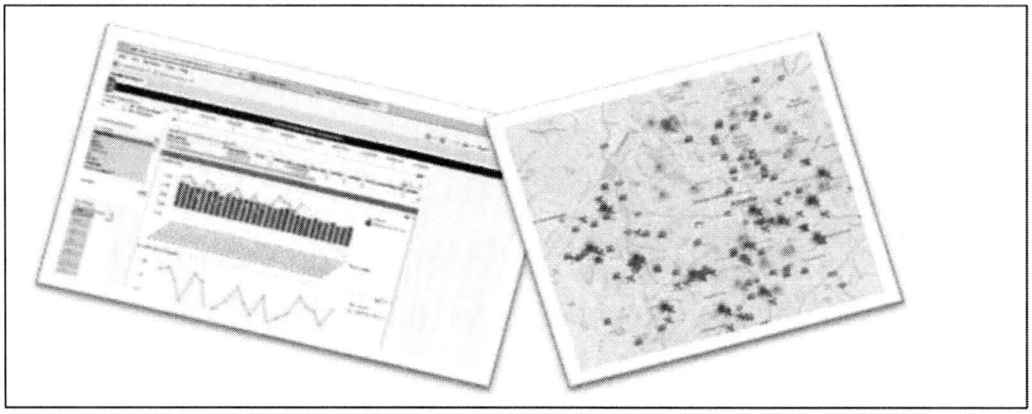

Figure 6. Sewer monitoring system predicts overflows and alerts staff before they occur

Results to Date

The results achieved to date on this project have included:

- Program pilot delivered on schedule and on within budget.
- 30% reduction in sewer overflows and spills using predictive analytics thus reducing fines and improving customer service levels
- Lowering of O&M costs for collection systems.
- More effective prioritization of capital spending to meet consent decree commitments

Case Study 2: Thames Water Plc, London, United Kingdom, Intelligence Hub (IHub)

Thames Water Utilities Ltd (TW) is the largest provider of water and wastewater services in the UK, serving 15 million customers in the London and Thames Valley region, and operating 102 water treatment plants and 350 sewage works. TW is amongst the first movers on the intelligent use of real time data to monitor its assets as an efficient way to deliver the quality of service customers expect. As part of a long-term vision to embrace the Internet of Things to transform decision-making on the frontline, the utility company selected the Eight 2 O Alliance led by MWH now part of Stantec to develop actionable insights on its Water Analytics platform.

A state of the art analytics platform has been developed that receives, cleanses and blends together both structured and unstructured information from thousands of sensors and many software systems such as SCADA, Water Modelling, GIS, Enterprise Resource Planning (ERP) and Enterprise Asset Management (EAM). IHub created a single view of its operating systems and assets, such as pipe and treatment facilities **(Figure 7).**

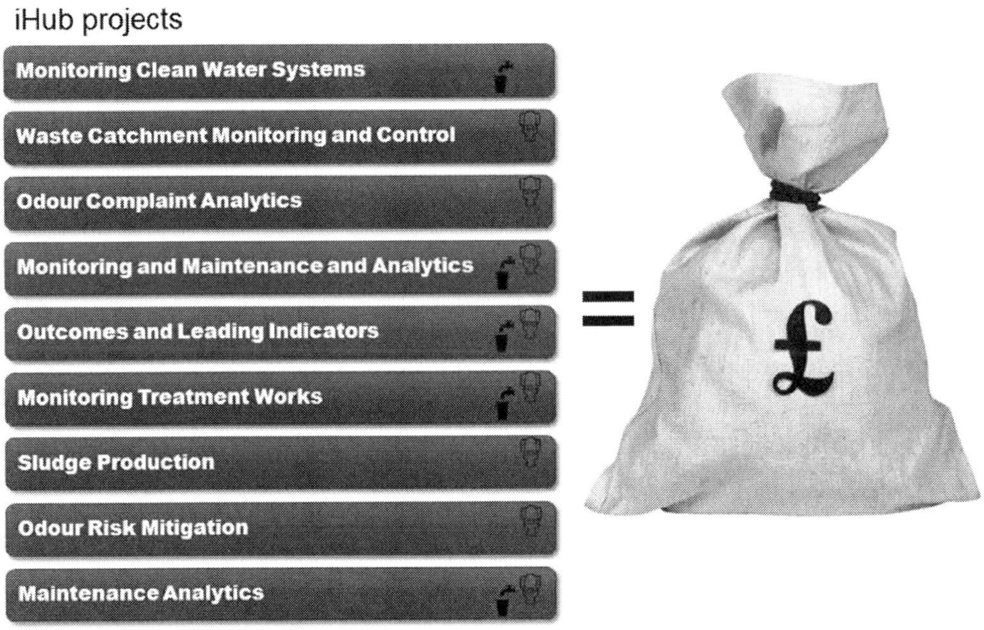

Figure 7. IHub Creates a Single View of Thames Water's Operating Assets and Key Performance Indicators.

The platform, which has been operational since April 2015, utilizes recent developments in cloud computing, real-time big data storage, cleansing and management, and predictive analytics to serve insights that support day-to-day operational decision making, as well as longer-term planning. The data is distributed to Thames Water's operational staff and management through real-time dashboard views that are available over the internet, allowing users to proactively respond to varying operating conditions, such as demand fluctuations, incoming weather fronts, and equipment failure.

Thames Water staff save significant time investigating and searching for data across source systems. As an example, the platform analyses data from five source systems to detect the risk of sewage spilling out of the network. It then shows it on a single, map-based views along with other contextual information needed to take action such as geolocation, affected customers, pump configuration, the next point of failure, flooding history, distance to nearest pollution point, asset condition, and ongoing work. With real-time cleansing, the platform also performs noise filtering on incoming alarms from ultrasonic sensors, highlighting data errors and communications failures.

New teams and new more effective working practices have emerged as a result of the right information being available at the right time to the right people. These include a centralized plant production planning process supported by a production optimization algorithm that calculates the most cost-effective routes to meet water demand.

The Intelligence Hub platform enabled significant (TOTEX) benefits over a five-year period for just one third of the TW asset base. This equates to a 25 to 1 ROI. These include increased production and headroom from low-cost sources, leveraging economies of scale and optimizing plant performance; measurable reliability improvements through proactive maintenance; reduced environmental compliance costs due to pollution prevention measures; increased efficiency in field work co-ordination; and intelligent use of storage capacity and tailored pumping regimes to reduce energy expenditure.

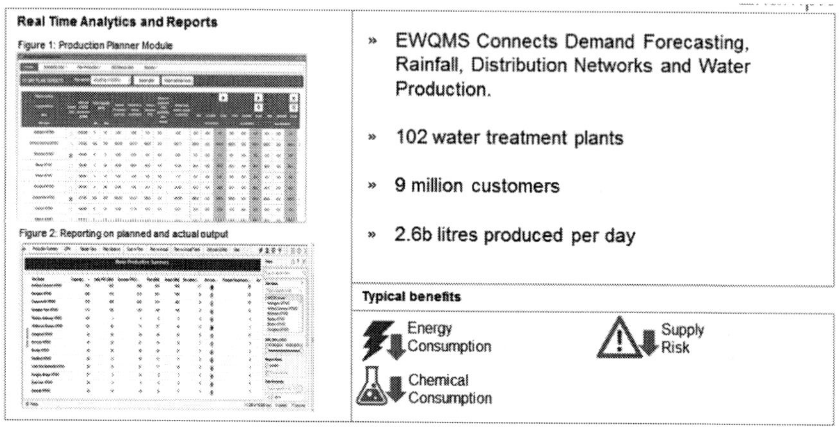

Thames Water has now embarked on developing new insights and rolling them out to new regions. The platform has been unique in its ability to work with whatever quality of data was available, cutting across siloed source systems. Unlocking a treasure trove of data is transforming how Thames Water view, understand and manage their assets to deliver the best possible service to their customers.

The results achieved to date on this project have included:

A 25 to 1 Return on Investment Resulting in Significant TOTEX Savings over a five year period:

- Capital avoidance
- OPEX reduction (Energy, Chemicals,)
- Leakage reduction:
- Productivity:
- Data and IT Storage Costs

CONCLUSIONS AND LESSONS LEARNED

When properly implemented using an intelligent platform approach, utilities have an unprecedented opportunity to take a leap forward in dramatically enhancing their capital planning efforts, improving operations and maintenance and service delivery to customers. Some of the key Lessons learned from our current efforts include:

1. **Management View**: More effective ways are emerging in managing competing utility demands
2. **Market Perspective**: Connected products and devices are transforming your relationship between utilities, regulators and customers
3. **Technology Perspective:** Expanded Automation/Big SCADA is key to achieving greater success
4. **Business Process Perspective**: Big Data and Business analytics deliver tangible results in weeks/months rather than years

In Conclusion, leveraging the Intent of Things, Big Data, Cloud Computing, Business Analytics, Utilities of the Future and Smart Cities can produce very positive results for utilities today

In order to apply this new approach that leverages existing infrastructure to the maximum benefit using modern methods of integration and data management to solve complex business problems, the approach to Intelligent Water must also include linkage with experts that understand this approach. While applications may be developed by programmers, it is important for expert teams to vet data correlations so that spurious results, which can be an artifact of "big data" analyses, do not occur. This requires a rigorous technical review of the information that is developed, and in addition to integrating systems and data, it is critical to ensure that the relationships are genuine with an understanding of the fundamental correlations. The analytical methods, underlying data quality and interpretation of the output must be understood for consistency with fundamental principles and when an expert team is engaged in this process, it is possible to make continual improvements to the platform and approach, extending the benefits to utilities.

Big Pipes plus Low Flow equals Cost Savings – Startup and Lessons Learned

Jon Schommer, EmNet; Oluwole McFoy, BSA; Mike Quinn, GHD; Tim Ruggaber, EmNet

ABSTRACT

In spring 2016, the Buffalo Sewer Authority (BSA) began operation of the City's first two Real-Time Control (RTC) facilities at Bird Avenue and Lang Avenue. These sites were implemented as a part of BSA's CSO regulatory requirements and the updated LTCP. Each site consisted of a chamber with two gates that stored and released wastewater in the existing combined sewer trunkline based on real time conditions in the trunkline. As part of the LTCP implementation, the BSA selected the two sites as demonstration projects to assess the effectiveness of all aspects of RTC and to inform the implementation of future RTC sites. RTC utilizes existing in-line storage capacity within the collection system to maximize the conveyance to the treatment plant and reduce the frequency and volume of overflows. The desktop analysis of RTC in the BSA system referenced in the updated LTCP, concluded that local reactive RTC could provide an approximately 15-20% reduction in annual overflow volumes (BSA LTCP, 2012).

This paper presents the results of the post construction monitoring analysis conducted during the three-month period following startup from March-May in 2016. During the post construction monitoring period both sites showed significant success with 13.3 MG of overflow prevented. The Bird RTC site prevented 12.1 MG or 64% of total potential overflow volumes at the Bird Sewer Patrol Point (Bird SPP) and the Lang RTC site prevented 1.2 MG or 48% of total potential overflow volumes at the Lang SPP. The success of the BSA LTCP is measured by the reduction in CSO activations and as such, control of Sewer Patrol Point (SPP) activations upstream of the CSOs is critical. That said, in terms of potential activation events, the Bird RTC site is estimated to have prevented 74% or 14 out of 19 events; and the Lang RTC site is estimated to have prevented 44% or 4 out of 9 events, based on the collected rainfall and flow data. The design and control logic approach implemented for the Bird and Lang RTC sites will be expanded upon in ongoing and future RTC projects throughout the City's collection system. The benefits these RTC projects provided to Buffalo Sewer Authority customers can be applied to many other collection systems in the US and beyond.

KEYWORDS
Real Time Control, Combined Sewer Overflow, In-Line Storage, Real Time Decision Support System, Smart Water Infrastructure, Smart Water

INTRODUCTION

Background

In 1900 the City of Buffalo, New York was the eighth largest city in the United States and by 1950 the population had increased to 580,000 people. During that time the City added a new large conveyance pipes network to serve the burgeoning population. However, over the last 60 years the population has had a net decrease of over 50%. This population decrease was mirrored by a decrease in the City's industrial base. The Buffalo Sewer Authority found that the resulting decreased load in the collection system meant that the large, shallow pipes in the system potentially had millions of gallons of unused storage capacity making them ideal for in-line storage use.

A thorough evaluation of BSA's collection system for in-line RTC storage was conducted. BSA identified in-line RTC locations along eight major trunklines (Figure 1) that were more than half empty even during the peaks of the largest storm events. Two inline storage locations, the Bird RTC site and the Lang RTC site, were selected as initial sites for implementation and would serve as examples for future implementations (BSA LTCP, 2012).

The Bird RTC site is located in a 6 foot 6 inch wide by 9 foot 9 inch tall egg shaped brick sewer built in the late 1800's in an area with very shallow bedrock and, as identified in the RTC evaluation studies, can potentially provide up to 1.0 MG of storage volume.

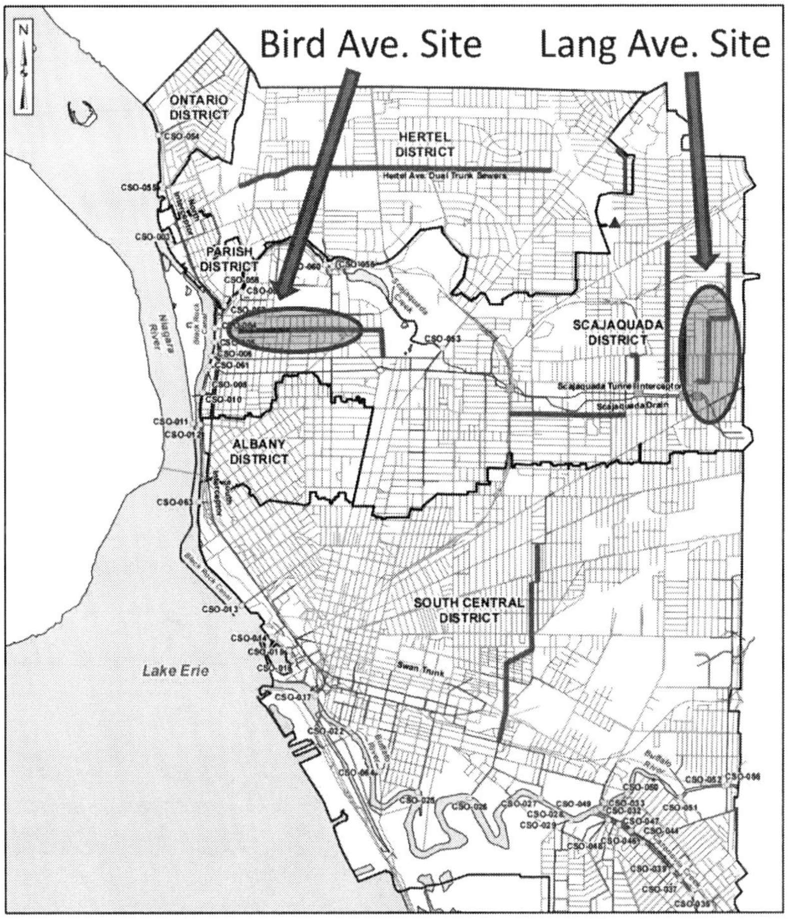

Figure 1. Inline RTC Storage Sites in the BSA Collection System (BSA LTCP, 2012)

Conversely, the Lang RTC site, located in an 8 foot 6 inch circular reinforced concrete pipe, is in an area with relatively deep bedrock, and can provide up to 0.84 MG of storage volume. The juxtaposition of the conditions at these two locations was chosen so as to cover a range of conditions expected to be encountered during construction of the remaining RTC in-line storage locations.

System Design

The system design of the RTC sites involved a diversely skilled project team that ensured the structural design and the gate control logic design would be most effective at regulating system storage capacity. Both the Lang and Bird RTC sites were designed for robustness and constructed with multiple fail-safes. Two smaller gates (5 foot x 5 foot for Lang and 4 foot x 4 foot for Bird) were built into the control chamber weir wall (Figure 2). Two gates were chosen because they were easier to install and maintain at a lower cost than one, larger gate. Furthermore, a smaller gate can be more easily modulated to provide an exact outflow, which is necessary for the control logic. Lastly, the two gates provide a level of redundancy. In the event that one gate fails closed, a majority of the flow could pass through the functioning gate with the remaining flow overtopping the emergency weir in the chamber. The chamber weir height was established to prevent basement backups upstream of the new chambers.

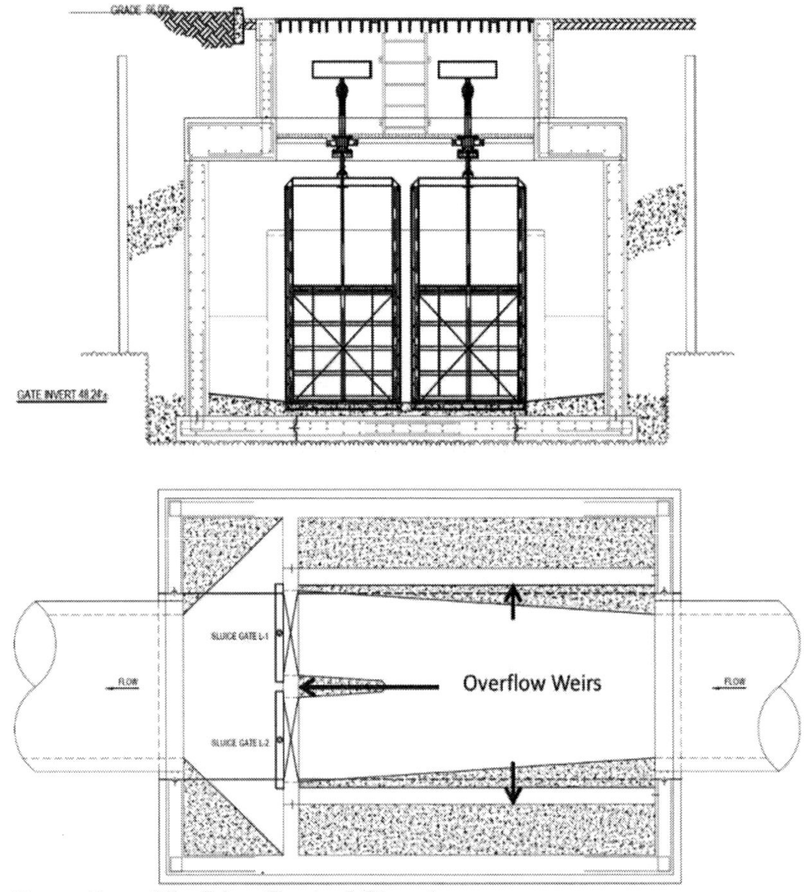

Figure 2. Configuration of In-Line Control Structures

Both sites were designed to convey the peak capacity of the respective sewer trunklines through the gates and over the chamber weir. This corresponds to a maximum flow of 500 cfs at Lang RTC site and 270 cfs at Bird RTC site. To get the necessary weir length in a constrained area, a multi-directional weir was implemented for each control site.

The control logic of the gates is the key to enabling efficient storage upstream of the gates. During dry weather, the gates at each site are normally in the open position except when exercised automatically after long periods of dry weather to ensure they are working properly. When the depth of flow indicates that the flow has increased to approximately twice the normal dry weather flow, both gates modulate the flow — one gate closes completely (called the lag gate), and the other one (called the lead gate) closes to the height of the water surface. The lead gate then modulates to keep approximately twice the normal dry weather flow going through the gate at all times while maintaining the upstream level in the chamber at or below the height of the chamber weir.

For both gates, the triggering depth initializes gate closing, and the gate opening curve determines how much the gates close at a given depth above the triggering depth. The trigger depth was initially set to just below the depth on the structure that correlated to the depth at the Sewer Patrol Point (SPP) being just below the weir for each site (as calculated by the SWMM model). During the post-construction analysis, the triggering depth was adjusted for Bird RTC and Lang RTC sites based on how the actual collection system performed during storm events. The trigger depth at the control site was tuned so that it was indicative of the potential overflow at the SPP. Once the trigger depth is reached, the gates begin to throttle back flow to hold water in the storage area. The goal is to send as much water downstream as possible during wet weather while maintaining the level below the SPP weir. This was done using the gate opening curves. These curves determined the gate position based on the depth upstream of the gate. In the current system configuration, there is no feedback to the control site about what is occurring at the SPP. Rather, the trigger depth and gate opening curves were tuned to achieve a certain depth at the SPP by analyzing how the system behaved during actual storm events.

Table 1 shows the original and adjusted values for trigger depth over time for the Bird RTC site. The trigger depth value at Bird RTC was first changed to four feet, which is much higher than what is expected in a typical storm. This was done to check the chamber performance without modulating gates. As the gate performance was monitored in the field it was found that the dry weather flow at Bird RTC was much lower than predicted in the model. This meant that the value initially set based on the model needed to be modified to prevent overflow at the SPP weir when the gate was activated. After monitoring the successful performance of the Bird gate at this trigger depth, it is likely that further calibration of the trigger depth can free up more storage by bringing the downstream depth at the SPP closer to the maximum depth of the SPP weir.

Table 1. Bird RTC Site Trigger Depth Adjustments

Parameter	Original	Chamber Check 2/1/2016	Final Trigger Depth 3/9/2016
Trigger Depth (ft)	0.75	4.00	1.7

Table 2 shows the original and adjusted trigger depth at the Lang RTC site. The triggering depth was adjusted to 1.2 feet based on depth data collected after activating the site. At all storms in the monitoring period, the Lang RTC gates triggered when the depth at the SPP was at or above 80%.

Table 2. Lang RTC Site Trigger Depth Adjustments

Parameter	Original	Final Trigger Depth 3/18/2016
Trigger Depth (ft)	1.55	1.2

A gate opening curve was used in the control logic to regulate flow out of the storage area without forcing overflow at the SPP. At the Bird RTC site, the original values set in the curve resulted in the expected performance. At the Lang RTC site, after adjustments in the trigger depth, there were still small amounts of volume being passed through the gate during closure that caused small overflows at the SPP weir when there was still capacity upstream of the gates. The values in the gate opening curve were then adjusted slightly on 3/28/2016 and 5/27/2016 to increase the gate sensitivity to depths that were observed to trigger the slight SPP overflow. Since the adjustments, the gates have been working as expected.

When the depth upstream of the gates approaches the height of the control site's overflow weir, the lead gate begins to open more to allow all excess flow to pass through the storage area and to keep the depth upstream of the weir as close as possible to the height of the weir without exceeding it. If the lead gate is fully open and not able to pass all the excess flow, the lag gate will then be modulated accordingly. This will only occur during severe storms.

Dewatering at the RTC location is a continuous process. The gates will continue to allow approximately twice the dry weather flow to pass through them until the flow drops below this threshold. At this point, both gates will open fully, in preparation for the next event.

METHODOLOGY

To conduct the post-construction analysis, data from 3/10/2016-5/20/2016 was used from flow and level sensors placed upstream of the control gate and downstream at the SPP. Prior to this, pre-construction monitoring was performed at both RTC sites to better understand the relationship between flow at the site and the SPP as well as to establish a value for typical dry weather flow used in the control logic to trigger the gates (Figure 3).

Once the RTC site was implemented, upstream depth was used to trigger the gate closing. The depth at the gate was used with the BSA SWMM model to create the volume curves (see Figures 4 and 5) that calculate the volume retained once the gates were closed. The curve equations correlated depth upstream of the chamber weir to volume stored using the SWMM model by totaling the available storage upstream of the gate at depths increasing in 1 foot increments. The equations used to calculate the volume at each depth are displayed in Figures 4 and 5.

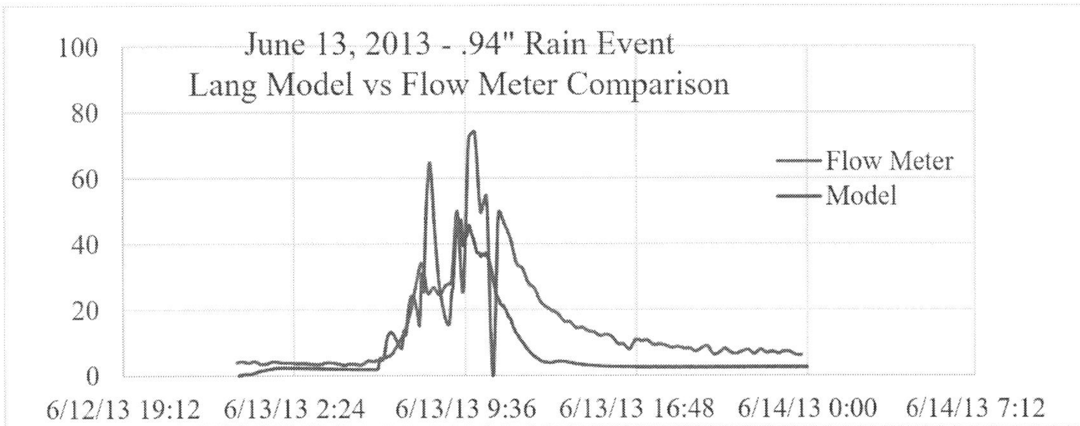

Figure 3. Example of results from preconstruction analysis comparing measured flow to modeled flow at the Lang site

The volume curves were then used with data for depths in the chamber that triggered gate closure to calculate the potential prevented overflow volume during gate activation. The control logic was designed to close the gates when local conditions upstream of the gates indicate a potential for SPP overflow, and the system was later tuned to achieve this objective. Therefore, additional volumes stored at depths greater than the trigger depth are the best estimates of prevented overflow volumes.

The actual overflow volume for the three-month period was calculated using sensor data that monitored the depth at the SPP weir. The measured depth at the SPP was used with the Francis rectangular weir flow equation to calculate the flow over the weir and convert to volume. This value for actual overflow was then added to the potential prevented overflow volume (stored volume) to determine the total potential overflow volume. The total potential, actual, and a potential prevented (stored) overflow volumes are displayed in Table 3.

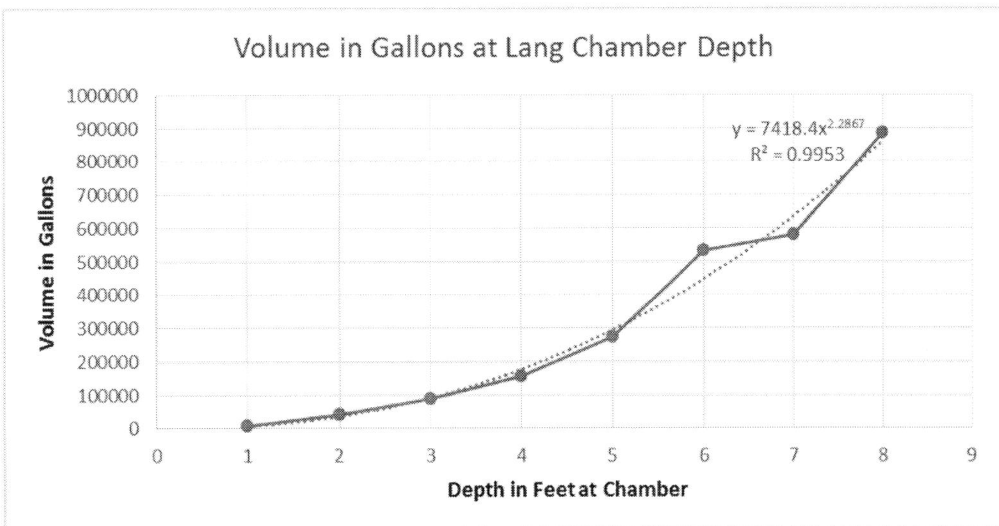

Figure 4. Volume Curve in gallons at Lang RTC Chamber for Depth in Chamber

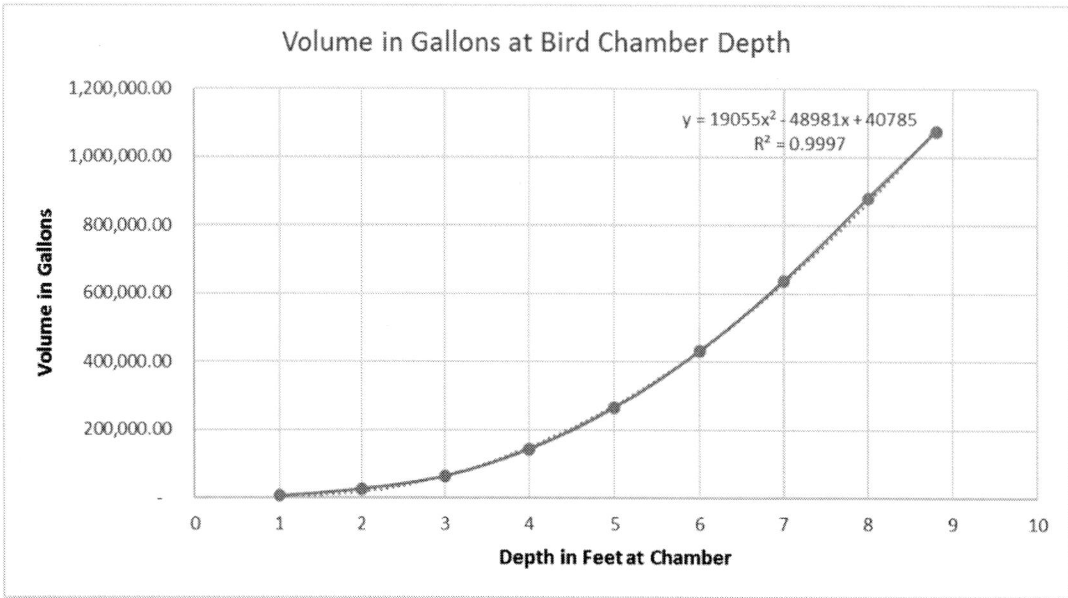

Figure 5. Volume Curve in gallons at Bird RTC Chamber for Depth in Chamber

RESULTS

After monitoring the RTC performance at each site from March-May, the values for actual overflow and potential prevented overflow in Table 3 were calculated. The chamber at the Lang RTC site was activated during a storm event a total of 9 times and the Bird RTC site was activated 19 times. Overall, the system worked exactly as designed and prevented up to 13.3 MG of potential overflows.

Table 3. Summary of RTC Chamber Performance

	Total Potential Overflow (MG)	Actual Overflow (MG)	Stored Volume (MG)	Percentage of Overflow Prevented	Number of Events with All Overflow Prevented	Number of Events with Actual Overflow
Lang RTC Site	2.5	1.3	1.2	48%	4	5
Bird RTC Site	19.0	6.9	12.1	64%	14	5

The values for potential prevented overflow, actual overflow, and potential percentage prevented per storm for Lang and Bird RTC sites are displayed in Tables 4 and 5. Daily precipitation data is also shown for each storm event as measured at the respective site. For events where there were multiple activations over the course of a day, only the precipitation just before and during the event is included. There were two storm events that triggered no response by the gate at the Lang RTC site, the first occurred on 3/16/2016, and was corrected in the calibration when the trigger depth was reduced to 1.2 feet. The second was on 5/2/2016 due to the gate being inadvertently left in

Table 4. Lang RTC Site Performance

Storm #	Date	Times	Daily Precip. (in.)	Peak rainfall intensity (in/hr)	Trigger Depth (ft.)	Max Depth (ft.)	Volume Stored (Gal.)	Actual Overflow (Gal.)	Total Potential Overflow Volume (Gal.)	% Prevented
1	3/10/2016	15:05-17:10	0.79	0.16	1.54	3.02	72,972	597,233	670,205	10.9%
2	3/24/2016	7:10-7:35	0.22	0.16	1.18	1.26	1,753	-	1,753	100.0%
3	3/24-25/2016	23:05-3:05	0.45	0.28	1.23	3.75	140,477	690,682	831,159	16.9%
4	3/28/2016	6:20-7:10	0.38	0.24	1.22	1.36	3,296	-	3,296	100.0%
5	3/31/2016	11:25-19:30	0.50	0.16	1.21	7.34	696,307	1,319	697,626	99.8%
6	4/11/2016	5:10-8:20	0.40	0.36	1.2	2.4	43,665	-	43,665	100.0%
7	4/11/2016	13:50-18:25	0.19	0.16	1.19	4.3	197,341	2,863	200,204	98.6%
8	4/26/2016	4:15-7:00	0.43	0.32	1.21	3.13	89,331	-	89,331	100.0%
9	3/16/2016	13:05-13:40	0.22	0.36	-	1.44	-	27,988	27,988	0%
No Trigger	5/2/2016	7:35-9:10	0.57	0.32	-	1.91	-	435,363	435,363	0%

Table 5. Bird RTC Site Performance

Storm #	Date	Time	Daily Precip. (in.)	Peak rainfall intensity (in/hr)	Trigger Depth (ft.)	Max Depth (ft.)	Volume Stored (Gal.)	Actual Overflow (Gal.)	Total Potential Overflow Volume (Gal.)	% Prevented
1	3/10-11/2016	6:20-4:35	0.71	0.24	2.19	8.18	890,230	1,665,282	2,555,512	34.8%
2	3/14/2016	4:35-13:25	0.28	0.20	2.15	7.97	837,240	-	837,240	100.0%
3	3/16/2016	11:50-19:50	0.12	0.20	2.13	7.97	837,891	-	837,891	100.0%
4	3/23/2016	20:45-22:50	0.14	0.04	1.95	3.23	63,647	-	63,647	100.0%
5	3/24/2016	6:15-13:00	0.20	0.12	1.75	7.98	849,922	-	849,922	100.0%
6	3/24-25/2016	22:35-10:15	0.47	0.52	1.76	8.23	914,719	2,484,271	3,398,990	26.9%
7	3/28/2016	5:50-13:00	0.28	0.28	1.96	8.02	855,599	-	855,599	100.0%
8	3/28/2016	15:55-19:30	0.10	0.08	1.7	4.82	234,804	-	234,804	100.0%
9	3/31-4/1	7:20-1:00	0.66	0.16	1.76	8.13	888,443	793,610	1,682,052	53%
10	4/1/2016	4:05-7:20	0.14	0.12	1.69	4.24	163,239	-	163,239	100.0%
11	4/6-7/2016	16:50-1:40	0.21	0.16	1.7	7.87	822,926	-	822,926	100.0%
12	4/11/2016	1:45-12:55	0.57	0.40	1.7	8.02	860,996	-	860,996	100.0%
13	4/11-12/2016	12:55-00:55	0.18	0.20	4.74	8.08	652,316	-	652,316	100.0%
14	4/26/2016	00:50-2:45	0.13	0.08	1.7	3.44	85,193	-	85,193	100.0%
15	4/26/2016	4:00-9:05	0.31	0.40	2.13	7.99	842,994	-	842,994	100.0%
16	4/26/2016	9:05-15:00	0.09	0.16	5.75	8.15	518,120	3,706	521,826	99.3%
17	5/1/2016	4:15-11:00	0.29	0.16	1.73	7.94	840,094	-	840,094	100.0%
18	5/2/2016	7:00-16:35	0.49	0.28	1.74	8.28	928,354	1,955,648	2,884,002	32.2%
19	5/13/2016	4:20-6:20	0.14	0.16	1.7	3.25	70,279	-	70,279	100.0%

manual mode after a generator testing. The May 2nd overflow is not included in the analysis values.

By the end of the monitoring period, the Lang site was operating at as designed—keeping the depth at the SPP weir a few inches below overflow level throughout the storm. The Bird RTC site successfully stored large volumes during storm events and there is potential for improvement in establishing a trigger value that will begin storage closer to when the weir at the SPP is about to overflow. In both sites, there was opportunity to further optimize performance by coordinating controls with other areas in the system and using data from the downstream SPP.

DISCUSSION

Operation Examples

Figures 6 through 9 demonstrate how the Lang RTC gates and Bird RTC gates function under different system conditions. The red line represents the depth in the chamber, the blue line represents the depth at the SPP weir, and the green and purple lines represent the percent closure of Gate 1 and 2 respectively. Different points of interest are highlighted on the graph using lines with letter labels. Figure 6 shows the typical process of the control gate—the control gate at Lang RTC closed when the depth upstream of the control chamber reached the triggering value (Point A), the depth upstream of the gate increased drastically once the control gate closed while the depth at the SPP weir did not exceed the overflow level (Point B). At the end of the storm the lead gate discharged the stored volume without increasing the downstream depth at the SPP Weir (Point C).

Figure 7 highlights the effectiveness of the control gate system at quickly preventing overflow at the SPP Weir. Before the gate closed, the rate of depth increase at the Lang SPP weir was acute and followed the rate of depth increase upstream of the gate (Point A). Once the gates were closed, the rate at the SPP Weir decreased drastically so that the weir depth began to steadily decrease as the gates remained closed and the depth behind the gate increased (Point B). At the end of the storm the lead gate discharged the stored volume without increasing the downstream depth at the SPP Weir (Point C).

Figure 8 from the Bird RTC site demonstrates the movement of the lead gate in order to keep the depth from going over the emergency weir in the chamber. When the depth upstream of the control structure was hovering around 8 feet which was the maximum emergency weir height, the lead gate opened up to allow more flow which was seen downstream at the Bird SPP weir (Point A). The lead gate closed as the depth in the chamber decreased and eventually the chamber was drained until a normal flow state occurred (Point B). At the end of the storm the lead gate discharged the stored volume without increasing the downstream depth at the SPP Weir (Point C).

Figure 9 demonstrates how the Bird RTC gates react to multiple storms happening in close series. The chamber filled up to maximum height then drained once the first storm event was finished (Point A). The chamber then filled up again due to another storm event that occurred before the first volume was drained completely while still preventing overflow in the chamber (Point B). At the end of the storm the lead gate discharged the stored volume without increasing the downstream depth at the SPP Weir (Point C).

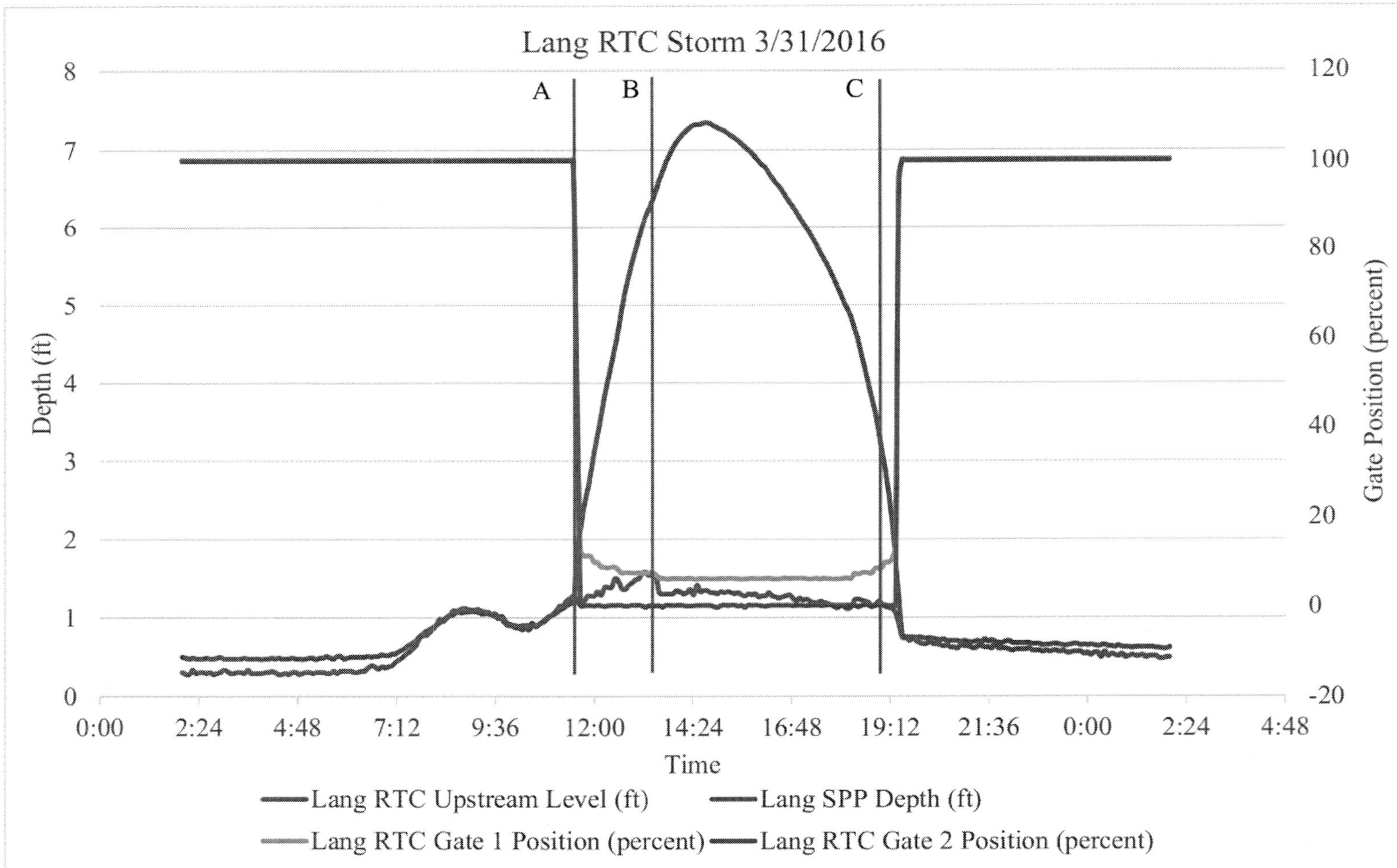

Figure 6. Lang RTC Gate Position and SPP Weir Depth March 31, 2016

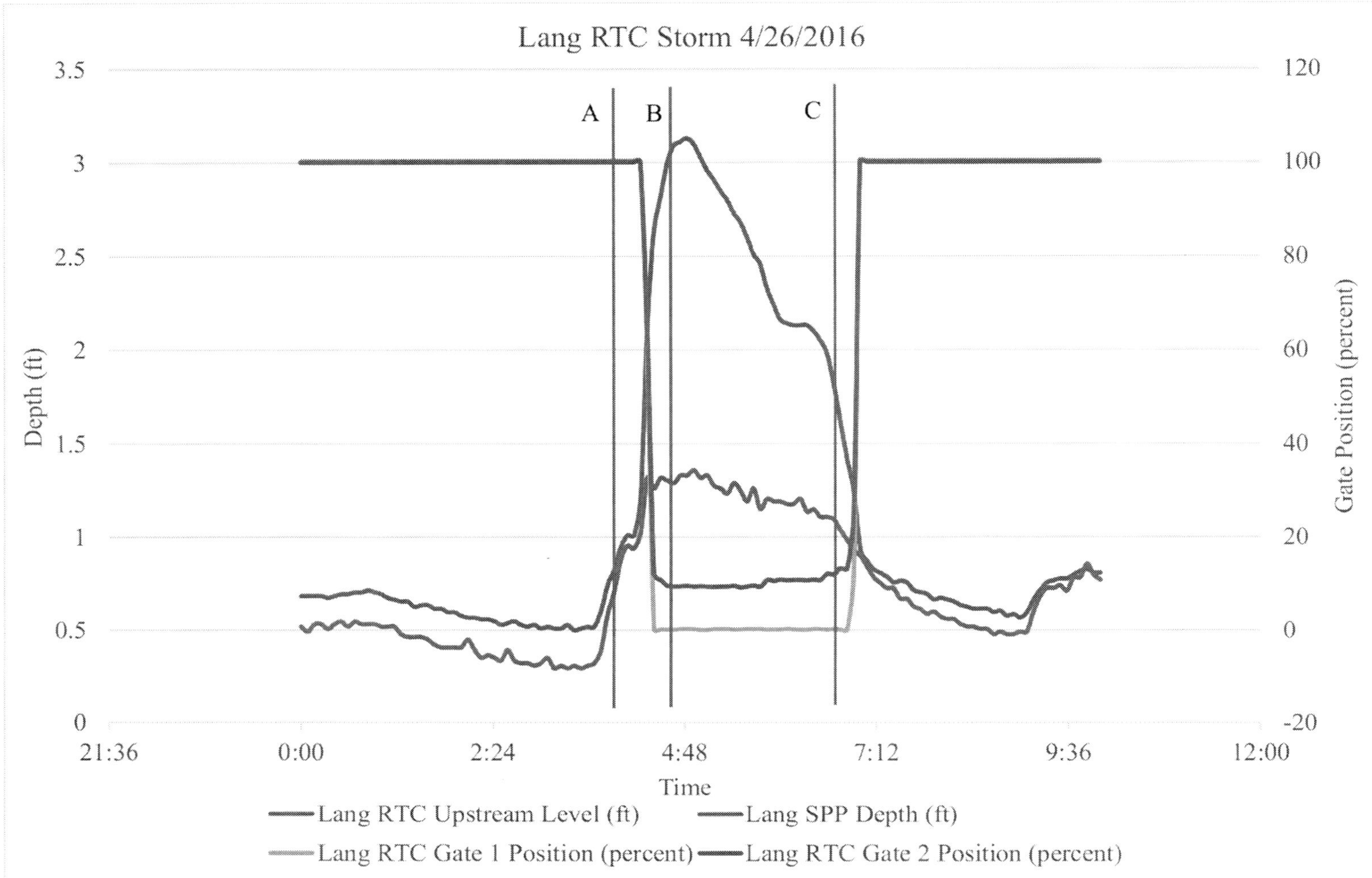

Figure 7. Lang RTC Gate Position and SPP Weir Depth April 26, 2016

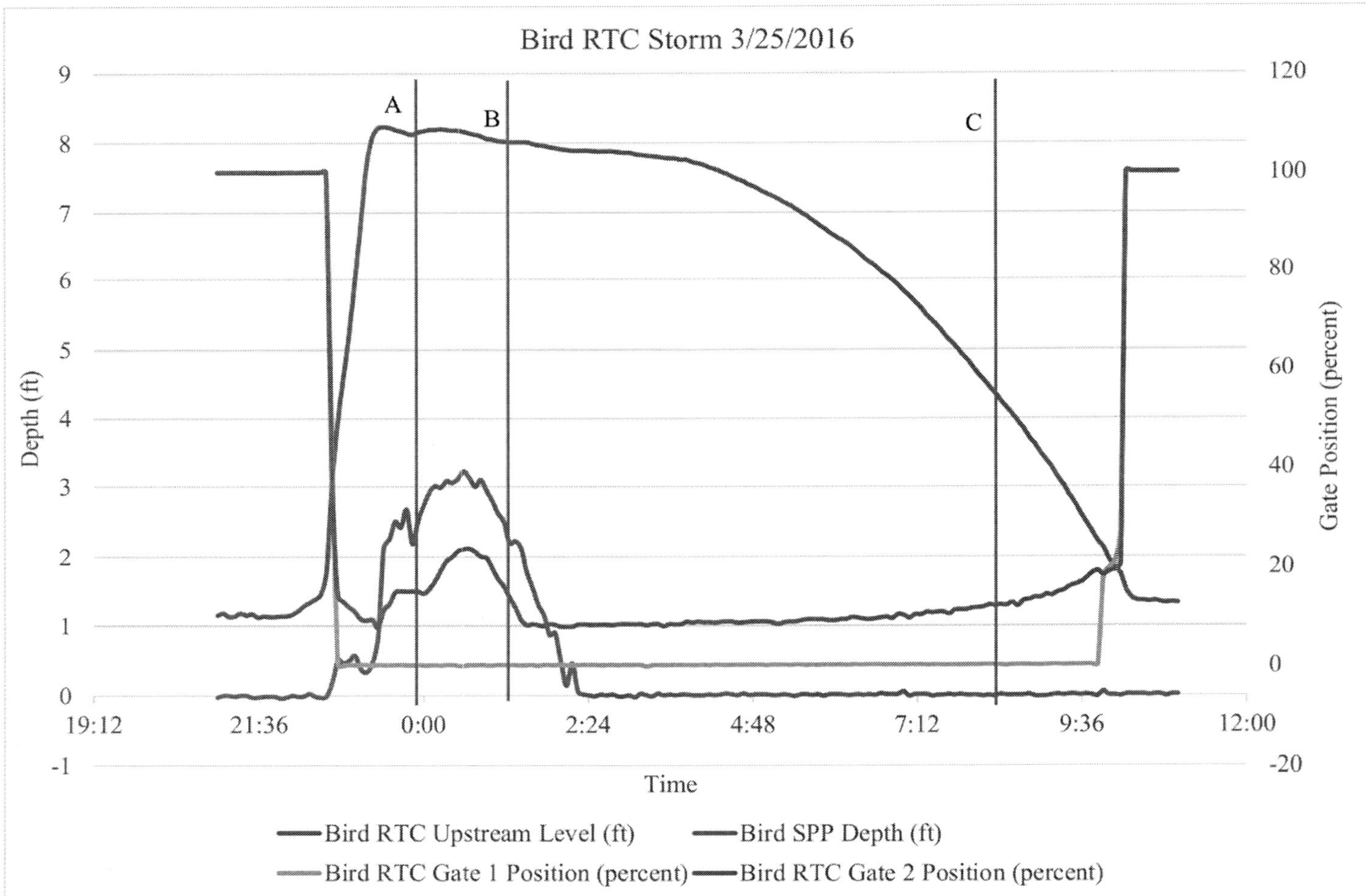

Figure 8. Bird RTC Gate Position and SPP Weir Depth March 25, 2016

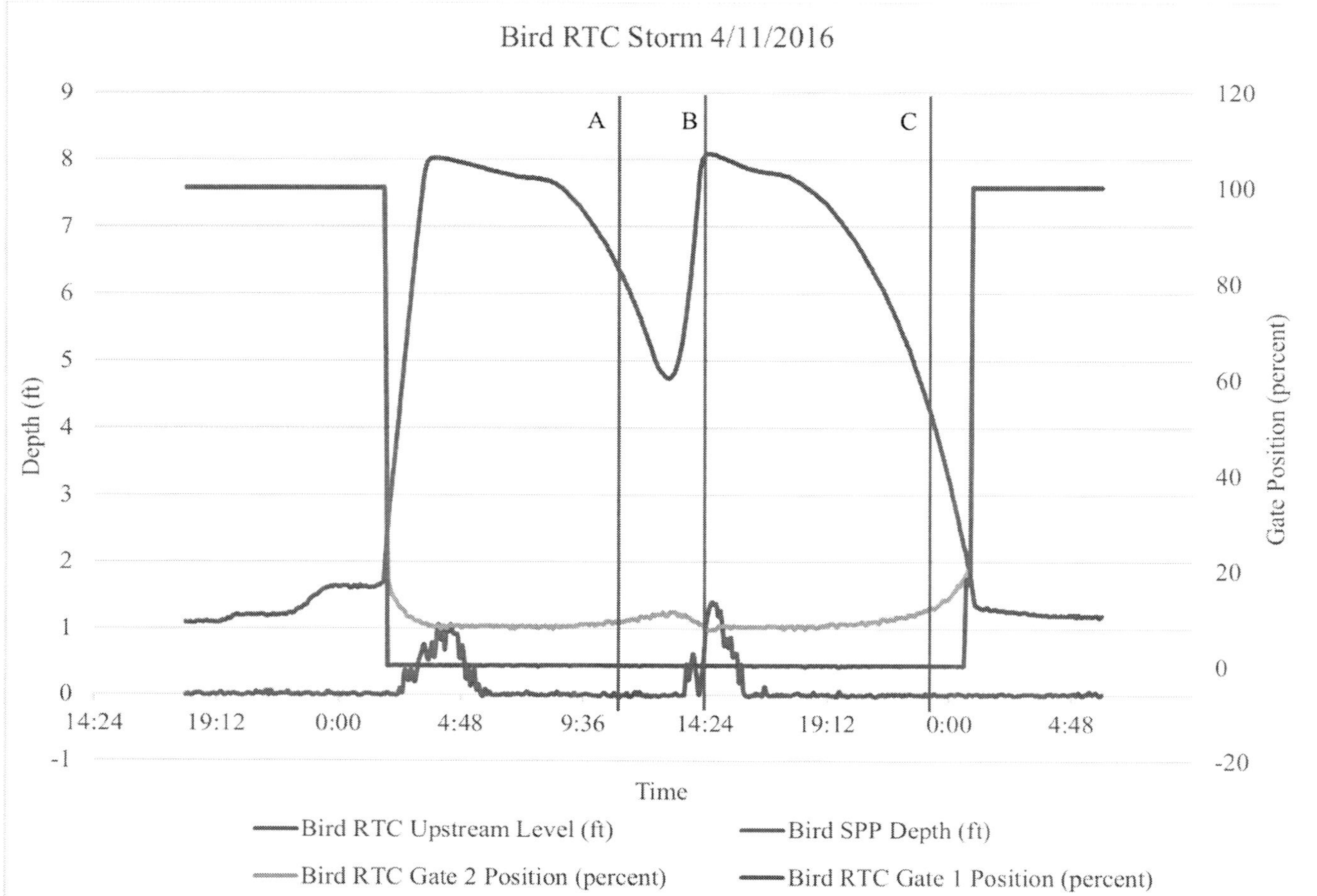

Figure 9. Bird RTC Gate Position and SPP Weir Depth April 11, 2016

In all of these scenarios, both gates responded to the control logic as designed—retaining potential overflow volume during the storm events and releasing to open up capacity for the next storm event. Even when storm events were in close succession, the chamber was effective in retaining most of the overflow volume as designed.

CONCLUSION

The results of the post construction analysis indicate a successful implementation of the RTC sites in the BSA collection system. The operation of the RTC sites was analyzed from March-May 2016. To summarize the results:

1. The Lang RTC site prevented:
   - up to 1.2 MG or 48% of total overflow.
   - up to 4 out of 9 events or 44% of the events.
2. The Bird RTC site prevented:
   - up to 12.1 MG or 64% of total overflow.
   - up to 14 out of 19 events or 74% of the events.
3. Both RTC sites worked as planned.

Next Steps

The successes of the Bird and Lang RTC sites indicate that there is more value that can be created from the RTC approach throughout the rest of the BSA collection system. The next steps moving forward following the project in this paper are to determine where more system-wide benefit can be reached by coordinating the control of the gates with conditions at the SPP, in the interceptor, or elsewhere in the system.

The BSA is currently beginning the construction phase of another RTC chamber upstream of the Lang RTC site at Hazelwood Ave. The new site will provide more storage capacity in the trunkline during large storm events. The Hazelwood RTC site will use a coordinated control logic that takes into account depth at the SPP and data coming from the Lang RTC chamber. According to our modeling, the addition of this RTC site will further reduce activations at the Lang SPP and potentially provide benefit to parallel trunklines as well.

The other locations for RTC sites proposed in the BSA RTC Evaluation Study are currently being explored for similar opportunities. In addition to the volumes retained upstream in each of these locations, the communication and coordination between the chamber locations open up a large potential to more effectively reduce overflow volumes with efficient use of existing assets. By building a coordinated network of RTC sites utilizing market-based optimization, more capacity can be created in the system in the areas that need it most, when they need it most.

REFERENCES

Buffalo Sewer Authority (2012) *Long Term Control Plan – FINAL,* Buffalo, NY

THE ULTIMATE INFRASTRUCTURE RECYLCING: PUTTING THE HOLLYWOOD HILLS ABANDONED SEWER TRUNKLINE TO GOOD RE(USE)

Authors

Adel Hagekhalil, P.E, City of Los Angeles

Ali Poosti, P.E., City of Los Angeles

Fernando Gonzalez, P.E., City of Los Angeles

Rowena Lau, P.E., City of Los Angeles

Kwasi Berko, P.E., City of Los Angeles

Javier Dennis, P.E., City of Los Angeles

Gary Skipper, P.E., Brown & Caldwell

Cleve Lee, P.E., Brown and Caldwell

Abstract

As part of its commitment to sustainability and efficiency, the City of Los Angeles, with the help of Brown & Caldwell, assessed the practicality of reactivating or repurposing abandoned sewer pipes. This study investigated the alternative uses of the 1.45-mile long continuous stretch of abandoned sewer ("Hollywood Hills Sewer") that is categorized as in good condition. The use alternatives analysed consisted of: wastewater conveyance, wastewater storage, stormwater conveyance/storage as a component of a stormwater quality project, or no project at all. The study found that repurposing the Hollywood Hills for stormwater quality improvement showed needed benefits to both stormwater conveyance and water quality, as well as providing the City with millions of dollars in cost savings by repurposing existing abandoned infrastructure.

Keywords

Collection Systems, Sustainability, Sewer, Modelling, Design Storm, Aging Infrastructure

Overview

The City of Los Angeles' (City) provides wastewater collection, conveyance, treatment and disposal services for over 6 million people. Each day, the City treats over 350 million gallons per day of wastewater generated by residential, commercial, and industrial customers. These flows are conveyed through over 6,700 miles of sewer pipelines. As the City's infrastructure ages and the population continues to increase, this extensive sewer and stormwater collection system must be continuously enlarged, replaced, rehabilitated, and repaired. In a city as densely populated as Los Angeles, making needed underground improvements can become a very expensive task and an inconvenience to residents and commuters. As part of its commitment to sustainability and efficiency, the City is currently assessing the practicality of reactivating or repurposing previously abandoned sewer pipes. This study investigated a number of alternative uses for the Hollywood Hills abandoned sewer.

The Hollywood Hills Sewer is a continuous 1.45-mile-long stretch of currently abandoned sewer ranging in sizes from 48" to 60" diameter. It was constructed of tile-lined reinforced concrete pipe in 1953, placed into service, and then abandoned in 1967. In its place, flows were consolidated into twin, parallel 42" trunk sewers referred to as the La Cienega/San Fernando Valley Relief Sewer (LCSFVRS), a major trunk sewer conveying valley flows through the Hollywood Hills to the Hyperion Treatment Plant.

Due to its relatively short period of service, the Hollywood Sewer was not permanently abandoned (e.g. slurry filled or other sediment backfill). The sewer was instead capped at the ends and left in place, leaving the option to reactive the sewer in case a need presented itself.

The alignment of the LCSFVRS (black) and Hollywood Hills sewer (blue) is shown in Figure 1.

Scope

The purpose of the study was to evaluate use alternatives for the Hollywood Hills abandoned sewers in order to put valuable infrastructure to beneficial use. It was initially determined that the best possible use categories for the Hollywood Sewer were: wastewater conveyance, wastewater storage, stormwater conveyance, as a component of a stormwater quality project, or no project at all.

For wastewater projects, the MIKE URBAN hydrodynamic model was used to identify beneficial sewer projects surrounding the Hollywood Sewer alignment. For stormwater projects, the Ballona Creek Watershed Management Group developed a preliminary list of regional projects that will be evaluated in the Enhanced Watershed Management Plant (EWMP) for the Ballona Creek, which encompasses the Hollywood Hills alignment. Regional projects near the Hollywood Hills alignment were then identified for further evaluation. The various use alternatives were narrowed down to the following four (4) alternatives, which the study analysed:

1. No Project (abandon)
    a. Abandon but maintain
    b. Permanently abandon
2. Wastewater Conveyance to Relieve the LCSFVRS
3. Wastewater Storage to Attenuate Peak Wet-Weather Flow (PWWF) in the LCSFVRS
4. Stormwater Quality
    a. Integrated Pipeline for Plummer Park BMP
    b. Integrated Pipeline for Poinsettia Park BMP
    c. Stormwater Silo for Inline Storage for both BMP projects
    d. Stormwater Silo with a Green Street Retrofit

Figure 1. LCSFVRS and Hollywood Hills Sewer Alignment

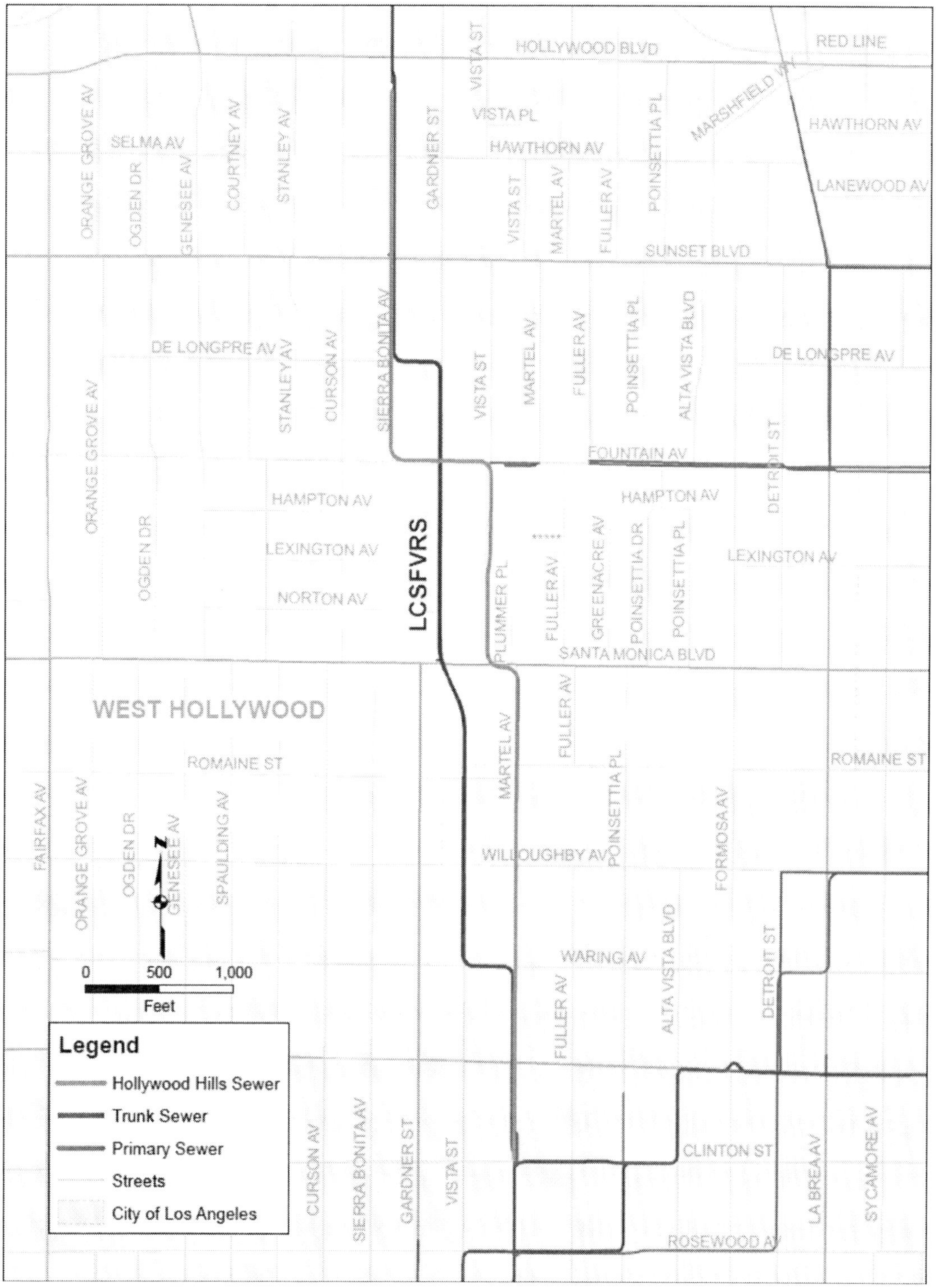

Alternative Description

Alternative 1a

Alternative 1a would leave the Hollywood Hills sewer abandoned, but will maintain the structural condition of the sewer by rehabbing the sewer, if needed. The alternative allows for reactivating the sewer or repurposing the sewer if a beneficial use is identified in the future. The alternative beneficially utilizes zero (0) percent of the Hollywood Hills sewer, but leaves the potential to reuse the sewer at a future date.

Alternative 1b

Alternative 1b would permanently abandon the Hollywood Hills sewer by backfilling the sewer with slurry or other sediment backfill. The alternative would make it unfeasible to reactive the sewer in the future. The alternative beneficially utilizes zero (0) percent of the Hollywood Hills sewer.

Alternative 2

Alternative 2 would connect the upstream and downstream points of the Hollywood Sewer to the LCSVFRS. The two sewers would be connected upstream near the intersection of Hollywood Blvd and Sierra Bonita Ave. The existing flows in the LCSFVRS would be split evenly between the two sewers. The two sewers would then reconnect downstream near the intersection of Clinton St and Martel Ave where flows would recombine into the LCSFVRS. The alternative beneficially utilizes 100 percent of the Hollywood Hills sewer.

Alternative 3

Alternative 3 would utilize the Hollywood Sewer for wet-weather storage. Similarly to Alternative 2, the upstream and downstream ends of the Hollywood Sewer would be connected to the LCSFVRS at Hollywood Blvd and Sierra Bonita Ave (upstream), and Clinton St and Martel Ave (downstream). However, a passive or actively controlled weir gate at the upstream connection would control flow into the Hollywood Hills sewer. During a wet-weather event, some LCSFVRS flows would be diverted to the Hollywood Hills sewer to be temporarily stored. Once the storm passes, stored flows would then be allowed to flow back into the LCSFVRS through a downstream gate. The alternative has the potential to store up to 0.9 million gallons of wastewater, but the actual storage capacity would be reduced through the use of flap gates to store wastewater in segments of sewer pipes as to not pressurize the sewer. The alternative beneficially utilizes 100 percent of the Hollywood Hills sewer.

Alternative 4a

The Plummer Park BMP was identified as one of the regional projects in the Ballona Creek EWMP near the Hollywood Hills sewer alignment. The project would construct a below-ground retention and infiltration basin under the parking lot of Plummer Park. Alternative 4a would utilize the Hollywood Sewer as a local stormdrain by installing new stormwater inlets. Stormwater flows would then be conveyed to Plummer Park through a new connection along Vista, just south of Fountain Ave. Additional dry-weather flows can be diverted from the Laurel Canyon and La Brea storm drains, which also parallel the Hollywood Sewer, to maximize the capacity of the Plummer BMP. Flows in excess of the Plummer Park BMP would be diverted to the Laurel Canyon storm drain. The alternative beneficially utilizes 46 percent of the Hollywood Hills sewer. Figure 2 shows the location of the Hollywood Hills sewer, storm drains, and Plummer Park.

Figure 2. Alternative 4a – Integrated Pipeline to Plummer Park BMP

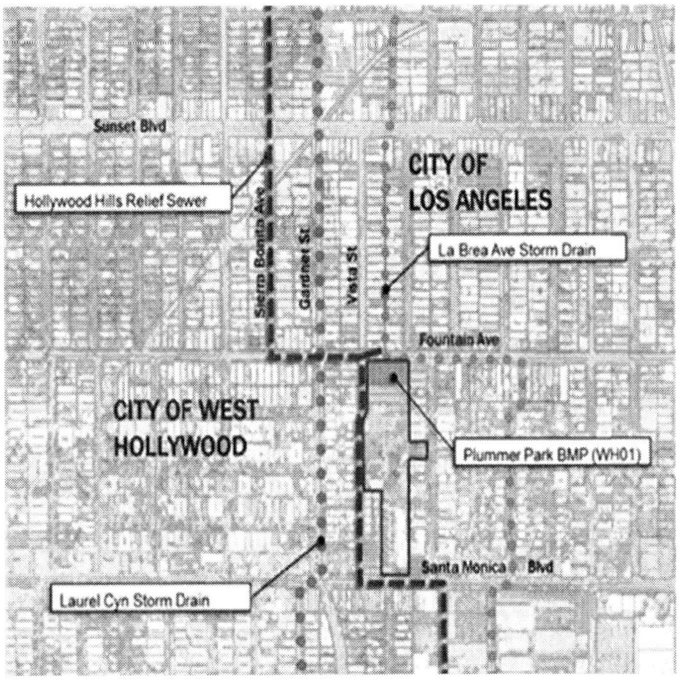

<u>Alternative 4b</u>

The Poinsettia Park BMP, just south of Plummer Park, was also identified as one of the regional projects in the Ballona Creek EWMP near the Hollywood Hills sewer alignment. The project would construct a below-ground retention and infiltration basin under the softball diamonds and open space. Alternative 4b would utilize the Hollywood Sewer as a local stormdrain by installing new stormwater inlets. Stormwater flows would then be conveyed to Poinsettia Park through a new connection along Romaine St from Martel to Gardner St. Additional dry-weather flows can be diverted from the Laurel Canyon and La Brea storm drains to maximize the capacity of the Poinsettia BMP. Flows in excess of the Poinsettia BMP would also be diverted to the Laurel Canyon storm drain. The alternative beneficially utilizes 66 percent of the Hollywood Hills sewer. Figure 3 shows the location of the Hollywood Hills sewer, storm drains, and Poinsettia Park.

Figure 3. Alternative 4b – Integrated Pipeline to Poinsettia Park BMP

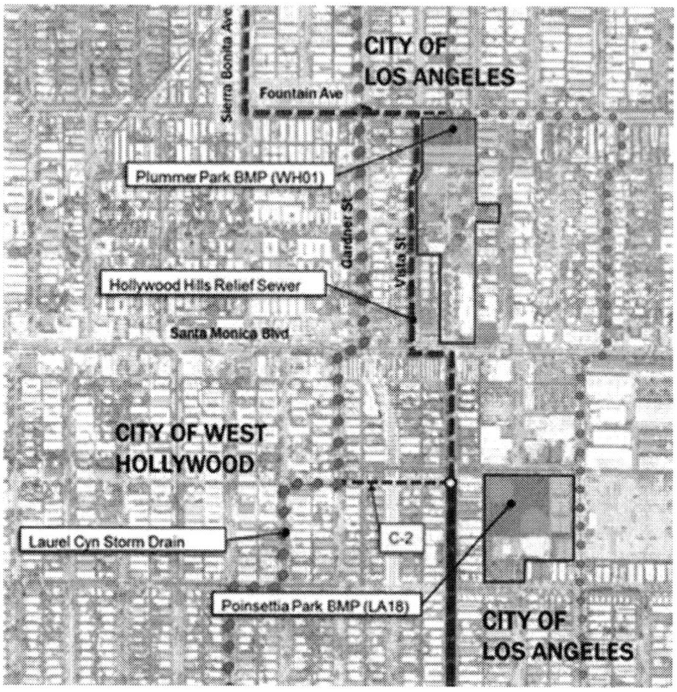

Alternative 4c

Alternative 4c is a combination of Alternatives 4a and 4b with the addition of utilizing the Hollywood Sewer as additional storage for both the Plummer and Poinsettia BMPs. Like the previous two alternatives, stormwater flows would be conveyed to Plummer Park and Poinsettia Park through the same proposed connections. Through installation of flap gates throughout the Hollywood Sewer, stormwater could be stored ranging from 6,600 to 27,500 cubic feet. Additional dry-weather flows can be diverted from the Laurel Canyon storm drain to maximize the capacity at both BMP. Flows in excess of the Plummer and Poinsettia BMPs would also be diverted to the Laurel Canyon storm drain. The alternative beneficially utilizes 66 percent of the Hollywood Hills sewer.

Alternative 4d

Alternative 4d a similar to Alternative 4c with a slight modification. Instead of installing new catchment inlets to the Hollywood Hills sewer, the alternative would instead incorporate bioswales or flow-through planters to provide pretreatment prior to stormwater entering the Hollywood Hills sewer. The alternative beneficially utilizes 66 percent of the Hollywood Hills sewer. Figure 4 shows a rendered example of green street infrastructure.

Figure 4. Alternative 4d – Example of Green Street Addition

Methodology Used

Physical Condition

The physical condition and integrity of the Hollywood Hills abandoned sewer was initially unknown. It was determined that an assessment was needed to inspect and evaluate the structural condition of the abandoned pipe prior to expending City resources to determine how this pipe might be repurposed. Inspection was conducted using a remotely operated closed circuit television (CCTV) unit.

Hydraulic Condition

The MIKE URBAN hydrodynamic modelling software was used to investigate hydraulic conditions surrounding the Hollywood Hills Sewers and to determine whether there is a need for hydraulic relief in the area. A sewer is considered to be in need of hydraulic relief if modelling results show a peak dry-weather flow (PDWF) d/D of 0.75, or a sewer overflow during a 10-year storm event. MIKE URBAN was also used to model changes to the hydraulic conditions when utilizing the Hollywood Hills Sewer for conveyance. The PDWF, PDWF d/D, and surcharge level during the PWWF for the "no project" alternative were compared to the wastewater conveyance alternatives to quantify the measure of relief. For the storage alternative, a "best case" relief scenario was estimated by "shaving" the PWWF by the total potential storage volume of the Hollywood Hills Sewer (approximately 0.9 million gallons). The PWWF flow corresponds to a citywide 24-hour, 10-year storm event modelled in MIKE URBAN.

Cost

Cost estimates were created for each alternative taking into account the design, construction, contingency, and cost escalation.

Each alternative was evaluated on the basis on level of water conveyance need, level of conveyance relief offered, level of water quality need, and level of water quality benefit offered, cost, and beneficial reuse.

Findings

Physical Condition
The physical condition of the Hollywood Sewer via CCTV inspection showed the pipe to be in good structural condition requiring minor rehabilitation and/or repair. The defects noted from the inspection included missing tiles, sediment build up at the downstream end of the alignment, and a short segment where the crown of the sewer pipe had been demolished to allow for the construction of a storm drain pipeline.

Hydraulic Condition

Alternative 1a & 1b (do nothing)
Analysing the existing sewer collection system using MIKE URBAN showed that the LCSFVRS, parallel to the Hollywood Hills Sewer, will reach a PDWF d/D of 0.58 by year 2050, as shown on Figure 5. Modelling the year 2050 hydraulic conditions during a 10-year storm event shows that the LCSFVRS will surcharge, as shown on Figure 6. However, no wet-weather related sewer overflows would occur.

Figure 5. Year 2050, PDWF d/D

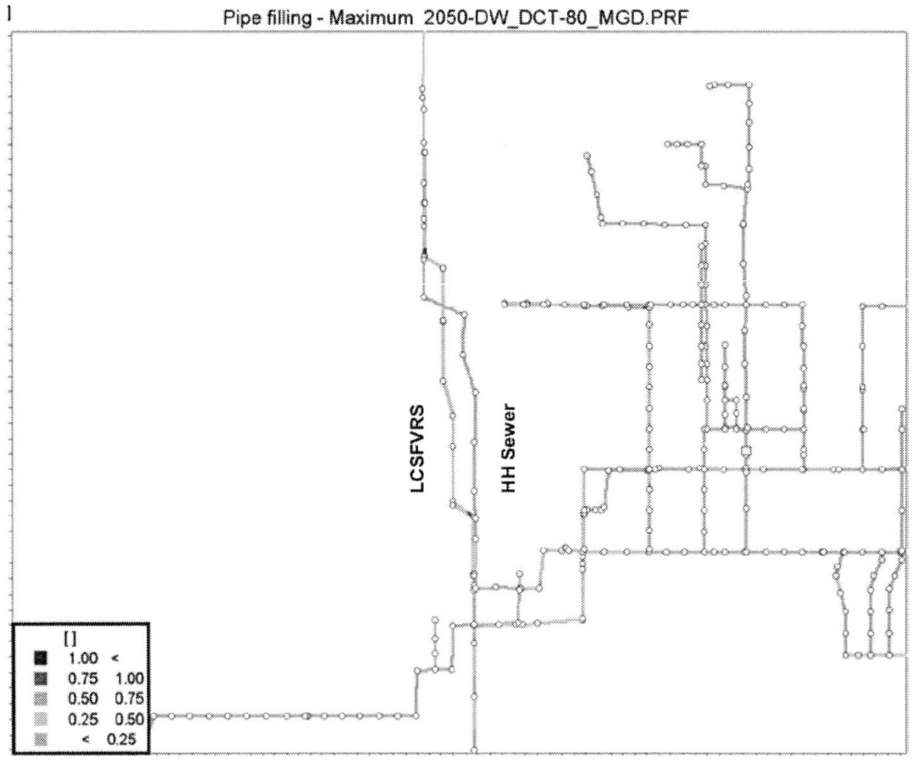

Figure 6. Year 2050, PWWF Surcharge Conditions

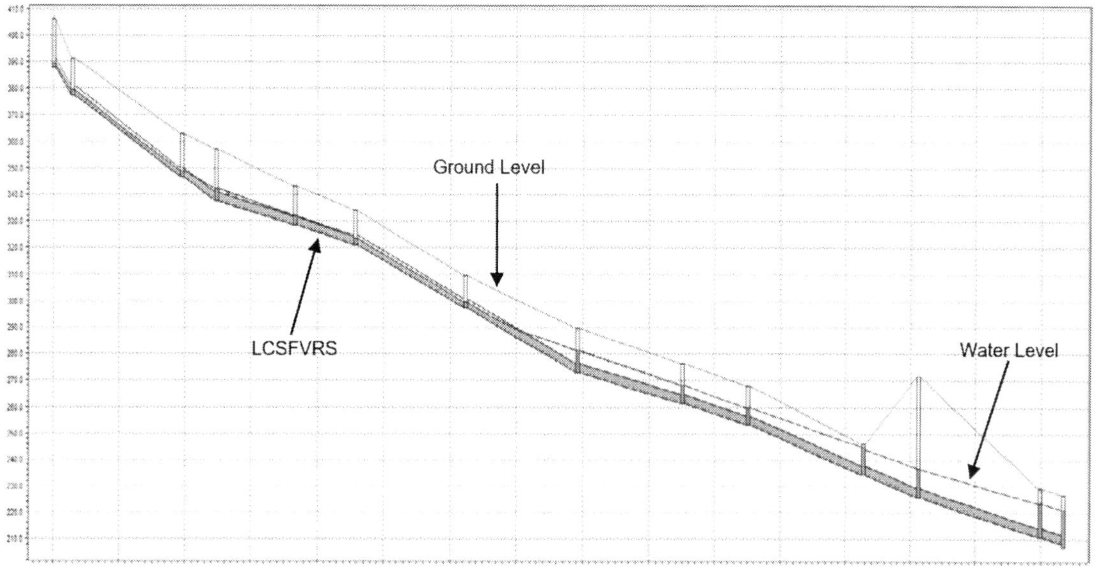

Additionally, historical odor complaint and flow gauging records were investigated, which showed problematic history of odor complaints and high surcharged levels during storm events along the LCSFVRS. In 2006, a carbon scrubber was installed along the upper portion of the LCSFVRS to mitigate odor issues; however, high surcharge levels still exist during wet-weather conditions.

Alternative 2 (Wastewater Conveyance)

Alternative 2 would connect the Hollywood Hills Sewer to the LCSFVRS and split flows evenly (50/50). The two sewers would reconnect at the termination of the Hollywood Sewer, and flows would return to the LCSFVRS. Modelling results showed Alternative 2 favourably reducing the LCSFVRS PDWF d/D from over 0.50 to 0.31, as shown on Figure 7. Alternative 2 would also reduce the year 2050 PWWF d/D from surcharged conditions to 0.57, as shown on Figure 8. However, the Hollywood Sewer parallels a small portion of the LCSFVRS and therefore, there is no relief throughout the rest of the LCSFVRS.

Figure 7. Year 2050, PDWF d/D

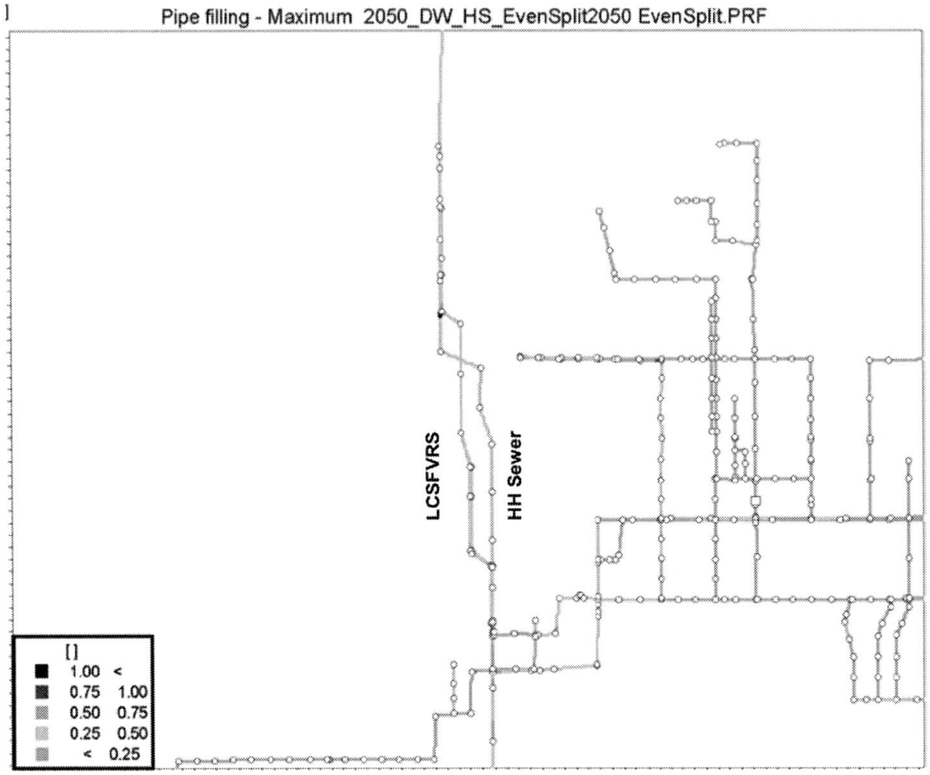

Figure 8. Year 2050, PWWF Surcharge Conditions

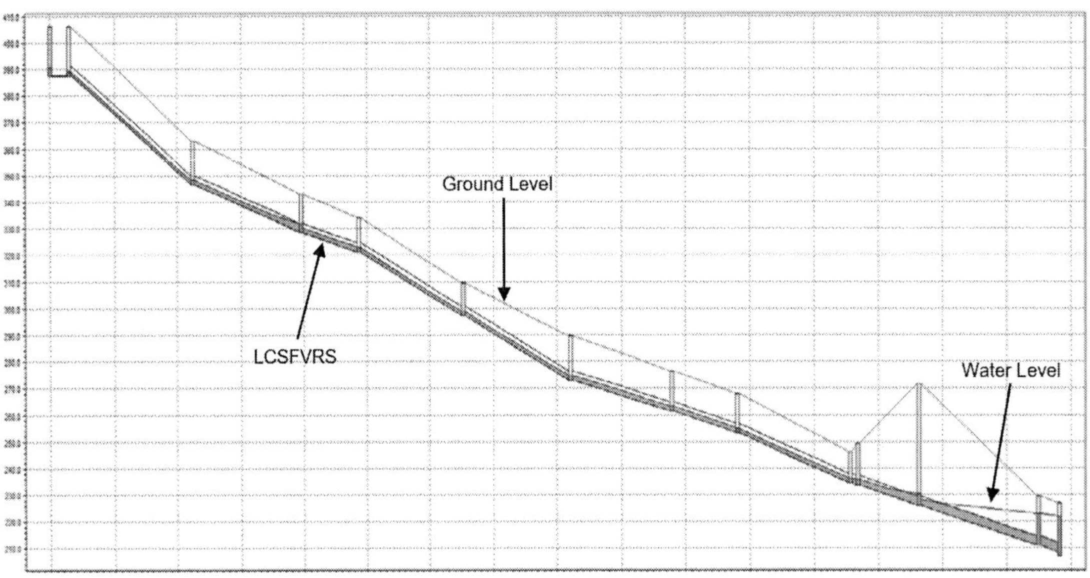

Alternative 3 (PWWF storage)
Alternative 3 (PWWF storage) did not require modelling using MIKE URBAN. Instead, the "best case" scenario was estimated by taking the PWWF pattern along the LCSFVRS and "shaving" the peak by the total potential storage volume of the Hollywood Sewer (i.e. 0.9 million gallons). In the best case, storage would reduce the LCSFVRS surcharged conditions by 9 inches, as shown on Figure 9.

Figure 9. Year 2050, Wet-Weather Flow and d/D - "Best Case" Reduction

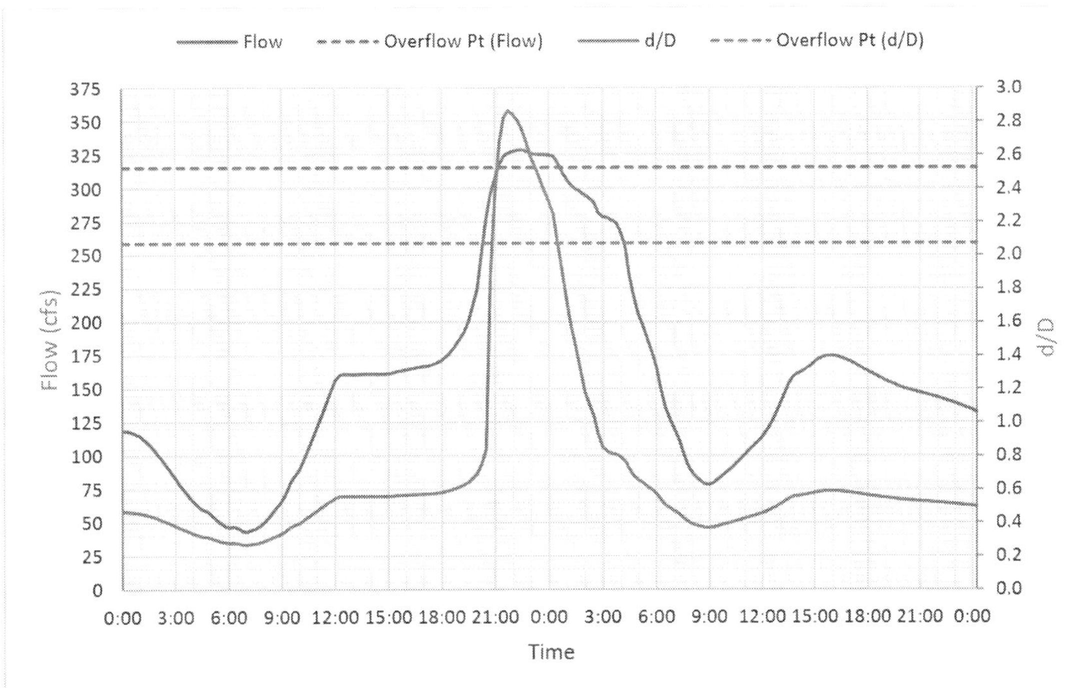

Alternative 4a (Stormwater Quality)
Alternative 4a would utilize the Hollywood Sewer to convey stormwater flows to the Plummer Park BMP, which would retain and infiltrate flows. The project would help mitigate water quality problems in the Ballona Creek by diverting dry-weather runoff that is polluted with high levels of metals, bacteria, E. Coli, and Zinc. The Plummer Park BMP is not sized to treat wet-weather flows and therefore, the alternative would not address the high levels of zinc and E. Coli typically found in wet-weather flows.

Alternative 4b (Stormwater Quality)
Alternative 4b would utilize the Hollywood Sewer to convey stormwater flows to the Poinsettia Park BMP, which would retain and infiltrate flows. The project would help mitigate water quality problems in the Ballona Creek by diverting dry-weather runoff that is polluted with high levels of metals, bacteria, E. Coli, and Zinc. The Poinsettia Park BMP is sized to treat the 85th percentile

design storm runoff volume and therefore, would address high levels of pollutants found in both dry- and wet-weather runoff.

Alternative 4c (Stormwater Quality)
Alternative 4c would utilize the Hollywood Sewer to convey stormwater flows to both Plummer Park and Poinsettia Park BMPs, which would retain and infiltrate flows. The project would help mitigate water quality problems in the Ballona Creek by diverting dry-weather runoff that is polluted with high levels of metals, bacteria, E. Coli, and Zinc. The Poinsettia Park BMP is sized to treat the 85th percentile design storm runoff volume and therefore, would address high levels of pollutants found in both dry- and wet-weather runoff. The runoff storage offered by the alternative (i.e. 6,600 to 27,500 cubic feet) is negligible compared to the volumes of wet-weather runoff and therefore, offer little added benefit.

Alternative 4d (Stormwater Quality)
Alternative 4c would utilize the Hollywood Sewer to convey stormwater flows to both Plummer Park and Poinsettia Park BMPs, which would retain and infiltrate flows. Additionally, the alternative would construct bioswales or flow-through planters that would provide treatment to all runoff captured by the Hollywood Sewer. The project would help mitigate water quality problems in the Ballona Creek by diverting dry-weather runoff that is polluted with high levels of metals, bacteria, E. Coli, and Zinc. The Poinsettia Park BMP is sized to treat the 85th percentile design storm runoff volume and therefore, would address high levels of pollutants found in both dry- and wet-weather runoff. Wet-weather events exceeding the 85th percentile storm would also benefit from water quality improvements through the bioswales or flow-through planters. The runoff storage offered by the alternative (i.e. 6,600 to 27,500 cubic feet) is negligible compared to the volumes of wet-weather runoff and therefore, offer little added benefit.

Summary and Cost

Each alternative was evaluated on the basis on level of water conveyance need, level of conveyance relief offered, level of water quality need, and level of water quality benefit offered, cost, and beneficial reuse. These benefits are summarized on Table 1.

Table 1. Summary of System Needs and Alternative Benefits

Alt	Alternative Description	Level of Conveyance Need	Level of Conveyance Relief	Level of Water Quality Need		Level of Water Quality Mitigation		Beneficial Reuse (%)
				Dry Weather	Wet Weather	Dry Weather	Wet Weather	
1a	Abandon and Maintain	Low	None	None*	None*	None*	None*	0**
1b	Permanently Abandon	Low	None	None*	None*	None*	None*	0
2	Wastewater Conveyance	Low	Medium	None*	None*	None*	None*	100
3	Wastewater WW Storage	Low	Low	None*	None*	None*	None*	100

Alt	Alternative Description	Level of Conveyance Need	Level of Conveyance Relief	Level of Water Quality Need		Level of Water Quality Mitigation		Beneficial Reuse (%)
				Dry Weather	Wet Weather	Dry Weather	Wet Weather	
4a	Conveyance to Plummer Park BPM	Medium	High	High	High	High	Low	46
4b	Conveyance to Poinsettia BMP	Medium	High	High	High	High	Medium	66
4c	Conveyance to Plummer & Poinsettia BMP, Storage	Medium	High	High	High	High	Medium	66
4d	Conveyance to Plummer & Poinsettia Park BMP, Storage, Treatment	Medium	High	High	High	High	High	66

Note:
*No wastewater quality needs or benefits were identified
**Does not reuse any portion of the Hollywood Hills, but allows future reuse

Planning level (Class "O") project costs were estimated for each alternative. It was assumed that the design cost would be 10 percent of the construction level cost, and a 30% estimating contingency was used. The total alternative project cost are shown in Table 2.

Table 2. Planning Level (Class "O") Project Cost

Alternative	Alternative Description	Total Design and Construction Cost
1a	Abandon and Maintain	$ 0
1b	Permanently Abandon	$ 0.98 M
2	Wastewater Conveyance	$ 7.72 M
3	Wastewater WW Storage	$ 8.35 M
4a	Conveyance to Plummer Park BPM	$ 5.20 M*
4b	Conveyance to Poinsettia BMP	$ 6.70 M*
4c	Conveyance to Plummer & Poinsettia BMP, Storage	$ 6.87 M*
4d	Conveyance to Plummer & Poinsettia Park BMP, Storage, Treatment	$ 7.98 M

Note:
*Cost estimate based on itemized cost of Alternative 4d

Conclusions

The evaluation of alternatives showed that the Hollywood Sewer provided benefits to both the wastewater collection system and stormwater quality initiatives. However, sewer modelling showed there was a minor need for hydraulic relief of the LCSFVRS along the alignment of the Hollywood Hills sewer. Projecting both dry-weather and wet-weather flows to the year 2050 shows the LCSFVRS can convey flows without exceeding the City project trigger limits. Therefore, alternatives to relief the wastewater collection were dismissed.

Repurposing the Hollywood Hills for stormwater quality improvement projects showed benefits to both conveyance and water quality. Additionally, a high need for conveyance of stormwater and water quality improvements were identified. There is a need to convey stormwater flows to the proposed Plummer Park and Poinsettia Park BMPs to undergo treatment, and the Hollywood Hills sewer is positioned to both capture stormwater flows and convey them to both BMP locations. Of the water quality improvements alternatives (Alternatives 4a-4d), Alternative 4c and Alternative 4d benefitted both the proposed Plummer Park and Poinsettia Park BMPs, and therefore, addressed the greatest stormwater improvement needs. The major difference between the two alternatives is that Alternative 4c would provide treatment to stormwater runoff up to the 85th percentile storm event, while Alternative 4d would provide treatment to all runoff captured by the Hollywood Hills sewer at an additional cost of approximately $1.1 million. Due to degree of impairment of the Ballona Creek and mandates to improve water quality (e.g., Regional Water Quality Control Board mandate for compliance of Metals and Bacteria TMDLs by year 2021), it was concluded that the benefits offered by Alternative 4d outweighed the additional cost. Additionally, the project would beneficially reuse 5,128 feet or 66 percent of the Hollywood Sewer, providing the City with millions of dollars in cost savings by repurposing existing abandoned infrastructure.

NOT IN MY BACKYARD, ANYMORE: A SUCCESSFUL BACKYARD SEWER REPLACEMENT PROGRAM

Authors: Richard E. Underhill, P.E., PACP, Greeley and Hansen; Kelvin Coles, P.E., ENV SP, Greeley and Hansen, Belinda Wilson, Virginia Beach Department of Public Utilities

ABSTRACT

Purpose
To present the process of designing and constructing a successful backyard sewer replacement project in a well-established, older neighborhood. To discuss the obstacles, challenges, and public concerns that needed to be addressed to obtain homeowner buy-in of the project and to provide a safe construction environment.

Benefits
The paper will share lessons learned in the design and construction phases of the project and provide suggestions to others on how to gain project acceptance from multiple stakeholders.

Status of Completion
The construction project is in progress and will be substantially complete by the date of the conference.

Narrative
Backyard sewers are common in many older sewers in American cities. The sewers in many residential subdivisions built in the 1950s were installed in the backyards of residences, along property lines or easements, because it was easier to put them there than in the street. Thus avoiding construction traffic disruption as new houses were built. In many older neighborhoods that originally had septic tanks and drain fields, it was easier to connect the lateral in the back of the house to a backyard sewer. This all seemed like a good approach back in the day. But, as any Collections System Operations and Maintenance supervisor will tell you, back yard sewers are difficult to maintain.

Eliminating back yard sewers often involves constructing a new main line sewer in the street in front of the house and routing the sewer lateral around the house, from the back to the front. To do this, the designer must resolve many issues. Over the years home owners make improvements to their properties such as constructing fences, sheds, and other structures in the their back yards and by planting trees and shrubs along the property line. These amenities inhibit maintenance crew access and make manholes and cleanouts hard to find. Older sewers that were poorly constructed and have open, leaky joints contribute to higher wet weather flows due to infiltration and inflow (I/I). Open joints can also contribute to root intrusion and cave-ins due to soil migration, both of which can cause sewer blockages. Poor access makes clearing blockages with conventional truck mounted equipment difficult if not impossible. Sewer access problems, homeowner complaints, and the need for I/I reduction drove the City's decision to eliminate the back yard sewers in the project area.

Homeowners also make improvements to their property that inhibit routing new laterals around the house, such as garden and flowerbed hardscapes, ornamental structures, privacy fences, decks, driveway and sidewalk extensions, and trees and shrubs. In the project area, all of these things had occurred over the years, and the property owners took pride in how their yards looked. The neighborhood had been constructed in the mid 1950's in a previously wooded area that contained many large white oak trees, some with diameters of 4 or more feet, hence the name of the neighborhood, White Oak. Because of the age of the neighborhood many trees that were planted years ago were also big and well dispersed about the properties.

Another consideration in design of the new house laterals is the minimum slope required by the local building code, usually about 1%. This can become an issue at the end of the main line sewer where it is shallower and long lateral runs may be necessary. Maintaining vehicular and pedestrian access to properties during construction is an absolute requirement. Traffic control plans can become extensive and require several phases to construct the project.

Our approach to addressing the project constraints and considerations discussed above included:

- Inspecting and photographing properties to document the existence of features that could impact lateral re-routing.
- Having discussions with property owners during the field inspections to hear their concerns.
- Sending maps of the proposed lateral re-routing to each property owner with a form requesting their acceptance and any information on abandoned septic tanks or underground fuel oil tanks.
- Conducting a public meeting at a nearby venue to describe the project and to get feedback from property owners.

The project is currently under construction, with few complaints, and will be completed in the next several months.

KEYWORDS

Backyard Sewers, Public Participation, Mitigation

INTRODUCTION

Overview

The City of Virginia Beach, Virginia is located in southeastern Virginia in the part of the state referred to as Hampton Roads. The city is bounded by the Chesapeake Bay to the north, the Atlantic Ocean to the east, North Carolina to the south, and the Cities of Norfolk and Chesapeake to the west. The city is one of the largest in Virginia with a total area of 1,287 square kilometers (497 square miles) and an estimated current population of about 456,000 people, which is about 27% of the total population in the Hampton Roads metropolitan area (1.7 million). The city is a thriving resort area with 59.5 kilometers (37 miles) of beaches, and includes three major

Department of Defense bases. Visitors to the city almost equal the resident population, averaging about 450,000 per month.

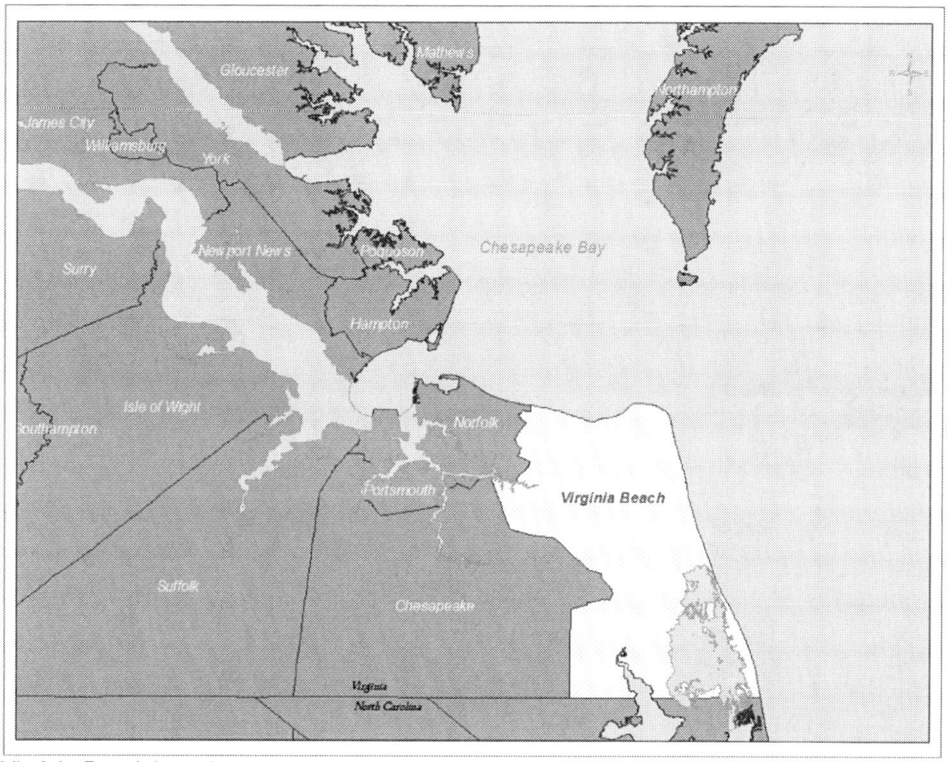

Virginia Beach Location

Virginia Beach Oceanfront

The City's sewage collection system consists of about 1,947 kilometers (1,210 miles) of gravity sewers ranging size from 10.16 centimeters (4 inches) to 91.44 centimeters (36 inches). The area is relatively flat; therefore, sewer run lengths away from a central sewage pump station are limited by the practical depth of the pump stations and the need to maintain minimum sewer slopes. As a consequence, there are about 405 sewer basins in the city, each served by a sewage pump station.

Project Area

The project presented in this paper is located in the White Oak neighborhood, which is served by Virginia Beach Sewer Basin pump station 417. A map of the sewer basin is shown to the right. White Oak is a relatively older neighborhood in the city, with the majority of the 50 residential structures built in the early 1950's. There are many large white oak trees scattered through the neighborhood which is probably how it got its name. Over the years, the various homeowners have made improvements to their property including hard and vegetative landscaping, drive way extensions, fences, gardens, walkways, decks and building additions. For the most part, the home owners take pride in the appearance of their properties which results in an overall attractive looking neighborhood. Some photographs of the neighborhood before the construction project are shown below.

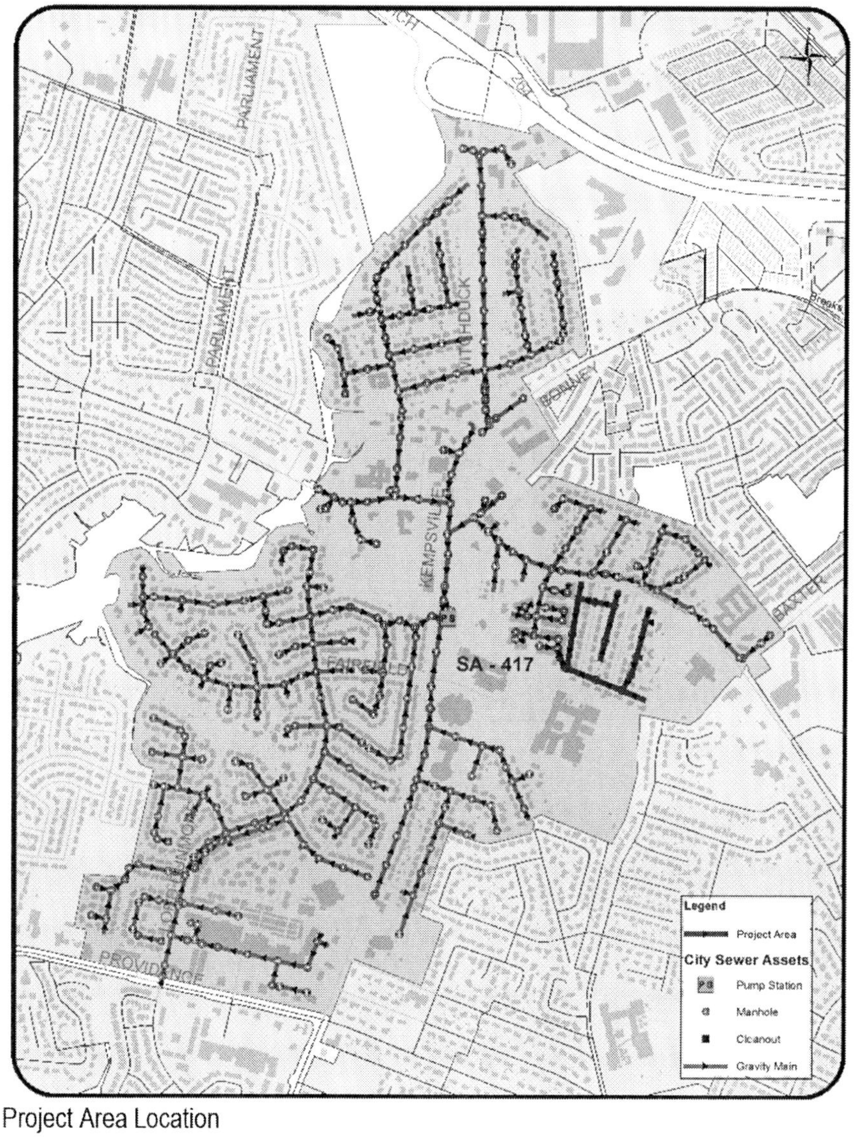

Project Area Location

White Oak Drive (New Pavement Overlay)

White Oak Drive (Large Trees)

Englewood Drive (Landscaping)

Wellston Court (Landscaping)

Gravity Sewer System
The original homes were served by on-site septic tank / drain field systems, which were abandoned when a developer installed a vitrified clay pipe and cast iron pipe gravity collection system in the mid 1950's. The developer installed most of the main line sewers along the back property lines of the residences, with one connecting sewer located in a city street right of way. This was done because it was easier to connect to the house sewers formerly going to septic tanks in the back yards and it avoided disruption to existing streets and traffic access. Homes built after 1955 were connected directly to the new sewers. This seemed to be a good idea at the time, but has become a problem for maintenance sixty years down the road. Fences, sheds and other amenities constructed by the homeowners along the back of their lots and trees and shrubs that have been planted or have emerged over the years have made access to the backyard sewers practically impossible. The photos below show some of the existing conditions along the backyard sewer routes.

Backyard Sewer Along Fence Line

Backyard Sewer (Fences and Trees)

Side Yard Sewer (Trees and Fences)

The tables below present the characteristics of the existing collection system.

Sanitary Sewer Collection System Summary

Category	Characteristic
Installation Year	1955
Gravity Main (LF)	3,858
No. of Manholes	13
No. of Public Main Line Cleanouts	7
No. of Laterals	50

Gravity Main Characteristics

Material	Diameter			
	6" (LF)	8" (LF)	10" (LF)	Total (LF)
Clay	1,167	1,892	0	3,059
Iron	0	692	107	799
Total	**1,167**	**2,584**	**107**	**3,858**

The figure below shows the locations of the existing mainline sewers in the neighborhood.

White Oak Neighborhood Backyard Sewer System

The sewershed that the White Oak neighborhood is part of was designated as a Sanitary Sewer Evaluation Study (SSES) basin based on field investigations and flow monitoring by the City. The results of SSES field investigations that were conducted in the White Oak neighborhood are shown in the figure below. Based on CCTV inspection, many of the existing sewer segments had a NASSCO PACP grade of 4 or 5 (based on the quick score method). The existing sewer conditions are shown in the figure below, which shows the sewer segments with a PACP grade of 4 or 5 as red lines. The poor sewer conditions and a history of service calls for blocked house sewers further justified the sewer replacement project. Replacing the old clay sewers with new PVC sewer pipes will reduce infiltration and inflow from the project area.

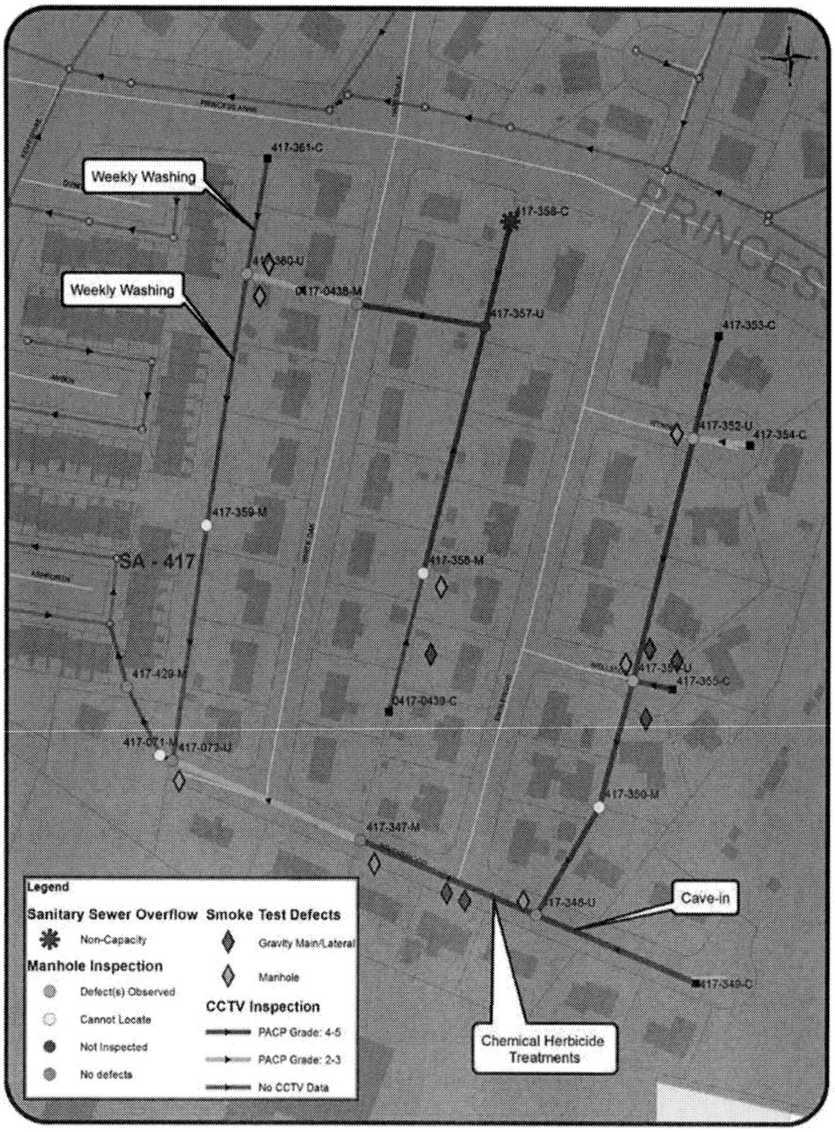

Existing Sewer Conditions

Four alternative sewer alignments were considered, ranging from no action to replacement of all gravity sewers. The selected alternative was replacement of the sewer system at a new tie-in location. The new 954 m (3,130 linear foot) neighborhood sewer system consists of 20.32, 25.40, and 30.48 cm (8, 10 and 12 inch) PVC pipe. The new sewer system is shown below.

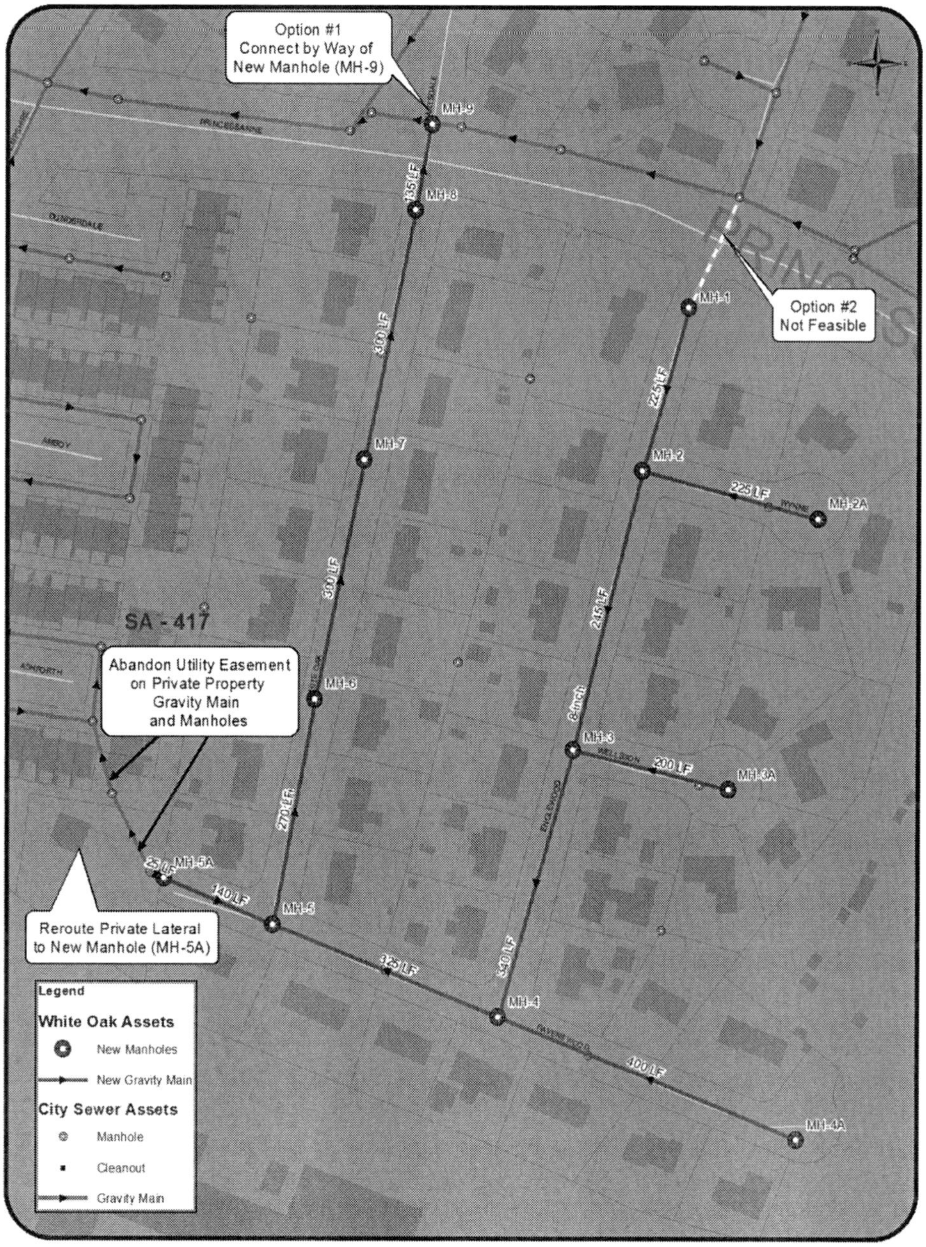

New Gravity Sewer System

Water System
The homes in the White Oak neighborhood were originally served by individual water wells. Cast iron water mains ranging in size from 10.16 to 15.24 cm (4 to 6 inches) were installed in 1972. The characteristics of the water distribution system which existed prior to this project are presented in the tables below.

Water Distribution System Characteristics

Category	Characteristic
Installation Year	1972
Water Main (LF)	3,013
No. of Service Lines	50
No. of Water Meters	50
No. of Fire Hydrants	3
No. of Isolation Valves	12

Water Main Characteristics

Material	Diameter		
	4" (LF)	6" (LF)	Total (LF)
Cast Iron	691	2,332	3,013

The City's water distribution system model and hydrant tests were used to check the fire flow adequacy of the existing looped system and it was determined to be inadequate due to high pipe velocities (>11 fps). This and the age and material of the pipe drove the decision to replace the water mains with a new looped system as part of the sewer replacement project.

The original and new water distribution systems are shown in the figures below. The new water lines are 4, 6, and 8 inches in diameter. The new system will meet fire flow requirements while not exceeding the City's maximum desired water flow velocity of 2.44 meters per second (8 feet per second).

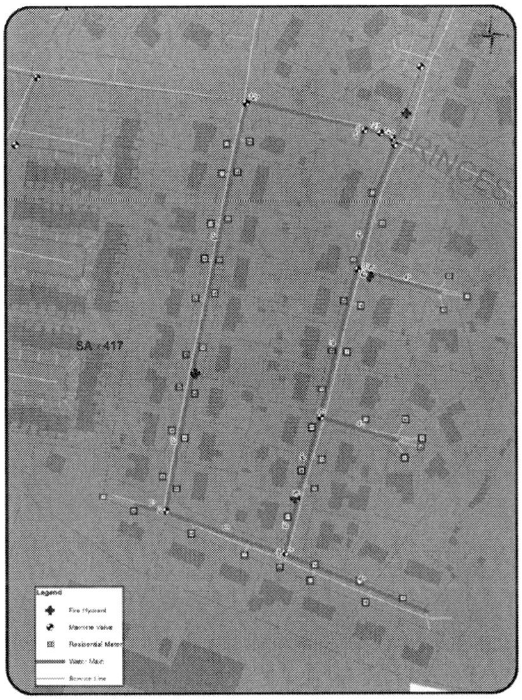

Existing Water System

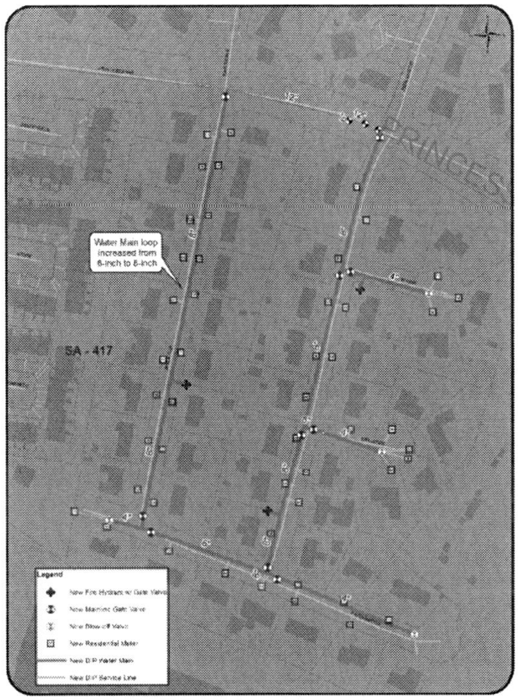

New Water System

METHODOLOGY

Field investigations: The field investigations that were conducted as part of the preliminary engineering phase included:

- Topographic survey of the neighborhood which included all structure and out building lines, property lines, trees and shrubs, driveways and sidewalks, and site improvements. Information on trees included the diameter and type of tree. Some of the older trees had diameters of around 1.22 m (4 feet). Property owners were requested to mark the location of the sewer lateral at the back of their houses for planning the new lateral tie-in locations.

- Subsurface utility investigations included utility markings by the Virginia Miss Utilities organization. Existing utilities included water and sewer mains, storm drains, gas mains and communications cables. Utility test holes were dug in areas of potential utility conflicts with new sewer and water mains.

- Soil corrosivity evaluation consisted of taking soil samples along the proposed water main route and analyzing the soil corrosiveness in the laboratory. The soils were found to be corrosive, and polyethylene wrapping of the ductile iron water mains was recommended and included in the design.

- Soil borings to assess new pipe and manhole bedding requirements. Soil borings to the along the proposed sewer main routes were made at about every 152.4 m (500 linear feet).

Project Constraints

There were a number of project constraints that needed to be resolved in order to design the new water and sewer mains as follows:

- Tapping the City water main: The existing water mains tied into a City water main located in a major four lane divided roadway. In order to avoid traffic disruption, the City required the contractor to make the new water main taps at night when traffic volume was low. The contractor used a detailed traffic control plan that was approved by the City as part of the design package.

- Maintaining water and sewer service during construction: The new water mains were constructed on the opposite side of the street from the existing mains and water service was transferred to the new main as it was advanced up the streets. The plan for the suggested sequence of construction was to install the sewer and water lines on one street at a time to avoid traffic disruption. The contractor opted to install the water system first and make all new connections prior to installing the sewer. The sewer mains were installed in the street with laterals up to the property lines. Connections to the existing backyard sewers remained in order to maintain service while the new sewers were being constructed. When all of the new sewers were in place, the contractor began rerouting the service laterals to the new street sewers.

Sewer Constructed After Water Line (White Oak Drive)

Water Main Removed to Install New Sewer (Englewood Drive)

Small Backhoe Used for Lateral Installation

Along several reaches of the new sewer, the existing sewer was located in the same area as the new sewer and had to be removed. A sewer bypassing plan was developed to allow removal of the existing sewer pipe and installation of the new pipe. The contractor had initiated bypassing, but determined that the day time, dry weather flows we so low that he could temporarily plug upstream existing sewers, let them surcharge and then remove sections of pipe to allow the new sewer to be installed in its place. The existing sewer was about the same elevation of the new sewer, so at the end of each day, the contractor made a temporary connection to maintain flow.

Bypassing for Sewer Removal / Replacement

Existing Sewer Removed and Replaced in Same Trench
(Ravenswood Drive)

- Traffic Control Plan: The City required that a detailed traffic control plan with detours and signage be part of the design project. This requirement added to the design cost. For the most part, the contractor was able to control traffic in the neighborhood without setting up detours. This deviation was approved by the City inspector. Not using the approved traffic control plan resulted in a cost savings for the contractor.

- New Sewer Lateral Routing: Since the sewer laterals along the backyard sewers were located in the back of the houses, the laterals needed to be routed either around the house to the street or the plumbing under the house needed to be rerouted to allow the lateral to exit from the front of the house. Locating the new laterals on each property required an extensive process that involved home owner participation as follows:

- The first step was inspecting and photographing properties to document the existence of features that could impact lateral routing.
- Discussions were held with property owners during the field inspections to hear their concerns.
- Maps of the proposed lateral routing plans were sent to each property owner with a form requesting their acceptance of the proposed routing. They were requested to sketch-in any information on the location of abandoned septic tanks or underground fuel oil tanks on the map and return it to the designer. An example of the lateral routing maps is shown below.

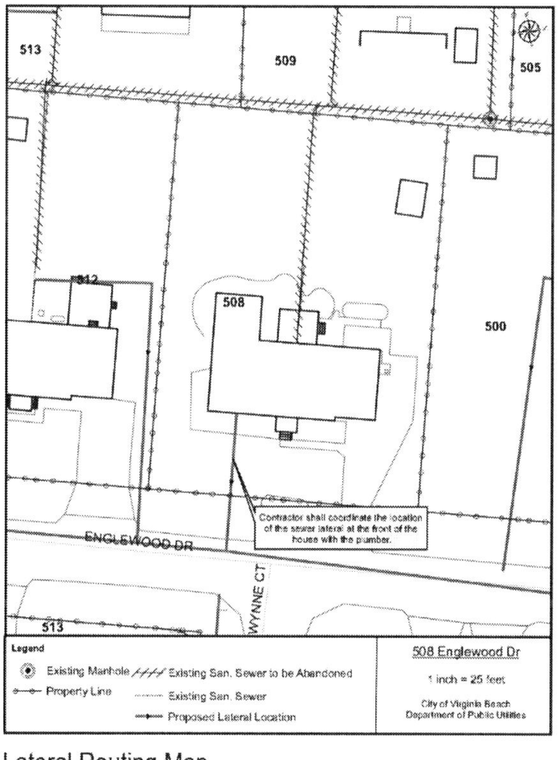

Lateral Routing Map

- A public meeting was conducted at a nearby elementary school to describe the project to the residents and to get feedback from property owners. Some property owners requested changes to the lateral routing to avoid hardscaped improvements. One property owner was particularly proud of his lawn which was an exotic type of Australian grass, Zoysia Grass. The design included the requirement that the contractor restore any of the grass that was disturbed in front of his house with the same type of grass. That made the home owner very receptive to the project.

Property with Zoysia Grass

- Subsequent follow up as the new sewers were constructed. At one house, it was determined that the lateral could not be routed around the house as shown in order to avoid damage to two large Oak trees. We met the home owner in the field and proposed an alternate routing on the other side of the house that he approved.

Proposed Lateral Would Impact Big Trees

- In another case, a new home owner was concerned about the condition of the drain piping under their house (in the crawl space) and proposed that the pipes under the house be rerouted to have the lateral exit at the front of the house and connect to the street sewer. The bid form for the project included a line item for the contactor's plumber to turn around piping under houses. So, this change was easily accommodated prior to construction of the new sewer along the street where the new home owner was located.

- House lateral minimum slopes. Another consideration in design of the new house laterals was the minimum slope required by the local building code, 1%. This could become an issue at the end of the main line sewer where it is shallower and long lateral runs may be necessary. The new mainline sewer has one tie-in to an existing manhole in the four lane street. This set the beginning invert elevation of the new sewer. Once the preliminary layout and grades of the new mainline sewers were determined, each proposed sewer lateral was checked to make sure that the lateral could be routed around the house and be at a depth of about 0.61 m (2 feet) where it would tie in to the existing house lateral. A spreadsheet was developed to make the calculations. It included the proposed invert of the mainline sewer connection at each house, the proposed lateral length, and the calculated depth of at the lateral tie-in while maintaining minimum slope. We were successful! The last house at the end of the line where the new sewer was shallowest had a lateral depth of slightly deeper than 0.61 m (2 feet).

RESULTS

As of the date of this conference, all of the new sewer and water lines had been constructed and all tie-ins to existing sewer laterals were well under way. The next step will be to CCTV the new sewers for acceptance by the City. The existing roads in the neighborhood will be milled and a new asphalt surface coat will be installed.

The project was constructed with minimal disruption to existing landscaping, trees, shrubs, and other physical features. The photographs below show how impacts were minimized and mitigated.

Lateral Installed Between Existing Features

Zoysia Grass Sodding After Water Main Constructed

New Water Service Avoided Landscaping

Conflicts with Post Design Additions Avoided

DISCUSSION

As the construction project progressed, the contractor made changes to the sequence of construction which saved him time and money, and did not result in water and sewer service disruption or major traffic access issues. Changes included avoiding detours and signage and extensive bypass pumping. This demonstrates that even though certain aspects of a design are necessary to get approvals by the owner's agencies, the contractor can often figure out an alternative way to get the job done that works to his advantage.

CONCLUSIONS

Doing extensive preliminary field investigations and working closely with property owners both during design and construction are important to a successful backyard sewer replacement project. A backyard sewer replacement program can be successful if the following elements are part of the design and construction process:

- Gaining an understanding of the history of the existing utilities and how they came to be where they are.
- Conducting thorough subsurface utilities locating.
- Locating and naming all major topographic features during the topographic survey.
- Doing detailed field reconnaissance and making contact with property owners to assess lateral routing constraints.
- Working with homeowners to get project buy-in including consent letters and public information meetings.
- Being flexible during construction to address changing conditions and to address new concerns by property owners.

This project was constructed with very minimal complaints by the area residences. The project will likely increase the value of their properties by having new, sustainable sewer and water systems.

REFERENCES

None.

End of Manuscript.

Evaluation of High pH Infiltration Water and Recycled Bedding Material

Keith Hobson, PE, BCEE, FOX Engineering Associates, Inc.

ABSTRACT

The purpose of this evaluation was to determine the cause and source of high pH (alkaline) wastewater in the CBI Southlands sanitary sewer in Council Bluffs, Iowa. This newly installed 48-inch sanitary sewer in south Council Bluffs filled with a high pH wastewater and the first discharge of this wastewater to the City wastewater treatment plant created a resultant rise in the pH of the total wastewater stream. A recycled crushed concrete material was used for the bedding support during the pipe installation.

A number of water and experiment samples and tests were conducted to identify the alkaline material and source potential. The potential that something in the pipe bedding could cause such an increase in pH and calcium through infiltration through leaking joints into the piping was explored. The results of testing are presented that shows that the calcium levels were over 2,000 mg/l as $CaCO_3$ in the distilled water. This is similar to the process that occurs in a lime softening water treatment plant.

Even though the recycled concrete had the same structural strength and pipe lateral support as natural rock, it did not prove to provide the same long lasting support. Based on the testing and confirmation of high alkalinity wastewater the sewer line was completely removed and replaced. The new installation was completed with new bedding material of natural rock to avoid this situation.

KEYWORDS

High alkalinity wastewater, recycled concrete, pipe bedding, lime softening chemistry, and sanitary sewer.

INTRODUCTION

A new sanitary sewer, pumping station, and force main were installed in Council Bluffs, Iowa in 2012 to serve a new industrial park in the southern part of the city. The site is also within one mile of the Missouri River and pipe excavation was in sand and loess material. Dewatering was required during construction to lower the water table in the area. This sanitary sewer portion consists of 48-inch diameter fiberglass-reinforced, polymer mortar pipe made by HOBAS and 21-inch diameter PVC pipe. The pipe was installed with bedding material consisting of recycled crushed concrete to simulate crushed rock.

After installation this sewer was idle for several months after leakage testing. During this period a certain amount of ground water had entered the pipe. When this water was discharged to the City wastewater treatment plant it created a resultant rise in the pH of the total wastewater stream at the plant creating a major concern. Figure 1 below illustrates the rise in pH of the plant influent from an average of about 7 to over 9 with the new sewer discharge. The new sewer discharge was terminated until the source of this high pH wastewater could be determined and treated. This newly installed sanitary sewer from the CBI Southlands Industrial site in south Council Bluffs had apparently filled with a high pH wastewater.

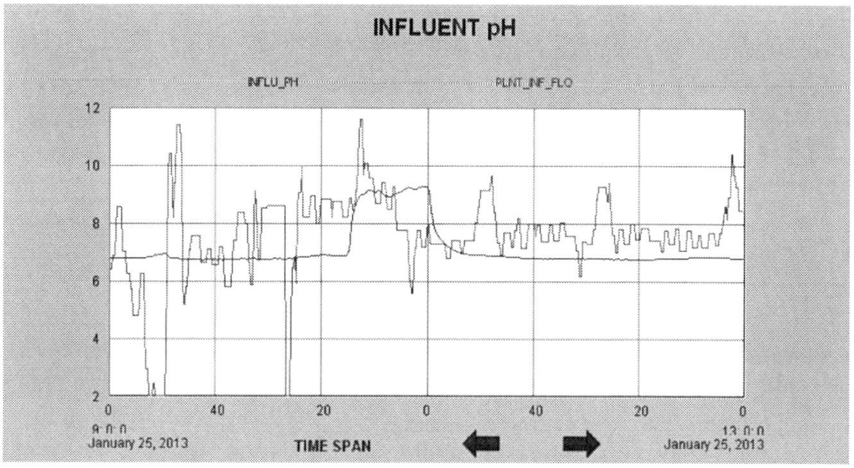

Figure 1. Council Bluffs Water Pollution Control Plant influent flow and pH levels during the first sewer discharge on January 25, 2013.

FOX Engineering was hired to investigate this situation and determine the cause and potential treatment. The purpose of this evaluation is to determine the cause and source of high pH (alkaline) wastewater in the CBI Southlands sanitary sewer and determine appropriate treatment methods and prevent further contamination.

METHODOLOGY

In order to evaluate the source of the high pH wastewater, we held discussions with a number of site personnel and visited the area to check on potential sources. Discussions were held with city wastewater staff and city engineering staff, city water works staff, sewer installing contractor staff, construction observers, geotechnical engineers, and area industries. We then reviewed the site of the sewer and potential water and contamination sources. We requested a number of water and experiment samples and testing to be conducted to identify the alkaline material and source potential. We also researched a number of papers and reports on related alkaline sources.

Based on the sewer layout, topography, and sewer tributary area a number of potential sources were identified for the alkaline contamination including:

- Potential Source #1 – Area industries and construction site
- Potential Source #2 – Illegal discharge from hauler into area manhole
- Potential Source #3 – Groundwater infiltration and surface water inflow

These are evaluated in the following sections.

SOURCE EVALUATION

Industrial Site Observation

FOX staff visited the industries in the area that were connected to the sewer to observe construction activity, water sources, and chemical storage within the buildings that had been connected to the sewer. A record of the potable water supplies to the sites was also reviewed to compare to the amount of sewage discharged. Based on this review of operations and chemical storage, there was not sufficient quantities of any chemicals present to produce the high levels of pH and alkalinity in the wastewater pipeline.

Water Chemistry Review

The measured pH in the wastewater contained in the sewer line was 12.5. As indicated earlier a number of water tests were conducted to identify the source chemicals. These samples were taken at various manholes throughout the sewer line and tested for typical inorganic parameters as presented in Table 2. A complete organic toxicity test was also completed on one sample to check for other potential contaminants. However, no potential toxic levels of organic compounds were found in the testing. Additional tests were then performed on constituents believed to contribute to the alkalinity.

Potential causes of high pH water in the sanitary sewer that were explored as part of this evaluation include the following:
1) High pH Due to Lime Softening Chemistry within Pipe Bedding
2) High pH Due to Strong Base Addition Due to Waste Dumping

A more detailed discussion is included below.

1) <u>High pH Due to Lime Softening Chemistry within Pipe Bedding</u>

a. Background – Lime Softening in Drinking Water

Lime softening is a conventional drinking water treatment process that involves the removal of calcium and magnesium (total hardness) by increasing the pH of the water and subsequently precipitating calcium as calcium carbonate and magnesium as magnesium hydroxide. In general, the calcium hardness is removed (precipitated and settled) by raising the pH to 10.3, while the magnesium hardness requires a higher pH of around 11. The pH is raised through the addition of some form of lime which may be either calcium oxide (quicklime, CaO) or calcium hydroxide (hydrated lime, $Ca(OH)_2$. After lime addition, the raw water is mixed and the white hardness precipitate is formed, flocculated and settled in a matter of hours. Because the precipitates are very slightly soluble, some hardness remains in the water typically on the order of 50-85 mg/l as CaCO3 for drinking water.

The total hardness of a water is the concentration of multivalent cations in water such as calcium, (Ca^{2+}), Magnesium (Mg^{2+}), iron, manganese, strontium, barium, zinc, aluminum and other ions. Total hardness is typically defined as simply the sum of calcium and magnesium hardness, since the other multivalent ions are generally not present in significant quantities. The total calcium (i.e. hardness) measured in the sanitary sewer samples was much higher than would be expected from a typical source of groundwater infiltration which raised a red flag and question regarding the calcium source. This issue is addressed in the following section.

b. Lime Softening in Recycled Concrete Bedding

The chemistry which occurs in a conventional drinking water plant with lime softening treatment was observed to be present in the samples collected from the sanitary sewer. The samples exhibited a very high pH and extremely high calcium (i.e. hardness) levels which are well above (2-5 times) the anticipated hardness levels in fresh/shallow groundwater that could infiltrate the pipe.

Table 1. Typical Hardness Levels in Water

Water Type	Total Hardness (mg/l as $CaCO_3$)	Notes
Soft	0 to 75	
Moderate	75 to 150	Typical Softened Drinking Water Quality
Hard	150 to 300	Typical Fresh Water/Shallow Ground Water ≈ 15-375 mg/l Typical Council Bluffs Raw Sewage ≈ 200 – 400 mg/l
Very Hard	>300	Iowa Ground Waters/Aquifers ≈ 300-1000 mg/l

The potential that something in the pipe bedding could cause such an increase in pH and calcium through infiltration through leaking joints into the piping was explored. In order to confirm or deny this theory, a sample of the recycled concrete pipe bedding was collected and distilled water was poured into the sample and allowed to set for 24 hours. After that time, there was a white precipitate in the sample. The liquid was poured off and sent to a laboratory for testing and the results are presented in the second column in Table 2 below. The results are also illustrated in Figure 2. The calcium levels were over 2,000 mg/l as $CaCO_3$ in the distilled water which shows that there is a strong potential for ground water in contact with the recycled pipe bedding to become saturated with calcium as it flows through the bedding. This results in a gain in alkalinity and increase in pH. This is similar to the process that occurs in a lime softening water treatment plant.

The fact that the measured sewer samples had elevated calcium, but not elevated magnesium as in lime softening required additional analysis. The distilled water soaked in the recycled concrete bedding had an elevated magnesium level, yet there was virtually no magnesium in the sewer samples. As a potential explanation, it was previously noted that calcium carbonate and magnesium hydroxide floc are quite different. The calcium carbonate floc is comprised of dense crystals that are negatively charged and are benefited by the use of a coagulant aid for settling. Calcium carbonate ($CaCO_3$) removal is aided by a coagulation/flocculation process to grow the particles to induce settling. A mechanism for growing the $CaCO_3$ particles would not be present within the pipe bedding and it is theorized that a portion of the fine $CaCO_3$ particles would make their way through the bedding and into the piping. Magnesium hydroxide floc, on the other hand, is similar to a coagulant aid such as alum (aluminum sulfate) which has positively charged sites that can neutralize the charges of particles that allow them to clump together. It is again theorized that the Magnesium Hydroxide particles might be more easily retained within the pipe bedding and would not pass through into the sewer pipe. The other theory is that the shallow ground water had a very low Magnesium level to begin with.

Testing results also showed that the recycled concrete significantly added potassium and sodium to the water which was also observed to be high in the sewer samples. This is another indicator of the potential for groundwater into the pipeline to be the source of high pH water.

Table 2. Measured Hardness Levels in Water

Chemical	Original Sample Collected 1-25-13 (Lift Station-MH Bypass)	Distilled Water w/ 24-Hr Soak in Recycled Concrete Bedding	MH#2 Sample Collected 02-07-13 (Lift Station-MH Bypass)	Pipe Seepage Sample @ Sta 575+66 Collected 2-09-13
Total Alkalinity, mg/l as CaCO3		2190	685	875
Total Calcium, mg/l	568	2,221	237	329
mg/l as CaCO3	1,420	5,553	593	823
Chlorides, mg/l	103	273	58	27
Hardness (Total), mg/l as CaCO3	*>1,420 est.*(1)	*5,850 est.*(1)	*594 est.*(1)	822
Iron, mg/l		107	0.6	0.04
Total Magnesium, mg/l		74.3	0.39	0.02
mg/l as CaCO3		297.2	1.56	0.08
Manganese, mg/l		3.38	0.02	n.d.(2)
pH	12.5	12.4	12	12.2
Potassium, mg/l	71.5	365	63.8	26.2
Sodium, mg/l	102	437	80.2	29.7
Sulfates, mg/l	14.7	8	10	30
Total Dissolved Solids (TDS), mg/l		4,980	1722	1,972
Total Suspended Solids (TSS), mg/l		1,071	274	7

(1) Total hardness estimated from the sum of the calcium and magnesium.
(2) n.d.= not detected

Several other studies (Thiele Geotech 1996 and Steffes 1999) have been conducted on recycled crushed concrete material and similar observations were made regarding reaction of the material with ground water to produce high pH water and resultant precipitation of calcium carbonate. Therefore it was thought that the high pH wastewater could be produced from the recycled concrete bedding material.

2) High pH Due to Strong Base Addition Due to Waste Dumping

The potential for the pH to be raised due to the dumping of a strong base was explored, since only a strong base would be capable of raising the pH to levels above 12. A list of strong bases was considered and the most common commercially available strong bases outside of hydrated lime were considered to be sodium hydroxide and potassium hydroxide.

The potential for the pH to be raised due to the dumping of a strong basic solution such as Sodium Hydroxide (NaOH) or Potassium Hydroxide (KOH) was explored since these strong bases would be capable of raising the pH to the measured levels and also due to the presence of sodium and potassium in the samples. Based on results of the Rothberg, Tamburini and Winsor Model for Corrosion and Water Chemistry (RTW3.0 Model), it was estimated that a dosage of over 500 mg/l of sodium hydroxide into ground water would be required to increase the pH of a typical groundwater from 8 to 12.5. This dosage would result in a sodium concentration of over 280 mg/l which is approximately three (3) times the measured values, so there is not enough sodium in the water to indicate that NaOH is responsible for raising the pH.

The potential for KOH dumping was explored and similar results were determined indicating that over 500 mg/l of KOH would be required as well. This dosage would result in a potassium concentration of over 350 mg/l which is approximately five (5) times the measured values, so there is not enough potassium in the water to indicate that KOH is responsible for raising the pH either.

Therefore, chemical analysis showed that the concentrations of the two strong bases of sodium hydroxide or potassium hydroxide were not high enough to cause the observed pH.

This ruled out the likelihood of a large chemical dump to cause the high wastewater pH issue.

CBFS - Sewer Effluent
12002.03

1 1/2" Pipe Bedding from Conreco

2,000g sample
innundated w/ 500g distilled water

date/time	ET (min)	pH
2/5/13 10:12	0	7
2/5/13 10:13	1	11.9
2/5/13 10:42	30	12.7
2/5/13 11:42	90	13.4
2/5/13 12:42	150	12.8
2/5/13 13:42	210	12.7
2/5/13 14:42	270	12.9
2/5/13 15:42	330	12.8
2/5/13 16:42	390	12.8
2/6/13 10:12	1440	12.9

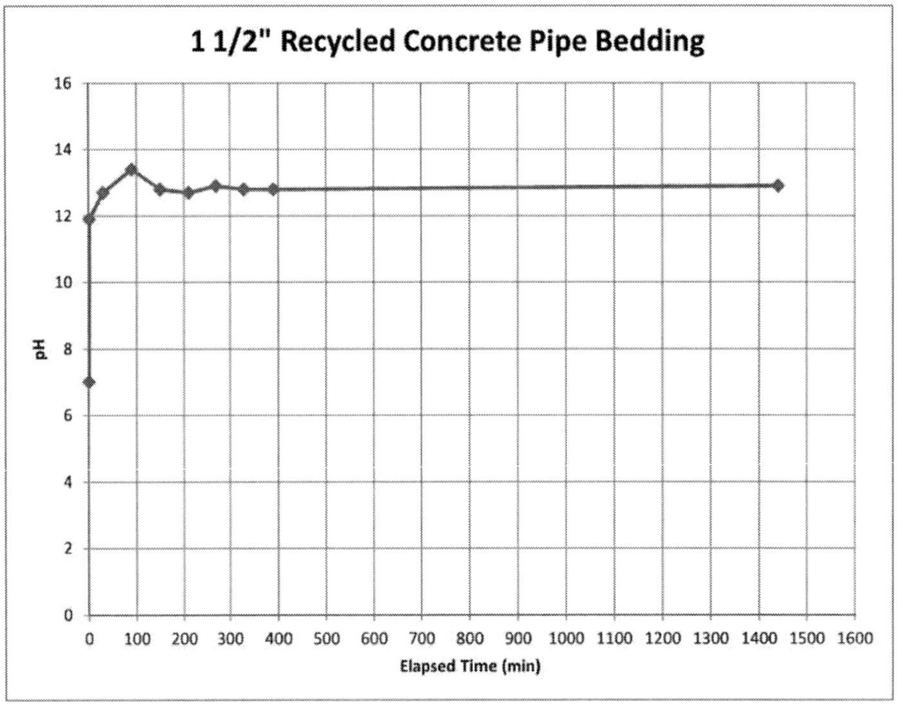

Thiele Geotech Inc

Figure 2. Results and graph from lab testing of distilled water and recycled concrete pipe bedding showing elapsed time and pH.

DISCUSSION AND CONCLUSION

The observations of the wastewater characteristics and the water testing indicated that the recycled crushed concrete bedding was reacting with the groundwater and producing a high pH water that infiltrated the sewer and precipitated out calcium carbonate. This created not only a high pH wastewater but also one high in total dissolved solids (TDS) and total suspended solids (TSS). In order to discharge this waste to the city the contractor was required to neutralize this waste with acid to bring the pH down to 9 before discharging. A procedure was developed for removing the wastewater, neutralizing with acid in fracking tanks and then discharging the equalized waste to the city for further treatment.

Even though the crushed concrete material was adequate for strength and abrasion as a bedding material and of correct gradation, the material was not deemed acceptable for long term use. In addition to producing the high pH water the bedding material also degraded from the chemical softening process and the leaching of calcium carbonate. Therefore it would lose its structural integrity over time.

The recycled crushed concrete bedding material needed to be replaced. In order to correct this situation the complete sewer and bedding material was removed and the pipe was re-laid with natural crushed rock bedding material.

For sewer pipe bedding material in groundwater conditions it is recommended that recycled crushed concrete material be avoided as a suitable bedding material. Even though there are benefits to using a reclaimed product and reducing natural resource use, it does not provide adequate long term pipe support and its use creates additional high pH (alkaline) wastewater.

REFERENCES

Thiele Geotech Inc. (August 20, 1996) *Recycled Aggregates in Infrastructure Construction* prepared for Nebraska State Recycling Association, Omaha, Nebraska.

Steffes, Robert (September 1999) *Laboratory Study of the Leachate From Crushed Portland Cement Concrete Base Material* produced for the Project Development Division, Iowa Department of Transportation, Ames, Iowa.

Superstorm Sandy Aftermath: Providing Resiliency for Critical Infrastructure With a New Stormwater Collection System

Anthony Brozier, P.E.[1*], Michael Vecchio, P.E.[1], Lina Posso, P.E.[1], Scott Davis, P.E.[1], Albie Perry, P.E.[2], Steve Biuso, P.E.[3]

[1] HDR, Mahwah, New Jersey
[2] Passaic Valley Sewerage Commission, Newark, New Jersey
[3] AECOM, Piscataway, New Jersey
[*]Email: anthony.brozier@hdrinc.com

ABSTRACT

The Passaic Valley Sewerage Commission (PVSC) wastewater treatment plant consists of a 140-acre facility located adjacent to Newark Bay in the city of Newark, New Jersey. The facility treats the wastewater and sludge from over two million residents from both New Jersey and New York, and is the fifth largest wastewater treatment plant in the United States.

In October of 2012, the PVSC facility was flooded and rendered inoperable for several days as a result of storm surge generated by Superstorm Sandy which impeded the facility to fully treat the incoming flows and to process sludge from other facilities that rely on this facility for sludge treatment and disposal. The New Jersey Department of Environmental Protection (NJDEP) subsequently identified the PVSC facility as a critical component of New Jersey's infrastructure and recommended that PVSC implement measures to protect the facility from similar future storm events.

In August of 2014, the Federal Emergency Management Agency (FEMA) awarded a $260 million Public Assistance grant to PVSC to build comprehensive mitigation measures that would protect the facility. To make the facility more resilient from future storm surge and rainfall events, PVSC chose to design and install two perimeter floodwalls to protect the entire plant from future storm surges. This included reconfiguring the existing stormwater collection system to convey stormwater within the limits of the new floodwalls to one of several new pump stations. A Joint Venture (JV) team of AECOM and HDR was tasked with finalizing the design concepts for the floodwall, stormwater collection system and pump stations, and a standby power facility, with each becoming a separate design-bid-build contract.

The stormwater collection system was designed to convey stormwater flows generated from the 100-Year, 24-Hour storm event, and a small amount of floodwall overtopping associated with the 500-Year storm surge event, while protecting critical facilities within the floodwall by ensuring the maximum allowable flooding elevation is at least 0.5 feet below the finished floor elevation (FFE) of all buildings. Design alternatives were provided to PVSC, with their preferred alternatives being selected for detailed design and construction. The methodology, design approach, and selection process are presented in this document.

Currently, the JV is assisting PVSC in contracting design professionals to complete the detailed design, and will be assisting in managing the design and construction over the next several years.

KEYWORDS

Critical Infrastructure, Superstorm Sandy, Resiliency, Storm Surge Flooding, Rainfall Induced Flooding, Collection System Improvements

INTRODUCTION

The PVSC wastewater treatment plant is the largest wastewater treatment plant in New Jersey, and the fifth largest in the United States. The plant serves over 1.5 million residents from New Jersey, spread out over five counties and 48 separate municipalities, plus receives trucked in liquid waste and barged sludge from NJ and NY.

Figure 1 shows the site location map, and Figure 2 shows the municipalities and counties served by the PVSC wastewater treatment plan facility.

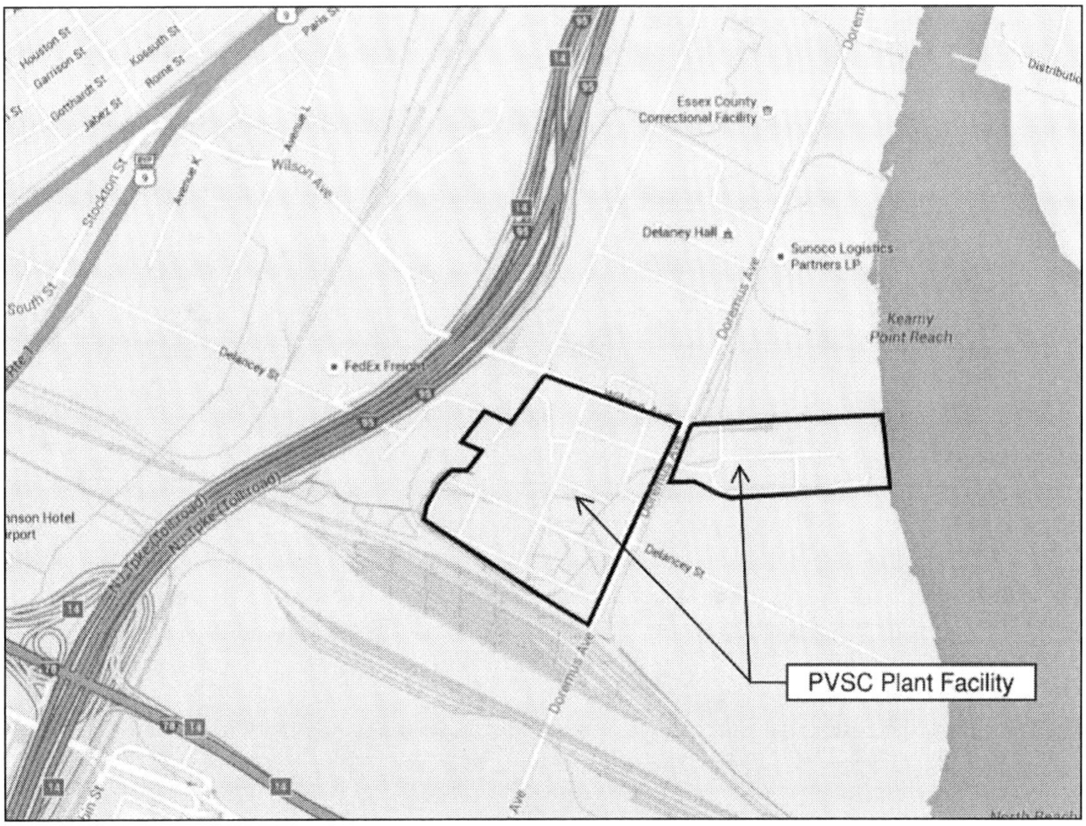

Figure 1 – PVSC Site Location Map

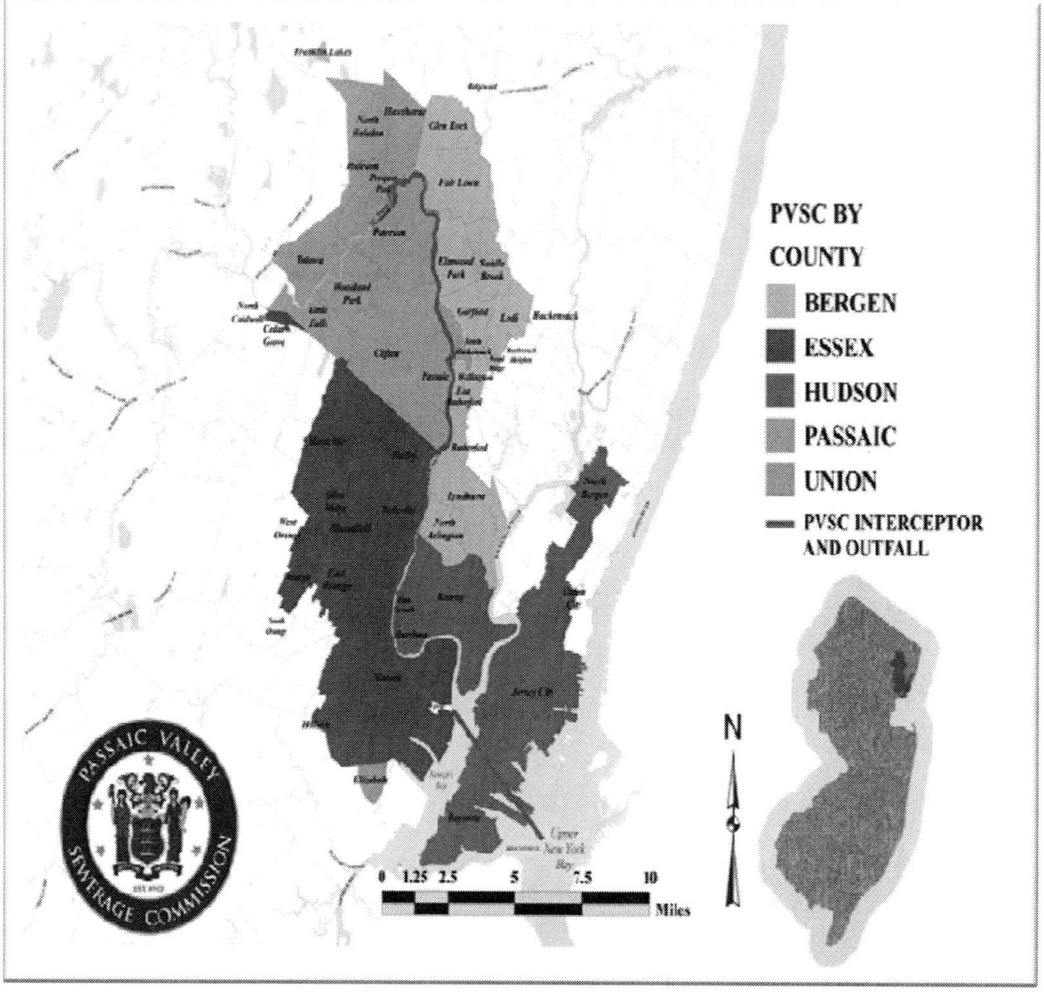

Figure 2 – PVSC Service District Map

On October 29, 2012, Superstorm Sandy made landfall along the coast of New Jersey. The winds and atmospheric pressure changes generated by the storm created a storm surge that pushed ocean water inland, inundating the 140+ acre PVSC wastewater treatment plant facility.

Figure 3 shows the track that Superstorm Sandy took as it approached landfall in New Jersey.

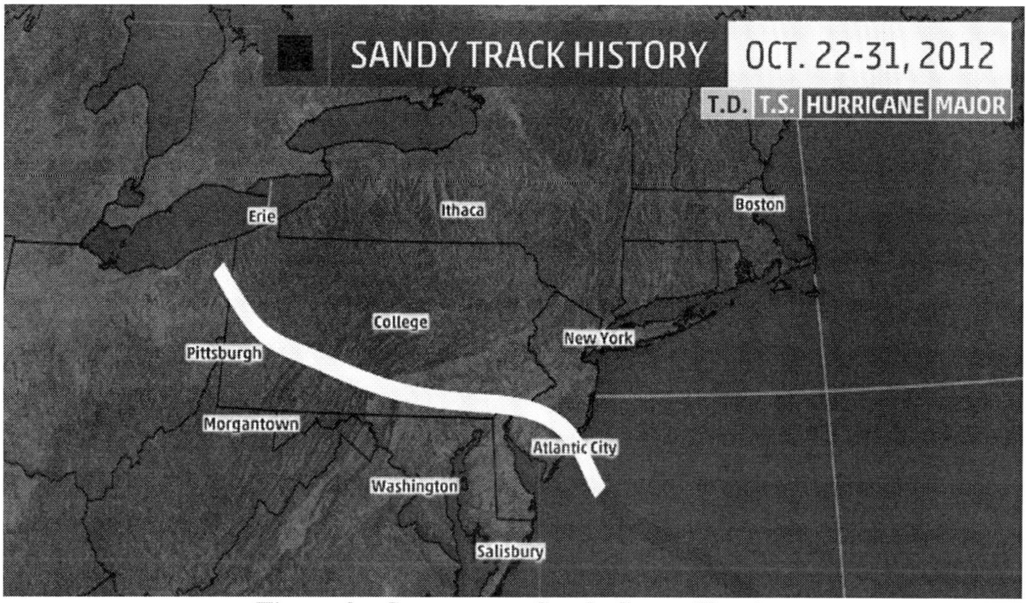

Figure 3 – Superstorm Sandy Storm Track
Source: Weather Channel (weather.com)

The saltwater from the storm surge flooded the site's buildings, tunnels and equipment, resulting in significant damage throughout the facility. In addition, PSE&G, the company which provides power to the PVSC facility, experienced an outage that lasted for days following the storm. The combination of these impacts rendered the facility inoperable for several days, also impacting Newark Bay and customer facilities that import sludge to PVSC for treatment.

Figure 4 shows some examples of the flood damage caused at the PVSC facility.

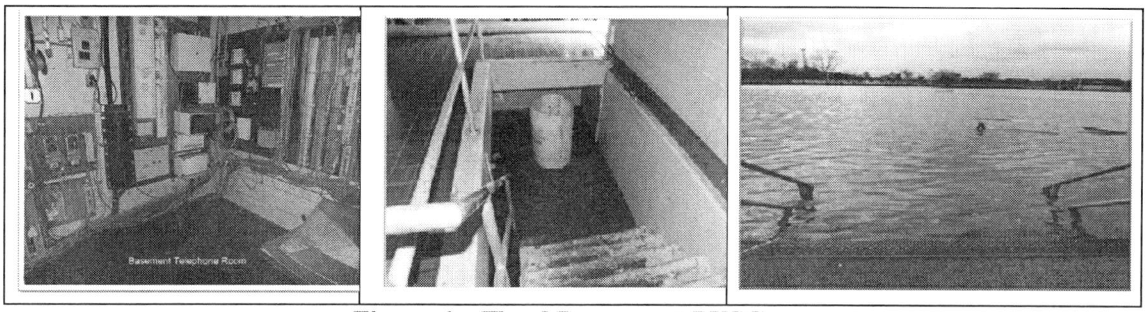

Figure 4 – Flood Damage at PVSC

The New Jersey Department of Environmental Protection (NJDEP) subsequently identified the PVSC wastewater treatment plant as a critical component of New Jersey's infrastructure and recommended that the facility implement measures to protect it from future storm events. In August of 2014, the Federal Emergency Management Agency (FEMA) awarded a $260 million Public Assistance grant to PVSC to build comprehensive mitigation measures that would protect the facility.

To protect the facility, PVSC chose to design and install a perimeter flood wall to protect the entire wastewater treatment plant from future storm surges. Since the PVSC facility is separated by a roadway, two separate floodwalls would be required. Figure 5 shows an aerial photograph of the PVSC facility with the approximate location of the floodwalls shown highlighted in yellow.

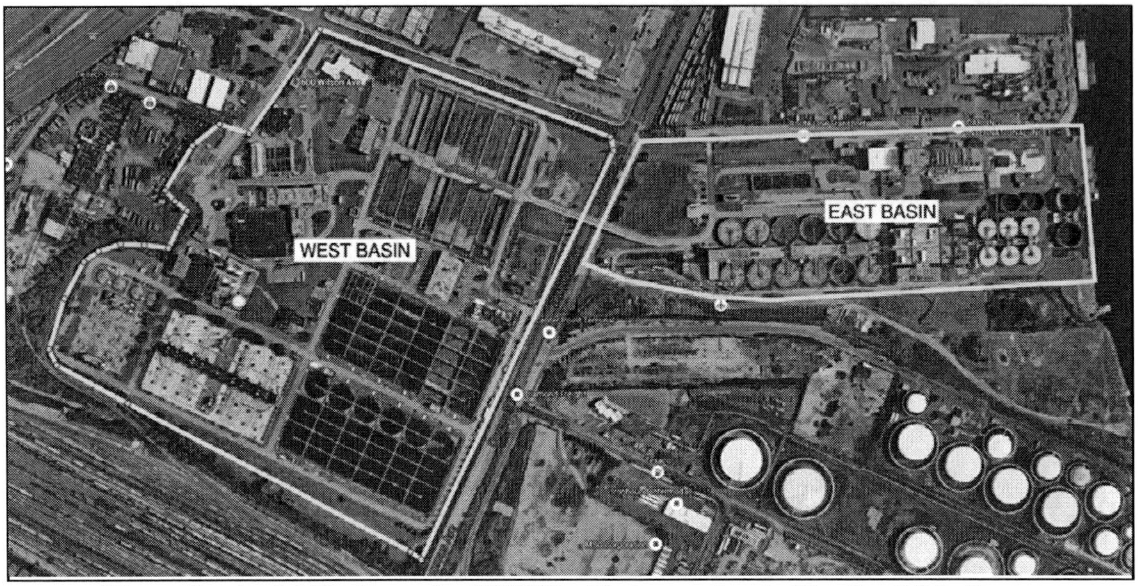

Figure 5 – Aerial Photograph of PVSC with Approximate Floodwall Limits
Source: Google Maps

In addition, the existing stormwater collection system would need to be reconfigured so the outfalls would no longer be impacted by tidal influences. As a result, the reconfigured stormwater collection system would only receive stormwater from rainfall events, with the water being directed to one of several new stormwater pump stations that will pump the stormwater over the floodwall to Newark Bay.

The existing stormwater collection system pipes consisted primarily of Reinforced Concrete Pipe (RCP) or Ductile Iron Pipe (DIP). The DIP was found in pipes 24" in diameter or smaller, while RCP was used for larger pipe sizes. Due to the geotechnical conditions of the soil and the proximity to Newark Bay, the pipes are provided with pile supported pipe cradles along their length, except where the pipes are located above another crossing utility or tunnel. The cradles provide pipe support and deflection control for the collection system.

The pipe cradles at the site are of two basic types – wood pile supported or concrete filled steel pipe pile supported. The pipe cradles constructed from wood timbers are supported on two (2) wood piles having a design capacity of 25 tons each. The pipe cradles constructed of concrete are supported on two (2) concrete filled pipe piles having a design capacity of 60 tons each. The storm manholes and catch basins were generally found to be supported on three (3) piles, with some exceptions where larger structures required additional piles.

METHODOLOGY

Design of the stormwater collection system first required the development of Design Criteria, which would determine the parameters to which the collection system would be designed, and the appropriate methodology to be used.

Prior to recommending design criteria, an initial investigation of the existing stormwater collection system was done to gain an understanding of the pre-construction drainage conditions. This was accomplished through on-site investigation, review of available data (surveys, as-builts, record drawings, etc.), and performing a preliminary hydrologic and hydraulic analysis of the existing collection system. Design Criteria recommendations were provided by the JV to PVSC, and through subsequent discussion, the final design criteria were agreed upon.

Using the Design Criteria, multiple hydrologic and hydraulic models were developed for each of the two separate sections (East and West Basin) of the PVSC site, with each model meeting the design criteria. Each model represented a unique stormwater collection system design alternative. For each alternative, Class IV cost estimates were prepared. The alternatives were ultimately presented to PVSC, after which PVSC selected the preferred stormwater collection system alternative for the East and West Basin.

Development of Design Criteria

PVSC determined that the stormwater collection system should be designed to handle a 100-Year, 2-Hour rainfall event (9.04" rainfall depth) during a coincident 500-Year storm surge coastal event. The reasons for this were fourfold:

- Using the 100-Year, 24-Hour rainfall event was found to be consistent with state regulatory requirements.
- Preliminary hydrologic and hydraulic analysis of the existing stormwater collection system based on surveyed data and record drawings showed that some portions of the system appeared to be designed for a storm similar to the 100-Year, 24-Hour rainfall event.
- Designing the existing collection system to the 500-Year, 24-Hour rainfall event would require all pipes to be upsized significantly, resulting in conflicts with other existing subsurface utilities and tunnels. This would create operational impacts to the wastewater treatment plant that PVSC preferred to avoid if possible.
- The size of pumps and wet wells that would be required to meet the peak flows of the 500-Year, 24-Hour rainfall storm event would be too large to fit within the available area.

Two external inflows to the stormwater collection system were identified during the design:

- Consistent with the floodwall design, an overtopping rate of 0.01 cubic feet per second per linear foot (cfs/LF) was applied to a 1,400 LF section of the eastern floodwall, adjacent to Newark Bay. This overtopping rate was associated with the water surface elevation encountered in Newark Bay according to FEMA data for the 500-Year storm surge coastal event.
- Currently, the drainage area for PVSC property to the western side of the facility flows to an on-site pump station which contributes to the existing collection system. The pump curves of the existing pump were obtained, and showed that the maximum pumping rate

was approximately 3.3 cfs. This flow rate was factored into the model as a worst-case scenario. However, the future goal of PVSC is to decommission the pump station, and create an alternative route for the creek flows to reach Newark Bay without having to enter the limits of the proposed floodwall.

No additional external inflows were considered, since the stormwater collection system model design was performed in a scenario that assumed the floodgates were all closed. The top of the floodwall was identified as the tailwater condition in the Design Criteria and used in the modeling and design of the collection system.

To provide PVSC with an added factor of safety, the stormwater collection system was designed to allow some minor short-term flooding that would not adversely impact the operation of the facilities. The modeling and design of the stormwater collection system was performed to maintain a minimum freeboard of six inches below the first floor elevations (FFE) of all buildings within the limits of the floodwalls.

The method for discharging stormwater captured in the collection system is via proposed pumps located within new pump stations on each side of the facility. The west side pumps would discharge into Jasper Creek, which flows to Newark Bay, while the east side pumps would discharge directly into the bay.

Modeling Methodology
Hydrologic and hydraulic models were developed to show how the proposed stormwater collection system would respond to the aforementioned Design Criteria loading conditions. The models were constructed using Storm and Sanitary Analysis (SSA) 2014 by Autodesk. SSA allows users to take the pipe networks and drainage area catchments created in AutoCAD Civil 3D and import them directly into the SSA model.

The United States Department of Agriculture's Technical Release #55 (TR-55) hydrologic method, which utilizes simplified procedures for calculating storm runoff volumes and peak rates of discharge for small urban watersheds was used for this design. The hydrographs developed with the models were created using the Soil Conservation Services (SCS) method. According to SCS rainfall distribution maps, the PVSC site falls under Type III rainfall distribution. The Type III hydrograph was applied to simulate the 100-Year, 24-Hour rainfall event.

Hydrologic Modeling Input Data
Using the most up-to-date aerial photography, the cover types for the PVSC were identified as either ">75% Grass Cover" or "Impervious Cover". The USDA Web Soil Survey showed that the site consisted entirely of HSG D-Type soils.

Using the topographic survey provided, drainage areas to each individual catch basin within the PVSC site was then delineated. Time of concentration (TOC) for each individual drainage area was determined. Due to the site being very heavily developed and impervious, and the presence of numerous catch basins, minimum TOC values were used based on the size of the drainage areas. TOC of 10 minutes was used for drainage areas greater than 0.5 acres, and 6 minutes was used for drainage areas smaller than 0.5 acres.

Hydraulic Modeling Input Data

The stormwater collection system was then modeled. The proposed stormwater collection pipes consisted of either Reinforced Concrete Pipes (RCP) or Steel Reinforced Polyethylene Pipe (SRPE). The SRPE was used in lieu of High-Density Polyethylene Pipe (HDPE), which is not rigid enough to use in conjunction with piles and pile caps. The SRPE was used for pipe sizes between 30" and 60" in diameter where it was found to be more cost-effective than RCP. Manning's n-values for SRPE and RCP are 0.011 and 0.013, respectively.

In addition, hydraulic loads from external inflows associated with floodwall overtopping were included. Consistent with the floodwall design, an overtopping rate of 0.01 cubic feet per second per linear foot (cfs/LF) was applied to a 1,400 LF section of the eastern floodwall, adjacent to Newark Bay. This overtopping rate was associated with the water surface elevation encountered in Newark Bay according to FEMA data for the 500-Year storm surge coastal event.

Additional Modeling Assumptions

In performing the H&H modeling, the following assumptions were made:

1. The stormwater collection system pipes to remain in place are all clean, intact, and with no significant blockages. The client agreed with this assumption based on their knowledge of the existing collection system.
2. The model does not account for infiltration of flow. The design consultant is to perform an evaluation of the existing pipes remaining to verify this.
3. Open tankage (clarifiers, chambers, etc.) were determined to be non-contributors of flow to the collection system. The client indicated rainfall to these tanks did not flow to the collection system. Open tankage areas were not modeled as contributing to the collection system in any way.
4. The existing inlets are not limiting in capacity. The client did not indicate that this was an issue during previous rainfall events.

DESIGN APPROACH

Several collection system designs were considered for each side of the project site. Since the site is bisected by a local road, two separate floodwalls were designed, one for portion to the east of the road (East Basin) and one for the portion to the west of the road (West Basin). As a result, collection system alternatives had to be designed for each basin, each of which met the aforementioned Design Criteria. An analysis of each alternative was presented to PVSC, after which the preferred alternatives were selected.

The overall design process included the following steps:

1. Development of Alternatives
2. Verification of Alternatives (Evaluate Freeboard Conditions)
3. Selection of Preferred Alternatives

Development of Alternatives

For each basin, several collection system design alternatives were prepared.

Using the aforementioned SSA program, each alternative was modeled, and run for the scenario described in the Design Criteria. Based on the storm event that is run, the SSA program will indicate which pipes will be surcharged, and which storm manholes and catch basins will flood. The figure below shows an example of what the SSA program displays once a storm event is run. Surcharged pipes are highlighted red, and flooding storm manholes or catch basins are shown with blue dots.

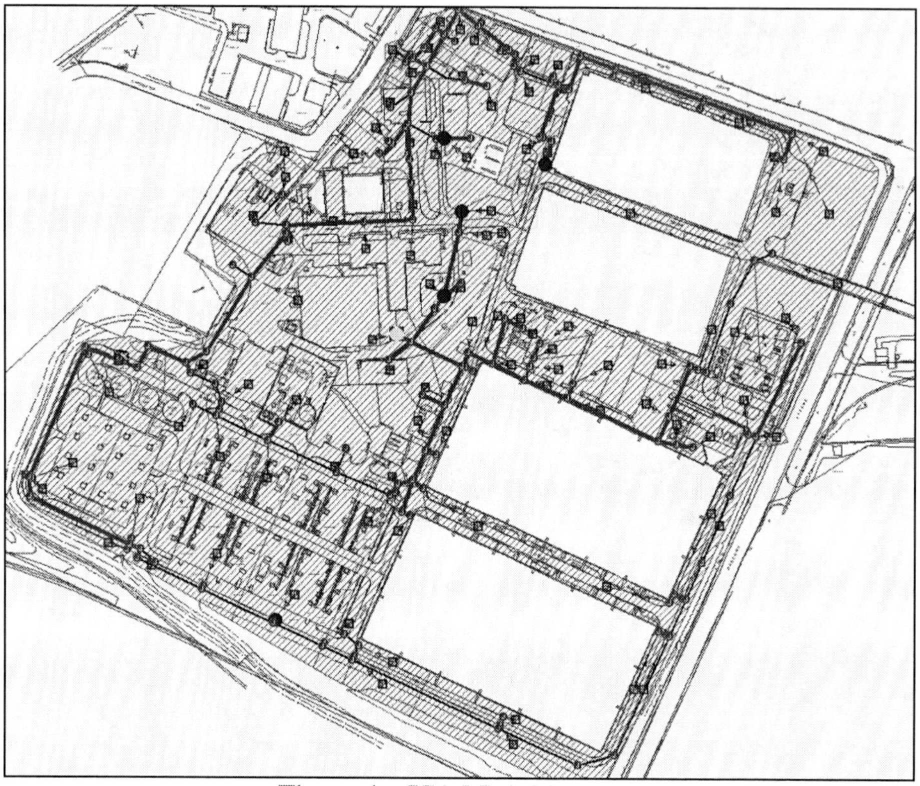

Figure 6 – SSA Model Display

Once the SSA models for each alternative were created, each model had to be verified to confirm that the freeboard criterion was being met.

Verification of Alternatives (Evaluate Freeboard Criterion)

Using the survey provided in AutoCAD Civil 3D, the linework showing the location of the proposed floodwall for each basin was added in. Using the topographic data, the locations of all "low spots" were identified within each basin. These are the locations where stormwater would travel when catch basins and storm manholes begin to flood during a large rainfall event.

For each low spot identified, the FFE information for each building was used to determine which buildings would be flooded first. Six inches from the building FFE was then subtracted to determine the maximum allowable flooding elevation for each low point. This will allow for verification that the collection system capacity is such that the freeboard criterion is not being violated. Closed polylines were drawn at each low point's maximum allowable flooding elevation, and stage-storage relationships were developed using the topographic data. This provided the maximum allowable flood storage volume for each low point that could occur without violating the freeboard portion of the Design Criteria.

For some of the low spots, there were no buildings nearby that were in danger of being impacted by flooding of the collection system. In these areas, flooding was connecting with adjacent low spots long before reaching the freeboard elevation of nearby buildings. The point of interconnection of the low spots was set as the maximum allowable flooding elevation. During modeling of the alternatives, these areas were shown to be among the last to reach the maximum allowable flooding elevation.

The following figures depict the closed polylines that were drawn at each low point's maximum allowable flooding elevation. This shows the limits of where flooding could occur while ensuring six inches of freeboard at all buildings. These figures do not represent actual flooding conditions.

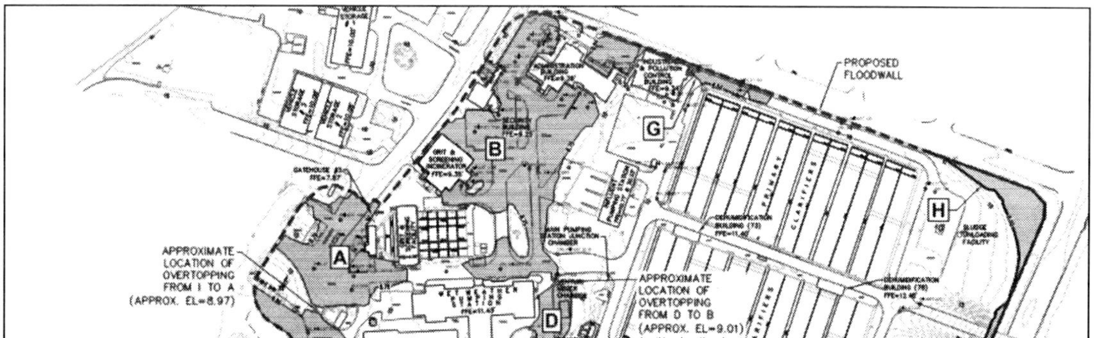

Figure 7 – Maximum Allowable Flooded Area Figure – West Basin

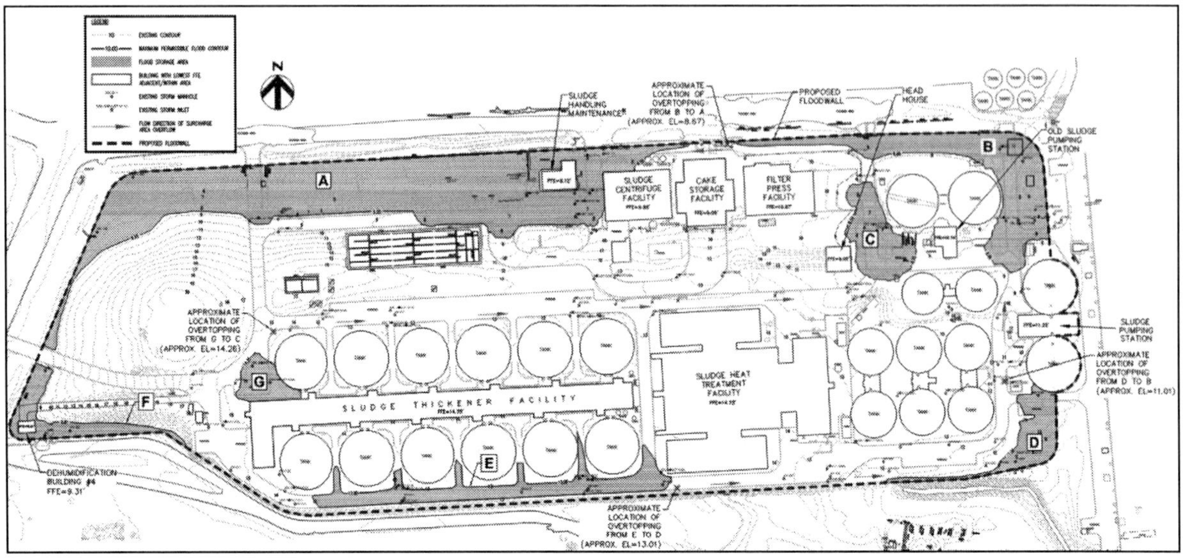

Figure 8 – Maximum Allowable Flooded Area Figure – East Basin

Once the stage-storage relationships had been developed, the relationships were combined with surcharge flooding predicted by the SSA models to evaluate whether each model met the freeboard criterion.

To determine how much flooding is occurring at each storm manhole or catch basin in a given SSA model, the user can click on the structure, and the total flooded volume is highlighted in units of acre-inches.

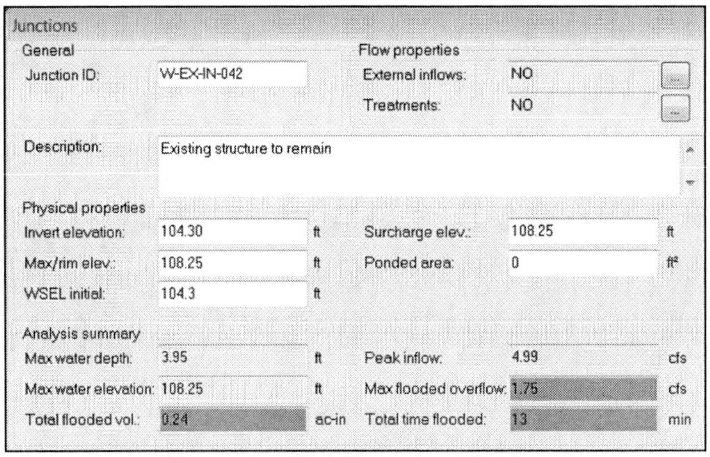

Figure 9 – SSA Junction Display

Using the topographic data, each storm manhole and catch basin was evaluated to determine which low spot any surcharged water would flow to. When each model was run, it was verified that the cumulative flooded volume reaching each of the low spots did not exceed the maximum allowable flooding calculated when the stage-storage relationships were developed. If the maximum was exceeded, further design was performed to upsize additional storm pipes and/or change pipe inverts until each model was verified as meeting the freeboard criterion.

Selection of Preferred Alternatives

The alternatives for each basin were presented to PVSC during a design workshop. During the workshop, each alternative was described in detail, focusing on the distinguishing features of each alternative, and their pros/cons. Tabular comparisons of pipe/structures to be replaced, as well as the Class IV cost estimate totals for each alternative were provided. The JV provided recommendations for each basin. At the conclusion of the workshop, PVSC selected their preferred alternative based on the information provided.

West Basin Collection System

PVSC selected the alternative that proposed to keep the largest amount of existing pipes and drainage structures in place. As shown in Table #1, the preferred alternative (Alternative #3) proposed to keep 25% of existing pipes in place, with 39% requiring new pipes without requiring new pile supports. Under this alternative, 41% of the catch basins and manholes were proposed to be kept in place. This compared favorably to Alternatives #1 and #2 which proposed the removal and replacement of all 12,000-plus LF of existing pipe, in addition to approximately 150 catch basins and manholes.

In addition, the preferred alternative had the lowest estimated construction cost, approximately $1.2 million less than the next closest alternative.

Table 1 – Comparison of West Basin Alternatives

	West Alt. #1	West Alt. #2	West Alt. #3
Total LF of All Pipes	12,641	12,586	12,252
LF of All Pipes to Remain	0	0	3,065
LF of All New Pipes (Needs New Pile Installation)	3,830	3,850	3,544
LF of All Pipes to be Replaced & Maintain Inverts	1,125	1,233	4,814
LF of All Pipes to be Replaced & Modify Inverts	7,686	7,503	829
Total # of All Junctions (CB & MH)	151	150	146
# of Junctions Remaining In Place	0	0	61
# of New Junctions or Junctions Being Replaced	151	150	85

East Basin Collection System

Similar to the West Basin Collection System, PVSC selected the alternative that proposed to keep the largest amount of existing pipes and drainage structures in place. As shown in Table #2, the preferred alternative (Alternative #5) proposed to keep 74% of existing pipes in place, with 7% requiring new pipes without requiring new pile supports. Under this alternative, 80% of the catch basins and manholes were proposed to be kept in place.

Of the alternatives presented this had the lowest estimated construction cost, approximately $1.4 million less than the next closest alternative.

The table below compares the East Basin Alternatives that were considered.

Table 2 – Comparison of East Basin Alternatives

	East Alt. #1	East Alt. #2	East Alt. #3	East Alt. #4	East Alt. #5
Total LF of All Pipes	**7,475**	**7,606**	**7,769**	**7,899**	**7,632**
LF of All Pipes to Remain	4,404	3,960	4,425	3,960	5,668
LF of All New Pipes (Needs New Pile Installation)	2,338	2,263	2,611	2,556	1,228
LF of All Pipes to be Replaced & Maintain Inverts	483	1,316	483	1,316	507
LF of All Pipes to be Replaced & Modify Inverts	250	67	250	67	228
Total # of All Junctions (CB & MH)	**106**	**103**	**108**	**105**	**106**
# of Junctions Remaining In Place	62	59	64	61	85
# of New Junctions or Junctions Being Replaced	44	44	44	44	21

CONCLUSIONS

The conceptual design process implemented for the PVSC facility resulted in the identification of locations where flooding was most likely to occur, and more importantly, how much flooding could occur in each location without violating the freeboard criterion. As a result, the design of the collection system could be focused on proposing changes to small portions of existing collection system in these areas only. The result was that many of the design alternatives allowed for significant portions of the existing stormwater collection system to remain in place.

PVSC ultimately preferred the alternatives which proposed the least amount of existing pipe and structure replacement. The benefits of selecting these alternatives included:

1. Lower anticipated construction costs when compared to the other alternatives
2. Fewer open excavations during construction
3. Fewer road closures and detours required during construction
4. Lower potential for damage to other subsurface utilities crossing over/under the collection system piping during construction
5. Potential for fewer impacts and/or temporary shutdowns to critical operations at the facility during construction

ACKNOWLEDGEMENTS

The authors gratefully acknowledge the cooperation of our Joint Venture partners at AECOM, with whom HDR has had the privilege of working with to finalize the design concepts that will ultimately protect the PVSC facilities for years to come. We also acknowledge the cooperation of the staff at PVSC, in allowing us the opportunity to distribute this manuscript, and more importantly, the opportunity to design this unique stormwater collection system.

REFERENCES

Federal Emergency Management Agency. (2015). Three Years After Sandy: Recovery and Resiliency at Passaic Valley Sewerage Commission [Press Release, Number SRO NJ NR 062]. Retrieved from https://www.fema.gov/news-release/2015/10/09/three-years-after-sandy-recovery-and-resiliency-passaic-valley-sewerage

Google Maps. (2016). 657 Doremus Avenue, Newark, NJ 07105. Retrieved from https://www.google.com/maps/place/675+Doremus+Ave,+Newark,+NJ+07105/@40.7109976,-74.1337,687m/data=!3m2!1e3!4b1!4m5!3m4!1s0x89c253dbbdf1d833:0x8d74d35aaa5b72fa!8m2!3d40.7109976!4d-74.1315113

Joint Venture (2016), Stormwater Collection System Basis of Design Report

Soil Conservation Service (SCS), Urban Hydrology for Small Watersheds, Tech. Release 55, Washington, DC. 1986. Retrieved from http://www.wcc.nrcs.usda.gov/water /quality/common/tr55/tr55.pdf

State of New Jersey, Governor Chris Christie. (2014). *New Jersey Receives Largest Public Assistance/Mitigation Grant Ever Issued by FEMA For Post-Sandy Infrastructure Hardening* [Press Release]. Retrieved from http://nj.gov/governor/news/news/552014/approved/20140821b.html

Bangor's Version of an Integrated Wet Weather Plan

T. Kate Mignone[1*], John Theriault[2], Gregory Heath[1]

[1]AECOM, Chelmsford, Massachusetts
[2]City of Bangor Engineering Department, Bangor, Maine
*Kate.mignone@aecom.com

Bangor's Version of an Integrated Wet Weather Plan

ABSTRACT

The City of Bangor, Maine is currently conducting wet weather planning under a Federal Consent Decree (CD) that was entered in November 2015. Remedial measures required under the CD relate to the city's Capacity, Management, Operation, and Maintenance (CMOM) Program, its Municipal Separate Storm Sewer System (MS4), and to the development of Phase 2 Combined Sewer Overflow (CSO) Controls. The main wet weather planning activities focused on the sewer system evaluation survey (SSES) required as part of the CMOM program and the development of Phase 2 CSO controls to further reduce remaining overflows.

EPA's policy on Integrated Wastewater and Stormwater Planning provides a framework for communities that have multiple, costly Clean Water Act compliance requirements and want to develop implementation plans that achieve desired water quality benefits in an affordable manner. During CD negotiations Bangor worked with United States Environmental Protection Agency (EPA), United States Department of Justice (DOJ), and Maine Department of Environmental Protection (DEP) to explore the use of integrated planning. Bangor prepared an outline for an Integrated Plan so that all Clean Water Act requirements, including MS4, SSES, CSO planning, and wastewater treatment plant (WWTP) projects, could be prioritized and scheduled in one document. EPA, however, informed Bangor that doing so would result in ***all*** projects in the Integrated Plan, even WWTP projects not envisioned to be within the scope of the CD, being subject to CD dates with stipulated penalties if any dates were missed. Bangor did not desire to expose its rate payers to more enforcement under the CD than necessary and the city was forced to abandon its plan for a formal Integrated Plan.

Having decided not to conduct formal Integrated Planning, the City was forced to develop necessary capital projects separately under its SSES and CSO LTCP planning activities. However, Bangor was able to negotiate with EPA to allow all SSES and CSO LTCP projects to be scheduled under the umbrella of the updated LTCP. This enabled Bangor to effectively integrate its SSES and CSO LTCP capital projects into one schedule to facilitate prioritizing CWA requirements to achieve water quality benefits in an affordable manner.

To date, Bangor has conducted substantial planning activities in support of its SSES and CSO LTCP requirements. CSO LTCP and SSES capital projects were jointly presented in an integrated implementation schedule in the draft LTCP, which was submitted in January 2017.

KEYWORDS: Integrated Plan, LTCP, CSO, SSES, financial capability analysis

INTRODUCTION

The City of Bangor is located on the Penobscot River in Penobscot County, Maine. See Figure 1. With a 2010 population of 33,039, Bangor is Maine's third largest City.

As with other older cities and towns throughout Maine and New England, Bangor's wastewater collection system contains combined sewers, and Bangor's CSOs are designed to regulate the amount of flow that can be intercepted and conveyed to Bangor's Wastewater Treatment Plant (WWTP) for treatment prior to discharge to the Penobscot River.

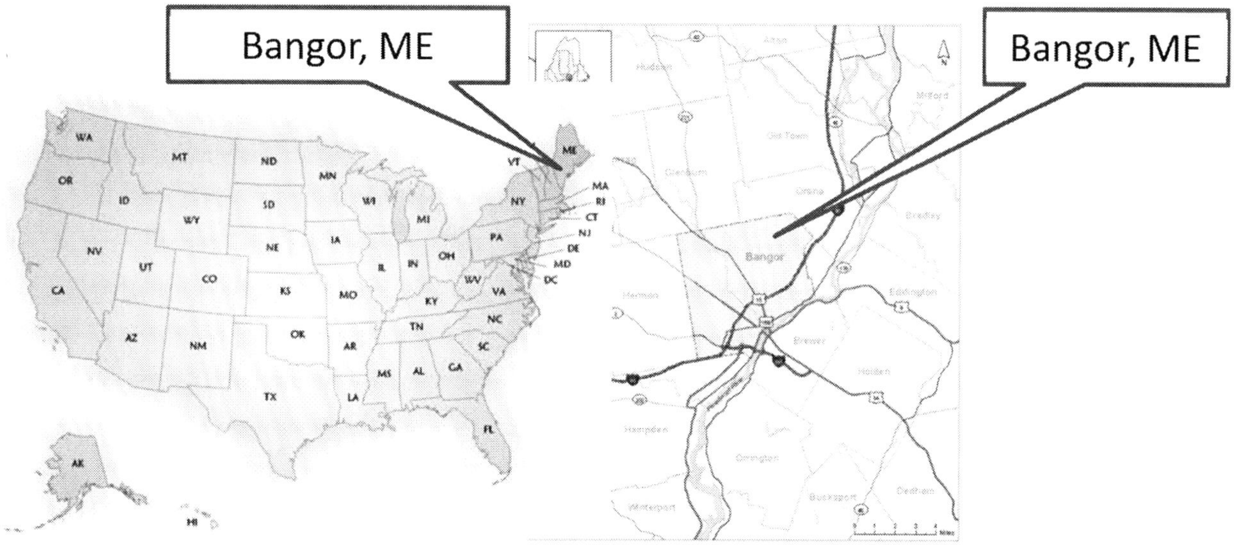

Figure 1. Location Plan

The City of Bangor, Maine is currently conducting wet weather planning under a Federal Consent Decree (CD) that was entered in November 2015. Remedial measures required under the CD relate to the city's Capacity, Management, Operation, and Maintenance (CMOM) Program, its Municipal Separate Storm Sewer System (MS4), and development of Phase 2 Combined Sewer Overflow (CSO) Controls. MS4 activities are limited to implementation of the city's revised 2013-2018 Illegal Discharge Detection and Elimination (IDDE) Plan, which was approved by EPA prior to completing negotiation of the CD. CMOM activities include development of geographical information system (GIS) deliverables, a Sewer System Asset Management Preventative Maintenance and Sewer Maintenance Improvement Plan, and a sewer system evaluation survey (SSES). Since GIS deliverable activities focus on enhancing existing GIS mapping and the Sewer System Asset Management Preventative Maintenance and Sewer Maintenance Improvement Plan focuses on closed circuit television (CCTV) inspection and cleaning requirements in the sewer system, the main wet weather planning activity under the CMOM Program is focused on the SSES. Bangor had already completed the development and implementation of Phase 1 CSO controls, so the CD focuses on additional Phase 2 controls to further reduce remaining overflows.

Collection System

Bangor's wastewater infrastructure includes a network of collector and interceptor sewers, pump stations, force mains, and siphons, all leading to a secondary WWTP. The collection system includes both a combined sewer system (CSS) and a separate sewer system (SSS).

Following are the main features of the Bangor collection system and shown on Figure 2:

- Five City owned Pump Stations
- Nine permitted CSOs
- 157 miles of collector sewers
- 9 miles of interceptor sewers
- Four interceptor sewers: Penobscot, Penobscot East, Kenduskeag, and Kenduskeag East

Approximately 24% of the collector sewers are combined, down from 58% at the start of the CSO abatement program in the early 1990s.

WWTP

Bangor's original WWTP began operation in 1966 as a primary plant. It was upgraded to provide secondary treatment in 1992. In response to recommendations contained in the 1993 CSO Facilities Plan, further improvements were completed in 1998 to accommodate higher rates of wet weather flow including the addition of a CSO Bypass around the secondary treatment facilities, the operation of which is described below. The application for the CSO Bypass was approved by EPA in 1996.

The secondary WWTP employs the Trickling Filter/Solids Contact (TF/SC) process of using both an attached-growth tricking filter and suspended-growth activated-sludge process in series. The WWTP includes raw sewage pumping, preliminary treatment (screening and grit removal), primary treatment, secondary treatment, and seasonal disinfection (chlorination/dechlorination). The secondary portion of the WWTP was designed for an average daily design flow of 68 mega liters per day (MLD) (18 million gallons per day (MGD)) and a peak wet weather flow of 114 MLD (30 MGD). During wet weather up to an additional 49 MLD (13 MGD) can receive treatment (sedimentation and disinfection) at the WWTP.

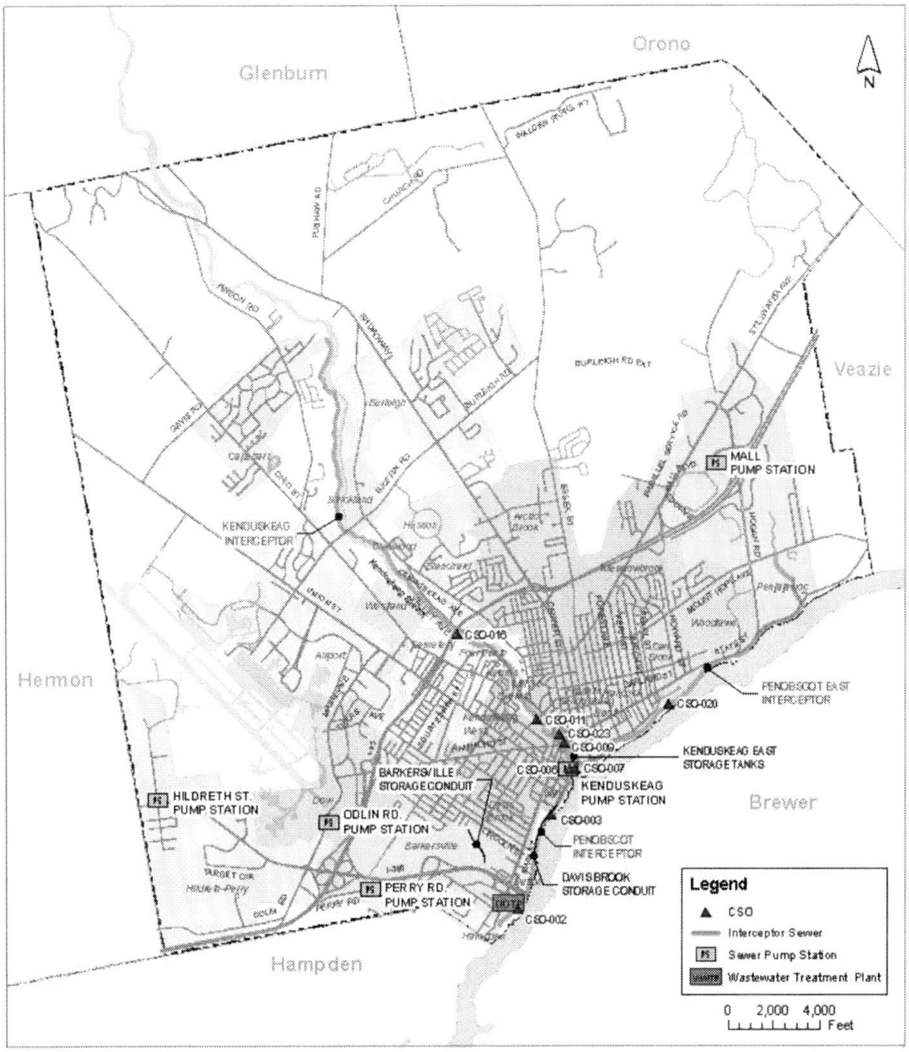

Figure 2. City of Bangor Collection System Features

Inter-Municipal Services

In addition to providing wastewater collection and treatment to Bangor, the City also provides treatment services to portions of the Town of Hampden to the South and to the Town of Hermon to the West. Hampden conveys flow to Bangor via three town-owned pump stations, and Hermon via two town-owned pump stations.

MS4 System

The City of Bangor small MS4 consists of approximately 3,400 catch basins, 155 kilometers (km) ((96 miles) of piping, and 243 km (151 miles) of ditches that drain to six small urban impaired streams, a major tributary stream, and a river. The City has an active stormwater program supported by a stormwater utility and participates in a regional stormwater working group. In the last permit year (July 2015 – June 2016), the City tracked over 75 best management practices (BMPs), conducted over 140 construction site inspections, and removed approximately 19,722 liters per day (5,210 gallons per day) of illicit discharges to its system.

PREVIOUS CSO CONTROL ACTIVITIES

The current CSO planning effort is the second comprehensive evaluation of Bangor's CSO system. Previous CSO control activities are described below.

1991 Consent Decree

The first major action pertaining to CSO abatement was a 1991 Consent Decree (CD) that Bangor entered into with the Maine Department of Environmental Protection (DEP) entered into with EPA. The 1991 CD superseded a 1987 consent agreement between Bangor and DEP that primarily addressed the requirement for secondary treatment at Bangor's then primary WWTP but also included a requirement to conduct a Sewer System Evaluation Survey (SSES) of the collection system.

A key requirement of the 1991 CD was the preparation of a CSO Facilities Plan which ultimately established a timeframe for the plan's recommended CSO abatement projects. A draft of the plan was submitted to EPA and DEP in 1992. A final report and accompanying implementation schedule were subsequently submitted to EPA in 1993. This document became known as the 1993 CSO Facilities Plan and was approved in 1994.

1993 CSO Facilities Plan

The 1993 CSO Facilities Plan was the first comprehensive evaluation of Bangor's CSO system. In addition to recommending a long-term program of CSO abatement measures, the plan:

- Developed the initial hydrologic/hydraulic model of the collection system using the EPA Storm Water Management Model (SWMM)
- Established baseline conditions
- Documented ambient water quality and CSO characteristics, and the impact of CSO discharges on area receiving waters
- Evaluated a wide range of alternative control measures based on their ability to reduce CSO discharges and provide water quality benefits
- Established 80% as the target level of CSO control
- Recommended a series of abatement projects that included storage, treatment and sewer separation
- Established a phased implementation schedule with key milestone dates

While specific projects and milestones were modified through processes consistent with both the 1991 CD and MEPDES Permits, the recommendations from the 1993 CSO Facilities Plan has formed the basis of Bangor's CSO abatement program up to the present. The modifications to the plan that did occur over time were the result of Bangor's adoption of an adaptive approach for implementation of CSO control projects. This allowed Bangor to use lessons learned from earlier projects, and more detailed field information, to modify CSO control technology and design configurations. A Final Report on CSO Abatement Program was submitted to EPA in April 2009, documenting all of the measures taken by Bangor to satisfy the requirements of the 1991 CD.

2010 Section 308 Requests for Information

Following Bangor's submittal of the Final Report on CSO Abatement Program, EPA issued a series of Requests for Information (RFI) under Section 308 of the Clean Water Act (CWA) beginning in May 2010. In addition to requests for information on the operation of the collection system, the key CSO-related requirement of the RFI was the preparation of a Phase 2 LTCP by September 2012.

2012 Draft Phase 2 LTCP

A draft Phase 2 LTCP, consisting of a series of technical memoranda (TM), was prepared to document evaluations of alternatives to meet higher levels of CSO control than the original 80% level of control contained in the 1993 CSO Facilities Plan. This document included:

- An update of the collection system model
- Baseline characterization
- Level of control evaluations
- Development of a decision-making framework and costing methodology
- Identification and evaluation of feasible CSO controls
- Affordability evaluations and a financial capability analysis
- Public participation summary

A major focus of the 2012 draft Phase 2 LTCP was the evaluation of additional control measures required to increase the level of CSO control to EPA-prescribed annual overflow frequencies of four, two, one and zero in a typical rainfall year.

2015 Consent Decree

Following the Section 308 RFI process and development of the draft Phase 2 LTCP, Bangor, EPA, and DEP entered into the 2015 CD. In addition to requiring preparation of the 2017 Phase II LTCP, the 2015 CD requires Bangor to (among other things) implement a Capacity, Management, Operations and Maintenance (CMOM) Corrective Action Plan and a Sewer System Asset Management Preventive Maintenance and Sewer Maintenance Improvement Plan (MIP), including a phased Sewer System Evaluation Survey (SSES).

During CD negotiations Bangor worked with EPA and DEP to explore the use of integrated planning in accordance with the EPA June 2012 framework to address its MS4, SSES, and CSO planning activities, and the integrated scheduling of capital projects expected to result from those planning activities. Bangor prepared an outline for an Integrated Plan so that all its Clean Water Act requirements could be accounted for and prioritized in one document. In addition to the MS4, SSES, and CSO planning activities included in ongoing CD negotiations, Bangor also sought to include WWTP capital projects in the Integrated Plan. Bangor's intent in proposing to do so was to ensure that WWTP needs would be properly sequenced and scheduled with the MS4, SSES, and CSO LTCP projects in an affordable and environmentally beneficial manner. EPA, however, informed Bangor that doing so would result in ***all*** projects in the Integrated Plan, even WWTP projects not envisioned to be within the scope of the CD, being subject to CD dates with stipulated penalties if any dates were missed. Bangor did not desire to expose its rate payers to more enforcement under the CD than necessary and the city was forced to abandon its plan for a comprehensive Integrated Plan.

2017 PHASE II LTCP

The purposes of the 2017 Phase II LTCP are summarized as follows:

- Assess progress made to date on the continuing abatement of Bangor's CSOs
- Ascertain current baseline conditions in comparison to those presented in the 1993 Facilities Plan
- Evaluate higher levels of CSO abatement consistent with the requirements of the 2015 CD
- Coordinate, and integrate to the extent practical, the findings of the CSO abatement evaluations and findings from the SSES investigations

As noted above, both the 2017 Phase II LTCP and SSES investigations are required under the 2015 CD. As part of CD negotiations EPA and DEP agreed that Bangor could integrate projects to be implemented into a single schedule to be contained in the 2017 Phase II LTCP. This was considered an important outcome of CD negotiations since, as described above, Bangor could not proceed with a fully integrated planning effort without expanding the reach of the CD to include additional compliance dates and stipulated penalties. At the time of CD negotiations, Bangor envisioned that being able to integrate at least CSO and SSES capital projects into a single schedule would help all parties to see the magnitude of work to be faced by the City, and to appropriately account for budget and schedule considerations in a practical and affordable manner.

System Characteristics

The City has been reporting annual CSO activity using data collected by flow meters installed at each CSO outfall since 2012. Table 1 contains a summary of CSO activation and volume data obtained from 2012 to 2015. Based on the meter data, four of the CSOs make up the vast majority of the City's typical annual CSO volume. These four major CSOs are Barkersville,

Table 1. Annual CSO Frequency and Volume for Years 2012 to 2015 Based on System-Wide Flow Monitoring

	2012		2013		2014		2015	
Annual Precipitation, centimeters (inches) of rain	**107.4 (42.3)**		**99.8 (39.3)**		**111.5 (43.9)**		**90.2 (35.5)**	
Location	**OF/yr.(1)**	**Vol.(2) (ML) (MG)**	**OF/yr.(1)**	**Vol.(2) (ML) (MG)**	**OF/yr.(1)**	**Vol.(2) (ML) (MG)**	**OF/yr.(1)**	**Vol.(2) (ML) (MG)**
Barkersville	17	49.6 (13.1)	11	34.8 (9.2)	22	147.6 (39.0)	10	14.8 (3.9)
Davis Brook	28	88.6 (23.4)	23	64.4 (17.0)	32	100.7 (26.6)	18	82.5 (21.8)
Kenduskeag West	10	75.0 (19.8)	6	12.1 (3.2)	7	45.4 (12.0)	3	35.2 (9.3)
Kenduskeag East	8	30.3 (8.0)	4	5.7 (1.5)	5	29.5 (7.8)	3	15.1 (4.0)
Hammond St.	12	6.1 (1.6)	3	1.1 (0.3)	3	3.4 (0.9)	1	N/A(3)
Meadowbrook	21	12.5 (3.3)	12	2.6 (0.7)	15	5.3 (1.4)	5	2.6 (0.7)
Cemetery	0	0.0	0	0.0	0	0.0	1	N/A(3)
Carr Brook	3	0.8 (0.2)	4	0.0	7	0.0	3	1.1 (0.3)
Central St.	7	0.4 (0.1)	8	1.1 (0.3)	7	0.4 (0.1)	5	0.4 (0.1)
Systemwide	--	**263.5 (69.6)**	--	**121.5 (32.1)**	--	**332.4 (87.8)**	--	**151.8 (40.1)**

1. OF/yr. = number of overflows per year
2. Vol. (ML) (MG) = annual volume in mega liters (ML) and million gallons (MG). Values in MG in parentheses.
3. Data not available

Davis Brook, Kenduskeag West, and Kenduskeag East. Together these four CSOs resulted in 95.7% of the City's CSO volume from 2012 to 2015.Inspection of Table 1 shows that, in general, there are more frequent overflows and higher volumes in the wetter years (2012 and 2014) than in the dryer years (2013 and 2015). This is most apparent at the four overflows that produce the larger overflow volumes: Barkersville, Davis Brook, Kenduskeag West, and Kenduskeag East. While the annual overflow frequency was relatively high at the Meadowbrook CSO, annual overflow volumes at this location were relatively low.

CSO Model Update and Baseline Conditions

As part of the 2017 Phase II LTCP, the SWMM model was refined and recalibrated based on the additional flow meter data collected since the 2012 Phase 2 LTCP Progress Report. Table 2 contains a comparison of the system-wide CSO volume distribution based on the meter data from Table 2 and two annual SWMM runs from the updated model: 2014 and the typical year (1974). The 2014 annual run represents a recent, relatively wet year. The 1974 typical year rainfall was reviewed against the past 40 years of rainfall data and was determined to be a reasonable representation of typical (long-term average) year rainfall characteristics.

Table 2. Comparison of System-Wide Distribution of Annual CSO Volume Between Meter Data and Model Predictions

CSO Outfall	Percent of Annual CSO Volume		
	Flow Meter Data (Average for 2012 to 2015)	2014 SWMM Model Run	Typical Year SWMM Model Run (1974)
Barkersville	28.4%	46.2%	36.3%
Davis Brook	38.7%	42.8%	45.0%
Kenduskeag West	19.3%	8.9%	15.8%
Kenduskeag East	9.3%	1.4%	1.2%
Hammond St.	1.6%	0.0%	0.0%
Meadowbrook	2.6%	0.5%	0.7%
Cemetery	0.0%	0.0%	0.0%
Carr Brook	0.2%	0.1%	0.7%
Central St.	0.3%	0.1%	0.2%

Inspection of Table 2 shows that the relative system-wide CSO volume distribution between meter data and model predictions is reasonable. Of the four largest outfalls, Davis Brook and Barkersville are the two largest contributors to the City's annual CSO discharge, followed by Kenduskeag West and Kenduskeag East which rank third and fourth, respectively. Among these four largest outfalls, inspection of Table 2 shows that meter results indicate that a greater percentage of volume discharges from Kenduskeag East as compared to model predictions. In the wetter year (2014) volume from both Kenduskeag West and Kenduskeag East is predicted to shift to the two largest volume outfalls: Barkersville and Davis Brook. In the typical year (1974) volume from the Kenduskeag East outfall is predicted to shift to the Barkersville and Davis Brook outfalls. While there are differences among the system-wide CSO volume distribution between meter data and model predictions, the general trends are correct: the four largest outfalls contribute the greatest annual volumes, especially in the case of the two largest volume outfalls at Barkersville and Davis Brook.

Once the SWMM model was refined and the calibration check completed, the model was used to establish baseline values for system-wide CSO activations and volumes based on the typical year. Baseline values for activation frequency, overflow volume, and peak flow are presented for each CSO outfall in Table 3.

Table 3. Baseline Activation Frequency, Overflow Volume, and Peak Flow by Outfall for the Typical Year (1974)

CSO Outfall	Activations(1)	CSO Volume (ML) (MG)(2)	Peak CSO Flow Rate (MLD) (MGD)(3)
Barkersville	19 – 20	137.8 – 175.3 (36.4 – 46.3)	418.3 (110.5)
Davis Brook	19	152.9 – 216.5 (40.4 – 57.2)	616.6 (162.9)
Kenduskeag West	8 – 9	75.3 – 76.5 (19.9 – 20.2)	339.9 (89.8)
Kenduskeag East	14	4.5 – 5.7 (1.2 – 1.5)	83.3 (22.0)
Hammond St.	0	0.0	0.0
Meadowbrook	12	3.4 (0.9)	22.0 (5.8)
Cemetery	0	0.0	0.0
Carr Brook	5	3.4 (0.9)	489.5 (129.3)
Central St.	7	1.1 (0.3)	7.2 (1.9)

1. Annual activations. Low end of range corresponds to a WWTP capacity of 162.7 MLD (43 MGD) and high end to 136.3 MLD (36 MGD).
2. Annual volume in mega liters (ML) and million gallons (MG) in parentheses. Low end of range corresponds to a WWTP capacity of 162.6 MLD (43 MGD) and high end to 136.3 MLD (36 MGD).
3. Highest peak flow predicted by the SWMM model in the typical year in MLD and MGD in parentheses.

As shown in Table 3, ranges of values are included in the table to reflect the current peak WWTP capacity of the WWTP versus the rated capacity. The rated capacity of the WWTP is 162.7 MLD (43 MGD) based on the 1993 CSO Facilities Plan. However, the current raw sewage pumps can reliably deliver 136.3 MLD (36 MGD) to the primary clarifiers, and up to 147.6-151.4 MLD (39-40 MGD) under certain conditions (e.g., optimal wet well level, condition of wear rings, etc.). Thus, baseline values for overflow frequency and volume were calculated based upon this range of peak wet weather WWTP capacities: a low of 136.3 MLD (36 MGD) to a high of 162.7 MLD (43 MGD). The higher values in Table 3 represent the lower WWTP capacity while the lower values in the table represent the higher WWTP capacity.

Alternatives for CSO Control

CSO controls are typically grouped into categories that range from simpler, less costly source control measures to more complex and costly treatment and storage facilities. A list of CSO controls identified for consideration in Bangor is presented in Table 4. This table lists CSO control measures grouped by the following categories: Source Controls, System Optimization, Conveyance Enhancements, Treatment, and Storage.

Table 4. CSO Control Measures Considered for Bangor

Source Controls
• Best Management Practices/Nine Minimum Controls
• I/I Reduction/Sewer Rehabilitation
• Sewer Separation
• Green Infrastructure
System Optimization
• Weir Adjustment
• Bending Weirs/Control Gates
• Real Time Controls
Conveyance Enhancements
• Parallel Relief Interceptors
• Pump Station Expansion
Treatment (Satellite and Centralized at WWTP)
• Vortex Separation
• Retention Treatment Basin (RTB)
• High Rate Clarification (HRC)
• WWTP Wet Weather Capacity Expansion
• Disinfection in conjunction with treatment technologies
Storage
• In-System/Conduit
• Off-Line Tank
• Tunnel

Alternatives Evaluation Process

The alternatives evaluation process began with a first-level or fatal flaw analysis to screen from consideration those measures not suited to Bangor. The remaining alternatives were then advanced for a more detailed evaluation and screening process.

One important consideration in the process of evaluating CSO control alternatives for Bangor was to acknowledge the types of controls that had already been implemented. Selecting CSO controls that would be compatible with existing controls, and with which the City was already familiar, would ease the burden of taking on additional CSO control facilities. The following are CSO controls that Bangor has implemented from the 1993 CSO Facilities Plan:

- All of the source control measures in Table 4 except green infrastructure
- Weir adjustments
- WWTP expansion
- In-system conduit storage
- Off-line tank storage

In general, and in particular for the largest CSO outfalls, Bangor felt that storage would be preferred over satellite treatment as it is a relatively simple process to operate. The stored volume is drained back to the WWTP following the storm where it receives full secondary treatment and seasonal disinfection. Bangor currently owns and operates three satellite CSO storage facilities and has a level of comfort with the technology. The satellite treatment alternatives under consideration would provide a lesser level of treatment than secondary treatment at the WWTP. Satellite treatment locations that would include disinfection would present both siting and operational challenges as each facility would require its own chemical storage and feed building along with the controls necessary to optimize the process and meet effluent limitations.

Following the first-level or fatal flaw analysis, the alternatives listed in Table 5 remained for further evaluation. Some alternatives were set aside due to space limitations, complexity of operation, or other considerations particular to Bangor and its current wastewater system operations. Another cause for alternatives to be set aside during this fatal flaw screening was that there were no further opportunities to implement the measure.

In addition to listing alternatives for further consideration, Table 5 identifies CSO controls currently in place following implementation of projects recommended in the 1993 CSO Facilities Plan.

Because of the relatively large volumes of overflow discharged from the four largest CSO outfalls - Barkersville, Davis Brook, Kenduskeag West and Kenduskeag East - storage was considered to be the primary control measure for those locations along with RTB and vortex treatment plus disinfection. Sewer separation, sewer replacement/rehabilitation and GI were only considered as supplemental controls for these outfalls.

For the smaller outfalls, sewer separation and sewer replacement/rehabilitation were the primary control measures considered, along with GI where the application of such could be cost-effectively included on a case by case basis.

In order to include CSO elimination in the evaluation process, costs for system-wide sewer separation were developed for both the four largest CSO outfalls and the smaller outfalls in order to view and compare a full range of CSO control levels.

Table 5. CSO Control Alternatives for Further Evaluation

CSO Outfall	Current Control	Storage/ Additional Storage[1]	Sewer Separation[2,3]	Treatment[4] (RTB or Vortex)	Sewer Replacement/ Rehabilitation	Green Infrastructure (GI)
Barkersville	5.3 ML (1.4 MG) Conduit	Yes	Yes	Yes	Supplement	Supplement
Davis Brook	4.5 ML (1.2 MG) Conduit	Yes	Yes	Yes	Supplement	Supplement
Kenduskeag West	--	Yes	Yes	Yes	Supplement	Supplement
Kenduskeag East	4.5 ML (1.2 MG) Tank	Yes	Yes	No	Supplement	Supplement
Hammond Street	--	No	Yes	No	Yes	Supplement
Meadowbrook	--	No	Yes	No	Yes	Supplement
Cemetery	--	No	Yes	No	Yes	Supplement
Carr Brook	--	No	Yes	No	Yes	Supplement
Central Street	--	No	Yes	No	Yes	Supplement

1. Storage tank or linear storage conduit.
2. Public inflow only (e.g., catch basins and drains); does not include private sources.
3. As (1) a stand-alone control (2) supplement to other control or (3) for CSO elimination/100% control.
4. Including seasonal disinfection.

The next step in the alternatives development and evaluation process was sizing of the control facilities for the range of possible levels of control. This was performed for the control measures remaining after the initial screening process at the four largest outfalls. The range of level of control was a high of 8 annual activations to a low of 2 annual activations with 6 and 4 activations included as mid-points.

Sizing of the alternatives was based on volume to size storage tanks or conduits and on peak flow rate to size flow-through treatment facilities (RTB or vortex, plus disinfection). Baseline SWMM model results were used to determine the volume and flow rate values used for sizing alternatives for the range of control levels. Overflow volumes and peak flows used to size alternatives for the four largest CSO outfalls are presented in Tables 6 and 7, respectively.

As shown in these tables, a range of values is included for some entries which relates to the range of peak wet weather capacity at the WWTP as discussed above. The lower values represent the WWTP with a peak wet weather capacity of 162.7 MLD (43 MGD) whereas the higher values represent the WWTP at 136.3 MLD (36 MGD). Because Kenduskeag East and Kenduskeag West are hydraulically upstream of the Kenduskeag Pump Station, peak flows and

Table 6. Overflow Volumes [1] Used to Size Alternatives for the Four Largest CSO Outfalls

CSO Outfall	Level of Control, Overflows per Year			
	8	6	4	2
Barkersville	4.2 – 5.7 (1.1 - 1.5)	5.3 – 12.1 (1.4 - 3.2)	11.4 – 13.6 (3.0 - 3.6)	13.6 – 17.8 (3.6 - 4 .7)
Davis Brook	4.9 – 6.8 (1.3 - 1.8)	5.7 – 10.2 (1.5 - 2.7)	9.5 – 14.4 (2.5 - 3.8)	20.1 – 24.2 (5.3 - 6.4)
Kenduskeag West	0.8 (0.2)	1.1 (0.3)	3.0 – 3.8 (0.8 - 1.0)	11.7 – 14.8 (3.1 - 3.9)
Kenduskeag East	0.4 (0.1)	0.4 (0.1)	0.4 (0.1)	0.8 (0.2)

1. Volumes in mega liters (ML) with million gallons (MG) in parentheses

Table 7. Peak Flows [1] Used to Size Alternatives for the Four Largest CSO Outfalls

CSO Outfall	Level of Control, Overflows per Year			
	8	6	4	2
Barkersville	134.8 – 149.1 (35.6 - 39.4)	145.4 – 162.0 (38.4 - 42.8)	179.1 – 179.8 (47.3 - 47.5)	189.6 – 199.(50.1 - 52.7)
Davis Brook	52.2 – 71.5 (13.8 -18.9)	69.7 – 76.8 (18.4 - 20.3)	95.0 – 99.6 (25.1 - 26.3)	124.5 – 127.9 (32.9 - 33.8)
Kenduskeag West	30.7 (8.1)	61.3 – 92.7 (16.2 - 24.5)	140.4 – 142.7 (37.1 - 37.7)	165.4 – 166.9 (43.7 - 44.1)
Kenduskeag East	7.9 – 14.0 (2.1 - 3.7)	17.0 – 18.2 (4.5 - 4.8)	18.2 – 18.5 (4.8 - 4.9)	20.8 – 24.6 (5.5 - 6.5)

1. Peak flows in mega liters per day (MLD) with million gallons per day (MGD) in parentheses

volumes are less impacted by WWTP capacity and a very small variation in peak flow or volume, or single value, is shown.

Using the values from Tables 6 and 7, control measures for the four largest CSO outfalls were sized for the range of control levels. The following sizing criteria were generally used:

- **Storage Conduits:**
  - Largest conduit cross-section was 3.66 meters (m) by 3.66 m (12 ft. by 12 ft.); multiple (parallel) conduits were used when that size was exceeded.
  - Conduits were not considered viable for Kenduskeag West or East (CSOs 006 and 007) due to availability of land and location of outfalls.

- **Storage Tanks:**
  - Storage tanks were assumed to be 5.5 m (18 ft.) deep and to have a length-to-width ratio of 2:1.
  - Storage tanks were the only control measure considered for Kenduskeag East (CSO 007).

- Because of their close proximity and discharge characteristics, a single storage tank was also considered for combining Kenduskeag West and East (CSOs 006 and 007).

- **RTB:**
  - RTBs were sized based on a Surface Overflow Rate (SOR) of 1,408 liters per day per square meter (LPD/m^2) (4,000 gallons per day per square foot (gpd/sf)) at peak flow, were assumed to be 5.5 m (m) (18 ft.) deep, and to have a length-to-width ratio of 2:1.

- **Vortex:**
  - Units were sized based on a SOR of 5,283 LPD/m^2 (15,000 gpd/sf), and would be 10.7 m (35 ft.) deep or less; units with calculated diameters larger than 10.7 m (35 ft.) would result in multiple units with an equivalent surface area plus the required flow distribution structure.
  - Vortex was deemed to be unsuitable for Kenduskeag West or East (CSOs 006 and 007) due to the potential for downstream solids deposition in the Davis Brook Conduit from the foul sewer which is required to be in operation when the vortex unit is receiving flow.

- **Disinfection Building (for RTB or Vortex)**
  - Sized at 18.3 m by 18.3 m (60 ft. by 60 ft.) to house chemical storage and feed equipment and building support functions.

Initial Cost Estimating

Following sizing, cost estimates were developed for each alternative control facility. The bases for the estimates were costs obtained from recent similar CSO abatement projects in Maine and, in the case of vortex, vendor-supplied estimates. The preliminary costs at this stage in the evaluation did not take into account all of the site-specific cost elements, such as land purchases or contamination mitigation, but were used to develop a generalized trend to establish the knee-of-the-curve (KOTC) or the cost-effective level of control.

Initial cost estimates are shown in Table 8. The costs span the range of control levels (8, 6, 4 and 2 overflows per year) and WWTP capacities 136.3 MLD to 162.7 MLD (36 MGD to 43 MGD). The higher costs correspond to the WWTP at 136.3 MLD (36 MGD) and the lower costs to the WWTP at 162.7 MLD (43 MGD). Also included in Table 8 is an allowance to modify the WWTP so that it could reliably achieve a peak flow rate of 162.7 MLD (43 MGD). This allowance would not be required with the 136.3 MLD-based (36 MGD-based) facilities since the satellite facilities would be sized based on the lower WWTP capacity.

These costs were used to establish the appropriate level of CSO control.

Table 8. Initial Cost Estimates for CSO Control Alternatives at Four Largest CSO Outfalls (1)

	Level of Control, Overflows per Year			
	8	6	4	2
Barkersville				
Storage Conduit	\$12 - \$13	\$13 - \$20	\$19 - \$22	\$21 - \$25
Storage Tank	\$9	\$9 - \$18	\$15 - \$22	\$20 - \$26
RTB and Disinfection	\$15 - \$17	\$16 - \$17	\$17	\$16 - \$17
Vortex and Disinfection	\$15 - \$17	\$16	\$19	\$21
Davis Brook				
Storage Conduit	\$11 - \$15	\$14 - \$18	\$17 - \$22	\$27 - \$30
Storage Tank	\$9 - \$11	\$10 - \$16	\$15 - \$27	\$29 - \$35
RTB and Disinfection	\$6 - \$8	\$8 - \$9	\$9	\$10 - \$11
Vortex and Disinfection	\$8 - \$10	\$10	\$11 - \$12	\$11 - \$15
Kenduskeag West and East				
Storage Tank	\$3	\$3 - \$4	\$9	\$1 9 - \$23
WWTP[2]	\$3	\$3	\$3	\$3

[1] Over range of 136.3 MLD (36 MGD) to 162.7 MLD (43 MGD) peak wet weather WWTP capacity
[2] Estimate of CSO treatment only to reliably attain 162.7 MLD (43 MGD)

Determination of Level of Control

The preliminary cost estimates presented in Table 8 were plotted to determine if a cost-effective level of control was evident. The average cost for each of the viable control measures at each location was used to develop the plot (Figure 3). Two lines were plotted, one for the WWTP at 136.3 MLD (36 MGD) (top) and the other with the WWTP at 162.7 MLD (43 MGD) (bottom). Based on inspection of Figure 3, four overflows per year was selected as the cost-effective level of control as indicated by the inflection point, or knee-of-the-curve, under either WWTP capacity. The incremental costs that would be incurred by moving to the next higher level of control, four overflows per year, would be both more costly and less cost-effective.

It should be noted that in addition to assessing the cost-effective level of control using average costs for each control measure and at each location, cost-effectiveness was also assessed on a site-specific basis using costs for each individual control measure. This site-specific assessment served as a check and supported the conclusion drawn from inspection of Figure 3 that four overflows per year was the cost-effective control level.

Based on this conclusion, subsequent evaluations and comparisons of CSO control alternatives were based on the four overflows per year level of control.

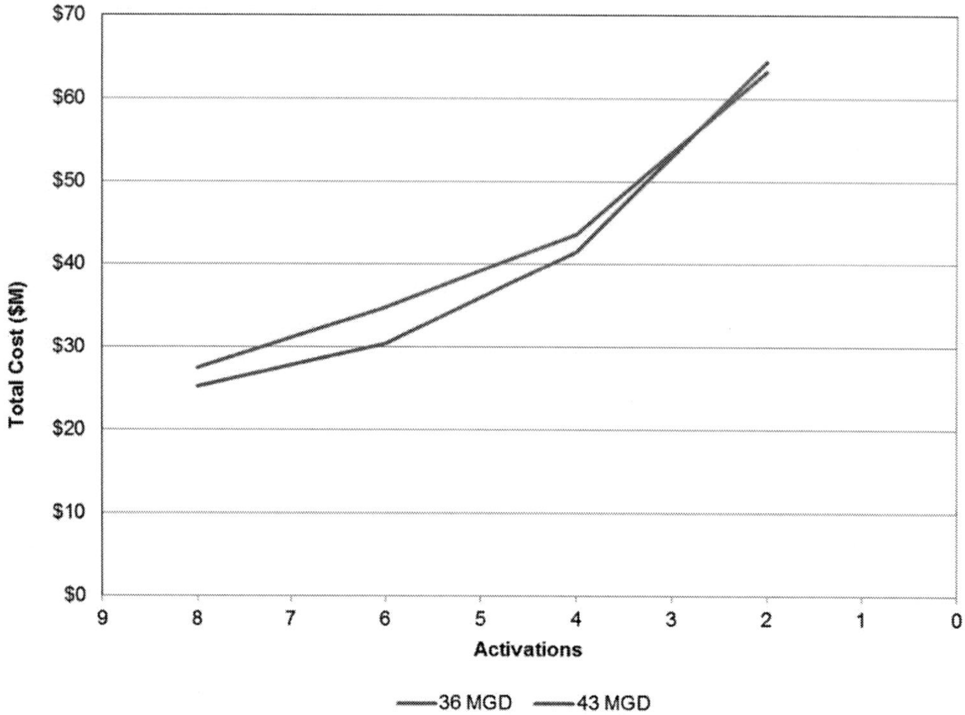

Figure 3. Cost Versus Overflow Frequency for CSO Alternatives for the Four Largest CSO Outfalls

Full Sewer Separation Alternative

Full sewer separation of all the CSO outfalls to achieve zero overflows in a typical year was evaluated. Costs for sewer separation were based on constructing new storm drains and replacing sanitary sewers as necessary. Based on review of the CCTV data, it was assumed that approximately 40% of the sanitary sewers would need to be replaced due to their structural condition, presence of excessive I/I, or to maintain minimum velocity. Costs for new storm drains would include new catch basins and new 305 millimeter (mm) (12 inch) catch basin laterals. The resulting cost estimate for public source sewer separation was $76 million.

Public source sewer separation plus private inflow removal would theoretically denote 100% CSO control. In other EPA Region 1 CSO communities where private inflow removal was pursued, it accounted for 50% of the sewer separation costs. Therefore, the estimated cost for the full sewer separation alternative would approach an estimated $150M, or roughly twice the estimated $76M cost estimate for public source sewer separation.

Based on the high cost of the full sewer separation alternative, potential resistance to private property inflow removal, and concerns related to future stormwater regulations, the full sewer separation alternative was not considered further. Instead, public source sewer separation for the smaller outfalls was carried forward to achieve a target level of control of 4 overflows per year. Sewer separation for the smaller outfalls could also include green infrastructure where feasible.

Additionally, targeted sewer separation for both the larger and smaller CSO outfalls will be considered on a case-by-case basis as opportunities arise, such as local and state roadway projects or as part of private development projects.

Detailed Evaluation of Remaining CSO Control Alternatives

The next step in the CSO alternatives evaluation process was a non-monetary and economic, or monetary, evaluation of the remaining alternatives. The remaining CSO control alternatives for the four largest CSO outfalls included storage (both conduit and tank) and RTB treatment with disinfection for the Barkersville and Davis Brook outfalls, and a single storage tank for consolidated Kenduskeag East and West outfalls. CSO control for the smaller CSO outfalls - Hammond Street, Meadowbrook, Cemetery, Central, and Carr Brook – will consist of sewer replacement/rehabilitation and/or sewer separation. Green infrastructure may be part of the CSO control plan for these outfalls when site-specific conditions are suitable.

Non-Monetary Assessment. The non-monetary assessment of remaining CSO control alternatives was done in a comparative manner using a series of evaluation criteria and qualitative ratings. Evaluation criteria used were:

- Regulatory compliance
- Health and safety requirements
- Regulatory uncertainties / expandability
- Water quality impacts
- Constructability
- Ease of operation
- Public / community acceptance
- Ease of implementation
- Visual / aesthetics
- Process reliability

Rating factors used were:

- Highly favorable
- Favorable
- Neutral
- Unfavorable
- Highly unfavorable

As noted above, the only alternative for the two Kenduskeag CSO outfalls (East and West) remaining in the detailed evaluation process was a single storage tank. Accordingly, the comparative non-monetary analysis was performed only to compare storage (conduits or tanks) versus RTB treatment for the Barkersville and Davis Brook CSO outfalls. As expected, the storage alternatives (conduit or tank) rated higher than RTB treatment and disinfection in virtually all of the evaluation categories.

Economic Assessment. Cost estimates for the four largest CSO outfall alternatives remaining in the detailed evaluation were prepared using a variety of methodologies including:

- Review of quantities and components included for each control alternative. Material costs for various components were obtained from suppliers when applicable
- Review of completed CSO control facility costs within the State of Maine based on a unit cost per liter (gallon) stored or treated
- Use of cost curves, which were adjusted for this Phase II LTCP based on ENR

These methodologies were also used to develop the preliminary cost estimates for establishing the targeted level of control of four activations in the typical year as discussed above.

The cost estimates developed for the detailed evaluation of remaining CSO alternatives include appropriate contingencies and allowances for engineering, land acquisition, and SRF loan administrative costs.

Estimated costs for the four largest CSO outfall alternatives are presented in Tables 9 and 10. Table 9 corresponds to a WWTP capacity of 136.3 MLD (36 MGD) and Table 10 to a WWTP capacity of 162.7 MLD (43 MGD).

Table 9. Estimated Cost for Alternatives for the Four Largest CSO Outfalls with WWTP Capacity of 136.3 MLD (36 MGD)

CSO Outfall	Storage Conduit	Storage Tank	RTB Treatment with Disinfection
Barkersville	$21.5 M	$20.4 M	$16.7 M
Davis Brook	$22.0 M	$22.0 M	$12.1 M
Kenduskeag West and East	--	$9.0 M	--

Table 10. Estimated Cost for Alternatives for the Four Largest CSO Outfalls with WWTP Capacity of 162.7 MLD (43 MGD)

CSO Outfall	Storage Conduit	Storage Tank	RTB Treatment with Disinfection
Barkersville	$19.4	$17.3	$16.7
Davis Brook	$17.4	$14.6	$10.6
Kenduskeag West and East	--	$9.0	--
WWTP[1]	$3.0	$3.0	$3.0

1. Estimate of costs to reliably attain peak capacity of 162.7 MLD (43 MGD); does not include costs related to improvements to influent screens, grit handling, etc., that would be eventually required at the WWTP regardless of peak wet weather capacity.

The results from the non-monetary and economic assessments presented above were used to select preferred alternatives.

Preferred Alternatives

As noted, only one alternative remained in the detailed evaluation of remaining CSO control alternatives for Kenduskeag West and East. A storage tank was determined to be the only alternative. Also, there was virtually no difference in the Kenduskeag tank size regardless of WWTP capacity at 136.3 or 162.7 MLD (36 or 43 MGD). As shown in Table 9, the estimated cost for the Kenduskeag West and East storage tank is $9 million.

For the Barkersville and Davis Brook outfalls, storage was preferred over RTB treatment and disinfection based on the non-monetary assessment. While RTB treatment and disinfection did have a lower estimated cost than storage, the cost difference was not considered significant enough in light of the differences in the non-monetary assessment. As discussed previously, Bangor currently owns and operates three satellite CSO storage facilities. Their level of comfort with this CSO control technology, coupled with the comparatively simple operation of storage versus treatment, and avoidance of the need for disinfection chemicals were prime factors in making the decision to select storage over treatment.

For Barkersville, conduit storage is more costly than tank storage at both WWTP flow rates: 136.3 and 162.7 MLD (36 and 43 MGD). Therefore a storage tank was selected as the preferred configuration.

For Davis Brook, conduit and tank storage is nearly equal in cost with the WWTP at 136.3 MLD (36 MGD) but tank storage is lower in cost with the WWTP at 162.7 MLD (43 MGD). Therefore tank storage was also selected as the preferred storage configuration for the Davis Brook outfall.

For both the Barkersville and Davis Brook outfalls the larger tank size, corresponding to a WWTP capacity of 136.3 MLD (36 MGD), was selected as the placeholder tank size. This conservative assumption can be re-visited at the time of design should plans to upgrade the WWTP to reliably handle 162.7 MLD (43 MGD). Alternatively, the larger storage tank size could be retained to provide a factor of safety and/or to increase the level of annual CSO control. As shown in Table 9, the estimated cost for the Barkersville storage tank is $20.4 million and the estimated cost for the Davis Brook tank is $22.0 million.

Public source sewer separation, with elements of green infrastructure incorporated into separation projects as deemed feasible, was selected as the preferred alternative for control of the smaller CSO outfalls. No separation activities targeted at private sources is planned. Sewer rehabilitation/replacement work has been included as part of this effort, to reflect the findings of SSES investigations that the City has completed to date. Based on review of CCTV data, it was assumed that, in addition to construction of new storm drains, approximately 40% of the sanitary sewer in a given CSO tributary area would need to be replaced due to structural condition, excessive inflow and infiltration, or due to minimum velocity issues following separation. In addition, the costs include installation of new 305 mm (12 inch) catch basin laterals. Based on these assumptions, the estimated cost for control of the smaller CSO outfalls is $11.5 million.

The estimated cost for Phase II LTCP preferred alternatives is $62.9 million, composed of the following components:

• Barkersville storage tank -	$20.4 million
• Davis Brook storage tank -	$22.0 million
• Kenduskeag West and East storage tank -	$ 9.0 million
• Sewer separation for smaller CSO outfalls -	$11.5 million
TOTAL	**$62.9 million**

The estimated cost to upgrade the WWTP to reliably provide peak flow treatment of 162.7 MLD (43 MGD) is $3 million. If this cost were added to the $62.9 million total above, the total estimated cost attributable to the Phase II LTCP would be $65.9 million.

INTEGRATING PHASE II LTCP AND SSES RECOMMENDATIONS

SSES activities included development of a Phase 2 Scope of Work, flow isolation of two sub-sections (Airport and Dow) as shown on Figure 4, encompassing 24,385 m (80,000 ft.) of sewer, and CCTV inspection of 8,840 m (29,000 ft.) of pipe. The areas selected for CCTV inspection were based on high infiltration rates, greater than 370,400 mega liter per day per meter of diameter per meter of pipe (4,000 gallons per day per inch-mile (gpd/in-mi)). Recommendations for rehabilitation work were developed based on the flow isolation and CCTV work performed, and are included in the schedule in the January 2017 Phase II LTCP.

In accordance with the 2015 CD, the Phase II LTCP includes the cost effectiveness analysis for removal of sources of I/I identified through SSES, including, at a minimum, a discussion of which public sources of I/I would be more cost-effective to remediate than to transport and treat. After review of the Phase 2 SSES data, including the CCTV and flow isolation work completed, recommendations on sewer rehabilitation were developed. The recommended rehabilitation work addresses both structural and inflow/infiltration issues.

Phase 2 SSES recommendations were based on a cost effectiveness analysis. The cost effectiveness analysis considered whether it would be more cost effective to convey and treat the inflow and infiltration at the WWTP or if it would be more cost effective to rehabilitate the existing pipe. For Bangor, the threshold where convey and treat is more cost-effective than rehabilitation is at a value of $2.90/liter ($11/gallon). This was calculated based on five years of WWTP flows; WWTP operating expenses, and DEP recommended interest rates. This means that any rehabilitation option that costs less than $2.90/liter ($11/gallon) would be deemed cost effective.

Each rehabilitation cost represents the estimated construction cost for the noted rehabilitation method without an allowance for engineering and contingencies. With both the estimated rehabilitation cost and the estimated removable infiltration rate known, the unit cost of infiltration removed expressed in dollars per liter (gallon) was calculated for each source by dividing the rehabilitation cost by the removable infiltration. Infiltration rates shown are estimated removable infiltration rates. It was assumed that recommended rehabilitation work

Figure 4. City of Bangor Collection System and SSES Subsection Field Work Completed

would remove 100% of the infiltration entering the system through manhole walls and bases and 50% of infiltration through main pipelines.

The types of rehabilitation evaluated include the following:

- Root treatment (chemical or cutting)
- Cleaning/CCTV
- Cutting protruding services
- Lateral lining
- CIPP lining
- Short liners

- Full pipe replacement
- Grouting of joints
- Manhole rehabilitation

The SSES recommendations were kept separate for public sources and private sources. It was assumed that, based on the cost effectiveness analysis and SSES recommendations, the associated costs from public sources would be input into the financial capability analysis (FCA) and the implementation schedule. However, during the SSES field investigations, it was found that a significant portion of the inflow and infiltration was coming from private sources. However, with the amount of public mainline SSES rehabilitation work identified in the Airport and Dow subsections, and in other subsections, the City decided to focus its private source inflow and infiltration control efforts on three of the largest sources as identified below:

1) During the Phase 2 SSES field work, it was noted that there were numerous laterals that were no longer in service due to the buildings they served having been demolished. The City committed to capping laterals no longer in service in the Dow and Airport subsections.

2) In addition, a significant source of private inflow (approximately ½ of the total) was attributable to a water line bleeding into a sewer manhole. The City worked with the Bangor Water District to remove this significant inflow source.

3) Through flow isolation work, the City became aware that a large inflow and infiltration contributor is the Air National Guard. In an effort to quantify the amount of inflow and infiltration from the Air National Guard, the City anticipates installation of a flow meter and review of flow monitoring data at this location for an extended period of time over the next five years.

Other SSES Results

During the field investigations, several locations were noted to have deficiencies during flow isolation, but were not identified as significant sources of infiltration during the CCTV inspections nor would they have been cost effective to rehabilitate based on the cost effectiveness analysis. However, these pipe segments were noted to have structural deficiencies. Rehabilitation recommendations were included and the costs associated with this work were included in the FCA.

The total estimated cost for implementation of the cost-effective SSES Phase II rehabilitation measures, including pipe segments noted to have structural deficiencies as noted above, is $1M.

FINANCIAL CAPABILITY ANALYSIS

The costs associated with the Phase II LTCP, along with other non-CSO wastewater system needs under the Clean Water Act (CWA), were evaluated in what is referred to as a financial capability analysis (FCA). The FCA allowed the City to assess its ability to pay for the preferred Phase II LTCP and other wastewater system needs using metrics established by EPA. The FCA established that Bangor would experience a "medium" financial burden as a result of these additional wastewater costs, based on application of EPA's FCA guidance. More importantly, the FCA also concluded that sectors of Bangor households are already experiencing a "high" financial burden and that additional wastewater costs would add to their existing financial stress. As such, care will need to be taken to evaluate all wastewater system needs and to prioritize projects in a manner that will achieve the greatest water quality benefit while still being affordable.

The most immediate major capital project identified was the Davis Brook CSO storage tank. While this Phase II LTCP project could be considered affordable at this time, the impact of future projects would need to be carefully evaluated while taking into consideration all of the other identified and yet to be identified WWTP and SSES-related needs of the City. The Phase II LTCP implementation schedule described below took this financial situation into consideration in its development.

SEQUENCING AND SCHEDULING

The 2015 CD requires Bangor to complete the CSO abatement work called for in the Phase II LTCP by December 31, 2031, less than 15 years from the submittal date of the plan itself. There are also intermediate submittal dates included in the CD for the preparation of Statements of Work (SOWs) for plan implementation: June 30, 2021 and June 30, 2026. However, because of the high cost of the preferred plan as summarized in the LTCP, and based on the findings of the FCA, this schedule would cause an undue burden on all of the Bangor ratepayers and even more so for the already financially stressed segment of the population. A time extension will likely need to be sought from EPA as plan implementation progresses and the financial impact of the initial project or projects can be more fully assessed. Therefore, the implementation schedule presented in Table 11 focuses on the initial projects with a FCA reassessment phase following each major project.

Table 11. Implementation Schedule

Action	Date	Remarks
1. Submit Phase II LTCP	Jan. 2017	Jan 31, 2017 per 2015 CD
2. Comment and Response Period	February-May 2017	Based on receipt of EPA and DEP comments by March 1, 2017
3. Phase 1 - Davis Brook Storage Tank		
• **Design (18 months)**	Aug. 2017–Jan. 2019	
• **Bidding/Construction (25 months)**	March 2019-March 2021	
• **Start-up and Commissioning (3 months)**	April 2021 to June 2021	
4. Preparation of SOW for LTCP and SSES Phase III analyses	June 2021	June 30, 2021 per 2015 CD

Action	Date	Remarks
5. Phase 2 - Barkersville CSO Storage Tank	***Could follow Kenduskeag West and East as Phase 3***	
• Design (18 months)	TBD	Date will be based on results of revised FCA
• Bidding/Construction (25 months)	TBD	Date will be based on results of revised FCA
• Start-up and Commissioning (3 months)	TBD	Date will be based on results of revised FCA
6. Preparation of SOW for LTCP and SSES Phase IV analyses	June 2026	June 30, 2026 per 2015 CD
7. Phase 3 - Kenduskeag West and East CSO Storage Tank	***Could precede Barkersville as Phase 2***	
• Design (18 months)	TBD	Date will be based on results of revised FCA
• Bidding/Construction (25 months)	TBD	Date will be based on results of revised FCA
• Start-up and Commissioning (3 months)	TBD	Date will be based on results of revised FCA
8. Phase 4 - WWTP Capacity Expansion (Optional)	***Need for Phase 4 WWTP Expansion to be determined through PCM and other considerations***	
• Design (18 months)	TBD	
• Bidding/Construction (25 months)	TBD	
• Start-up and Commissioning (3 months)	TBD	
9. Abatement of Smaller CSO Outfalls	Continuous small projects over implementation schedule	Detailed schedule to be developed based on related infrastructure improvements, SSES-related needs and other considerations
10. Post Construction Monitoring (PCM)	Following start-up of each major facility	Depending upon scheduling included in SOWs; will determine need for WWTP capacity expansion
11. Financial Capabilities and Affordability (FCA) Update	Following start-up of each major facility	And included in SOWs depending upon scheduling

SUMMARY AND CONCLUSIONS

The primary purpose of this paper is to demonstrate how Bangor, Maine was able to effectively integrate the scheduling of capital projects from separate CSO and SSES planning efforts required under its CD into one comprehensive schedule included in its Phase II LTCP. After learning that the purported benefit of integrated planning (i.e., being able to schedule ***all*** wastewater and stormwater projects so that Clean Water Act requirements could be accounted for and prioritized in one document) would result in ***all*** projects in the Integrated Plan being subject to CD dates with stipulated penalties, Bangor was able to successfully negotiate the integration of scheduling for SSES and CSO LTCP capital projects into a single schedule in the LTCP.

Other lessons learned during conduct of this work include the following:

- Allow adequate time for flow monitoring. Bangor allowed for two spring seasons (high groundwater) for SSES flow monitoring. Due to rapid snowmelt near the start of the first season, followed by a prolonged drought extending throughout the rest of the first season and on into the second season, Bangor only experienced high groundwater for a short period despite the lengthy monitoring program. The drought also impacted the number of suitable storm events for collection system model calibration. If Bangor had selected a shorter flow monitoring period, adequate data would likely not have been obtained.
- Tailor the level of analysis to the size of the community and/or wastewater system. Bangor used a level of flow monitoring and collection system modeling that was appropriate for its collection system size and for the number and magnitude of overflows remaining following implementation of Phase I CSO controls.
- Data collection and organization of data is imperative. Bangor has a comprehensive GIS system that interfaces with the asset management IT pipes for CCTV inspection and maintenance. During this project the City and a subcontractor completed SSES inspections, which required coordination of naming convention of newly located manholes, field data and final data storage. It was important for the City and the subcontractor to have open dialogue on the field inspections as the City needed to manipulate and use the data moving forward and not the subcontractor.
- Favor alternatives that include technologies already in use in the community. While due consideration was given to a full range of CSO control technologies, it was recognized that Bangor has three off-line storage facilities and off-line storage became a favored technology during Phase II LTCP evaluations.

It is hoped that Bangor's innovative approach for integrating multiple CD-driven program requirements into a single implementation schedule will be of benefit to other communities and agencies facing multiple Clean Water Act compliance requirements.

REFERENCES

Phase II 2017 Long Term Control Plan, City Of Bangor, Maine, January 2017

Odor Control Management for Combined Sewer Overflow Structures

Ray F. David, PhD, PE
Greeley and Hansen, LLC

ABSTRACT

The provision of enhanced protection against wet weather flow events while improving our ability to minimize nuisance concerns continues to become a more pressing concern. This issue will become more prominent and visible as communities continue to expand and utilize areas surrounding treatment facilities and their related appurtenances. An area that many utilities are increasing their efforts is combating odors related to combined sewer overflow (CSO) events. In an attempt to reduce combined sewer overflows into receiving waters, utilities are building tunnels and storage tanks for these normally high intensity, short duration flow events. These events result in large volumes of runoff water mixed with residential, commercial, and industrial sewage. This combination water during wet weather events poses potential odor concerns due to its storage and transmission. The transfer of water to and from the storage structure combined with the storage phase during a wet weather event can liberate hydrogen sulfide gas from the liquid phase and into the air phase. Presented are case studies from two cities outlining their approaches to odor control management within their CSO systems.

The City of Lafayette, Indiana, is in the process of designing a new diversion structure composed of a 5.9 million, underground, CSO Storage Tank. The approximately 300-foot by 150-foot rectangular CSO Storage Tank was designed to best utilize the general site layout constraints and account for the geotechnical and structural requirements deemed appropriate for the site. As the area surrounding the location of the CSO Storage Tank is undergoing various exciting stages of development, a decision was similarly made to adjust the odor control strategy in complimentary stages. In the near-term, the decision was to employ Air Ventilation and Dispersion using a ventilation strategy with strobic fans to evacuate the nuisance odors from the tank and the surrounding area. As the area becomes further developed and a greater population surrounds the CSO Storage Tank, a proposed enhancement of the odor control strategy is the implementation of a targeted, granular activated carbon unit.

The District of Columbia Water and Sewer Authority (DC Water) is implementing a $2.6 billion Long Term Control Plan (DC Clean Rivers Project). A primary concern for this system is that many of the related infrastructure is located in residential areas of DC. The system is anticipated to release tunnel air at the upstream termini adjacent to historic row homes when the air would be displaced by rapid tunnel filling and long-term stack effects. Specific focus was placed on the system when it is empty and full to properly size odor control units. The technology selected was radial flow granular activated carbon units to meet the odor treatment needs. Additional design considerations included providing vents to prevent nuisance noise and employing gravity dampers to inhibit fugitive emission when the tunnel is empty.

The selection of an odor control management strategy is dependent on the size of a given treatment and collection system, space and budgetary constraints, and the level of odor treatment desired. With a fluid nature of land development that effectively maximizes our land use capability, it is imperative we prioritize our management of CSO flows and related odors to provide the best communities for all.

Keywords: Combined Sewer Overflows, Odor Control, Granular Activated Carbon

INTRODUCTION

An important consideration for many cities, town, and villages across the country and the globe, is: how do we continue to provide enhanced protection against wet weather flow events while improving our ability to minimize nuisance concerns? The prevalence of ever-increasing extreme weather events, including wet weather events (Goswami et al. 2006; Easterling et al. 2000), necessitates careful planning to address the surge of rainwater as well as its accompanying effects.

One notable area that many utilities are increasing their efforts is combating odors related to combined sewer overflow (CSO) events. These CSO events are objectionable for various reasons. Combined sewers, as the name suggests, are collection systems for both sewage and surface runoff due to precipitation (*i.e.* wet weather) events that are transported to a wastewater treatment plant (WWTP). During high wet weather events, the deluge of precipitation can overwhelm the capacity of the combined sewer and/or WWTP resulting in CSO events where a relief structure facilitates a portion of this combined stormwater and sewage to be discharged into a receiving stream of water untreated. The negative effects of releasing untreated sewage and urban surface pollution contained within runoff water into receiving waters may result in oxygen depletion due to the high organic loads, reduction in photosynthesis due to high levels of turbidity, increase in metal concentrations, and increase in pathogenic microorganisms (Passerat et al. 2011). In an attempt to reduce combined sewer overflows into receiving waters, many utilities are constructing storage structures (*e.g.* tunnels and tanks) where flow can be diverted prior to reaching the WWTP during these high intensity, short duration flow events. While this addresses the concerns of a liquid overflow event, this event poses odor issues due to the large storage volume and turbulent transmission.

Two cases studies are provided indicating the plethora of opportunities available to manage CSO-originating odors and the unique characteristics of each system that necessitate creative approaches. The D.C. Clean Rivers Project is the implementation of a long-term control plan (LTCP) that includes a series of underground storage and conveyance tunnels, shafts, diversion chambers, overflow facilities, and a dewatering pump station located at D.C. Water's Blue Plains Advanced Wastewater Treatment Plant. The Greenbush CSO Project is associated with the City of Lafayette, Indiana's CSO LTCP with one of the four phases the construction of a 5.9 MG CSO storage tank near an existing CSO diversion structure.

CASE STUDY NO. 1 – D.C. CLEAN RIVERS PROJECT (DISTRICT OF COLUMBIA)

Methodology

The D.C. Water LTCP consists of a series of underground storage and conveyance tunnels with all pieces related to the Blue Plains Advanced Wastewater Treatment Plant. These tunnels range in size up to 23-feet in diameter, extend over 13-miles in length, and are located approximately 100-feet underground. The project is to be completed in multiple divisions and once completed, the tunnel system will reduce combined sewer overflows by 96% overall, and by 98% in the Anacostia River alone.

The approach to this project included the completion of an exhaustive hydraulic modeling analysis of the entirety of the existing sewer system. The design of CSO-related infrastructure including tunnels, storage tanks, and related odor management strategies were subsequently preformed with an ultimate goal to reduce flooding events in the susceptible project areas while also eliminating the release of nuisance rogue odors and emissions. To achieve these goals, the project involved the development of design criteria for the Ventilation, Odor Control, and Maintenance Access for the entire system. The approach to this task included evaluating similar CSO systems throughout the Country to identify the positives and shortfalls of each facility from the operational, design, and construction perspectives of the municipalities. The ability to communicate with the owners and operators of the facilities, specifically in terms of ventilation, odor control, and maintenance, provided invaluable in identifying design parameters for each of the shaft locations to prevent negative impacts on the surrounding areas by the presence of the CSO infrastructure.

Results/Discussion

One unique characteristic of designing an odor control strategy for a CSO structure is the inherent surges of both water and air that require appropriate control and treatment. When a significant wet weather event necessitates the use of a given CSO structure, large quantities of air will be exhausted from the tunnel at high rates that parallel with the rush of water. This release of air is short in duration and during heavy rainfall where it is not conducive to being outdoors. Prior to this short release event, the tunnel or CSO structure will have been near empty or completely empty. As such, the air would not be typical CSO air. The odorous air associated with a CSO event and the anaerobic processes the produce these odors would not have impacted the empty headspace in the tunnel or CSO structure. Furthermore, this immense surge of air, both volume and rate, makes it impractical to provide odor control. A similar odor control strategy of typically not providing odor control for the large rate of air released as the tunnel fills was found in other tunnel systems including Milwaukee, Portland, and Richmond(Greeley and Hansen 2008). Other programs have not experienced problems using this approach. Based on these factors, an explicit odor control strategy is not proposed for the large rates of air to be exhausted during tunnel filling.

To combat the release of rogue emissions when the tunnel is nearly or completely empty, which is anticipated to be the majority of the time, the maintenance of the CSO tunnel under slightly negative pressure is recommended. During these periods of active odor control, the strategy would be to convey air from the CSO tunnel to odor control systems such as carbon absorption. Adjustable intake dampers were employed at shafts to allow air to enter the structure and adjustable to allow fine tuning of the system to prevent short circuiting of air. The required design capacity of odor control facilities at various sites resulted in ventilation rates of 1 air change every 4, 2, and 1 hour. The evaluation of similar tunnel systems indicated sizing the system to provide one air change every two hours (*i.e.* 0.5 air changes per hour) is recommended (Greeley and Hansen 2008). The treatment of odors when the tunnels are full presents distinctive issues since this water prevents odor control facilities at lower ends of the tunnel from drawings air from upper ends of the tunnel. This may thus present conditions for selective shaft locations to become odorous. This was addressed by designing space for small, skid-mounted odor control facilities that service these selective shaft exclusively.

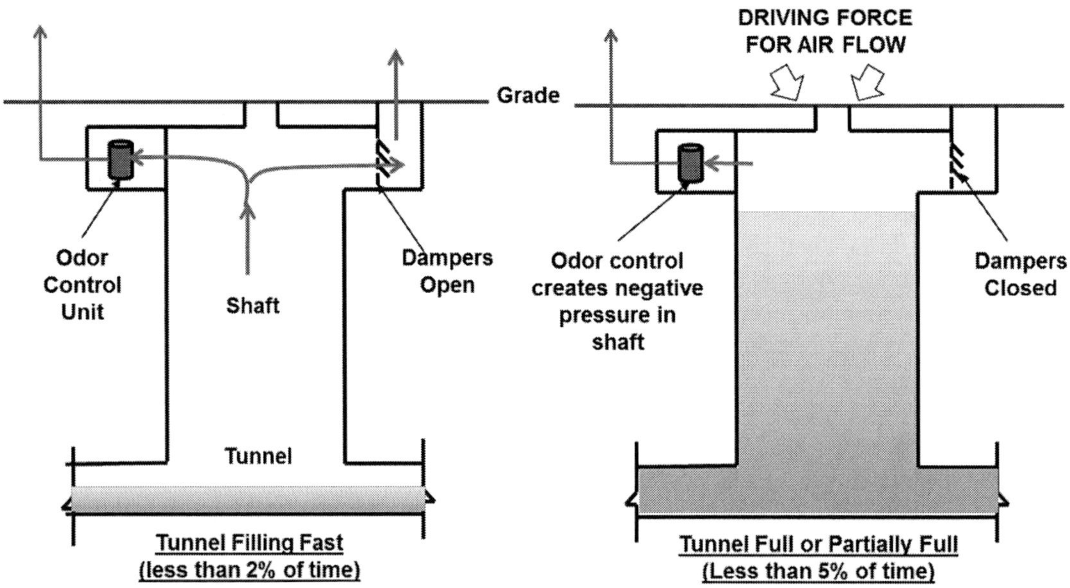

Figure 1: Approach for odor control during tunnel filling (left) and when the tunnel is full (right).

CASE STUDY NO. 2 – GREENBUSH COMBINED SEWER OVERFLOW PROJECT (LAFAYETTE, INDIANA)

Methodology

The City of Lafayette's Combined Sewer Overflow Long Term Control Plan has similarities and distinctions in comparison to the D.C. Clean Rivers Project. The associated Greenbush CSO Project required the planning, designing, and constructing of a 5.9 MG CSO storage tank in the area near an existing CSO diversion structure. Similarly, this Greenbush CSO Storage Tank is needed to store and convey wet weather flow to the Lafayette Waste Water Treatment Plant (WWTP) when the WWTP has limited capacity thereby improving the water quality in the receiving body of water (*i.e.* Wabash River).

Municipalities that have implemented CSO storage facilities have typically encountered limited odor issues due to the dilution (*i.e.* rainwater and runoff) of sewage when associated with wet weather flow conditions. Issues with nuisance odors are more predominant with dry weather flow conditions. The immediate area surrounding the proposed Greenbush CSO Storage Tank location is in various stages of development. As such, the current conditions in the area require a different approach to odor control in comparison to anticipated future conditions. In stark contrast, many of the locations associated with the D.C. Clean Rivers Project were in densely populated areas so the strategies and approaches varied to reflect those dissimilarities.

In the phased approach, the CSO Storage Tank will employ air ventilation and dispersion in the immediate term to allow for nuisance odors to be exhausted from the structure. This includes the use of strobic fans to disperse the air from within the CSO Storage Tank. As the area becomes more developed, the implementation of an active odor control technology is desired to alleviate odors associated with organic and inorganic substances, with hydrogen sulfide and ammonia anticipated to be the target gasses for odor control.

Results/Discussion

Hydrogen sulfide concentrations generated in the Greenbush CSO Storage Tank are likely to be in the low (<10 ppm) to mid-range (10-50 ppm). Air Ventilation and Dispersion is recommended for ventilation in the near term to control site-related odors. As future development in the area continues, there may be a desire for additional odor control at the site. An ultimate decision on the type of unit should be aided by onsite odor sampling to determine odor concentration. Additionally, the large volume of the Greenbush CSO Storage Tank makes odor control decisions more complex. Based on current information, it is recommended that granular activated carbon (GAC) odor control units be utilized for future treatment. The application of GAC is appealing because of its ability to treat the anticipated odors, ease of maintenance and operation, and relatively low capital cost. While other units (Table 1) may have slight advantages in a parameter such as a slightly lower operation cost, the totality of the parameters indicate GAC would likely be an appropriate technology. Manufacturers recommend a maximum velocity through the carbon units of 60 feet per second or less to prevent breakthroughs and short circuiting. Piping was designed to accommodate the Air Ventilation and Dispersion approach but also consider the potential expansion to a GAC unit.

Table 1: Summary of a few available odor control technologies potentially applicable for CSO structures

Technologies	Footprint	Capital Cost	Operating Cost
Air Ventilation and Dispersion	Small	Low to medium	Low to medium
Granular Activated Carbon (GAC)	Medium	Low to medium	Medium to high
Chemical Scrubbers	Small to medium	Medium to high	Medium
Biofilters	Large	High	Low
Odor Masking Agents	Small to medium	Medium to high	Medium

CONCLUSIONS

As communities continue to expand and the buffer between treatment-related activities and residential, commercial, and industrial areas shrink, some undesirable aspects of the treatment processes may become a greater nuisance.

The transfer of water to and from the storage structure combined with the storage phase during a wet weather event can liberate hydrogen sulfide gas, along other odors, from the liquid phase and into the air phase. The release of these malodorous gases produce offensive odors and can severely corrode structures and materials when oxidized (Fan; et al. 2001) thus reducing the lifetime of purchased equipment. An efficient odor control management strategy for CSO structures is required to: provide storage and conveyance of wet weather flows, minimize or eliminate overflow events into receiving bodies of water, and provide suitable working and living conditions for these communities. Each community has its own unique treatment and collection system, and the ability to tailor an appropriate treatment approach is essential in achieving a successful outcome.

ACKNOWLEDGEMENTS

The author would like to acknowledge and thank Carlton Ray of D.C. Water and Justin Carl, John Cassidy, and Edward Shea of Greeley and Hansen for their work on the D.C. Clean Rivers Project. The author would like to thank and acknowledge Brad Talley of the City of Lafayette and Scott Hall of Greeley and Hansen for their work on the Greenbush Combined Sewer Overflow Project.

REFERENCES

Easterling, D. R., J. L. Evans, P. Ya Groisman, T. R. Karl, K. E. Kunkel, and P. Ambenje. 2000. "Observed Variability and Trends in Extreme Climate Events: A Brief Review." *Bulletin of the American Meteorological Society 425* 81 (3): 417–25.

Fan;, Chi-Yuan, Richard Field;, William C. Pisano;, James Barsanti;, James J. Joyce;, and Harvey Sorenson. 2001. "Sewer and Tank Flushing for Sediment, Corrosion, and Pollution Control." *Journal of Water Resources Planning and Management* 127 (3): 194–201.

Goswami, B. N., V. Venugopal, D. Sengupta, M. S. Madhusoodanan, and Prince K. Xavier. 2006. "Increasing Trend of Extreme Rain Events Over India in a Warming Environment." *Science* 314 (5804): 1442–45.

Greeley and Hansen. 2008. "DC Clean Rivers Project: Anacostia River Projects Facility Plan Operational Plan and Hydraulic Design."

Passerat, Julien, Nouho Koffi Ouattara, Jean-Marie Mouchel, Vincent Rocher, and Pierre Servais. 2011. "Impact of an Intense Combined Sewer Overflow Event on the Microbiological Water Quality of the Seine River." *Water Research* 45 (2): 893–903.

Minne Lusa Stormwater Conveyance Sewer Tunnel and Detention Basin Forebay Hydraulics

Ken Christison, PEng, NHC
Janet Strickland, PE, AECOM
Tom Demlow, PE, NHC
Eitan Tsabari, PE, City of Omaha

ABSTRACT

The design process for a new multi-facetted project can be complex and require a multi-disciplinary team. For the Minne Lusa Stormwater Conveyance Sewer Tunnel and Detention Basin Forebay consideration of flow capacity, hydraulic transients during filling of the tunnel, air management, and sediment deposition were all required. The objective of hydraulic design effort was to optimize the design for capacity, operations and maintenance, constructability, surge control and capital cost. AECOM and Northwest Hydraulic Consultants (NHC) addressed these design challenges with a suite of design tools including desktop hydraulic design, 1D numerical modeling, hydraulic transient analysis, and physical modeling. The overall effort demonstrated an efficient use of available design tools for developing a safe and robust project.

The objectives of the project were to develop a design to significantly reduce untreated combined overflows in Omaha by conveying up to 56.63 m^3/s of stormwater via a tunnel to a proposed stormwater detention basin utilizing two existing basins, and provide water quality benefits for the stormwater prior to discharging to the Missouri River. This design approach provided the necessary hydraulic information to develop a very efficient and robust design to divert stormwater out of the combined sewer system, reducing CSO's and alleviating flooding problems in northeast Omaha. This approach to evaluating complex urban stormwater sewer system and flood control hydraulics is a method that can be used in many flood prone cities.

KEYWORDS

Combined Sewer Overflow, Hydraulic Design, Sediment Transport, Transient, Physical Modeling

INTRODUCTION

The Minne Lusa Stormwater Conveyance Project is located in the east central portion of the Minne Lusa Basin in Omaha Nebraska to provide conveyance of separated stormwater to the expanded Storz West and Pershing Detention Basins and subsequently, the Missouri River. The project was developed to reduce Combined Sewer Overflow (CSO) volume and number of CSO occurrences. In addition, the Pershing Detention Basin would provide water quality benefits for the stormwater prior to discharging to the Missouri River.

The Minne Lusa Stormwater Conveyance Sewer includes an approximately 1859.28 m long, 4.27 m diameter tunnel followed by a 73.152 m length of 4.27 m x 4.27 m box conduit. The inflow to the tunnel from the Paxton Blvd. and John Creighton Blvd. (JCB) stormwater conveyance sewers (collectors) is supplied through a drop shaft followed by a de-aeration chamber fitted with air release vents.

An outlet control structure is located at the terminus of the stormwater conveyance sewer and inlet to a sediment detention basin, and includes four 0.46 m x 0.91 m low level orifices to permit sediment flushing and a 22.86 m long weir crest to accommodate high flows to the Storz West Basin. The maximum design inflow is approximately 56.63 m^3/s and is based on a 10-year, 24-hour storm event.

APPROACH

To support the design effort an iterative approach was followed using 1D numerical modeling, transient analysis, and physical modeling. Each design tool contributes in a unique way to the design. The 1D modeling develops the overall size of the project and prescribes modifications based on well-established empirical relationships while the transient effort identifies possible unsteady flow conditions that drives the project geometry. Physical modeling helps develop the details to help optimize performance while providing a higher level of assurance that the overall project performs as intended.

1D NUMERICAL MODELING

A 1D numerical model was adopted to guide the overall design of a circular conduit considering free-surface flow, pressurized flow, and sediment transport. For the free-surface flow alternative, the tunnel would operate as a circular open channel at peak flow, venting air up through the open air core and up through vents in the deaeration chamber and other locations along the tunnel alignment.

For the pressure flow alternative, the tunnel operates as a circular open channel at low flows, transitions through to pressure flow at moderate flows (less than design flow), and under pressure at maximum design flow capacity. All air venting in the pressure flow alternative would occur at the drop shaft and de-aeration chamber, vents along the tunnel, or at the exit portal.

Preliminary-level design effort started by defining an envelope of potential slopes in order to: a) identify potential tunnel size/roughness combination(s); b) define potential constraints to the vertical alignment of the tunnel; c) establish boundary conditions for numerical and physical models; and d) design tunnel, drop shaft, transitions, deaeration chamber, discharge structure, and detention basin. Figure 1 illustrates one of many iterations that were considered by adjusting tunnel diameter, roughness, and slope.

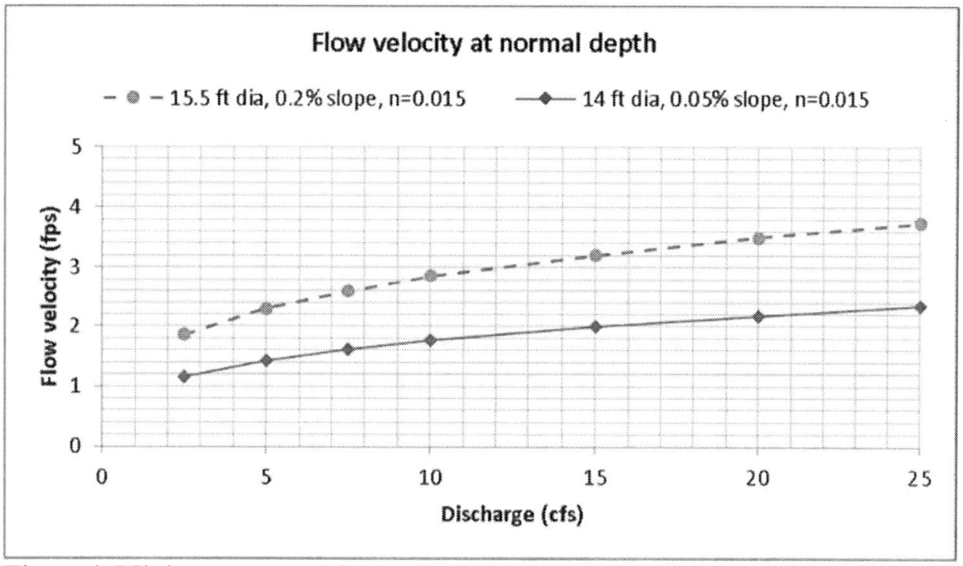

Figure 1. Minimum normal flow velocity in free surface (i.e. open channel) tunnel (4.72 m diameter, 0.2% slope). Pressure tunnel alternative (4.27 m diameter, 0.05% slope).

The design of the tunnel, inlet, and discharge structures must also consider system maintenance concerns. Sediment flushed from City streets into the storm sewer system has collected in the stormwater conveyance system in the past, and is expected to be a key issue to address in the future. Most of this material is medium to fine grained sand, typically used by the City in winter to treat streets. Figure 2 illustrates one of many iterations that were considered to identify flushing characteristics of various tunnel geometries.

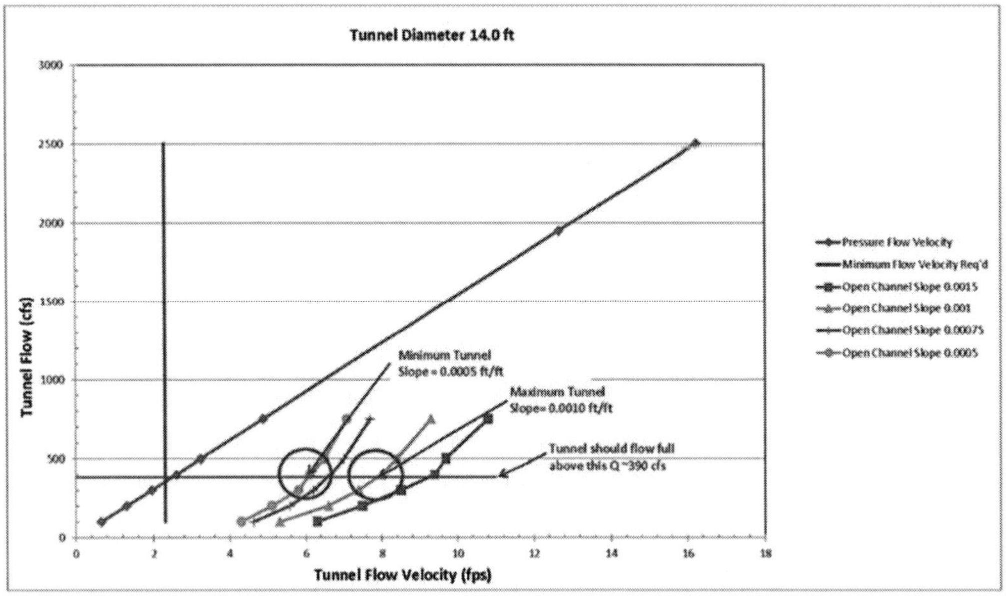

Figure 2. Flow regime for 4.27 m diameter tunnel, various slopes

For sediment flushing consideration, evaluations were directed at lower flows, since the higher flows will provide flow velocities in excess of the minimum required to transport the expected sands and silts through the tunnel to the Storz West detention basin.

Based on the results of the analyses, NHC determined that all tunnel diameters, slopes and roughness values considered would not impede the transport of sediment through the tunnel under virtually all discharges.

TRANSIENT ANALYSIS

A hydraulic transients computer model of the Stormwater Conveyance Sewer was developed and performed for the 10-Year, 24-Hour Design Storm Event and a 70.79 m^3/s Storm Event using the ITM (Illinois Transient Model) software. The transient analysis was based on the tunnel conveyance system drawings that were developed by AECOM as part of the preliminary design efforts and included the prediction of hydraulic grade line elevations in the tunnel and maximum water levels at the Intake/Drop Shaft with and without entrained air in the fluid.

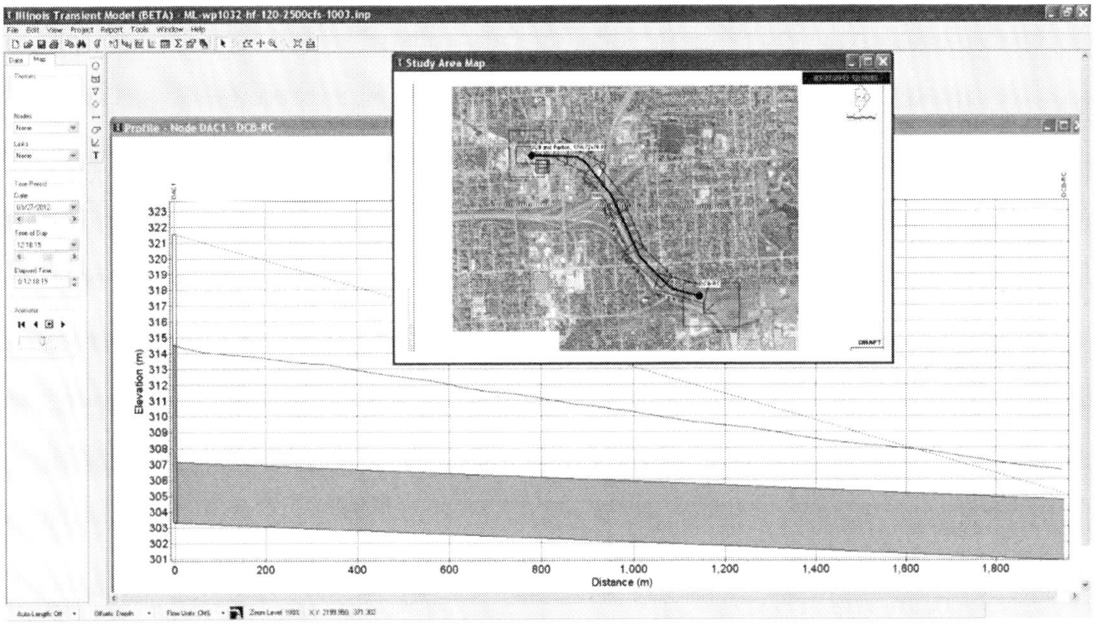

Illinois Transient Model (ITM) Software

Figure 3. ITM (Illinois Transient Model) software

The results of the transient analyses for the Stormwater Conveyance System showed that the water level at the Intake/Drop Shaft is predicted to exceed the invert elevation of the Paxton and JCB Sewers, which could cause these sewers to surcharge during the 10-Year Design Storm Event.

These same results also show that the maximum fill and drain rates at the Intake/Drop Shaft are predicted to be 0.70 m/sec and 0.40 m/sec, respectively, during the 10-Year Design Storm Event. Although the water level at the Intake/Drop Shaft is predicted to not overtop the drop shaft, it is possible that any air that is vented from the drop shaft at these velocities could briefly release water from the shaft.

The maximum outflow at the Discharge Control Box is predicted to be 69.80 m^3/s and will occur for a duration less than 5 seconds. The oscillatory nature of the HGL in the tunnel just before the peak of the storm event is predicted to create rapid surges in the tunnel, which in addition to creating negative pressures in the tunnel, also creates small sloshing waves at the outlet. These small waves briefly raise the water level at the outlet above that which would be associated with the peak inflow rate, which accounts for the predicted outflow rate exceeding the maximum inflow rate at the Intake/Drop Shaft.

The transient analysis identified that to prevent surcharge of the Paxton and JCB Sewers during the 10-Year Design Storm Event, it is necessary to modify the Deaeration Chamber as follows:

- With a weir crest elevation of 304.80 m at the Discharge Control Box, the length of the Deaeration Chamber need only be 16.46 m, however, the top of the Deaeration Chamber should be raised to the same elevation as the top of the Intake/Drop Shaft. It would not be necessary to change the width of the Deaeration Chamber as long as a plan area of 70.23m^2 is provided.

With the above modifications in place, the results of the transient analyses showed that the maximum water level at the Intake/Drop Shaft is predicted to not exceed the invert elevation of the Paxton and JCB Sewers and the hydraulic grade line is predicted to not exceed the ground elevation along the tunnel alignment.

The 914 mm diameter air vents shown on the tunnel plan and profile drawings should be installed along the tunnel to provide for the release of air that may be trapped in air pockets that are created as the tunnel fills.

The results of the transient analyses show that the HGL is predicted to briefly drop below the crown elevation of the tunnel and create negative pressures in the Stormwater Conveyance Sewer. However, the installation of the above recommended air vents will provide for vacuum relief in the event that negative pressures are created in the tunnel during the storm event. To be conservative, the tunnel segmental lining should be designed to withstand full vacuum pressure in case negative pressures do occur even with the air vents in place.

A sensitivity analysis showed that the above recommended surge control will prevent surcharge of the Paxton and JCB Sewers for high and low friction loss in the tunnel and with and without entrained air in the fluid.

Additional transient analyses were performed to study the response of the Stormwater Conveyance Sewer to a storm event with a peak flow that is approximately 26 percent greater than the peak flow for the Design Storm Event (56.07 m^3/s). With the above recommended surge control in place, the results of the transient analysis for the 70.79 m^3/s storm event show that the maximum water level in the Intake/Drop Shaft is predicted to exceed the invert elevation of the Paxton and JCB Sewers as well as the crown elevation of the JCB Sewer. If the City of Omaha wishes to prevent surcharge of the Paxton and JCB Sewers under the 70.79 m^3/s Storm Event, it will be necessary to significantly increase the plan area of the Deaeration Chamber to 1053.52 m^2.

Alternatively, a tunnel diameter sensitivity analysis showed that by increasing the tunnel diameter sufficiently to prevent full pressurization of the tunnel (i.e., install an 5.49 m diameter tunnel in place of a 4.27 m diameter tunnel), it is possible to pass the 70.79 m^3/s Storm Event through the Stormwater Conveyance System without modifying the original Deaeration Chamber design while at the same time preventing the occurrence of negative pressures in the tunnel, eliminating the possibility of surcharging the Paxton and JCB Sewers, and preventing any release of water from the tunnel through the Intake/Drop Shaft.

PHYSICAL MODELING

A 1:12 scale physical hydraulic model study was conducted to document and develop the 60% Stormwater Conveyance Sewer design. The physical model optimized the initial hydraulic design (desktop) effort, 1D modeling, and the transient analysis conducted previously. Evaluation of the initial design revealed that the design did not accommodate tunnel pressurization at sufficiently low discharges. Modifications to the tunnel, box conduit and outlet control structure were developed in the model to facilitate pressurization of the tunnel at lower flows. In addition, modifications were developed in the model to increase sediment trapping in the detention basin and improve sediment transport through the tunnel.

Drop Shaft and Collector

Evaluation of the drop shaft and upstream collector conditions conducted under the initial design configuration determined that free-surface flow conditions were present in the Paxton collector when operated at 56.63 m^3/s. This met design criteria and confirmed that the design developed to meet transient conditions was acceptable for steady state and lower flow operation.

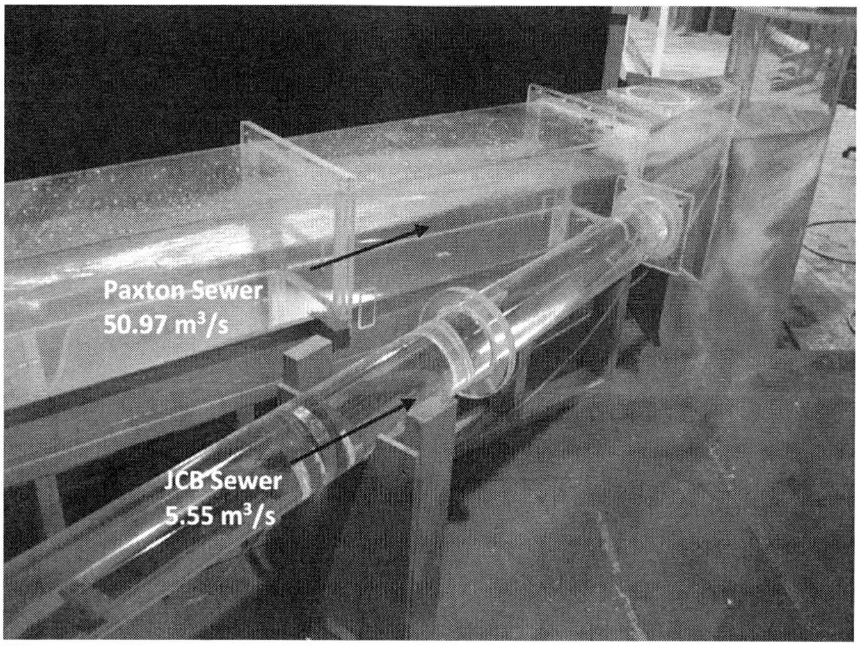

Figure 4. Photograph of the drop shaft and upstream collectors.

The recommended design test results confirmed that the Paxton collector could pass 50.97 m^3/s (with the JCB collector passing 5.66 m^3/s). In addition, the Paxton collector had sufficient capacity to pass 56.63 m^3/s, confirming that the collectors could be combined further upstream without causing the system to back up. The collectors are capable of passing a steady state 70.79 m^3/s event with HGL levels near design limits and therefore are likely to surcharge during transient conditions typically present during operation.

Tunnel, box conduit and outlet control structure

Modifications to the tunnel, box conduit and outlet control structure were developed in the model to facilitate pressurization of the tunnel at lower flows. In addition, modifications were developed to reduce sediment deposition within the tunnel and improve the potential for sediment to settle out downstream of the outlet control.

Figure 5. Photographs of outlet control structure modifications.

The final recommended design, as developed in the model included the following modifications:

- Tunnel invert raised by 0.31 m above the original design concept along the entire tunnel;
- Outlet control structure weir crest raised by one foot;
- Width of box conduit reduced from 4.27 m to 3.05 m;
- 0.61 m by 0.61 m fillets installed along the bottom corners of the 3.05 by 4.27 m box conduit;
- Width of outlet control structure reduced from 7.62 m to 3.05 m;
- Transition between the tunnel and the 3.05 m x 4.27 m box conduit increased 3.05 m;
- Outlet control structure moved upstream so that upstream face meets the toe of slope along the west (left) side of the detention basin;
- A 0.46 m by 3.05 m opening was included along the bottom of the end wall of the outlet control structure;
- 0.46 m by 0.91 m low level orifice added along the east (right) side of the outlet control structure; and
- 1.07 m high, 2.74 m long sediment diversion wall installed at east (right) side of the outlet control structure just downstream of the low level side orifice.
- Top of de-aeration chamber raised to El. 321.56 m; and,
- 45 degree fillets installed upstream of the tunnel entrance.

Detention Basin

Modifications to the detention basin were developed in the model to promote the deposition and collection of sediment to facilitate maintenance while maintaining design water levels over the control weir and backwater impacts on the tunnel.

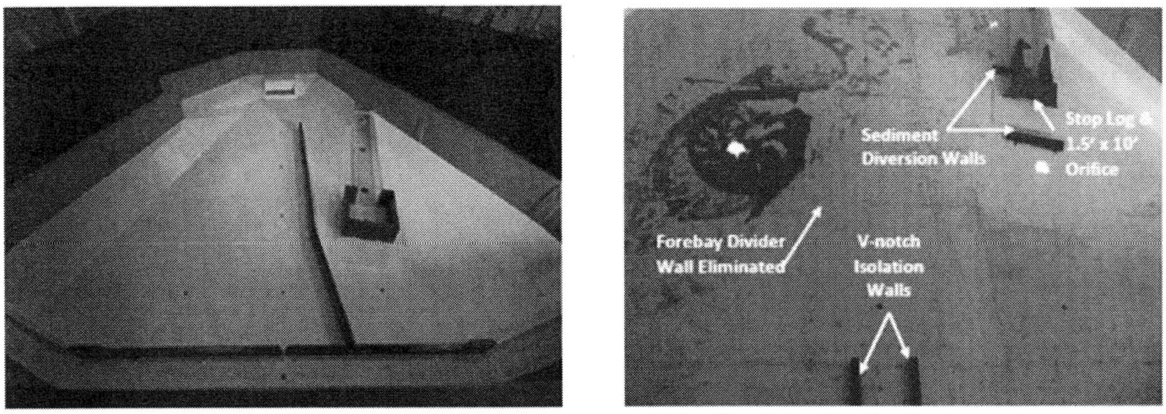

Figure 6. Photograph of the outlet control structure within the detention basin.

The final recommended design within the sediment basin, as developed in the model included the following modifications:

- Downstream weir lowered;
- Detention basin diversion wall removed;
- Three baffle blocks installed in the floor of the outlet control structure;
- Sediment diversion wall installed 6.10 m downstream of the outlet control structure end wall;
- Isolation walls installed on either side of the V notch to control sediment deposition;

To facilitate sediment removal the tunnel invert was raised by 0.31 m and fillets were added along the bottom of the box culvert. Any sediment that settled within the transition from the 4.27 m diameter tunnel to the box culvert was transported through the transition for all flows greater than approximately 8.50 m^3/s. The recommended V notch isolation walls were effective in minimizing the amount of sediment passing downstream resulting is sediment exiting the tunnel to deposit in the basin before the weir.

SUMMARY AND CONCLUSIONS

The multifaceted design effort using hydraulic design, 1D, transient, and physical hydraulic modeling was a successful and efficient approach to develop and document the complicated Minne Lusa tunnel system. Leaning on the strengths of this individual tools a design that satisfied capacity and stability requirements, minimized construction and maintenance costs was attained while the providing assurance that the highlest level of engineering analysis was adopted.

Using Computational Fluid Dynamics Modeling to Improve Flow Partition in a Combined Sewer Diversion Structure under Supercritical Flow

Yovanni A. Cataño-Lopera, PhD, PE, ENV SP, Stantec, Boston, MA.
David Bedoya, PhD, PE, ENV SP, Stantec, Boston, MA
Nicholas Anderson, P.Eng., Stantec, Boston, MA

ABSTRACT

Bottom slot racks are used in combined sewer overflow (CSO) diversion structures to convey dry and wet weather flows through an underdrain conduit to the intercepting sewer. Once the capacity of the underdrain sewer system is reached, excess flows are conveyed to facilities, such as storage tanks and deep storage tunnels, or directly to the CSO outfall. Often a bottom weir or splash plate is placed on or near the overflow sewer to maximize flows into the underdrain and minimize splash-over flows.

The design of these structures commonly relies on semi-empirical formulae, designer experience, and/or one-dimensional (1D) conveyance models. These approaches are not able to account for the effects specific geometric configurations have on localized head losses caused by such things as flow recirculation cells and air entrainment. This, in turn, decreases the amount of flow into the underdrain conduit.

In this study, the hydraulic behavior of a diversion structure was initially evaluated within InfoWorks ICM against flow meter data. The structure was comprised of an access manhole chamber, bottom slot rack, splash plate, a steep inflow pipe, underdrain, and overflow conduits. The 1D conveyance model did not fully capture the complex flow behavior at the diversion structure, which was augmented by high inlet velocities inducing a "flow leap" over the splash plate (or overflow weir) even before the capacity of the underflow was exceeded. The highly unsteady and irregular free-surface patterns at the interior of the structure cannot be accurately represented by conventional 1D conveyance models. This limitation of 1D hydraulic models may result in instances of gross miscalculation of offline storage systems devised to handle excess flows.

A computational fluid dynamics (CFD) three-dimensional model was used to evaluate the flow hydraulics at the diversion structure under several discharge rates representing free and pressurized flow conditions. This was compared against flow meter data. The CFD model was used to evaluate alternative geometric configurations of the splash plate within the CSO regulator structure to maximize flows through the slot rack and minimize overflows due to flow leaps. The resulting design is also intended to be represented in conveyance models more accurately and thus enhance their ability to calculate the flow partition between underflow and overflow conduits.

KEYWORDS

Combined Sewer Overflow, Passive Control, Bottom Slot, Numerical Modeling, CFD, Design.

INTRODUCTION

Hydraulic structures to control flooding are used extensively in urban and rural water conveyance infrastructure, such as weirs, relief canals and underdrains, flow diversion structures, and storage tanks. The flows at or in the proximity of these structures are typically highly turbulent and often characterized with changes between open channel (free surface) and pressurized conditions. Such turbulent flows frequently involve complex interactions between air and water (Lopes, P., 2013) as in the case of manholes with multiple in/out pipes, stepped spillways, and flow diversion structures. The latter structures are typically composed by an entrance manhole and inflow, overflow, and underflow conduits. The main design challenge of the diversion structure is the allowance of overflows only after underdrain capacity is exceeded, while minimizing head losses that reduce the underdrain flow capacity. A second challenge is the prevention of significant backwater effects.

In urban sewer infrastructure, debris often enters underground sewers through defects in catch basins and manhole covers. These objects may cause flow obstruction and increase the likelihood of street flooding. For this reason, bottom slot racks are typically used in combined sewer overflow (CSO) diversion structures to prevent major objects from entering underdrain conduits draining to the intercepting sewer.

Under extreme rainfall events, the conveyance capacity of the underdrain/downstream collection system may be exceeded and excess flows are conveyed to storage tanks, deep storage tunnels, or directly to the CSO outfall via the overflow conduit. In other cases, when the inflows are supercritical, as is the case of pipes in steep hills, an elevated plate is placed somewhere between the inflow and overflow sewers to act as a barrier to minimize the flow splash (or flow-leap), and in turn reduce overflows. The plate, also called a splash plate, can be curved or flat and placed at different elevations with respect to the bottom rack.

Historically, practitioners make use of laboratory tests or semi-empirical equations to determine suitable splash plate dimensions and configurations. In either case, it's extremely difficult to accurately model the hydraulic performance of diversion structures that include this type of structures when using one-dimensional (1D) conveyance models. Such models do not fully capture the complex flow behavior associated with highly unsteady and irregular free-surface patterns at the interior of the structure. Generally, 1D models assume flat water levels at a specific node based on incoming flows and backwater conditions. These limitations may result in significant divergence between modeled and measured overflows and potentially result in miscalculation of offline storage systems devised to handle excess flows.

In recent years, high resolution computational fluid dynamics (CFD) simulations have been used with great success in the study of environmental flows. Ramamurthy et al. (2007) applied a CFD based free-surface turbulence model to obtain flow parameters, such as pressure head distribution, velocity distribution, and water surface profiles, of a bottom slot located in an open channel. This study demonstrated that CFD modeling can be used to predict free surface flow characteristics with reasonable accuracy.

The open-source CFD software OpenFOAM is equipped with a well-designed C++ library that allows the numerical simulation of various engineering applications. Through its object-orientated structure and the open-source code concept, OpenFOAM is flexible and can be adjusted to very specific environmental fluid flow problems (Schulze and Thorenz, 2014). Bayón-Barrachina et al. (2015) used OpenFOAM to analyze flow partition at a South Valencia sewage collection system diversion. Wang et al. (2015) used OpenFOAM to study elevated total dissolved gas (TDG) formation under various spillway regimes.

OpenFOAM has been used extensively in coastal and ocean engineering applications, including flooding in urban coastal areas (Gallien et al., 2014). Several authors have explored the applicability of OpenFOAM in low-head dams (López-Egea et al., 2015), hydraulic jumps (Bayón-Barrachina, and López-Jiménez, 2015), and sediment scouring around hydraulic structures (Kazuyuki and Sato, 2015).

In the present study, OpenFOAM was employed to analyze the complex hydraulics within the diversion structure for several flow discharges representing underdrain free and pressurized flow conditions. The effect of the location of the splash plate on the flow partition were analyzed. The results indicate the current configuration of the splash plate is prone to induce undesired overflows. A second configuration, in which the splash plate was moved to the leading edge of the overflow conduit, was investigated and resulted in reduction of overflows while increasing underflows.

The main objective of this study is to advice engineers in the potential of CFD approaches to help understand the hydraulics associated with this type of structures and improve its representation when using 1D collection system models. The secondary objective is to propose a fully 3D CFD-based method to model the hydraulics of a typical collection system structure using the open-source platform OpenFOAM. The use of freely available open-source codes allows a continuous community-based improvement of the model, and avoids the necessity of costly software license fees.

METHODOLOGY

A detailed description of the field conditions, available measured data, and the numerical model used to simulate existing and alternative configuration of the CSO diversion structure is presented below.

Field Conditions and Measurements

Collection of flow monitoring data on the inflow and underflow pipes was required to calibrate and validate the collection system 1D model. Two flow meters were deployed near a diversion structure (the subject of this study). The location of flow meters M1 (inflow) and M2 (underflow), located upstream and downstream of the diversion structure, are shown in Figure 1. The diversion structure receives storm and sanitary flows from a catchment area of about 3.4 km^2 (830 acres) composed primarily of industrial and residential land uses. The majority of pipes tributary to the diversion structure are circular with diameters ranging from 20 cm to 152 cm and slopes ranging from 0.5% and 30%.

The inflow meter was placed at a distance of approximately 30 conduit diameters to ensure measurements were not affected by the local diversion structure hydraulics. The underflow meter was installed about 3 m downstream the diversion structure. The flow measurement uncertainty is estimated to be ±5% of the measured flow discharge.

Figure 2 shows plan and side views of the configuration of the diversion structure composed by inflow, underdrain and outfall sewers, access manhole, bottom slot rack, and splash plate.

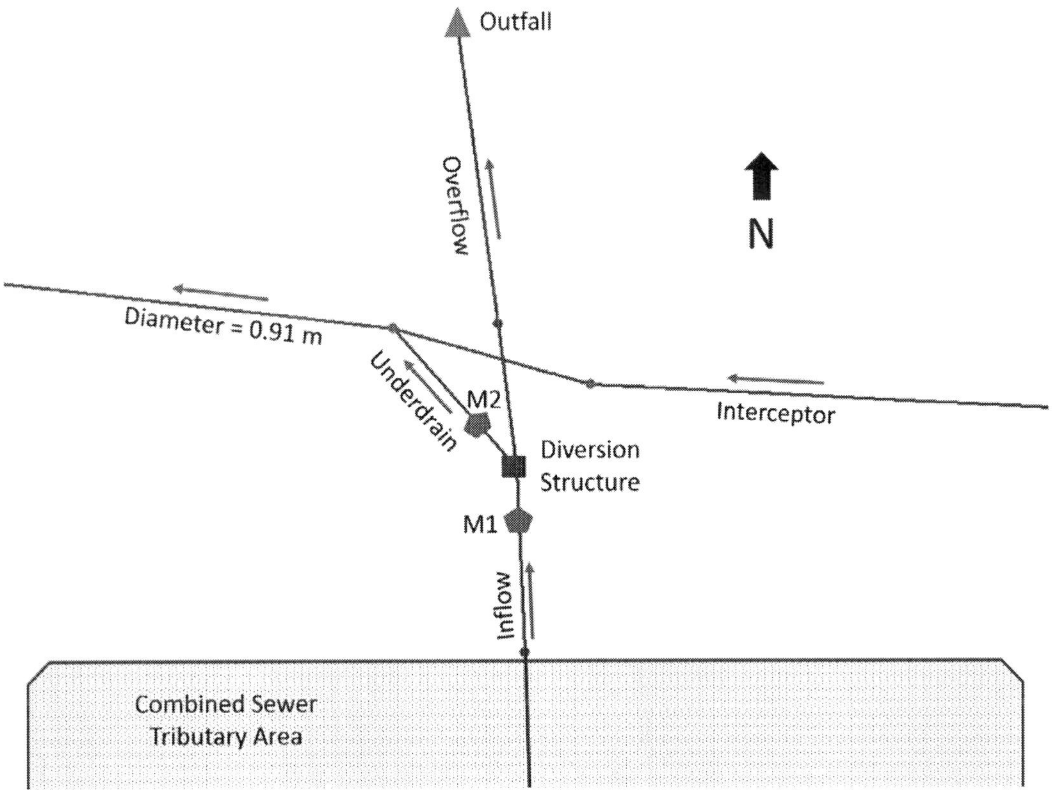

Figure 1. Hydraulic model schematics of CSO diversion structure, inflow, underdrain, and overflow sewers. The location of flow meter on the inflow conduit is shown.

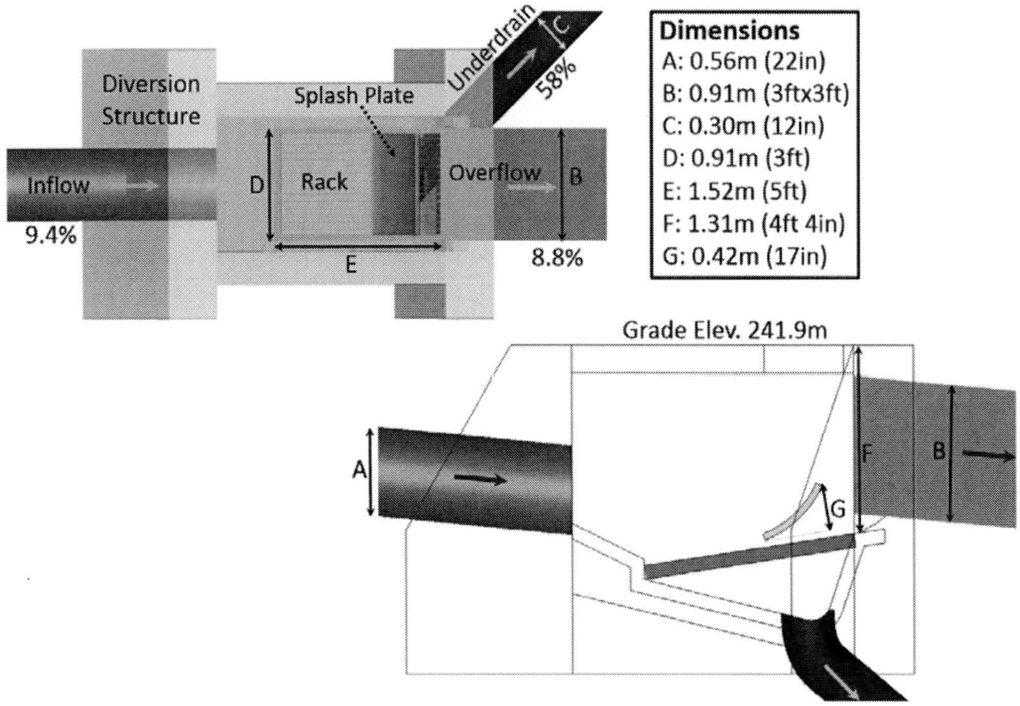

Figure 2. Plan and profile views of 3D rendering of the diversion structure with bottom slot rack and splash plate.

Assuming a Manning's roughness coefficient of 0.015 for the inflow and underdrain conduits, the full flow capacity is estimated as 1.34 m^3/s and 0.67 m^3/s, respectively. Table 1 summarizes the flow meter characteristics and recording period that was available for analysis. Figure 3 shows recorded inflow and underdrain hydrographs during the 01/03/2015 storm event, the largest storm event recorded during the period. During this storm event, the maximum inflow is 0.47 m^3/s while the peak underflow is 0.44 m^3/s, both smaller than the underdrain flow capacity. The walls of the diversion structure are made of aged masonry brick, and the inflow and underdrain sewers are made of concrete.

Table 1. Description of flow meter locations near the diversion structure.

Flow Meter	Start Date	End Date	Conduit Dimensions (cm)	Conduit Slope (m/m)	Peak Inflow (m^3/s)	Remarks
Meter 1	10/21/2014	01/08/2015	55.9	0.09	0.47	Inflow Sewer
Meter 2	01/03/2015	01/19/2015	30.5	0.58	0.44	Underdrain Sewer

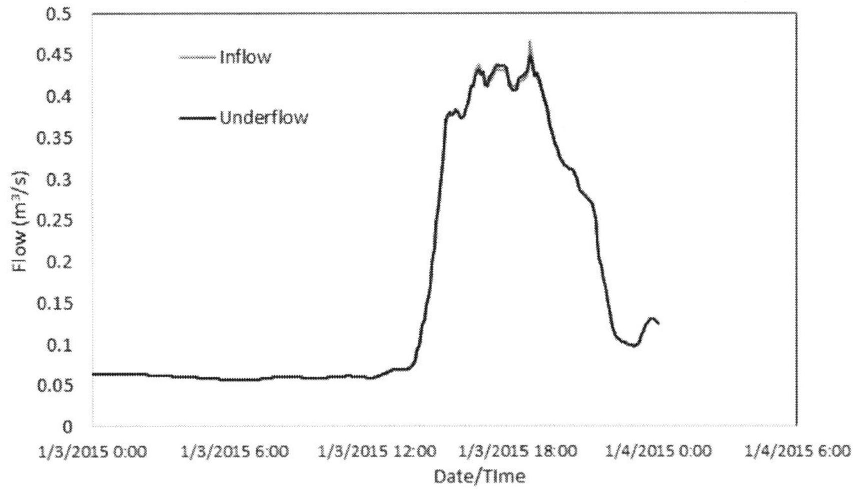

Figure 3. Recorded underdrain hydrograph during the January 3, 2015 storm event.

The calibrated 1D model was used to simulate the flow partition at the diversion structure during other storms events outside the period with available record data. During some of these storm events, the model did not estimate overflows even though overflows were visually documented in the field. The 1D model fails to reproduce documented overflows primarily due to its inability to accurately account for the complex flow dynamics induced by the interaction of the inflow and geometric configuration of the diversion structure.

At the interior of the diversion structure, the flow is highly turbulent and exhibits high deformation of the free surface. It is suspected that due to the high incoming velocities incoming flows "leap" over the splash plate even when there is available flow capacity in the underdrain conduit. To confirm the latter is the case, a 3D numerical model of the diversion structure and inflow and outflow conduits was developed.

Numerical Model

High resolution numerical simulations were conducted using a 3D non-hydrostatic model coupled with a RANS-based κ-ε turbulence closure scheme implemented in the control-volume open-source code OpenFOAM. The numerical approach used in this study resolves the interface between two (e.g., air and water) or more immiscible fluids by solving a single set of momentum equations and tracking the volume fraction of each fluid throughout the domain using the volume of fluid (VOF) model.

Typical applications include the motion of large bubbles in sludge flows (Taha and Cui, 2006), highly turbulent dambreak two-phase flow propagation, and the steady or transient tracking of any liquid-gas interface, such as free surface flows (Ferrari et al., 2008). Each fluid considered in the simulation is treated as a separate phase, and in each control volume, the volume fractions of all phases sum to unity. The fields for all variables and properties are shared by the phases and represent volume-averaged values as long as the volume fraction of each phase is known at each location.

The PIMPLE algorithm, which is a mixture of the pressure implicit with splitting of operators (PISO) and semi-implicit method for pressure-linked equations (SIMPLE) algorithms, was used to couple pressure and velocity. At each iteration, the pressure distribution is calculated based on the local velocity field and volume fraction values to satisfy both momentum and continuity equations. Therefore, if the volume fraction value in a cell is equal to 0, this cell is occupied by the air phase, and the momentum equation is solved for air. Depending on the velocity and pressure values in surrounding cells, the PISO scheme is used for correction of pressure in this cell.

The Euler blended Crank-Nicholson method was used to discretize the temporal term. This is a first-order, bounded implicit method. The convective terms were discretized using Gauss upwind, which refers to the standard finite volume discretization of Gaussian integration. The Laplacian terms used Gauss linear corrected interpolation to calculate values from cell centers to face centers. Due to the transient nature of the problem, an adaptive time-stepping technique based on the Courant number was employed. A maximum Courant number of 0.75 was adopted in this study.

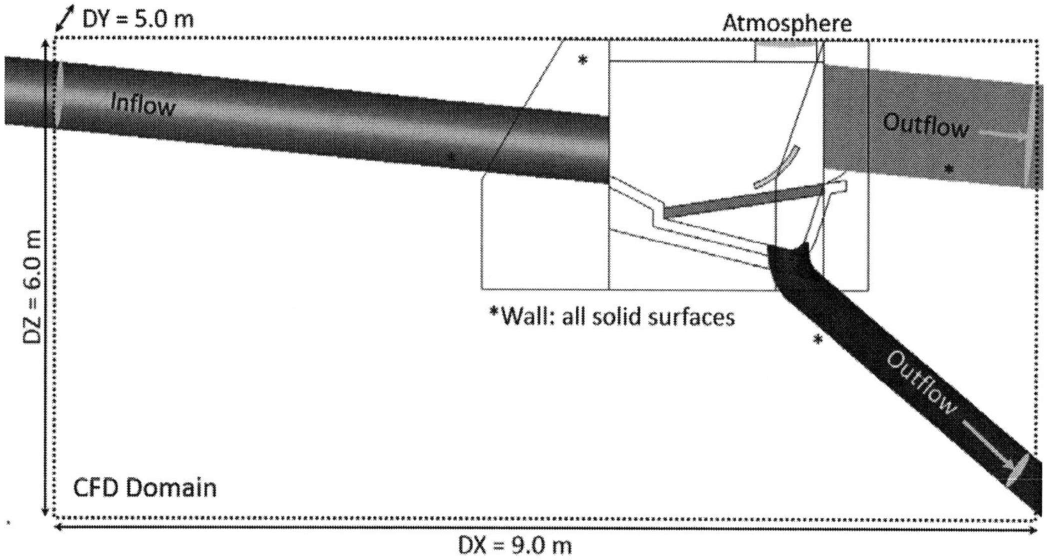

Figure 4. Configuration of diversion structure and inflow and outflow conduits. The box encloses the extents of the numerical model.

The model used a body-fitted mesh consisting of hexahedral elements. The dashed box in Figure 4 exhibits the extent of the computational domain. As noted, the domain encapsulates a limited region of the entire diversion system shown above in Figure 2 due to the highly computational demands in terms of memory and simulation times. The CFD model domain covers an outer bounding box of dimensions 9.0 m x 5.0 m x 6.0 m. A mesh size sensitivity analysis was conducted for which mesh independency results were attained for a computational grid contains approximately 346,000 elements with average cell sizes ranging from 1 to 6 cm. Further mesh refining was required to fully resolve each individual bar defining the bottom slot rack and splash plate for which cell elements range from 0.5 cm to 2 cm.

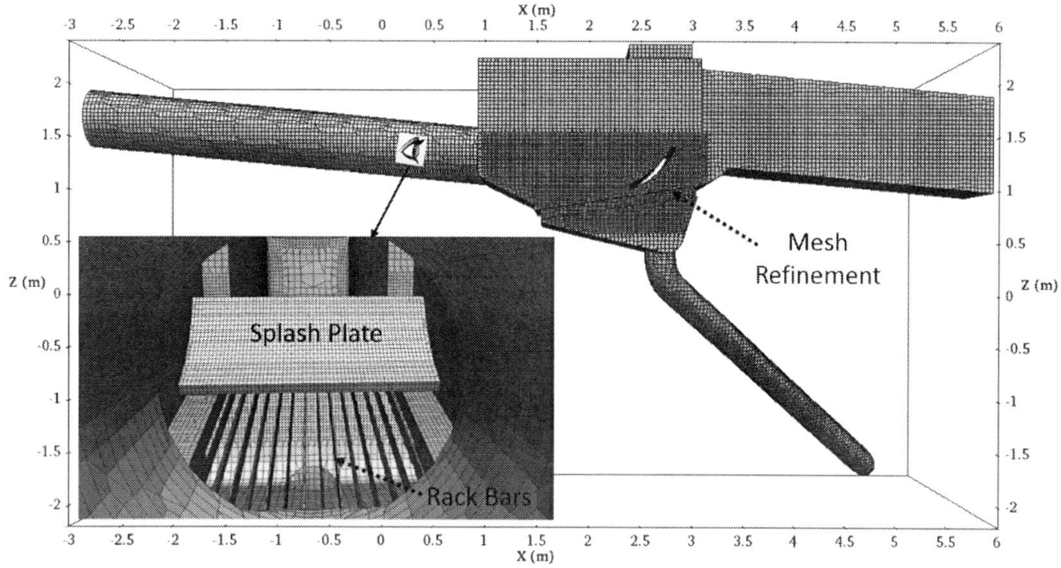

Figure 5. Numerical grid used in to resolve the diversion structure, including the bottom slot rack and splash plate.

Simulations were conducted on an 8-core cluster for at least 70 seconds until steady state conditions were achieved. The model automatically calculates flow discharges through the inflow, overflow, and underdrain conduits.

The following are the boundary conditions applied to the numerical model:

- Inlet: Constant flow rate and turbulence intensity at inlet were set at 1%-5% and applied at the inflow sewer inlet for the bottom area of the inlet representing water phase. The upper region of the inlet was set as air with 0 inlet velocity, and a 0 static pressure (representing atmospheric conditions).
- Outlet: Continuative, zero (0) static pressure head boundary condition was set at the outlets of both underdrain and overflow sewer conduits. Zero pressure and velocity gradients were imposed as well. It is assumed that no flow restriction occurs either on the underdrain or receiving interceptor (Figure 1).
- Walls: Interior walls of structures were modeled as no-slip boundary conditions and high-Reynolds rough wall function. No fluid was allowed to pass through this face. A roughness height corresponding to concrete was used to characterize the walls.
- Top of domain: Atmospheric boundary condition was set at the top of the access manhole. Air or water can freely leave the domain with zero gradients.

RESULTS

The numerical model was run during the period covering the peak underflow recorded during the 1/3/2015 storm event (e.g., 0.44 m^3/s), and no overflow was calculated. Next, the 3D numerical model was run for a set of eight inflows ranging from 0.25 m^3/s to 2.50 m^3/s to investigate the flow partition between underdrain and overflow with the splash plate positioned in the current field conditions. Simulated underflows and overflows are listed in Table 2. Overflows appear for inflows barely below 0.75 m^3/s and increased progressively for increasing inflows.

The inflow is supercritical under both open and closed conduit conditions due to the relatively large sewer slope (i.e., 9.4%). The flow structure within the diversion structure is rather complex, and due to its incoming speed, the inflow bounces off the bottom and walls of the structure, making it highly turbulent with a highly irregular free surface. Thus, some flow pockets detach from the bulk of the flow and leap over plate, resulting in overflow augmentation.

Figure 6 shows surface water levels within and near the diversion structure for an inflow of 1.0 m^3/s. The flow exits the inflow pipe and free-falls into the diversion structure exhibiting high velocities in excess of 5 m/s. The highly complex flow structure and irregular free surface configuration at the interior of the diversion structure can be seen. A small average overflow of 0.08 m^3/s occurs as the momentum of the inflow induces flow leaps and causes the water levels to overtop the invert of the overflow conduit. Pockets of flow are pointed along the overflow sewer in Figure 6. Even though overflows are simulated, the underdrain flow capacity has not yet been exceeded, an undesired condition.

Simulations show that for a range of transitional inflows, between about 1.5 m^3/s and 2.0 m^3/s, the splash plate in its current position prevents some of the flow splash from reaching the overflow conduit. On the other hand, for higher flows (i.e., > 2.0 m^3/s) both rack and splash plate become completely drawn. Therefore, internal flow and free surface flow irregularities are dampened compared to the transitional flows case, and the maximum head driving the magnitude of the underflow is controlled primarily by the upstream invert of the overflow conduit.

The computational domain is meshed with hexahedral elements using the snappyHexMesh mesh generation toolset accompanying OpenFOAM. This toolset requires an existing mesh block, an OBJ/STL file defining the interior surfaces of the flow domain (i.e., the interior of the diversion structure and inlet/outlet conduits), and edge features describing sharp corners. The numerical grid is shown in Figure 5.

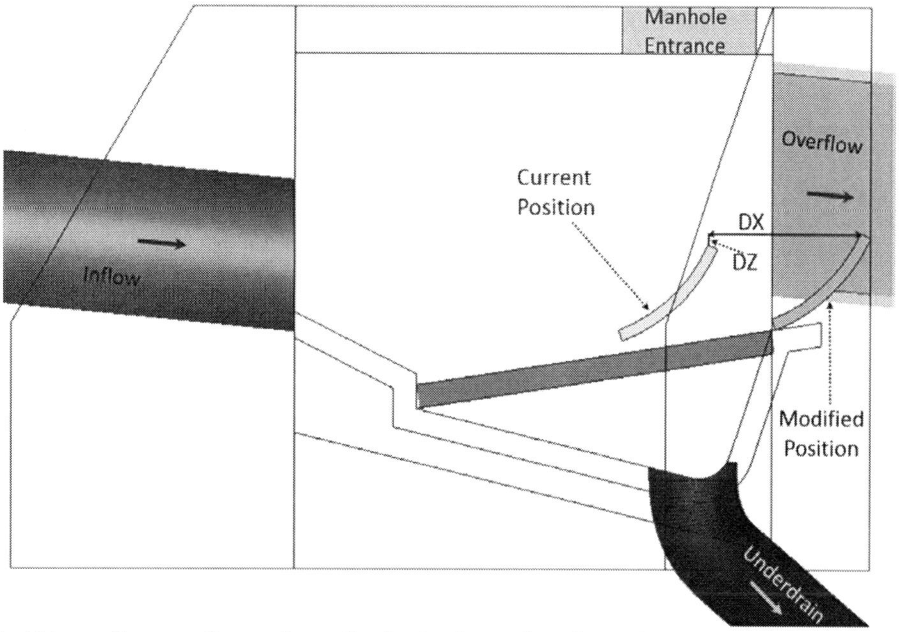

Figure 6. Alternative configuration of splash plate; the plate sits at the leading edge of the overflow conduit (DX = 0.70 m, DZ = 0.05 m).

Table 2. Simulated overflows for different inflows with Splash Plate under Current and Modified Positions

	Current Position		Modified Position		
Inflow (m3/s)	Underflow (m3/s)	Overflow (m3/s)	Underflow (m3/s)	Overflow (m3/s)	% Overflow Reduction
0.25	0.25	0.00	0.25	0.00	0
0.50	0.50	0.00	0.50	0.00	0
0.67	0.67	0.00	0.67	0.00	0
0.75	0.746	0.0004	0.75	0.00	100
1.00	0.92	0.08	1.00	0.00	100
1.50	1.30	0.20	1.498	0.002	99
2.00	1.42	0.58	1.59	0.41	30
2.50	1.67	0.83	1.98	0.52	37

To reduce undesired overflows, the position of the splash plate was altered to rest over the invert of the overflow conduit. The invert of the plate was moved by 0.70 m and 0.05 m in the horizontal and vertical directions, respectively. Figure 6 shows the current and modified locations of the splash plate.

A set of eight simulations were carried out to investigate the flow partition between overflow and underflow under the modified splash plate configuration. Columns 3 and 4 in Table 2 list the resulting underflow and overflow distributions. The simulation results indicate that the new splash plate configuration attains a higher level of overflow control. Note that no overflows occur under this configuration for an inflow of 0.75 m^3/s, resulting in a net overflow reduction of 100% with respect to the current field configuration. Figure 7 shows that even as the water levels within the diversion structure are higher than the invert of the leading edge of the overflow sewer, the presence of the splash plate prevents the overflow. Overall, flow reduction decreases with increasing inflows, varying from 100% to 30% for inflows of 0.75 m^3/s and 2.0 m^3/s, respectively.

DISCUSION

The numerical results verified that the current field configuration of the splash plate is non-optimal as overflows occur even when there is still underdrain flow capacity. An alternative splash flow configuration was evaluated with the 3D numerical model. The simulation results indicate the modified splash plate configuration induces significant overflow reduction for the range of inflows tested. Additional splash plate configurations, including modifications to the plate's height and orientation with respect to the inflows, could be investigated to further reduce overflows. The addition of a bottom weir at the leading edge of the overflow conduit could also be investigated; however, this is left for future studies.

The 1D model is not able to accurately estimate the reduction in overflows induced by the modification of the splash plate without relying on model recalibration using experimental data. Although it is possible to mimic the revised splash plate position by setting a weir with crest elevation equating that of the top of the splash plate at the leading edge of the overflow pipe, there is not direct way to account for the elimination of the initial position of the plate. On the other hand, the results of the 3D numerical simulation can be used to recalibrate the 1D model and enhance its reliability at estimating flow.

Refined 3D numerical simulations can be conducted at a fraction of the cost of expensive flow monitoring campaigns. Any alternative splash plate configuration can then be directly implemented in the field. It is recommended to frequently monitor and maintain the diversion structure (e.g., removal of blocking debris and objects) to maximize the underdrain flow capacity and thus reduce overflow occurrences.

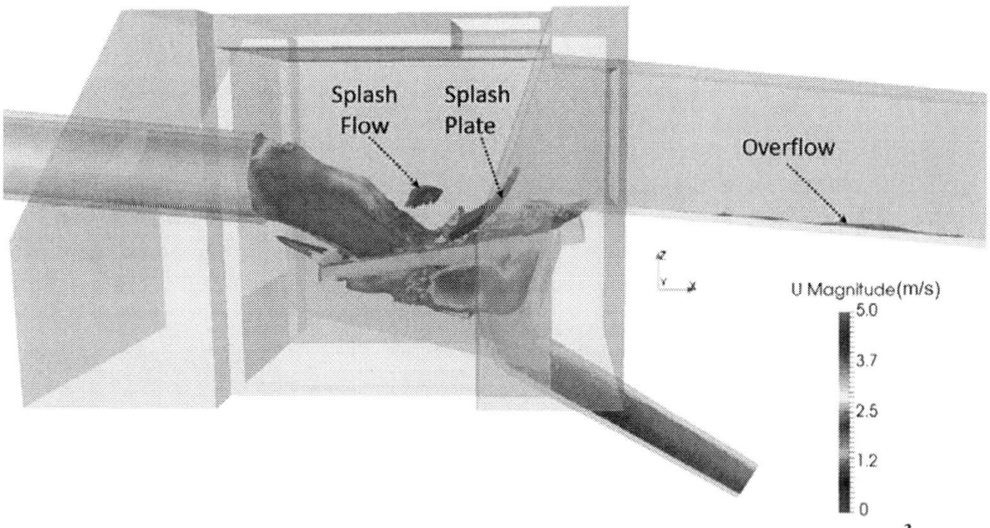

Figure 6. Simulated flow velocity magnitude overflow for the case of Inflow of 1.0 m^3/s (22.8 MGD) with splash slate in current position.

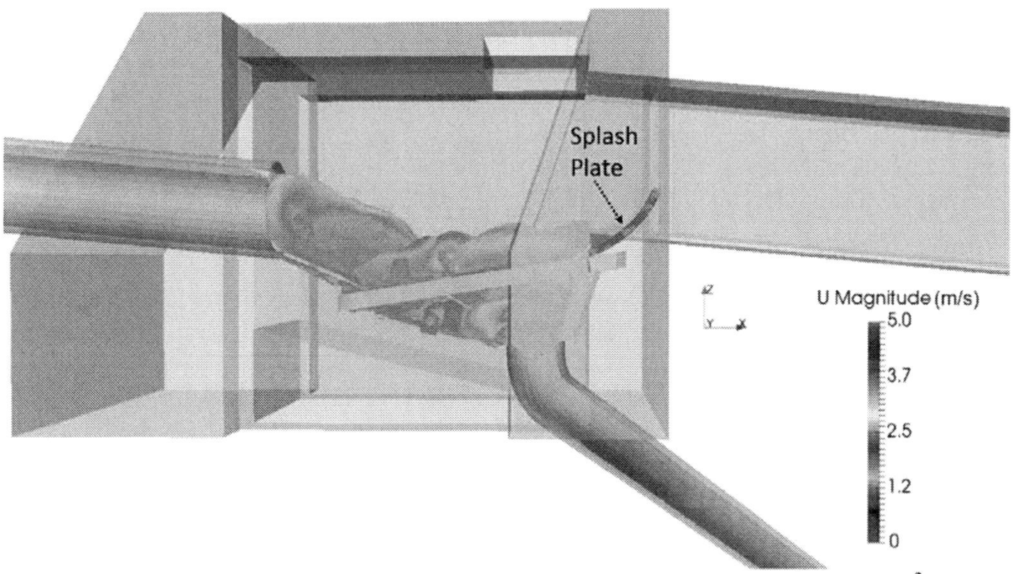

Figure 7. Simulated flow velocity magnitude overflow for the case of Inflow of 1.0 m^3/s (22.8 MGD) with splash slate in modified position.

CONCLUSIONS

This study used a combination of measured flow data and 1D and 3D numerical models to analyze the hydraulics of a CSO diversion structure equipped with a bottom slot rack and splash plate to minimize overflows. Simulation results allowed identification of flow-pathways underneath the splash plate that resulted in undesired augmentation of overflows. Under the current field configuration of the splash plate, the numerical simulations show splash-over flow induced by the fast approaching inflow due to the steepness of the inflow pipe. A second configuration of the splash plate, in which the device was placed at the leading edge of the overflow sewer, was investigated. Simulation results indicate the second configuration is better suited for reducing overflows without causing backwater effects on the upstream sewer system.

The developed CFD model proved useful to evaluate potential geometric modifications of the splash plate to minimize overflows. The results of the 3D numerical simulation can be used to recalibrate the 1D model and enhance its reliability at estimating flow partition between underflow and overflow. The simulation results can be used to develop stage-discharge relationships and average flow discharge coefficients that can be readily implemented in collection system 1D models. Refined 3D numerical simulations can be conducted at a fraction of the cost of often expensive flow monitoring field campaigns.

ACKNOWLEDGEMENTS

This study was made possible by the sponsorship of Stantec.

REFERENCES

Bayón-Barrachina, A., and López-Jiménez, P.A., (2015). "Numerical analysis of hydraulic jumps using OpenFOAM." Journal of Hydroinformatics, Vol. 7 (4), pp. 662-678.
Bayón-Barrachina, A., Vallés-Morán, F.J., and López-Jiménez, P.A. (2015). "Numerical analysis and validation of South Valencia sewage collection system diversion." 36th IAHR World Congress, The Hague, The Netherlands.
Ferrari, G., Politano, M., and Weber, L., 2009. "Numerical simulation of free surface flows on a fish by-pass." Computers and Fluids, Vol. 38 (5), pp. 997-1002.
Gallien, T., Sanders, B., and Flick, R. (2014). "Urban coastal flood prediction: Integrating wave overtopping, flood defenses and drainage." Coastal Engineering, Vol. 91, pp. 18-28.
Kazuyuki O, and Sato, T., (2014). "Experimental and numerical study of the local scour caused by sediment releasing through a dam gate." Journal of Japan Society of Civil Engineers, Vol. 3, pp. 184-190,
Lopes, Pedro, (2013). "Free-surface Flow Interface and Air-Entrainment Modeling using OpenFOAM." PhD Dissertation, Universidad de Coimbra, Portugal (URL: Pedro Lopes_ Free-surface flow interface and air-entrainment modeling using OpenFOAM_FINAL.pdf).
López-Egea, M., Nistor, I., Towsend, R., Paudel, B., and Sullivan, P., (2015). "An experimental and numerical study of submerged hydraulic jumps forming at low-head dams." E-proceedings of the 36th IAHR World Congress, 28 June – 3 July, The Hague, the Netherlands.
Ramamurthy, A., Qu, J., and Vo, D. (2007). "Simulation of Flow past an Open-Channel Floor Slot." J. Hydraul. Eng., 133(1), pp. 106–110.
Schulze, L., and ThorenzI, C., (2014). "The Multiphase Capabilities of the CFD Toolbox OpenFOAM for Hydraulic Engineering Applications." CHE 2014, Hamburg - Lehfeldt&Kopmann (eds), Bundesanstaltfür Wasserbau ISBN 978-3-939230-32-8.
Taha, T., and Cui, Z.F., 2006. "CFD modelling of slug flow in vertical tubes." Chemical Engineering Science, Vol. 61, pp. 676-687.
Wang, Y., Politano, M., Laughery, R., and Weber, L., (2015) "A Model development in OpenFOAM to predict spillway jet regimes." Journal of Applied Water Engineering and Research, Vol. 3 (2), pp. 80-94.

You Can Dig It, But Can You Operate It? Proactive CSO Tunnel Operational Planning

Jessica Bastin, PE[1], John Trypus, PE, LPG[1], Jeff Hansen[1], Emily Damian[1], Christopher J. Ranck, PE, BCEE, D.WRE, ENV SP[2*]

[1]Citizens Energy Group, Indianapolis, Indiana
[2]Arcadis, Indianapolis, Indiana
*Email: chris.ranck@arcadis.com

ABSTRACT

In 2011, Citizens Energy Group began construction of the Dig Indy Deep Rock tunnel system, which includes 47 km (29 miles) of tunnels with a finished internal diameter of 5.5 meters (18 feet). This system is the critical part of Citizens' combined sewer overflow (CSO) consent decree, which is required to be fully operational by the end of 2025. Due to the phased implementation of tunnel design and construction, Citizens initiated a focused, proactive operational planning to ensure that there are no loose ends between multiple design contracts and construction phases. The operational planning included:

- Development of a SCADA strategy to connect all drop shaft sites to the treatment plant network via fiber optic cable;
- Development of an instrumentation strategy to reliably monitor the tunnel as part of surge mitigation;
- Development of a conceptual tunnel dewatering strategy;
- Physical modeling of drop and ventilation shafts;
- Development of an air management tool to evaluate forward air induction, air pocket pressurization, and air pocket migration based on current research.
- Development of a tunnel ventilation strategy based on the physical modeling and air management tool. The strategy focuses on routing air to working and retrieval shafts that are larger than the tunnel diameter;
- Odor Dispersion analysis for both H_2S and ammonia for the end shaft of the Eagle Creek tunnel branch that will be located in a city park; and
- Tours of established CSO tunnel systems and collaboration with other design and operations professionals.

The work in developing the Dig Indy operational plan is derived both from highly technical sources, as well as boots-on-the-ground practical experience. Each element of the plan has helped to minimize Citizens' long-term risk in operating the tunnel system.

KEYWORDS: Tunnel, CSO, Surge, Air Movement, Odor Control

INTRODUCTION

The City of Indianapolis is a combined sewer overflow (CSO) community with approximately 130 outfalls distributed over the following six major receiving streams: White River, Fall Creek, Pleasant Run and Bean Creek, Pogues Run, and Eagle Creek. The watersheds for each receiving stream have significant areas within and upstream of the Indianapolis combined sewer area. Figure 1 presents the Indianapolis combined sewer system area, interceptors, and treatment plants. Wastewater is treated at two locations, the Belmont and Southport Advanced Wastewater Treatment (AWT) Plants, which discharges to the White River.

THE GRAPHIC ON THIS PAGE WAS NOT AVAILABLE ON THE E-MEDIA VERSION OF THIS PAPER.

Figure 1: Indianapolis Combined Sewer Outfalls and Combined Sewer Area

On August 26, 2011, the wastewater collection and treatment system assets were transferred from Indianapolis to Citizens Energy Group and CWA Authority, Inc. was formed for the wastewater system. In this manuscript, all activity relevant to the Dig Indy tunnel system is referred to as Citizens' work, including work performed before August 26, 2011 by Citizens' predecessor, Indianapolis Department of Public Works.

Figure 2 presents the overall tunnel alignment. The tunnel segments that make up the Dig Indy system are in various stages of design and construction, with the mining operations and the tunnel lining complete for the Deep Rock Tunnel Connector (DRTC) and Eagle Creek branches. The two completed branches are in orange and red in Figure 2.

Figure 2: Dig Indy Overall Tunnel Alignment

At the time of this manuscript, mining is 50% complete for the Lower Pogues Run branch, which is shown in yellow. Design is complete for the White River branch, which is shown in light blue on Figure 2, and design is nearly complete for the Fall Creek branch, which is shown in green. Finally, design for the Pleasant Run branch is approximately 50% complete, which is shown in dark blue.

Figure 3 presents the tunnel boring machine breakthrough for the Deep Rock Tunnel Connector. The tunnel system is the primary infrastructure required by the Indianapolis Long-Term Control Plan (Indianapolis, 2006), and was extended as part of modifications to the Indianapolis consent decree (Ranck et. al, 2012) for a required design criteria of 946,000 cubic meters (250 MG) of combined sewage for treatment at the Southport AWT Plant.

Figure 3: Dig Indy Tunnel Breakthrough – Deep Rock Tunnel Connector

Citizens has implemented a plan for construction award of all remaining segments (Citizens, 2016), and has recently awarded the contract. With phased implementation of the tunnel system, it was necessary for a holistic operational plan of the tunnel system to be developed by Citizens. The following sections of this manuscript discuss the methodology in developing the operational plan, the resulting operational plan, and next steps for Citizens.

METHODOLOGY

This section discusses the operational planning approach utilized by Citizens for each aspect of the operational plan. Specific recommendations and results are presented in the following section. The methodology for our specific technical items are discussed in this memorandum:

SCADA connectivity for drop shaft sites, Shaft level monitoring for inflow control, Dewatering of the tunnel, Ventilation of the tunnel, Physical modeling of shafts, Odor control at shaft sites, and Tours of established tunnel systems. Each is discussed in their respective subsections.

SCADA Connectivity

The tunnel operation plan will be largely dependent on the use of inflow gates to control the volume in the tunnel and mitigate surge conditions. For this reason, a reliable connection between tunnel inflow gates and the SCADA system is of high importance. The SCADA system will allow operators to remotely monitor and control the tunnel, ensuring that it works in harmony with the rest of the collection system. Therefore, in choosing a medium for transmitting SCADA signals, an emphasis must be placed on the reliability of the system during these events (Citizens, 2016). While radio telemetry was initially planned to connect the shaft sites, the first phase of operational planning identified the opportunity to utilize fiber optic cable to connect the shaft sites to the control center at the Belmont AWT Plant (Citizens, 2014). Following the first phase of operational planning, Citizens performed a business case analysis for fiber optic cable, and has developed the performance specification for the fiber optic network. Installation of the fiber in the DRTC and Eagle Creek tunnel segments will begin after the lining operation is completed.

Shaft Level Monitoring

With the shaft sites connected to the control center at the Belmont AWT Plant, the next aspect of operational planning is where will the signal come from to close inflow gates at shaft sites and what technology will be used at the monitoring location or location(s). Candidate locations include the pump station shaft, intermediate working shafts, and drop shafts. As part of the its tunnel surge modeling work (Ridgway et. al, 2011 and Rowe et. al, 2015), Citizens identified that a single tunnel shaft, an intermediate working shaft in the Fall Creek branch could be adequate for control decisions on inflow gates. However, Citizens will also utilize additional monitoring locations for reliability and redundancy.

Tunnel Dewatering

From Citizens' amended consent decree, the tunnel system design criteria requires that 946,000 cubic meters (250 MG) of combined sewage can be reliably stored and dewatered during a wet-weather event or multiple events, provided sufficient precipitation has occurred (Ranck et. al, 2012). Since Citizens has proactively committed to construct approximately 110,000 cubic meters (290 MG) of storage through value engineering initiatives (Citizens, 2016), Citizens has significant flexibility in developing a strategy to dewater the tunnel to maintain available storage, energy efficiency, and pump operation.

Tunnel Ventilation

Air management has become a recent area of focus for tunnel operations due to geyser events in a handful of tunnel systems (Wright et. al, 2011). Citizens' plan for ventilation of the tunnel during wet-weather events has evolved since the initial DRTC design and surge modeling in

2010 (Ridgway et. al, 2011). Significant numerical and physical modeling activity has taken place, as well as comparing to current research and ventilation approaches for other tunnel systems.

As shown in Figure 2, the Eagle Creek tunnel branch will end with a drop shaft to collect CSO flow and a ventilation shaft to deaerate the CSO flow as well as the tunnel branch. The Iowa Institute of Hydraulic Research (IIHR) evaluated vortex and plunge drop shafts for the upstream end of the Eagle Creek tunnel. Their evaluation included several simulations of potential geysering from the 8' ventilation shaft (IIHR, 2016). Due to the potential operational risks of ventilation, a spreadsheet tool was developed to post-process surge model output and evaluate the potential for forward air induction, air pocket migration, air pocket pressurization, and shaft ventilation (Citizens, 2016). The spreadsheet tool allows Citizens staff to periodically evaluate tunnel ventilation without needing a proprietary study. Table 1 presents the dimensionless numbers utilized in Citizens' tool.

Number	How it's Calculated	What it Means
Vol*	Pocket / D^3	Ratio of air pocket to tunnel size
Q*	Q / √(g * D^5)	Surge Screening - Q* > 0.5 expected surge issues
D*	Dvent / D	Ratio of vent shaft restriction to tunnel size
V*	V / √(g * D)	Ratio of air pocket velocity to tunnel size
H*	Air pressure / D	Ratio of air pocket pressure to tunnel size
P*	P / [pgD * (A^2/(Ao * (A- Ao)))]	Ratio of surge pressure to tunnel size and size of the bore in front of the air pocket. A is the tunnel cross-sectional area and Ao is the area of water ahead of the bore
Va / Vw	(-1 + √a) / (a - 1)	Ratio of inducted air velocity to water velocity

Table 1: Dimensionless Number Summary for Air Management Tool

The above numbers are derived from equations presented in research papers for forward air induction (Brocard et. al, 2012), air pocket migration (Benjamin, 1968 and Vasconcelos and Chosie, 2013), and surge pressures due to trapped air pockets (Wright et. al, 2011). The calculations are based on tunnel geometry and flow, depth, and velocity. To confirm the findings' of Citizens internal tool, a follow-up study was performed with Auburn University (Auburn, 2016) to evaluate the potential for geysering in the Eagle Creek ventilation shaft and the three utility shafts in the DRTC.

Odor Control Evaluation

As shown in Figure 2, tunnel ventilation may occur at shafts in or near public property. Because of this an odor modeling evaluation was performed for the Eagle Creek ventilation shaft that has been constructed in Ross Claypool Park (Citizens, 2015). Both hydrogen sulfide (H_2S) gas and ammonia were evaluated. After reviewing industry standard approaches to calculate H_2S emissions and the conservativeness of the assumptions used in those approaches, the equation used in the EPA design manual document (US EPA 1985) was selected as being appropriate for the Eagle Creek shaft. This method has been utilized in the Wastewater Aerobic/Anaerobic Transformation in Sewers (WATS) concept and was confirmed by utilizing a second method based presented in Influence of Wastewater Constituents on Hydrogen Sulfide Emission in Sewer Networks (Yongsiri et. al, 2005). The dispersion evaluation was subsequently performed using AERMOD, a steady-state Gaussian plume model that simultaneously simulates pollutant concentrations from a single source or a variety of sources, and is the appropriate tool for evaluating air discharges from the Eagle Creek shaft (US EPA, 2005).

Site Tours

Citizens staff performed site tours of the Tunnel and Reservoir Program (TARP) in Chicago, IL and the Monroe County Pure Waters tunnel system in Rochester, NY. Additional site tours are planned for 2017.

RESULTS

This section discusses Citizens' results for each of the subsections discussed in the methodology section, and represents the details of the operational plan developed to date (Citizens, 2016).

SCADA Connectivity

Table 2 presents the initial business case evaluation for the use of fiber optic cable vs. radio telemetry. As shown in the table, the average total capital cost estimate for fiber is approximately 7% higher than the cost for radio telemetry. However, Citizens would have full ownership of the fiber optic network connecting the drop shaft sites to its SCADA system. While the two options have similar capital costs, the added resiliency of fiber during storm events, and the potential for future application make fiber cable the better choice.

The connectivity between tunnel components and the SCADA system is critical in a successful tunnel operating plan. In choosing a transmission medium, a number of criteria were considered; the reliability of the signal is of the upmost importance, however, potential for future collaboration with other utilities and possible revenue streams was weighed as well. These added benefits of fiber cable out-weigh the marginal cost increase, and therefore make it the most viable solution for Citizens (Citizens 2016). Following the acceptance of the business case, Citizens has conceptually planned the fiber network, developed the performance specifications, and will proceed with installation of the fiber in constructed tunnel segments after the lining is completed.

	Radio/Phone Signal	Fiber Cable
Capital Costs:	- $25K/ site = $925,000	- 52 km cable @ $18.93 / m =$991,500
Reliability	- Possibly jeopardized during storm events due to heavy traffic - Equipment susceptible to weather conditions and geographic obstructions	- Provides dedicated service that is not shared by other users - Not impacted by weather. Once installed, no new geographic obstructions
Operation and Maintenance	- Easily accessible if in need of repair - Costs will increase over time due to provider fees	- Sound construction with all splices placed at surface will facilitate access for testing and future connections - No provider fees exist
Ownership and Security	- Service will be owned by 3rd party provider and Citizens will share ownership of service with public network	- Citizens will have private network and not rely on 3rd party
Data Constraints	- Limited capacity. May not meet future needs?	- Unlimited capacity. Will not limit future upgrades on sending/receiving side.
Connectivity and Operation	- Signals need to be converted from Ethernet to radio telemetry	- Connects directly to programmable controllers - Supports localized Wi-Fi to assist in on-site data collection
Potential for Funding and Collaboration		- Other utilities can benefit from fiber network. Also creates potential to expand to other sewer networks and pump stations.

Table 2: Business Case Evaluation for SCADA Connectivity (Citizens, 2016)

Shaft Level Monitoring

From a review of established tunnel systems, Citizens recommend that monitoring for control decisions such as inflow gate closure and tunnel dewatering, that a primary contact method (pressure transducer or bubbler tube) with physical float switches as redundancy is used to determine critical elevations. For the DRTC Screen & Grit Shaft, Citizens will be installing a bubbler tube for monitoring to assess dewatering decisions such as the number of pumps to use.

In the long term, the level data in the downstream end of the tunnel and the Fall Creek Working shaft will be supported by ultrasonic level measurement in the upstream end of each of the 5 tunnel branches. Based on the surge modeling performed in 2010, inflow gate closure would occur when an elevation of 171 meters (560 feet) is reached in the Fall Creek Working Shaft (Ridgway et. al, 2011). The inflow gates could reopen when the HGL recedes to the crown of the tunnel at the working shaft, which is approximately 155 meters (508 feet). Water level monitoring of the Fall Creek Working shaft is a critical aspect of the proposed tunnel operating plan. An accurate and reliable measurement in this shaft will allow CEG to use the tunnel storage up to its capacity while preventing surge conditions from developing (Citizens, 2016). For redundancy, the next closest large shaft, the DRTC Retrieval Shaft, may also be monitored for inflow gate closure.

The phased implementation of the tunnel system provides Citizens with flexibility in implementing the level monitoring approach. Before the Fall Creek shaft is constructed, Citizens will monitor the DRTC Screen & Grit Shaft with a bubbler tube and evaluate its effectiveness in the near term for gate closure and the long term for tunnel dewatering. This same contact instrument can then be replicated in the Fall Creek shaft, with redundancy through double redundant physical float switches intended to read elevations of 167 meters (550 feet) and higher; this will allow monitoring of the critical level that triggers inflow gate closure 170 meters (560') as well as higher elevations for emergency monitoring, such as 15 meters (50 feet) from ground. Should the initial operation of the DRTC or Fall Creek tunnel identify the need for additional redundancy in level monitoring for surge control purposes, CEG could consider a redundant method such as an air bubbler encased in a tube due to its ability to withstand extreme velocities that may occur. An alternative for a redundant contact application would be a submersible pressure transducer intended to read elevations 167 meters and higher. The available above ground space or near surface space in this shaft should be confirmed during final design to allow for the installation of a contact application (Citizens, 2016).

In addition to global control of all inflow control gates based on an elevation of 171 meters (560 feet) reached in the central monitoring location in the tunnel, localized control for individual drop shafts will be performed by float switches installed in the ventilation shafts to prevent a localized overflow from the tunnel or adit. For the localized shaft monitoring, the close elevation does not need to be set as low as 171 meters (560 feet), an elevation closer to the ground could be utilized. Additional localized control for two drop shafts should also be considered:

- DRTC CSO 008 – opening at higher elevations to utilize the existing CSO outfall for passive surge overflow, with closure when the river stage at the nearby Stout USGS gaging station reaches an elevation of 204 meters (669 feet) to prevent river intrusion to the tunnel; and
- Lower Pogues Run Dropshaft 3 – closure to prevent creek intrusion from the nearby Pogues Run Creek (Citizens, 2016).

Tunnel Dewatering

Citizens' initial operational planning (Citizens, 2014) identified several key aspects of tunnel dewatering for further refinement in the Phase II plan. The key aspects are as follows:

- When the Dig Indy Tunnel is completed, approximately 28 l/s (.63 MGD) of baseline infiltration can be expected based on the allowable infiltration limit for the tunnel. This could be addressed be intermittent operation of a single Deep Rock Tunnel Pump Station (DRTCPS) pump in the middle of the night; and

- Using Citizens' collection system hydraulic model (Rubchinskaya et. al, 2015), it was determined that the largest storm events can deliver enough combined sewage to allow for up to 6 days continuous dewatering at 4 m^3/s (90 MGD); and
- There is the opportunity to balance flow at both AWT plants to preserve capacity for tunnel dewatering; and
- During and after wet-weather events, Southport will have flow capacity to receive flows from the DRTCPS in excess of 4 m^3/s (Citizens, 2014).

The above four bullets illustrate that there is a lot of flexibility for Citizens in operating both the DRTCPS and the AWT plants during a wet-weather event. Figure 4 presents the approximate storage curve for the tunnel system based on the Dig Indy alignment. The curve could serve as a reference for dewatering operations, as the # of pumps and VFD adjustments could be based on the HGL in the Screen & Grit Shaft. For example, when an HGL near 153 meters (501 feet) is observed after CSOs have stopped discharging, there will be approximately 946,000 cubic meters (250 MG) to pump out (Citizens, 2016).

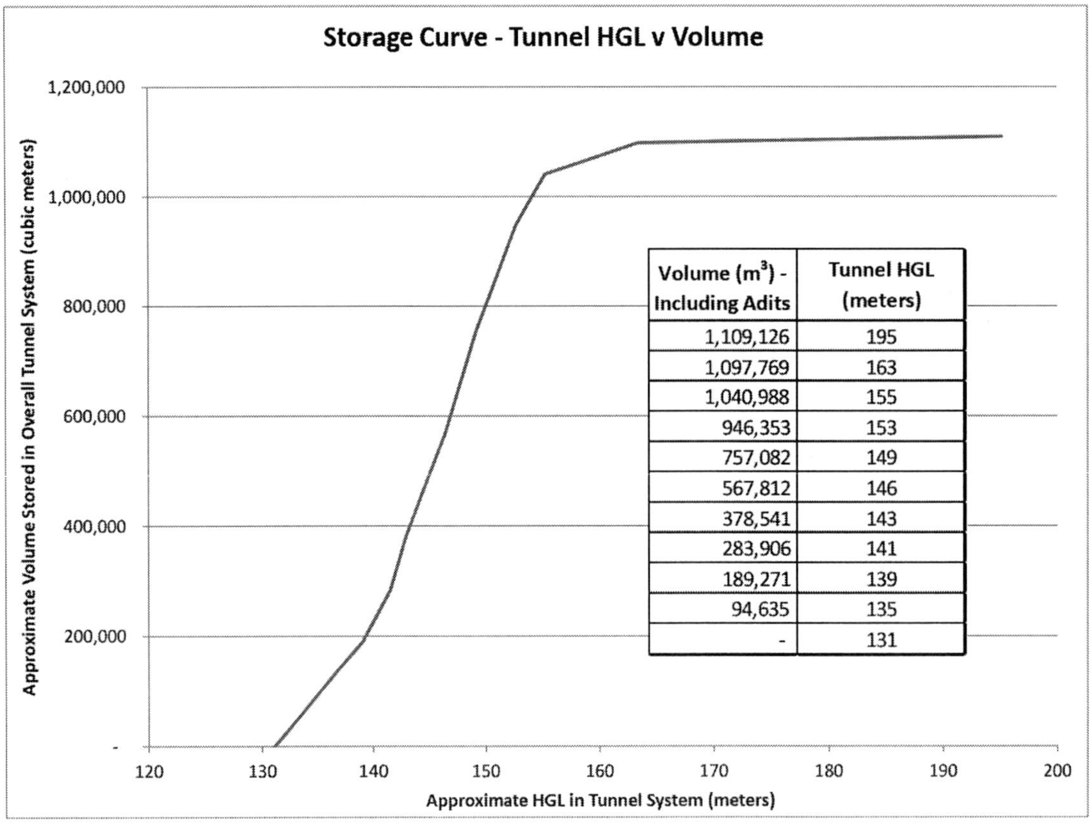

Volume (m^3) - Including Adits	Tunnel HGL (meters)
1,109,126	195
1,097,769	163
1,040,988	155
946,353	153
757,082	149
567,812	146
378,541	143
283,906	141
189,271	139
94,635	135
-	131

Figure 4: Tunnel Storage Curve

It should be noted that a consistent flat HGL in the Dig Indy Tunnel will only occur after the tunnel has stopped filling. While the tunnel is receiving flow, the HGL will have positive slope as flows move towards the Screen & Grit Shaft. Based on the above, the conceptual tunnel dewatering plan for CWA Authority is as follows:

- Turn on one dewatering pump to remove accumulated groundwater when the following is observed:
    - HGL 135 meters (445 feet), and the forecasted rainfall for the next 24 hours is one inch of rain or greater.
    - HGL 139 meters (456 feet);
- During the event, Gradually initiate the dewatering operation via VFDs and number of pumps:
    - HGL 139 meters (456 feet) – 1 pump
    - HGL 143 meters (469 feet) – 2 pumps
    - HGL 153 meters (501 feet) – 3 pumps;
- During the event, consider operating only two pumps when the equalization basin at the Southport AWT Plant is full;
- Following the event, consider operating four pumps as flow conditions at Southport are receding to dry-weather; and
- Maximize the secondary treatment capacity at the Belmont AWT Plant to the fullest extent practical to preserve additional capacity at Southport (Citizens, 2016).

The above conceptual plan should minimize the number of start/stop cycles for the dewatering pumps. Determining when to increase or decrease the number of pumps on during the event can also be based on the rainfall forecast for the remaining of the event.

Tunnel Ventilation

As part of the IIHR's physical modeling of the Eagle Creek drop shaft, the effects of an air pocket released from the downstream DRTC tunnel was simulated to test the 2.4 meter (8 foot) ventilation shaft, which is 44% of the diameter of the tunnel. Figure 5 presents two snapshots of the experiment showing the predicted geyser. IIHR evaluated numerous variations of drop shaft flow, surcharge conditions in the tunnel, and geyser mitigation alternatives such as an orifice restriction in the ventilation shaft. IIHR was not successful in developing a configuration change to mitigate geyser potential if an air pocket is released upstream when the Eagle Creek branch is full (IIHR, 2016). Based on this, Citizens began evaluating alternatives to prevent air migration from the DRTC to the Eagle Creek branch.

The geysering evaluation also suggested high risks of venting the DRTC from the three 2.4 meter (8 foot) utility shafts, which is why Citizens' is currently evaluating the best means of closing off the utility shafts to mitigate geysering.

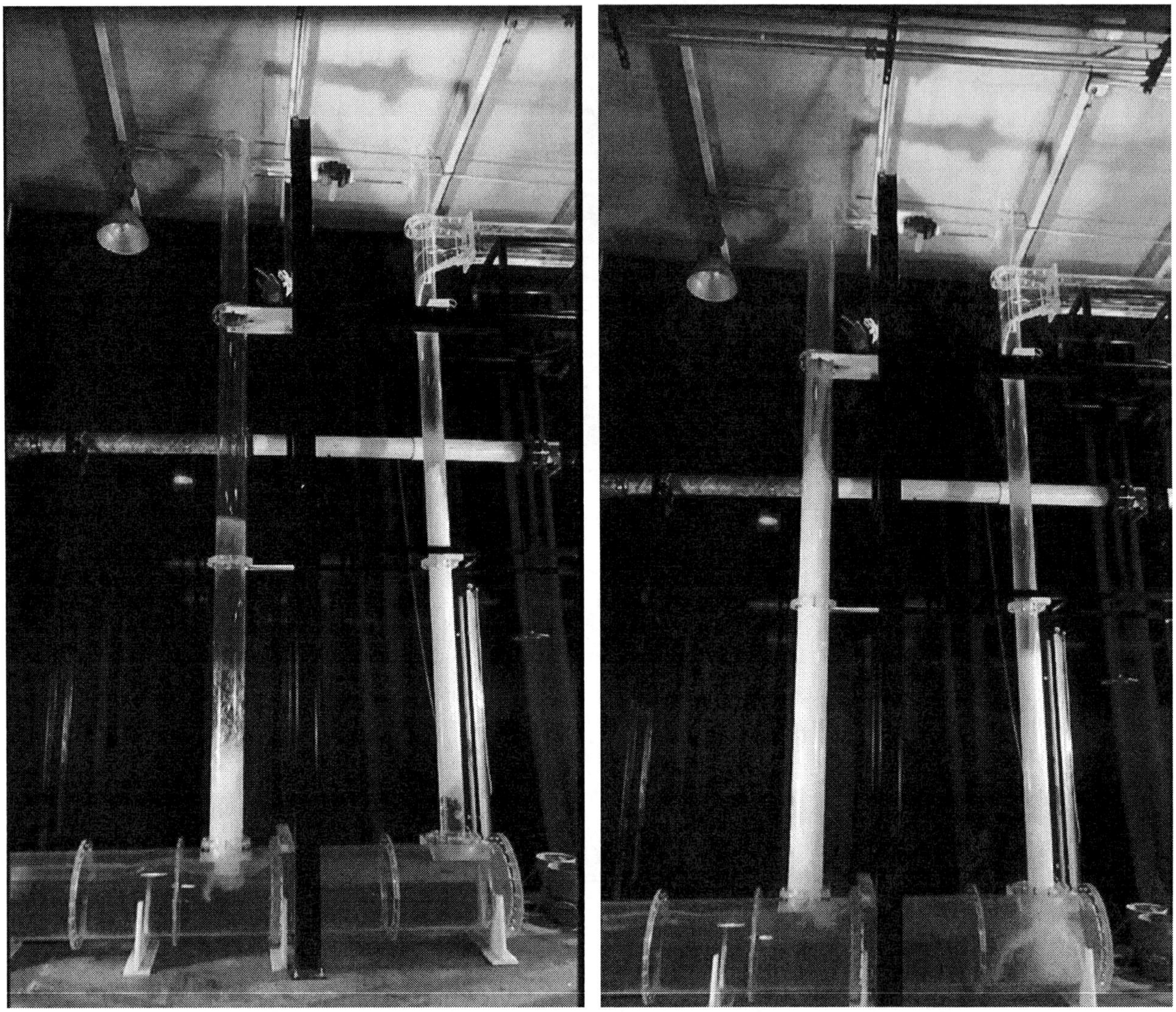

Figure 5: Simulated Geyser in Eagle Creek Ventilation Shaft (IIHR, 2016)

For an initial evaluation of the three utility shafts, Citizens applied its spreadsheet tool to updated surge model output. Figure 6 presents a profile of the downstream DRTC tunnel from the surge model (Rowe et. al, 2015) showing a trapped air pocket adjacent to two of the three utility shafts. Based on published air migration research (Vasconcelos and Chosie, 2013), if the utility shafts were closed, the pocket would migrate to the DRTC retrieval shaft in 50 to 100 minutes. During filling of the tunnel, peak velocities will briefly prevent the air pocket from migrating upstream (Benjamin, 1968). Since the retrieval shaft is larger than the tunnel diameter, the geyser risk at the retrieval shaft would be much lower that at the smaller utility shafts.

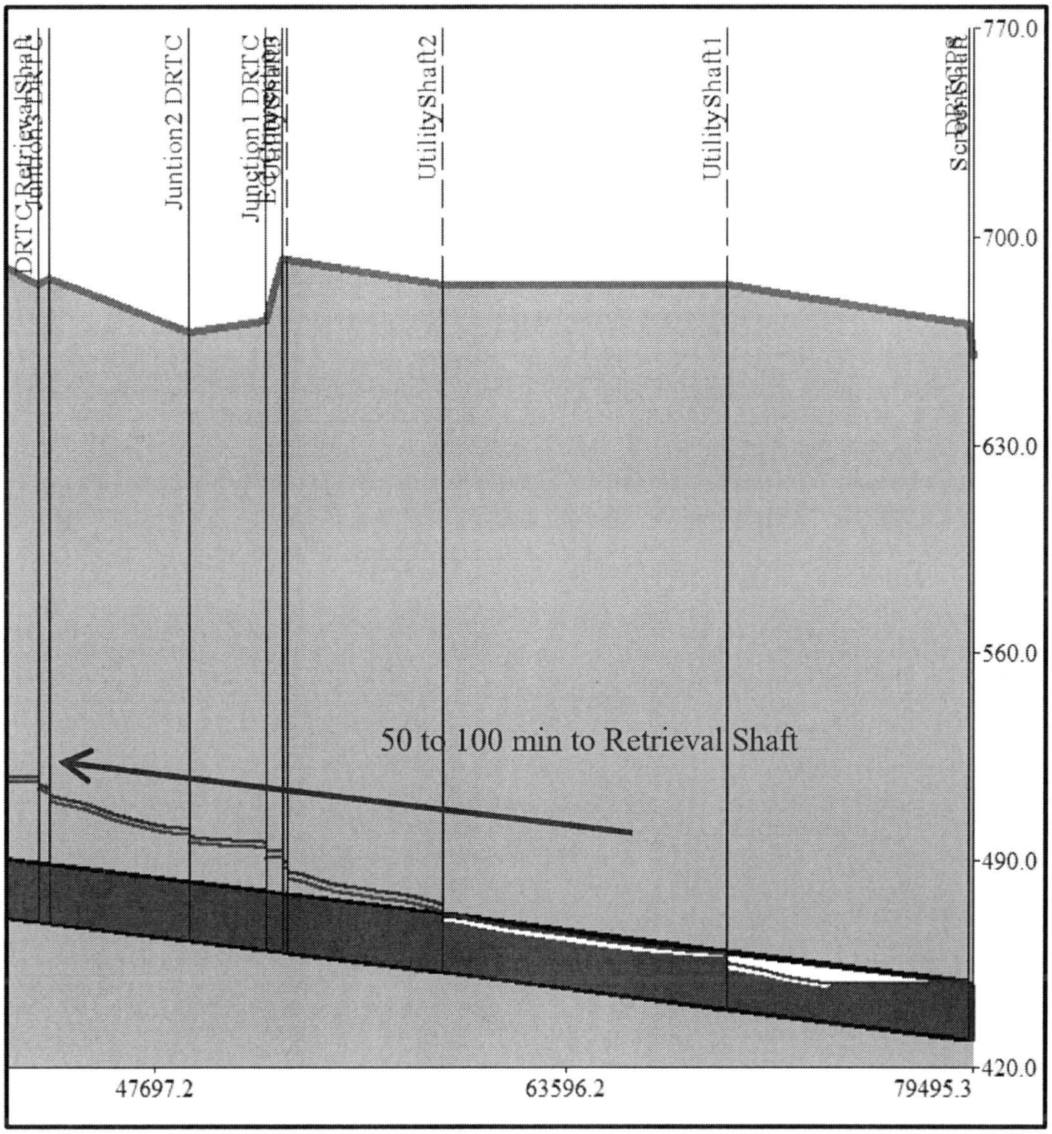

Figure 6: Simulated Upstream Migration of Air Pocket

A key finding of the spreadsheet tool analysis is that the pocket depth in at the leading edge of the pocket in Figure 6 is approximately 1.2 meters (4 feet). As the pocket migrates upstream towards the bifurcation junction with the Eagle Creek branch, the pocket will spread longitudinally and the depth will be reduced. For Citizens, this means a partial bulkhead of 1.5 meters (5 feet) would be adequate to keep the air pocket away from the Eagle Creek branch.

Another key finding from the spreadsheet tool analysis is that if the pocket is not immediately vented at the utility shafts, there is the risk for surge pressures due to the trapped pocket. Based on published research (Wright et. al, 2011), the maximum surge pressure was estimated to be higher than the ground elevation, but less than the specified strength of the tunnel liner. This instantaneous pressure means enough energy could lift the lid off a utility shaft or place additional stress on the liner if the utility shafts are closed. Since the maximum estimated pressure is below the strength of the liner by a factor of six, closing the utility shafts represents a lower risk to Citizens. The analysis also identified the potential for negative pressures as the compressed air pocket expands (Citizens, 2016).

Another key finding based on published air induction analysis (Brocard et. al, 2012) is that during filling of the tunnel, forward air induction up to 21 m^3/s (730 cfs) can be expected. This means that ventilation will be necessary at the downstream Screen & Grit shaft.

A follow-up evaluation was performed by Auburn University to confirm or refine Citizens' findings from its spreadsheet tool analysis. The evaluation utilized the Hydraulic Analysis for Stormwater Tunnels, version 2 (HAST2) program to evaluate transients, air movement, and air pressurization Figure 7 presents a profile of HAST2 for the DRTC tunnel branch. The profile shows both a pipe-filling bore and a trapped air pocket (Auburn, 2016).

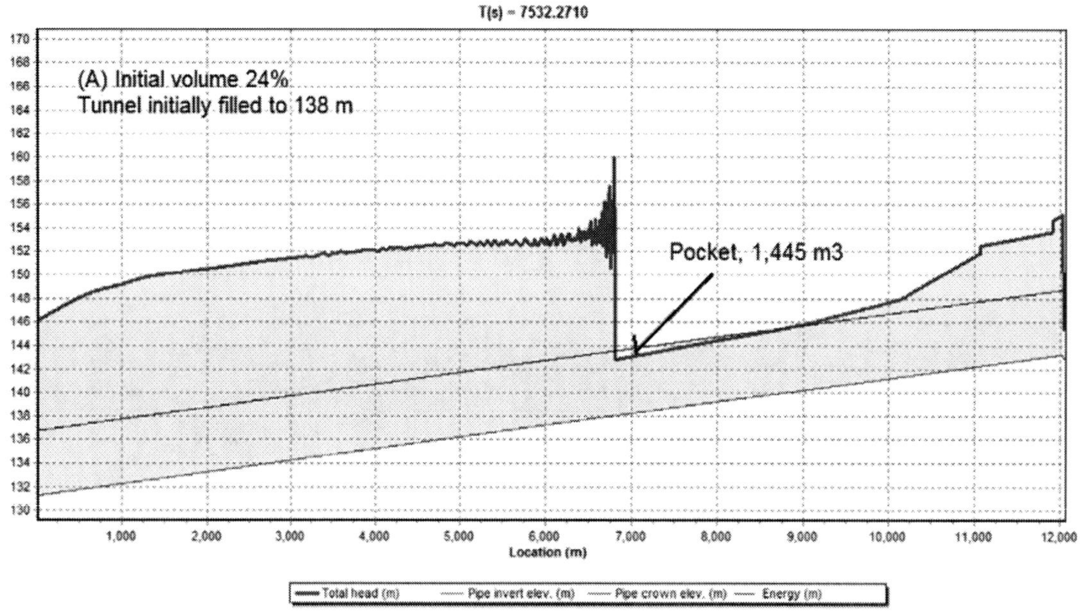

Figure 7: HAST2 Model Profile of DRTC, July 18, 1996 event (Auburn, 2016)

The HAST2 evaluation confirmed the potential geyser risk at the DRTC utility shafts due to the potential volumetric release of air at each shaft. The HAST2 evaluation also confirmed the potential geyser risk at the Eagle Creek shaft, but identified the risk as moderate due to the timing of when the Eagle Creek tunnel would be filling v. when an air release would be occurring. Compared to the estimated air pocket depth of 1.2 meters from the spreadsheet tool, a leading edge pocket depth of 0.5 to 1 meters was calculated at the Eagle Creek bifurcation. Finally, the HAST2 evaluation determined that the maximum surge pressure estimated with

Citizens' spreadsheet tool was conservative – transient pressures in excess of the ground elevation are not expected (Auburn, 2016).

Based on the analysis summarized above, the conceptual narrative for the ventilation approach of the Dig Indy tunnel system is as follows:

- Fully dewater the tunnel at the end of wet-weather events to minimize the presence of combined sewage in the tunnel during dry-weather;
- Preferentially vent the tunnel during wet-weather events at retrieval and working shafts that are equal or larger to the tunnel diameter;
- Allow for ventilation at the DRTC Screen & Grit Shaft for air inducted forward during filling of the tunnel. Air flow rates are expected to reach 21 m^3/s (730 cfs);
- An air pocket is expected to be trapped upstream of the Screen & Grit Shaft during peak filling conditions. The pocket is expected to be held in place briefly during peak filling conditions but will migrate upstream after peak flows subside;
- Allow the air pocket to migrate to the DRTC Retrieval shaft and vent there;
- Close the DRTC Utility shafts from venting during wet-weather events to minimize the risk of geysering. Sealing the DRTC Utility shafts will prevent geysering but may expose the liner to transient pressures within the strength of the liner; and
- Utilize a partial bulkhead at the crown of branch tunnels that terminate without a retrieval shaft to minimize the risk of geysering. This applies to the Eagle Creek and Pogues Run branches (Citizens, 2016).

Figure 8 presents a picture of the tunnel bifurcation during construction of the liner. The Eagle Creek branch is on the left, while the main branch of the DRTC is on the right. Based on the above analyses, a 1.5 m (5 foot) depth for the partial bulkhead to mitigate geysering was carried forward, which is illustrated in the figure. The partial bulkhead, inflow gate control, and dewatering pump station represent the overall geyser mitigation for Eagle Creek.

Figure 8: DRTC & Eagle Creek Bifurcation Junction with Partial Bulkhead Illustration

Odor Control Evaluation

As stated in the methodology section, the AERMOD analysis was for a single source: the Eagle Creek ventilation shaft. Figure 9 presents the shaft location within Ross Claypool Park. A key consideration for this shaft site is that the ventilation shaft is located near the western extent of Citizens' easement for the tunnel, with a private facility, Whitecase Logistics, located adjacent to the easement. This means that the critical receptor in the air dispersion analysis would be a person standing in Whitecase's parking lot close to the shaft.

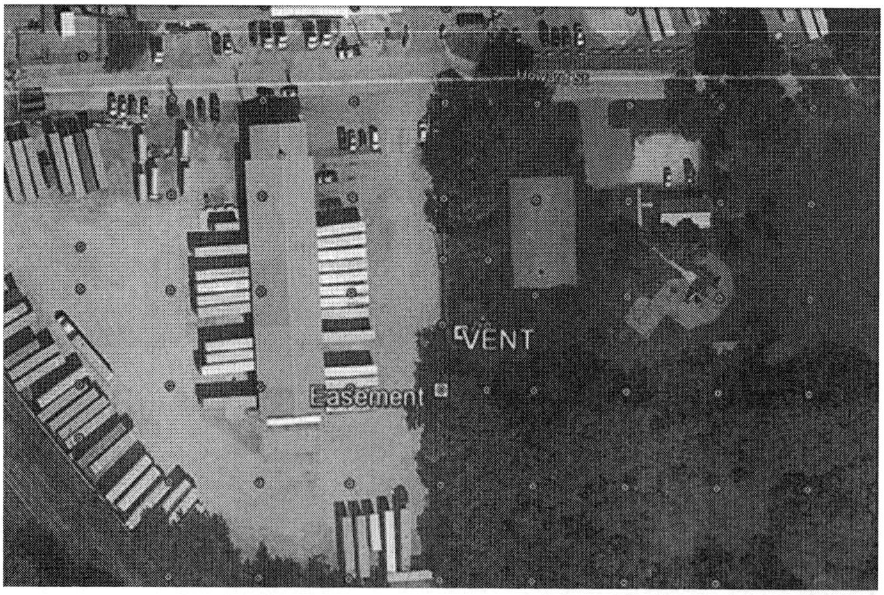

Figure 9: Eagle Creek Shaft Site & Receptor Grid (Citizens, 2016)

The dispersion evaluation for H_2S and ammonia was performed based on a maximum ventilation velocity of 5 m/s (17 fps) based on Citizens' surge modeling (Rowe et. al, 2015), and estimated emission rates for H_2S and ammonia of 5 g/hr (.012 lb/hr) and 232 g/hr (.51 lb/hr), respectively. The AERMOD evaluation utilized local meteorological data provided by the Indiana Department of Environmental Management (IDEM) for the years 2008-2012, which included upper air data from the nearby Indianapolis International Airport (Citizens, 2015).

Based on a review of odor thresholds in Indiana and other adjacent states, H_2S thresholds of 10 ppb over a 1-hour timeframe and 30 ppb over a 30-mintue timeframe were evaluated, as well as a health protective level of 140 ppb for ammonia. The analysis showed that there is moderate risk of detectable odors for a brief period of time during a wet-weather event when a potential receptor is in the immediate area of the shaft. For example, a person standing at the western property of Ross Claypool Park effectively has no opportunity for odor dispersion – the odor leaving the shaft reaches the receptor. Additional AERMOD simulations were performed to identify odor mitigation alternatives, which included an orifice or nozzle restriction on the 2.4 meter (8 foot) ventilation shaft, or adding a ventilation stack on top of the shaft. These modifications would allow the air released from the tunnel to rise above a potential receptor (Citizens, 2015).

Based on the odor dispersion analysis, Citizens' approach to mitigating potential odor risks is:

- Collecting H_2S data at the Eagle Creek shaft during initial operation to confirm or refine the findings of the analysis;
- For other tunnel ventilation shafts, siting the shafts in areas where the immediate vicinity can be fenced off where possible;
- In response to observed odor, use a orifice or nozzle restriction at the top of the shaft to increase air velocities – this facilitates dispersion above the height of a potential receptor; and
- In response to observed odor after the installation of a below ground restriction, install a narrow vent stack on top of the shaft. A diameter of 0.6 meters (2 feet) will allow for additional dispersion as well as raising the vented air above any potential receptors. This vent structure can resemble a lamppost and can be incorporated into the architectural features of the shaft site (Citizens, 2015).

Site tours

Citizens' two site tours in 2015 were valuable in both what was observed, and the development of relationships for future knowledge exchange. From the two site tours, Citizens was able to observe the following aspects of established tunnel operations:

- Control room configuration and staffing;
- SCADA and level monitoring installations;
- Site access and safety;
- Priming of the system before wet-weather events begin; and
- Sediment and floatables management.

Based on the experience in 2015, Citizens is planning additional site visits and networking with utilities operating combined sewer tunnels.

SUMMARY AND CONCLUSIONS

Citizens Energy Group has developed its conceptual tunnel operational plan to mitigate operational risks before the tunnel system is commissioned. At the summary level, Citizens' operational plan is as follows:

1. The tunnel pump station can expect to see flow within 2 hours of rainfall beginning – both AWT Plants will need to be ready to respond;
2. Tunnel dewatering does not need to begin at its full capacity to satisfy Consent Decree requirements;
3. For surge mitigation and protection from high river stages, the conceptual operational scheme for the inflow control gates is:
    a. Fall Creek Working Shaft < 170 meters (560 feet) – Open
    b. > 170 meters (560 feet) – Close
    c. < 155 meters (508 feet) – Reopen
    d. > 198 meters (650 feet) – CSO 008 open for passive overflow
    e. High Water – All gates reopen;
4. The inflow gates may close simultaneously or sequentially from downstream to upstream.
5. Ventilation will be expected at all retrieval shafts to minimize the risk of geysering. Ventilation will not be performed at any shaft that is smaller than the tunnel diameter

As Citizens continues to implement its operational plan, next steps include the startup of the DRTC and Eagle Creek tunnel branch. This represents the 2017 conditions and is Version 1.0 of the Operational Plan. Version 1.0 will include the O&M documents for the DRTC, DRTC Pump Station, and Eagle Creek. Citizens will be bringing additional tunnel branches online in 2021 and 2025, which represents versions 2.0 and 3.0 of the overall operational plan.

REFERENCES

Auburn University (2016). *Simulation of the rapid filling of the DRTC tunnel system with HAST2 model – Final Report.*

Benjamin, T. B. (1968). Gravity currents and related phenomena. *J. Fluid Mech.*, 31(2), 209–248.

Brocard, D., McMasters, M., and Thomas, D. (2012). "Air Flow in Sewers." Water Environment Federation Technical Exhibition and Conference, New Orleans, LA.

Citizens Energy Group (2014) *Tunnel Operational Plan Phase I Memorandum.*

Citizens Energy Group (2015) *Odor Modeling Evaluation – Eagle Creek Extension Shaft Memorandum.*

Citizens Energy Group (2016) *Draft Tunnel Operational Plan Phase II Memorandum.*

Citizens Energy Group (2016) *Dig Indy Tunnel System Completion Plan Announcement.* http://www.citizensenergygroup.com/Our-Company/News-Financials/News-Releases/DigIndy-Tunnel-System-Completion-Plan-Announced

Indianapolis DPW (2006) *Raw Sewage Overflow Long-Term Control Plan and Water Quality Improvement Report.*

Iowa Institute of Hydraulic Research (2016). *Eagle Creek Drop and Vent Shaft Physical Model Study.*

Ranck, C; Sherman, D; Jacob, M; Nielsen, S; Reichlin, D; Masbaum, B. (2012) *Behind the $740M: Building a Technical Consensus for Indianapolis Consent Decree Amendment No. 2;* Water Environment Federation Technical Exhibition and Conference (WEFTEC) 2012, New Orleans, LA.

Ridgway, K; Ranck, C; Morgan, J; Kluzinger, D. (2011) *The Indianapolis Tunnel Surge Analysis: Debugging Future Operational Issues in the Present;* ASCE EWRI Annual Congress, Palm Springs, CA.

Rowe, S; Ridgway, K; Ranck, C; Rubchinskaya, E; Trypus, J. (2015) *Riding the Wave – Updating Transient Models Throughout Design;* ASCE EWRI Annual Congress, Austin, TX.

Rubchinskaya, E; Sutton, D; Ranck, C; Rowe, S. (2015) *Model Meets World: Guiding an Evolving Integrated Model;* Water Environment Federation Technical Exhibition and Conference (WEFTEC) 2015, Chicago, IL.

US EPA. (1985) *Odor and Corrosion Control in Sanitary Sewerage Systems and Treatment Plants*, Office of Research and Development: Washington DC. EPA/625/1-85/018

US EPA. (2005) *Revision to the Guideline on Air Quality Models*: Federal Register/Vol. 70, No. 216.

Vasconcelos, J. G. and C. D. Chosie. (2013). *Kinematics of Entrapped Air Pockets Within Stormwater Storage Tunnels;* EWRI Annual Congress, Cincinnati, OH.

Wright, S.J., Determan, K.V., and Vargas, S.M. (2011). *Pressure Transients due to Compression of Trapped Air in Rapidly Filling Combined Sewer Overflow Storage Tunnels.* EWRI Annual Congress, Palm Springs, CA.

Yongsiri, C; Vollertsen, J; Hvited-Jacobsen, T. (2005) *Influence of Wastewater Constituents on Hydrogen Sulfide Emission in Sewer Networks.* ASCE Journal of Environmental Engineering, (2005)131:12(1676).

Mind Over Model: Optimization of Gate Closure for the South Hartford Conveyance and Storage Tunnel

Lawrence Soucie[1*], Dominique Brocard[1], Kate Mignone[1], Andrew Perham[2]

[1]AECOM, Chelmsford, Massachusetts
[2]Metropolitan District, Hartford, Connecticut
*Larry.Soucie@aecom.com

ABSTRACT

Hydraulic modeling is frequently used for planning and design of deep tunnel conveyance and storage systems. However, care must be taken to check that the model results are reasonable and that the end result is the best possible design. This paper illustrates some examples of how human "thinking" can and should be used when reviewing model results. These principles are illustrated using the design of the South Hartford Conveyance and Storage Tunnel (SHCST) in Hartford, CT. This project includes a South Tunnel and a North Tunnel. The South Tunnel is under construction while the North Tunnel is currently still in the planning phase. However, the South Tunnel must work as part of an integrated system with the North Tunnel, and therefore the analyses described in this paper included both tunnels. During wet weather, CSO / SSO flows will be diverted into the tunnel when the capacity of the collection system is exceeded. Hydraulic modeling was used to design and optimize the control logic for the gates that regulate flows into the tunnel. Hand calculations of the initial model results indicated discrepancies which were resolved by decreasing the length of the tunnel segments and time step used in the model. The model was subsequently applied to determine the optimal gate closure based on 54 year continuous simulations. Hydraulic modelers often try to make the model completely self-contained, so that the gates in the model automatically open and close based on calculated parameters in the model, such as the hydraulic grade line (HGL). Analysis of the model results showed that extreme storms in the period of record that caused the tunnel to fill would have been predicted by modern weather forecasting days in advance. As a result, human "Predictive Control" could be used to further optimize the operation of the tunnel by closing the CSO gates earlier than the pre-programmed set-points used in the hydraulic model. Another aspect of the tunnel design that involved imposing "human thinking" over raw model results was sediment calculations. Sediment modeling was performed to estimate the distribution and quantity of sediment in the tunnel over time. There is a lot of uncertainty in sediment modeling, and therefore the results were checked in multiple ways, including using two different sediment models. Through careful thinking about the model results and how the tunnel will actually be operated, an optimal tunnel size and operating strategy was developed for the SHCST.

KEYWORDS: Hydraulic Modeling, Tunnel storage, CSO control, SSO control, tunnel conveyance and storage

INTRODUCTION

Background

The South Hartford Conveyance and Storage Tunnel project (SHCST) is part of the Hartford, CT Metropolitan District's (MDC) program to comply with state and federally mandated clean water regulations. Hartford is the capital of Connecticut, with a population of approximately 125,000 persons. The Connecticut River, which flows from north to south from near the Canadian border into Long Island Sound, forms the city's eastern boundary (Figure 1).

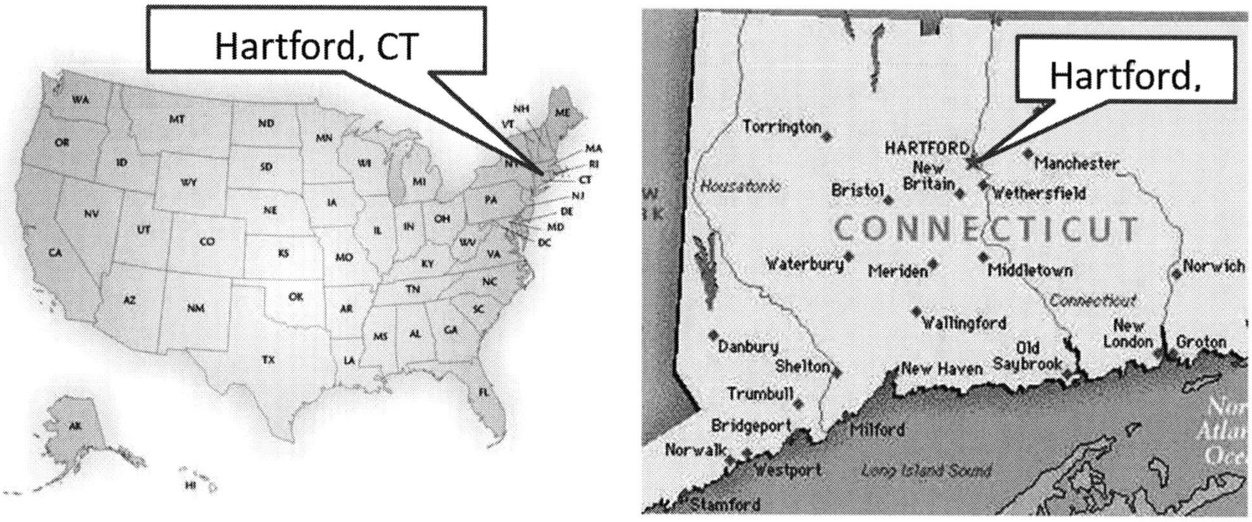

Figure 1. Location Plan

Figure 2 shows the conceptual layout of the tunnel system used for Preliminary Design. There is a South Tunnel and a North Tunnel. The South Tunnel is under construction while the North Tunnel is currently still in a planning phase. However, the South Tunnel must work as part of an integrated system with the North Tunnel, and therefore the analyses described in this paper included both tunnels.

During wet weather, CSO / SSO will be diverted into the tunnel when the capacity of the collection system is exceeded. Diversion structures will be constructed at each CSO and SSO to intercept overflows and divert them to near-surface consolidation sewers. These, in turn, will discharge to vortex drop shafts which will convey the flow in a controlled manner to the deep rock tunnel. Once flow reaches the downstream tunnel pump station, it will be pumped to the Hartford Water Pollution Control Facility (HWPCF).

Mignone et.al. (2016) summarize the functional design of the SHCST. The main components of the tunnel system include:

- 5.49 meter (18-foot) diameter, deep rock tunnel with 0.1% slope, 6,645 meter (21,800 feet) length for South Tunnel (final design) and 9,327 meter (30,600 feet) length for North Tunnel (planning phase)

- Consolidation conduits (0.61 to 1.68 meter (24 to 66 inches) in diameter, over 2,134 meters (7,000 feet) long
- Vortex drop shafts, seven in South Tunnel and eight in North Tunnel
- 2.19 cms (50 mgd) tunnel dewatering pump station
- Odor control at all potential air release points

Final design has been completed for the South Tunnel, and portions have been bid and are under construction. The layout of the North Tunnel will likely be revised in later design phases.

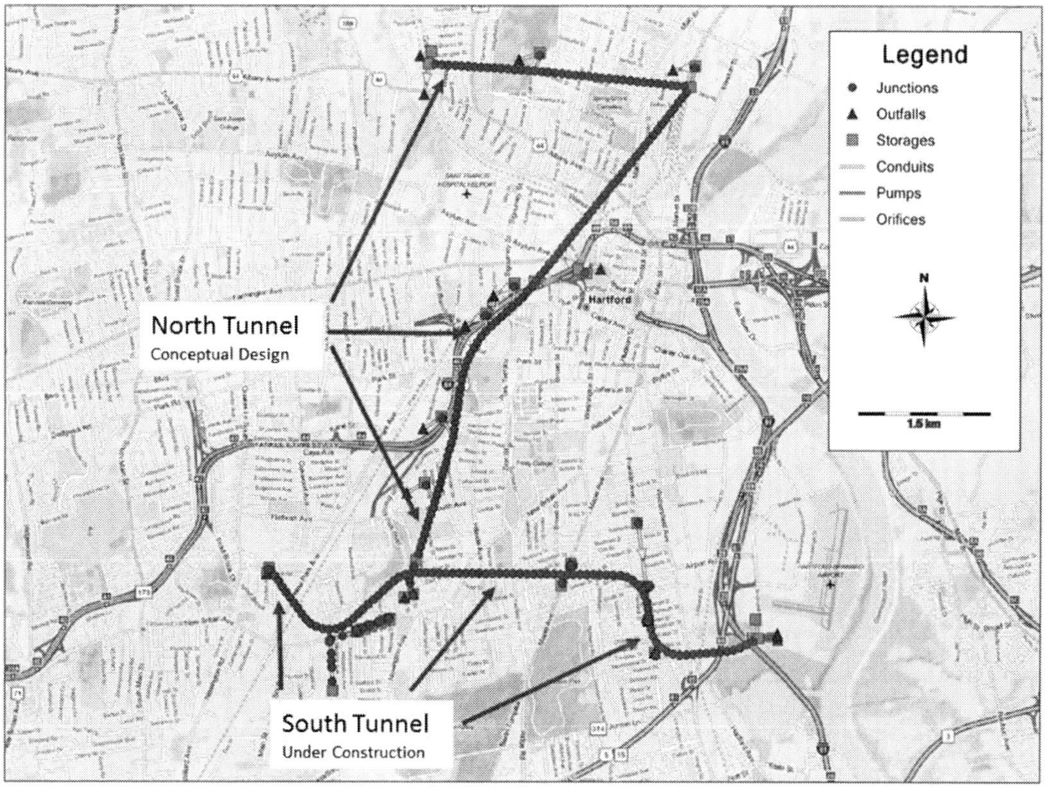

Figure 2. Conceptual Layout of South Hartford CSO Storage Tunnel

Overflow Control Design Criteria

Design criteria for each overflow varies depending on either the characteristics of the overflow (CSO or SSO) or the receiving water to which the overflow discharges (more or less sensitive). Overflow control objectives are one of the key design criteria for the tunnel, and specific performance measures were defined. These performance measures were assessed through the use of collection system models and a long term model simulation based on a 54-year long precipitation record between 1958 and 2011. Further information on the collection system modeling approach is provided below.

SSOs are required to be eliminated. The performance measure for SSO control is that no overflow can occur based on simulation of the 54-year period of record. There are two levels of control for CSOs. A design storm with an 18-year return frequency was selected for CSOs tributary to sensitive receiving waters because it would result in no more than two gate closures during the 54-year period of record. The performance measure for the remaining CSOs, referred to as 1-year CSOs, is that less than one overflow per year on average (i.e., less than 54 overflows in the 54-year period of record) can occur. These performance measures are summarized in Table 1, while the characteristics of the 1-year and 18-year design storms are summarized in Table 2.

Table 1. Tunnel Overflow Performance Measures

SSOs	CSOs Tributary to Sensitive Receiving Waters (18-year return frequency)	1-Year CSOs
No overflows allowed	Two overflows per 54-year period of record	Less than 54 overflows in the 54-year period of record

Table 2. Design Storm Characteristics for Initial Tunnel Sizing

Storm Event	Date	Total Depth of Rainfall During Storm (cm)	Peak 1-hr intensity (cm/hr)	Storm Duration (hrs)
1-Year Storm	October 7, 1951	6.1	1.83	22
18-Year Storm	May 24, 1989	12.4	4.01	16

Objective

As noted above, the tunnel is designed to store the overflow, convey it to the HWPCF, and then treat it when there is available capacity. The tunnel is designed to provide a specified level of service. This requires that gates be installed that are set to close at specified set-points to limit flow into the tunnel from the CSO that discharge to non-critical areas, thereby reserving capacity for the SSOs and CSOs in sensitive areas.

Hydraulic modeling is an important tool in planning, design, and optimization of tunnel systems. However, care must be taken to carefully check the results and not blindly accept the model predictions. The first part of this paper describes the various modeling tools and how they were applied. The second part of the paper discusses modeling of the gates and the initial model results, how the results were checked, and how the model was modified based on the checking. Once it was confirmed that the model produced reasonable results, the model was applied to determine the optimal gate closure based on 54 year continuous simulations. The hydraulic model can be completely self-contained, so that the gates in the model automatically open and close based on calculated parameters in the model, such as the hydraulic grade line (HGL) at a particular point in the tunnel. However, analysis of the model results showed that extreme storms in the 54-year period of record that caused the tunnel to fill would have been predicted by modern weather forecasting days in advance. The third part of this paper discusses how human "Predictive Control" could be used to further optimize the operation of the tunnel by closing the CSO gates earlier than the pre-programmed set-points based the hydraulic model. Another component of the tunnel design that involved imposing "human thinking" over raw model results was sediment calculations. Sediment modeling was performed to estimate the distribution and quantity of sediment in the tunnel over time. There is a lot of uncertainty in sediment modeling, and therefore the results were checked in multiple ways. The fourth part of this paper discusses the sediment modeling, including how the results were checked, and the final results.

HYDRAULIC MODELING TOOLS

Various hydraulic models were used to support the design of the tunnel system, and these are briefly described below.

System-Wide Hydraulic Model

A system-wide hydraulic model of the collection system, using the Stormwater Management Model (SWMM), was used to generate SSO and CSO inflow hydrographs for the 54-year period of record. The period of record flows represent predicted flow over existing or proposed SSO/CSO control points, typically weirs or high pipe outlets. These overflow control points represent the interface point between the system-wide hydraulic model of the collection system and the tunnel system model.

Tunnel System Hydraulic Model

A detailed model of the combined South and North Tunnel system was developed using the Personal Computer Storm Water Management Model (PCSWMM) modeling software. PCSWMM is distributed by Computational Hydraulic Inc. (CHI) and is a graphical user interface for the USEPA SWMM model. This model included both the deep rock tunnel and the consolidation conduits. A schematic of the model was shown in Figure 2. This model was used to develop gate controls and adjust tunnel diameters to meet the overflow performance measures in Table 1.

Sediment Deposition Model

An analysis of sediment deposition was conducted to estimate the quantity and distribution of sediment build-up in the tunnel system. These analyses were conducted using the InfoWorks CS collection system modeling software, which is distributed by Innovyze. The SWMM model of the tunnel system was imported into InfoWorks CS, which was subsequently configured to simulate sediment.

HYDRAULIC MODELING OF GATES

The tunnel system is designed to intercept, store, and convey SSO and CSO to the HWPCF, where it will be treated when there is available capacity. As was noted in Table 1, there are three types of discharges: SSOs, CSOs in sensitive areas, and CSOs in non-sensitive areas. This requires that gates be installed at the inlets to the tunnel that are set to close at specified set-points to limit flow into the tunnel from the CSO areas, thereby reserving capacity for the SSOs.

A typical gate chamber is shown in Figure 3. Each gate chamber will be constructed in close proximity to an associated drop shaft and will limit CSO and SSO into the tunnel system based on measured water level in the tunnel. Flow in excess of tunnel system capacity will be excluded from the tunnel system using an electric motor-actuated slide gate. This slide gate closure will cause flow to back up in the associated consolidation conduit and overflow to receiving waters. Since slide gate closure is critical for protecting the tunnel system, redundant electric motor-actuated slide gates will be provided.

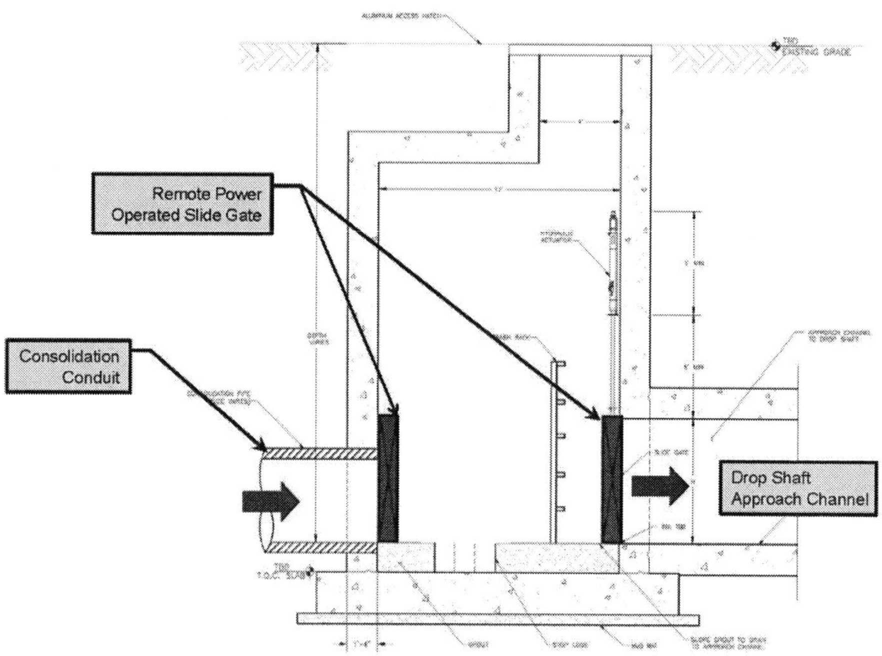

Figure 3. Typical Gate Chamber

The initial sizing of the tunnel and design of the gate control system utilized the design storms summarized in Table 2. The tunnel sizing and gate control system were subsequently refined based on the 54-year continuous simulations. The tunnel configuration used for the current analysis was a diameter of 18 feet for the South Tunnel and a diameter of 17 feet for the North Tunnel with a 0.1% slope for both tunnels.

The basic procedure followed for the initial design of the gate control system was:

1. Run the model for the 1-year design storm and note the maximum predicted HGL at the downstream pump station connection to the HWPCF.

2. Set the 1-year CSO gates to close at the HGL determined in Step 1. This process allows the 1-year storm to be captured by the tunnel. CSO volumes greater than the 1-year storm elevation will be diverted to the existing CSO discharge. This process reserves tunnel capacity for the SSOs and CSOs in sensitive areas.

3. Run the model for the 18-year design storm and note the maximum predicted HGL at the downstream pump station connection to the HWPCF.

4. Set the gate elevations for the CSOs in sensitive areas to close at the 18-year design storm elevation determined in Step 3. This process reserves capacity for the SSOs, which can continue to enter the tunnel after the 1-year and 18-year CSO gates have closed.

One of the limitations of model simulations is that they may be affected by numerical instability. To assess the impact of potential numerical instability, the model was run with 3-second, 1-second, and 0.5 second time steps. Table 3 shows the initial model simulation results. The continuity error is calculated internally and is the sum of the inflows divided by the sum of the outfalls. A negative continuity error indicates more water exited the system than entered. A low continuity error means the model has converged to a stable solution. The continuity error for the 18-year design storm model run with the 3-second time step model run was higher than the 1-second or 0.5 second model runs, which indicates the model results for the 3-second run may be inaccurate.

The maximum amount of water stored in the tunnel was calculated based on the volume into the tunnel minus the volume out. This is illustrated in Figure 4 for the 1-year design storm. The cumulative volume into the tunnel is the sum of all the inlets and the cumulative volume out is the volume pumped at the connection to the HWPCF. The amount of water stored is equal to the difference, which is 244,605 m^3 (64.61 MG) for the 1-year storm. Similar calculations were made to determine the stored volume for the model runs summarized in Table 3.

Table 3. Sensitivity of Volume Stored in Tunnel to Time Step Used in Model for Initial Assessment

Time Step (sec)	1-Year Design Storm		18-year Design Storm		Tunnel Volume Comparison for 18-year Design Storm		
	Peak HGL (m)	Continuity Error (%)	Peak HGL (m)	Continuity Error (%)	Maximum Volume Stored in Tunnel (m^3)	Inclined Cylinder[1] Volume Check (m^3)	Percent Difference (%)
3	-38.44	-0.26	-33.35	-6.24	361,986	337,644	6.7
1	-38.45	-0.24	-33.26	0.29	361,948	338,780	6.4
0.5	-38.51	0.35	-33.95	1.40	361,456	329,164	8.9

1 Hand calculations using inclined cylinder volume formula were used to check model results

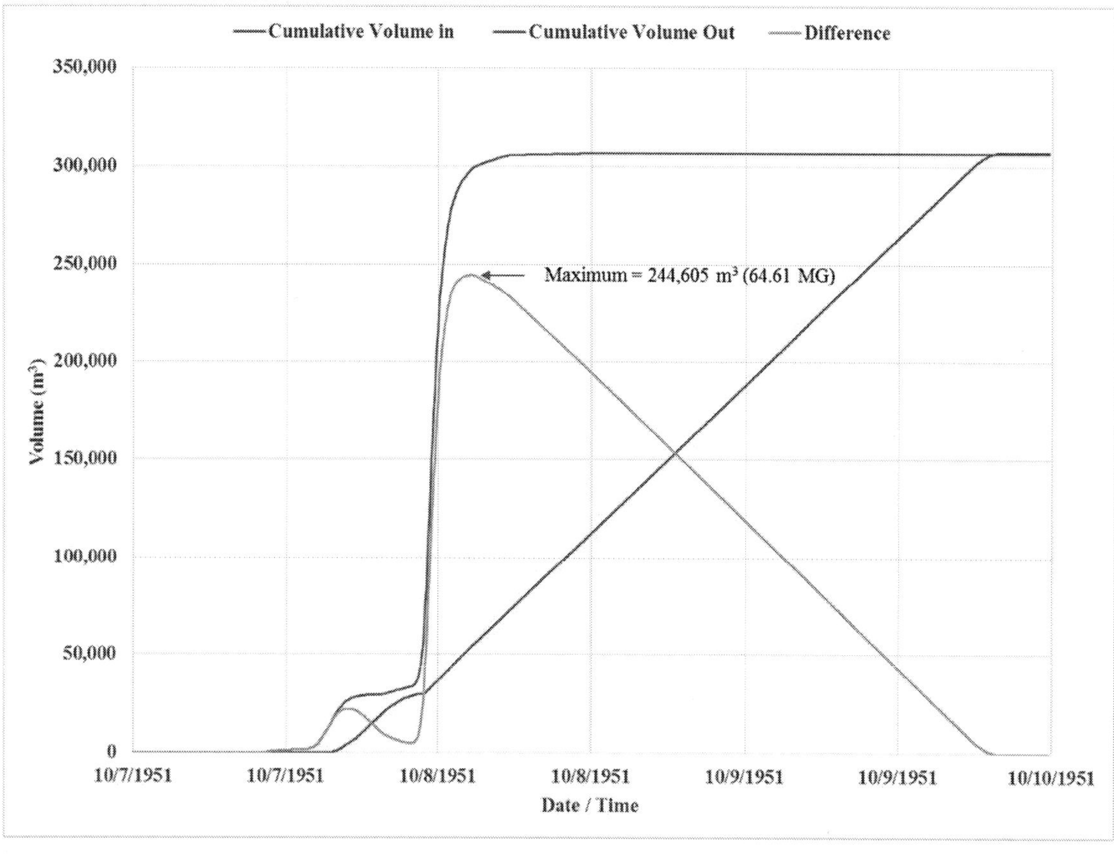

Figure 4. Example of Calculation of Volume Stored in Tunnel for 1-year Design Storm

In general, the predicted volumes stored in the tunnel are within about 500 m^3 (0.14 MG) for all three time step simulations, suggesting that the simulated volumes are not sensitive to the time step used in the simulations. However, the difference in peak elevation is 0.69 meters (2.3 feet) between the 1 second and 0.5 second time step model runs, which may indicate the simulated peak elevations are sensitive to the time step.

Since the tunnel is essentially a large, inclined cylinder, the results from the mass balance volume calculations should be reasonably close to hand calculations performed with the inclined cylinder formula. The peak elevations predicted by the model were used to compute the volume in a partially filled inclined cylinder, which was then used to assess the accuracy of the model results. This is illustrated in Figure 5. These results are summarized in Table 3 and indicate that the volumes predicted by the inclined cylinder calculations are about 6 to 9 % lower than the volume calculations based on cumulative volume in minus cumulative volume out. This is considered too great a difference and indicates the predicted peak HGL may be too low. Additional investigations were performed to determine the reason for this outcome.

In most collection system models, the HGL versus flow rate (which causes CSOs) is the most significant relationship. For a tunnel system, what matters most is the HGL versus volume correlation. Collection system models need to be configured correctly to simulate the important phenomena. The initial model configuration has model nodes only at the inlet connections and at the downstream pump station. This results in relatively long pipe lengths ranging from 331 meters (1,086 feet) to 2,361 meters (7,746 feet). To further investigate the level of discretization used in the model, the pipe lengths were subdivided into 76 meter (250 feet) segments and the analysis was repeated. The results are summarized in Table 4.

Table 4. Sensitivity of Volume Stored in Tunnel to Time Step Used in Refined[1] Model Assessment

Time Step (sec)	1-Year Design Storm		18-year Design Storm		Tunnel Volume Comparison for 18-year Design Storm		
	Peak HGL (m)	Continuity Error (%)	Peak HGL (m)	Continuity Error (%)	Maximum Volume Stored in Tunnel (m^3)	Inclined Cylinder[2] Volume Check (m^3)	Percent Difference (%)
3	-38.21	-1.81	-31.62	-1.07	356,573	351,766	1.3
1	-38.19	-0.63	-31.56	-0.26	356,686	351,985	1.3
0.5	-38.19	-0.55	-31.61	-0.22	356,535	351,803	1.3

1 Model refined by splitting the tunnel links into 76 meter (250 feet) segments
2 Hand calculations using inclined cylinder volume formula were used to check model results

The continuity errors decrease as the time step is reduced, which is expected. These results indicate the 0.5 second time step produces the most accurate result.

In general, the predicted HGL for the 18-year storm are 1.7 to 2.3 meters (5.7 to 7.7 feet) higher for the refined model than the comparable elevation in the initial configuration. As a result, the inclined cylinder volume calculations are higher and are within 1.3 % of the volumes calculated based on the cumulative volume in minus cumulative volume out calculations. These results suggest that the refined model is more accurate than the initial model configuration with the long tunnel segments. Further refinement may increase the accuracy further. However, this model was to be run for numerous 54-year continuous simulations, with each run taking approximately 45 hours. Since the maximum volume stored calculations are within 1.3% of the inclined cylinder volume calculations, the level of accuracy was considered adequate to support the application of the model. All additional analyses were performed using the refined version of the model with a 0.5 second time step.

This discussion demonstrates why it is important carefully review model results. The initial model configuration used very long tunnel segments, which appears to have resulted in water level predictions that were lower than would be predicted based on hand calculations using an inclined cylinder formula. Since the water levels predicted by the model were critical for the design of the gate control strategy, it was important that the model results were reasonable. Subdividing the tunnel segments into 76 meter (250 foot) segments and using a time step of 0.5 seconds resulted in model predications that were consistent with inclined cylinder hand calculations, and were therefore believed to be more accurate. Tunnel segments of 152 meters (500 feet) and 30.5 meters (100 feet) were also investigated, but the 76 meter (250 foot) segments produced the best results.

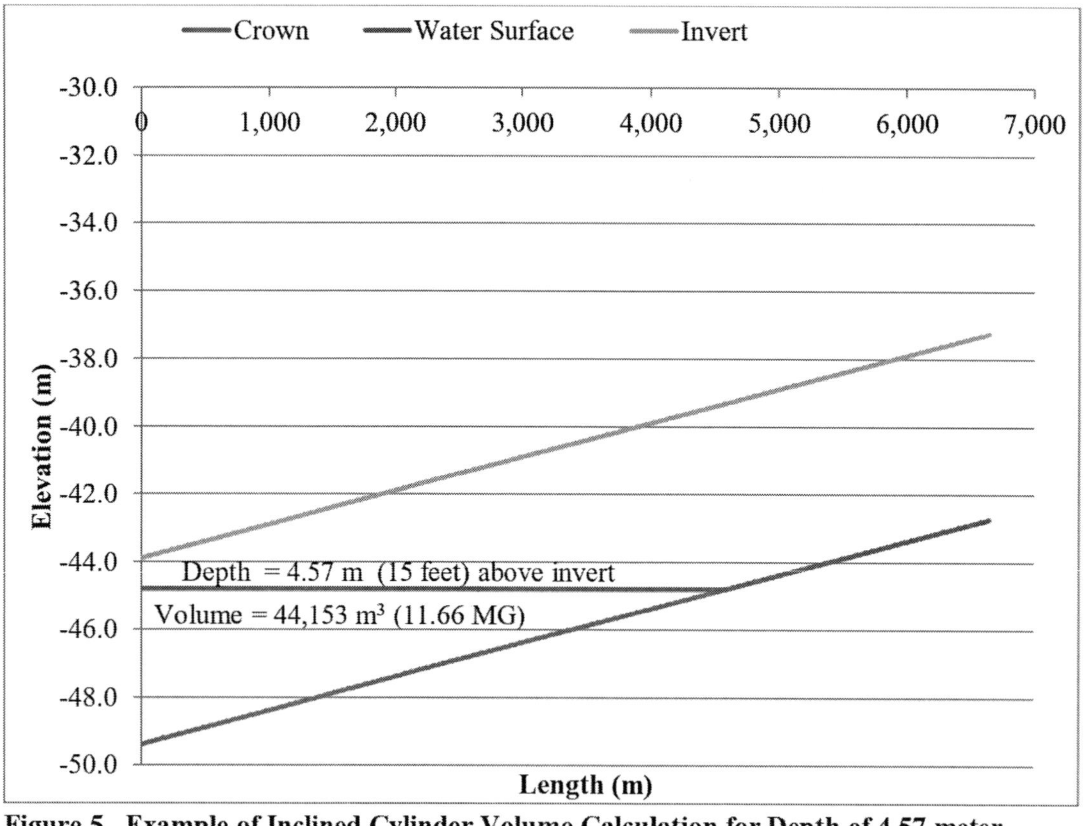

Figure 5. Example of Inclined Cylinder Volume Calculation for Depth of 4.57 meter (15 feet) in 5.49 meter (18-feet) Diameter Tunnel

GATE CONTROL STRATEGY

The refined model described above was used to design the gate control strategy. Three control system strategies were initially considered:

- Passive control
- Direct measurement control
- Predictive control

The passive control strategy would involve the use of fixed weirs at the inlets to the tunnel and at an emergency overflow at the downstream end. This option was eliminated because of the need to provide different levels of service at the various inlets. It was also determined that installation of a passive emergency overflow at the downstream end of the tunnel system was infeasible. This alternative was not considered further. Both direct measurement and predictive control are discussed below.

Direct Measurement Control

Direct measurement control, as used for the SHCST system, involves measuring water levels at the downstream pump station, and then controlling the upstream gates based on the set points to achieve the desired level of control. The gate control set points were initially determined based on the 1-year and 18-year design storms as described above. The model was then run for the 54-year period of record (POR) which included the period from 1958 to 2011. Since the 1-year design storm was used to set the 1-year gate closure level, it was anticipated that there would be on the order of 54 activations during the period of record simulation. However, the POR simulation indicated that there were 76 1-year gate activations. When the 1-year gate control elevation was raised in an attempt to reduce the number of 1-year gate activations, the number of 18-year gate activations increased since there was less storage available in the tunnel for the 18-year CSOs. A detailed evaluation of the 1-year design storm was conducted to determine why using it to set the 1-year gate levels resulted in more than the expected number of 1-year overflows during the POR simulation.

Figure 6 shows the predicted cumulative volume pumped at the pump station and the corresponding water level at the pump station for the 1-year design storm. At the time of the peak water level elevation of -38.19 m (-125.3 feet), the pumps had extracted 53,508 m^3 (14.1 MG). The model was configured to shut the 1-year gates at an elevation of -38.19 m (-125.3 feet) and the model was run for the 18-year storm. As shown in Figure 7, the simulated HGL reached the 1-year gate closure level at hour 17 compared to hour 40 for the 1-year design storm. Because of the shorter pump run time, the pumps had only extracted 20,001 m^3 (5.3 MG) from the tunnel. This is about 33,334 m^3 (8.8 MG) less than the 53,508 m^3 (14.1 MG) for the 1-year design storm at the peak HGL. The reason for this difference is the shape of the 1-year design storm hydrograph. The 1-year storm begins with low rainfall rates, during which time the pumps are able to keep up with the incoming flows. As a result, the pumps are able to remove more water during the 1-year design storm than during the 18-year design storm. These results show that the 1-year design storm is not a suitable surrogate event for determining the elevation at which to close the 1-year CSO gates. It was determined from these model simulation results that

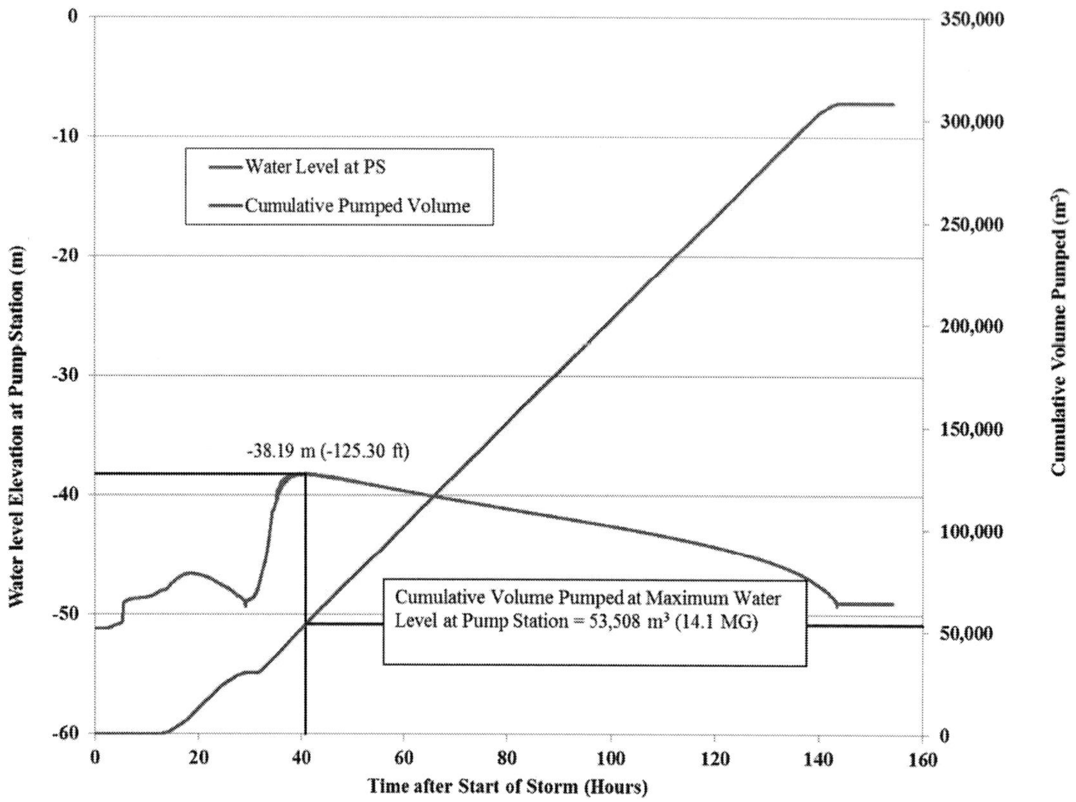

Figure 6. Water Levels at Pump Station and Cumulative Volume Pumped for 1-Year Storm

an iterative approach using the 54-year POR would be necessary to size the tunnel and establish the 1-year and 18-year gate closure levels.

Multiple 54-year POR simulations were run with various 1-year CSO gate closure elevations in order to determine the 1-year gate closure elevation that would result in 53 overflows during the period of record. The 1-year gate closure elevation determined from this iterative POR approach was higher than the 1-year gate closure level determined from the 1-year design storm due to the shape of the 1-year design storm. The higher 1-year gate control level resulted in less storage being available for the 18-year CSOs. As the POR analysis progressed, it became apparent that it wasn't feasible to meet performance goals for the 18-year CSOs with a direct measurement control approach without increasing the size of the tunnel. Accordingly, evaluations of a simple predictive control approach were conducted.

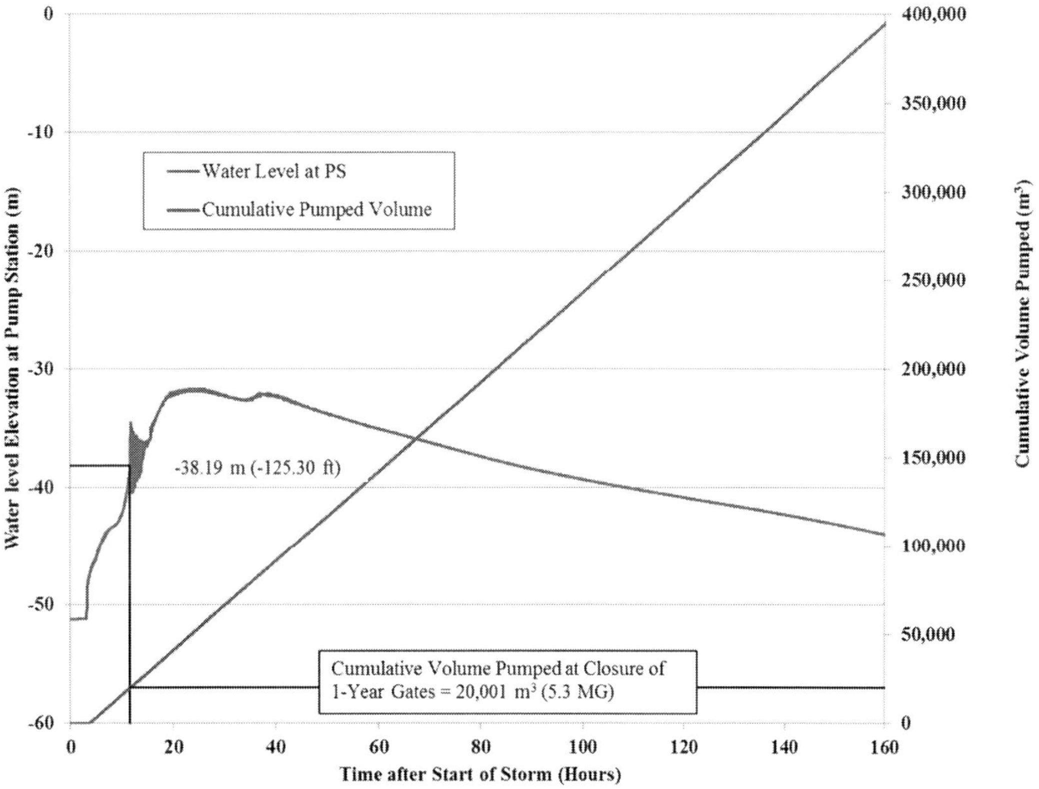

Figure 7. Water Levels at Pump Station and Cumulative Volume Pumped for 18-Year Storm

Predictive Control

Traditional predictive control strategies use weather forecasting along with rainfall-overflow volume relationships derived from hydrologic/hydraulic models, plus level monitoring, to assess available tunnel volume. Instead of a more complex, traditional predictive control system, a simple predictive control system was developed and modeled in order to meet the specified overflow performance measures. Essentially, the operators would watch the weather forecast. When very large and readily predictable storm events were forecasted (i.e., predicted total rainfall depth greater than a threshold value), operators would indicate this to the control system. This could be as simple as clicking a selection on the appropriate Supervisory Control and Data Acquisition (SCADA) screen. With this entry, the slide gates associated with the 1-year CSOs would close at a lower downstream tunnel water level than normal, preserving tunnel volume for the 18-year CSOs associated with sensitive receiving waters and the SSOs. This would not impact the key performance measure for the 1-year CSOs of less than one overflow per year on average because the gates would have closed during these large storms even if based solely on downstream tunnel level. Although this control approach would not increase the number of overflows from the 1-year CSOs in the

54-year period of record, it would increase the overflow volume from the 1-year CSOs during large events when the 1-year CSO gates would close sooner than the pre-programmed set points. But, by closing at a lower than normal downstream tunnel water level, overflow performance measures for the 18-year CSOs associated with sensitive receiving waters would be met in all but the very largest storm events in the 54-year period of record. In addition, performance measures for the SSOs would be met for all storms in the 54-year period of record.

To investigate the practicality of simple predictive control, the predictability of storms that would cause closure of the 18-year gates was assessed. This assessment consisted of first assuming tunnel diameters for the North and South tunnel segments. For the South tunnel segment, a constant diameter of 5.49 m (18 feet) was used. For the North tunnel segment, diameters of 4.88, 5.18, and 5.49 m (16, 17, and 18 feet) were assessed. For each tunnel system size (16N, 18S; 17N, 18S; and 18N, 18S) the 1-year gate closure elevation required to achieve less than 54 1-year CSO discharges in the 54-year POR was determined. Using this 1-year gate closure elevation, the corresponding number of 18-year gate closures and the storms that resulted in these closures were determined. An assessment was then conducted to determine whether each storm that caused an 18-year gate closure could have been predicted. In order to meet the performance criterion for number of 18-year gate closures in the 54 year POR, the analysis had to result in fewer than three 18-year gate closure storms that could not be predicted.

The results of this iterative assessment are presented in Table 5. In summary, it was determined that both the North and South tunnel segments would have to be 5.49 m (18-feet) in diameter. It was also determined that, for large, predictable storm events, the 1-year CSO gates should close when the water level elevation at the pump station reaches -39.32 m (-129.00 feet.) and the 18-year CSO gates should close when the water level elevation at the pump station reaches -34.29 m (-112.50 ft.). Finally, it was determined that the 1-year CSO gates should re-open when the water level elevation reaches -45.69 m (-149.89 ft.). These results will need to be updated for the new boundary conditions for the North Tunnel, as the North Tunnel design progresses.

Key observations from inspection of Table 5 are summarized below:

- With the 5.49 m (18 ft.) South Tunnel and 4.88 m (16 ft.) North Tunnel configuration, a total of 15 18-year CSO gate closures were predicted. Of these, at least six were judged to be "unpredictable storms", for which the 18-year CSO gates would have closed even if simple predictive control was used.

- With the 5.49 m (18 ft.) South Tunnel and 5.18 (17 ft.) North Tunnel configuration, a total of 11 18-year CSO gate closures were predicted. Of these, at least four were judged to be unpredictable storms for which the 18-year CSO gates would have closed. Thus, the 5.49 m (18 ft.) South Tunnel and 5.18 (17 ft.) North Tunnel configuration also failed to meet the performance requirement for 18-year CSO gate closures.

- With the 5.49 m (18 ft.) South Tunnel and 5.49 m (18 ft.) North Tunnel configuration, seven 18-year CSO gate closures were predicted. Of these, only two were judged to be associated with unpredictable storm events. Thus, the 5.49 m (18 ft.) South Tunnel and 5.49 m (18 ft.) North Tunnel configuration would meet the performance requirement for the 18-year CSO gates.

- If simple predictive control was not used, the performance requirement for the 18-year CSO gates would not be met, even with the 5.49 m (18 ft.) South Tunnel and 5.49 m (18 ft.) North Tunnel configuration.

Table 5. Results of Simple Predictive ControlAnalysis

Tunnel Diameters	1-year gate Closure Level[1]	Number of 18-year CSO Gate Closures	18-year CSO gate Closure Storms	Assessment of Storm Predictability	No. of Unpredictable Storms
18S / 16N	-125.0	15	Oct 24,1959 Sep 12,1960 Oct 03, 1979 Jun 05, 1982 Aug 09, 1982 Aug 11, 1985 May 24, 1989 Sep 16, 1999 June 17, 2001 Aug 21, 2004 July 27, 2005 Oct 14, 2005 Sep 06, 2008 Aug 28, 2011 Sep 08, 2011	No- Long Duration Yes - Hurricane No - Tornado Yes – Flood No-Summer Storm No - Thunderstorm Yes - Big Storm Yes – Hurricane Yes- Big Storm No –Thunderstorm No – Thunderstorm Yes -Tropical Storm Yes- Tropical Storm Yes – Hurricane Yes – Hurricane	6
18S / 17N	-127.0	11	Oct 24, 1959 Sep 12,1960 Oct 03, 1979 Jun 06, 1982 Aug 11, 1985 May 24, 1989 Sep 16, 1999 Aug 21, 2004 Oct 14, 2005 Sep 06, 2008 Sep 08, 2011	No – Long Duration Yes - Hurricane No – Tornado Yes - Flood No - Thunderstorm Yes - Big Storm Yes - Hurricane No – Thunderstorm Yes - Tropical Storm Yes -Tropical Storm Yes – Hurricane	4
18S / 18N	-129.0	7	Oct 24, 1959 Sep 12, 1960 Aug 11, 1985 May 24, 1989 Oct 15, 2005 Sep 06, 2008 Sep 08-, 2011	No – Long Duration Yes – Hurricane No - Thunderstorm Yes - Big Storm Yes -Tropical Storm Yes - Tropical Storm Yes – Hurricane	2

Notes: 1. Level resulting in 53 closures of the 1-year CSO gates

Perspective on Predicted Tunnel System Performance

There is a certain degree of uncertainty in the predicted tunnel system performance presented above. Factors that should be taken into account with respect to the 54-year period of record results are:

- The 54-year period of record modeling results show long periods with no predicted gate activations for CSOs associated with sensitive receiving waters. For example, for the 18S / 18N configuration, no activations of these gates are predicted for the 25-year period from 1960 to 1985. Then, five activations are predicted in the next 25-year period. This means that it will be important for project stakeholders to note that an assessment of the system's ability to meet overflow performance measures cannot be made until many years of operation have passed.

- Only one rain gage was available for the 54-year period of record rainfall data. Many high-intensity summer storm events picked up by that single gage, and applied uniformly over the study area, likely only impacted a limited portion of the study area. Accordingly, high-intensity summer storm events may be less likely to cause the tunnel system to fill than predicted.

- The system-wide hydraulic model does not take into account inflow restrictions. For example, catch basin inlet grates have a limited capacity to admit high rates of runoff during high intensity storm events. Thus, it is possible that peak flows entering the tunnel may be less than modeled.

- The results are based on a historical period of record which will not re-occur in the future. In particular, global climate change has been predicted to result in a larger number of intense storms. Thus, some degree of conservatism is warranted in the design.

As indicated in the bulleted text above, in some cases, the inherent uncertainty in the analyses performed leads to a conservative (over-designed) condition. This is considered appropriate given the need to attain regulatory compliance with a costly built solution. It also indicates that optimization of tunnel system controls, 1-year CSO gate closure elevations in particular, can likely be improved over time based on actual operating experience.

The simple, predictive control approach described above is an example of not blindly accepting pre-programmed gate control set points based on model output and then applying them for all situations. Instead, operators inject "human" thought into the control system, and over-ride the normal control gate set points based on modern weather forecasts. In so doing, they are able to operate the tunnel system more efficiency than would otherwise be the case using only pre-programmed set points.

SEDIMENT DEPOSITION INVESTIGATIONS

Another aspect of the tunnel design that involved imposing "human thinking" over raw model results was sediment calculations. Sediment modeling was performed to estimate the distribution and quantity of sediment in the tunnel over time. The SWMM model used for the hydraulic analyses does not support sediment calculations. Therefore, the SWMM model was imported into the InfoWorks CS modeling software. InfoWorks CS is a collection system modeling software distributed by Innovyze which includes three sediment transport algorithims:

- KUL
- Ackers-White
- Velikanov

The KUL methodology did not contain sufficient information to apply it for the SHCST, and therefore it was not used.

The Ackers-White sediment model is the default sediment model used in the InfoWorks CS Software. This algorithm computes the maximum concentration of sediment that can be held within the flow. This is defined as the "carrying capacity". If the actual concentration is greater than the carrying capacity, then deposition occurs, and if the actual concentration is less than the carrying capacity, then bed erosion occurs. Erosion is assumed to occur instantaneously, while the rate of deposition is a function of the settling velocity.

The Velikanov sediment model uses two total suspended solids (TSS) concentrations in the water (C_{min} and C_{max}). If the flow concentration is below C_{min} then erosion occurs to achieve Cmin if possible. If the flow concentration is above C_{max} then deposition occurs to achieve C_{max} if possible.

The sediment analysis was performed using both the Ackers-White and Velikanov algorithms, and the variation in the results provides an indication of the uncertainty that is inherent in these calculations.

Model Setup

A schematic of the InfoWorks CS model is shown in Figure 8. Sediment transport in the North Tunnel System may impact the South Tunnel System, and therefore the two tunnels were analyzed as part of an integrated system. The InfoWorks model was based on the refined version of the SWMM model with 5.49 m (18 feet) diameter tunnel reaches subdivided into 76 m (250 ft.) segments. Flow inputs to the tunnel were extracted from the SWMM model and were added at the corresponding location in the InfoWorks model. The pump station at the HWPCF was configured the same as in the SWMM model.

The InfoWorks software allows two sediment fractions to be simulated, such as organic (volatile fraction) and inorganic (grit). Table 6 provides a summary of the sediment characteristics used in the modeling. Information from previous studies indicated the average TSS were 142 mg/L. Based on typical composition of untreated domestic wastewater (Metcalf & Eddy, 1991), the TSS concentration of 142 mg/l was divided into an organic fraction of 106 mg/l and an inorganic

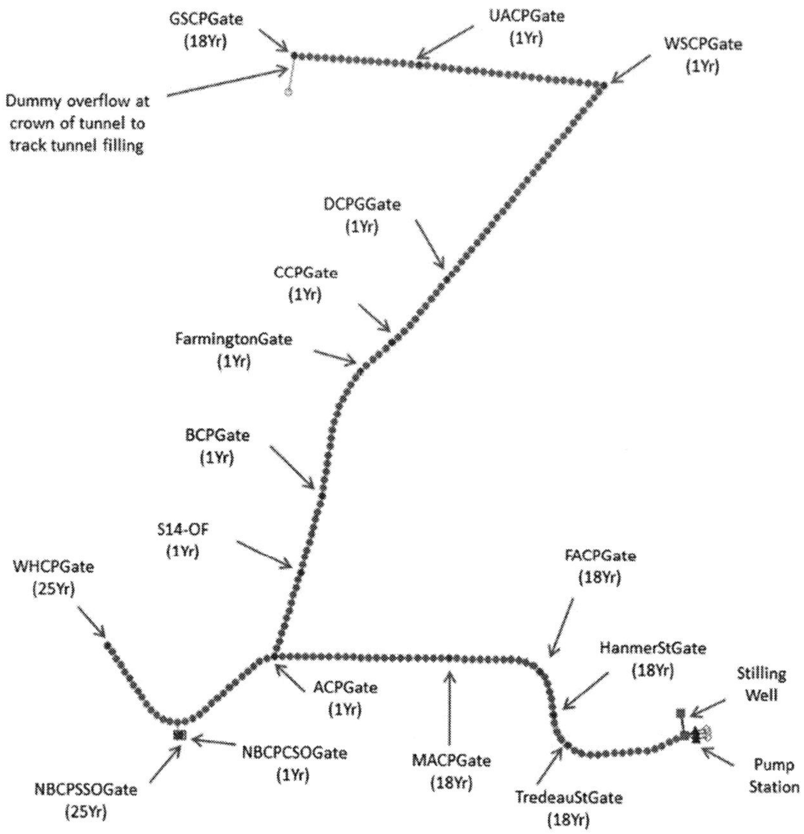

Figure 8. Schematic of InfoWorks CS Sediment Model

Table 6. Summary of Sediment Characteristics for Settlement Modeling

Sediment Fraction	CSO	SSO	Particle Size	Specific Gravity
Inorganic	36 mg/L	None	0.75 mm	2.6
Organic	106 mg/L	130 mg/L	0.04 mm	1.4

fraction of 36 mg/l. For flows entering tunnel dropshafts with CSO flows, both fractions were applied. However, for SSO inflows, only the organic fraction was used. An inorganic fraction was included in the TSS from CSOs to reflect roadway constituents (i.e. sand and grit) that could wash off and become part of the solids matrix in CSOs. Other parameters required for the sediment modeling include the median particle size, and specific gravity of the inorganic (grit) and organic (sanitary solids), and these were obtained from CIRIA (1996).

Sensitivity Testing

Experience has shown that sediment modeling can be sensitive to the time step used in the model. One of the ways that the results of sediment models can be assessed is by computing the mass balance. This involves extracting from the simulation results the mass of TSS entering the tunnel, the mass leaving the system through the pump station, and the mass that accumulates (deposits) in the tunnel. Multiple model runs for the year 2002 were performed using both the Ackers-White and Velikanov sediment modeling algorithms with various time steps. The mass in for the year 2002 were the same for all of the model runs and was equal to 237,324 kg for the organic fraction, and 80,222 kg for the inorganic fraction (total 317,546 kg). The model results for the mass out, accumulation, and mass balance are summarized in Table 7.

As can be seen from the results in Table 7, a much greater fraction of the TSS is accumulating in the tunnel for the Ackers-White sediment model than for the Velikanov. The model results indicate the total mass balance errors generally decrease as the time step decreases for the Ackers-White model, and increase for the Velikanov model. Generally, a lower time step results in an improved mass balance. It is not known why the opposite occurred for the Velikanov model. In this respect, the sensitivity results were inconclusive. Both sediment models were used for the full analysis to provide an estimate of the uncertainty inherent in these types of analyses.

Table 7. Results of Sediment Sensitivity Analysis

	Time	Mass Out			Accumulation			Mass Balance		
Sediment Model	Step (sec)	Organic SF1 (kg)	Inorganic SF2 (kg)	Total (kg)	Organic SF1 (kg)	Inorganic SF2 (kg)	Total (kg)	Organic SF1 (%)	Inorganic SF2 (%)	Total (%)
Ackers-White	60	158,118	368	158,486	42,767	70,250	113,018	15.4	12.0	14.5
Ackers-White	30	168,109	197	168,305	45,219	84,187	129,406	10.1	(5.2)	6.2
Ackers-White	10	175,786	221	176,007	49,271	73,312	122,583	5.2	8.3	6.0
Ackers-White	0.5	177,541	216	177,757	50,279	71,099	121,378	4.0	11.1	5.8
Velikanov	60	220,881	68,155	289,036	8,352	21,558	29,911	3.4	(11.8)	(0.4)
Velikanov	30	225,928	65,282	291,210	6,697	18,734	25,431	2.0	(4.7)	0.3
Velikanov	10	226,541	47,376	273,916	8,222	28,424	36,646	1.1	5.5	2.2
Velikanov	0.5	227,227	47,113	274,341	7,942	28,704	36,646	0.9	5.5	2.1

Sediment Model Results

Model simulations were conducted over the most recent 10 years for which flow data were available (2002 – 2011), and the results are summarized in Table 8. Long term simulations are warranted for sedimentation modeling as accumulation occurs gradually and erosion tends to occur during larger storms.

It can be noted from the results in Table 8 that the net accumulation based on the Ackers-White sediment model for the second 5-year period (years 6 through 10) was approximately half of the accumulation for the first 5-year period, suggesting that the rate of sediment accumulation may decrease. In contrast to the Ackers-White sediment model, the rate of sediment accumulation based on the Velikanov sediment model is roughly the same for the two 5-year periods.

Table 8: Predicted Sediment Accumulation and Volume

Year	Component	Sediment Model Results	
		Ackers-White	Velikanov
End of Year 5	Mass In (kg)	2,478,900	2,478,900
	Accumulation (kg)	651,900	288,500
	Percentage of Mass Inflow Predicted to Accumulate	26%	12%
	Bulk Density (kg/m^3)	1,595	1,662
	Total Volume of Sediment (m^3)	409	174
	Mass Balance (%)	5.8	2.0
End of Year 10	Mass In (kg)	5,006,300	5,006,300
	Accumulation (kg)	973,800	586,700
	Net Accumulation (kg)	321,900	298,200
	Percentage of Mass Inflow Predicted to Accumulate	19%	12%
	Bulk Density (kg/m^3)	1,640	1,650
	Total Volume of Sediment (m^3)	594	356
	Mass Balance (%)	5.4	2.0

Assuming a porosity of 30%, the bulk density of the sediment was calculated for the various simulations, and results are presented in Table 8. In general, the bulk density is on the order of 1,600 kg/m^3 (99.7 lb/ft^3). The mass balance errors were about 5.5 % for the Ackers-White sediment model and 2% for the Velikanov sediment model. A positive mass balance error indicates that more TSS entered the tunnel than accounted for by the mass existing at the pump station and the mass that accumulated in the tunnel. Thus, the actual accumulations in the tunnel could be marginally higher. However, the mass balance errors are considered within acceptable limits for this type of analysis.

Figures 9 and 10 shows the sediment depth distribution in the South tunnel after 5 and 10 years for the Ackers-White and Velikanov models, respectively. The sediment depth for the North Tunnel is similar. In general, higher sediment accumulations are predicted to occur at the downstream end of the South Tunnel. This follows logically since water will back up at this location and provide the greatest opportunity for sediments to deposit. The sediment modeling also predicts that the depth of sediment is lower just downstream of the S-19 and S-21 Gate connection (refer to Figure 8). This is believed to occur because the North Tunnel connects at this location, and the additional flow from the North Tunnel reduces the sediment deposition. The maximum depth of sediment after 10-years is predicted to be less than about 0.3 meters (one foot) for both the Ackers-White and Velikanov sediment models.

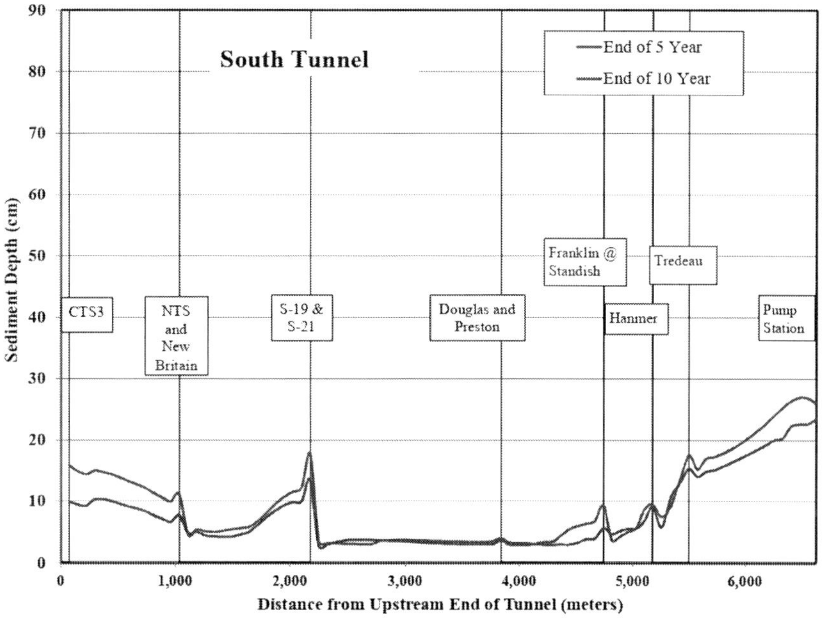

Figure 9. Predicted Sediment Depth for Ackers-White Sediment Model

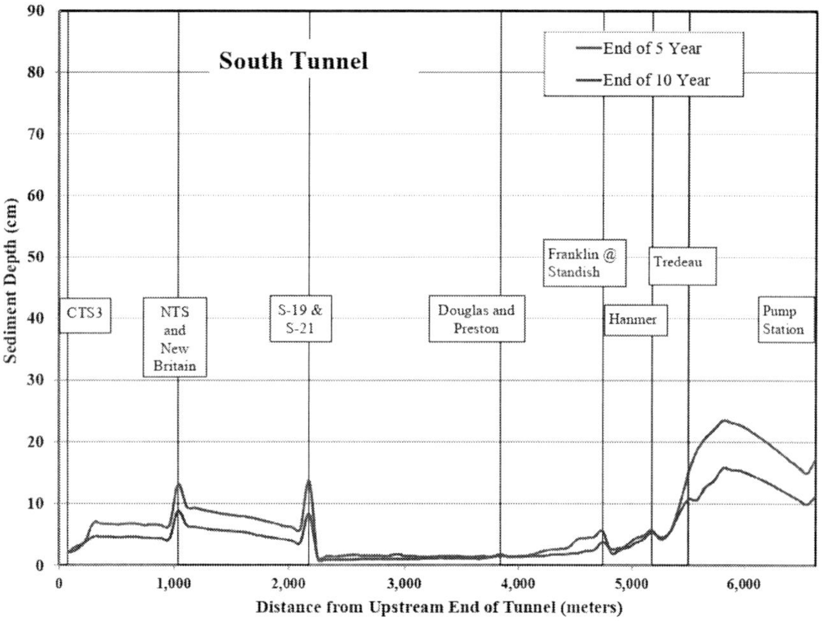

Figure 10. Predicted Sediment Depth for Velikanov Sediment Model

As these results show, there is a lot of uncertainty with modeling of the sediment. The sediment models produce results, and the challenge then becomes how to gain confidence in the predictions. Some of the tools that were employed for the SCHST sediment modeling include time step sensitivity testing, comparing results from two different sediment modeling algorithms, and mass balance analyses. Based on these analyses, the general magnitude of the potential sediment deposition was assessed and is believed to be reasonable.

CONCLUSIONS AND RECOMMENDATIONS

Hydraulic modeling is an important component in the planning, design, and optimization of deep tunnel systems such as the SHCST. However, the model predictions have to be checked, and challenged, in order to gain confidence in the results. Some of the ways this was done for the SHCST include:

Compare Model Results for Different Time Steps

Numerical models can be unstable. In general, the longer the time step, the faster the model runs and the more susceptible the results are to numerical instability. The instabilities may be difficult to detect in a 54-year simulation. Time step sensitivity testing was performed on shorter runs which indicated a 0.5 second time step produced the most accurate result.

Recommendation: An assessment of the time step should be made during model development to confirm the model has time step independence, which means the model results are not significantly affected by the selection of the time step used in the model.

Compare Volume Stored with Hand Calculations

The maximum volume of water stored in the tunnel was computed based on a mass balance analysis using flow volume into the tunnel minus flow volume exiting the tunnel. The peak hydraulic grade predicted by the model was then used to compute the volume of water stored in the tunnel using hand calculations based on an inclined cylinder formula. The initial results indicated a significant discrepancy (6 to 9 %) for the initial model configurations which utilized relatively long tunnel segments. The discrepancy was reduced to 1.3 % by subdividing the tunnel reaches into smaller segments.

Recommendation: Model results should be checked against hand calculations whenever possible.

Recommendation: When applying SWMM for large diameter, relatively flat pipes, such as a tunnel, care should be taken to avoid excessively long segments. If necessary, long pipes should be subdivided.

Compare Design Storm and Continuous Simulation Results

Initial sizing of the tunnel was performed using design storms selected from the historical period of record. Gate control and tunnel sizing for a 1-year storm should have, in theory, resulted in about 54 overflows over a 54-year period. This was not the case for the SHCST because the shape of the 1-year design storm used for the SHCST was not typical. The storm started out slow and stopped, which allowed pumps to dewater the tunnel. When the storm picked up again, it was essentially a new storm with a smaller volume than a 1-year design storm.

Recommendation. Confirm design storm simulation results with continuous simulations whenever possible.

Use Predictive Control to Further Optimize Tunnel Performance

The 54-year period of record simulations resulted in too many CSOs in sensitive areas, which are only allowed to overflow on average once every 18-years or twice in the 54-year period of record. Iterative model simulations indicated that it was not possible to meet the CSO control objectives with a pre-programmed gate control system without increasing the size of the tunnel. However, many of the storms that resulted in CSOs in sensitive areas could have been predicted days in advance by modern weather forecasts. As a result, human "Predictive Control" could be used to further optimize the operation of the tunnel by closing the CSO gates earlier than the pre-programmed set-points based the hydraulic model.

Recommendation: Don't rely solely on model results for design and operations. Look for opportunities to inject human thought, such as the application of Predictive Control, into the process.

Check Sediment Calculations Carefully

Sediment calculations are inherently uncertain. Some of the tools that were employed for the SHCST sediment modeling to check the results included time step sensitivity testing, comparing results from two different sediment modeling algorithms, and mass balance analyses. The time step testing established that a 0.5 second time step was appropriate. The results from the different sediment modeling algorithms established the range of variability in these types of analyses and the mass balance analyses provided a measure of the accuracy. Based on these analyses, the general magnitude of the potential sediment deposition was assessed and was determined to be minimal.

Recommendation: When performing sediment calculations, use multiple sediment models, conduct time step testing, and check the mass balance.

It is hoped that the conclusions and recommendations presented above will be valuable to others facing the design of similar tunnel systems to control wet weather overflows. Each tunnel design should be carefully considered in terms of the site-specific overflow performance targets and design criteria.

REFERENCES

AECOM in association with Black and Veatch, February 2013. *South Hartford Conveyance & Storage Tunnel, Hartford Connecticut. Final Basis of Design Report.*

CIRIA, 1996, *Report 141: Design of Sewers to Control Sediment Problems*, Construction Industry Research and Information Association, London.

Hartford MDC and CDM Smith, December 4, 2014. *Long-Term Combined Sewer Overflow Control Plan 2012 Update*, The Metropolitan District, Hartford, Connecticut.

Metcalf & Eddy, Inc. 1991, *Wastewater Engineering: Treatment, Disposal, and Reuse*, Third Edition, McGraw-Hill, Inc. New York.

Mignone, T. K, S. Craig, G. Heath, and J. Sullivan (2016), *Functional Design for the South Hartford CSO Storage Tunnel*, Presented at Water Environment Federation Collection Systems Specialty Conference, May 2, 2016.

Innovative Parabolic Depth Sensor Allows Agencies to Detect Growing Blockages, Conduct Micro-metering and Measure CSOs

Patrick Stevens, PE, ADS
Kevin Enfinger, PE, ADS

ABSTRACT

Dozens of sewer agencies in the United States and Australia have found important uses for an innovative depth sensor that relies on the combination of traditional ultrasonic depth technology with the parabolic microphone seen on the sidelines of football games. People will recall from high school geometry that the parabola has the distinctive characteristic of directing any sound energy approaching it to the mathematical focal point of the parabola. This allows the sideline microphone to hear every time Payton Manning yells 'OMAHA' before a snap.

It is possible to make the same sideline parabolic microphones operate as a speaker and transmit focused sound back to Payton Manning's replacement. All it takes to get the device to operate in both modes is to place both a speaker and a microphone at the focal point of the parabola. This is exactly what ADS Environmental Services has done with its Long Range Depth (LRD) sensor and its stand-alone version, ECHO.

The result is a technology that is mounted inside the manhole riser section and can measure distance to the water surface as far as 6.1 m (20 feet) below. A depth calibration conducted from the surface will allow the device to measure precise sewer flow depth measurements.

KEYWORDS

Depth-only measurement, parabolic depth sensor, blockage detection, micro-metering.

BACKGROUND

Whisper Dishes

The best way to discuss this new technology is to describe the similarity to Whisper Dishes. Whisper Dishes appear in science museums and many outdoor public facilities. Whisper Dishes consist of two large parabolic reflectors that face each other and can be separated across a room, across a street or even farther. Figure 1 is a schematic showing how they are arranged.

Figure 1 Principle behind a pair of Whisper Dishes

Figure 2 is a close-up view of a Whisper Dish in Dallas, Texas. Two people placing their heads in the circular frame on the opposing dishes can literally hear each other whisper. At the same time ambient sounds are minimized at the listener's position.

Figure 2 Example of a Whisper Dish reflector from Dallas, Texas

The science behind the Whisper Dish is a parabolic curve and its unique relationship with the focal point of the parabolic curve. We have all seen these parabolic reflectors in action on the sidelines of sporting events. Figure 3 shows an audio technician on the sidelines of a football game to listen in on what is being said on the field. Ornithologists use the same technique to record the songs of birds in far-away nests. Satellite TV antennas on the roofs of homes are also parabolic reflectors. The important reflective property of the parabolic curve is that any sound originating at the focal point is focused in a very directional 'beam'. Conversely any sound approaching the companion Whisper Dish is reflected to the same focal point, amplifying it greatly.

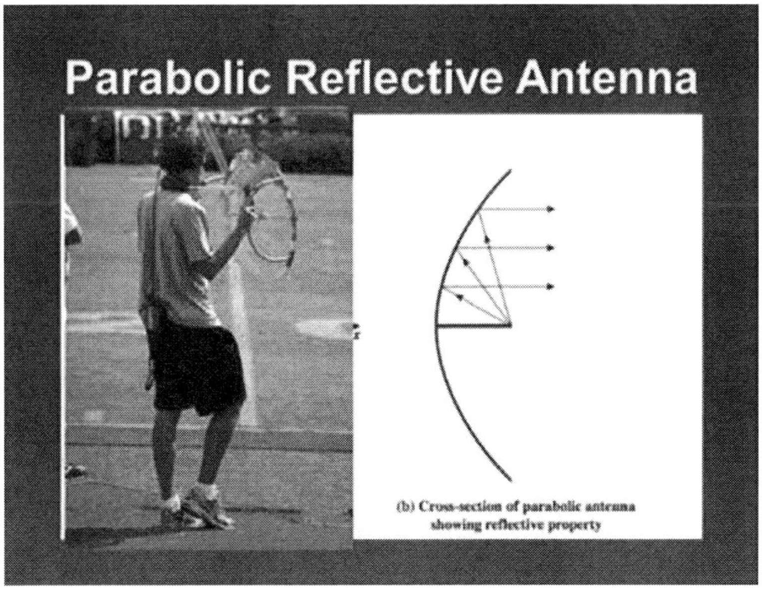

Figure 3 Physical principle of focal point and parabolic curve.

The Technology

ADS Environmental Services has developed a unique acoustic depth sensor that simultaneously operates as both sides of the Whisper Dishes to accurately measure the distance to a water surface. Figure 4 is a schematic view of the sensor showing 1) the horizontally-mounted and highly-directional transducer, 2) the parabolic reflector/housing, 3) the directional acoustic path of the ultra-sonic sound leaving the sensor. The design refocuses and amplifies the returning ultrasonic echo onto the transducer, allowing it to recognize very weak reflections and calculate distances to the water surface of up to 6.1 m (20 feet).

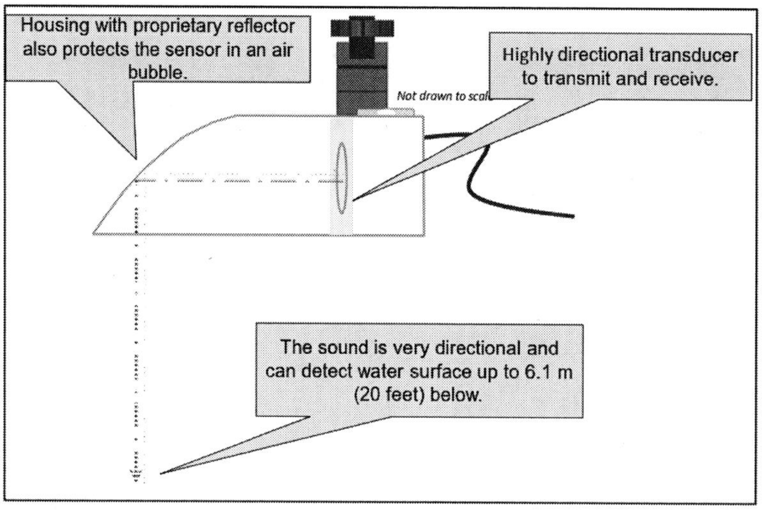

Figure 4 Operating features of the acoustic Long Range Depth sensor.

Figure 5 displays a view of the reflective housing from the underside and it is seen that the large diameter transducer is mounted in the horizontal position at the focal point of the parabolic reflector. The sensor face is vertical and prevents the collection of condensation on the face. The dead zone common to all ultrasonic sensors

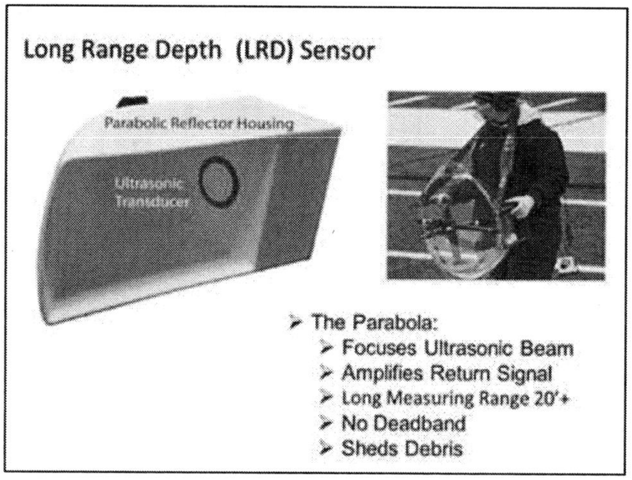

Figure 5 View of the housing with precision parabolic reflector.

occurs in the distance between the sensor face and the parabolic reflector, allowing the sensor to measure the distance to a water surface right up to the bottom of the sensor.

The Long Range Depth (LRD) sensor shown in Figure 5 is a sensor that connects to an ADS Triton meter as one of several sensor types that it can operate. There is a stand-alone version of the technology called the ECHO and it is displayed in Figure 6. The LRD and ECHO mount at the top the manhole and but the ECHO contains its own battery and wireless modems.

The focused beam will 'paint' a target of around 25 cm (10 in.) at a distance of 3.1 m (10 feet) from the sensor face.

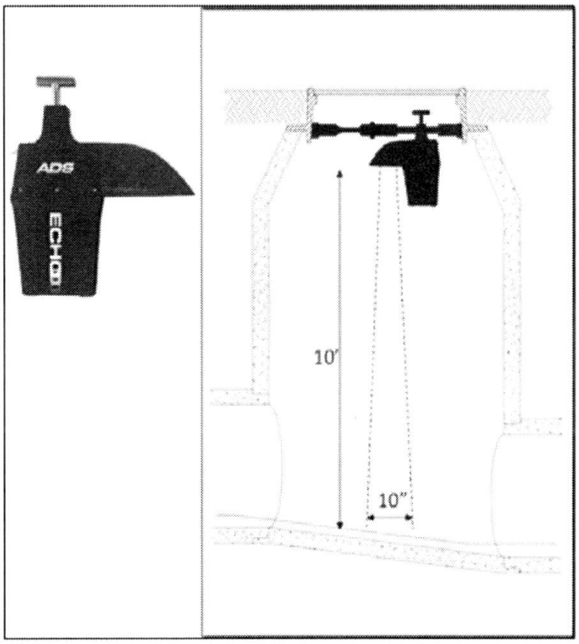

Figure 6 Narrow acoustic 'beam' will paint a small target.

The 'shower curtain rod' type of mounting allows the sensor to be installed directly above most invert channels. Figure 7 demonstrates the flexibility of installation positions.

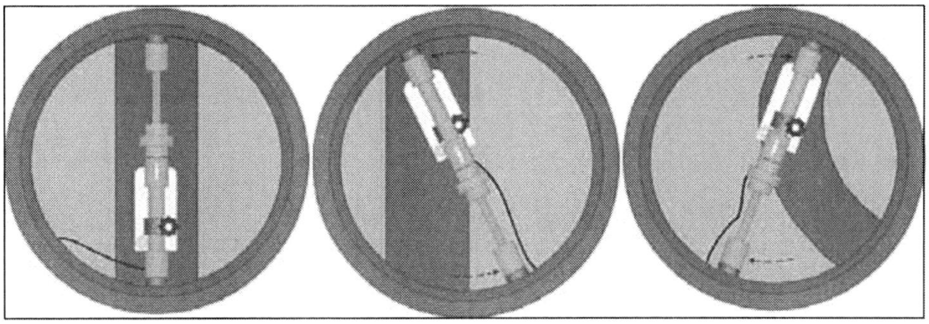
Figure 7 Flexible mounting system allows sensor to be above invert in various locations.

The ECHO also is equipped with a pressure sensor that can measure surcharge up to 2.5 meters over the sensor. During surcharge a bubble forms inside the housing and keeps the sensor from being fouled so that it immediately becomes active as when the water level drops below the sensor.

Five alarm levels can be set to notify the operator of any level of pipe depth or manhole surcharge elevation, including an overflow above the sensor as shown in Figure 8.

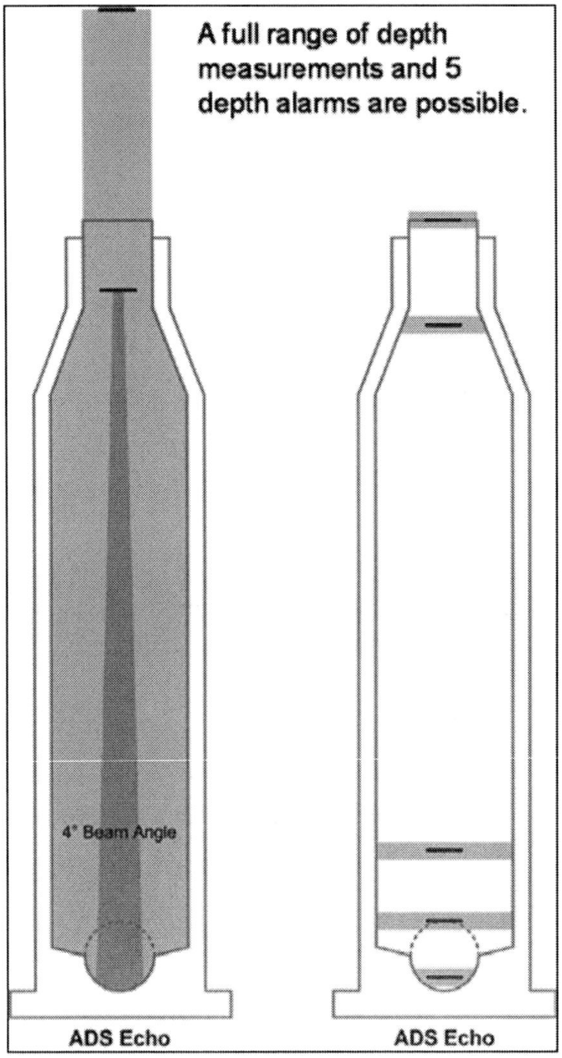

Figure 8 Pressure sensor allows the ECHO to detect an overflow from the manhole in which it is installed.

Blockage Detection

The Water and Sewer Department in the City of Murfreesboro Tennessee had implemented a regularly-scheduled cleaning program at several locations that routinely experienced blockages. As a test of the LRD technology, two LRD sensors were installed upstream and downstream of the known blockage location.

Figure 9 is an aerial view of the blockage location on the yellow 450 mm (18 inch) sewer flowing from east to west. The LRD sensors were installed in the blue-circled manholes upstream and downstream of the blockage area.

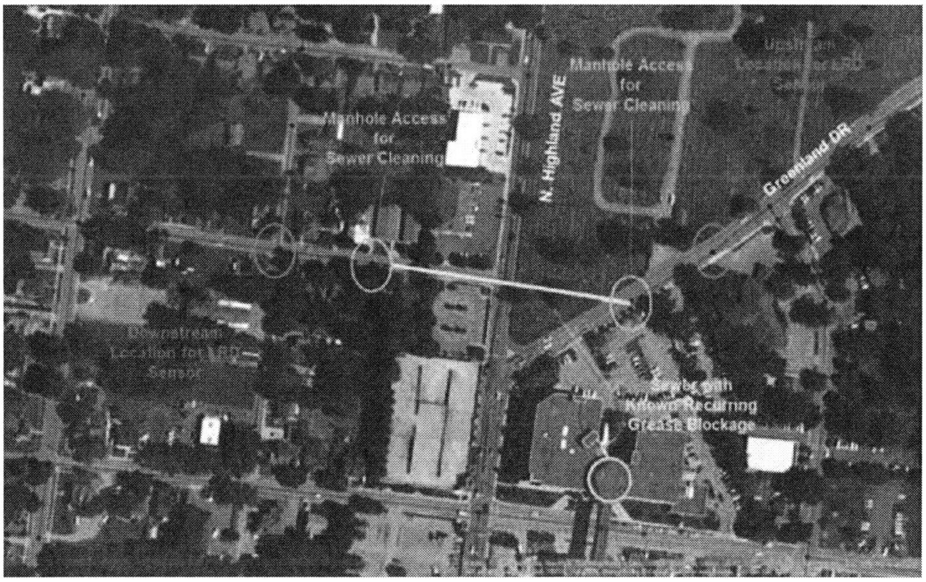

Figure 9 Aerial view of yellow 457 mm sewer line that experienced regular blockages.

Figure 10 is a view of the upstream LRD installation with a distance of 4.8 m to the water surface of the 457 mm sewer. Figure 11 is a view of the downstream installation with a distance of 6.7 m (22 feet) above the water surface. Note that there was a side sewer penetrating almost half the manhole diameter at the downstream site and the sensor was able to obtain an accurate measurement.

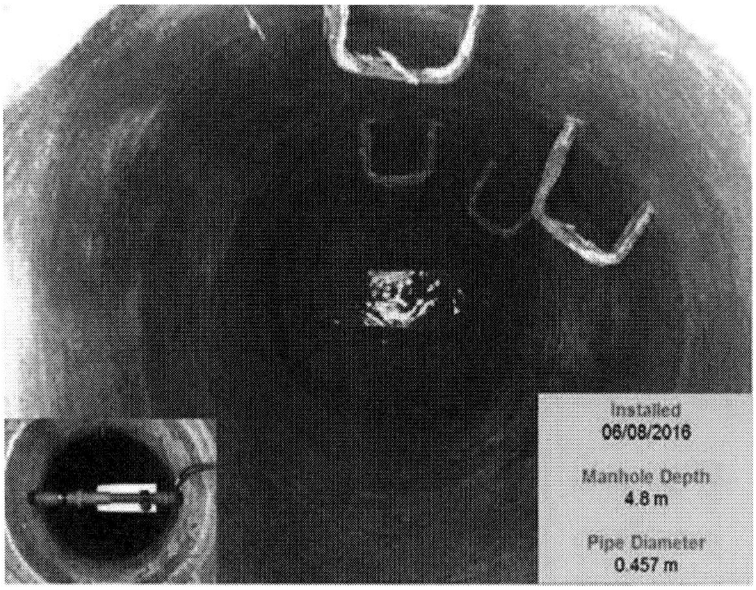

Figure 10 View of upstream sewer 4.8 m below sensor.

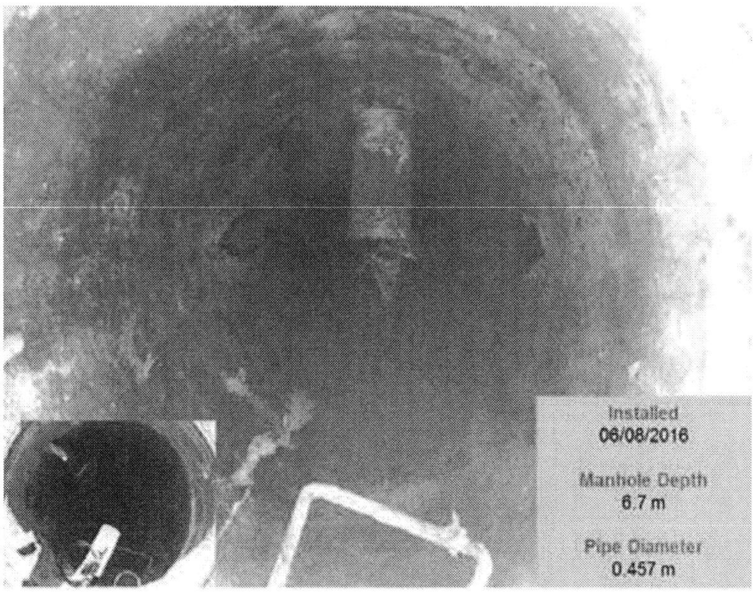

Figure 11 View of downstream sewer with water 6.7 meters below sensor. Note side sewer penetrating into manhole.

On 22 June, ten days after the 10 June installation, the two LRD sensors simultaneously recorded a blockage. Figure 12 displays the two depth hydrographs and both reflect the pattern we would expect. The upstream sensor recorded increasing depths leading to a surcharge on the 23rd.

The pattern recorded by the downstream meter is a relatively constant flow that would be expected downstream of an orifice-type blockage.

Although no alarms were set up during this test, the ECHO is able issue alarms for up to 5 user-selected depths. Had alarms been set, one alarm certainly would have been at full pipe and that alarm would have occurred around noon on Saturday, 23 June. But if we study the pattern of the upstream meter we can see the pattern began to take on a different shape shortly after the early-morning hours of Friday, 22 June.

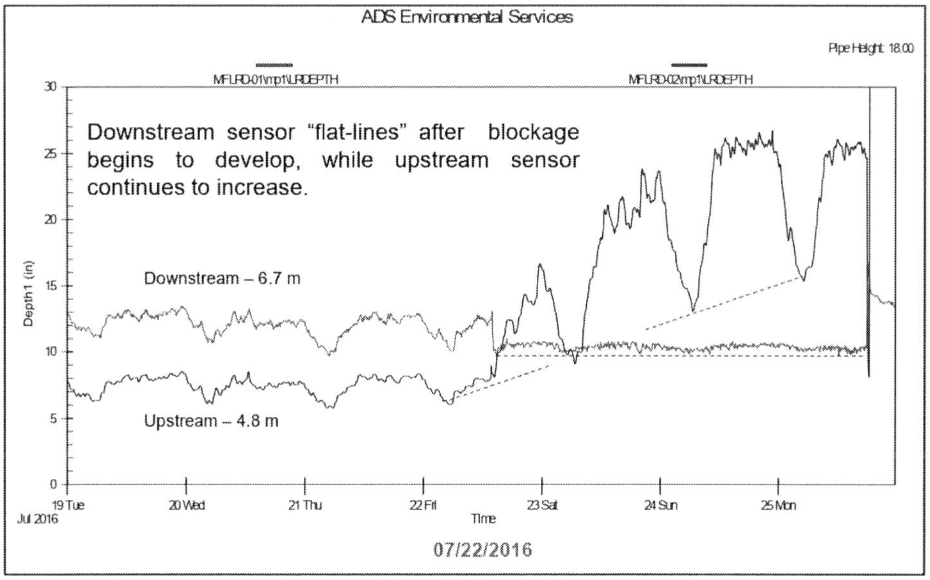

Figure 12 Depth hydrographs of both upstream and downstream sensors, each identifying the blockage that occurred on Friday 22 June.

Micro-metering

There is a lot of 'buzz' in our industry concerning micro-metering. Depending on who is using the term, it can refer to 1) the principle of looking for Rainfall Dependent Infiltration Inflow (RDII) in small and smaller meter basins, 2) a metering device that can measure shallow flows in small sewersheds (micro-basins) or 3) the practice of sequential metering in which a second or third round of metering is conducted upstream of the original meter. The LRD and ECHO sensors are showing promise as a tool to obtain a yes/no indicator of RDII sources in the second or third round of such metering.

Figure 13 is an example one of four small and similarly-sized sewersheds that typically would not be metered by conventional flow meter because of the cost. This small sewer shed is two blocks long and contains 335 meters of sewer. Practitioners are looking for a low cost way to achieve a yes/no indicator in branches in a study basin.

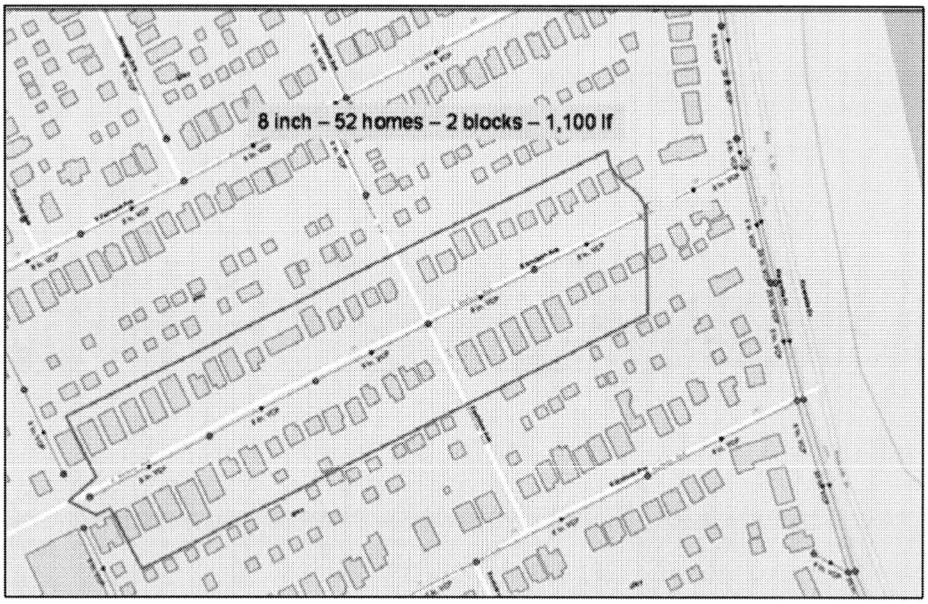

Figure 13 Example of a small 602 m. sewershed that typically would not be metered.

Figure 14 displays the depth data collected for the four micro-basins and is apparent that only one of the four small basins warranted additional metering or SSES work.

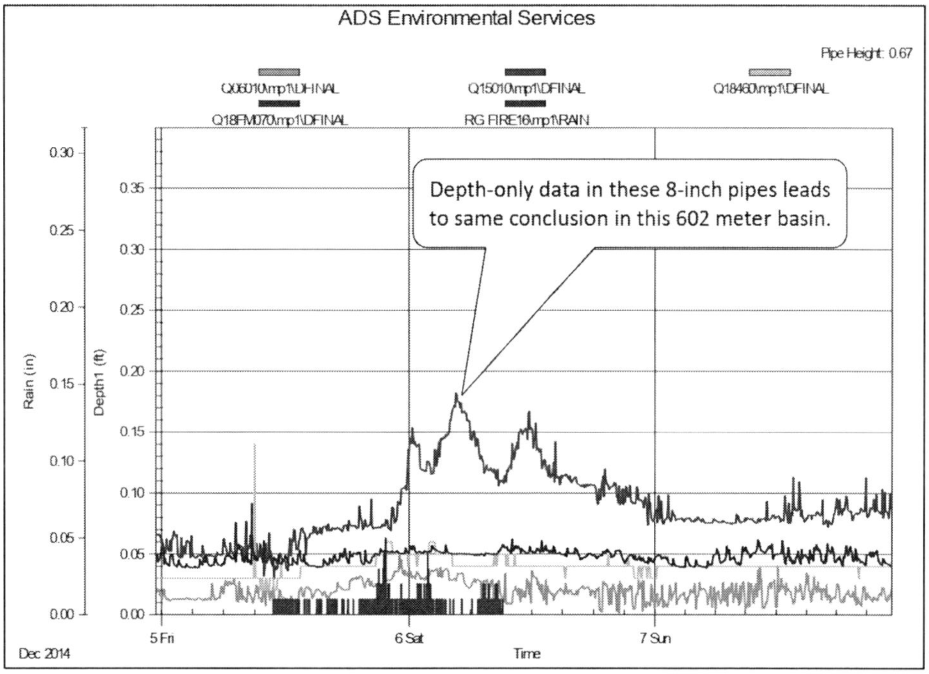

Figure 14 Depth data by itself is able to determine which of the four micro-basins experiences RDII.

Of course, looking for RDII with depth-only data suffers from certain weaknesses. The first problem occurs if the sewers being studied become surcharged during rain events. But in theory this would be known during the first round of metering that depth-only metering would be ruled out.

The second weakness appears if the sewers being studied have a wide range of slopes. A flat sewer could exhibit a large depth change during a storm whereas a more-leaky sewer with a steeper slope may exhibit a small change and appear to be 'good'. The issues can be noted and overcome with good field work during the installation of depth sensors.

CSO Applications

The two critical depths that CSO managers need for public notification or to calculate overflow volumes are the depth on the upstream side of the overflow weir and the water level in the receiving water. Figure 15 shows an ECHO sensor mounted in the manhole above the regulator to determine if the elevation is above the weir elevation and then to use a weir equation to calculate overflow volume.

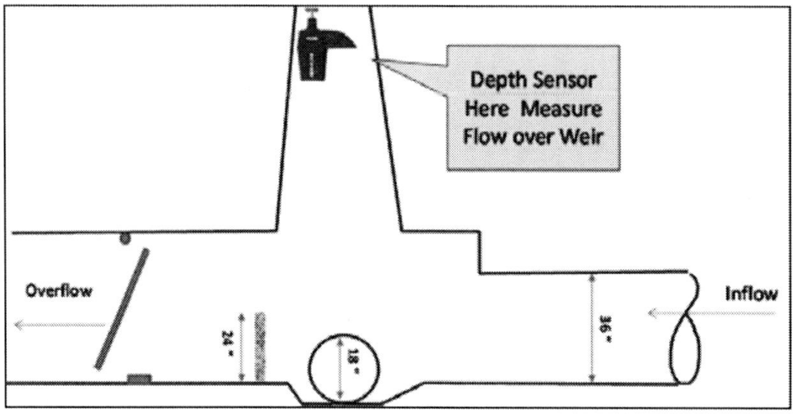

Figure 15 ECHO sensor measuring water surface elevation in a CSO regulator.

Figure 16 is the LRD sensor measuring elevation of the receiving water outside of the CSO regulator. By comparing the two elevations an operator can determine if an overflow is active.

Figure 16 LRD sensor measuring elevation of receiving water body.

CONCLUSIONS

The parabolic 'Whisper Dish' technology employed by the LRD and ECHO sensors is a new and innovative addition to the Collection System Management tool box. It offers the opportunity for managers to detect blockages in such a way that scheduled maintenance visits to 'hot spots' can be replaced by on-demand cleaning conducted only when the data indicate that a blockage is developing.

Its low cost combined with wireless communication and multi-year battery life makes it an economical way to monitor known hot spots as well as overflow alarms and measurements to quantify flow at CSO regulators. Managers can also use the sensor to measure tidal or river elevation to help sort out if an overflow occurred.

Summary of Current and Future Wet Weather Regulations in the United States

Santtu Winter, PE, CH2M
Bill Kreutzberger, CH2M

ABSTRACT

Since the passage of the Clean Water Act of 1972 (CWA), significant progress has been made in improving the quality of the waters in the United States (US). Over the past 25 years, a significant portion of these efforts focused on reducing pollutants in wet weather discharges, such as combined sewer overflows (CSOs), sanitary sewer overflows (SSOs), and municipal separated storm sewer systems (MS4s), in addition to the reductions in pollutants accomplished from wastewater treatment plants (WWTPs) over the nearly 45 years since the CWA was enacted. The policies and regulations surrounding wet weather discharges are complicated and time consuming to understand. This manuscript provides an overview of the current state of these requirements, as well as offering some thoughts on the futures direction of these requirements. An understanding of wet weather requirements is a key component in being able to effectively and efficiently manage these systems while reducing the adverse impacts of these discharges.

The CWA of 1972 and its subsequent amendments is the primary federal law that drives water pollution control activities in the United States, including those targeted towards wet weather discharges. As described in the CWA §101.(a), "The objective of this Act is to restore and maintain the chemical, physical, and biological integrity of the Nation's waters." The CWA sets the lofty goal of completely eliminating the discharge of pollutants into navigable waters, and further stipulates that "the discharge of any pollutant by any person shall be unlawful," unless the discharge is authorized under a permit in accordance with the CWA. The far-reaching and comprehensive CWA is the enabling statute for all federal wet weather requirements and establishes the goal towards which all wet weather programs must strive—namely, the control of all pollutant discharges into US waters.

A number of regulations, policies, and guidance addressing wet weather discharges have been established following the passage of the CWA. This manuscript provides an overview of these regulations, which include the:

- CSO Control Policy (1994)
- Stormwater Phase I and Phase II Rules
- Blending and bypass requirements
- SSO requirements, including the proposed SSO rule in 2001

In addition to discussing requirements, this manuscript discusses the various enforcement actions that have been undertaken by the United States Environmental Protection Agency (EPA) related primarily to CSO and SSO discharges, but expanding more recently to address discharges from MS4s. The background, current state, and trends in these enforcement actions, such as requirements to consider environmental justice and green infrastructure in alternatives analysis, are discussed.

Lastly, this manuscript discusses potential future regulations or regulatory activities that may shape the future of wet weather requirements. Examples include increasing attention being given to enforcement actions against MS4 dischargers, and efforts to increase the affordability of implementing wet weather programs through integrated wastewater and stormwater planning and renegotiation of CSO and SSO consent decrees. Several case studies are presented on how municipalities, both within the US and internationally, have applied unique wet weather planning approaches such as integrated planning to comply with regulatory requirements.

KEYWORDS: Wet weather, regulations, CSO, SSO, stormwater, blending, bypass, integrated planning, climate change

INTRODUCTION

Since the passage of the Clean Water Act of 1972 (CWA), significant progress has been made in improving the quality of the waters in the United States (US). Over the past 30 years, a significant portion of these efforts focused on reducing pollutants in wet weather discharges, such as combined sewer overflows (CSOs), sanitary sewer overflows (SSOs), municipal separated storm sewer systems (MS4s), and wastewater treatment plants (WWTPs). The regulations surrounding wet weather discharges can be complicated and time consuming to understand. This technical memorandum provides an overview of these regulations, as well as references to documents where additional information can be found. An understanding of wet weather regulations is a key component in being able to effectively and efficiently reduce the adverse impacts of these discharges.

SUMMARY

Table 1 summarizes the various wet weather discharges and relevant regulations discussed in this document. Figure 1 illustrates a timeline of the development of the regulations addressing the various wet weather discharges. The following text presents of summary of the regulatory developments.

CLEAN WATER ACT OF 1972

The CWA of 1972 and its subsequent amendments (see Figure 1) is the primary federal regulation that drives water pollution control activities, including those targeted towards wet weather discharges, in the United States. As described in the CWA §101.(a), "The objective of this Act is to restore and maintain the chemical, physical, and biological integrity of the Nation's waters." The CWA sets the lofty goal of eliminating the discharge of pollutants into navigable waters. The CWA further stipulates that "the discharge of any pollutant by any person shall be unlawful," unless the discharge is authorized under a permit in accordance with the CWA. The far-reaching and comprehensive CWA is the ultimate regulatory driver for all wet weather regulations and establishes the goal towards which all wet weather programs must strive—namely, the control of all pollutant discharges into US waters.

A concise summary of the CWA is provided in *Clean Water Act: A Summary of the Law* (Copeland, 2010), an excerpt of which is provided below:

The principal law governing pollution of the nation's surface waters is the Federal Water Pollution Control Act, or [CWA]. Originally enacted in 1948, it was totally revised by amendments in 1972 that gave the act its current shape. The 1972 legislation spelled out ambitious programs for water quality improvement that have since been expanded and are still being implemented by industries and municipalities.

The [CWA] consists of two major parts, one being the provisions which authorize federal financial assistance for municipal sewage treatment plant construction. The other is the regulatory requirements that apply to industrial and municipal dischargers. The act has been termed a technology-forcing statute because of the rigorous demands placed on those who are regulated by it to achieve higher and higher levels of pollution abatement under deadlines specified in the law. Early on, emphasis was on controlling discharges of conventional pollutants (e.g., suspended solids or bacteria that are biodegradable and occur naturally in the aquatic environment), while control of toxic pollutant discharges has been a key focus of water quality programs more recently.

Prior to 1987, programs were primarily directed at point source pollution, wastes discharged from discrete sources such as pipes and outfalls. Amendments in that year authorized measures to address nonpoint source pollution (stormwater runoff from farm lands, forests, construction sites, and urban areas), now estimated to represent more than 50% of the nation's remaining water pollution problems.

Under this act, federal jurisdiction is broad, particularly regarding establishment of national standards or effluent limitations: Certain responsibilities are delegated to the states, and the act embodies a philosophy of federal-state partnership in which the federal government sets the agenda and standards for pollution abatement, while states carry out day-to-day activities of implementation and enforcement.

To achieve its objectives, the act embodies the concept that all discharges into the nation's waters are unlawful, unless specifically authorized by a permit, which is the act's principal enforcement tool. The law has civil, criminal, and administrative enforcement provisions and also permits citizen suit enforcement.

During the early years of CWA program implementation, the focus was on funding and implementation of improvements in publicly owned treatment works (POTWs) to achieve secondary treatment and water quality based treatment requirements. The US Environmental Protection Agency (EPA) recognized the significant potential impacts from CSO and urban stormwater in early funding guidance (USEPA, 1980) and well as the early CSO and urban stormwater needs assessments (USEPA, 1979). Urban stormwater was identified as an important source of pollutants in many of the area-wide water quality plans developed across the United States in the 1970's and EPA established the Nationwide Urban Runoff Program in 1978 in coordination with the US Geological Survey and published a comprehensive report documenting the extent of pollution from stormwater runoff (EPA, 1983).

Despite the early recognition of the significance of pollutants discharged from wet weather sources, little implementation of control programs was initiated in the 1980's with the exception of lawsuit-driven CSO control programs such as the Milwaukee Pollution Abatement Program and the Chicago Tunnel and Reservoir Plan which was initiated as a flood control initiative in the mid-1970's. The Water Quality Act of 1987, which amended the CWA, included a framework for permitting and controlling stormwater from municipal and industrial sources. While EPA began initial efforts to implement CSO and SSO control in many locations in the late 1980's and early 1990's, there was not a comprehensive framework nor consistent approach between EPA Regions and between states for this effort. The following sections further describe the development of the frameworks that are in place today.

Table 1. Summary of Wet Weather Discharge Categories and Current Regulatory Requirements

Wet Weather Discharge	Current Regulatory Requirements	Trends and Possible Future Requirements
CSOs	Implement nine minimum controls, which includes technology-based requirements such as the maximizing flow to the WWTP, maximizing system storage, and public notification requirements. Develop and implement a long-term control plan (LTCP) that details what control measures are required to reduce CSO discharges so that remaining discharges do not cause or contribute to waterbody impairment. Focus of EPA enforcement actions through issuing of consent decrees to major dischargers. Relevant federal regulations: CSO Control Policy (EPA, 1994), published in the Federal Register, Vol. 59 No. 75, April 19, 1994	Further modification of receiving water quality standards or their interpretation, as called for in section III of the CSO Policy and potentially addressed in Total Maximum Daily Load (TMDL) planning for urban runoff pollutants Continued enforcement actions by the EPA Re-opening of consent decrees to change CSO control measures Increasing consideration for integrated planning as a way to extend timelines for implementing CSO control programs Development of exit strategies from consent decrees as communities complete required reduction efforts
SSOs	Wastewater NPDES permit holders must "at all times properly operate and maintain all facilities and systems of treatment and control (and related appurtenances)" (CFR 40 122.41(e))– how this is done and what it means for the wastewater collection system has not been codified at the federal level All SSOs are a violation of the CWA, regardless of the cause Proposed SSO rule developed in 2001 but was never finalized, it clarified what properly operating and maintaining the collection system means, required implementation of a CMOM	Eventual release of SSO rule likely in combination with updated blending regulations Continued enforcement actions by the EPA Re-opening of consent decrees to change CSO control measures Increasing consideration for integrated planning as a way to extend timelines for

	program, monitoring/reporting/ public notification, and a framework for raising a defense for unavoidable discharges. Under this rule all wastewater collection systems (with 1 mile or more of pipe) would have required an NPDES permit. Some states implemented their own SSO regulations, including California, Wisconsin, North Carolina, and South Carolina. These state regulations require permits for wastewater collection systems, and are largely based on the proposed 2001 SSO rule from the EPA. On a federal level SSO regulation is primarily driven by enforcement actions by the EPA and state regulators, which has typically included the implementation of a CMOM program. Relevant federal regulations: Standard NPDES permit language to properly operate and maintain wastewater facilities, and the duty to mitigate overflows.	implementing CSO control programs Development of exit strategies from consent decrees as communities complete required reduction efforts
WWTP	Bypasses (intentional diversion of waste stream from any portion of a treatment facility) are a violation of all permits unless specifically authorized by the NPDES permit. Blending (diversion of waste stream around secondary treatment, typically occurring during wet weather, and re-combined with treated effluent prior to disinfection and discharge, where final discharge meets effluent limits) regulations are undefined at the federal level. Some states allow it under certain conditions, others do not. Proposed blending rule issued by the EPA in 2003, received significant comments and was withdrawn. Proposed revised blending rule issued by the EPA in 2005, but was not finalized. Would have allowed blending under specific circumstances and would need to be approved by regulators each permit renewal.	Eventual finalization of blending rule, possibly in combination with SSO regulations Court case currently underway in the Washington D.C. Circuit Court that would apply the 8th Circuit Court's Iowa League of Cities ruling nationwide, allowing blending in all states. Individual permit renewal applications and re-negotiations

	Recent court case allows blending in states within the Eighth Circuit Court (Iowa League of Cities versus the EPA) Relevant federal regulations: Bypass regulations as defined in 40 CFR 122.41	
MS4	Implementation of stormwater management program, which includes six minimum control measures: public education and outreach; public participation/ involvement; illicit discharge detection and elimination; construction site runoff control; post-construction runoff control; and pollution prevention/good housekeeping. Must be implemented in all "urbanized areas" as delineated by the Bureau of the Census, as well as for construction sites that disturb one acre or more. Relevant federal regulations: NPDES Permit Application Regulations for Storm Water Discharges (a.k.a. MS4 Phase I Stormwater Rule), 40 CFR Parts 122, 123, and 124, published in the Federal Register Vol. 55, No. 222, 11/16/1990 NPDES Regulations for Revision of the Water Pollution Control Program Addressing Storm Water Discharges Final Rule (a.k.a. MS4 Phase II Stormwater Rule), 40 CFR parts 9, 122, 123, and 124, published in the Federal Register Vol. 64, No. 235, 12/8/1999	• Increased emphasis on federal enforcement actions for compliance with stormwater regulations • Possible resurrection of proposed stormwater rule revisions • Possible enforcement to support TMDLs

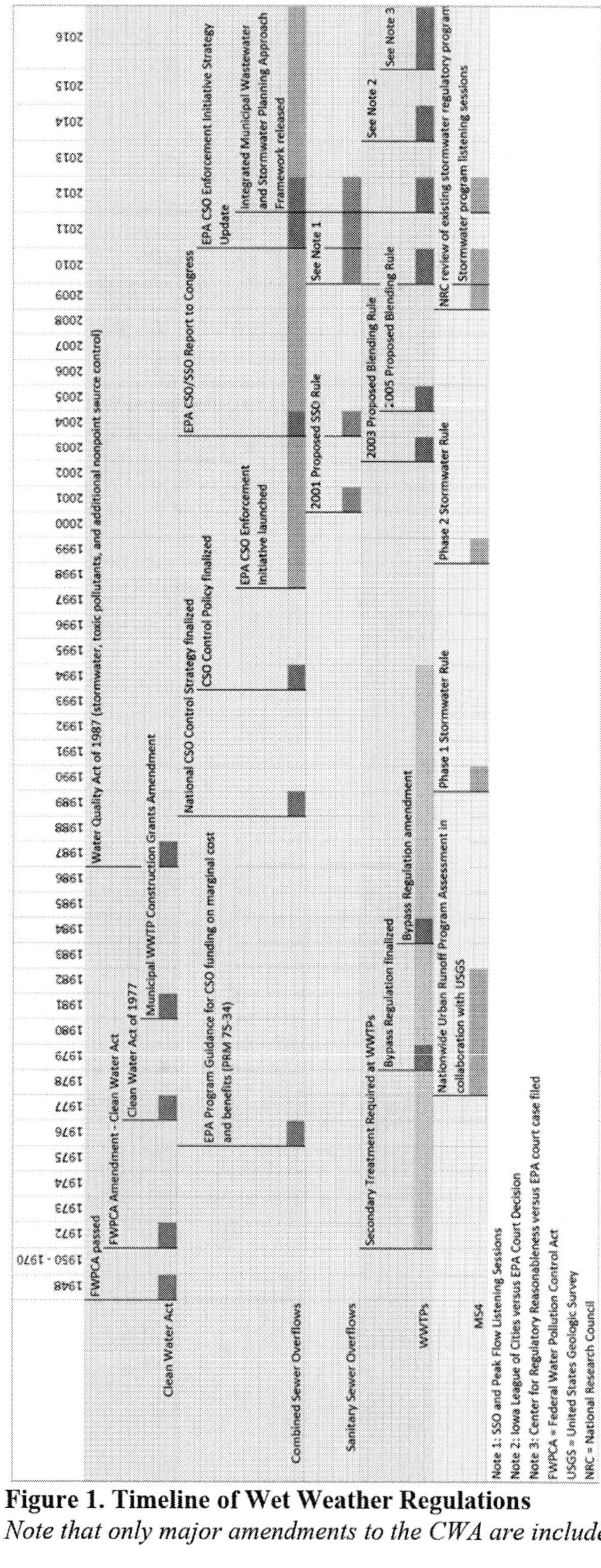

Figure 1. Timeline of Wet Weather Regulations

Note that only major amendments to the CWA are included in the figure.

COMBINED SEWER OVERFLOWS

A description of CSOs is provided in EPA's *Report to Congress on the Impacts and Control of CSOs and SSOs* (EPA, 2004), as excerpted below.

Two types of public sewer systems predominate in the United States: combined sewer systems (CSSs) and sanitary sewer systems (SSSs). CSSs were among the earliest sewer systems constructed in the United States and were built until the first part of the 20th century. As defined in the 1994 CSO Control Policy (EPA [1994]), a CSS is:

A wastewater collection system owned by a state or municipality (as defined by Section 502(4) of the [CWA]) that conveys domestic, commercial, and industrial wastewaters and storm water runoff through a single pipe system to a publicly-owned treatment works.

During wet weather events (e.g., rainfall or snowmelt), the combined volume of wastewater and storm water runoff entering CSSs often exceeds conveyance capacity. Most CSSs are designed to discharge flows that exceed conveyance capacity directly to surface waters, such as rivers, streams, estuaries, and coastal waters. Such events are called CSOs.

A CSO is defined as: the discharge from a CSS at a point prior to the POTW treatment plant.

Current Combined Sewer Overflow Regulations

EPA's regulatory approach regarding discharges from Publically Owned Treatment Works (POTWS) have been interpreted for CSO discharges primarily through the CSO Control Policy (EPA, 1994), which elaborated on the EPA's earlier released CSO Strategy (EPA, 1989) by clarifying what actions CSO communities must take to comply with the CWA with regard to their CSO discharges. As described in both the CSO Strategy and CSO Control Policy, the objectives of the national effort to reduce CSO discharges are to accomplish the following:

1. Ensure that CSOs are only caused by wet weather (i.e., dry-weather CSOs do not occur).
2. Bring all CSO discharges into compliance with technology-based and water quality-based requirements of the CWA.
3. Minimize the water quality, aquatic life, and human health impacts of CSOs.

To achieve these objectives, the CSO Control Policy establishes two key requirements: (1) implementing the nine minimum controls, and (2) developing and implementing long-term control plans (LTCPs). As described in EPA's *Guidance for Nine Minimum Controls* (EPA, 1995), implementing the nine minimum controls "can reduce CSOs and their effects on receiving water quality, do not require significant engineering studies or major construction, and can be implemented in a relatively short period," and include activities like controlling solid and floatable materials in CSOs and implementing a program to notify the public when CSOs occur. LTCPs contain a set of CSO control projects that, once implemented, "will ultimately result in compliance with the requirements of the CWA" (EPA, 1994). LTCPs must also include post-construction compliance monitoring programs that should be "adequate to verify compliance with water quality standards and protection of designated uses." Once an LTCP project is complete, its maintenance and management become an additional component of the NPDES permit for the system along with the nine minimum controls. The relationship between these requirements is described in the flow chart presented in Figure 2.

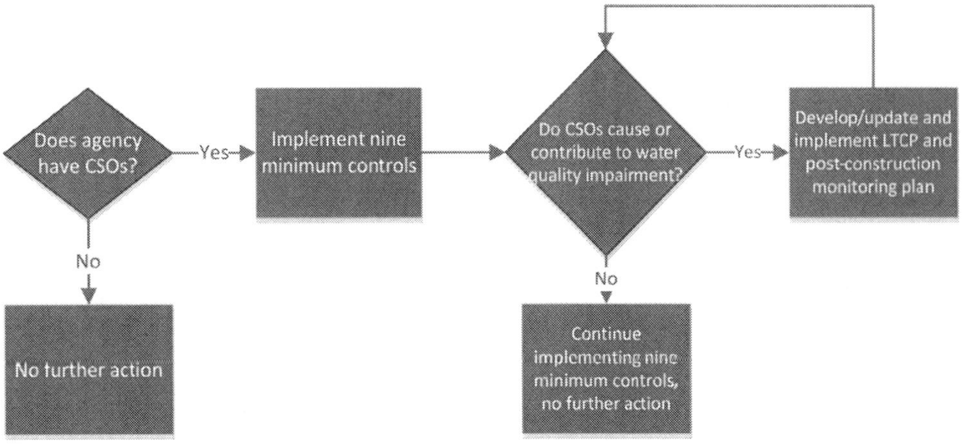

Figure 2. Summary of Requirements in the CSO Policy

The EPA CSO Control Policy provides two approaches for determining whether the improvements in a LTCP contribute to violations of the water quality standards of the CWA: the presumptive and demonstrative approaches, which are summarized below; most agencies have selected the presumptive approach to determining compliance:

Presumptive approach—CSO discharges are "presumed to meet an adequate level of control to meet the water quality-based requirements of the CWA" (EPA, 1994) if they are reduced to one of the following levels:

1. "No more than an average of four overflow events per year, provided that the permitting authority may allow up to two additional overflow events per year…
2. The elimination or the capture for treatment of no less than 85% by volume of the combined sewage collected in the CSS during precipitation events on a system-wide annual average basis; or
3. The elimination or removal of no less than the mass of the pollutants, identified as causing water quality impairment through the sewer system characterization, monitoring, and modeling effort, for the volumes that would be eliminated or captured for treatment under [item #2] above" (EPA, 1994).

Demonstrative approach—"A permittee may demonstrate that a selected control program, though not meeting the criteria specified [under the presumptive approach described above] is adequate to meet the water quality-based requirements of the CWA. To be a successful demonstration, the permittee should demonstrate each of the following:

1. The planned control program is adequate to meet water quality standards and protect designated uses, unless water quality standards or uses cannot be met as a result of natural background conditions or pollution sources other than CSOs;

2. The CSO discharges remaining after implementation of the planned control program will not preclude the attainment of [water quality standards] or the receiving waters' designated uses…

3. The planned control program will provide the maximum pollution reduction benefits reasonably attainable; and

4. The planned control program is designed to allow cost effective expansion or cost effective retrofitting if additional controls are subsequently determined to be necessary…" (EPA, 1994).

State regulations vary from state to state, but must at a minimum be as stringent as the federal requirements. State water quality standards regulations are also very important in determining the actions in LTCPs that are needed to protect water quality standards.

CSO control requirements are implemented through the National Pollutant Discharge Elimination System (NPDES) permit program. Many municipalities with CSOs have also entered into consent decrees or orders with the EPA and/or state environmental regulators, as discussed in Section 3.2. These decrees/orders contain requirements for implementing specific CSO reduction measures by certain dates and allow for monetary penalties if compliance is not achieved. CSO consent decrees/orders also often address SSOs and their reduction.

In summary, compliance with CSO control requirements ultimately demands that the receiving waters meet water quality standards and designated uses. Regardless of whether a CSO community takes the presumptive or demonstrative approach to CSO discharges, they must ultimately prove, through the post-construction compliance monitoring program, that the implemented CSO control projects result in receiving water compliance with water quality standards and designated uses or that non-compliance is not the result of the CSOs.

U.S. Environmental Protection Agency Combined Sewer Overflow Enforcement Initiative

The EPA's Office of Enforcement and Compliance Assurance (OECA) has undertaken numerous enforcement actions aimed at increasing compliance with the CSO Control Policy, and ensuring that municipalities are making progress with reducing CSO discharges. A summary of the EPA's enforcement actions related to CSOs is presented in the EPA Office of Inspector General's *EPA Needs to Track Whether Its Major Municipal Settlements for Combined Sewer Overflows Benefit Water Quality* (EPA, 2015a), an excerpt of which is provided below:

In 1998, OECA began a National Enforcement Initiative (CSO initiative) to focus enforcement action on cities that were not complying with their discharge permits and CSO policy. Since the initiative began, the EPA said it has targeted communities where CSOs have caused the most contamination of surface water. In OECA's 2011 strategy for this initiative, the EPA set a goal, which said OECA would address all large CSO communities by 2016. In their 2011 strategy, the agency identified 204 large CSO communities (those with more than 50,000 residents), and the number of large CSO communities grew to 213 by 2015. The EPA considers a CSO community "addressed" when the system is under an enforceable order, or is on a schedule to develop a long-term control plan that the EPA accepts.

Since 1998, the EPA has addressed 92 percent (196) of the 213 large CSO communities nationwide. Enforcement actions can include administrative orders and consent decrees. When the EPA uses consent decrees as enforcement tools, the agency, typically the state, and the U.S. Department of Justice negotiate and enter into a judicial consent decree with the municipality to bring about compliance.

Figure 3 presents the status of the EPA's effort to address large dischargers of CSOs through the CSO enforcement initiative.

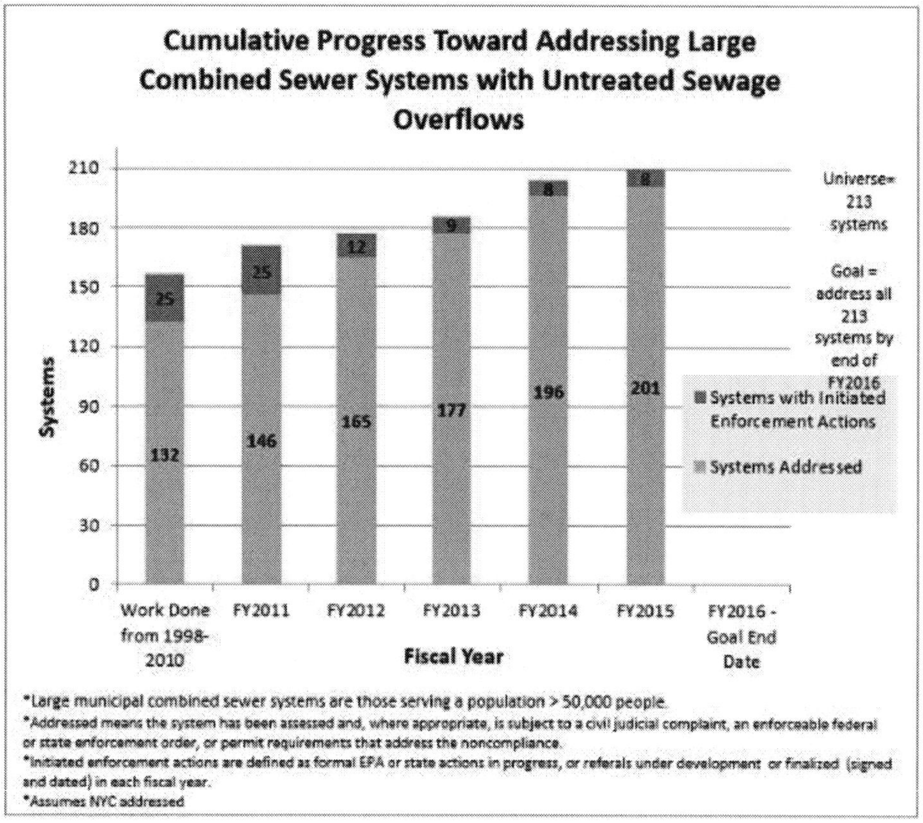

Figure 3. EPA progress in addressing CSOs (EPA, 2016a)

Recent trends in CSO consent decrees have been a topic of interest at the National Association of Clean Water Agencies Law Seminar. A presentation by Linda Benfield of Foley and Lardner, LLP, described three recent trends with CSO consent decrees (Benfield, 2015):

- Green infrastructure (GI)—Recent consent decrees are requiring assessing and considering GI as a means to achieve CSO reductions. The EPA has also released a number of guidance documents related to GI, including guidance on incorporating GI into CSO plans, a GI Strategic Agenda, and guidance on substituting GI control measures for planned gray infrastructure control measures related to CSOs. In general, GI is used to reduce the size of gray infrastructure, but it does not typically eliminate the need for it.

Evaluating gray infrastructure versus GI must involve comparing environmental and socioeconomic benefits, as well as environmental justice concerns.

- Environmental justice—The EPA recently released a draft of the Environmental Justice 2020 Action Agenda Framework, which "will help [the EPA] advance environmental justice through its programs, policies and activities" (EPA, 2015b). One goal the framework is to "deepen environmental justice practice within [EPA] programs to improve the health and environment of overburdened communities" through compliance and enforcement. For example, the City of Harrisburg, Pennsylvania, consent decree from February 2015 includes a requirement to consider the impact of alternatives on environmental justice populations when selecting CSO controls.

- Reopenings—Discussed in the following section.

Possible Future Combined Sewer Overflow Regulatory Trends

Reducing CSOs was the second major effort by the EPA after passage of the CWA, following the national effort to implement secondary treatment at all WWTPs. Initial policy setting the stage for CSO reduction have been around for more than 25 years, and agencies have been striving towards reducing CSOs for several decades. CSO requirements themselves are unlikely to change in the foreseeable future; rather, future regulatory activities are likely to focus on continued enforcement actions or on clarifying and further encouraging integrated planning approaches for CSOs and other CWA requirements. Following are these anticipated trends, and a few others:

- Continuing enforcement actions—As described in Section 3.2 the EPA is still expecting to address approximately 12 public agencies with CSOs (as of 2015) with either a federal consent decree or other enforcement action.

- Reopening consent decrees—Most public agencies with CSSs have finished developing LTCPs and are now in the midst of implementing their CSO reduction programs. As the true costs of these programs are being realized, many agencies are looking for ways to extend regulatory milestones to reduce rate increases or to identify other means of reducing the costs of their programs through reopening and renegotiating consent decrees. These reopenings can be driven by a desire to take a different approach to reduce program costs and/or address financial affordability concerns, leading to investigating approaches like integrated planning (see Section 7.2).

- Considering integrated wastewater/stormwater planning—One approach municipalities are considering to control costs, address multiple regulatory requirements and to factor affordability into their requirements is integrated planning. This allows them to look at their system as a whole and develop regulatory deadlines that can be achieved. This approach is discussed in greater detail in Section 7.2.

- Finishing LTCP implementation and terminating consent decrees—A few public agencies have finished implementing their LTCPs and are seeking to terminate (or have successfully terminated) their consent decrees. More and more agencies are expected to be reaching this phase in the coming years, and what happens when LTCPs are fully implemented is somewhat uncertain. What happens if receiving waterbodies are still impaired? If agencies took the presumptive approach, do they need to prove that discharges do not cause remaining violations of water quality standards? Will the EPA require additional consent decrees to implement further CSO reductions?

SANITARY SEWER OVERFLOWS

A description of SSOs is provided in EPA's *Report to Congress on the Impacts and Control of CSOs and SSOs* (EPA, 2004), as excerpted below.

Since the first part of the 20th century, municipalities in the United States have generally constructed [separated sanitary sewer systems] (SSSs). For the purposes of this Report to Congress, an SSS is: a municipal wastewater collection system that conveys domestic, commercial, and industrial wastewater, and limited amounts of infiltrated groundwater and storm water, to a POTW.

SSSs are not designed to collect large amounts of storm water runoff from precipitation events. Areas served by SSSs often have [an MS4] to collect and convey runoff from rainfall and snowmelt.

Untreated or partially treated discharges from SSSs are commonly referred to as SSOs. SSOs have a variety of causes including blockages, line breaks, sewer defects that allow excess storm water and groundwater to overload the system, lapses in sewer system operation and maintenance, inadequate sewer design and construction, power failures, and vandalism. An SSO is defined as: an untreated or partially treated sewage release from a SSS...

Current Sanitary Sewer Overflow Regulations

Like all other point-source discharges of pollutants into the nation's waters, SSOs are illegal unless they are allowed by an NPDES permit. The following subsections describe several current regulatory requirements related to SSOs, as well as the EPA's proposed SSO rule from 2001.

All NPDES permits for WWTP discharges and/or CSO discharges must contain specific language that requires permittees to "at all times properly operate and maintain all facilities and systems of treatment and control (and related appurtenances)" (CFR 40 122.41(e)). What constitutes a properly operated and maintained collection system was something that the EPA sought to clarify when developing the capacity, management, and operations and maintenance (CMOM) program requirement in the proposed SSO rule (see Section 4.1.2).

NPDES permits must also contain language requiring permittees to report SSOs to the state regulator and local public health agency—this is the primary means through which the EPA can track which agencies are discharging large amounts of SSOs and where to pursue enforcement actions. Reporting requirements vary significantly by state.

The EPA estimates that nearly 5,000 satellite wastewater collection systems (i.e., systems that convey sewage to a regional wastewater conveyance and treatment provider) are in the United States (EPA, 2004). If these collection systems do not have CSOs, then they are not required to have an NPDES permit. As such, they are not required to report SSOs to state regulators (in most states), and EPA enforcement actions to reduce SSOs against these public agencies are relatively uncommon. However, SSO discharges are not authorized under the CWA, and third parties can sue the operator of the collections system with the unpermitted discharge.

The need to clarify what a "properly operated and maintained" wastewater collection system is, as well as the fact that satellite agencies do not need NPDES permits, pushed the EPA to undertake rulemaking to address SSOs (EPA, 2010). In the early 2000s, the EPA appeared poised to establish a national standard for regulating SSOs by developing a proposed SSO rule in 2001 and releasing several guidance documents in the years following. The proposed SSO rule required all owners of sanitary sewer systems, including "satellite" systems, to obtain an NPDES permit for their collection systems. The NPDES permits would contain three general permit conditions (Strifling, 2003):

- Implement a capacity, management, operations and maintenance (CMOM) program
- Prohibit discharges from the collection system prior to the WWTP, subject to certain defenses for unavoidable discharges.
- Adhere to reporting and recordkeeping requirements related to SSOs, as well as required public notification.

Conditions under which permittees would be allowed to establish affirmative defense were divided into the following two categories (Strifling, 2003):

- SSOs caused by severe natural conditions, where:
    - The violation was caused by a condition such as a hurricane, tornado, widespread flooding, earthquakes, or tsunamis
    - No feasible alternatives were available.
    - The NPDES or local authority was notified within 10 days of the discharge.
- SSOs caused by factors other than severe natural conditions, where:
    - The cause of the SSO was identified.
    - The SSO was exceptional, unintentional, temporary, and caused by factors beyond the reasonable control of the permittee.
    - The SSO could not have been prevented by the exercise of reasonable control.
    - The NPDES or local authority was notified within 10 days of the discharge.
    - The permittee took all reasonable steps to stop, and mitigate the impact of, the SSO as soon as possible.

The proposed rule laid out the components of a CMOM program. The EPA describes the CMOM approach in the *Guide for Evaluating CMOM Programs at Sanitary Sewer Collection Systems* (EPA, 2005), which is excerpted below.

CMOM programs incorporate many of the standard operation and maintenance activities that are routinely implemented by the owner or operator with a new set of information management requirements in order to:

- *Better manage, operate, and maintain collection systems*
- *Investigate capacity constrained areas of the collection system*
- *Proactively prevent SSOs*
- *Respond to SSO events*

The CMOM approach helps the owner or operator provide a high level of service to customers and reduce regulatory noncompliance. CMOM can help utilities optimize use of human and material resources by shifting maintenance activities from "reactive" to "proactive"–often leading to savings through avoided costs due to overtime, reduced emergency construction costs, lower insurance premiums, changes in financial performance goals, and fewer lawsuits. CMOM programs can also help improve communication relations with the public, other municipal works and regional planning organizations, and regulators...

In CMOM planning, the owner or operator selects performance goal targets, and designs CMOM activities to meet the goals. The CMOM planning framework covers operation and maintenance (O&M) planning, capacity assessment and assurance, capital improvement planning, and financial management planning. Information collection and management practices are used to track how the elements of the CMOM program are meeting performance goals, and whether overall system efficiency is improving.

The proposed rule was withdrawn prior to being finalized pursuant to a memorandum issued by the White House. In this memorandum the "newly-inaugurated President George W. Bush directed all administrative agencies to withdraw all pending documents for review by his newly appointed agency heads" (Strifling, 2003). Although a revised rule was perceived to be imminent for a number of years following the 2001 rule's withdrawal, no revised rules have been proposed by the EPA. The EPA held a series of SSO and peak-flow listening sessions in 2010, which gathered input from stakeholders on how the EPA should move forward.

With the exception of SSOs occurring in states that have implemented wastewater collection system permitting requirements (see Section 4.1.3), there is no framework to issue NPDES permits for SSOs, and as such all SSOs are considered a violation of the CWA regardless of the cause of the discharge.

Although there is no national rule clarifying how SSOs should be regulated, several states developed state laws regulating the discharges, including California, North Carolina, South Carolina, and Wisconsin (Bradley, 2011). In addition, many communities took it upon themselves to prepare for the eventual finalization of SSO rulemaking, which was highly expected to include a requirement to implement a CMOM program. These communities chose to adopt many of the components in a CMOM program, which are typically good utility management practices in themselves.

The EPA has been successful in requiring the implementation of the general content of the proposed rule through consent decrees. These SSO consent decrees, which are sometimes combined with CSO consent decrees as applicable, typically require implementing a CMOM program in line with the proposed 2001 SSO rule. In this fashion, implementing CMOM programs has become relatively commonplace among larger utilities.

EPA has not moved expeditiously to finalize a national framework regulating SSOs, partially due to a lack of consensus in the agency regarding concepts promoted by the regulated community. This actually gives EPA flexibility in negotiating consent decrees or other administrative orders since the starting point for negotiation is that all SSOs are considered prohibited under the CWA (if they reach the water of the US). The lack of a national framework also allows for a wide array of public agencies that could be subject to SSO consent decrees. If the SSO rule was finalized, then enforcement actions could only be initiated against agencies that are not complying with their wastewater collection system NPDES permit, as opposed to against any agency that has SSOs.

The EPA has indicated that taking enforcement actions against large SSO discharges will remain a priority. Figure 4 presents the status of the EPA's enforcement initiative related to SSOs, which indicates that the EPA expects to address an additional 106 agencies, in addition to the 83 with enforcement actions underway, with SSOs in the coming years.

Possible Future Sanitary Sewer Overflow Regulatory Trends

In 2010, EPA indicated that it was continuing to evaluate how to move SSO rulemaking forward, likely in coordination with regulations addressing blending at WWTPs (see Section 5.0). These two issues are connected because reducing SSOs in wastewater collection systems often results in an increase in flow to WWTPs. Whether or not blending is allowed at WWTPs, and under what conditions, could have a significant impact on how public agencies address SSOs. At some point, these rules likely will be finalized, but no timetable has been released by the EPA. As discussed in Section 4.1.4 the EPA is expected to continue addressing large SSO dischargers via consent decrees, and SSO activities in the coming years are likely to focus on enforcement action initiation by the EPA and implementation of existing SSO reduction programs. These consent decrees will be subject to the same trends and anticipated future trends discussed in Section 3.2.

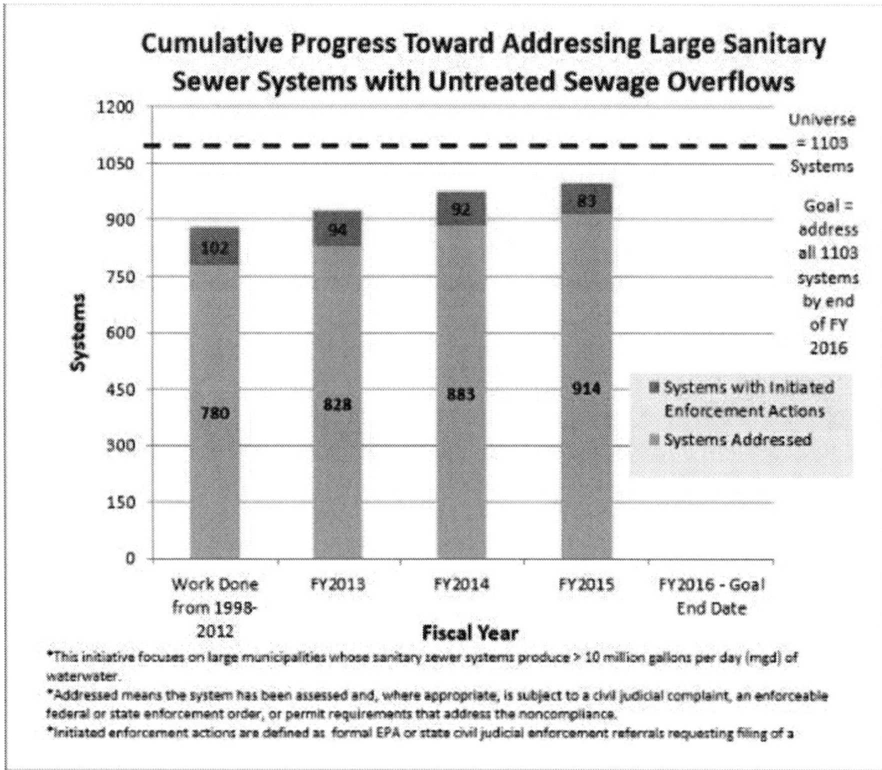

Figure 4. EPA progress in addressing SSOs (EPA, 2016a)

WET WEATHER IMPACTS TO WASTEWATER TREATMENT PLANTS

Wet weather can create several challenges to WWTP operation and performance. For example, diluting sanitary sewage with stormwater runoff can impact the effectiveness of biological secondary treatment processes, and bottlenecks within the WWTP can result in discharges of untreated or partially treated sewage—or, in some cases, can pose a danger to staff and/or property. Bypassing or blending flows within a WWTP are two strategies used to address these challenges, and they are the primary focus of wet weather regulations as they relate to WWTPs. These issues are further discussed in the remainder of this section.

Current Wastewater Treatment Plant Wet Weather Regulations

EPA regulations define a bypass as an "intentional diversion of waste streams from any portion of a treatment facility" (40 CFR 122.41[m]). Figure 4 illustrates the concept of a bypass, which shows flow being discharged without biological secondary treatment and disinfection (i.e., bypassing secondary treatment and disinfection processes). The resulting discharge does not meet permit limits.

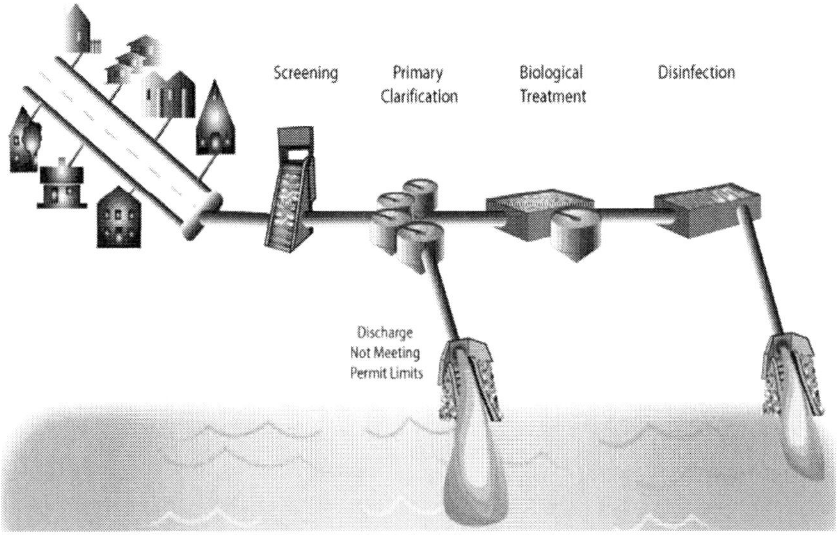

Figure 4. Bypass Scenario at a WWTP (EPA, 2010)

The CWA prohibits bypasses, except under specific conditions (CFR 122.41):

- The bypass was unavoidable to prevent loss of life, personal injury, or severe property damage.
- No feasible alternatives to the bypass, such as using auxiliary treatment facilities, retaining untreated wastes, or performing maintenance during normal periods of equipment downtime, were available.
- The permittee submitted notice to the regulator of the bypass.

Of the criteria listed above, the "no feasible alternatives" criteria has been challenged numerous times by permittees (EPA, 2000). Court rulings have upheld a narrow interpretation of this term, and clarified that constructing additional treatment capacity and/or storage, although in many cases quite expensive, can be considered a feasible alternative. However, many public agencies have been able to demonstrate no feasible alternatives, and have been allowed bypass.

For bypasses caused by "exceptional incidents" such as natural disasters, permittees may establish an affirmative defense to protect the permittee from enforcement action and citizen lawsuits. These bypasses are called "upsets" (CFR 122.41).

The term "blending" is used to describe situations where peak wet weather flows are diverted around a treatment process (typically biological secondary treatment) and recombined with secondary effluent before disinfection and discharge. The diverted flow may or may not receive some form of treatment in lieu of biological secondary treatment, such as chemically enhanced or high-rate clarification. The blended effluent meets permit limits at the point of discharge. Figures 5 and 6 illustrate the two variations of blending (with and without treatment of the diverted flow).

Regulations are not consistent as to whether blending is allowed in a WWTP. The following subsections summarize the EPA's efforts to establish a national policy for blending, as well as a recent court case that may have national implications for blending.

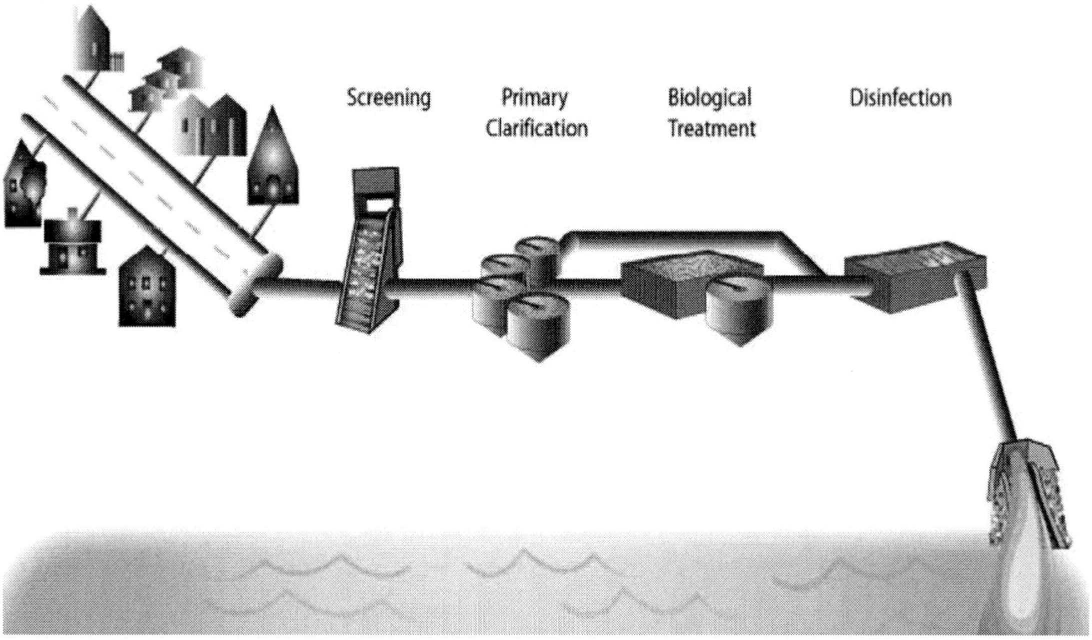

Figure 5. Blending without Treatment Scenario at a WWTP (EPA, 2010)

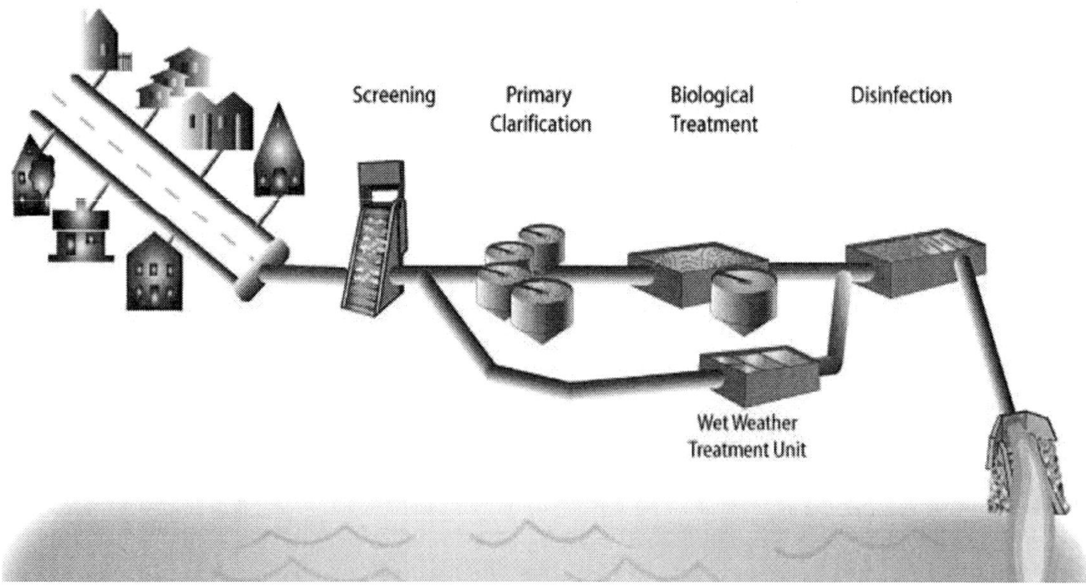

Figure 6. Blending with Wet Weather Treatment Scenario at a WWTP (EPA, 2010)

In November 2003 the EPA released a draft rule governing the use of blending flows within a WWTP. This draft rule considered blending a bypass (see Section 5.1.1), which would be prohibited unless the following six criteria were met (EPA, 2003):

1. The final discharge meets permit limits
2. The permittee's NPDES permit defines the treatment scenario when blending is allowed
3. Flow diverted around secondary treatment receives at least the equivalent of primary clarification
4. Blending is only allowed when the secondary treatment unit is being fully utilized
5. Monitoring requirements
6. The collection system must be properly operated and maintained

The EPA received significant public comment on the proposed policy "due to the concerns about human health risks," and in 2005 the EPA "indicated that after consideration of the comments, the [EPA] had no intention of finalizing the 2003 proposal" (EPA, 2005b). As a result, municipalities and state environmental regulators remain uncertain as to whether blending should be allowed.

Subsequent to the withdrawal of the 2003 proposed blending rule, the Natural Resources Defense Council (NRDC) and the National Association of Clean Water Agencies (NACWA) provided the EPA "with their joint proposal recommending further action that the Agency should take regarding the blending issue. The NRDC/NACWA recommended approach includes an interpretation of the bypass regulation that is significantly different from the November 7, 2003, proposal, in that it would clarify that the bypass provision would apply, in all instances, to wet weather diversions at publicly owned treatment works treatment plants serving separate sanitary sewers" (EPA, 2005b).

The 2005 proposed blending rule specified that blending "cannot be approved if feasible alternatives are identified" and described a "no feasible alternatives analysis process" (EPA, 2005b). The process required the permittee to complete a comprehensive utility analysis that assesses current conditions within the WWTP and collection system and also assesses ways to reduce peak wet weather flow volumes at the WWTP. The alternatives to be evaluated must include expanding capacity of treatment processes at the WWTP, constructing storage at the WWTP and/or in the collection system, and reducing infiltration and inflow. The 2005 proposed blending rule also would have required the permitting authority to conduct a "rigorous review" for all permit renewal requests that seek to continue blending.

The EPA indicated that most comments received on the draft policy were positive, but the policy was never finalized. Since the release of the 2005 proposed blending rule, the EPA has conducted a number of listening sessions regarding peak flows at WWTP and SSOs (see Section 4.2).

A recent court case from the US Eighth Circuit Court of Appeals has continued the debate over whether blending is allowed. The result of the case—the Iowa League of Cities v. the EPA—was that the EPA's "statutory authority only permits the [EPA] to set 'effluent limitations based upon secondary treatment' for the 'addition of any pollutant to navigable waters' from 'a point source'" (Hall et al., 2014). In other words, the court ruled that the EPA did not have statutory authority to set effluent limitations within a WWTP, and thus could not interfere with how a municipality chooses to meet effluent limitations (as long as they provide secondary treatment). The EPA chose not to appeal the case, and as a result, the court's ruling only applies to states within the US Eighth Circuit Court District (Arkansas, Iowa, Minnesota, Missouri, Nebraska, North Dakota, and South Dakota). Further, as a result of this case, blending is allowed by permittees within these states, provided that permit limits are met at the point of discharge. However, the legality of blending remains uncertain in all other states.

Possible Future Wastewater Treatment Plant Wet Weather Regulatory Trends

At some point, EPA likely will release and finalize a revised blending rule. When this rule will be issued in draft form is uncertain, but the ruling of the U.S. Eighth Circuit Court of Appeals increases the sense of urgency felt by the EPA to issue a national blending rule. Furthermore, a lawsuit against the EPA is underway that seeks to force the EPA to apply the ruling from the Iowa League of Cities case nationally. Of key concern to the EPA is that, without a national blending rule, a permittee could blend under any condition as long as the effluent discharged meets permit limits.

Given the joint listening sessions for peak-flow impacts on WWTPs and SSOs in 2010, as well as the interconnectedness of SSOs and peak flows at WWTPs (reduced SSOs could result in increased peak flows at the WWTP), national rules for these issues will likely be advanced together.

MUNICIPAL SEPARATED STORM SEWER SYSTEMS

The EPA defines an MS4 as "a conveyance or system of conveyances that is:

- Owned by a state, city, town, village, or other public entity that discharges to waters of the U.S.,
- Designed or used to collect or convey stormwater (e.g., storm drains, pipes, ditches),
- Not a combined sewer, and
- Not part of a sewage treatment plant, or POTW" (EPA, 2016b)

The EPA defines stormwater as "storm water runoff, surface runoff, street wash waters related to street cleaning or maintenance, infiltration (other than infiltration contaminated by seepage from sanitary sewers or by other discharges) and drainage related to storm events or snow melt" (EPA, 1990).

Current Municipal Separated Storm Sewer System Regulations

The regulation of stormwater discharges from MS4's was initiated with the Water Quality Act of 1987, which amended the CWA to require the EPA to permit stormwater discharges. Before this amendment, stormwater runoff regulation was considered too difficult to implement, because of the variability in flows, lack of control by the owner for stormwater discharges (Pomeroy and

Curtis, 2015), perceived lack of water quality impacts from stormwater in comparison to sewage, and the lack of an administrative framework to deal with the large numbers of outfalls.

The passage of the Water Quality Act of 1987 resulted in the development of NPDES regulations governing the discharge of stormwater runoff: the 1990 Phase I Stormwater Rule and the 1999 Phase II Stormwater Rule. These rules were modifications to the existing permitting regulations. The EPA estimates that the regulated area from the Phase I and Phase II rules is only 4 percent of the land area in the United States, but over 80 percent of the total population (Pomeroy and Curtis, 2015). The Phase I and Phase II rules are summarized as follows (Copeland, 2012):

- Phase I Stormwater Rule (1990)—Applies to MS4s in urban areas that serve populations greater than 100,000 people.
- Phase II Stormwater Rule (1999)—Applies to MS4s in urban areas with populations less than 100,000 people.

The Phase I and Phase II rules also addressed stormwater discharges from commercial, industrial, and construction activities, in addition to discharges from MS4s. This document focuses only on the portions of the rules that govern the discharge of stormwater runoff from MS4s.

MS4 Phase I and Phase II permits contain a number of requirements, the primary one being implementing a stormwater management program (SWMP). A SWMP for Phase I municipalities typically includes the following components, but varies by NPDES permit:

- Ongoing program for mapping and documenting MS4 discharges
- Coordination mechanisms among departments to eliminate barriers to compliance with the permit
- Public involvement process
- Program to prevent and control the impacts of runoff from development, redevelopment, and construction sites
- Program to implement structural stormwater controls to prevent or reduce impacts of stormwater runoff to receiving water bodies
- Source control program for existing development
- Illicit connection and discharge detection and elimination program
- Ongoing operations and maintenance program
- Education and outreach program

MS4 permit requirements vary from permit to permit, but all must meet minimum requirements established by the CWA. Requirements vary depending on site-specific issues and can be influenced by total maximum daily load (TMDL) wasteload allocations and other factors.

One feature of the current MS4 regulations is the lack of specific performance metrics. Some other regulated discharges have specific categorical performance metrics, such as the number of CSO events discharged per year, which can be used to assess whether regulations are being met. MS4 NPDES permits do not typically have these requirements, and permittees are often allowed to set their own goals for stormwater-related pollutant reduction. While some criticize that this allows permittees to set very low targets for improvement, others applaud that the permittee is afforded the flexibility to define and achieve the level of pollutant control that makes sense in their unique situation.

Possible Future Municipal Separated Storm Sewer System Regulatory Trends

In 2009, the National Research Council of the National Academy of Sciences released a report that recommended EPA to adopt a "watershed-based permitting system encompassing all discharges— stormwater and wastewater—that could affect waterways in a particular drainage basin, rather than individual permits that do not account for cumulative conditions from multiple sources in the same watershed" (Copeland, 2012). As a result of this report, EPA has been considering various possible regulatory changes that would "embrace the report's recommendation to adequately control all sources of stormwater discharges that contribute to waterbody impairment" (Copeland, 2012).

A summary of recent EPA activities related to MS4 regulations is summarized in *Stormwater Permits: Status of EPA's Regulatory Program* (Copeland, 2012), which is excerpted below:

"In early 2010 EPA held a series of listening sessions across the country as part of a process seeking public comments on potential considerations for regulatory changes. The agency also sent survey questionnaires to property owners and developers, municipal sewer system authorities, state regulators, and EPA regional offices to obtain their input. Some industry groups reportedly criticized possible expansion of the current program, saying that EPA's authority to regulate stormwater does not extend to regulation of post-construction discharges. Some states also said that EPA lacks the technical knowledge to regulate stormwater across the nation, while states with comprehensive regulations, such as Florida and Maryland, demonstrate that regulation is best done at the state and local level, because of locational differences in stormwater discharges. A number of commenters urged EPA to ensure that performance standards designed to reduce storm runoff be flexible so that communities can create requirements appropriate to their stormwater needs. Cost is a key issue raised by states and municipalities that are concerned about the possibility of a [sic] mandatory retrofit requirements that would impose a significant economic burden on cities."

Copeland went on to describe several EPA revisions to schedule dates for promulgation of updates to the regulations. As of the writing of this document, no further updates have been issued by EPA regarding major revisions to stormwater rules. These rules will likely be updated at some point in the future, but the timing of this update is unknown.

In January 2016 the USEPA published draft revisions to the Phase II Stormwater Rule that responded to a court case from 2009, in *Environmental Defense Center, et al. v. EPA,* 344 F.3d 832 (9th Cir. 2003). This draft revision addressed inadequate public notice and opportunity to request a hearing for small MS4 general permits, as well as permitting authority reviews of best management practices used by MS4 permittees to achieve pollutant reduction to the maximum extent practicable.

EPA and state regulators have two other means they can use to implement increasingly stringent requirements on MS4s apart from revisions to the stormwater rules —enforcement actions and implementation of TMDLs.

EPA has increased its enforcement actions addressing MS4s, including increasing amounts of EPA audits of MS4 programs and issuing consent decrees focused solely on enforcing MS4 stormwater permit requirements. A recent MS4-focused consent decree for Rockford, Illinois, included requirements to implement standard operating procedures for activities in the stormwater management program, such as street sweeping, monitoring and sampling stormwater, training city staff, and inspecting and maintaining the stormwater system.

TMDLs are developed for specific pollutants being discharged into specific receiving waterbodies to address waterbody impairment that exists despite implemented CWA-required technology-based discharge limits. A key component of a TMDL is developing wasteload allocations that establish how much of a given pollutant each identified pollutant discharger is allowed to discharge. These wasteload allocations can then be incorporated into NPDES permits to require further pollutant reduction efforts.

Many TMDLs identify MS4 discharges as a significant source of pollutants. The incorporation of the wasteload allocations in those TMDLs into MS4 NPDES permits results in additional pollutant removal requirements, such as requiring the installation of specific stormwater best management practices. TMDLs can also be a way to require specific performance metrics or milestones into agencies' stormwater management programs.

An example of how TMDLs can drive increased stormwater regulation is the Chesapeake Bay TMDL, which targets the reduction of nitrogen, phosphorus, and sediment across multiple states in the Chesapeake Bay basin. This TMDL is resulting in more specificity in MS4 permit requirements, such as the requirement to implement "specific best management practices or develop TMDL Plans or Pollutant Reduction Plans" (Pennsylvania Department of Environmental Protection, undated).

OTHER REGULATORY TRENDS

The following subsections describe other regulatory trends that do not fall under the regulations associated with the specific discharges described above.

Climate Change

Climate change is expected to significantly impact wet weather discharges and associated programs focused on reducing the impacts of those discharges. The EPA has supported considering climate change when developing and implementing wet weather programs, but the agency has not required this consideration in NPDES permit or consent decree requirements.

The EPA conducted a study in 2008 that evaluated the potential impacts of climate change on CSOs in the Great Lakes and New England regions (EPA, 2008). This study summarized estimated CSO frequency increases and made recommendations on mitigating those impacts when designing CSO reduction facilities. Several other scholarly research projects have been conducted more recently to assess the impact that climate change may have on CSO and SSO frequency and volume. The research generally agrees that climate change poses a risk to achieving and maintaining regulatory compliance for wet weather discharges, but the impacts are typically very site specific and need more focused research.

The EPA has also developed a climate risk assessment tool for water utilities, called the Climate Resilience Evaluation and Awareness Tool (CREAT), "to assist drinking water, wastewater, and stormwater utility owners and operators in assessing risks to utility assets and operations" (EPA, 2012). This tool can be used to help assess the potential impacts of climate change on a utility and help to prioritize where investment should be made to mitigate risk.

Regardless of whether public agencies consider climate change when developing their wet weather programs, climate change is expected to increase many regulated wet weather discharges. Public agencies that are subject to wet weather regulations based on actual performance may be particularly impacted by climate change from a regulatory compliance perspective. For example, the State of Washington requires CSO discharges to be reduced to no more than an average of one event per outfall per year. This requirement must continue to be met, regardless of what impact climate change has on CSO discharges. Other CSO communities use a more flexible "typical year" approach, where CSO's must be reduced to a given frequency for a predefined typical year. If a CSO has more than the allowed number of events in a year, but it can be proven that the rainfall during the year was greater than the pre-defined typical year, then the public agency remains in compliance with their NPDES permit.

Integrated Municipal Stormwater and Wastewater Planning Approach Framework

Public agencies that manage urban wet weather discharges in the United States are facing major financial burdens: developing and implementing expensive capital programs to reduce CSOs and SSOs, adhering to increasingly stringent requirements to reduce pollutant discharges from their MS4s, adhering to pollutant reductions required by TMDLs, and meeting more stringent effluent limitations. These requirements result in major regulatory obligations placed on public agencies to develop and construct water quality facilities, often pushing financial budgets beyond their capacity.

The ultimate goal to which CSO, SSO, wastewater treatment, and stormwater regulatory requirements strive towards is meeting the goal of the CWA—eliminating pollutant discharges to navigable waters, to the extent that they prevent attainment of the fishable-swimmable goals of the act. Individual regulations may have intermediate requirements, such as reducing CSOs to a set threshold, but in the end, they all strive towards the same goal. As the primary enactors of CWA requirements, public agencies must see their CWA-related responsibilities as improving water quality and attaining state-designated uses, rather than simply meeting current NPDES permit or consent decree requirements—this is a key concept that allows integrated planning to fulfill its potential.

The EPA released the *Integrated Municipal Stormwater and Wastewater Planning Approach Framework* (Integrated Planning Framework; EPA, 2012) in 2012 to allow municipalities to balance CWA requirements in a manner that addresses the most pressing public health and environmental protection issues first. The Integrated Planning Framework describes the overarching principles that the EPA "will use in working with municipalities to implement an integrated approach to meet [a municipalities'] wastewater and stormwater program obligations under the CWA." Following are the four overarching principles:

- Maintain existing regulatory standards.
- Balance CWA requirements in a manner that addresses the most pressing public health and environmental protection issues first.
- Ensure that integrated plans are voluntary.
- Allow for innovative technologies, such as GI, which may be fundamental aspects of municipalities' plans.

The Integrated Planning Framework further describes the scope of what type of projects may be included in an integrated plan: separate sanitary systems, combined sewer systems, MS4s, WWTPSs, source water protection efforts, and/or nonpoint source control.

The Integrated Planning Framework also describes what elements an Integrated Planning Framework approach should include:

1. Description of water quality, human health, and regulatory issues to be addressed
2. Characterization and description of the wastewater and stormwater systems
3. Public and stakeholder involvement
4. Process to develop, evaluate, select, and implement alternatives
5. Plan for how to measure the success of implemented projects
6. Process for how new projects or modifications will be made

As public agencies enter into the implementation phase of their LTCPs and other wet weather-related programs, they are beginning to realize the true cost and financial impact of these programs. Financial affordability has been a major challenge facing the agencies implementing these programs, and many agencies are looking for ways to ease the financial burden of complying with regulations and making sure that invested funds are used to address the most pressing problems. Integrated planning is one approach that could both ease the financial burden on public agencies and increase the human and environmental benefits from implementing wet weather programs.

For integrated planning to be successful, both the permittee and regulatory agencies must see integrated planning as a benefit. For the permittee, this could mean a water quality program that places less financial burden on its rate-payers while achieving compliance with regulatory requirements. For the regulatory agencies, this could mean an enforceable schedule that results in higher or faster achievement of human health and environmental benefits than would be achieved through implementation of the each program separately. Without this, permittees may be disinclined from pursuing integrated planning, because they are just committing to a more expensive capital program than without integrated planning, and regulatory agencies cannot claim success with enforcing the CWA and achieving the environmental and human health improvements they are responsible for shepherding. A hypothetical example of the successful outcomes of integrated planning is presented in Figure 7.

Integrated planning has huge potential benefits for both permittees and regulatory agencies—it presents an opportunity to take significant steps towards achieving the ultimate goals of the CWA.

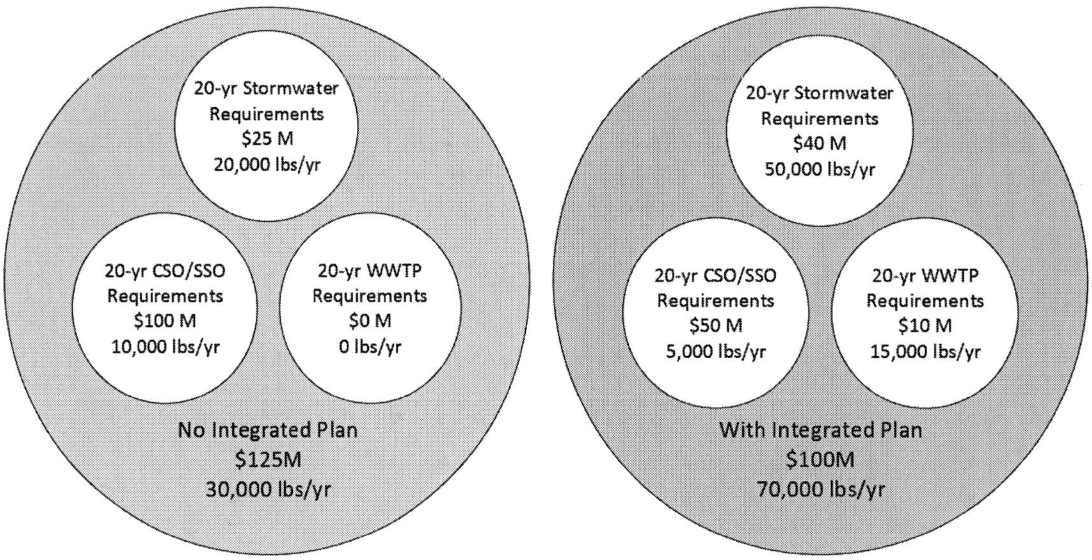

Figure 7. Hypothetical Example Illustrating Potential Benefits of Integrated Planning

Costs represent life-cycle cost of projects devoted to the stated system. Pounds per year are removal amounts of a pollutant from the implemented projects. All values are hypothetical and used for illustrative purposes.

Integrated planning also poses risks to public agencies. Being a relatively uncommon approach, regulators may not know how to review and approve these plans. Integrated planning may also result in increased regulatory responsibility for public agencies. For example, if an integrated plan recommends investing more in stormwater rather than on other wet weather discharges, then those recommendations could be included in NPDES permits or consent decrees, increasing the amount of regulatory requirements on the agency.

REFERENCES

Benfield, L.E. 2015. *Enforcement & Consent Decree Developments: Part 1 – Wet Weather Enforcement*. November.

Bradley, P. 2011. *Wet Weather NPDES Permitting (Including SSOs): Case Studies from Around the Country*.

Copeland, C. 2010. *Clean Water Act: A Summary of the Law*. Prepared for the Congressional Research Service. April.

Copeland, C. 2012. *Stormwater Permits: Status of EPA's Regulatory Program*. Prepared for the Congressional Research Service. July.

Hall, J., G. Cohen, and P. Rosenman. 2014. *Eighth Circuit Vacates EPA's Illegal Rules Prohibiting Blending and Bacteria Mixing Zones: Implications of the Decision and the Practical Consequences*.

Legal Information Institute (LII). 2016. Affirmative Defense definition. Available at https://www.law.cornell.edu/wex/affirmative_defense. June 6. 2016.

Pennsylvania Department of Environmental Protection (DEP). Undated. *Municipal Stormwater: Program Updates*. Available at http://www.dep.pa.gov/Business/Water/PointNonPointMgmt/StormwaterMgmt/Stormwater/Pages/default.aspx#change.

Pomeroy C. and J. Curtis. 2015. *Stormwater Primer*. Presented at the NACWA 2015 Law Seminar. November.

Strifling, D. 2003. *Sanitary Sewer Overflows: Past, Present, and Future Regulation*. Marquette Law Review Volume 87 Issue 1 Article 6. 2003.

U.S. Environmental Protection Agency (EPA). 1979. *1978 Needs Survey: Cost Methodology for control of Combined Sewer Overflow and Stormwater Discharge*. February.

U.S. Environmental Protection Agency (EPA). 1980. *Program Requirements Memoranda: Municipal Wastewater Treatment works Construction Grants Program*. January.

U.S. Environmental Protection Agency (EPA). 1983. *Results of the Nationwide Urban Runoff Program: Volume 1*. Available at https://www3.epa.gov/npdes/pubs/sw_nurp_vol_1_finalreport.pdf

U.S. Environmental Protection Agency (EPA). 1989. *National Combined Sewer Overflow Strategy*. Available at https://www3.epa.gov/npdes/pubs/owm0356.pdf. August 10.

U.S. Environmental Protection Agency (EPA). 1990. *National Pollutant Discharge Elimination System Permit Application Regulations for Storm Water Discharges; Final Rule*. Available at https://www.epa.gov/sites/production/files/2015-10/documents/stormwater_phase1_rule.pdf. November 16.

U.S. Environmental Protection Agency (EPA). 1994. *Combined Sewer Overflow (CSO) Control Policy*. Available at https://www3.epa.gov/npdes/pubs/owm0111.pdf. April 19.

U.S. Environmental Protection Agency (EPA). 1995. *Combined Sewer Overflows Guidance for Nine Minimum Control Measures*. Available at https://www3.epa.gov/npdes/pubs/owm0030.pdf. May 1.

U.S. Environmental Protection Agency (EPA). 2000. *Enforcement Alert: Clean Water Act Prohibits Sewage 'Bypasses'*. April.

U.S. Environmental Protection Agency (EPA). 2001. *Proposed Rule for National Pollutant Discharge Elimination System (NPDES) Permit Requirements for Municipal Sanitary Sewer Collection Systems, Municipal Satellite Collection Systems, and Sanitary Sewer Overflows.* January.

U.S. Environmental Protection Agency (EPA). 2003. *Request for comment on proposed policy: National Pollutant Discharge Elimination System (NPDES) Permit Requirements for Municipal Wastewater Treatment Discharges During Wet Weather Conditions.* Published in the Federal Register Vol. 68, No. 216. November 7.

U.S. Environmental Protection Agency (EPA). 2004. *Report to Congress: Impacts and Control of CSOs and SSOs.* August.

U.S. Environmental Protection Agency (EPA). 2005a. *Guide for Evaluating Capacity, Management, Operation, and Maintenance (CMOM) Programs at Sanitary Sewer Collection Systems.* January.

U.S. Environmental Protection Agency (EPA). 2005b. *Request for comment on proposed policy: NPDES Permit Requirements for Peak Wet Weather Discharges from Publicly Owned Treatment Works Treatment Plants Serving Separate Sanitary Sewer Collection Systems*. Published in the Federal Register Vol. 70, No. 245. December 22.

U.S. Environmental Protection Agency (EPA). 2008. A Screening Assessment of the Potential Impacts of Climate Change on CSO Mitigation in the Great Lakes and New England Regions. February.

U.S. Environmental Protection Agency (EPA). 2010. *Sanitary Sewer Overflows and Peak Flows Listening Sessions*. PowerPoint presentation.

U.S. Environmental Protection Agency (EPA). 2012. Climate Resilience Evaluation and Awareness Tool Version 2.0 A Climate Risk Assessment Tool for Water Utilities. December.

U.S. Environmental Protection Agency (EPA). 2015a. *EPA Needs to Track Whether Its Major Municipal Settlements for Combined Sewer Overflows Benefit Water Quality.* September.

U.S. Environmental Protection Agency (EPA). 2015b. *Draft EJ 2020 Action Agenda Framework.* June.

U.S. Environmental Protection Agency (EPA). 2016a. *National Enforcement Initiative: Keeping Raw Sewage and Contaminated Stormwater Out of Our Nation's Waters.* Web. 6 June. 2016.

U.S. Environmental Protection Agency (EPA). 2016b. *Stormwater Discharges from Municipal Sources*. Web. 13 June. 2016.

United States of America and the State of Illinois, Plaintiffs, v. City of Rockford, Illinois, Washington, Defendant. In the United States District Court for the Northern District of Illinois, Western Division. Consent Decree. Civil Action No. 3: 15cv50250. Filed October 8, 2015. Available at https://www.epa.gov/sites/production/files/2015-11/documents/cityofrockford-cd.pdf.

Washington State Department of Ecology (Ecology). 2012. *Phase I Municipal Stormwater Permit*. August 1.

Using Robust Asset Management Technology to Tackle Inaccessibility in Dallas

H. Steed[1], S. Mandhle[1] [1]Arcadis US, Inc., TX.
*Email: sacheev.mandhle@arcadis.com

ABSTRACT

In the absence of inspection data due to accessibility issues, statistical probability models were used to leverage the condition of inspected pipes to forecast condition of inaccessible pipes within the Dallas Water Utilities interceptor system. The combination of forecasted condition scores with risk based asset management allowed for the prioritization of selected interceptors for rehabilitation and replacement projects. Arcadis provided DWU with a complete program that allows for input criteria to be amended to reflect current needs of the system, and create budgetary planning information with the goal of reducing overall risk within the network.

KEYWORDS: NASSCO, GompitZ, RRPS, CIP

INTRODUCTION

How do you proactively characterize risk within a system that you can't inspect? We faced that situation in Dallas. The standard approach, closed-circuit television (CCTV) and manhole inspections, for a Dallas Water Utilities (DWU) project was utilized initially with limited success on hard to access interceptor sewers. More expensive and sophisticated technology (laser and sonar) was next utilized with mixed results, which still left over 20% of the large diameter interceptor pipe in the system off limits to inspection. The costs for access improvements would be prohibitively significant even with the highest risk pipe. These challenges spurred the team to find a solution that would produce reasonable remaining useful life for pipe yet to be inspected. Ultimately, predictive condition modeling, in combination with a customizable capital improvement planning tool and access improvements for the highest risk pipes, provided the needed results.

Many utilities have the equipment, staffing and programs to inspect, evaluate and prioritize repair and replacement programs for their wastewater collection systems with the exception of larger diameter interceptors. Large diameter interceptors are a small percentage of the total wastewater system and require specialized equipment and staffing. Inspecting and evaluating these large interceptors also comes with numerous challenges including access (along creeks and waterways), physical design challenges (offset manholes, curves, distances between access points), and flows (depth, velocities, obstructions). Dallas Water Utilities undertook a program to address these challenges, evaluate technologies and develop a sustainable program for their interceptor system. This program included a phased plan to inspect, evaluate, prioritize and develop a capital program.

DWU operates and maintains over 4,000-miles of wastewater pipelines, of which 231.2-miles are wastewater interceptors larger than 24-inches. DWU is equipped and regularly inspects the majority of their system; however the larger size interceptors require specialized equipment that has limited regular inspections and evaluation. These interceptors are mostly concrete pipelines 15 to 84 years in age with the majority being installed over 50 years ago.

METHODOLOGY

Inspections

In September 2010 the multi-phased project began with a desktop review of GIS information, maintenance records and the wastewater master plan to refine the system risk profile and prioritize areas for inspection. This desktop risk refinement identified substantial GIS updates and data modifications that significantly changed the system risk profile and prioritization. Considering inspection budgets and prioritization evaluations, an initial 91 miles of interceptors were identified for Phase 1 field inspection and evaluation.

DWU's consultant team, conducted scouting activities to locate manholes and evaluate access challenges. The triage approach would utilize the lowest cost inspection technology (zoom camera) to find the "good" pipe and if necessary recommend a CCTV or Laser/Sonar/CCTV follow-up inspection. The zoom camera inspection would also recommend field maintenance activities necessary for inspection with other technologies (cleaning, manhole rehabilitation, access improvements). Below, Table 1 summarizes the anticipated use of technology for the inspection phase of the project.

Table 1: Inspection Technology Plan

Inspection Method	Scoped	Considerations
Zoom Camera	66.0 miles	Lowest inspection cost, can be deployed to difficult locations, manhole inspection and pipeline inspection, horizontal and vertical curves limit views, no concerns for stuck/lost equipment, cannot be used with offset manholes, may not view entire pipe segment
CCTV	18.1 miles	Medium inspection costs, skid and float deployment, consider hazards for deployment in pipe segment, proven technology and condition scoring, requires nearby truck access
Laser/Sonar/CCTV	6.5 miles	Highest inspection costs, detailed measurements of pipe corrosion and sedimentation, CCTV information and condition scoring, robust robot can travel upstream and downstream from single deployment, requires large truck access and larger manhole openings

Standardized Inspection Scoring

Following the physical interceptor inspection, the condition assessment phase began. The purpose of this phase of the work is to accurately determine the condition of the existing pipe per nationally accepted standards. Then based on the standard, and good engineering judgment, the likelihood of failure or remaining useful life can be estimated. The condition assessment and condition grading of the defects and overall pipe ratings is performed in accordance with the National Association of Sanitary Sewer Companies (NASSCO) Pipeline Assessment Certification Program (PACP) standard. For the success of the DWU inspection and condition assessment program, it is essential that a standard approach is followed to define the existing condition of the inspected assets. This standard approach allows the whole team from inspectors to planning staff to design engineers to effectively communicate the pipe condition.

NASSCO developed the PACP in 2001 in conjunction with the Water Research Centre (WRc) of the United Kingdom to standardize the defect coding and condition grading of sewer pipe inspection in the United States. The standard also includes guidelines regarding proper positioning of the CCTV camera in the pipe along the pipe axis, speed of advancement of the camera, and standard database format, among other things. CCTV operators are trained by NASSCO approved trainers and are required to be re-certified on a recurring interval.

A condition grade is a pre-defined value between 1 and 5 for each defect observed with implications as shown below in Table 2 for structural defects. The PACP includes both Structural and Operations & Maintenance condition grades.

Table 2: Structural Condition Grade Implications

Condition Grade	Condition	Defect Description
1	Excellent	Minor defects
2	Good	Defects that have not begun to demonstrate
3	Fair	Moderate defects that will continue to deteriorate
4	Poor	Severe defects that will become grade 5 within the foreseeable future
5	Immediate Attention	Defects requiring immediate attention

Structural defects are those that directly impair the structural condition of the pipeline such as joint separations, joint offsets, cracks, fractures, broken and collapsed pipe, and eroded inverts. Concrete pipe corrosion has unique defects which include the following in ascending order of significance: roughness increased, surface spalling, exposed aggregate, missing aggregate, and steel reinforcing visible / corroded. Structural defects are those that may only be addressed through rehabilitation, replacement, or spot repairs. On the other hand, operational defects include a range of conditions that can either directly affect the performance of the sewer or are indicators of potential structural defects. Operational conditions include debris, sedimentation, grease, intrusions (root or service laterals), pipe sags, hydraulic / capacity problems, and infiltration. Operational defects are those that generally can be addressed by maintenance.

Risk

In order to use prioritize pipe for rehabilitation and replacement projects, there needed to be a system to identify the most critical pipes. The introduction of a risk system based on the likelihood of failure and the consequence of failure allowed for the desired prioritization. The following sections discuss the components of risk in greater detail.

LoF

Likelihood of failure (LoF) includes the consideration of two different failure modes, physical and performance. For this project, we have defined the physical condition to be equivalent to the structural condition. From this point forward, the term structural will be used to describe the physical condition of the pipe. Structural and performance criteria are broken out into separate components, as shown in Figure 1 .

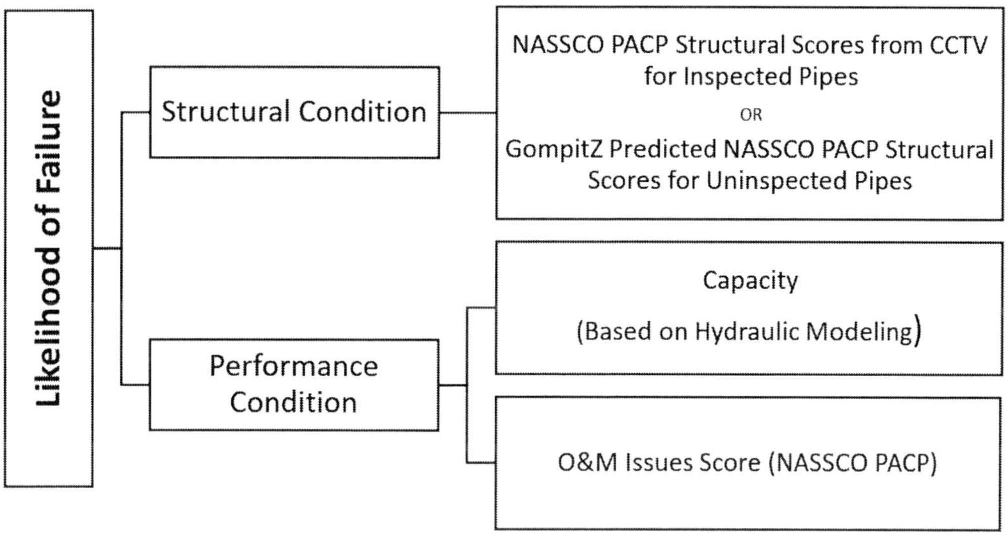

Figure 1: Likelihood of Failure Criteria

The LoF score is comprised of both structural condition and performance condition scores. The sum of the two scores is equal to the LoF score, as shown in the following equation.

$$Performance\ Score + Structural\ Score = Likelihood\ of\ Failure\ Score$$

This approach results in LoF scores ranging from a minimum of 2 to a maximum of 10. In order to get a score of 2, both the performance and the structural scores would have to be 1, or in other words, it would have to be in near perfect condition. A score of 10 would indicate that the interceptor is failing in both areas. Physical condition is based solely on the structural assessment, with scores ranging from 1 (very good) to 5 (very poor). The pipe that was able to be inspected using CCTV was assigned a value based on the condition seen in the actual pipe, based on NASSCO PACP condition scoring standards. For the segments that were unable to be inspected, predictive condition modeling was used to produce scores.

Statistical Modeling

Given the extensive portion of the DWU interceptor system which was not inspected it, became necessary to "forecast" or model the condition of the uninspected pipe based on past pipe inspections. InfraPLAN was retained to apply the GompitZ model and to provide DWU with an estimate of the physical condition for a pipe of interest at any year by using the inspection information from the sample of previously inspected pipe.

GompitZ considers the current condition state of a pipe at a particular time and then predicts when a pipe will likely transition to the next highest condition state. The model represents the transition between condition states using Markov Chain and Matrices of probabilities. A Markov chain is a random model that identifies a sequence of possible events and the likelihood of each event occurring is only dependent upon the state reached in the previous event. Essentially, it suggests that the probability that one event occurs is only influenced by the previous state rather than the whole sequence in its entirety.

Further, GompitZ calls for the Markov chains to be organized within time dependent matrices. This set up guarantees that a pipe will either stay in the same condition or to move to the next. GompitZ is a combination of a Gompertz function and a Probit model. A Gompertz function displays slow growth at the beginning and end of a time period while a Probit model only allows for the dependent variable to take two values. Applying these statistical tools to the given GIS dataset allowed for predictive modeling to provide a forecast of the condition for the remainder of the Dallas system. However, before the model was able to produce results, the input data needed to be analyzed to ensure accuracy.

The first step to this process was to review and ensure that the pipe data, such as material and diameter, were drawn from inspection records. However, there were some discrepancies and missing entries that required attention. Inspection identification numbers were incorrectly assigned, duplicate inspections existed, multiple inspections occurred for the same pipe segments, and there were discrepancies between the pipe and inspection data. In order to address the issues, Arcadis reviewed their own records to identify any errors and then consulted with DWU to come to a resolution. The main attributes that needed review were pipe length, material, and year of installation.

Pipe length data was given by three different sources: GIS length, length of pipe from the inspection report and the length surveyed from the inspection report. Pipe material data was provided by the inspection report and when inconsistencies occurred, Arcadis and DWU collaborated to select the correct option. DWU also provided the year of installation and Arcadis clarified data for entries that were missing. Other data provided by the inspection report that were considered for the GompitZ model include: inspection date, condition grade and environmental factors (pipe capacity, slope/turbulence).

Each pipe also has a value corresponding to its respective condition score, flow value, velocity, and slope. The condition scores, as mentioned earlier, were provided by the inspections and the values were determined using the standard PACP methodology. Velocity for each pipe was determined using Manning's Equation which has several inputs such as diameter and slope. Using Manning's equation the Froude number was calculated to characterize the flow in any given pipe. Locations where the Froude number changed from a value greater than one to a value less than one experienced a change from supercritical flow to subcritical flow, or rather a hydraulic jump. Pipes that had flow undergoing hydraulic jumps were flagged and incorporated into the model. Slopes were determined using GIS data and manual calculations when certain data was unavailable. Upon completion of this step, statistics were able to be produced to better understand the data.

Results from the statistical analysis from InfraPLAN were presented as graphs, charts, and tables in order to completely assess the validity of the data. At first glance, the data seemed to be accurate, but upon closer look it became apparent that several pieces of data did not fit within expectations. Analysis of this data for concrete pipes raised doubt due to the observation that there is old pipe that is scored as good condition pipe. This finding led to further investigation of the inspection logs and CCTV videos for the pipes of interest and the concerns proved valid as many of the pipes were in much worse condition than initially scored during the first round of review. Revisions were made during another round of review to the scores in the database to reflect the true condition of the pipes.

Additional findings were made while observing more of the statistical presentations for PVC pipes. After analysis of the data, it was apparent that the trend is unrealistic as the condition appears to improve over time. Areas of high concern were the younger pipes that had high condition scores. After investigation of these pipes, the scores were once again revised to reflect the true condition of the pipes. It also became apparent that most of the pipes were rated as a 3 and very few of the pipes were coded as a 2 or 4. These findings are not surprising as the defects that coincide with 2 or 4 are more difficult to identify during the inspections, so it is far more likely to see the other scores represented more frequently. Once the data was completely revised, it became necessary to consider all the potential causes for the deterioration of a pipe.

There are many factors that could affect the condition of a pipe over time and in order to have a model that includes those factors, it is necessary to fully understand them. These factors are called covariates, or a variable that has potential to predict the outcome of a study. During the modeling stage there were several covariates that were initially considered to identify any correlation between the pipe condition and that particular covariate. Below are several of the covariates that were studied:

- Pipe Material: Within the Dallas system there are several different pipe materials so each material was considered individually in order to determine if any correlation between the pipe condition and type of material exists. The system includes materials such as concrete, PVC, VCT, brick, fiberglass, HDPE, RTP and wrapped steel pipe. Some of the materials are not represented in large quantities within the system and for modeling purposes it was necessary to create groups. The materials and the abbreviations used in the analysis of the project is listed below in Table 3. The materials were grouped into 4 main categories and Figure 2 illustrates the breakdown.

Table 3: Pipe Material Abbreviation Key

Abbreviation	Material
CONC	Concrete
PC	Precast Concrete
PCCP	Pre-stressed Concrete Cylinder Pipe
RC	Reinforced Concrete
PVC	Polyvinyl Chloride
ELSE	Any other material not previously listed (e.g., steel, clay)
MULT/UNK	Multiple and/or Unknown

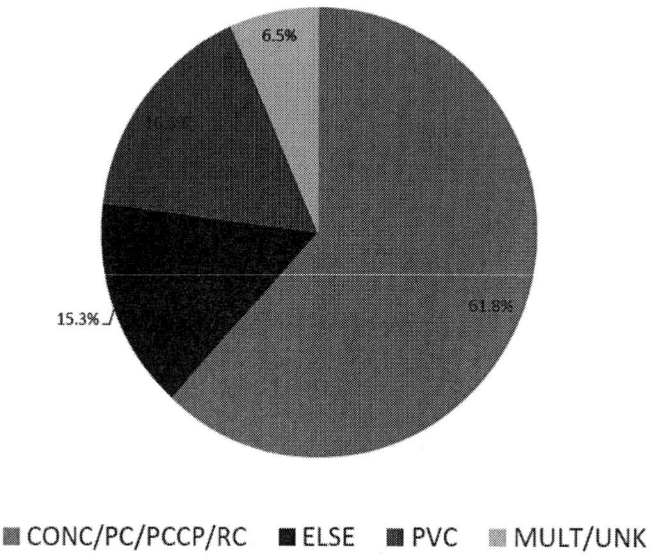

Figure 2: DWU Pipe Material Grouping

- Discharge of Force Main: Force mains are pressurized pipes that are typically used for moving water when gravity is not the best options. The pipes of interest for this study were those that had connecting force mains 16-inch diameter and larger, given the force mains are typically conveying older wastewater without much oxygen available until it reaches a possibly turbulent discharge point and connection to the interceptor system. The initial results pf the GompitZ modeling showed a low correlation for this covariant and thus was not included.

- Near Junction of Large Diameter Pipes: As corrosion can be most evident just near the junction structures, consideration was given to pipes upstream and downstream of the junction structures. Specifically, a junction is defined as a location where there is at least one downstream pipe to two upstream pipes which are both large diameter. The study of this covariate included analysis of pipes downstream (DS), upstream (US), upstream and downstream, and not near junctions. From the given dataset there was no trend apparent for any of the scenarios and so this covariate was not included into the predictive model. However, as more data is collected in the future, it may be necessary to reconsider and assess the inclusion of junction data into the model.
- Near Hydraulic Jump: Using the flow data and slope information in the GIS database areas were located where supercritical flow transition to subcritical flow. This is characteristic of a hydraulic jump. Locations of hydraulic jumps were studied and explored for possible correlation to the condition of the pipes. The hydraulic jump data proved to be significant and seemed to contribute as an indicator to the condition. As more data is provided regarding slope and flow it would be prudent to reassess the continued importance of this covariate.
- Evidence of Inflow /Infiltration: Pipes were identified with observed inflow and infiltration (I/I) defects from the condition assessment. It was challenging to see any significant trend that would suggest an influence on pipe condition due to I&I. As previously stated, once more data is collected it is very possible that this factor could have much more significance, but with the limited data it did not make sense to include I&I in the model. Based on the results, there is no confirmation that I/I is the sole contributor to the increased average condition score because the pipes were also older which could imply that age effected the rating.

Once the errors in the data were reconciled and the covariates were considered, GompitZ provided results. The model's output resulted in the probability of a pipe being rated each of the PACP scores for any given year. Further, the model not only projected these probabilities for every pipe in the system but also for each year until 2100. Below, Table 4 shows an excerpt from the dataset that shows the probability of two pipes being any of the ratings for a five year span. Additionally, there is an aggregated state which is essentially an expected value and is determined by using a weighted average method. The aggregate state is the PACP score that best represents a pipe at an age or year of interest. The formula for the aggregated state is listed below, followed by the excerpt from the table.

$$\textit{Aggregated Value} = p1(1) + p2(2) + p3(3) + p4(4) + p5(5)$$

Table 4: Probability Example for Two Pipes

ID	Year	p1	p2	p3	p4	p5	Length	aggregated state
01000000006M-Central	2016	0.120	0.013	0.353	0.008	0.505	0.215	3.765
01000000006M-Central	2017	0.119	0.013	0.354	0.007	0.508	0.215	3.772
01000000006M-Central	2018	0.117	0.013	0.355	0.005	0.510	0.215	3.777
01000000006M-Central	2019	0.116	0.012	0.356	0.004	0.511	0.215	3.782
01000000006M-Central	2020	0.115	0.012	0.357	0.004	0.512	0.215	3.786
01000000010M-01000000006M	2016	0.120	0.013	0.353	0.008	0.505	0.116	3.765
01000000010M-01000000006M	2017	0.119	0.013	0.354	0.007	0.508	0.116	3.772
01000000010M-01000000006M	2018	0.117	0.013	0.355	0.005	0.510	0.116	3.777
01000000010M-01000000006M	2019	0.116	0.012	0.356	0.004	0.511	0.116	3.782
01000000010M-01000000006M	2020	0.115	0.012	0.357	0.004	0.512	0.116	3.786

After the modeling and inspections conducted during this project, expected useful life has proven to be much longer than the initial values used. In short, it is apparent that the current assessment of the DWU system indicates pipe is in far better condition than was predicted using the initial desktop study. Based on its probabilistic method, the model predicts that the average structural score of the concrete pipe for the entire Dallas system is 2.98. In Table 5, the average score for each of the next 10 years if no improvements are made to the system are shown.

Table 5: GompitZ Predicted Average System Structural Scores 2016 - 2025

Year	Average System Structural Score (Including GompitZ)
2016	2.98
2017	3.00
2018	3.02
2019	3.03
2020	3.05
2021	3.06
2022	3.08
2023	3.09
2024	3.11
2025	3.12

The NASSCO PACP inspection scores were completed in December 2015 and after receiving the GompitZ projected 2016 scores for the uninspected sections of the pipes, it was possible to estimate the scores for the system as a whole as shown in Table 6.

Table 6: Structural Condition Results Summary

Structural Condition	Segments	Miles	% Miles
1	1077	53.28	23%
2	602	32.03	14%
3	2438	115.80	50%
4	301	17.72	8%
5	180	12.36	5%
Total	4598	231.19	100%

To further refine the likelihood of failure, performance condition is also analyzed, as there is more than one failure mode for a pipe. Performance condition criteria includes the ability to meet current capacity demands, the ability to meet future capacity demands, and the O&M score that was recorded in the field. The capacity scoring is based on the criteria that DWU has used in the past, which uses both current (2004 model) and future (2050 projected) capacity information in order to determine the capacity score for the pipe.

The O&M scores were previously explained and essentially, the scores were based on maintenance issues (defects) observed at the time of inspection which negatively impact the performance of the pipe, namely water level. Since only a fraction of the pipes were able to be physically inspected, O&M scores are not available for all segments. In the case when no PACP data was available, the pipes were assigned the average O&M score for the entire system which is 1.5. In Table 7 it can be seen that over 60% of the system does not have O&M data, as O&M data is only available for pipes inspected during this project.

Table 7: System O&M Scores

System O&M Score	Segments	Miles	% Miles
1	811	51.10	22.10%
1.5	3180	148.70	64.32%
2	501	25.83	11.17%
3	68	3.42	1.48%
4	34	2.04	0.88%
5	4	0.09	0.04%
Total	4598	231.19	100.00%

In the above table, it is very noticeable how many of the interceptors within system did not have O&M scores available. To look further into trends of actual inspected pipe with O&M scores, Table 8 was created to examine the breakdown of the field-recorded scores.

Table 8: Inspection O&M Scores

Inspection O&M Score	Segments	Miles	% Miles
1	811	51.1	61.95%
2	501	25.83	31.32%
3	68	3.42	4.15%
4	34	2.04	2.47%
5	4	0.09	0.11%
Total	1418	82.48	100.00%

Approximately 82 miles of the interceptor system had O&M scores from the field inspections. Of those 82 miles, more than 60% had an O&M score of 1, indicating that the vast majority of the pipes had either minute O&M defects or none at all. Less than 3% of the inspected pipes received scores above a three, indicating that very little of the system had serious O&M defects that might progress into failure modes. Additionally, the 2.13 miles of pipe with these high O&M scores were coded with defects indicating high water levels, which is a capacity concern rather than a maintenance issue. It is important to keep in mind that many of the lower level defects (i.e., scores from 1-3) could be issues that can be resolved through maintenance operations such as: grease or sediment accumulation, ragging, encrustation, minor obstructions and the presence of rocks.

The other component of the performance condition is the hydraulic modeling data. Hydraulic modeling information was extracted from flow data from 2004; earlier risk models also used the data to represent the system. The 2004 data was then used to project 2050 flows. Based on 2004 and projected 2050 hydraulic data, an overall hydraulic performance score was created. The criteria is listed below in Table 9.

Table 9: Hydraulic Performance Criteria

Criteria	Score				
	1	2	3	4	5
Current Capacity (2004 Model)	q/Q_{full} =0.65	q/Q_{full} =0.85	q/Q_{full} =1.00	q/Q_{full} >1; Surcharge predicted at manholes within 3' of ground surface	q/Q_{full} >1; Predicted overflows
Future Capacity (2004 Model for 2050)	q/Q_{full} =1.0	N/A	N/A	q/Q_{full} >1; Surcharge predicted at manholes within 3' of ground surface	q/Q_{full} >1; Predicted overflows

Each pipe was evaluated based on 2004 data showing the current (at the time) capacity and the projected 2050 capacity. A conservative scoring approach was used, which adopted the highest score in either the current or future capacity was used as the overall hydraulic performance score. Unfortunately, not all pipes were evaluated in the 2004 hydraulic modelling. For pipes that do not have data from the model, the average hydraulic score for the entire system, a score of 1.9, was assigned. Table 10 shows that a third of the system earned a performance score of 1 and indicates that a large portion was assigned the average score.

Table 10: Hydraulic Performance Scores based on 2004 Model

Hydraulic Performance	Segments	Miles	% Miles
1	1480	74.05	32.03%
1.9	2074	98.04	42.41%
2	388	19.06	8.25%
3	487	26.51	11.47%
4	19	2.19	0.95%
5	150	11.33	4.90%
Total	4598	231.19	100.00%

It should be noted that due to the lack of system wide hydraulic modelling information and the age of the available information, these scores are provisional. These scores are provided to demonstrate how likelihood of failure is influenced by more than structural condition alone. Updated hydraulics are vital to establishing a more reliable measure of performance and should be incorporated into the condition scoring data as soon as it is available. This will not only update current scores, which are over ten years old at this time, but also, hopefully, provide data where there is currently none. Over forty percent of the system is lacking hydraulic modelling data completely, and, thus, was assigned a score of 1.9.

After the components of likelihood of failure, performance and structural condition, were evaluated and scored, they were added together to produce the likelihood of failure, an indicator with a minimum score of 2 and a maximum score of 10. A score of 2 would imply that the asset received the minimum score of 1 for both performance and structural condition, indicating that the asset was in near perfect condition. A score of 10 would identify an asset with a five in both performance and structural condition, which would be a pipe for which failure is imminent or has already occurred. Table 11 provides a summary of the likelihood of failure scores for the system.

Table 11: Summary of LoF Results

LOF	Segments	Feet	Miles	% of System
2	69	18,580	3.52	1.52%
3	951	242,693	45.96	19.88%
4	729	214,339	40.59	17.56%
5	2020	482,204	91.33	39.50%
6	389	108,008	20.46	8.85%
7	263	80,544	15.25	6.60%
8	90	37,356	7.07	3.06%
9	56	27,531	5.21	2.26%
10	31	9,412	1.78	0.77%
Total	4598	1,220,66	231.19	100.00%

The likelihood of failure scores here range from 2 to 10. Scores with non-integer values were rounded up. As shown in the table above, there was decent gradation of scores, with the majority of the scores falling in between 3 and 5. A distribution of the scores can be seen in Figure 3.

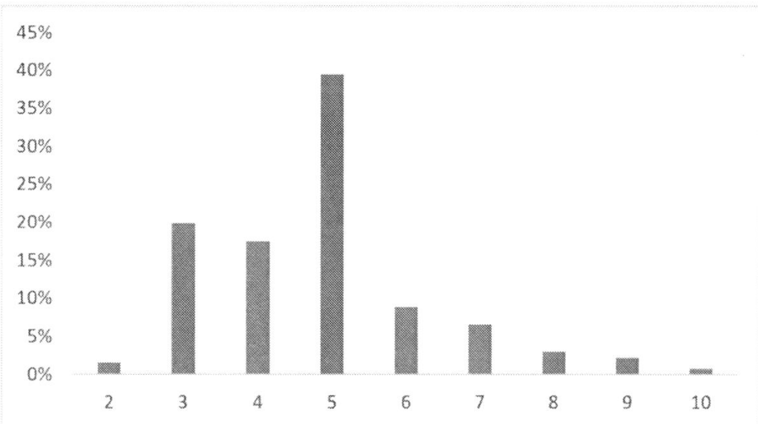

Figure 3: Distribution of System LoF Scores

The closer the likelihood of failure is to 2, the less likely the pipe is to fail. These scores have a favorable distribution in that the majority of the scores are concentrated at a score of 5. It is important to remember that these scores include the performance data that is old and incomplete.

This should be updated and the scoring exercise repeated to get a more accurate depiction of the system. With these current numbers, the system appears to be in fair condition. These scores will be combined with the consequence of failure, explained in the following section, to identify pipes and general locations with the highest risk to target for projects.

CoF

CoF scoring was developed with an emphasis on creating granularity and the size was the primary factor in determining the score. If there are critical infrastructure or sensitive areas intersected by the pipe then it will affect the score. Table 12 provides a comprehensive list of all the criteria which were incorporated into the scoring.

Table 12: CoF Scoring Criteria

CoF 1	CoF 2	CoF 3	CoF 4	CoF 5
Diameter <=12	Diameter >12 and <24	Diameter >=24 and <=72	Diameter >72 and <=96	Diameter >96
	Major Shopping 50ft	Hospital 100ft	Under Buildings	
	Arterial 60ft	Schools 100ft	Rail Road or Dart Rail 20ft	
		Industrial 50ft	Canals Rivers 100ft	
		Major Thoroughfare 60ft	Highway 60ft	
			Environmentally Sensitive Area 20ft	

The initial score is taken as the highest score that the pipe achieves based on any one of its characteristics. Each two additional occurrences of that maximum score will add an additional point to the base score to a maximum of 5. The equation of the CoF score is listed below.

$$CoF = Maximum\ Score + (Additional\ Two\ Occurrences\ of\ Max\ Score / 2)$$

In developing a new scoring methodology, the approach was to increase the gradation of scores throughout the system by increasing the size divisions and, effectively the scoring range, from three categories to five. This will also allow this CoF approach to be applied to smaller diameter pipe (i.e., non-interceptor). Previously for DWU, CoF had been scored using a 1-3 scale, with diameter as the primary determining factor of the final score. Secondary consideration was also given to interceptors with smaller diameters which intersected critical infrastructure or sensitive areas.

RESULTS

CoF Results

The CoF score can generally have a minimum score of 1 and a maximum score of 5; however, as all the pipes evaluated for this project were interceptors with a diameter 24 inches or greater, the minimum score possible for the purposes of this report is a 3. The maximum score remains a 5. A summary of the system CoF scores is presented below in Table 13.

Table 13: CoF Scores for DWU WW Interceptors

COF Score	Segments	Miles	% Miles
3	1689	83.44	36.09%
4	2727	130.54	56.46%
5	182	17.22	7.45%
Total	4598	231.2	100.00%

The consequence of failure for the system is generally high, with the majority of the pipes receiving a score of 4. This is not surprising as all the pipes assessed are interceptors. If any one of these fails, it will have a big impact purely from size alone. If you add other factors, such as intersecting critical infrastructure or sensitive areas, this increases even further. In the future, this CoF scoring could be applied to the entire system, as it has categories that would be applicable to smaller diameter pipe. If the entire system were included in the analysis, the numbers would have even more gradation.

Asset Management

DWU's interceptors support vital services to customers, and failure represents a risk to the organization from social, environmental or financial ramifications. Through risk analysis, DWU is able to compare the relative risks embedded in the system to make better resource allocation decisions. An evaluation is performed on the LoF (i.e. physical and performance condition considered together) and consequence of failure (CoF) of an interceptor. A risk score is calculated at DWU using the following formula:

$$Risk = Likelihood\ of\ Failure\ x\ Consequence\ of\ Failure$$

In this project, Arcadis focused on three areas while developing this approach to risk scoring. The main focus was to achieve gradation and granularity, or a better spread of discrete scores, in order for DWU to allocate resources efficiently to those pipes which need it most. Arcadis sought to accomplish this through CoF criteria such that assets could have scores at non-integer values ranging from 1-5. Secondly, Arcadis wanted to develop a holistic approach to risk which would give more consideration to several categories of consequence, including social, economic, and environmental, while also stratifying the LoF such that criteria are considered for the performance score. Finally, Arcadis wanted to deliver a tool that provided flexibility to the changing priorities over the years.

It is vital that the risk scoring scale truly represents the system even if it still categorized a majority of the system as high risk. Gradation is important for discerning between selecting one potential project from another, however it is crucial that high risk pipe is not overlooked. The current risk scoring system displays the high risk within the system while still allowing for a certain amount of gradation to assist in the selection and creation of projects by increasing the granularity of the CoF scores though incorporating a way for multiple criteria ranked at the same level of consequence to add additional points to the CoF. The CoF and LoF were multiplied together to attain the risk for each pipe segment. The segments were categorized by risk score and total miles of each risk category were categorized into five ranges according to the risk. The breakoff points for risk score categories are listed below in Table 14.

Table 14: Risk Categories and Corresponding Score Ranges

Risk	Min. Score	Max. Score
Highest	40	50
High	30	39
Medium	20	29
Low	10	19
Lowest	2	9

Using these risk category ranges, the risk scores were translated into their corresponding categories to give a better picture of the system as a whole. A summary of this exercise is presented in Table 15.

Table 15: DWU Interceptor Sewer Risk Summary

Risk	Miles	Percent of Total (mi)
Highest	0.57	0.24%
High	8.87	3.84%
Medium	105.65	45.70%
Low	87.77	37.96
Lowest	28.33	12.25%
Total	231.2	100.00%

As seen above, nearly half of the collection mains fall within the lowest and low risk categories, with an additional 46% falling within the medium risk category. Less than five percent of the collection mains are considered high or highest risk. This is good news, as, although quite a large portion of the system is considered medium risk, the highest risk pipes are few and easy to identify for projects.

Rehabilitation and Replacement Planning System

As an aid to the condition monitoring Arcadis developed a Rehabilitation and Replacement Planning System (RRPS) tool. All of the risk scores, based on the likelihood at the time of inspection, were inserted into the RRPS database in order to be utilized by the tool to identify potential projects and funding scenarios. The Arcadis Rehabilitation and Replacement Planning System (RRPS) is a software tool designed for distributed assets managed within a Geographic Information System (GIS). The system applies service level, CoF, LoF and replacement cost criteria in conjunction with budget alternatives to plan for the rehabilitation or replacement (R&R) of linear assets. Upon selecting the most desirable long-term budget scenario, assets can then be grouped into logical CIP projects to begin the design and construction process. The results of the selected long-term scenario and project grouping can be used to update the enterprise GIS to provide visibility of this information to the entire organization.

The CIP modeling process begins by loading the approriate assets inventory into the RRPS project database. In this project, the inventory is composed of the large diameter interceptors that are owned and maintained by DWU. After loading the pipe inventory, the asset cohorts should be defined. Cohorts are used to categorize the linear assets into groups having similar physical characteristics that would affect their expected useful life. Each cohort includes degradation curve information for the current pipe and for the replacement pipe.

Each cohort record requires the following minimum information:

- Material(s)
- Minimum Diameter (>Min Diameter)
- Maximum Diameter (<=Max Diameter)
- Minimum Year (>Min Year)
- Maximum Year (<=Max Year)
- Initial Pipe Equation Type
- Initial Pipe Equation Constant/Y-Intercept
- Initial Pipe Equation Exponent/Slope
- Replacement Pipe Equation Type
- Replacement Pipe Equation Constant/Y-Intercept
- Replacement Pipe Equation Exponent/Slope

The deterioration curves for current and replacement pipe are specified by the equation type of linear or exponential, the Y-intercept, and the slope/exponent values.

Exponential: *Condition* = *Y-intercept* * *e* $^{(Exponent\ *\ Age)}$

Linear: *Condition* = *(Slope* * *Age)* - *Y intercept*

The project database is configured to use exponential curves base on a condition value of 1 at installation (age 0) and condition value 5 at end of effective useful life (EUL). Therefore the Y-intercept is 1 and the exponent is defined by the equation: Exponent = ln (Condition / Y-intercept) / Age. For an EUL of 100 years the exponent is ln (5/1) / 100 = 0.01609

The project database has been configured with two cohorts, one for concrete and another for non-concrete material within the interceptor inventory. Concrete is defined to have an 80 year EUL and non-concrete uses a 100 year EUL. For most of the concrete interceptor assets, the cohort EUL is just a placeholder, because the results from the GompitZ analysis will override the cohort degradation curve with a pipe specific curve. The concrete cohort degradation curve is only used for pipes of this material that were not analyzed with the GompitZ model. All cohorts are defined to use replacement material having a 100 year EUL. The replacement degradation is used to predict the effect on system wide condition after each pipe is replaced.

Additionally, there is an R&R Cost table, shown below, that identifies pipe costs by diameter ranges. Each row represents the cost per linear foot for the range of diameters specified (> Min Dia and <= Max Dia). The project database is configured with the cost for inspection and three types of R&R costs (slip lining, cured-in-place pipe, and full replacement). Additionally, the cost for easements are defined for seven areas of the City: Northwest, North Central, Northeast, Central, Southwest South Central and southeast.

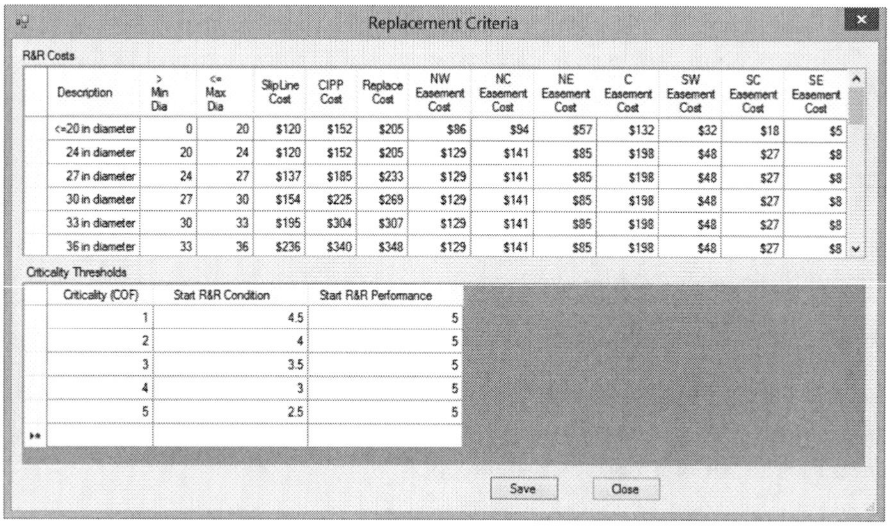

Replacement Criteria

R&R Costs

Description	> Min Dia	<= Max Dia	SlipLine Cost	CIPP Cost	Replace Cost	NW Easement Cost	NC Easement Cost	NE Easement Cost	C Easement Cost	SW Easement Cost	SC Easement Cost	SE Easement Cost
<=20 in diameter	0	20	$120	$152	$205	$86	$94	$57	$132	$32	$18	$5
24 in diameter	20	24	$120	$152	$205	$129	$141	$85	$198	$48	$27	$8
27 in diameter	24	27	$137	$185	$233	$129	$141	$85	$198	$48	$27	$8
30 in diameter	27	30	$154	$225	$269	$129	$141	$85	$198	$48	$27	$8
33 in diameter	30	33	$195	$304	$307	$129	$141	$85	$198	$48	$27	$8
36 in diameter	33	36	$236	$340	$348	$129	$141	$85	$198	$48	$27	$8

Criticality Thresholds

Criticality (COF)	Start R&R Condition	Start R&R Performance
1	4.5	5
2	4	5
3	3.5	5
4	3	5
5	2.5	5

Save Close

The Criticality Thresholds table stores the range of pipe structural condition and performance values within which a service may be provided. R&R will be considered if a pipe reaches either the condition or performance limit for its CoF. The project database is configured with a different range of structural condition values for each distinct criticality (integer values from 1 through 5). Each pipe in the model must therefore be assigned a criticality which represents the CoF. The project database is configured to start replacement at lower (better) condition for higher CoF pipes and wait for higher (worse) condition for pipes of lower CoF. Different performance scores can be defined, but this project database uses only the highest performance value of 5 for all CoF pipes.

Once all of the processes above are accomplished, the RRPS tool presents the assets graphically or in a tabular form so that projects can be aggregated in a logical manner. Graphic representation of the data allows the user to view the assets based on risk and visually group assets that are contiguous. The reasoning behind this is that these projects would be more sense to be performed if they are contiguous rather than scattered.

The R&R planning process utilized four software tools: Arcadis RRPS, GompitZ, Esri ArcGIS Desktop and Microsoft Access. The RRPS tool is used to develop long-term R&R budget scenarios and quantify changes to the system-wide risk and condition of the modeled assets. GompitZ is a statistical model used to define the structural degradation of each interceptor segment over time. Esri ArcGIS is used to manage the GIS data and define groups of interceptor assets for near-term projects. Microsoft Access is used for database management, system configuration and results reporting. RRPS uses an Access 2003 database configured as an Esri Personal Geodatabase (PGDB). All of these tools seamlessly work together and update to any edits or changes so that the robust features are always up to date.

SUMMARY AND CONCLUSIONS

The 231 miles of the Dallas interceptor sewers were evaluated for this project based on both physical inspection using NASSCO PACP guideline and GompitZ predictive modelling for those pipes which were unable to be inspected. After structural scores were assigned to each pipe, they were then evaluated based on O&M defects and hydraulic capacity to produce a performance score. These two components came together to create the likelihood of failure.

The consequence of failure is a measure of the severity of the effect on the system and the public if a particular asset were to fail. With discussion from Dallas, criteria were developed that were used to rank the segments based on their consequence of failure, ensuring that pipes which met more than one criteria were given more weight than those who might have only met one. The consequence of failure score and the likelihood of failure scores were then multiplied together to determine the overall risk score for the segment. The possible risk scores range from 2 to 50, but actual risk scores ranged from 6 to 45.

The risk categories include the following: Highest Risk (40-50), High Risk (30-39), Medium Risk (20-29), Low Risk (10-19), and Lowest Risk (2-9). Once the segments were distributed amongst the five ranges, it was easy to see, both visually and graphically, the current state of risk the system was experiencing as a whole. Figure 4 illustrates the distribution of the miles of pipe based on risk category. It is clear the majority of the system is in the medium to low risk range, with the medium risk range slightly greater than the low risk range by around twenty miles. It is important to keep in mind that these scores include a significant amount of modeled information when compared to actual inspection information; however, it is a good start to get a picture of overall risk of the system. The most promising news is that less than one percent of all the interceptors in the system were at the highest risk.

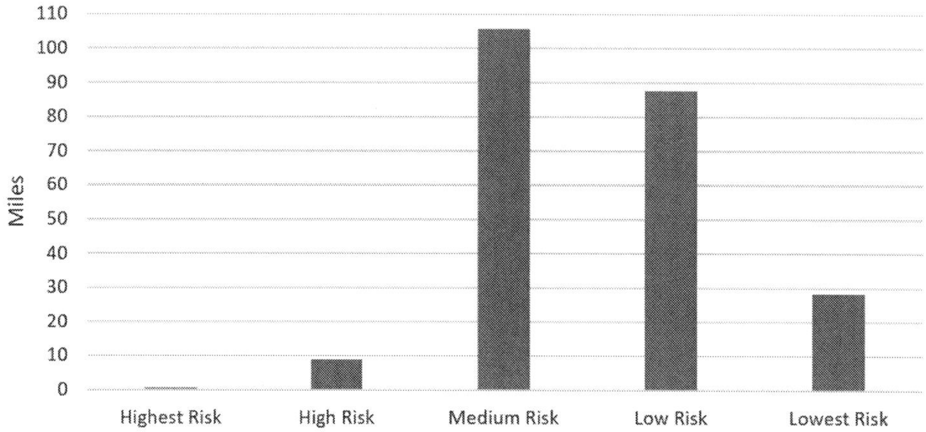

Figure 4: Miles of Pipe Classified by Risk Category

Based on the evaluation of risk for 2016, there is only a small portion of the system which is flagged for high or highest risk. This will simplify the project selection process, as it will be easier to narrow down potential projects based on where the highest concentration of high risk pipe is. The largest risk category is medium. These medium-risk interceptors will eventually progress to high risk. Considering the high number of segments, this could cause a large shift in the distribution of risk. This underscores the importance of monitoring the condition of the system and making risk mitigation and reduction, though rehabilitation or replacement, a priority now in order to reduce unplanned reactive costs in the future.

Finally, the risk system in conjunction with the RRPS planning tool allowed for the prioritization of pipes with inspection data. This prioritization assisted in the creation of rehabilitation and replacement projects that align with the DWU annual budget and with the desire to decrease risk within the system. Additionally, the prioritization of the pipes with GompitZ data allowed for the identification of pipes that required inspection. These suggested inspections would also be required to be considered for budgetary and planning purposes. Ultimately, DWU is left with a program that utilizes robust tools that can be altered to adhere to the current needs of their system and a lack of inspection data will not prevent a thorough evaluation of their network.

REFERENCES

Using Multi-Sensor Technologies in Large Interceptor Sewers to Develop Sensible Risk-Managed Capital Improvement Planning

Mark G. Wade, P.E.[1*]; Michelle Beason, P.E.[2]

[1]Providence Infrastructure Consultants, Lawrence, Kansas.
[2]National Plant Services, Long Beach, California.
*Email: mwade@providenceic.com

ABSTRACT

Large-diameter wastewater interceptor sewers are probably some of the most critical assets that are owned and operated by a utility. They are also and absolutely the most difficult to inspect and maintain. And the bigger they are, the harder they are to determine their remaining useful (or service) life (RUL). However, there is good news. During the past five years, technologies related to inspection, assessment and trenchless renewal of these monster wastewater conveyance assets have grown and improved dramatically. These changes come with the equal challenges of processing enormous quantities of detailed inspection data and other information regarding the condition of each of these very imposing assets. For wastewater conveyance systems, this can include information associated with its structural, hydraulic, and maintenance-related conditions.

This paper will present a summary of the technologies that are now available for such big-pipe inspection platforms and how these multi-sensor inspection (MSI) technologies can be combined to give utilities a true and reliable basis for estimating critical condition parameters such as structural integrity, hydraulic reliability, RUL, and cost-effective engineered solutions for their long-term operation. Specifically, the paper will highlight the methods that are now being used by system owners to utilize advanced MSI platforms to determine RUL for reinforced concrete pipelines (RCP), which we believe is the "holy grail" of large-diameter RCP wastewater interceptor management and operations. We will describe, in detail, those technologies now available on sophisticated MSI platforms which include conventional CCTV, laser scanning, sonar imaging, LiDAR, H2S monitoring, and pipe-penetrating radar (PPR). Finally, we will look at how these results from the information provided by MSI platforms for large-diameter wastewater pipelines can be directly applied to predictive models to identify relative risk as a utility asset based on likelihood of failure scoring.

KEYWORDS: Large-Diameter, Wastewater, Interceptor, Multi-Sensor Inspection, Assessment

INTRODUCTION

Historically the condition of most large-diameter wastewater conveyance systems has remained unknown due to their relative inaccessibility and the type of equipment or technologies available to perform necessary internal inspections to the level required for a full assessment. In the past, we have made extreme and often dangerous attempts to enter such systems to perform these requisite inspections using whatever means that made sense (or didn't make sense but we went ahead with them anyway) at the time including boats, rafts, pontoons, elaborate skate boards, and (yes) even motorized carts and other vehicles with combustion engines. We have even walked them end-to-end using SCBA, special footwear, and industrial-strength chest waders, all the time being tethered remotely from an access structure. However, a number of technology breakthroughs in recent years, and the evolution of robust robotic inspection platforms that can endure such harsh environmental conditions, have created new pathways to perform these assessments with none of the above dangerous and labor-intensive approaches. However, such technology comes with a significant cost. Even the largest cities and utilities do not have the financial means to own these systems due to their high cost of purchasing and maintaining them. Many of these MSI platforms are now available from larger utility contractors who routinely perform large-diameter pipeline inspections. For many of us who have endured the hardships of more conventional and manned-entry approaches we give a hearty "thank you" to everyone who has had a part in making such systems available and useful to those of us who inspect, engineer, and manage these critical assets.

DESCRIPTION OF MSI TECHNOLOGIES

The main purpose and strategy of using MSI platforms in large wastewater conveyance systems is that it offers a one-pass opportunity to utilize a wide range of inspection technologies in order to fully assess its condition. Transport systems for MSI technologies are usually designed around (a) crawler and (b) floatation platforms. The choice of a transport platform depends on a number of site conditions and factors including:

- Level and velocity of flow;
- Restrictions or options available for flow control;
- Diameter or dimensions of the conveyance asset;
- Access to the pipeline
- Available head space during high and low flow conditions;
- Anticipated debris that will remain in the pipeline during inspection;
- Depth of asset.

Generally, MSI platform systems combine traditional high-definition CCTV, sonar, and laser technologies to provide a fairly complete database of the condition of the pipeline being inspected. Depending on additional and optional equipment such as H2S monitors, inclinometers, 3-D LIDAR, and temperature sensors, the information gathered from MSI verses other more convention approaches to inspecting large-diameter structures can be used to better assess the condition of the pipeline both above and below the water surface. A conventional CCTV camera can be substituted for a side-scanning digital universal camera (DUC). The use of pipe penetrating radar (PPR) is also now available on MSI platforms and this offers the utility a means to measure wall thickness and locate and characterize voids outside of the pipe.

Figure A shows a few examples of typical MSI platforms that are commercially available for large-diameter pipelines.

Figure A: Example MSI Platforms

Inspection data for each asset, regardless of the inspection platform is typically collected in a NASSCO PACP code-based format for CCTV. PACP quick rating scores can be assigned for (a) structural, (b) O&M, and (c) overall. Although NASSCO encourages the use of these scores as a means to create a grade score from the user's own post-processing protocols and procedures, the use of the quick scoring and condition grading and scoring is not necessarily a sufficient tool to determine asset condition grades for large-diameter pipelines and, more importantly, calculating RUL. Rather, the use of additional MSI technologies such as DUC, LIDAR and sonar provides the necessary information to measure critical conditions such as (a) pipe configuration, (b) internal pipe dimension(s), (c) remaining wall thickness, (d) exterior voids, (e) pipe location and slope, (f) level, type, and volume of debris and settlement, (g) dimensional values of holes and exposed or offset joints, and (h) location of reinforcement. For RCP, in particular, data from MSI assessments can be significant and even critical in determining pipe wall loss. Understanding of the type of pipe, pipe size, asset age (year of construction), and depth of corrosion can also be used to more fully estimate the condition of each asset and ultimately its respective RUL. In essence, the tools that can be attached to an MSI platform opens up a new way of getting closer to fully characterizing a large-diameter sewer which means we can better predict its functional longevity and put renewal processes in place so it can "keep on ticking".

Although large wastewater conveyance systems vary widely in materials of construction, we will now focus our remaining attention on those that are classified as RCP since this class of large-diameter pipe generally represents the largest group of large pipe assets in a wastewater utility's inventory.

Summary Discussion of Hydrogen Sulfide and Reinforced Concrete Pipeline

Many large-diameter wastewater conveyance systems are constructed of materials that are susceptible to internal and exterior corrosion. In particular, these systems include unlined RCP. In many RCP-constructed wastewater interceptor sewers, internal corrosion of the wall surface can be easily observed using traditional technologies such as limited CCTV or zoom-style cameras from access structures to the asset. Even jetter cameras that can be installed on the head end of any conventional jet nozzle can be sufficient to discern major structural or maintenance-related issues. However, the use of MSI technologies can transform these limited inspection results for RCP to higher-level information that can accurately measure data such as corrosion due to atmospheric hydrogen sulfide (H2S). It can change the dynamic of characterizing the impact of atmospheric H2S and its corrosive effects that impact its remaining service life.

The earliest recorded information of manufactured RCP is 1904. However, the degree of wall loss within RCP systems, regardless of age, can vary significantly and may not necessarily represent a direct correlation to pipe age. This is why we observe some RCP interceptor systems that have remained in service for 80 to 100 years and exhibit no actual wall loss due to corrosion while newer systems exhibit the effects of corrosion that has fully compromised the structural integrity of the pipeline. Therefore, correlating the visual image of the CCTV inspection (and PACP codes) with the additional data that can be captured from a properly equipped and calibrated MSI platform (such as scans from LiDAR systems) is critical in determining the extent of wall loss in RCP.

Wall loss for any concrete manufactured pipeline, specifically large-diameter pipes, is crucial to measure as it determines if and when wall loss has reached the pipeline's steel reinforcement. The use of MSI to confirm this information is now considered foundational for the evaluation of most large-diameter assessments. Figure B shows an example of advanced wall loss that has penetrated the first row, or cage, of reinforcement. Once this occurs, corrosion can advance quickly through the steel bars and even penetrate further to the second cage. Therefore, a further discussion of the nature and chemistry of RCP and corrosion is important in order to understand the methodologies and procedures used from MSI data to determine existing condition (score), remaining useful life (RUL), and recommendations for future replacement, rehabilitation, and/or preventative maintenance.

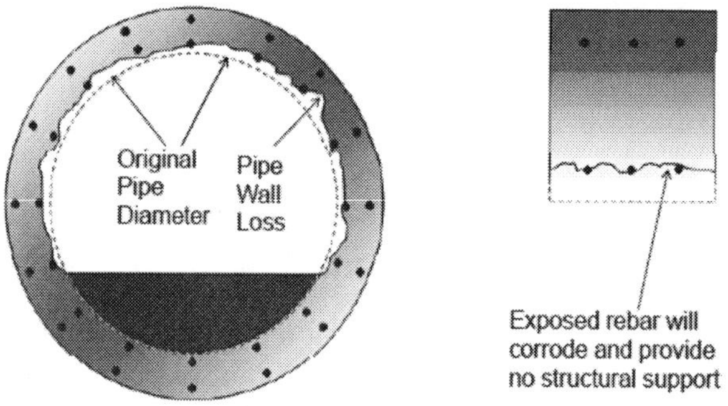

Figure B: Example of Wall Loss and Compromise of Steel Reinforcement in RCP

The most significant environmental impact on larger-diameter reinforced concrete pipelines is that of sulfuric acid formation within the pipeline itself. In advanced stages, corrosion and subsequent wall loss thickness due to the effects of sulfuric acid attachment can result in significant compromise to the structural integrity of the pipeline and even sudden and catastrophic failure. In order to better predict the remaining useful life of buried RCP infrastructure, it is important to also understand the biological process of concrete corrosion resulting from the presence of hydrogen sulfide in wastewater atmosphere.

The corrosion cycle within large-diameter concrete pipelines (unlined) can be best described by the illustration shown below in Figure C. In this illustration, aqueous H_2S is a product of biological sediment and wall build-up below the flow line. Once released into the wastewater interceptor headspace, it is attached to the wall surface and converted to sulfuric acid (H_4SO_4). Condensation that contains high concentrations of H_4SO_4 results in the corrosion in concrete and steel, which in turn results in observed and measured wall loss.

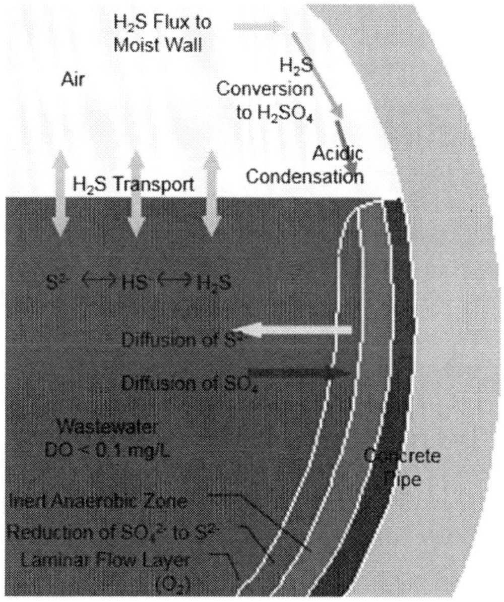

Figure C: Illustration of H2S Generation Cycle in Wastewater Pipelines

As Figure C shows, the factors affecting sulfide buildup in gravity sewers are complex. They consist of biological and physical processes in both the aqueous and the gas phases of the wastewater pipeline. The rate of each process varies (among other parameters) according to flow characteristics, temperature, and pH.

When initially installed, most major RCP interceptor sewers probably start off with a surface pH of 12-13, depending on the amount of cement in the mix design. Such a high pH level initially prevents the formation and growth of bacteria. Over time, however, the level of pH on the interior surface of the pipe wall declines due to the impact from CO_2 and H_2S. Combined, these two elements produce acids that can begin to attach the surface of the pipeline. In order for bacteria formation to begin, the pH level must be lowered to 9-9.5. Sometimes this takes several years but under more "optimal" conditions, it can take only a matter of months. The most common form of bacteria under these lower pH levels is a genus called Thiobacillus which has the unique ability to convert H_2S to sulfuric acid (H_2SO_4) in the presence of oxygen. This process always occurs above the waterline in a sewer system, particularly at the crown and wetted perimeter between the diurnal low and high water levels. As pipe ages, the level of pH continues to drop and new species of Thiobacillus evolve. This results in the production of increasing levels of sulfuric acid, thereby increasing the rate of corrosion to the pipe wall.

The major problem with H_2SO_4 is that it attacks the matrix of the concrete which is composed of calcium silicate hydrate gel, calcium carbonate from aggregates, and un-reacted calcium hydroxide. The primary product of RCP wall decomposition due to H_2SO_4 corrosion is calcium sulfate ($CaSO_4$), otherwise known as gypsum, a common material used to manufacture drywall board. Those involved with internal pipeline inspections often describe it as material with a white, pasty composition. It has been implied by some published information that a coating of gypsum paste somewhat protects the underlying sound concrete by providing a "buffer zone" through which freshly produced H_2SO_4 must penetrate. Interestingly enough, the zone between high and low diurnal flow levels keep washing away the constantly forming thin layer of gypsum in this region of RCP. This allows this section of the pipeline to experience higher-than-average corrosion rates. And oddly enough, high-pressure cleaning at frequent intervals can also exacerbate the situation and create wall surfaces that are more influenced to higher rates of corrosion.

The need to measure the effects of long-term corrosion due to the above described conditions can now be accomplished through the selection and use of those technologies that make sense and can be attached to an MSI platform. Once accomplished, we are on a pathway to determine those measurements and parameters that, combined with other asset information, lead to predicting the asset's particular remaining useful life.

Estimation of Remaining Useful Life

In general, RUL (also referred to as remaining service life or RSL) is the expected duration that a pipeline and ancillary access structures will perform satisfactorily based on its current structural condition and hydraulic performance. Remaining useful life, particularly for large-diameter wastewater interceptors, is generally considered the life span of any structural element within that pipeline from the current time of assessment to the time of its forecasted failure. However, the forecasted time of failure depends on more factors than just internal pipeline condition. It can range widely depending on the methodology used and assumptions chosen to calculate RUL.

For most new interceptor sewers, design service life (DSL) is the intended life span from the time of construction to the time of failure, based on the design criteria originally determined by the parties responsible for its construction. Currently, most interceptors are designed for a service life of 50-100 years. However, DSL is not necessarily a true indicator of the asset's deterioration rate or when the asset will actually fail due to the many factors that result in a very broad range of actual service life. As an example, there are case histories of RCP interceptor sewers that have required replacement or rehabilitation within 25-30 years of service life while others that show little evidence of deterioration after nearly 100 years of service life. It is, indeed, a paradox.

Admittedly, the determination of RUL for wastewater conveyance systems is not a fully developed science among utilities and consultants at this point in time. There are a few well-published articles and presentations on the topic, the best perhaps being the program underway by the City of Fort Worth called ICAP.

In general, NASSCO has encouraged the use of their PACP codes to develop appropriate scores that can then be leveraged to develop appropriate RUL estimates (or ranges). However, the use of PACP codes, although helpful for large interceptors, is a better tool to predict RUL for smaller diameter systems that commonly include pipe materials such as VCP. For larger-diameter systems, and RCP specifically, the use of LiDAR data in combination with high-definition CCTV video records is a better and more accurate methodology to estimate RUL. This approach is now routinely used by the authors to estimate RUL for nearly all of the interceptor assessment programs they undertake. Since mode-of-failure (i.e. likelihood of failure) for RCP is almost always structural in nature due to wall loss and corrosion of the asset's reinforcement, a combination of the following parameters are typically used when the proper MSI technologies are applied:

- Original installation date (and manufacturer, if available);
- Original pipe diameter;
- Original pipe wall thickness (including pipe class assumptions);
- Measured average and maximum wall loss (calculated from LiDAR);
- Observed corrosion (wall loss) consequence such as ribbing, exposed reinforcement, cracks, joint failures, etc. (observed from CCTV);
- Wastewater flow characteristics;
- Measured H_2S;
- Groundwater chemistry;
- Wastewater chemistry.

MSI technologies offer results that can be used in a number of ways to characterize the loss of pipe wall material. Since most decay curves available from scholarly journals and publications are limited (most being decay curves that are empirically derived), the ability to predict or forecast for RULs unique to each asset must be leveraged from MSI data. The above parameters for each pipe asset are then used to analyze and estimate the following:

- Average corrosion (wall loss) rate (inches/year);
- Maximum corrosion (wall loss) rate (inches/year);
- Years to first row of reinforcement failure (for pipelines greater than 30-inches);
- Remaining years to 50% wall loss;
- Remaining years to second row of reinforcement failure;

In the absence of more detailed information regarding a pipeline that may score high in condition (such as petrographic analysis of pipe coupons extracted at critical locations, RUL can be determined on the basis of maximum corrosion rates as a function of these corrosion estimates to both rows of reinforcement as well as the mid-point of the pipe wall.

Calculation of Corrosion Rate and Remaining Service Life

As mentioned previously, there are a limited number of publications on the matter of decay curves for wastewater conveyance systems. A characteristic decay curve is shown in the following figure. In this example, the separate curves represent RCP and brick interceptors.

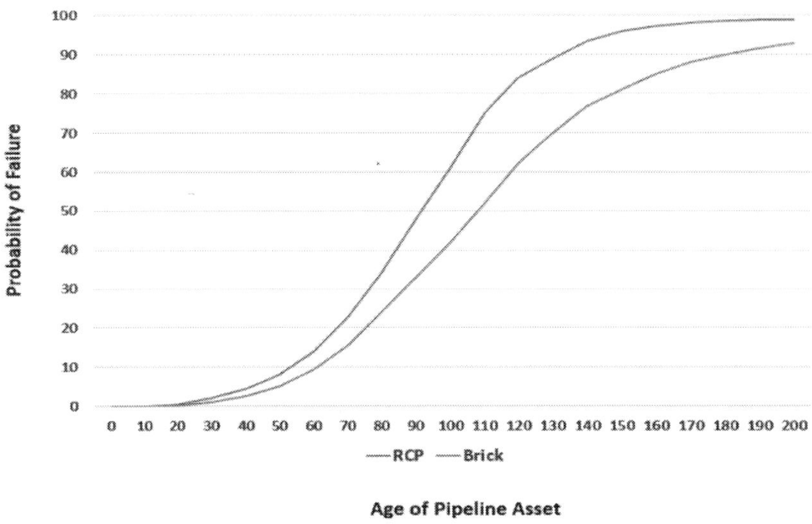

Figure D: Example Pipeline Decay Curve

Information in Figure D also represents multiple inspection times and resulting assessments that were used to construct these decay curves. This includes analysis of measured internal and external environmental and geotechnical conditions such as H_2S, petrography, and hydraulic modeling.

However, there is little agreement among practitioners on how such decay curves, when available, can be applied to a specific asset classification due to the multitude of variables, many which may or may not be monitored or measured using MSI technology. These variables include original materials of construction (these have changed through the years even in the manufacture of concrete pipe), manner of its original manufacture (sometimes referred to as rebar-creep), internal pH of the wastewater, unique microbiology within the "slime" layer of the pipeline, level and amount of settled sludge, velocity gradients, soil chemistry, soil structure within the original trench, groundwater, burial depth, type of pipe joint, and many more.

Therefore, in the absence of further and additional data regarding internal and external environmental conditions, rate of annual wall loss for each of the interceptor assets is generally modeled as a linear function over time (as opposed to some decay curves which show variable rates of corrosion during the lifetime of an RCP asset. In comparing a linear decay to that of the curves shown in Figure C allows for relatively conservative RUL values for the first 50 years of service life. With additional and future condition information available through data such as MSI technology, water and soil chemistry, petrographic analysis, and biology of debris and slime, the individual decay curves for those assets ranked highest in condition severity within a utility's asset inventory can be determined, thereby allowing for a more accurate and reliable failure prediction model.

The assumption that all pipelines are undergoing a linear model for wall loss also needs to include a brief discussion on additional assumptions:

- Laser Measurement – Different MSI platforms offer differing technologies and science to estimate wall loss in RCP. Laser types include 3D, 2D, Diode lasers, and ring lasers. In addition, there are serious engineering challenges in the interpretation and analysis of this data (some better than others, to be sure). To account for this, in certain situations the outcome of the laser data may need to be adjusted based on the visual observation from the CCTV and side-scan inspection records. This includes pipelines where extensive ribbing may not necessarily correlate to the maximum measurement from the laser scan, but more correlates to the average measurement obtained.

- Debris and Build-up – In certain situations, debris and solids will gather and accumulate in the pipe invert of larger pipelines and thus prevent a full and detailed assessment of the entire asset. This particularly happens where velocities are well below the pipe's natural scouring velocities. Therefore, estimates will need to be made on the overall condition of the entire pipeline based on information collected from the portion of the pipe that can be inspected. For some assets, this partial inspection may only include a simple view of the pipe within close proximity to an entry manhole, or from a zoom-style camera.

- Pipe Joints – The assessment of the entire pipe asset using MSI must include a "good look" at the pipe joint. However, some pipe assets may include wall build-up (or even high flow levels) that prevent a clear view of the pipe joints even with the most sophisticated MSI technologies. Only in the situations where the actual joint can be observed or where active groundwater infiltration can be documented, are PACP codes assigned to their condition. In general, however, it is the experience of the authors that the overall condition of most pipe joints can be determined from data derived from a program that leverages MSI technologies.

- Variability of Wall Loss Rate – A review of the data gathered from a number of interceptor assessment programs using MSI technologies has shown that the rates of wall loss among pipe assets of nearly identical age and location can vary significantly among RCP infrastructure. The results of one study, in particular, showed that wall loss rates ranged from 0.001 inches/year to 0.11 inches/year. In this study the MSI data showed that generally the older pipelines had a lower annual rate of wall loss than those of new age. Under these conditions, there may be justification enough to confirm wall loss rates and their associated causes through the use of petrographic analysis from selected pipe coupons and/or direct measurements using technology such as pipe penetrating radar.

Remaining Useful Life and Asset Prioritization

As discussed briefly, determining RUL only on the basis of an internal inspections is challenging as there are additional factors and conditions that one does not see from within the pipeline alone. However, knowing the age and diameter of the pipeline in combination with a clear image of the pipe wall and extent of wall loss, a good first estimate of RUL can be reasonably made.

There is significant debate within the pipeline assessment industry when the RUL for RCP is "zero" (i.e. no remaining service or useful life). For large-diameter RCP (36" in diameter and greater) there are two rows of steel reinforcement (cages or hoops). For RCP under 36" there is only a single row and it is typically located in the center of the pipe wall. By knowing the diameter, type, and class of RCP the following chart (Figure E) can then be created.

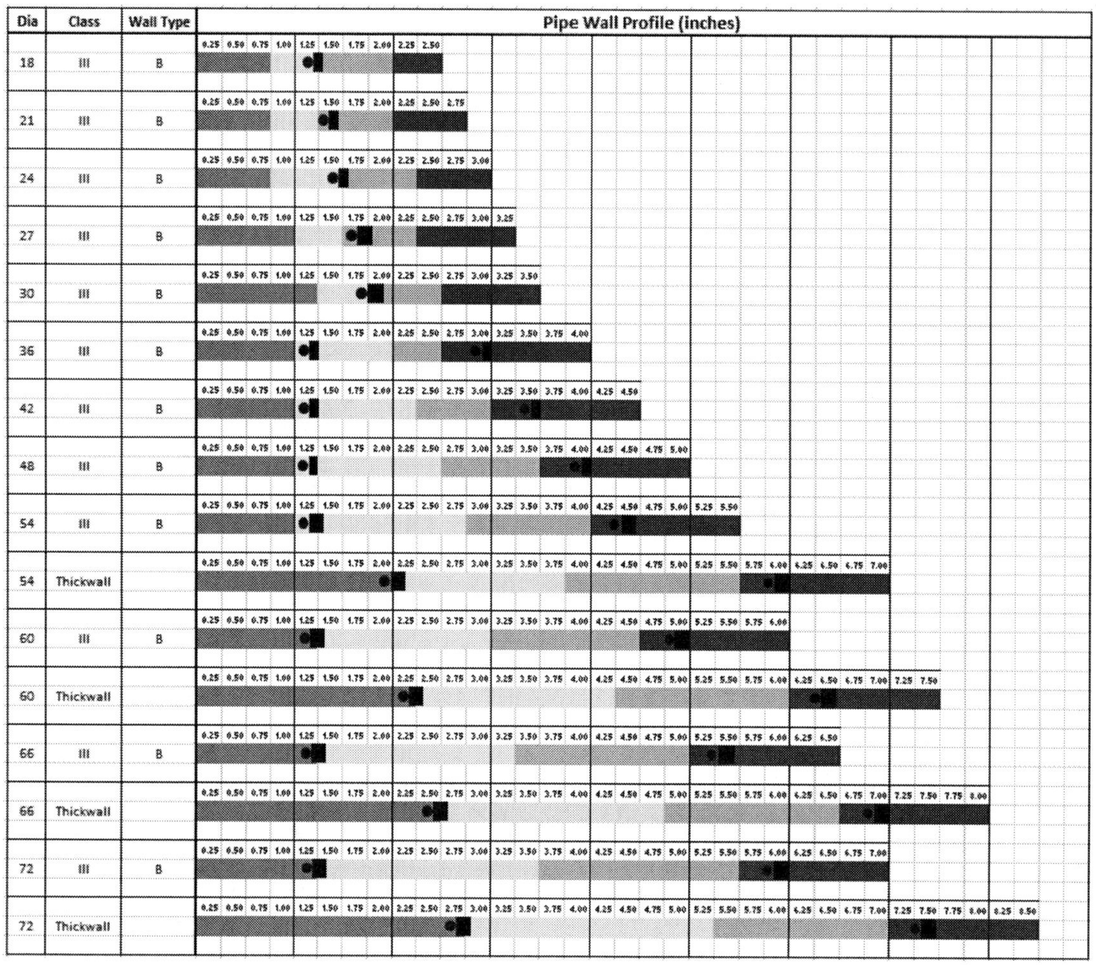

Figure E: Pipe Wall Thickness and Reinforcement Chart

A chart such as this is useful as it can help determine where a cage or row of reinforcement is located relative to the inner pipe wall surface at any point along its alignment when data from an MSI platform is used.

It can be argued that once the first (or inner) row of reinforcement is exposed to sulfuric acid due to corrosion, the RUL is approaching zero. However, if this same pipeline is structurally analyzed under normal site and burial conditions, there may only a modest decrease in its structural integrity with a safety factor above 1.0. This is why it is vitally important that utilities routinely observe wastewater RCP interceptors that are still fully functional even after the first row of reinforcement has been fully corroded (in the case of RCP 36" in diameter and greater). There is agreement among pipeline design engineers that the real structural strength of large-diameter RCP interceptors is the integrity of the outer (or second) cage of reinforcement.

To help direct the discussion toward a metric that might be considered in determining when RUL is zero, the authors have discovered that there is general consensus among a number of large utilities that three points of integrity (or special interest) exist within the RCP wall that can be used to establish RUL. These points included the following:

- Point A - Depth to the first row of reinforcement (double cage only);
- Point B – Depth to the mid-point of the pipe wall (single and double cages);
- Point C – Depth to the second row of reinforcement.

RUL for each of these three points can be determined by combining the MSI results, the information from the above chart, and estimating the asset's maximum point of corrosion rate (manhole-to-manhole, where available).

Where wall loss reaches one or more of these points within the wall profile, the RUL can then be assigned a value in years. The City of Fort Worth, as an example, has determined that a pipeline RUL is approaching zero (1-3 years) when the depth of corrosion has reached the mid-point of the pipe wall. The rationale is that this allows sufficient time for the City of Fort Worth to budget, plan, perform additional assessments, design, and either replace or rehabilitate that particular asset (often a 2-3 year period or longer) before it reaches the second row of reinforcement.
Once RUL has been determined, it is then possible to use this value along with other data to assign a structural or condition score. The chart in Figure D is color coded to show when the deterioration condition has reached a particular score reflected the following color scheme.

Table 1: Condition (structural) Score

Color	RUL (yrs)	Score	Re-inspection Metric
Blue	>30	1	20 Years
Green	10-30	2	10 Years
Yellow	4-10	3	Five Years
Orange	1-3	4	One Year
Red	<0	5	Replace/Rehabilitate

Notes:
- Score can be used as a basis to determine likelihood of failure for a risk analysis.
- Assignment of RUL varies depending on other parameters such as existing and future hydraulic conditions and O&M.

Of course, the selection of an appropriate RUL cannot be established without taking into consideration the PACP score as well as an overall grade score of the pipeline based on other factors (depth of normal flow, heavy wall build-up, condition of joints, protrusions, etc.). Using the combination of these factors, with corrosion-based wall loss as the primary factor, a prioritized list of all RCP assets pipe assets can be developed. An example of such a prioritized list using Point B as the criteria for setting RUL at zero is shown in Table 2.

Table 2: Summary of RUL (Point B Failure Criteria)

RUL	No. of Assets	Footage	% of System
0 (failure)	14	2,500	2.7
1-10 years	12	1,430	1.6
10-20 years	21	6,710	7.3
20-50 years	55	16,220	17.7
50-100 years	44	13,420	14.6
>100 years	181	51,580	56.2
Total	**327**	**91,860**	**100%**

In this example, of the 327 large-diameter assets included in the wastewater collection system, approximately 5% of them have reached a RUL approaching or reaching zero (years) with a condition score of 5 (poor or very poor).

Conclusions

MSI is a valuable tool to quantify the structural condition of our most critical large-diameter assets. Measurable data is obtained that can be combined with PACP, design, and operational conditions to evaluate RUL, and design the most appropriate rehabilitation necessary based on the results. Pipeline age, material, and location does not directly correlate to percent of pipe wall loss as many variables are at play. The most accurate method to gauge and evaluate the condition of our large pipeline assets is through regular MSI inspections.

City of Ottawa's Strategic Infrastructure Management Plan: Development and Application of a Triple-Bottom-Line-"Plus" Risk-Assessment Framework for Collection Systems Criticality Assessment and Needs Prioritization

Adrien Comeau, P,Eng., Stantec Consulting
Peter McKay, P,Eng., City of Ottawa, Canada.

ABSTRACT

Critical assets are generally those that are large, deep, have large flows, are difficult to access, and, inherently, come with a much higher cost of renewal. It is thus important to understand their implications on long-term financial planning within a Capital Improvement Program. The programming of overall asset management priorities must consider multiple factors, such as; integrated cross-asset class needs, timing constraints to conducting the work, as well as political factors in the decision-making process.

A key component of the City of Ottawa's Strategic Management Plan for its wastewater and stormwater infrastructure has been the development of a risk-based critical infrastructure assessment framework to serve as a decision-making tool in establishing and prioritizing investments in sewer infrastructure projects and operational needs. It also provides the basis for forecasting, justifying and maintaining, these investment needs within the City's overall Comprehensive Asset Management Program.

The risk-based criticality assessment framework developed by the City (Risk-based Criticality Model and guidelines for prioritization of the needs) is designed to look at infrastructure from a longer term strategic point of view and to serve as a key tool to justify the acceleration and/or advancement of the programming for inspection, maintenance, and/or repairs of those assets that are identified as being most critical and vulnerable to failure. Given the magnitude of overall system needs, it is not be possible or practical to undertake all identified needs at once. For this reason, the guidance provided for the prioritization of on-going condition assessment and contingency planning efforts plays a key role in the City's overall Strategic Management Plan.

KEYWORDS

Risk, Criticality, Asset Management, Renewal, Capital Improvement Programs

INTRODUCTION

In October 2012, City Council adopted a Comprehensive Asset Management (CAM) Program. The CAM Program outlines how the City of Ottawa's (the City) assets, valued at over $32B, are to be managed. The CAM Program founded on full life-cycle funding and identifies areas of continuous improvement, with the objective that the right infrastructure investments are made at the right time. It also recognizes that a risk-based approach to managing these assets is needed to develop an understanding of vulnerabilities, build resilience in the services provided to the community, and prioritize investments such that level-of-service impacts from asset deterioration are minimized and the high costs associated with failures of critical assets are avoided.

Similarly, the City's Service Excellence Plan is focused on customer needs and includes initiatives that leverage technology to not only achieve improvements in service, but to also maintain a focus on long-term sustainability though improved efficiency and cost savings for both the community and the City.

In keeping with the CAM Program and Service Excellence Plan, a key component of the City's Strategic Management Plan for its wastewater and stormwater infrastructure is the development of a risk-based critical infrastructure assessment framework to serve as a decision-making tool in establishing and prioritizing investments in sewer infrastructure projects and operational needs. It also provides the basis for forecasting, justifying and maintaining, these investment needs within the City's overall CAM Program.

The City has undertaken the development and implementation of the Strategic Management Plan for its wastewater and stormwater infrastructure in two phases; The first phase, which is the subject of this paper, is to develop a viable and effective strategy for facilitating the identification and prioritization of condition assessment and rehabilitation needs for sewers and small to medium culvert infrastructure. In the second phase, the outcomes of the first will be applied to life-cycle modeling and planning tools for on-going update of renewal programming and funding plans.

This first phase of development strategy thus includes the development of a risk-based criticality assessment framework that:

1. Identifies and defines the critical assets within the storm and wastewater collection systems as well as small to medium sized culverts (i.e. less than 3 m).

2. Defines the risk of failure of each asset (structural and operational) and ranks assets on the basis of total risk (based on criticality and condition)

3. Provides guidelines for prioritizing pre-emptive actions (i.e. inspection and rehabilitation needs) on the basis of the criticality and assessed risk of failure of the sewer and culvert infrastructure assets

4. Provides guidelines for inspection frequencies of for all sewers and culverts on the basis of the assessed risk of failure.

This paper describes the approach to the development and shares outcomes of the application a comprehensive triple-bottom-line based risk assessment framework for the strategic identification of critical assets and prioritization of rehabilitation, condition assessment, and preventative maintenances needs on the City's combined, sanitary and storm sewer networks (including drainage culverts).

APPROACH

The initial phase of the assignment included a review of Industry Best Practices and identified the approaches for the identification of critical sewer infrastructure implemented by several municipalities and utilities across Canada and the United States. During the review, it became apparent that many of the approaches developed have built upon a methodology that was originally developed by the UK's Water Research Centre (WRc) in the 1970's to meet demands that the water authorities were facing as a result of their aging infrastructure. The WRc approach considers many of the factors that form the basis of a Triple Bottom Line assessment of the consequences of infrastructure failure. Given that the availability of data and technology for the analysis of this data at the time was limited, the analysis was conducted at a high level to facilitate manual review.

Based upon an assessment of recent practices for risk and criticality assessment, it was apparent that there is currently no standard method for the creation of a Sewer/Culvert Criticality Model. It remains a somewhat customized process in which the agency selects and defines a range of criteria and scoring parameters that they feel best defines their own perspective of critical sewer/culvert criticality. Nevertheless, many of the agencies have adopted approaches that either employ the WRc method or have adapted a set of criteria to better reflect their priorities with respect to the traditional "triple-bottom-line (TBL)" factors of social, environmental and economic impacts.

The criticality evaluation adopted by the City of Ottawa expands on the WRc method by balancing weighted "Consequence Criteria" in four "Impact Categories", including:

- Social impacts,
- Environmental impacts,
- Economic impact to the community,
- Economic impact to the City (direct repair costs)

The criticality model thus incorporates the traditional triple-bottom-line (TBL) considerations for social, economic and environmental impacts (the cornerstones of sustainability when assessing the criticality of the assets) with a fourth, "Plus", dimension to the TBL based evaluation. This "plus" dimension of criticality is the consideration of direct replacement costs to the utility of infrastructure failures.

The selected "Triple-Bottom-Line Plus" approach to defining asset criticality and the criteria contained within each impact category are designed to be aligned with the City's CAM Program and Service Excellence Plan. In other words, the "Triple-Bottom-Line Plus" approach expands upon the WRc method by inherently considering sustainability from the perspective of minimizing overall costs to both the City and the community.

Figure 1 illustrates the overall framework for the Criticality and Risk Assessment model. The "Triple-Bottom-Line Plus" evaluation on the left side of the figure defines the consequence of failure for east asset in the system. Within the overall Risk-Based Decision Model, separate "likelihood of failure factors" are determined for each of the following failure scenarios:

1. Structural Failure
2. Operational Failure
    - o Hydraulic Performance Failure
    - o Odor Complaints and Corrosive conditions (sanitary and combined sewers)

Within the City's model, a structural failure scenario refers to the potential collapse of the asset resulting in either the formation of a sinkhole or flooding on the surface. Within the operational failure scenario, hydraulic performance failures and the likelihood of odors and corrosive conditions are considered. The likelihood of hydraulic performance failures refers to the asset's inability to maintain its hydraulic performance objectives and maintain its design level of service (i.e. for the public and the environment). The assessment of the likelihood of hydraulic performance failures also considers the likelihood of a sewer or culvert's hydraulic capacity being compromised due to hydraulic boundary conditions or conduit obstructions. For sanitary and combined sewer systems, the likelihood of experiencing nuisance odors and corrosive conditions is also assessed as a separate operational failure risk.

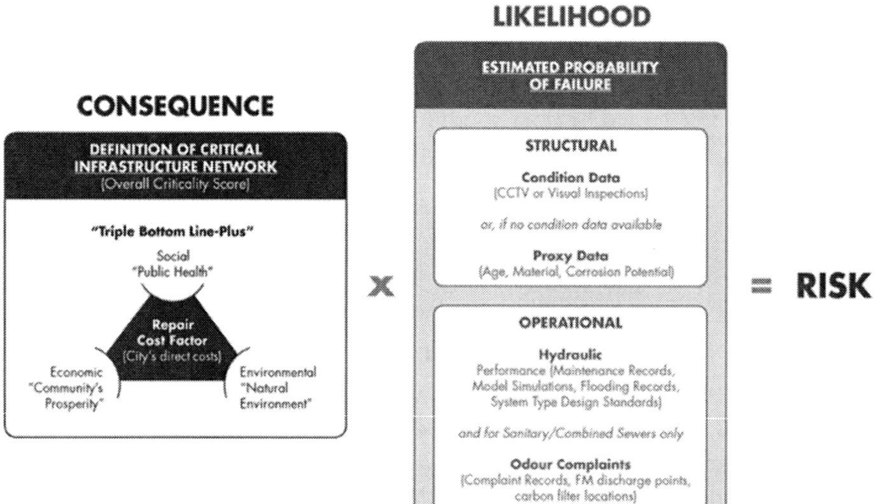

Figure 1: Triple-Bottom-Line -"Plus" Criticality and Risk Assessment Model

METHODOLOGY

The method consists of a three-step process to establish the relative risk of structural and operational failure associated with each sewer and culvert asset.

1. Definition of Critical Infrastructure Network
2. Definition of the Likelihood of Failure
3. Quantification of the Relative Risk of Failure

At each stage of development of the assessment framework and model, stakeholders from various City departments were consulted in order to develop consensus on the criteria and associated weightings that would be used to determine asset criticality (consequences of failure) and then incorporated asset condition to determine the likelihood of failure (structural and operational/functionality).

Definition of Critical Infrastructure Network

The critical sanitary, storm and culvert assets in the network first defined by assessing the consequence of failure of each asset. This is evaluated by using a series of impact criteria to establish an overall consequence of failure score for each asset based on location specific conditions as defined in GIS databases (e.g. population served, land use, road type, environmental sensitivity, river/rail crossings, etc …).

Within each of the four TBL "Plus" impact categories listed in **Table 1**, sub-criteria representing category specific consequences are identified for evaluation of each pipe segment and each culvert. Recognizing that there are some differences in the definition of consequences that can occur as a result of sewer or culvert failures, a distinct set of consequence criteria have been developed for each asset group.

The criticality of specific assets is determined by calculating an "overall criticality score" for each asset as depicted in **Figure 2**. The total scores derived as part of this process were then used to rank and categorize the level of criticality of each asset.

The scoring of the individual pipe segments is completed within GIS by means of spatial and database queries of various criticality indicators that are specific to each of the individual consequence criteria. With the assignment of a criticality level for each pipe, results are then represented in the form of colour coded thematic maps. The criticality assessment results are also available for querying within the underlying geo-database. The tool provides full flexibility for representing the sanitary/combined sewers, storm sewers and culverts on separate thematic maps or combined on the same map.

Definition of the Likelihood of Failure

Once the criticality model has established the critical infrastructure networks for each of the sanitary, storm and culvert systems, each pipe segment within the defined critical networks is then assessed to determine its likelihood of failure. The likelihood, or probability, of failure (structural or operational) were best estimated though direct observation (i.e. condition assessment and monitoring data). For assets lacking reliable condition assessment data, the likelihood of structural failure was inferred by using proxy data such as pipe age, type, and soil conditions. Similarly, where there were no direct observations available, the likelihood of operational failures was also estimated by using system models (hydraulic models for hydraulic performance as well as sulfide and H2S models to assess odor and corrosion likelihood).

Table 1: Consequence Criteria for Sewer and Culvert Networks

IMPACT CATEGORIES	Consequence Criteria for Sewers	Consequence Criteria for Culverts
SOCIAL (Impacts on Public Safety and Quality of Life)	Transportation Safety	Transportation Safety
	Property Flooding Impacts	
	Public Service Interruption	
	Odour Impacts	
ENVIRONMENTAL (Impacts on Natural Environment)	Spill to Environmentally Sensitive Area	Greenhouse Gas Emissions
	Inflow and Discharge of Contaminated Materials	Inflow and Discharge of Excessive Sediment or Contaminated Materials
	Contamination of G'water/Drinking Water Sources	
ECONOMIC - COMMUNITY (Economic Impacts on the Overall Community)	Disruption to Private Businesses or Public Services/Gov't Agencies	Transportation Impacts
	Transportation Impacts	
ECONOMIC - DIRECT REPAIR COST (Cost of Repair or Direct Cost to the City)	Diameter/Size (mm)	Diameter/Size
	Depth of Bury (m) /Ground Classification	Depth of Bury
	Access/Environmental Restrictions and Repair Technology	Access/Environmental Restrictions and Repair Technology
	Traffic Control Requirements	Culvert Length

Quantification of the Relative Risk

The risk of failure of each asset, relative to all the others, is then established as the product of its criticality score and its estimated likelihood of failure score. The likelihood of failure score is thus applied to the corresponding "overall criticality score" for each asset in the Criticality Model database in order to establish "overall risk scores". These "overall risk scores" are then used as the basis for defining priorities for further condition assessments and/or for life-cycle renewal (short and long term needs). From this assessment, standards for on-going inspection frequencies and other operation and maintenance activities can also be assessed and assigned on the basis of both criticality and risk.

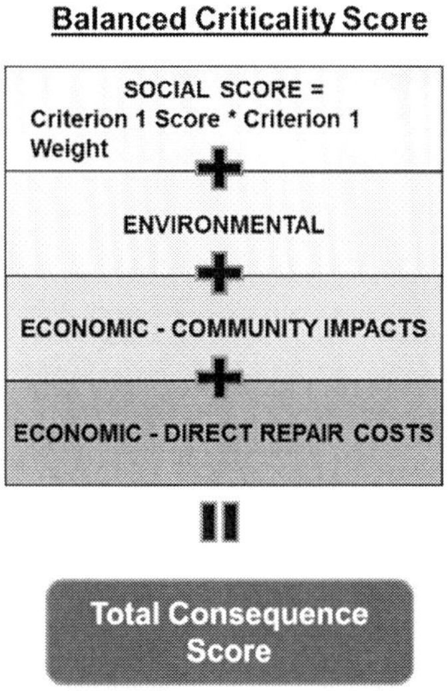

Figure 2: Overall Criticality Score Calculation

RESULTS

Based on established database queries and criteria using the most current information available within the City's databases, the model output and reporting framework, as illustrated for risk of structural failures in **Figures 3 and 4**, establishes a criticality and failure risk score for each and every asset in the network consisting of over 5656 km (3515 mi) of pipe. Ranking of results then identifies the most critical assets and provides prioritization of infrastructure upgrade and/or inspection needs.

Sanitary/Combined

Consequence of Failure \ Probability of Failure	1	2	3	4	5
5	260	182	155	94	111
4	567	321	194	100	88
3	1757	1394	684	307	193
2	5990	3838	2098	722	240
1	14779	9991	4894	1136	153

Storm

Consequence of Failure \ Probability of Failure	1	2	3	4	5
5	533	601	259	59	25
4	1087	1052	378	107	23
3	2755	2377	1053	198	18
2	6758	5742	2278	457	50
1	15490	10041	3120	674	38

Figure 3: Summary Risk Categories – Structural Failure

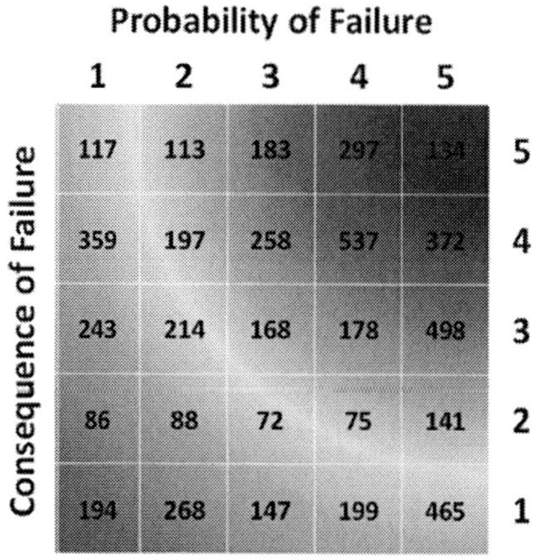

Figure 4: Summary Risk Categories – Structural Failure (All Culverts)

Prioritization of Pre-Emptive Actions for Structural Failures

The risk based criticality model allows the City to identify where the risks and the associated consequences of failure are the highest. This allows the City to prioritize inspection, maintenance and rehabilitation activities. **Table 2** and **Figure 5** were developed as guidelines for prioritizing further actions for individual assets on the basis of their overall structural failure risk scores. Based on the assessed level of risk, actions range from:

- Proceeding with the implementation repairs and contingency measures.
- Assessing and defining repairs required in the near-term.
- Planning for and scheduling future repair requirements.
- Monitoring the condition of the asset; additional guidelines are provided for the frequency of inspection based upon the current condition and consequences of failure of the asset.

Figure 5 depicts the proposed process for assessment, validation and application of the risk model results, while **Table 3** provides specific guidance on action planning as it relates to maintenance, contingency planning and inspection frequencies once the risk score is validated or accepted.

As a first step, the approach in **Figure 5** highlights the need to validate the actual condition of those assets with a risk score greater than 20. Therefore, if proxy data (i.e. the age and material of the asset) has been used to define the likelihood of failure, it is important that the actual condition of the asset is determined prior to assessing rehabilitation requirements. This will either lead to the re-classification or confirmation of the asset's level of risk, and hence a better determination of appropriate pre-emptive actions.

Table 2: Prioritization of Actions Based on Structural Failure Risks

Critical Infrastructure Risk Priority	Risk Score	Actions
First	> 20	Prioritize per Figure 5 Implement Repair and Contingency Measures
Second	> 16 to 20	Define Repair Requirements Monitor condition with Enhanced(1) Maintenance and/or Inspection (per Table 3 recommendations)
Third	> 9 to 16	Plan and Schedule Repairs Monitor condition with Routine(2) Maintenance and/or Inspection (per Table 3 recommendations)
Fourth	<= 9	Monitor condition (per Table 3 recommendations)

(1) Enhanced: Increased frequency as it relates to typical/industry standard requirements for assets in generally good condition
(2) Routine: Typical/industry standard requirements for assets in generally good condition

Once the risk scores are validated, **Table 3** then provides the basis for developing an overall prioritized action plan and overall cost for managing all the sewer assets in the network(s). **Table 3** provides further guidance on repair response actions for the full range of assets on the basis of their condition as well as the prioritization of those repair actions on the basis of their assessed criticality level (i.e. the combination of the two representing the risk level).

Based on the prioritized repair requirements, **Table 3** also provides corresponding recommendations on maintenance (enhanced vs. routine) and the need for enacting contingency plans, including preliminary guidance on the recommended frequency of re-inspection. Within the context of this report, routine maintenance refers to the City's current baseline conditions (i.e. established inspection and cleaning frequency) while enhanced maintenance refers to more frequent interventions (i.e. inspection and/or cleaning) as a result of operational or structural concerns. A modified set of inspection frequency guidelines were adopted for drainage culverts as shown in **Table 4**.

Based on the recommendations outlined within **Tables 2, 3 and 4**, it is possible to develop both near-term and long-term inspection program budget estimates for the sewer and culvert assets.

Table 3: Structural Risk of Failure – Recommended Inspection Frequency for Sewers

Condition Based Response		Risk Based Prioritization			Recommended Action Plan		
	Response	Estimated Remaining Service Life	Criticality Category	Repair Priority	Maintenance	Contingency Plans Pending Repair	Re-Inspection Frequency
5	Implement Repairs	0%	3, 4 or 5	1	High Priority Repair	Enact with Frequent Inspection	Annual[1][2]
			1 or 2	2	Lower Priority Repair + Enhanced Maintenance	Frequent Inspection	2 yrs [1]
4	Define Repairs	<= 20%	4 or 5	3	Prioritize Repair	Enact with Frequent Inspection	2 yrs
			1, 2 or 3	5	Routine Maintenance	Inspect Routinely	5 yrs
		> 20%	4 or 5	4	Enhanced Maintenance	Prepare to Enact with Frequent Inspection	3 yrs
			1, 2 or 3	7	Routine Maintenance	Inspect Routinely	5-10 yrs
3	**Plan and Schedule Repairs**	<= 20%	4 or 5	**6**	Routine Maintenance	Define Contingency Plans for Criticality Categories 3, 4 and 5	5 yrs
			1, 2 or 3	**12**			10 yrs
		> 20%	4 or 5	**8**			5-10 yrs
			1, 2 or 3	**13**			10 yrs
2	**Monitor Condition**	<= 20%	4 or 5	**9**	Routine but Lower Frequency Maintenance	Define Contingency Plans for Criticality Categories 3, 4 and 5	5-10 yrs
			1, 2 or 3	**14**			10 yrs
		> 20%	4 or 5	**10**			5-10 yrs
			1, 2, or 3	**15**			10-20 yrs
1	**Monitor Condition**	<= 20%	4 or 5	**11**	Routine but Lowest Frequency Maintenance	Define Contingency Plans for Criticality Categories 3, 4 and 5	5-10 yrs
			1, 2, or 3	**17**			10-20 yrs
		> 20%	4 or 5	**16**			10-20 yrs
			1, 2 or 3	**18**			20 yrs

(1) If voids in the pipe wall exist, re-inspection every 6 months is recommended.
(2) Upon inspection, may require continual or more frequent monitoring.

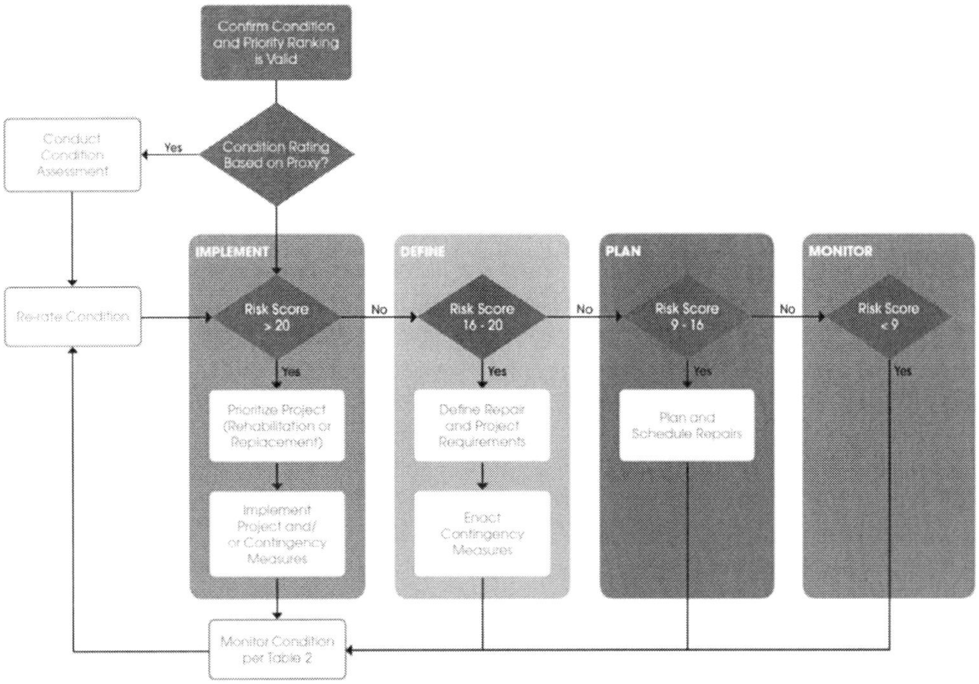

Figure 5: Response Plan for First Priority Items

Table 4: - Recommended Inspection Frequency (Culverts)

Structural Condition Rating	Criticality Category	Medium Culverts: Re-Inspection Frequency	Medium Culverts: Number of Culverts	Small Culverts: Re-Inspection Frequency	Small Culverts: Number of Culverts
5	4 or 5	Annual	167	2 yrs	339
	1, 2 or 3	2 yrs	43	2 yrs	1,061
4	4 or 5	2 yrs	318	5 yrs	516
	1, 2 or 3	5 yrs	55	5 yrs	397
3	4 or 5	5 yrs	227	5 yrs	214
	1, 2 or 3	5 yrs	75	5 yrs	312
2	4 or 5	10 yrs	115	10 yrs	195
	1, 2 or 3	10 yrs	45	10 yrs	525
1	4 or 5	10 yrs	65	10 yrs	411
	1, 2, or 3	10 yrs	24	10 yrs	499

Hydraulic Performance Failures

Actions related to addressing the risk of hydraulic performance failures are ultimately tied to the City's Basement Flooding Remediation Program as well as its Sewer Operations and Maintenance (O&M) Program. The results of the Hydraulic Performance Failures Risk Analysis can thus be applied in helping to prioritize design and construction projects stemming from the Basement Flooding Remediation Program. It can also be used to provide input to the O&M program, in particular where enhanced maintenance activities are to be undertaken.

As applied within this framework, the expected level of service is defined with systems modeling tools and/or the use of historical data (i.e. flooding records, historical design standards). These levels of service evaluations are typically done by the City as part of various systems assessment initiatives; such as the analyses completed for identifying and prioritizing flood prone areas within the Basement Flooding Remediation Program, Infrastructure Master Plans or Master Servicing Plans, as well as Operational Review Studies. The results of these evaluations can then be used as the basis for characterizing the likelihood of flooding at each asset within the network for subsequent application in the Risk-Based Criticality Model.

Specific data characterizing hydraulic boundaries and conduit obstructions are also important elements to consider and integrate in the assessment of expected level of service. Without hard data to characterize the level of conduit obstruction, however, the likelihood of reduced conveyance capacity serves only as a proxy to indicate where there is a possibility of reduced level of service, not the extent to which it is compromised.

Once study areas are prioritized, evaluated and a series of remedial measures are defined within the Basement Flooding Remediation Program, the asset criticality (and hence overall risk) can then be considered as an input to prioritizing the implementation of flood control projects within various flood prone areas. In addition, the results of the structural failure risk assessment can also be incorporated to further prioritize and/or integrate asset improvement needs.

Odor and Corrosion Potential (Sanitary and Combined Sewers)

Similar to a hydraulic performance failure that leads to a flood event or spill, an odor complaint or pipe corrosion represents a location specific impact that is often the result of systematic operating conditions as opposed to being caused by the impacted asset. As such, the root cause and definition of appropriate remedial measures must be evaluated from the perspective of overall system operation. Once higher likelihood odor and corrosion zones are prioritized for further study, evaluated and a series of remedial measures are defined, the asset criticality, and hence overall risk, can then be considered as an input to prioritizing the implementation of odor and corrosion control projects within the various study areas. In addition, the results of the structural failure risk assessment can also be incorporated to further prioritize and/or integrate asset improvement needs where pipe corrosion is problematic.

DISCUSSION

The most effective means of preventing an infrastructure failure incident is to significantly reduce or avoid the risk. The guideline recommendations provided in **Table 3** are risk management measures that are designed to prioritize actions to reduce failure risks and proactively address repairs where needed. This will ultimately minimize the number of sewer and/or culvert assets that are in such a condition that they require immediate repairs to avoid complete failure or emergency actions in response to a failure.

The Criticality and Risk Assessment Framework Model described herein is GIS based and is now fully integrated within the City's Asset Management and Programming software, where the consequence and risk scores are regenerated on an annual basis as assets are inspected, added and renewed. This now provides the City with a means to monitor the overall risk profile of the system as well as a better understanding of the unique challenges that each critical asset poses, whether it be the impact that assets (and specifically asset failures) have on citizen's lives operational, financial, or from a renewal perspective.

An understanding of the risk profiles for critical assets as well as the overall system also gives a better understanding of the non-financial planning requirements for these assets. The more critical assets are typically the most expensive assets to repair; typically representing asset replacement values and repair costs of more than $10M. For financial planning purposes, system owners need to be in a position where they understand both the near-term and long-term financial implications, as well as replacement timeframes, for critical assets.

The guidelines provided in **Table 3** allows for the application of the model scoring results to not only prioritize repairs, but also develop and budget for operational and failure prevention strategies that include prioritized preventative maintenance activities and condition assessment priorities. Higher risk assets will require more operational attention (e.g. inspection, maintenance) and thus require an increase in operational resources to handle the required inspection frequencies. Thus, the City must be strategic in selecting the level of service it will provide for various assets. The appropriate frequency and type of maintenance or inspection must be selected to ensure the assets remain in service, pose minimal risk to the community and minimize longer term investments. For the City, this means that acceptance of a lower level of service for lower risk assets is appropriate and can be justified within an overall risk management strategy.

Appropriate risk treatment not only includes enhanced maintenance and/or repairs, but also the development of contingency plans as part of the planning and implementation process. While the risk of failures can be significantly reduced, they can never be eliminated. Furthermore, is it not realistic to think that they can all be immediately addressed, especially if they entail complicated, long duration, or longer term repairs. As such, it must be recognized that, where risk elimination cannot be immediately addressed for high or significant risks, contingency planning measures should be enacted to manage the consequences and minimize the harm that could be caused. The appropriate set of contingency measures will vary from asset to asset and should be based on an evaluation of the key consequences of failure as identified within the criticality analysis.

CONCLUSIONS

The Criticality and Risk Assessment Framework Model provides City of Ottawa renewal planning engineers and managers with a tool to effectively understand and re-baseline risks, as well as expected service life, on the more critical assets within a very large wastewater and stormwater collection system. As needs are addressed and as further asset condition data is collected, the tool provides a basis for re-assessing risk and re-establishing critical priorities as conditions change. It has also enabled a greater appreciation of the impact asset failures have had on the general public, particularly for the higher risk assets.

Overall, the tool provides the basis for establishing renewal and operational priorities and, in turn, identifying and justifying different strategies and funding requirements for higher risk assets. Another key outcome of the initiative was the identification of the types of data that are required to assess the network risks (consequence and probability). The identification of this data (including gaps in the data) allowed the City to also derive strategies to both acquire and update these data sources.

The Strategic Management Plan is now focused on applying these results in the development of Life-cycle Modeling and Renewal Planning tools for the on-going update of asset renewal programming and funding plans. Similar risk assessments and strategic plans are now being developed for all the asset classes under City authority (e.g. watermains, roads, buildings, parks, structures). These assessments are being done so that a common frame of reference can be used to quantify and compare risk evaluations for each asset class in a manner that the City is then able to understand the relative importance of each asset class against the other classes. This will allow for a better more strategic formulation for funding allocations between asset classes and how projects can be funded when there are conflicting priorities.

REFERENCES

Federation of Canadian Municipalities and National Research Council, *Assessment and Evaluation of Storm and Wastewater Collection Systems: A Best Practice by the National Guide to Sustainable Municipal Infrastructure,* 2004

Heart, Simon, M. Cottingame, J. McLamarrah, *Dallas Water Utilities: A Risk-based Approach to Asset Renewal,* Proceedings of the Water Environment Federation, Utility Management Conference, 2008

National Cooperative Highway Research Program (NCHRP) and US Transportation Research Board (NCHRP Synthesis 371), *Managing Selected Transportation Assets: Signals, Lighting, Signs, Pavement Markings, Culverts, and Sidewalks*, 2007

Road Transportation Authority (RTA) for New South Wales (NSW), *Culvert Risk Assessment Guideline (V3.02)*, 2010

Water Research Centre (UK), *Sewerage Rehabilitation Manual, 2nd Edition*, 1986

Utilizing ArcGIS Online for Project Delivery on Sewer System Evaluation Study

Brian Villers[1], Joe Pavlick[2]*, Matt St. Pierre[3], Imad Salim[3], Tiffany Ashforth[4], Dan Porter[4], Ryan Pulis[4], Kevin Vander Tuig[3]

[1] Northeast Ohio Regional Sewer District, Cleveland, OH
[2] Water Resources and Coastal Engineers, Cleveland, OH
[3] Wade Trim, Cleveland, OH
[4] Brown and Caldwell, Independence, OH
*Email:Jpavlick@wrceinc.com

ABSTRACT
The Northeast Ohio Regional Sewer District (NEORSD) owns and operates regional interceptors, combined sewer overflow (CSO) regulators, storage tunnels, pump stations, and three wastewater treatment plants (WWTPs). NEORSD serves more than one million residents, as well as commercial/industrial customers in 62 communities covering 355 square miles (SM) near Cleveland, Ohio. The communities served convey and discharge wastewater to NEORSD interceptors, and are responsible for addressing Clean Water Act (CWA) issues such as Sanitary Sewer Overflows (SSOs), basement flooding, illicit discharges, and failing septic systems in their respective local sewer systems. The local community separate sewer systems comprise approximately 274 SM of the total NEORSD service area.

NEORSD has estimated the cost of addressing CWA issues in the local sewer systems at over $3 billion. NEORSD has developed a Member Community Infrastructure Program (MCIP) to help communities within NEORSD's service area address CWA issues. Funding is distributed through a grant program in which local communities can apply for assistance. To support the development of the MCIP, NEORSD has begun a series of local sewer system evaluation studies to identify existing CWA issues and propose solutions for implementation by the respective communities. The completed solutions, funded by the MCIP, will be monitored for performance and incorporated into NEORSD's MCIP strategy and program refinement.

Esri's ArcGIS Online (AGOL) tool is being used for the first of these studies known as the Heights Hilltop Interceptor - Local Sewer System Evaluation Study (HHI-LSSES). The AGOL tool is supporting project delivery from start to finish. Existing data from previous studies was loaded to AGOL for the development of priority study areas. New data collected from priority study areas and added to AGOL includes flow monitoring information, manhole inspections, smoke and dye testing, and pipeline inspections. AGOL is being used to analyze the information collected and to aid in the development and evaluation of alternatives. This paper details the AGOL framework and how it is being applied for the HHI-LSSES project.

KEYWORDS: Clean Water Act, Sanitary Sewer System, SSO, Illicit Discharges, Basement Flooding, Inflow/Infiltration, Sewer System Evaluation Study, Arc-GIS Online, Esri, Project Delivery, Common Trench Sewers, Clean Water Act, Water Quality, NEORSD, Lake Erie

INTRODUCTION

The NEORSD was created in 1972 to assume the operation and management of wastewater conveyance, treatment and disposal facilities serving the Cleveland metropolitan area. NEORSD serves all or part of 62 communities and over one million people in a 355-square-mile (920 square kilometers) service area. NEORSD is responsible for the operation and maintenance of three WWTPs and more than 300 miles (483 kilometers) of interceptor sewers. **Figure 1** shows the NEORSD combined (81 SM, 210 SK) and separate sanitary sewer (228 SM, 590 SK) and septic, if converted to sewer, (46 SM, 120 SK) service areas.

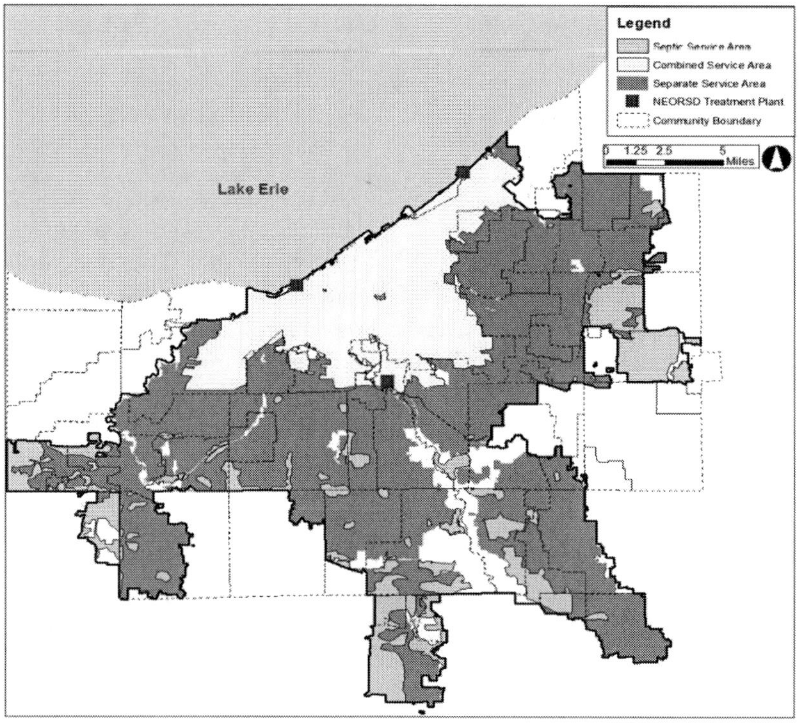

Figure 1. NEORSD Service Area

The District's 25-year Project Clean Lake Program is controlling CSOs under its 2011 Consent Decree, and is scheduled for completion in 2035. While significant progress has been made, CSOs remain one of many contributors to water quality degradation. Other CWA problems include basement flooding, Sanitary Sewer Overflows (SSOs), illicit discharges and failing septic systems that are located outside the combined sewer service area in separate sewer systems owned and operated by the local communities. The communities are responsible for maintenance and correction of performance deficiencies in their respective systems.

NEORSD conducted a preliminary analysis of CWA issues in the local community service areas using USEPA's integrated planning framework. In addition to SSOs, basement flooding and failing septic systems, the analysis confirmed earlier findings that common trench sewers are prevalent throughout NEORSD's local collection system areas, and are a major contributor of inflow and infiltration (I/I). The common trench systems feature storm and sanitary sewers that were constructed in the same trench between 1920 and 1970, primarily as a cost-saving measure. NEORSD common trench sewers are typically found in three manhole configurations, including Over/Under, Dividing Wall and Common Standard, as illustrated in **Figures 2** through **4**. The common trench systems are often leaky, and have been difficult to rehabilitate/separate with consistently acceptable performance results.

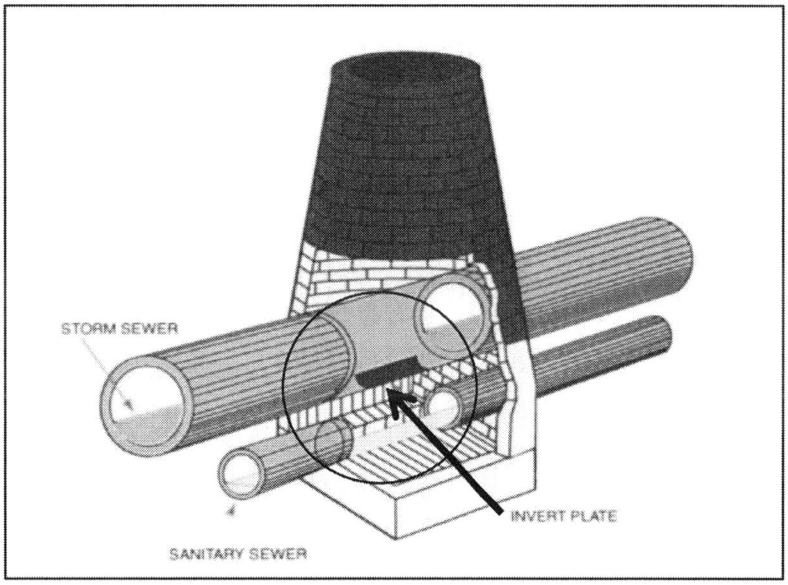

Figure 2. Over/Under Sewer in Common Trench Configuration

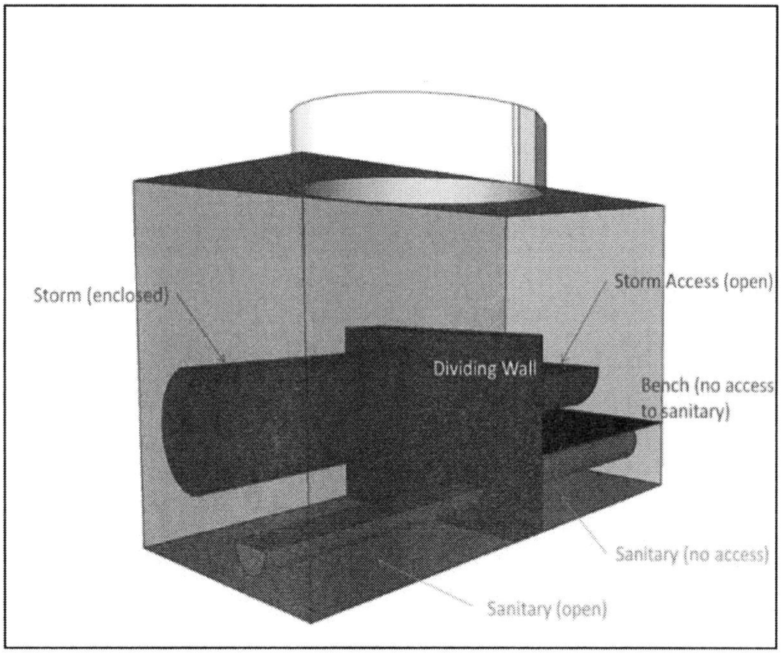

Figure 3. Dividing Wall Sewer in Common Trench Configuration

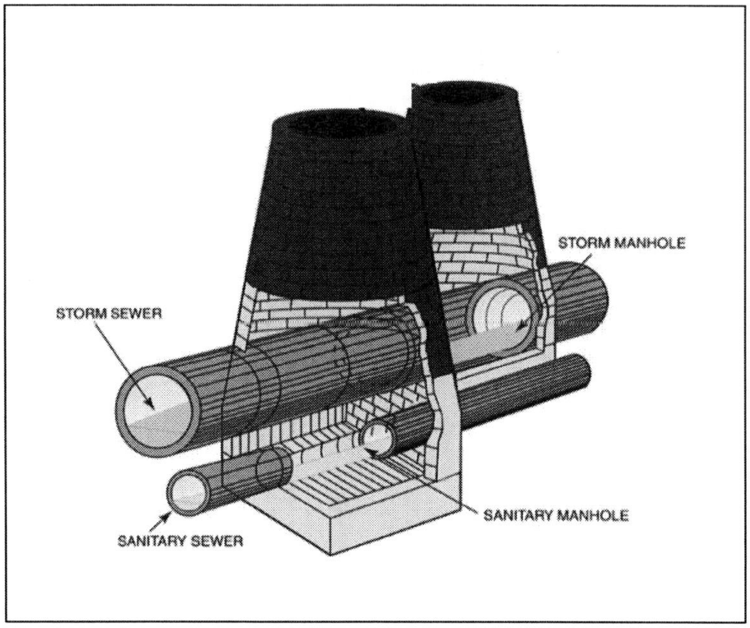

Figure 4. Standard Manholes in Common Trench Configuration

The analysis concluded that approximately $3 billion would be required to properly address separate sewer system CWA issues throughout the NEORSD service area. In response, NEORSD developed the Member Community Infrastructure Program (MCIP) to provide grant funding to local communities for projects that help solve pollution and basement flooding problems in their systems. NEORSD has also begun a series of Local Sewer System Evaluation Studies (LSSESs) to help identify significant CWA problems and causes, evaluate alternatives, and support implementation of solutions under the MCIP. The District's AGOL tool is serving as a key component of the LSSES projects.

METHODOLOGY

The first of the LSSES projects is underway in the Heights Hilltop Interceptor (HHI) service area. The HHI-LSSES is investigating CWA issues within a separate sanitary sewer area of approximately 50 square miles (130 square kilometers) immediately east of Cleveland. **Figure 5** identifies known CWA issues within the HHI-LSSES area.

The District's Geographic Information System (GIS) was used to optimize initial project organization, data collection and analysis, and is being set up to support alternatives development and analysis. The AGOL tool is being used to store and display all key project spatial information and field work efforts within a web browser.

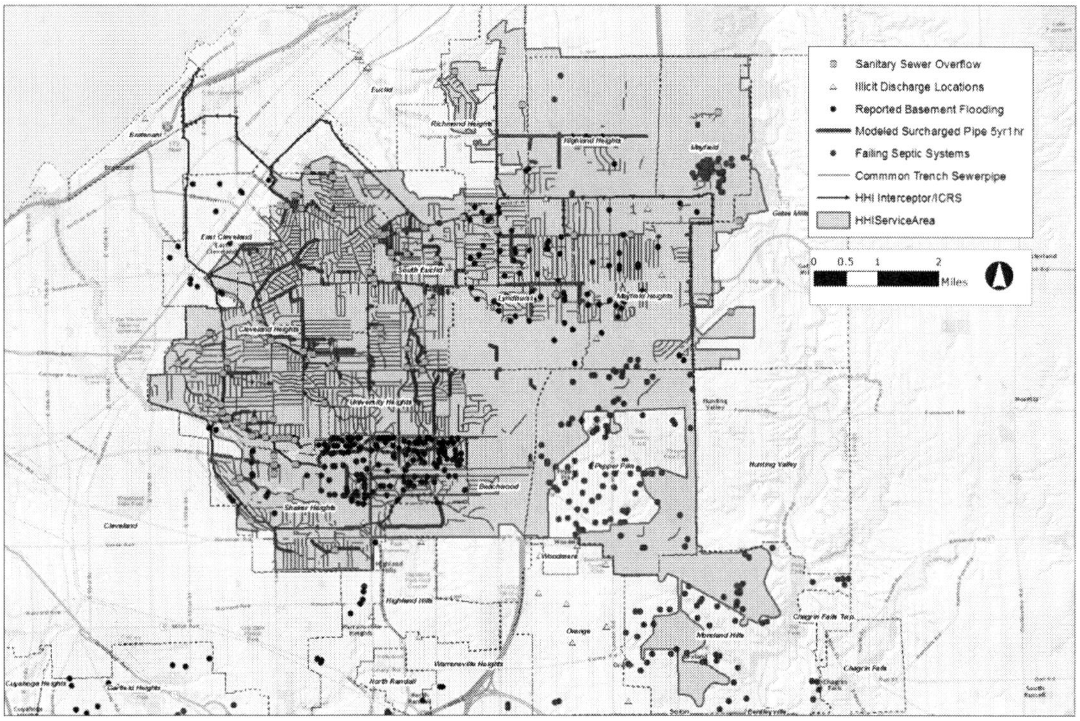

Figure 5. Known CWA issues within the HHI-LSSES area

NEORSD GIS Development

NEORSD has used GIS extensively for many years. In 2013, NEORSD implemented Esri's AGOL tool to broaden the reach and availability of GIS and related asset information across the organization. To date, the cloud-based mapping platform has been used to collect and store GIS data and publish targeted web maps for more than 100 internal workflows ranging from stormwater stream walk data collection to monitoring lawn mowing contractor activity. More recently, NEORSD broadened their use of the AGOL platform to include access by external consultants and contractors (approximately 200) working on NEORSD projects. AGOL has become the central data management platform for all data collection and visualization efforts on the District's internal and consultant-led projects.

NEORSD's long history using Esri GIS software made incorporating the use of AGOL as part of their overall enterprise GIS strategy a natural next step. NEORSD maintains an AGOL Organizational Account that includes individual user accounts assigned to both internal employees as well as external consultants and contractors. Access to GIS resources in AGOL is managed by NEORSD GIS Administrators using role-based permissions.

NEORSD publishes collection system GIS data for both its own assets as well as the tributary community assets as "feature services" in AGOL. Various feature services can be combined into web maps customized to specific workflows and project needs. Web maps are viewable by all project team members using a standard web browser without the need for specialized GIS software or the knowledge to use it. Web maps are primarily used for viewing by most SSES project users, but support editing directly in a web browser or via ArcGIS Desktop by designated editors and data managers. **Figure 6** provides an example of the HHI-LSSES project web map including locations of field activities. Access to AGOL maps is controlled using membership in project-specific Groups. A consultant/contractor is granted access to each project Group that they are participating in.

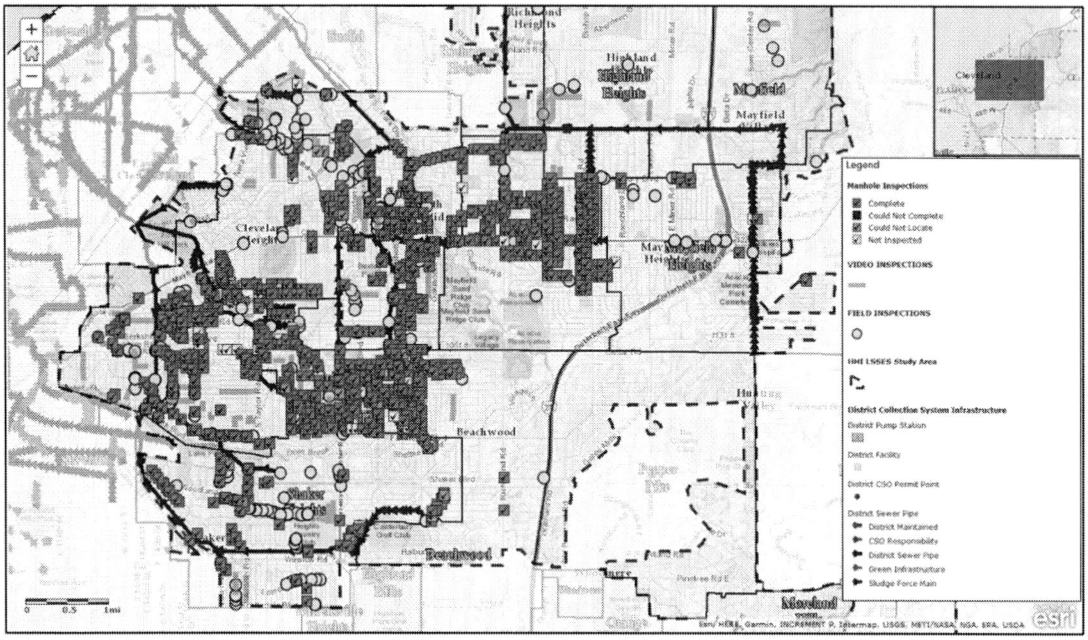

Figure 6. HHI-LSSES AGOL Web Map

Collector Application

One of the principal uses of AGOL on NEORSD projects centers around the ability for Esri's free cross-platform mobile applications to push data from field inspectors into AGOL for management and viewing. The primary application (app) used on NEORSD data collection efforts is called *Collector*, which is available on mobile devices running iOS, Android, and Windows operating systems. Digital inspection forms can be developed and published to AGOL allowing Collector App users to record data in the field. **Figure 7** shows an example of the HHI-LSSES project information displayed through the Collector App, and an example of the mobile inspection form. Photographs can be attached to inspection forms in the field, providing all project team members with near real-time access to observed field conditions.

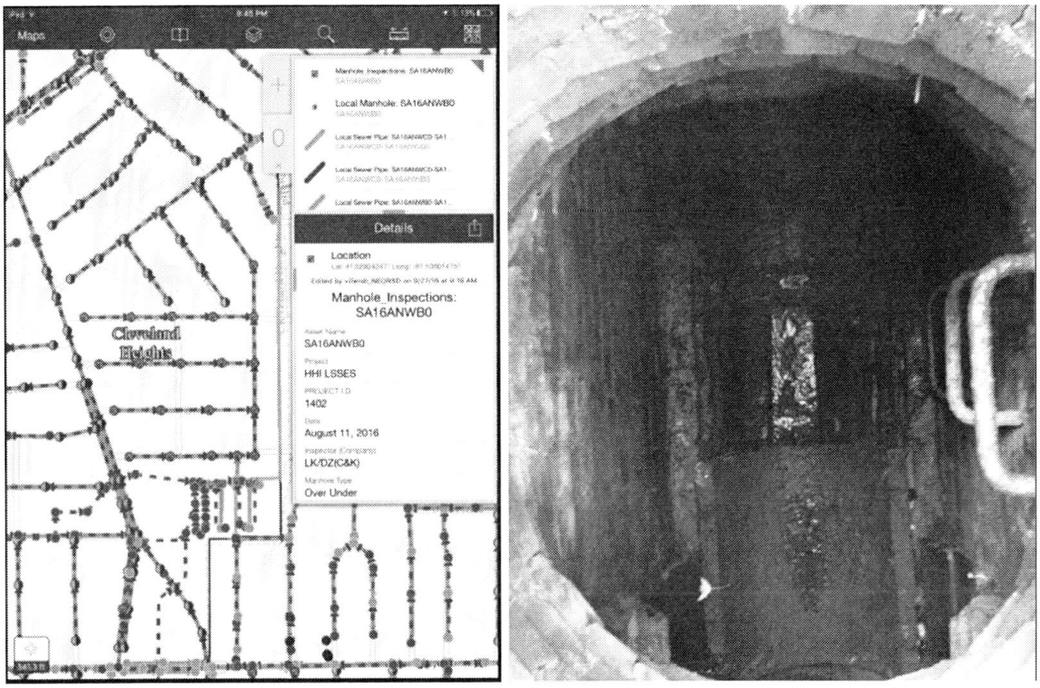

HHI-LSSES AGOL Mobile Inspection Form Photo saved to HHI-LSSES AGOL page

Figure 7. HHI-LSSES AGOL Collector App View

DISCUSSION

The anticipated data collection, analysis, collaboration and need for long-term storage of results made utilizing AGOL on the HHI-LSSES project an easy choice. AGOL has provided the following advantages:

- Real-time access to field data in the office
- Efficient field data collection augmented by photo/video capture
- Streamlined collaboration between team members
- Improved sewer system information quality by supporting efficient QA/QC
- Reduced potential for errors between engineers and field crews regarding field inspection locations
- More efficient and accurate interaction between project contractors and consultant team
- Efficient and accurate summation of field activities and results for progress reporting, alternatives analysis and project deliverables

Due to the size of the study area, a traditional SSES with detailed field assessment of the entire local sewer system was cost prohibitive for NEORSD's planning level effort. A cost-effective process was needed to focus field efforts in areas with the worst system performance, and to provide enough detail of field conditions for alternatives analysis. Areas were prioritized based on existing information about the study area obtained from previous reports, local engineers and community public service directors. This information helped prioritize where to investigate, and the field investigation required to quantify the CWA issues across the study area. The initial goal of the prioritization was to monitor flows and refine the existing model for 60% of the study area, micro-monitor 30% of the study area and perform traditional SSES field investigations in at least 15% of the study area. Approximately 40% of the HHI-LSSES area was found to have acceptable system performance at the beginning of the project. These areas were excluded from further consideration.

Collecting and organizing existing information within AGOL enabled identification and prioritization of problem areas. Information about the collection system was loaded from the District's GIS Enterprise system to AGOL. The collection system GIS includes information on pipe network, pipe attributes, constructed trench and manhole type, record drawings provided by local communities and locations of SSOs, suspected illicit discharge locations and septic systems. The collection system GIS is periodically uploaded to AGOL so users can access the latest information on the collection system.

Examples of data loaded to AGOL include existing model catchments with hydrology information, pipes with projected surcharging during the 5-year,1-hour District design storm and locations of reported basement backups. Interviews were conducted with community engineers and service directors to identify and verify locations of sanitary sewer capacity issues. These areas were identified in AGOL and descriptive information related to each issue was added in the attributes field.

Existing information was exported from AGOL and loaded to the desktop ArcMap application. Information on each CWA problem was organized for all existing model catchments. Using a weighted approach, SSOs, common trench sewers and basement flooding were evaluated to determine areas for model calibration flow monitoring and a first round of micro-monitoring and field inspection. These priority areas were loaded to AGOL and displayed for project team members to plan field work. **Figure 8** shows HHI-LSSES priority areas and how they were displayed in AGOL for the development of project activities. Red areas were identified as high priority areas with probability of multiple field activities and yellow areas were identified as lower priority but still receiving some evaluation in terms of flow monitoring, modeling and/or field investigation.

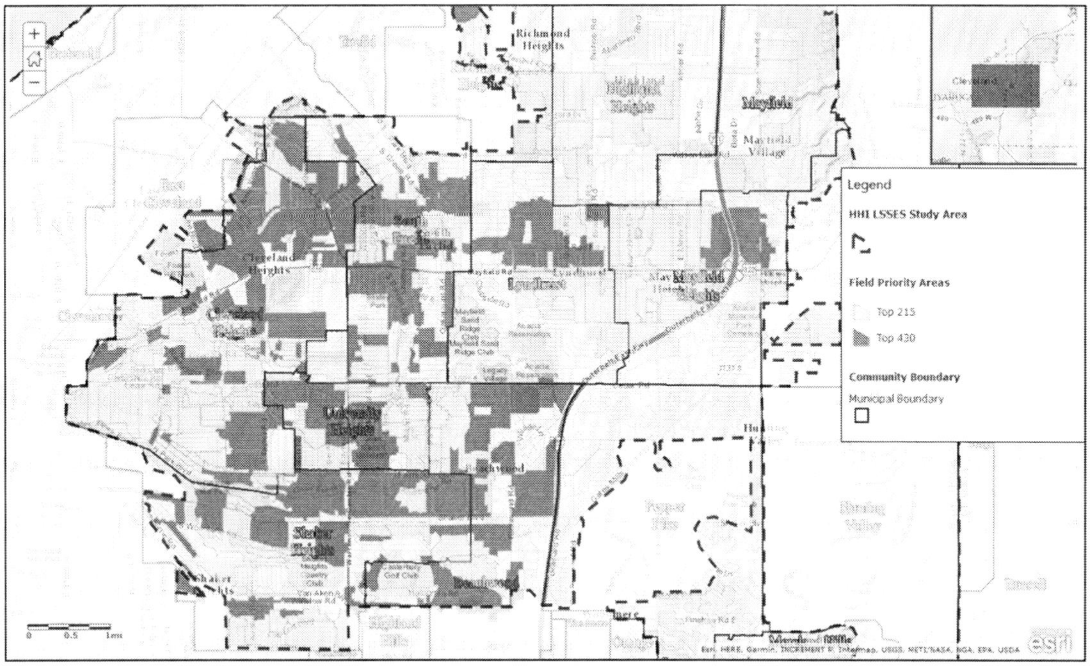

Figure 8. HHI-LSSES Priority Areas

Sanitary Sewer Overflow, Manhole Inspections and Mapping
Information collected during manhole inspections (depth to invert of each pipe, depth to top of wall (if applicable), and photos of the manhole) was entered into the manhole inspections layer on AGOL. . Manhole inspections were performed throughout the study to:

- Collect detailed information at known SSOs
- Verify system connectivity to update mapping
- Support model development
- Check structural condition

The District provided the project team with a list of documented HHI-area SSOs compiled from a report provided to the Ohio EPA as required under the NEORSD WWTP National Pollutant Discharge Elimination System (NPDES) Permit. The list consisted of 149 locations including the control status of each SSO: controlled (inactive for 5-year rainfall), active, unknown, physically eliminated, or not applicable. Additional SSO structures were encountered and categorized during field activities.

A field task to confirm the status for each SSO was completed using physical inspection and flow monitoring. The physical inspection included a standard National Association of Sewer Service Companies (NASSCO) Manhole Assessment and Certification Program (MACP) inspection of the SSO manhole. Additional information was collected including photographs, overflow height, weir dimensions and condition, pipe sizes and inverts, and other relevant comments. **Figure 9** shows an example of an SSO field sketch preserved in AGOL though the

Manhole Inspection layer. The field information was compared to available record drawings and previous studies. Based on observations of MACP inspections and monitoring, the team either confirmed status or recommended status changes for the Permit.

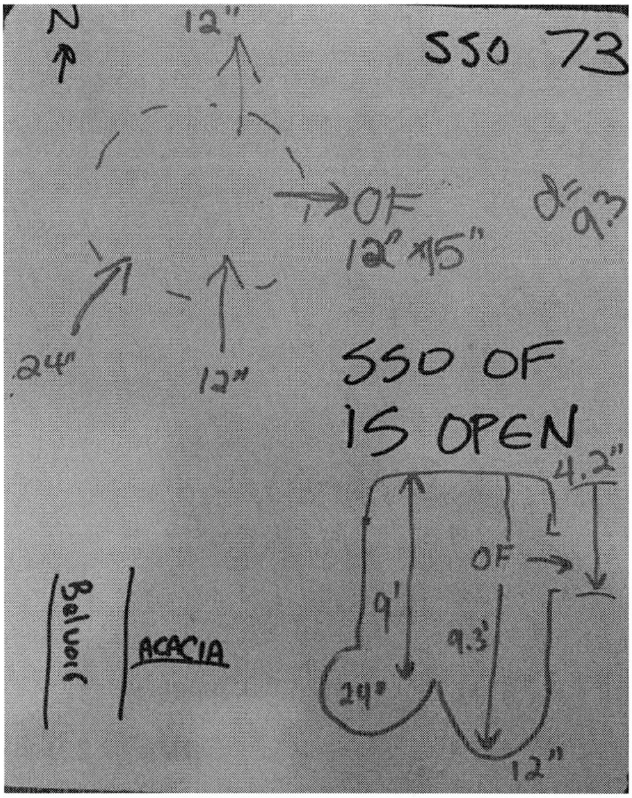

Figure 9. Example of SSO Field Sketch Preserved in AGOL

In addition to SSO inspections, manhole inspections were performed to collect information on the system for mapping and hydraulic model development. Throughout the mapping phase of the project, data inconsistencies were noted and investigated. Manhole findings were documented using the manhole inspection tool. Generally, information collected in non-SSO manhole inspections included observed pipe sizes, depth to pipe inverts, notes on any junctions and other noteworthy field information. All dividing wall manholes were inspected and dividing wall measurements recorded in a schematic to support hydraulic modeling.

Results of manhole inspections were used with record drawings from local communities to update mapping of common trench sewers within the study area. Quantifying common trench sewers and properly identifying the location of each type within the study area proved critical for numerous project tasks. AGOL was used to summarize field information and communicate areas targeted for in-depth investigation by NEORSD GIS analysts. Mapping of common trench sewers changed significantly as a result of the information review and field investigations.

System Investigations
During the course of field investigations, it is common to discover discrepancies between the existing system mapping and even record drawings in GIS with what is observed in the field. Common conflicts include differing trench configuration (e.g., common trench sewers recorded as separate trench), pipe diameters/materials, non-existent/buried manholes, and general pipe connectivity issues.

Although the District has a map layer used to officially communicate "Map Change Requests" to their GIS department, using it to note the observed field discrepancies before further investigation can be conducted could be premature. Discrepancies can often be investigated more in-depth in the office than in the field, and should not be automatically submitted with a modification to the GIS. To this end, the project team developed and implemented the "System Investigation" layer which both field crews and office staff can use to request further investigation from each other. Potential outcomes include a detailed field inspection of neighboring assets by field crews to clear up questions, investigation by office staff using record drawings and other previous inspection data, and ultimately a request to modify the GIS using the "Map Change Request" layer if the existing GIS data is incorrect.

Model Calibration Flow Monitoring
The project prioritization identified 60% of the area where flow monitoring was required to support model refinement and extension into the local collection system. The goal of the flow monitoring was to support model calibration of SSOs and calibrate depth and flow within the local collection system to determine capacity issues that may result in basement flooding. An added goal of the calibration flow monitoring was to determine the system performance and contribution of I/I from the separate trench and common trench systems throughout the study area.

A flow monitoring plan was developed to place 150 flow meters throughout the system. Based on the size of the study area, number of meters and desired model resolution, flow meters were placed with the goal of an average meter basin size of 150 acres. Concurrently, NEORSD staff were reviewing GIS connectivity based on newly available record drawings provided by the local communities. Record drawings were organized in AGOL using reference polygons drawn around referenced pipe segments. Updates to mapping and reference polygons were performed and posted daily to AGOL. The project team then used AGOL to verify connectivity of the collection system as updates became available. AGOL was used to determine proposed meter manholes, pipe sizes at proposed meter locations, CWA issues tributary to each flow meter and meter basin sizes.

Proposed meter locations were uploaded to AGOL and used to assist with meter site reconnaissance. The project team visited each proposed site with the flow monitoring contractor, ADS Environmental Services, to determine site suitability for access and hydraulic conditions. AGOL digital inspection forms were developed to collect and store information at each site as shown in **Figure 10.** Information collected includes:

- Manhole configuration
- Condition of common trench manhole, if applicable

- Suitability as a monitoring location
- If proposed location is found unsuitable:
    - Reason proposed site is unsuitable
    - Alternate target meter location
- Location and hydraulic conditions photos

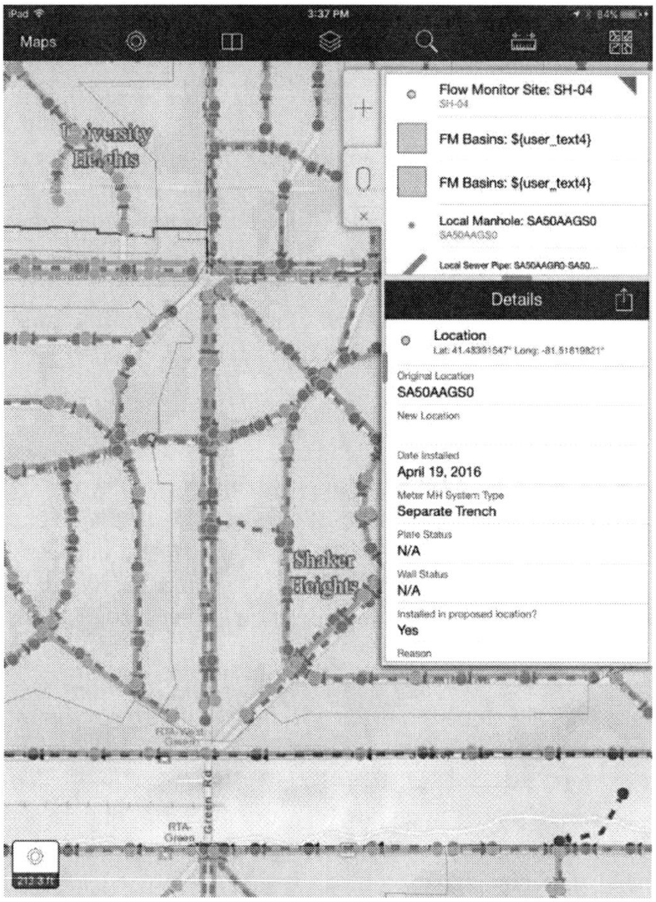

Figure 10. Flow Monitoring Site Mobile Inspection Form

The project team used the field reconnaissance results to detail changes from the proposed locations to the field confirmed locations, and then ADS used the information collected at each site to verify the correct meter manhole during installation.

Upon completion of the flow monitoring program, site installation sheets and final data files were uploaded to AGOL for permanent storage. Flow monitoring locations were then added to NEORSD's Archived Flow Monitoring layer, containing reference information on historical NEORSD flow monitoring. Based on model catchments, flow meter basins were developed and loaded to AGOL for data analysis.

Model Development

A project model was developed in InfoWorks ICM to evaluate local sewer capacity and determine SSO activation frequency. NEORSD's existing master model for the HHI system was developed in 2009 to support an operational evaluation/capacity analysis of HHI District interceptors and determine flow to the District's Easterly WWTP. During the study, the local system was modeled to the extent necessary to evaluate flows to NEORSD infrastructure. The HHI-LSSES project expanded the interceptor model to include local sewers within the priority areas. The modeling team used AGOL extensively to access system information including record drawings, SSO inspections, manhole investigations, dividing wall inspections and mapping of common trench systems. Periodic updates of the system as new information was received by NEORSD required modelers to be attentive to mapping changes within AGOL.

Micromonitoring

Micromonitoring aided assessment of the I/I problem areas throughout the study area. Micrometer locations were selected based on flows observed at model calibration flow meters. The calibration meter areas with the largest wet weather responses were further delineated using micromonitoring to focus field investigation efforts. The micromonitoring used 60 meters in rotation for an average of three weeks per installation, for five rounds of monitoring.

Similar to deployment of the calibration flow monitoring, micromonitoring locations were identified in AGOL for use by the project team. The locations were used by ADS to verify meter placement and by the project team to quantify I/I within the micrometer basins.

Work Order Management

The leakiest micrometer basins were identified for field testing to identify I/I sources. The type of field investigation selected for each area depended on the quantity of I/I and system type identified from the mapping. For example, the presence of over/under and dividing wall systems made smoke testing impractical because of the direct connections between the storm and sanitary sewer systems at the common manholes. A layer was developed in AGOL to summarize meter analysis and system type. The location and type of work was selected with locations highlighted. **Figure 11** shows the resulting HHI-LSSES RDII Analysis and Work Order Layer. From this layer, the type of field work performed in each area and the pertinent information about the work order area is identified.

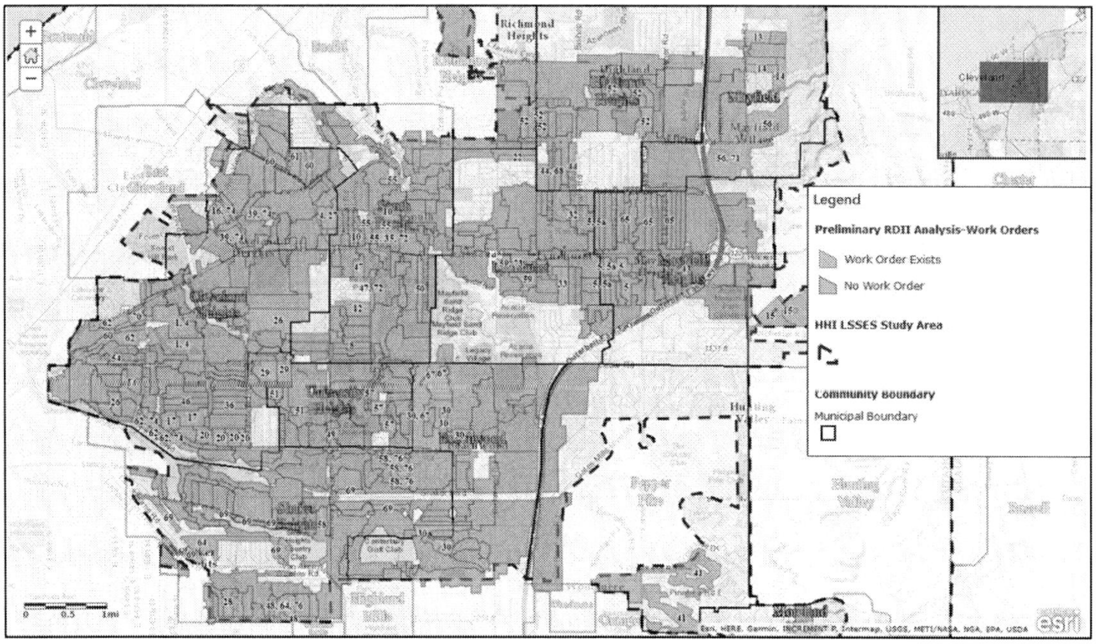

Figure 11. HHI-LSSES RDII Analysis and Work Order Layer

Dyed Water Testing

Two AGOL layers were set up to collect data for the dyed water flood testing, including Dye Testing Ground Level and Catch Basin Tests. The Dye Testing Ground Level activities used dyed water to flood storm sewer systems to identify and quantify cross-filtration between the storm and sanitary sewers. The Dye Testing Ground Level layer set points for each test and identified the limits of the test, system characteristics, testing procedures and equipment used. The Catch Basin Tests collected info for each specific catch basin tested during the dye test set up (often 4 to 6 catch basins per dye test). The Catch Basin Test points also identified if the catch basins were directly connected to the sanitary sewer. Both test set up points had a field for comments for the field inspector to describe specific inflow sources found during the test. Both points also permitted upload of pictures to document the severity of the info noted as well as set up locations.

Smoke Testing

AGOL was used to collect Smoke Testing data in a similar way. Smoke Test Set Up points were created for each set up location, and Smoke Test Defect Points were used to collect the defects. The Smoke Test Set Up points included the testing set up locations, limits of the test, crew and equipment information, and photos of the smoke test set up and equipment. The Smoke Test Defect Points included the location of the defects, type of defect, associated drainage area and photos of each defect.

PACP Inspection

Unlike the other field inspection efforts conducted on the project, NASSCO Pipeline Assessment and Certification Program (PACP) pipe inspections using CCTV technology have defined data collection standards and certified software packages are commonly used to log the information. As such, ArcGIS Online and Collector are not used to store the detailed CCTV inspection information. However, to track CCTV inspections in the same environment as the other field efforts, an AGOL GIS layer is populated by CCTV crews to indicate when a pipe has been inspected with basic details about the inspection (inspection date, inspector's name, and summary condition ratings). Once the CCTV data has been offloaded from the truck and reviewed, the GIS feature is populated with a link to the CCTV digital video file and a copy of the PACP inspection report is attached to it. Project team members then have access to the inspection information from the AGOL map without using special CCTV viewing software.

Water Quality Sampling

Water quality sampling is performed by NEORSD with input from the project team to provide a high-level assessment of stream quality within the study area, and identify potential sewage connections or other water quality impairments. Results of the stream sampling will be used to evaluate potential improvement alternatives. Pollutants of interest include E.coli, ammonia and phosphorus. The study area drains to several Lake Erie tributary streams including Kingsbury Run, Doan Brook, Dugway Creek, Nine-Mile Creek, Green Creek, Euclid Creek and the Chagrin River.

Dry weather stream sampling was performed throughout the study area to assess water quality, identify signs of failing septic systems and illicit discharges, and to identify stream reaches where unknown pollution sources may be impairing water quality. Samples were targeted in headwaters, downstream of suspected pollution sources such as failing septic systems, at community borders and at the project downstream extents. **Figure 12** shows the dry weather stream sampling locations. Wet-weather storm discharge quality samples are also being collected from the effluent of common and separate trench storm sewers to identify representative concentrations of pollutants of interest. **Figure 13** shows the locations and approximate drainage areas of the wet-weather storm sampling locations.

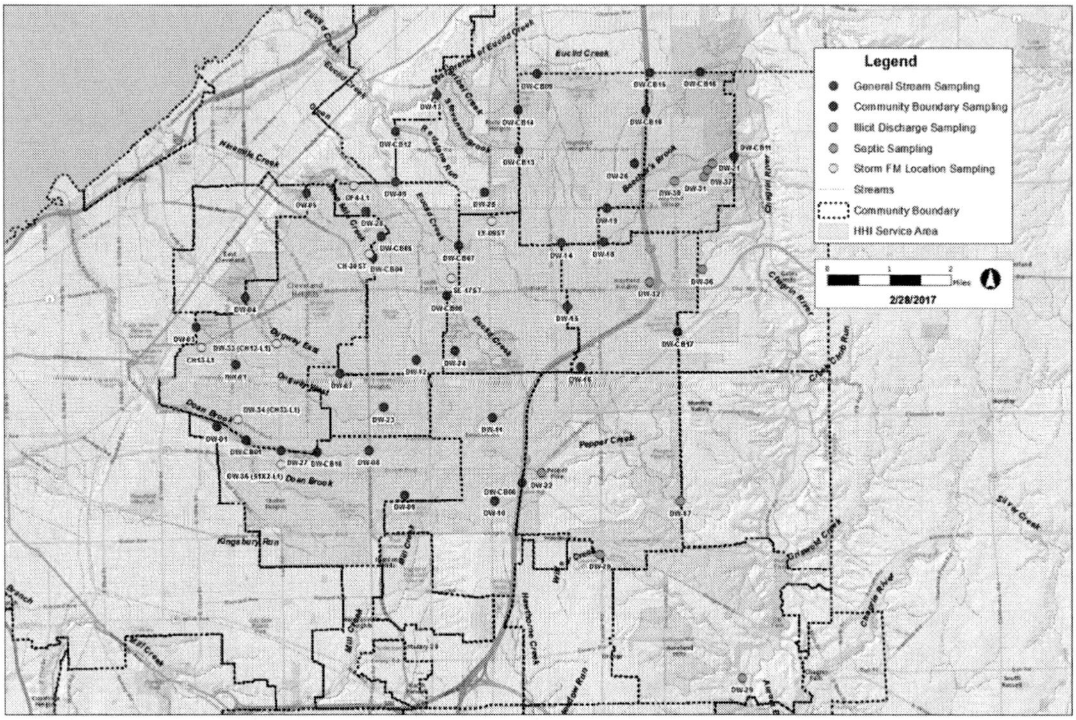

Figure 12. HHI-LSSES Dry Weather Water Quality Sampling Locations

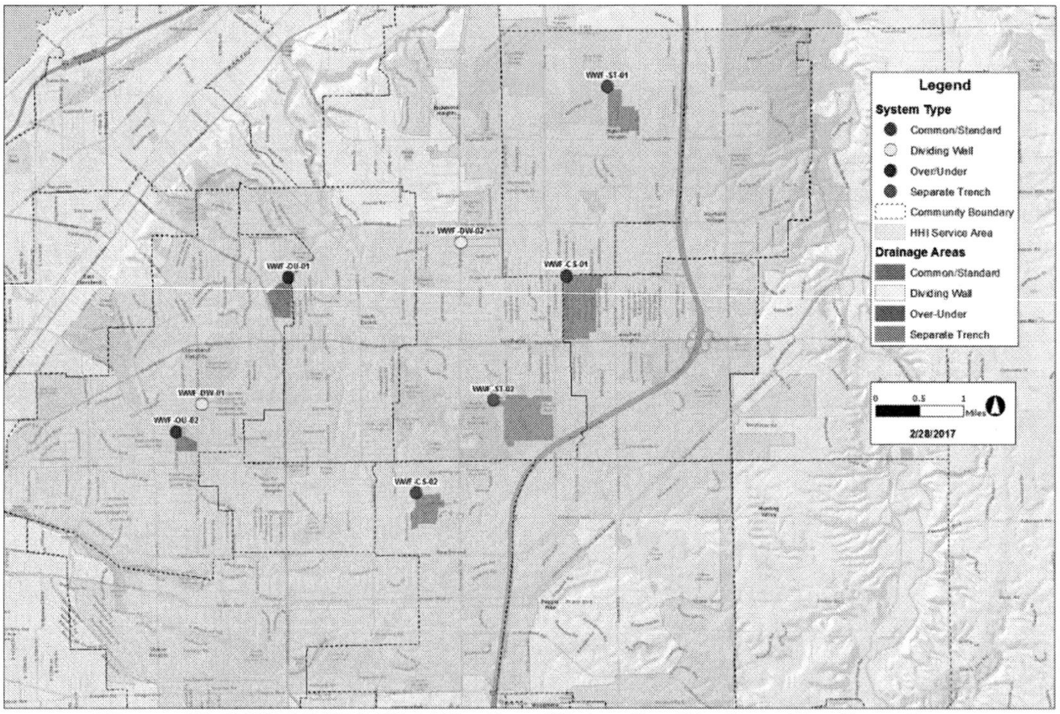

Figure 13. HHI-LSSES Wet Weather Water Quality Sampling Locations

Data Preservation
Preservation and cataloging of project data in a manner that makes the data readily available for future use has been a significant challenge for NEORSD. Typical methods to preserve data have included off-site storage, project hard drives provided by the consultant team and the use of non-GIS, cloud-based storage solutions. These options have produced mixed results.

Information developed and analysis performed for the current study is uploaded to AGOL as activities are performed. Maintenance activities with the NEORSD GIS Enterprise are ongoing as edits are needed to match GIS mapping to inspected field conditions provided by the project team. Upon completion of the project, the GIS mapping within the HHI-LSSES area will have undergone a significant upgrade compared to understanding of the local system before the start of the project. Project information collected and key deliverable alternatives and recommendations will be preserved with the HHI-LSSES project page. This information will be available for use by NEORSD and the local communities for future study and potential design of projects to remediate CWA issues.

CONCLUSIONS

AGOL has been an effective, versatile tool with an intuitive interface to consolidate information for delivery of the HHI-LSSES project. It has provided the framework for the project team to load information to a central repository that is used by all levels of the project team and owner's organization including project managers, engineers, data administrators, data collectors and data analysts. AGOL remains flexible enough for the development of new tools and is expected to be a critical tool as the project enters the alternative development and analysis phase. All data and results will be easily cataloged for long-term storage when the study concludes and projects are identified for further study or implementation to remediate CWA issues. Success of the HHI-LSSES AGOL has supported incorporation of AGOL on other design and planning projects including subsequent sewer system studies in other District service areas. The tools and workflows developed during HHI-LSSES are being used and continuously improved by the District and new project teams to improve efficiency using AGOL.

GOOD TO GREAT: A MORE DYNAMIC AND ROBUST LINEAR ASSET MANAGEMENT PROGRAM

AUTHORS

Paul Longo, P.E. – Dewberry Consultants, LLC
Jonathan Okafor, P.E. – Fairfax County Department of Public Works and Environmental Services

ABSTRACT

The Fairfax County Department of Public Works and Environmental Services, Wastewater Collection Division (DPWES, WCD) owns, operates and maintains over 3,400 miles of sanitary sewer collection lines spanning over 375 square miles. Unlike many other aging infrastructure systems, the Fairfax County sewer collection system is a truly separate system with no combined sewers. Over the past 25 years, DPWES has developed a linear asset management system to assist in operation and maintenance (O&M) decisions of their assets. Over the years, DPWES, WCD has been very successful in maintaining a high level of service with minimal sanitary sewer overflows (SSO). However, DPWES, WCD found that they were still struggling to quantify their risks, prioritize sewer segments within their sewer system and determine annual budgets required for their capital improvement program (CIP) and O&M program. In response, DPWES decided to develop a more dynamic and robust linear asset management system. The linear asset management system that has been developed for Fairfax County includes input stakeholders while developing Likelihood of Failure (LoF) and Consequence of Failure (CoF) models that analyze the risk associated with each pipe segment in the system

KEYWORDS

Asset Management, Risk Analysis, Sanitary Sewer, GIS, Condition Assessment

INTRODUCTION

The Fairfax County Department of Public Works and Environmental Services, Wastewater Collection Division owns, operates and maintains over 3,400 miles of sanitary sewer collection lines spanning over 375 square miles. Unlike many other aging infrastructure systems, the Fairfax County sewer collection system is a truly separate system with no combined sewers. Over the past 25 years, DPWES has developed a linear asset management system to assist in O&M decisions of their assets. Recently, WCD has decided to develop a more robust linear asset management system utilizing the most up to date technology available and strengthen the connection to the EPA Asset Management guidelines that have been issued.

The primary goal was to evaluate large diameter sewer segments in one of Fairfax County's five sewersheds (Arlington/Alexandria Sewershed) and develop a more robust linear asset management system to be utilized by DPWES, WCD staff. Once calibrated, this system is proposed to be implemented throughout the balance of the sewer system. The objectives were to develop a risk matrix to quantify risk and prioritize sewer segments; work closely with DPWES, WCD to develop CoF and LoF models to score sewer segments; standardize CCTV and multi sensor inspection data to the most up to date Version (6.0.1) of NASSCO PACP sewer condition inspection requirements; establish appropriate balance between optimal cost and desired level of service at acceptable risk level and to determine budgetary costs for the renewal and replacement of sewer segments. This prioritized list of sewer segments and scoring criteria are utilized by DPWES, WCD as a decision support tool within their overall asset management program. Testing and verification occurred post completion to verify risk factors.

METHODOLOGY

The process for developing a new linear asset management system for DPWES, WCD included several steps. The driving force for the initiation of this process was the desire for DPWES, WCD to have a GIS-based (ESRI (Redlands, CA) ArcMap 10.3) linear asset management program. The steps taken to develop the appropriate models are as follows:

Data Collection

The first phase in generating the LoF and CoF models that were used to prioritize the linear assets within Fairfax County was to gather the necessary geospatial data and determine the viability of using the data in detailed analysis. The analysis of the data included review of network connectivity, metadata and spatial data to determine data layers that could be used in a detailed analysis. In addition to the data provided by DPWES, WCD for Fairfax County, additional data sources were gathered from other utilities, the Department of Transportation and the USGS to be included in the analysis. In total, over 20 different data layers were used in the risk analysis. Figure 1 below is a representation of those data layers.

Figure 1 - Data Layers

Stakeholder Input

Once the data was collected, and determined viable for geospatial analysis, several stakeholder meetings were held to determine the risk criteria to be utilized. The stakeholder meetings were vital in capturing the institutional knowledge and preferences for the risk analysis.

Risk Model Development and Calibration

Utilizing the information gathered during the stakeholder meetings, CoF and LoF models were developed to calculate risk. The CoF and LoF models utilized several custom tools that were developed within ESRI ArcMap 10.3.1 (Redlands, CA). These tools were used to calculate the distance of sewer lines crossing streams or calculate the distance of sewer lines crossing major roadways. After the models were developed and tested, calibration of the models took place. The calibration utilized pipe segments where recent issues were noticed as a baseline for testing the model. Through the calibration of the models, weighting of specific criteria was adjusted to capture the actual field conditions of certain pipe segments. The final weighting was verified through meetings with key stakeholders.

The results of the CoF and LoF model were based on a 0-1,000 scale. Risk categories were used to determine the priority of each pipe segment. The risk categories used were low priority, regular monitoring, frequent assessment, medium priority, high priority and highest priority. An example of the risk matrix with risk categories is shown in Figure 2.

Consequence of Failure	0 - 199	200 - 399	400 - 599	600 - 799	800 – 1,000
800 – 1,000	Regular Monitoring	Frequent Assessment	Frequent Assessment	Highest Priority	Highest Priority
600 - 799	Regular Monitoring	Frequent Assessment	Frequent Assessment	Highest Priority	Highest Priority
400 - 599	Regular Monitoring	Regular Monitoring	Regular Monitoring	High Priority	High Priority
200 - 399	Low Priority	Low Priority	Low Priority	Medium Priority	High Priority
0 - 199	Low Priority	Low Priority	Low Priority	Medium Priority	Medium Priority

Likelihood of Failure

Figure 2 - Risk Matrix

Each pipe segment was then analyzed through the CoF and LoF models that were developed using Python Code within ESRI (Redlands, CA) ArcMap version 10.3.1. Each pipe segment was then graphed on the risk matrix as shown in Figure 3.

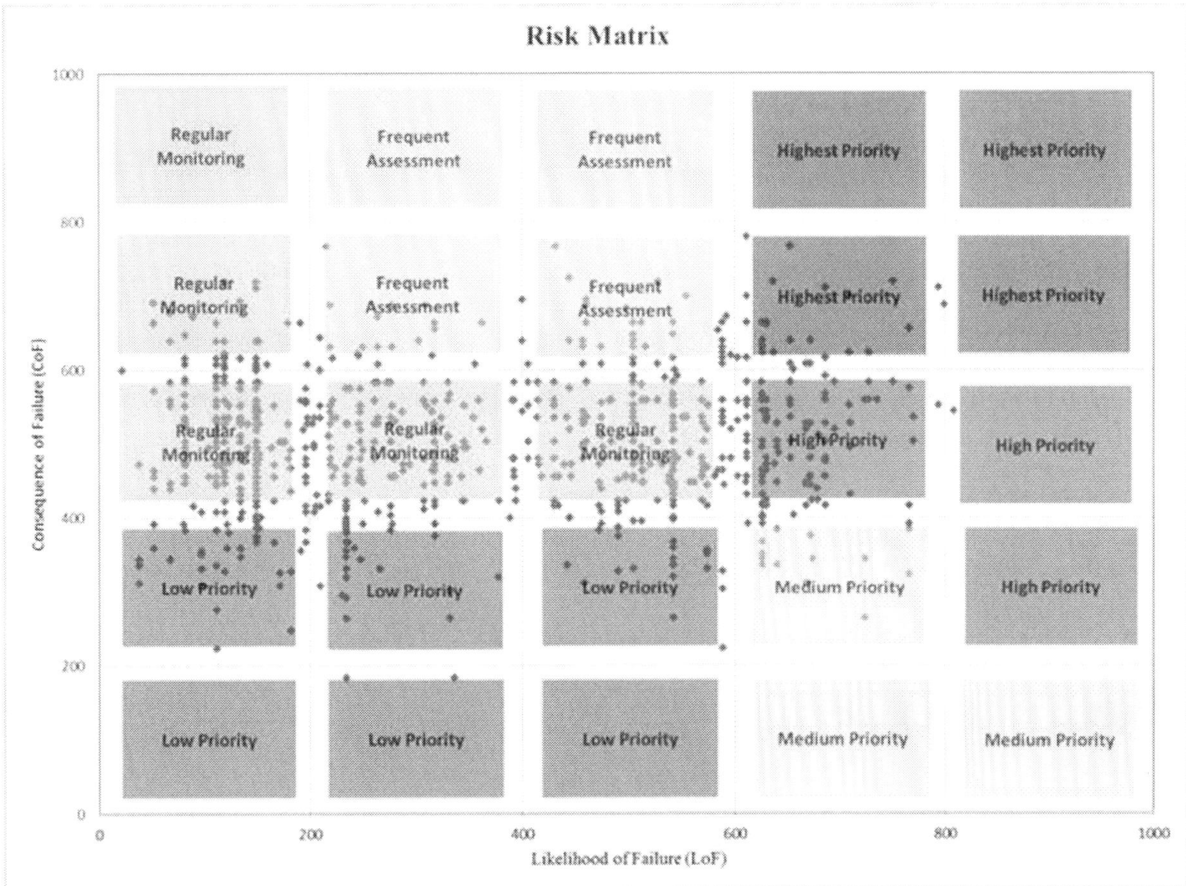

Figure 3 - Risk Matrix with Segments

In addition to the risk models developed, a cost model was developed for estimating the cost of the recommended improvement method e.g., replacement, rehabilitation, and/or repair costs for each pipe segment. Separate cost models were developed for each pipeline improvement method, and each model and subsequent pipe segment cost estimate includes three categories of costs:

Unit Costs (linear feet): based on pipe characteristics including pipe depth, pipe material, soil conditions, and pipe size

Lump Sum Costs: accounts for additional factors that may impact the cost of the pipe improvement e.g., stream crossings, bypass pumping, and locations relative to wetlands and Resource Protection Areas (RPAs)

Contingencies: includes mobilization/demobilization, soft costs, and other miscellaneous items

The overall Contingencies Cost included in the pipe improvement cost varies for each pipe segment, as it is based on a number of elements that are affected by the specific location of the pipe and for which a more detailed analysis of the situation is included. The Mobilization/Demobilization Contingency Cost represents the approximate cost for mobilization and demobilization to complete the recommended improvement work for an individual pipe segment. The Mobilization/Demobilization Contingency Cost used in each Cost Model is 5% of the sum of the Unit Costs and the Lump Sum Costs.

The Soft Costs Contingency represents the approximate costs for engineering, legal, and permitting fees that may be encountered to complete the improvement work. The Soft Costs Contingency used in each Cost Model is 10% of the sum of the Unit Costs and Lump Sum Costs. The Miscellaneous Costs Contingency, based on engineering judgment, includes additional costs that are identified after further detailed analysis of each pipe segment (e.g., location, situation, etc.) is performed. For example, if a pipe segment recommended for replacement crosses under a railroad and is located in a highway, a 25% contingency may be applied to account for access issues, permitting issues, and specialized design requirements that may be needed. The Miscellaneous Costs Contingency percentage is unique to each pipe segment.

RESULTS

Fairfax County DPWES, WCD has always provided its customers a high level of service with minimal impacts to public health and the environment. As with most municipalities that own and operate aging underground linear assets, over time, the status and condition of these assets become less certain. Fairfax County decided that by utilizing information that they have already collected, as well as incorporating new standards and practices, their linear asset management system will become more robust. In the end, the decision support tools developed for Fairfax County DPWES, WCD:

- Provides a list of prioritized sewer segments in its most critical sewershed;
- Provides DPWES, WCD with a better understanding of its overall risk;
- Establishes appropriate balance between optimal cost and desired level of service at acceptable risk level;
- Estimated costs for renewal and replacement of assets in its system that can be used for longer term CIP and O&M budgeting; and
- Provides a clear audit trail to continually demonstrate sustainability.

CONCLUSION

The implementation of a large scale linear asset management program takes the hard work and dedication of numerous key stakeholders and requires full support from upper level management. The process for developing these tools has brought forward issues that were not previously considered. This program was used as a pilot program for the implementation of a system wide asset management program. DPWES, WCD is currently in the process of implementing their system wide asset management program utilizing Innovyze's (Broomfield, CO) InfoMaster software.

EXTENDING THE SERVICE LIFE OF A BRICK & CAST IN PLACE CONCRETE SEWER

By: Bryon Livingston, P.E.[1], Andy Clements, P.E.[2]

[1]Black & Veatch, Sr. Project Engineer, Buried Infrastructure Practice, 8400 Ward Parkway, Kansas City, Missouri, 64114; PH (913) 458-3368; email: livingstonb@bv.com

[2] City of St. Joseph, Missouri, Asst. Director of Public Works, 1100 Fredrick Ave. St. Joseph, MO, 64501 PH (816) 271-4653; email: aclements@stjoemo.org

ABSTRACT

This paper will examine a case study illustrating the approach used by the City of St. Joseph, Missouri to evaluate the condition and extend the useful life of a 12 to 17 foot diameter concrete and brick combined sewer. The case study will include description of the methodology used to collect data on the condition of the pipeline, evaluation of methods to address the defects identified and how funding is an important part of the project.

PROJECT BACKGROUND

The Blacksnake Creek Sewer serves the drainage basin that covers the northern portion of the City and drains approximately 5300 acres. This system is one of the longest and largest sewer systems in the City. The oldest portion of the system is brick construction and dates back to about 1884. The remaining portions of the brick system were constructed in 1899 to 1909. The concrete pipe segments were constructed in the 1940's using cast in place methods.

As part of the Blacksnake Creek Stormwater Separation project, the need for a condition assessment of the existing Blacksnake Sewer pipeline became apparent since it will continue to be used in the conveyance system. The purpose of the inspection was to determine the current condition of the Blacksnake Sewer to identify defects and determine recommended repairs or rehabilitation to extend the useful life of the pipeline.

The sewer consists of a 15 ft x 12 ft double concrete box entrance, approximately 6,800 feet of 12 ft diameter cast in place concrete sewer; approximately 2,000 feet of 15 ft x 13 ft egg shaped brick sewer; and approximately 4,200 feet of 15 ft x 17 ft egg shaped brick sewer. The system also includes 30 manholes, the diversion structure, and the outfall structure and channel. The condition assessment was used to develop a plan for required repairs or rehabilitation as part of the combined sewer separation project.

FIRST STEP CONDITION ASSESSMENT

The large diameter pipe required the use of a multi-sensor inspection approach. A request for proposals was sent to three firms to determine the best approach with the most accurate equipment for the condition assessment. The results of the condition assessment would be used to develop a repair/rehabilitation plan to extend the useful life of the pipeline to be consistent with the new stormwater separation project.

The inspection of the Blacksnake Sewer was completed using CCTV, laser scanning to identify defects and provide a rating based on the National Association of Sewer Service Companies (NASSCO). Sonar profiling was available on the inspection platform but the depth of flow in the sewer did not exceed 12 inches so the sonar was not required. Since this is a combined sewer system the work was scheduled during the winter to avoid high flows from rain events. The inspection involved collecting information required to complete a condition assessment on approximately 13,000 linear feet of the existing Blacksnake Creek sewer system. The technicians used the standard rating of the structural defects and operation and maintenance defects that have been developed based on a Pipeline Assessment Certification Program (PACP). The rating score is a 1 to 5 scale with a "1" indicating excellent condition and "5" requires immediate attention.

RESULTS OF CONDITION ASSESSMENT

The inspection of the Blacksnake Creek Sewer did not identify any critical defects that would require immediate repair. However, there are 26 structural defects identified, that were rated a S4, for which repairs should be made in the next 5 years. These structural defects are multiple fractures, holes and missing bricks. The total length of continuous structural defects rated a S4 is about 2,600 feet. Also there are 33 additional structural defects rated S3 identified that include longitudinal cracks that are about 2,200 feet. These that do not appear to present an imminent potential to cause pipe failure but will continue to deteriorate and over time will become a potential for failure in the future.

DESIGN PROCESS

The design process began with an evaluation of the results of the condition assessment. It was determined to extend the useful service life of the existing Blacksnake Creek sewer pipeline to repair the structural defects that have been identified and rated S3 and S4 be scheduled for repair within the next 5 years. The estimated total length of these repairs is 4,800 feet or about 37% of the total length.

This portion of the project would be included with the overall work on the drainage basin and was submitted as part of the package for State Revolving Funding (SRF). The plans and specification will require approval from the Missouri Department of Natural Resources and special requirements for the SRF loan.

Rehabilitation alternatives of sprayed on cement mortar, slip lining, carbon fiber wrap and cured in place pipe were evaluated. These methods provide a fully structural rehabilitation and based on the observed defects a repair method would be more cost effective. The recommended repair methods for the cracks include the use of a hydro-active polyurethane grout or similar material that can be injected into the cracks and seal them from infiltration. The grout should expand when exposed to water before curing and adhere to the concrete or brick substrate. The areas with missing bricks or holes should be repaired with a concrete patch or other material suitable for applications in water. The material would have to be rapid setting, able to stop leaks and bond to damp surfaces.

PERMITTING REQUIREMENTS

The contractor will be required to obtain permits from the City of St. Joseph for working in the right of way and for possible road closures. Traffic control will be a requirement to ensure the public is not restricted by this work.

The contractor will be required to submit erosion control plans and Stormwater Pollution Prevention Program (SWPPP) plans before starting construction. Requirements for the SWPPP submittal will be included in the specifications. The design team will develop a draft SWPPP submittal obtaining key regulatory reviews, notices, permits etc. prior to the project's advertisement. This sample document will be incorporated into the specifications and made electronically available to the selected contractor for use in its submittal to the City and State. Construction will not take place near any bodies of water or sensitive areas. Therefore, traditional silt fences, haying, and storm water drain covers will likely be the extent of the erosion control plan. The plan will be reviewed as part of the submittal process to ensure compliance with any permitting requirements.

The contractor will be required to obtain necessary permission from Missouri American Water Company prior to start of work for access to fire hydrants. High pressure water jetting is required for cleaning and removing dirt material from cracks. Fire hydrants are located at couple of places adjacent to the Blacksnake Sewer.

DESIGN PARAMETERS

Various crack repair materials have been evaluated as listed in Table 1. In addition, the option was evaluated of providing a 0.20 to 1.5 inch thick liner of Quest Inspar. The liner would provide structural enhancements to the pipe surface. Since, the sewer pipe is in fair condition and there were no defects which identified requirement of structural integrity, this option is not needed and was not evaluated further.

The evaluation during the designed compared hydrophilic and hydrophobic grouts. Both grout materials are considered low viscosity to provide flow characteristics and are suitable for filling small voids. The results of the evaluation determined the hydrophilic grouts provided the desired characteristics for this project. The hydrophilic grout absorbs water while curing which was desired to address cracks that began over 10 mils but become smaller. The grout also cures to become a flexible foam that would provide for expansion of the concrete and bricks due to temperature changes and other factors. The hydrophilic grout also bonds tightly with damp surfaces.

Following evaluation and based on properties of the material, it was recommend to use DeNeef Hydro Active, Avanti AV-202-LV Multigrout LV, Prime Resins Prime Flex 900 XLV polyurethane based crack repair material listed in Table 1. The materials selected are more viscous and have good water sealing properties. The grout which will be used to seal the cracks have been specified to possess the following minimum characteristics:

1. Density of selected material should comply to ASTM D1622/ASTM D1638/ASTM D3574
2. Bond strength > 200 psi. The material should expand when exposed to moisture
3. Viscosity > 200 cps and should comply to ASTM D2196/ASTM 2196A
4. Tensile strength should comply to ASTM D3574
5. Elongation should comply to ASTM D1622/ASTM D638/ ASTM D3574

The crack will be sealed with an epoxy and then injected with the polyurethane material. The epoxy material shall be approved by the polyurethane manufacturer and suitable for application in the pipe with the existing temperature and moisture conditions.

Table1 – Crack Repair Material Properties

MANUFACTURER	PRODUCT NAME	HYDROPHOBIC/ HYDROPHILIC	TENSILE STRENGH	ADHESION PROPERTIES	COMPONENT	ELONGATION	VISCOCITY
DeNeef	Hydro Active	Hydrophillic	380 psi	250-300 psi	Polyurethane	400%	650-800cps
Prime Resins	PRIMEFLEX 900 XLV	Hydrophilic	450 psi	Seals active water leaks	Polyurethane	350%	250-350 cps @ 73° F
Avanti Grout	AV202 LV Multigrout LV	Hydrophilic	To Be Determined	Requires Primer	Polyurethane	400-600%	650-800 cps @ 72° F

PROCEDURES FOR REPAIR

Because the results of the inspection are about 3 years old, prior to the beginning of the work a visual inspection by the resident engineer, project engineer, City staff and contractor will be conducted. This preconstruction inspection will confirm the cracks and holes identified in the contract documents and provide information that will be used to repair "engineer directed cracks" and "engineer directed surface repair". As part of the bid documents alternate bids were taken that will allow the resident engineer to identify additional cracks and holes to be repaired in addition to those specified in the plans. This work will ensure the City is able to repair as many of the cracks and missing bricks as possible.

There are two different procedures applied depending on width of crack. For crack width smaller than quarter inch (considered an S3 crack) epoxy is applied directly on the crack surface after proper cleaning, removing dirt material and making 'V" groove. For cracks wider than quarter inch (considered an S4 fracture), polyurethane grout will be injected. For repair of fractures, the steps followed are:

- Drill holes in pipe wall at a 45 degree angle. Alternate positioning from left side to right side as you move along the crack. Drilling depth should be generally half the thickness of the pipe wall thickness.
- Install mechanical injection packers and secure in place at specified interval.
- The surface of the crack needs to be cleaned first using mechanical grinder or a wire brush.
- It is then flushed with clean water at minimum pressure of 250 psi to clean dirt and other contaminants out, open the crack up and to ensure enough water is present to activate the grout. Pressure should be slowly increased as necessary to allow the flush water to penetrate the crack. However, pressure should not exceed

2,500 psi. Flushing should continue until contaminate free water is flowing out of the crack.

- Apply a surface seal if necessary to contain the resin during injection. Hydraulic cement or wooden wedges may be used to reduce or stop the flowing water and seal the surface of the crack.
- Put a tip on the first packer and begin injecting polyurethane grout until grout material start flowing in second packer. Move the pump to next packer and continue the same procedure until entire stretch is sealed with grout material.
- Flush the pump and hoses with water or other material as specified by material supplier.
- Packer is to be pulled out of pipe surface and gap is filled with hydraulic cement.
- Excess grout material will be trimmed off from pipe surface to make it smoother.

The surface repair of locations with missing bricks will be accomplished using a reinforced concrete repair. The concrete material will be a one-component cementitous product that is compensated for shrinkage. Because the brick portion of the sewer was constructed using four layers the repairs planned for thickness that varied from 3 to over 9 inches.

Before working on any stretch of the Blacksnake Sewer for crack repair, the surface will be made dried by diverting the flow. This can be done by creating a temporary barrier with sand bags and pumping the flow to downstream manhole. The pipe flow entering the stretch directly into the sewer will be plugged temporarily.

The areas repaired will be inspected with CCTV to document and confirm the final condition of the repair. Also the locations and repairs made will be documented to provide a baseline of the condition of the pipe to monitor for future cracks.

CONSTRUCTION

The construction is planned for the summer of 2017. The project was bid and the results ranged from $1.4 million to $500,000. The results of the bidding provided a lower than anticipated bid cost but the low bidder was evaluated and found to be acceptable. Based on the bid form that used an Alternate 1 for engineer directed cracks and Alternate 2 as engineer directed surface repair the City will be able to repair additional locations.

As part of the inspection several laterals were observed that are plugged but have significant deposit build up. These locations will be cleaned and the deposit removed.

CONCLUSION

The determination to repair the cracks on this combined sewer was based on the need to reduce inflow and infiltration but also prevent further damage from the cracks. Sealing these cracks will prevent further deterioration that could result in loss of structural integrity.

The details of the repair alternatives were evaluated in a preliminary design and the use of a hydrophilic grout was selected. The grout will repair and seal the cracks along the alignment and help reduce the inflow and infiltration. The repairs to the defects should extend the useful life of the pipeline, but also provide a basis to monitor for future deterioration within the pipeline.

The repair alternatives were evaluated and based on the recommended repair of S3 and S4 cracks a cost estimate of about $2 million was developed. The currently available budget was very close to this amount and so the decision was made to proceed with the project. The available budget is not always close to the estimated cost of a project. With the excellent bid prices the City will be able to make the repairs necessary to extend the service life of the Blacknake Creek Sewer. The funding by the city for this work is incorporated as part of the overall stormwater project with SRF funding.

Future inspections are recommended in 5 years and this information will serve as a baseline to determine the condition of the concrete and brick pipeline.

Large Scale, Priority-Driven Field Data Collection through to Rehabilitation: How DeKalb County Department of Watershed Management Managed Multiple Assessment Contractors to Target and Coordinate Sewer Rehabilitation

Darren Eastall[1], Courtney Call Kennedy[2], Burhan Shaikh[2], Reggie Rowe[3]

[1] DeKalb County Department of Watershed Management
[2] CH2M
[3] HKA

ABSTRACT

DeKalb County Department of Watershed Management (DWM) is performing a large scale conveyance condition assessment and rehabilitation program under a United States Environment Protection Agency (EPA) Consent Decree. DWM has collected and coordinated vast quantities of condition assessment data from various contractors to meet tight program schedules. DWM's assessment work has involved from 15 to 20 assessment crews working simultaneously to perform combinations of six inspection technologies over approximately 1,300 km (~800 miles) of sewers and 27,000 manholes. Managing the data, prioritizing the asset conditions, and developing a rehabilitation plan was an enormous task, especially when overseeing and integrating hydraulic modeling capacity improvements with the various condition inspection techniques and contractors.

Using four key components (succinct data collection protocols, mandated quality control procedures, integrated data flow and robust decision logic) DWM automatically generated prioritized lists of assets needing either further inspection or corrective actions. The lists were confirmed by rehab review teams prior to assigning to design-build contractors. DWM realized the value of automating as many of the condition assessment business process steps as possible in order to efficiently process large quantities of condition assessment data and prioritize corrective actions. To date, DWM has collected over 10 Terabytes of media and 300 GB of inspection data. Processing this information by hand is not an option. DWM's automated tools have made processing the data with a small staff possible.

KEYWORDS

Data integration, condition assessment, Consent Decree, SCREAM, CCTV, quality control, corrective action

INTRODUCTION

DWM is performing a large scale conveyance condition assessment and rehabilitation program under an EPA Consent Decree. DWM has collected and coordinated vast quantities of condition assessment data from various contractors to meet tight program schedules. Managing the data, prioritizing the asset conditions, and developing a rehabilitation plan is an enormous task, especially when overseeing and integrating hydraulic modeling capacity improvements with the various inspection techniques and contractors.

This paper describes the data collection processes, decision logic, and coordinated software systems DWM uses to manage a variety of inspections and prioritize rehabilitation. DWM's approach is presented so that other utilities may benefit from their implementation experiences.

DWM'S INFRASTRUCTURE AND ORGANIZATION

DWM owns and operates approximately 4,300 km (2,600 miles) of sewers ranging from 15 cm to 183 cm (6 to 72-inch) diameter; 65 lift stations; and two treatment plants serving approximately 193,000 customers.

The assessment project is subdivided into two components – 48 priority areas known as the Priority Area Sewer Assessment and Rehabilitation Program (PASARP) and the remaining portion of DWM's system are known as the Ongoing Sewer Assessment and Rehabilitation Program (OSARP) areas. PASARP covers approximately 1300 km (30%) and 27,000 of the DeKalb County sanitary sewer system and the OSARP area covers the remainder of the County maintained assets. Figure 1 shows the collection and treatment system service areas and the 48 priority areas.

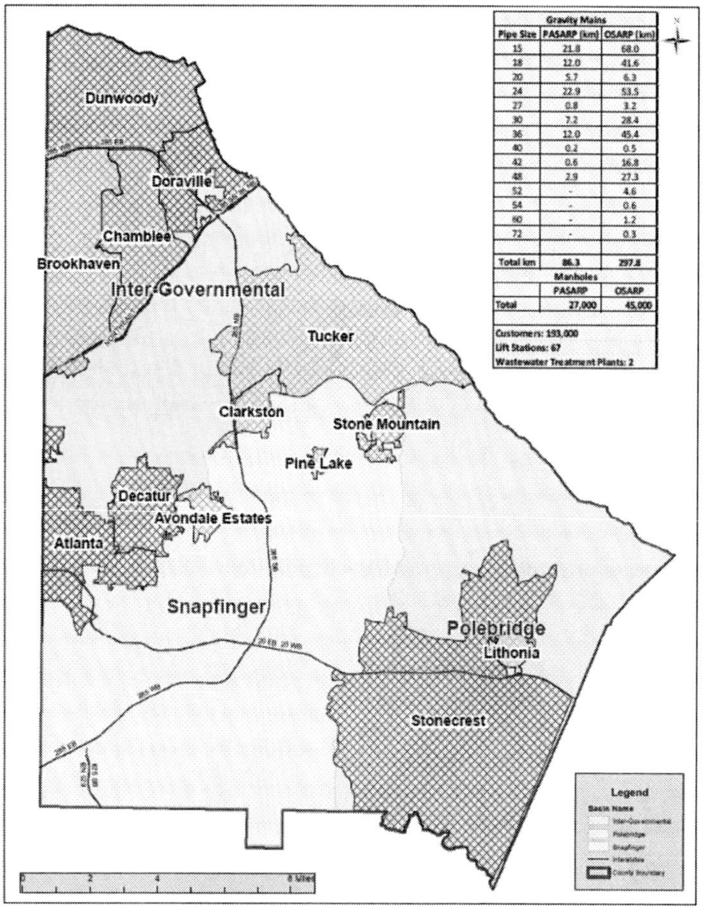

Figure 1: DWM's Wastewater Service Area.

Figure 2 displays the PASARP priority areas which were the primary focus areas of the EPA Consent Decree.

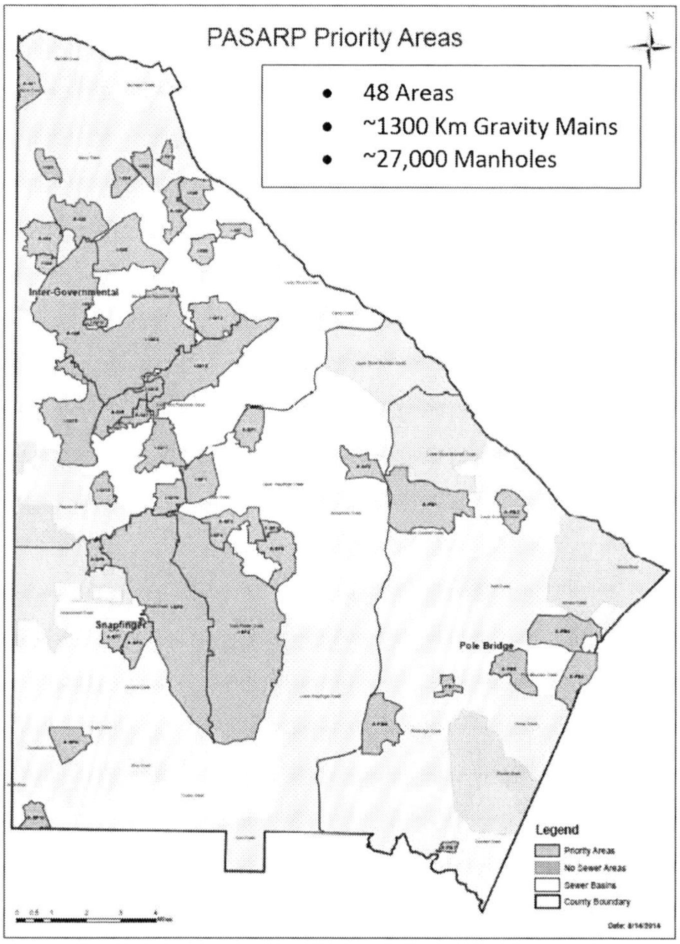

Figure 2: DWM's PASARP Priority Areas

DWM'S COORDINATED DATA COLLECTION APPROACH

DWM's assessment work has involved from 15 to 20 assessment crews working simultaneously performing combinations of six inspection technologies over approximately 1,300 km (~800 miles) of sewers and 27,000 manholes. To manage the complexities of this program, DWM developed a thorough data collection, integration and analysis approach.

DWM decided to use a tiered approach to inspect the pipes in the PASARP areas. Instead of performing resource-intensive CCTV inspections on all the pipes, all the pipes were pre-screened using the less expensive Tier 1inspection techniques. Tier 1 pipes with high condition scores (poor condition) were assigned to Tier 2, receiving subsequent CCTV inspections. The inspection techniques included:

- Acoustic inspections for all Tier 1 pipes.
- Smoke inspections for all Tier 1 pipes.

- CCTV inspections for pipes in Tier 2.
- TISCIT inspections for all large diameter pipes.

The methodology for moving from Tier 1 to Tier 2 is detailed later in this paper, but generally follows the logic shown in Figure 3.

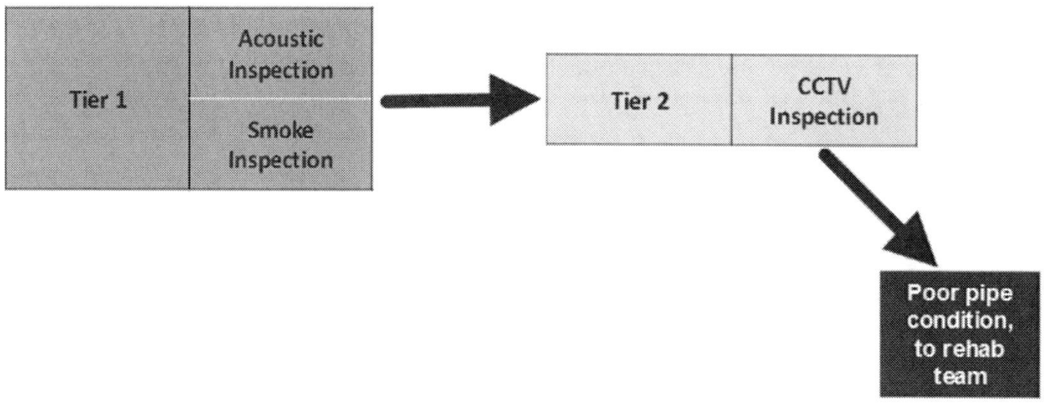

Figure 3: Tiered inspection logic

DWM knew that having a tiered approach to field inspection and subsequent data collection would help prioritize their assets with regards to condition and need for corrective action. DMW also needed to provide inspections that were properly conducted and that accurately documented the representative conditions of the pipes and manholes. Additionally, the program needed to carefully manage and standardize the data for easy and repeatable analysis. For instance, the SCREAM application enabled each asset's inspection data from the different field methodologies to be combined to provide a single condition score. The data handling process, from start to finish, required standardized data formatting protocols. Otherwise, the data would not have been useful to the semi-automated analysis processes set up by DWM.

Finally, DWM established data protocols and decision algorithms to screen the inspections and target assets in need of corrective action review. Four key components were initially developed which has made DWM's program successful:

1. Succinct data collection protocols
2. Mandated inspection quality control procedures
3. Integrated data flow processes for easy analysis
4. Robust decision logic matrix

Data collection protocols

DWM recognized early in their program the importance of creating succinct data collection protocols for each of the inspection techniques: smoke testing, acoustic testing, CCTV inspections, TISCIT, and manhole inspections. The protocols went beyond requiring the National Association of Sewer Service Companies' (NASSCO) PACP and MACP (Pipeline Assessment Certification Program and Manhole Assessment Certification Program, respectively) defect coding formats and included listing additional information and photos/videos to collect as well as procedures for transferring digital records and performing quality control. These protocols were necessary to confirm that the field contractors would provideg complete inspection data for DWM analysis. The following are example protocols that were used:

- **Smoke** testing protocols detailed how contractors used the ESRI collector application to collect smoke information. The collector forms included mandatory fields such as gravity main ID, diameter and smoke source.
- **Acoustic** testing protocols detailed the acoustic process for pipes with diameters below 38cm (15 inch) and required gravity main identification and length.
- **CCTV** inspection protocols detailed the key data requirements, for instance, gravity main identification, diameter, invert elevation, etc.

The protocols also described how contractors would collect and manage photos and videos. For example, snapshots of all significant defects were required to be taken and named using a smart naming convention. The protocols also detailed how the data was to be transferred to DWM. For example, the contractors delivered hard drives with both the databases and media files weekly.

Quality Control Procedures

Quality control (QC)_was an essential component of the data collection process. To encourage the field contractors to meet established data standards the contractor submittals would only be approved when the quality control was properly and successfully performed using tools developed both for the contractor and DWM's program management team. A quality control tool was developed for each inspection technique. With the help of an easy user interface, contractors imported their electronic submittals into the Microsoft-Access based tool. Figure 4 provides a screen capture of the tool's interface page. Contractors ran QC queries to check for a variety of potential problems, from unusual manhole depths to missing photos or data. The quality control checks helped contractors and DWM's program managers easily target problem inspections and make corrections. Additional quality control was performed by DWM through review of generally 10 to 15 percent of the randomly selected inspections.

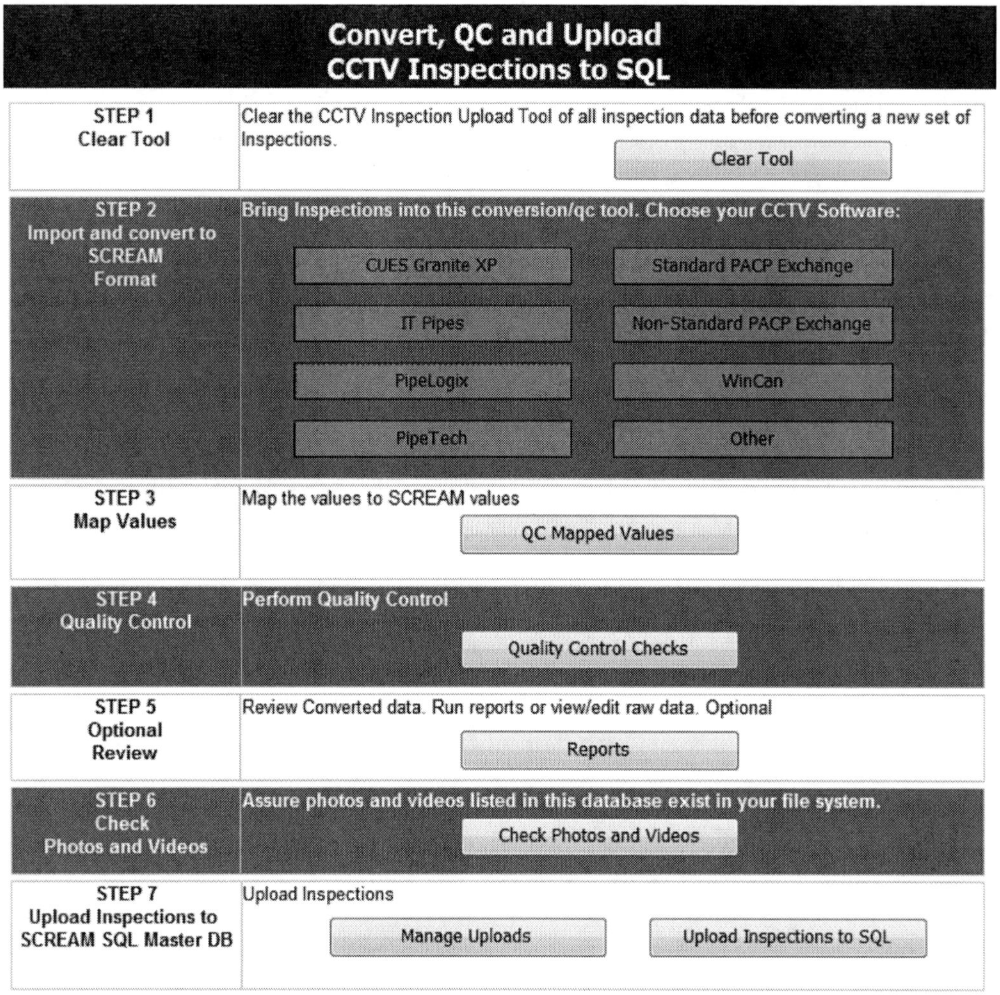

Figure 4: Example quality control interface

To confirm that the contractor performed the QC checks and resolved any important errors, DWM required the QC reports with any data submittal. Each of the contractors were given classroom training with hands-on case studies of how to handle the inspection data. DWM's emphasis on quality control has increased the quality of the field inspection data, as evidenced through less time spent on data issues during monthly progress meetings and fewer formal data submittal rejections.

Integrated data analysis methodology

The automated, integrated field-to-recommendation applications helped leverage the value of the inspection data more quickly, minimized data errors, prioritized rehabilitation projects, and save costs by more appropriately targeting contractor activities and the assets needing rehab.

The decision logic was automated through the use of several integrated systems. Figure 5 displays how data flows from the field through the quality control process to analysis using various software. First, CH2M's SQL Server-based System Condition Risk-Enhanced Assessment Model (SCREAM) processes all of the QC'd inspections through an automated upload procedure in the QC tool. More detailed discussions of SCREAM have been previously published (for example, Martin, L; Rowe, R; 2008). SCREAM produces condition scores for each inspection type as well as a combined condition scores, representing all of the inspection techniques used on each asset. For instance, smoke testing and CCTV scores were combined if performed on the same asset. The SCREAM scores were easily shared with DWM's InfoMaster software (Innovyze). InfoMaster's logic tables were customized to approximate SCREAM Next Step's logic, evaluating each asset's condition scores and risk consequence of failure (COF) score as well as inspection details, hydraulic capacity needs, and asset attributes to determine the next logical step for the asset.

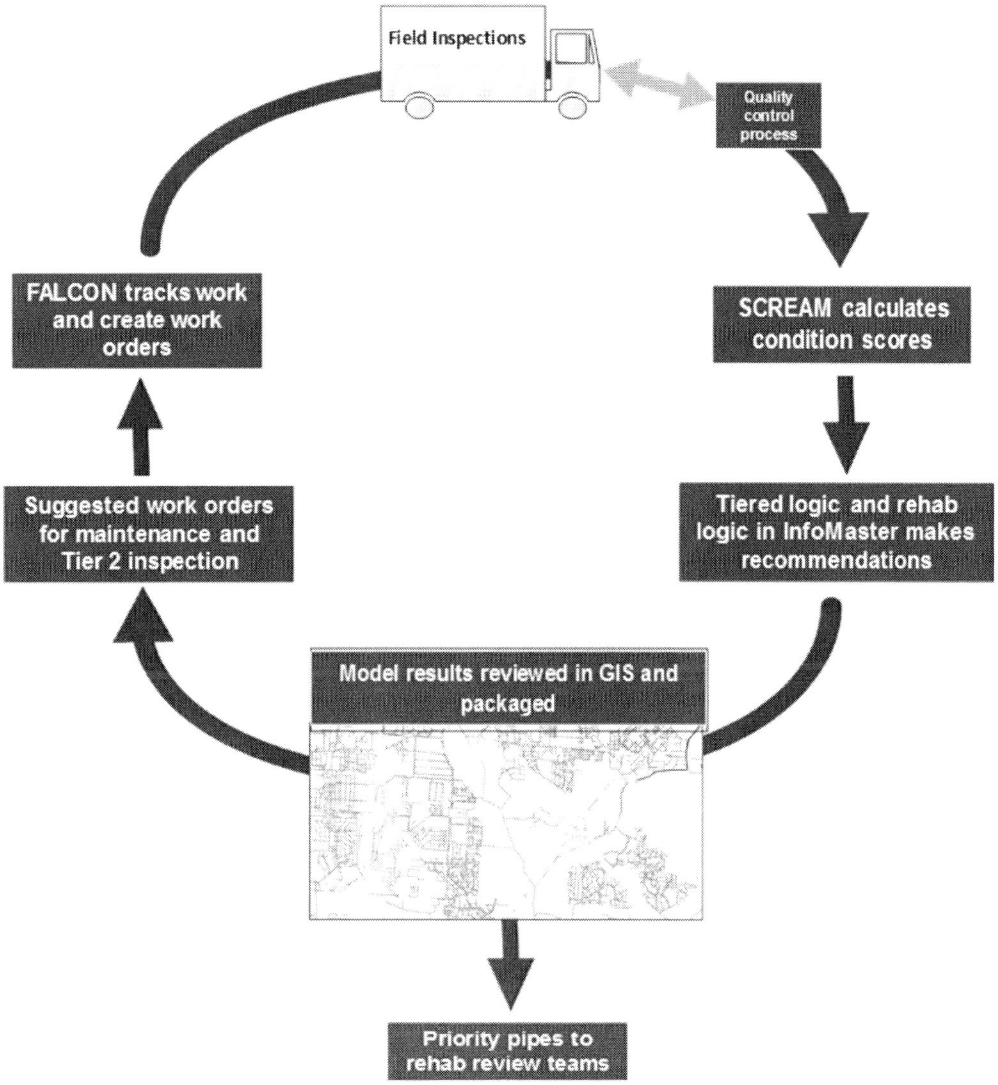

Figure 5: Data Flow Sequence through Software Applications

InfoMaster created a prioritized list of assets needing either CCTV inspection, emergency action, cleaning or priority rehab/repair work. DWM then used a custom workorder management system, FALCON, produced by CH2M as an interim tool until DWM's Cityworks computer maintenance management system could be installed. FALCON was used to track priority assets, and to help project managers create work orders to track contractor's work, invoices, performance metrics, and payments.

Decision Logic Flow Diagrams

DWM used two decision flow diagrams to make decisions about the tiered inspection approach and about which type of rehab was needed. The tiered flow diagram, shown in Figure 6, was used as assets were inspected. The logic determined which inspection technique to use. As the assets were inspected, condition scores were calculated for each of the inspection technologies. Additionally, combined scores were calculated to represent a blend of all the inspection scores. The tiered logic included pre-defined limits to the scores.

First tier inspections were quick inspections, using smoke and acoustic testing to target assets in need of more thorough inspection techniques. Second tier inspections were more intensive, using CCTV and manhole inspections. Assets that met certain criteria, were sent to DWM's rehab logic, Figure 7. Pipes entering the rehab logic potentially needing repair, rehabilitation or replacement. Included with the scoring criteria and decision logic were other factors like emergency needs that might be observed during condition inspection work and capacity upsizing requirements that were separately identified by DWM's hydraulic model runs.

The logic was created in InfoMaster and was designed to be similar to CH2M's SCREAM logic. The recommended corrective actions, coming out of the rehab logic were provided to the rehab teams as they begin their review.

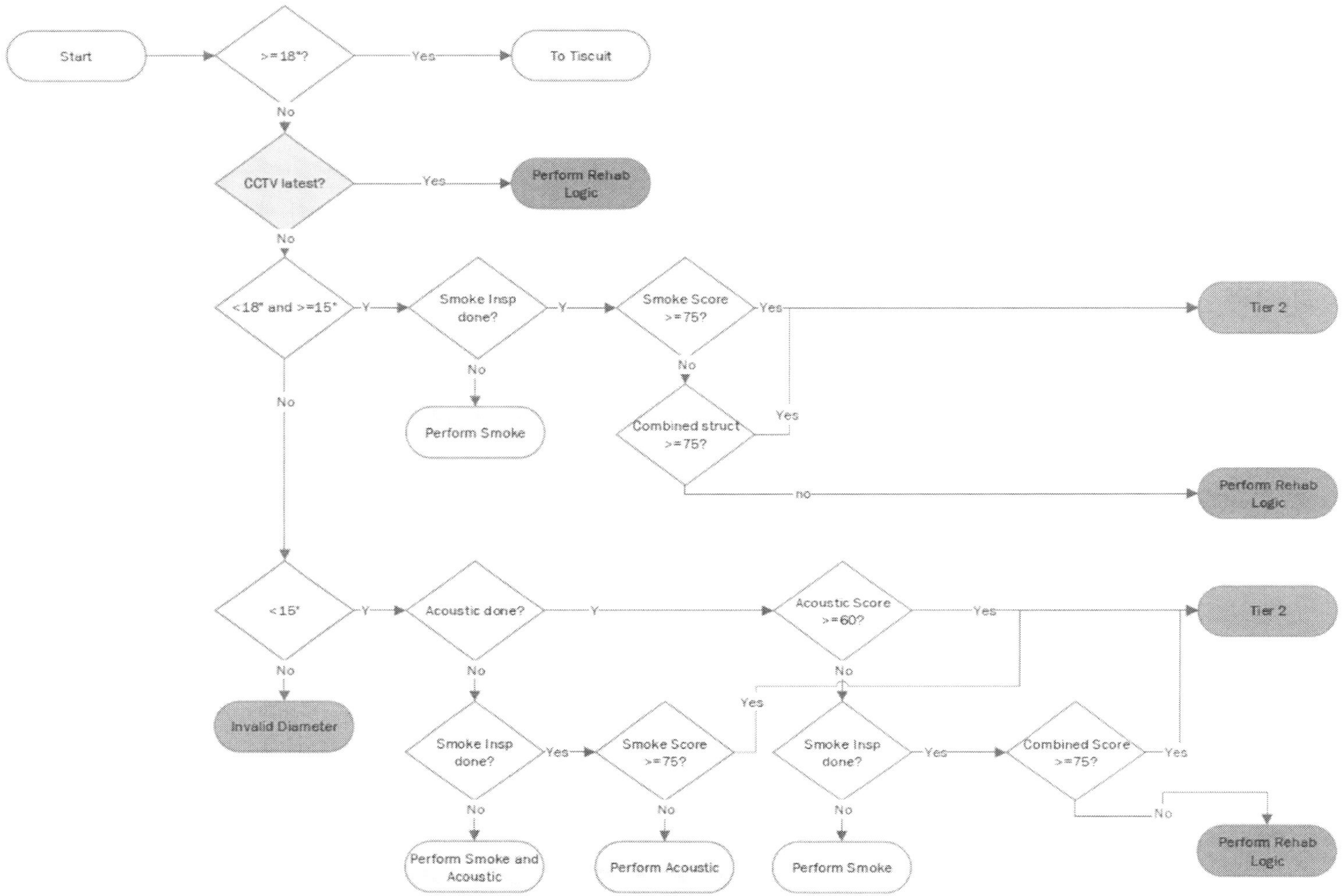

Figure 6: DWM's Tiered Decision Logic

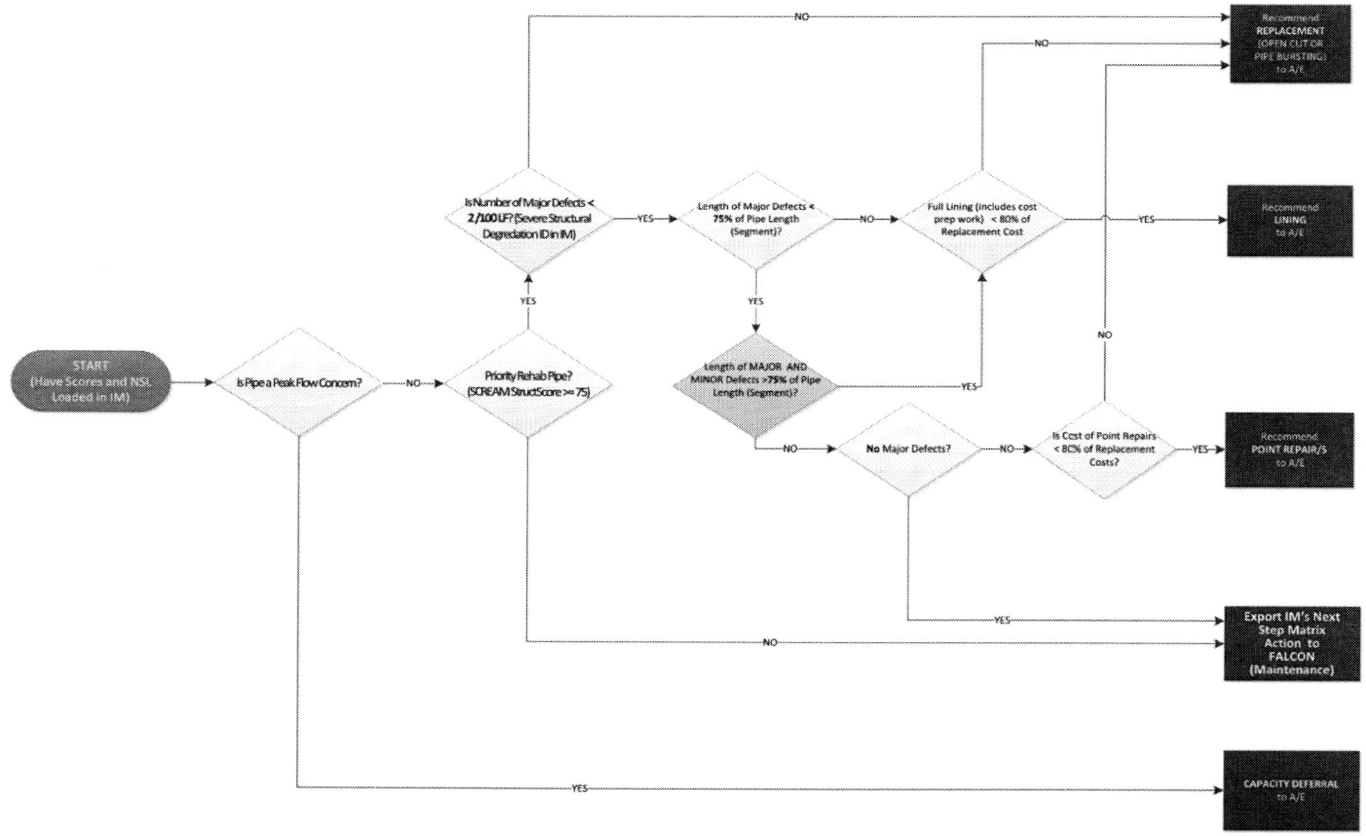

Figure 7: DWM's Rehab logic

STATUS OF WORK

DWM is approximately halfway through the assessment period which ends in December 2017. Approximately 55% of the priority system has been inspected. Rehab packaging and assignments to the design-build contractor have just initiated. The data submittal process has been progressively shortened and the contractor's submittals are cleaner with fewer QC problems as the contractors were encouraged to focus more on the importance of data quality. To date, DWM has collected 10 Terabytes of media and 300 GB of inspection data. Processing this information by hand is not an option. DWM's automated tools have made processing the data with a small staff possible.

Results of Tier 1 (Acoustic & Smoke Inspections) Assessment Leading to Tier 2 (CCTV):

To date DWM has performed a total of 17,832 Tier 1 assessments on gravity mains within the PASARP Areas. As Figure 7 illustrates, from the Tier 1 assessments there were 2,626 (15%) CCTV Inspection work orders that were issued and/or performed on the gravity mains. The leading factor for mains needing CCTV is the Acoustic score given by SCREAM to each main based on the inspection. SCREAM's scoring range is from 1 to 100 where 100 is the worst condition.

The breakdown on the factors leading up to CCTV inspection were 1991 (76%) based on an Acoustic score of 60 or more, 32 (1%) were based on a Smoke score of 75 or more and 603 (23%) were based on Combined Score (both Acoustic and Smoke) of 75 or more. These thresholds were agreed upon in early planning meetings.

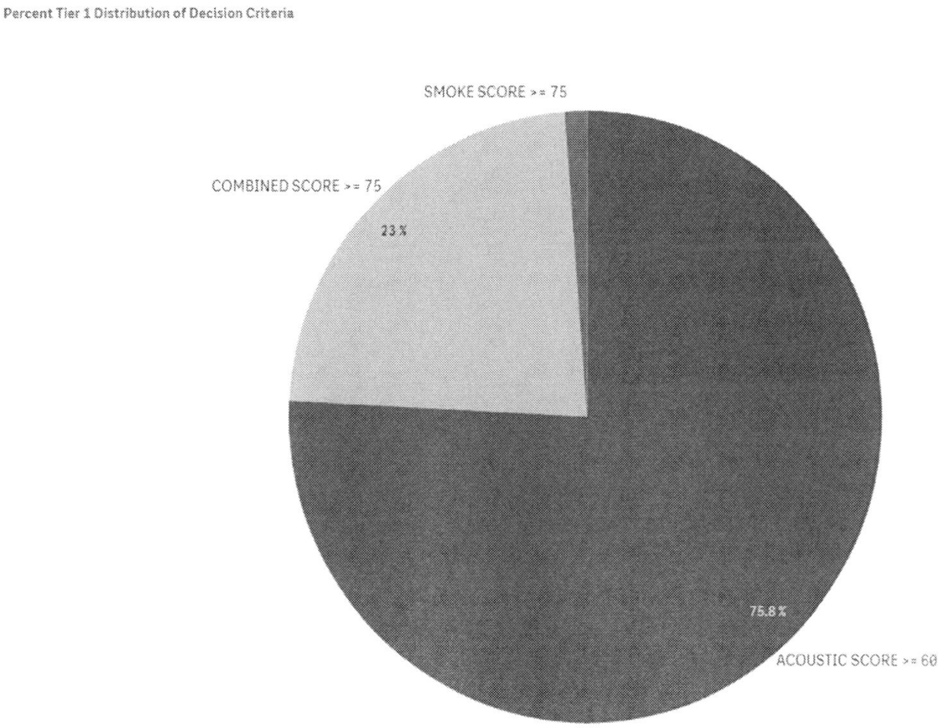

Figure 7: Percent Tier 1 Leading to Tier 2 Inspections

Result of Tier 2 (CCTV) Assessment Leading to Rehabilitation:

To date DWM has performed approximately 14,600 Tier 2 assessments on gravity mains within the both PASARP and OSARP Areas. The results of these assessments were classified into three categories as illustrated in Figure 8; Monitor (Re-CCTV between 2 years to 10 years), Assess (In-Complete Assessments based on Obstacle(s) in Pipe or Re-CCTV less the 2 years), or Rehab (Replace, Lining, or Point Repair). A breakdown of the Tier 2 inspection scoring results have been 9,523 (65%) as Monitor; 795 (5%) as Assess; and 4,298 (30%) as Rehab. The high trending percent needing Rehab might be due to the fact that DWM's work has been pre-screened and directed towards the parts of the sewer system which were prioritized to be of greater concern.

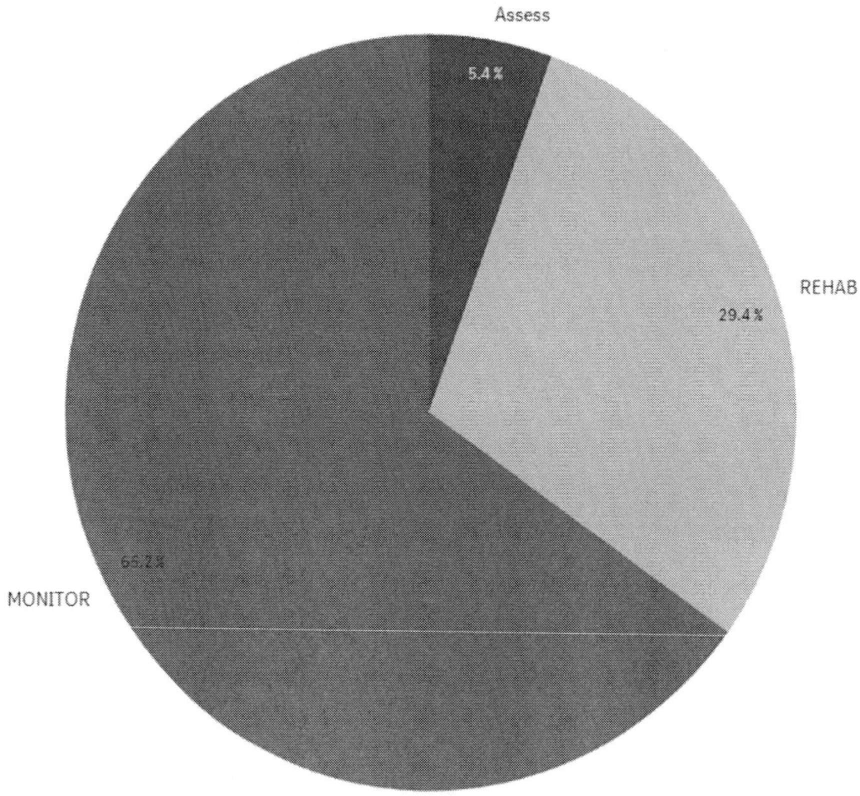

Figure 8: Breakdown of CCTV Scoring Results.

BENEFITS AND SIGNIFACANCE OF ROBUST DATA PLANNING AND IMPLEMENTATION

Using the steps and processes described above, DWM automatically generated prioritized lists of assets needing either further inspection or corrective actions, which were confirmed by Rehab Review Teams prior to assigning to the design-build contractor. DWM realized the value of automating as many of the condition assessment business process steps as possible in order to efficiently process large quantities of condition assessment data and prioritize corrective actions.

Large or small, other utilities can also integrate and automate their own software applications and platforms within their organization. What is common to DWM and other utilities, however, is the planning and design of how the condition assessment field data is applied and manipulated throughout the organization to more efficiently perform work. Processing the vast quantities of data and optimizing the rehab packaging according to geospatial proximity within the consent decree timeline would not have been possible without DWMs automated tools.

REFERENCES

Martin, L.; Rowe, R. (2008) *Sanitation District No. 1 of Northern Kentucky Selects gbaMS and CH2M HILL to Integrate Continuous Sewer Assessment Program Data*; gbaMS Annual Users Conference, Atlanta, Georgia; September.

Update on the City of Omaha's CSO Long-Term Control Plan

Jim Theiler, PE, City of Omaha, Nebraska
Tom Heinemann, PE, CH2M

ABSTRACT

The City of Omaha, Nebraska, is in the process of implementing their Long Term Control Plan (LTCP) as part of the City's *Clean Solutions for Omaha* – a major Combined Sewer Overflow (CSO) Program to bring the City into compliance with the requirements of the U.S. Environmental Protection Agency CSO Control Policy of 1994 (EPA CSO Control Policy). Taking a holistic and efficient approach, the City is looking at its CSO Program as an investment to improve water quality while achieving multiple community benefits. The LTCP must be implemented in a cost-effective manner that does not burden ratepayers. Omaha is not alone in this effort and is one of approximately 800 communities nationwide with combined sewer systems that must reach compliance with the EPA CSO Control Policy. The purpose of this paper is to provide general background on the development of the LTCP, an update on the implementation of the LTCP, and information on how those responsible for the implementation of the CSO Program utilize adaptive management along with a strong working relationship with the Nebraska Department of Environmental Quality (NDEQ) to keep the CSO Program on track in spite of the numerous challenges that have been encountered.

KEYWORDS

NDEQ, EPA, Omaha, combined sewer overflow (CSO), combined sewer system (CSS), Missouri River Wastewater Treatment Plant (MRWWTP), Consent Order, Missouri River, Papillion Creek, Long Term Control Plan (LTCP), retention treatment basin (RTB), Deep Tunnel System (DTS), affordability

INTRODUCTION

The City of Omaha, Nebraska, is located on the eastern border of Nebraska. The Missouri River is the eastern boundary of the City, with Council Bluffs, Iowa, located across the river to the east. The population of the wastewater service area is approximately 650,000. The City operates over 1,368 kilometers (km) (850 miles) of combined sewers in the eastern part of the City. The combined sewer service area is 11,655 hectares (45 square miles) in area, out of a total sewer service area of 86,247 hectares (333 square miles).

When the City's first CSO Permit was issued in 2002, there were 29 permitted CSO outfalls. Over the past 8 years, a net of three CSO outfalls have been deactivated. Another ten CSO outfalls are planned to be deactivated before completion of the LTCP in 2027. This will result in 16 remaining CSOs, overflowing at a greatly reduced frequency, volume, and duration to meet the requirements of the EPA CSO Control Policy. CSO outfall locations are shown in Figure 1.

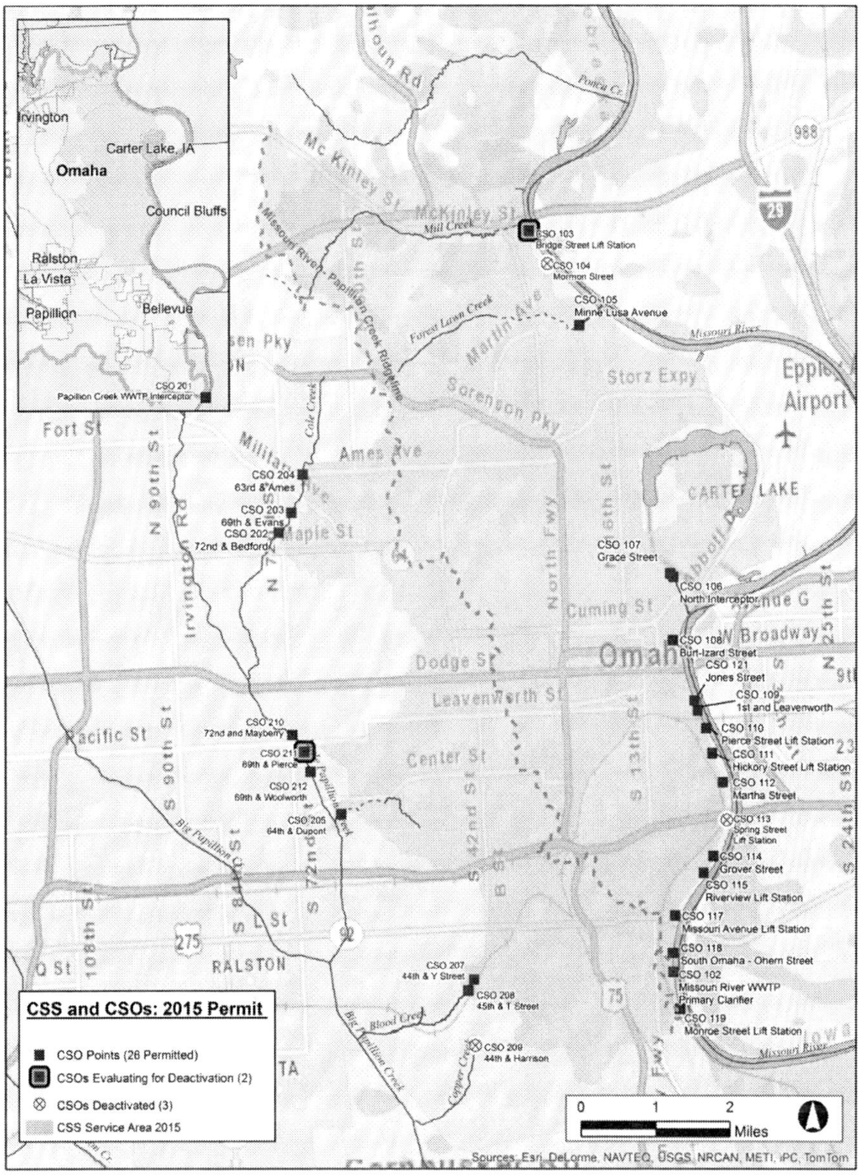

Figure 1. CSO Outfall Locations

Under "Existing Conditions," which represent the combined sewer system (CSS) as it was in 2002, it was estimated in 2007 that an average of 12,298,000 m^3 (3.25 billion gallons) per year of combined sewage overflows to receiving streams during approximately 70 CSO events. Of this volume, 8,892,000 m^3 (2.35 billion gallons) was estimated to overflow to the Missouri River and 3,406,000 m^3 (0.90 billion gallons) was estimated to overflow to the Papillion Creek system. The main pollutant of concern is *E. coli* bacteria.

In 2005, the Nebraska Department of Environmental Quality (NDEQ) notified the City in a letter of the requirement to submit a "Substantively Complete" LTCP by October 2007 and a Final

LTCP by October 2009, and to construct the selected controls by 2024. NDEQ also indicated that these dates would be included under an "enforcement mechanism" such as a Consent Order. Shortly after receiving the NDEQ letter, the City selected CH2M to serve as the Program Manager, in association with HDR Engineering and Lamp Rynearson & Associates, to develop a CSO LTCP. These firms formed the Program Management Team (PMT). The City divided its CSS into ten drainage basins, as shown in Figure 2, and selected a "Basin Consultant" for each basin to work with the City and PMT in evaluating and selecting CSO controls.

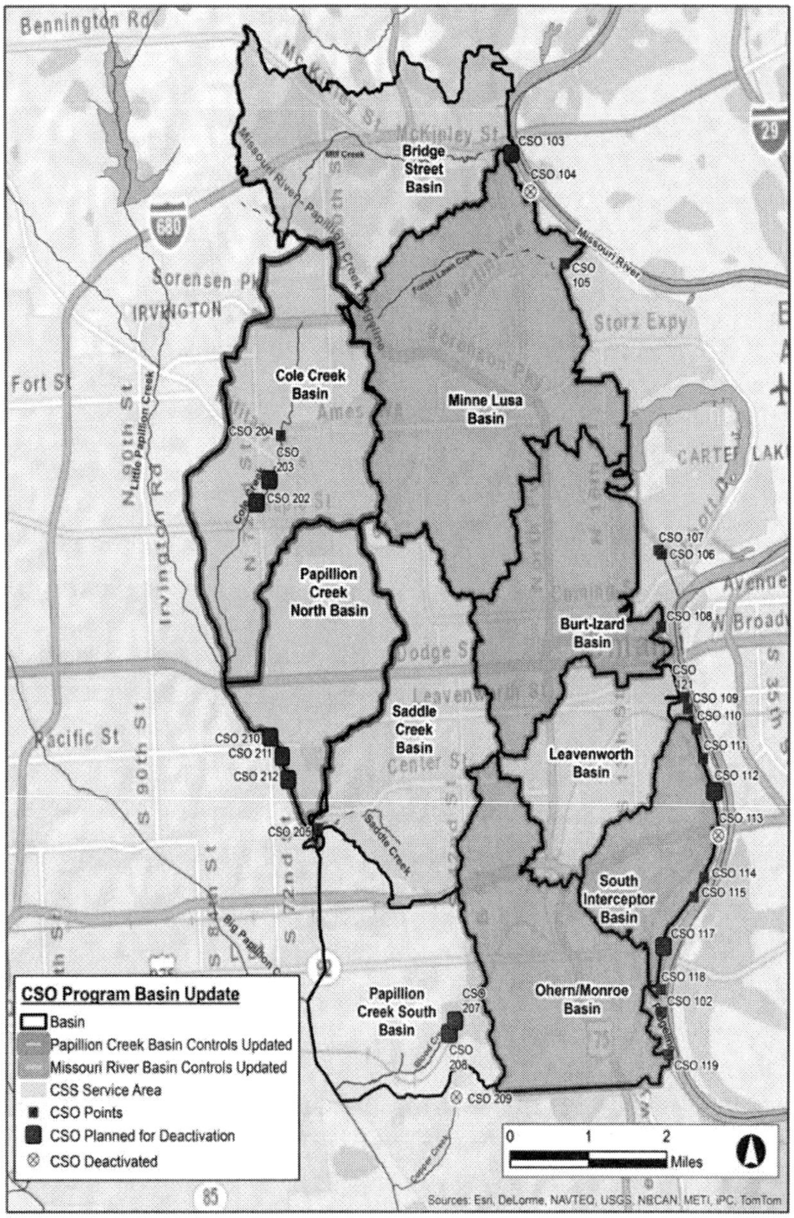

Figure 2. Combined Sewer System Drainage Basins

The resulting LTCP, completed in 2009 and approved by NDEQ in early 2010, identified 92 individual projects that would need to be completed by 2024, at a cost of approximately $1.66 billion in April 2009 dollars. The following components were included:

- Improvements to the Missouri River Wastewater Treatment Plant's (MRWWTP) treatment and delivery system to allow treatment of up to 567,800 cubic meters per day (m^3/day) (150 million gallons per day [mgd]) during wet weather.

- Extensive sewer separation and inflow reduction to decrease the flow of stormwater into the CSS.

- Two retention treatment basins (RTB) to provide high-rate treatment of combined sewage during wet weather.

- Two storage tanks.

- A Deep Tunnel System (DTS) that included a deep rock tunnel with a diameter of 5.2 meters (17 feet) and a length of 8.7 km (5.4 miles).

Implementation of the LTCP began in 2009 when design and construction of projects started in accordance with the schedule contained in the LTCP. The LTCP is not being implemented in accordance with the requirements of a federal Consent Decree, which many other CSO cities have. Instead, the City worked with NDEQ to develop an administrative Consent Order with the State of Nebraska that was signed in 2007. While providing NDEQ with the assurance that the City will implement the CSO controls, the Consent Order also provides flexibility for the City to adapt to changes as they inevitably occur. A key goal of the City in implementing the LTCP is to maintain their flexible Consent Order.

The City's CSO Permit and Consent Order required the City to update the LTCP by October 1, 2014. The initial effort by the City and PMT to update the LTCP was increased in 2011 when former Omaha Mayor Suttle asked the PMT to evaluate the potential of eliminating the CSO Deep Tunnel from the LTCP to reduce the cost of the Program. Current Mayor Stothert, who took office in 2013, challenged the PMT to work on project designs and evaluate technological solutions that are more cost effective, with the goal of reducing the overall cost of the CSO Program. The LTCP Update report, completed in October 2014, summarized efforts to improve the 2009 LTCP and reduce costs. The most significant change that was made to the CSO controls in the LTCP Update was to plan for storage tanks at CSOs 118 and 119 rather than including those CSOs in the DTS. As a result, the Deep Tunnel was shortened, the diameter was reduced from 5.2 meters (17 feet) to 4.6 meters (15 feet), and the capacity of the RTB was reduced from 196,800 to 83,280 m^3/day (52 to 22 mgd).

Figure 3 illustrates the main project categories of the LTCP following the 2014 Update and their proportional costs. The escalated cost of the CSO Program in the LTCP Update was $2.08 billion in December 2013 dollars. The LTCP Update provides for an overall volume capture of combined sewage that is comparable to the 2009 LTCP; however, it relies on volume capture as a means of compliance rather than on four CSOs. While providing for a volume capture in excess of that required by the EPA CSO Control Policy, it is predicted that the LTCP Update controls would result in no more than eight CSOs at any CSO outfall. In addition, the Program's water quality model predicts that the Missouri River downstream of the CSO outfalls will meet the water quality standard for *E. coli* after implementation of the LTCP. The CSO Program is sometimes referred to by the public as the "sewer separation program." While sewer separation is a significant component of the LTCP, only 18 percent of the CSS would be separated under the LTCP.

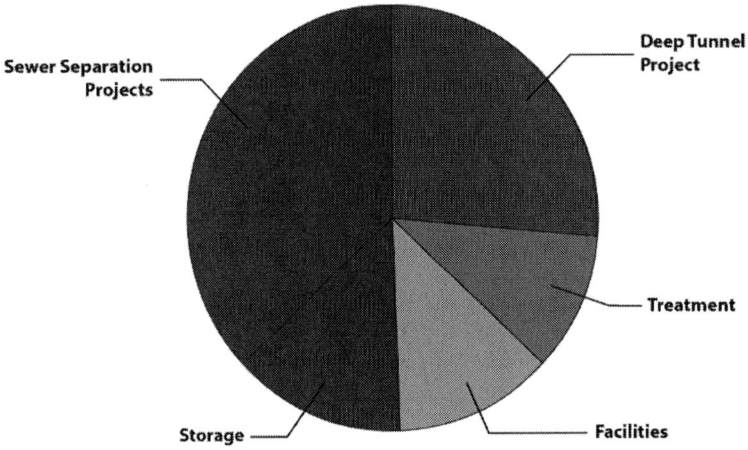

Figure 3. Cost Proportions for CSO Program Project Categories

Finalization of the LTCP Update did not mark the end of the City's efforts to reduce CSO Program costs. In addition, it is recognized that the LTCP is a plan, not a design document, and many of the factors that the LTCP is based on will continue to change over time. For these reasons, the City continues to use an adaptive management strategy for implementing the LTCP. This strategy is illustrated in Figure 4.

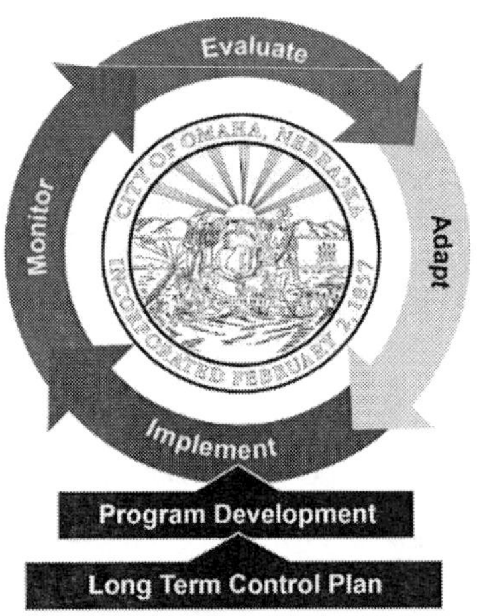

Figure 4. Adaptive Management Strategy for Implementing the LTCP

The LTCP Update notes the following specific areas of evaluation in the near term to further adapt the LTCP:

- Continued development and refinement of the City's computer model of the CSS.
- Green Infrastructure.
- Collection System Optimization.
- Operation, assessment, and optimization of the new wet weather treatment facilities at the MRWWTP and Saddle Creek CSO 205.
- Assessment of performance of implemented controls.
- Conceptual designs of storage tanks, and DTS.
- Assessment of affordability given the demands of other portions of the City's collection and treatment system.

The overall mission of the CSO Program is to meet the requirements of the EPA CSO Control Policy, and ultimately the Clean Water Act, in a fiscally responsible manner for the ratepayers and the community that is served. The Program continues to be executed by an integrated team comprising the City and PMT. The PMT firms currently consist of CH2M, HDR, and the City's Public Participation Facilitator (Lovgren Marketing Group). This integrated team works together to provide the necessary oversight for CSO Program implementation. More specifically, the purpose of this integrated team is to:

- Maintain consistency among the various CSO projects.
- Ensure that the individual CSO projects are not only quality projects themselves, but work together to meet the ultimate goals of the Program.
- Continually seek ways to achieve compliance at a reduced cost.
- Adapt the LTCP as needed to respond to changing conditions.

CURRENT STATUS OF THE CSO PROGRAM

Through adaptive management efforts, the LTCP has evolved, with the result that the original 92 LTCP projects have been reduced to a total of 74. Projects have been either removed, combined, or pooled. For example:

- Several sewer separation projects in the Burt Izard Basin have been removed from the LTCP because of their minimal direct benefit to CSO reduction; they may be eventually constructed to increase the level of service. Other funding sources might be used if these projects are constructed.

- Projects that originally had multiple phases have been accomplished under just the first phase, resulting in less overall construction work being necessary.

- Several rehabilitation-phase sewer separation projects have been pooled into a single Inflow Reduction Budget to allow optimized use of this funding for reduction of inflow into the CSS where needed.

Of the remaining 74 LTCP projects, as of the end of February 2017, 19 have been completed; 15 are under design, bidding, or construction; and 13 are on hold pending a technical re-evaluation and cost reduction efforts being conducted by the City and PMT in the Saddle Creek and Minne Lusa Basins. For a LTCP project to be considered complete, facility projects must be operationally complete (i.e., ready to be placed into service), and sewer separation projects must be substantially complete. The 74 LTCP projects have been further broken into 95 separate construction contracts.

The current Total Program Budget is $2.06 billion in June 2015 dollars, with a total of $512 million (27.4 percent) already been spent. In terms of construction value, the City has paid out $316 million, and:

- $240 million is budgeted for projects that are currently under bid or construction.

- $88 million is budgeted for projects that are currently under design.

- $319 million is budgeted for projects that are currently on hold and being re-evaluated.

The benefit of the work that has been done is significant. As shown in Figure 5, the City's collection system and water quality models estimate that *E. coli* loading from CSOs has been reduced by 26 percent, or 29 percent of the ultimate 90-percent reduction Program goal. The percent reduction in loading will continue to increase rapidly over the next several years as projects are completed. By approximately 2023, bacterial loading will have been reduced by an estimated two-thirds. At the same time, about one-half of the Program's budget will have been spent.

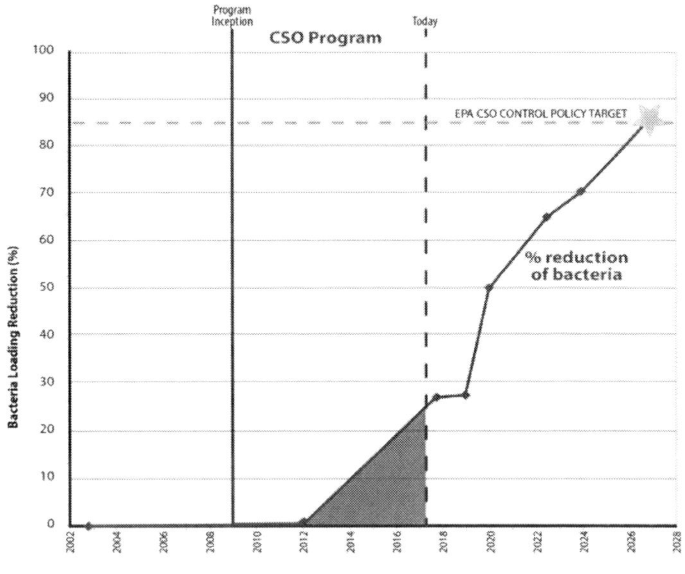

Figure 5. Projected Reduction in Bacterial Loading over Time

Following are examples of notable facility and sewer separation projects that have been completed or are under construction as part of the CSO Program.

MRWWTP Improvements Project

The MRWWTP Improvements Project was originally planned to be completed as a single design and construction project. However, effects of the 2011 Missouri River Flood resulted in it being split into three phases, named Schedules A, B1, and B2. This major project was included in the LTCP as an early-action project because of the significant reductions in the size and number of untreated overflows and the resulting *E. coli* loading to the Missouri River. The MRWWTP Improvements Project, along with the associated projects that are required to feed wet weather flows to the plant (listed later), will result in an estimated 52-percent reduction in the loading of *E. coli* to the Missouri River. Completion of Schedule A alone is estimated to reduce the loading by 26 percent.

Schedule A includes separate primary treatment of the high-strength wastewater from meat packing industries in the South Omaha Industrial Area (SOIA), and modifications to the plant's primary effluent lift station so that the high-strength wastewater can go directly to the secondary treatment system. The end result of the project is that this wastewater will no longer be present in CSOs 102, 118, and 119 along the Missouri River. Instead, it will go directly to secondary treatment. Schedule A was substantially complete in September 2014 at a cost of approximately $20 million.

Schedule B1 modified the plant to enable it to receive a peak-hour flow of 567,800 m^3/day (150 mgd) during wet weather and to send up to 290,950 m^3/day (64 mgd) of flow to secondary treatment. Schedule B1 included construction of a new headworks facility and modifications to preliminary and primary treatment facilities. The flow that is not sent to secondary treatment will receive primary treatment and disinfection prior to discharge through CSO 102. Disinfection facilities are being constructed as part of Schedule B2, as noted below. Due to the presence of significant concentrations of ammonia in the wastewater entering the plant at the beginning of wet weather events, an operational scheme needed to be developed and implemented whereby the plant's primary clarifiers are used for storage and dilution so that the concentration of ammonia entering the disinfection system is below 6 mg/L. This will provide for more effective chlorine disinfection. Schedule B1 was substantially complete in August 2016 at a cost of approximately $62 million.

Schedule B2 is currently under construction and will be completed in 2019. The bid award amount is approximately $52 million. Schedule B2 provides the disinfection facilities noted in the previous paragraph for CSO 102, along with a large odor control facility for primary clarifiers.

To allow the MRWWTP to treat flows as described above, several other significant LTCP projects had to be designed and constructed. They include the following:

- Leavenworth Lift Station Replacement Project. This project replaces an aging lift station and increases pumping capacity. It also allowed for the abandonment of two smaller lift stations, Pierce Street Lift Station and Hickory Street Lift Station. The new lift station has two wet wells, one for dry weather and one for wet weather. The dry weather pumping capacity is 34,096 m^3/day (7.5 mgd), and the wet weather pumping capacity is 204,574 m^3/day (45 mgd). This project is finished, but the lift station will not be placed into service until the new South Interceptor Force Main (SIFM) is completed.

- South Interceptor Force Main Project. The existing SIFM was constructed in the early 1960s and has remained in continuous operation since then. The current condition of the SIFM makes it unreliable for continued long-term use, particularly at the higher flow rates that need to be delivered to the MRWWTP during wet weather under the LTCP concept. A single SIFM project was envisioned in the LTCP; however, as with the MRWWTP Improvements Project, the 2011 Missouri River Flood caused the project to be broken into three design packages and, ultimately, two construction packages. The South and Central Segments of the SIFM include approximately 5,610 meters (18,400 feet) of 163-centimeter (64-inch)-diameter pipe from the new Leavenworth Lift Station south to the MRWWTP. The North Segment includes approximately 1,340 meters (4,400 feet) of 122-centimeter (48-inch)-diameter pipe from a connection point near the I-480 Bridge over the Missouri River to the new Leavenworth Lift Station. As described elsewhere in this paper, construction of the North Segment has met challenges; however, the force main should be placed into operation in early 2018.

- Ohern/Monroe Industrial Flow Area Sewer Separation. This early project removed industrial discharges from the CSS and directed them to the site of the new Industrial Lift Station, described below. This was a critical step in removing high-strength industrial wastewater from CSOs 118 and 119 during wet weather events.

- Ohern/Monroe Industrial Lift Station, Force Main, and Gravity Sewer. The new lift station provides pumping of separated high-strength flows directly to the MRWWTP through the SOIA Force Main and Gravity Sewer. The peak pumping capacity is 65,870 m^3/day (17.4 mgd). The lift station includes screens, but grit removal was constructed at the MRWWTP as part of Schedule A to reduce potential odor issues at the lift station.

- Monroe Street Lift Station. This existing lift station located south of the MRWWTP pumps wastewater from the Monroe Basin to the plant. As part of the LTCP's vision of maximizing wet weather treatment at the MRWWTP, this lift station will be rehabilitated and its capacity increased. The City is currently in the process of hiring a design consultant for this project.

Missouri Avenue (CSO 117) Sewer Separation

This significant sewer separation project is being designed and constructed in two phases. Phase 1 has been constructed, and Phase 2 is currently under design. The overall project will provide sewer separation to the 169-hectare (416-acre) Missouri Avenue sub-basin through a combination of new storm and sanitary sewers. Sanitary flows will be directed to an existing lift station located along the Missouri River, north of the MRWWTP. Stormwater will be conveyed to the river through the existing combined sewer, which will eventually be converted to a stormwater-only sewer. Phase 1 included construction of a stormwater treatment train within Spring Lake Park, including wetlands and a multi-use pond at the downstream end to provide detention of stormwater runoff, reducing downstream stormwater flow rates. The reduction in flow rate saved approximately $5 million by reducing the size of downstream sewers. The green elements making up the stormwater system in the park provide wildlife habitat diversity and are a significant benefit to the neighborhood. The pond replaces a lake that was originally established in the 1870s and was drained in the 1930s.

The project, which has received several awards, was completed through a partnership with the Nebraska Environmental Trust; Omaha Parks, Recreation, and Public Property; Keep Omaha Beautiful; the Spring Lake Park Neighborhood Association; and the Spring Lake Park Habitat Restoration and Preservation Team.

Other LTCP projects that incorporated significant cost-effective Green Infrastructure components include those within Adams Park, Elmwood Park, Fontenelle Park, and Westlawn Cemetery. Overall cost reduction is estimated to be over $15 million.

Current Situation

Nearing completion of the first 8 years out of the 18-year LTCP implementation period, the CSO Program has accomplished a lot, but it is now at a critical juncture where progress is being assessed and plans for the remainder of the Program are being refined. The remainder of this paper summarizes some of the challenges and changed conditions that have been addressed during implementation and describes a few that are being dealt with proactively now. A few lessons learned, and the future direction of the CSO Program as currently envisioned, are also described. The City retains a strong commitment to completing the CSO Program and to maintaining a good working relationship with NDEQ and EPA.

CHALLENGES AND CHANGED CONDITIONS

When implementation of the LTCP began in 2009, the City realized that the plan would need to be adapted to meet various challenges and changed conditions. This has indeed been the case. A few examples of events that have been successfully addressed thus far are provided below.

2011 Missouri River Flood

In 2011, the Missouri River experienced an historic flooding event, with the river reaching levels more than 1.8 meters (6 feet) higher than any flood elevation since construction of the mainstream dams along the river. Figure 6 is a photograph of a statue near Lewis & Clark Landing that is normally well above the river's surface level. The river was at flood stage for 104 days. The peak flow rate was measured at 6,120 m^3/s (216,000 cubic feet per second [cfs]) versus a prior record flow of 3,400 m^3/s (120,000 cfs) in 1960, and a summer average flow of 991 m^3/s (35,000 cfs).

Figure 6. 2011 Missouri River Flood

The 2011 flood resulted in significant impacts on the design and construction schedules of several projects. It was declared a *force majeure* event, and on May 12, 2012, a 3-year extension to the LTCP implementation period was granted to the City by NDEQ, with an amended Consent Order revising the completion date from October 2024 to October 2027. Due to the City's flexible Consent Order, this extension was negotiated between the City and NDEQ in a straightforward manner, without the need for action by federal courts.

The 2011 flood continued to impact LTCP projects well past the actual event. For example, in 2013, unstable river bank conditions along the site of the MRWWTP became apparent, as shown in Figure 7. These unstable conditions threatened the ability to construct new facilities called for in the LTCP and threatened existing infrastructure at the plant. Design and construction of a river bank stabilization project were required to mitigate these impacts.

Figure 7. Unstable River Bank Conditions

Program Optimization Review

In communicating with the public about the cost and scope of the Omaha CSO Program, it has often been referred to as an "unfunded federal mandate" that carries with it a significant cost to the ratepayers. Concerns have increasingly been expressed by elected officials and some stakeholders about the high cost of the Program and its impacts on industrial, commercial, and residential ratepayers. In 2016, the City retained the services of a third-party consultant to perform a CSO Program Optimization Review, working with City staff and the PMT to identify opportunities for cost savings and efficiencies.

The review was done in two phases. Phase 1 was a general assessment of the CSO Program. Program-related documents were reviewed, and a series of meetings and workshops were conducted with key staff members from the PMT and Public Works Department. The Program's work was divided into three fundamental areas: Regulatory Strategy, Planning/LTCP Optimization, and Program Delivery. Phase 2 consisted of two intensive workshops to develop strategies and tactics to optimize the CSO Program, and a subsequent session to review concepts for cost savings developed by the City and PMT, with the intent of identifying a specific cost-savings target.

The CSO Program Optimization Review resulted in an overall endorsement of the work being done by the City and PMT, along with recommendations for further optimization with the intention of reducing cost and increasing efficiency. The City and PMT are currently working on a number of action items that are based on the strategies and tactics identified in Phase 2 of the review. Most significantly, a Technical Assessment for Cost Savings (TACS) Task is being performed to evaluate alternatives for achieving the cost savings goal that was identified as part of the Assessment. The goal is to save 10 to 20 percent of the remaining $1.66-billion cost of the Program. Work on this Task will be completed this year.

Economic Inclusion

The City of Omaha's Economic and Inclusion Mission is:

> "To decrease unemployment and poverty and increase access and opportunity in economically distressed areas of the City"

The size of the CSO Program and location of many of the CSO projects has resulted in the Program being a focal point during implementation for economic inclusion efforts in the Omaha Metro area. A significant inclusion achievement early in the Program came with CH2M's hiring of KAI Design & Build, a minority-owned firm headquartered in St. Louis, to serve as a subconsultant for construction management work associated with the MRWWTP Improvements Project, Schedule B1. Proposals were received in April 2013 from three minority-owned firms in response to a Request for Proposals (RFP) issued by CH2M. An important aspect of KAI's successful proposal was their plans for benefitting the Omaha community by their presence, including opening an Omaha office and hiring locally for their positions. KAI has provided a full-time Quality Assurance Representative (QAR) and a full-time Administrative Assistant (AA), both hired locally. KAI's performance was excellent on Schedule B1, and they are continuing with their work on the $50-million Schedule B2 project, which will be completed in 2019. KAI is also providing "Value-Added Services," working with the PMT to enhance economic inclusion and diversity on the CSO Program.

In June 2014, the PMT was challenged by members of City Council to develop a formal Diversity Action Plan (DAP) on behalf of the CSO Program. The PMT developed the DAP, and subsequently a more specific Execution Plan that focused on four key areas: contracting, workforce development, youth outreach, and assignment of a PMT Diversity Coordinator. Following completion of the Execution Plan, a Diversity Task Force was organized by the City to discuss city-wide diversity issues in the context of the CSO Program, set measurable goals, identify specific action items to pursue, and develop a communication plan with City officials to highlight progress and successes.

In 2015, the PMT participated in the Mayor's Best Practices Council, established with the local collaborative program partners, to coordinate and enhance the economic inclusion effectiveness and sustainability, and with the City's Economic Inclusion Task Force. Since that time, the City has increasingly become the focal point for economic inclusion efforts city-wide, with the CSO Program being just one important aspect of the overall effort. The focal point for the City's efforts is the Human Rights and Relations Department. In addition, the Greater Omaha Chamber of Commerce has organized REACH, which is "a comprehensive approach to ensuring our certified small and emerging businesses (SEBs) are ready to bid and capitalize on new projects – and reach their strategic goals."

Working as part of the City's comprehensive efforts, following are key aspects of the CSO Program's Economic Inclusion:

- The PMT firms (CH2M and HDR) plus KAI Design & Build have transformed the original DAP and Execution Plan into an Economic Inclusion Plan (EIP) that addresses the key areas of Subcontracting, Workforce Development, and Youth Outreach. Enhancement of diversity is an important value of the firms, and the EIP includes copies of the Corporate Policies for each firm.

- The PMT firms and KAI, along with Lovgren Marketing Group (the City's Public Participation Facilitator) have developed an Economic Inclusion Team, led by an Economic Inclusion Officer who performs his work at no cost to the City.

- All CSO Design Teams are required to develop an initial construction phase SEB utilization memorandum at the 60 percent design phase of projects. This information is utilized by the City and PMT to communicate potential SEB activities to SEB contractors in an effort to maximize SEB participation. This 60 percent memorandum is updated with the 90 percent deliverables identifying the final percentage set-aside for SEB utilization for project construction. This percentage is included as a requirement in the Instructions to Bidders in the construction contract documents. From 2009 through 2016, SEB utilization on CSO projects has been 8.1 percent of the accumulated bid total.

- All bidders on CSO construction projects must complete and include with their bid, or have on file with the City, an EIP "which addresses the Contractor's plan to collaborate and partner with the City of Omaha to create a culture of economic inclusion on the project." An EIP template is included as an attachment to the Instructions to Bidders.

- In compliance with the City's recently established policy, all consultants working on or pursuing CSO work must have an approved annual Economic Equity and Inclusion Plan (EEIP) on file with the City.

Affordability

As noted in the 2014 LTCP Update report, the City's latest financial plan suggested that annual rate increases of about 9 percent would be necessary to fund the LTCP and other foreseeable wastewater collection, treatment, and capital expenses through 2027. These projections did not include impacts from new or stricter regulatory requirements that may be mandated. In May 2013, the University of Cincinnati completed a financial capability assessment for Omaha's wastewater enterprise fund. The report concluded that with the sewer rate increases adopted by City Council through 2018, the Omaha service area will be at a "Medium Burden" level as defined by EPA. However, the report also noted that some "Communities of Concern" already approach the "High Burden" threshold, and by 2027, the entire service area may be near "High Burden."

An updated financial capability assessment was conducted in late 2016 and early 2017 by the City's rate consultant, and the resulting report is currently nearing completion. The findings in terms of burden on the community are reportedly similar to the previous assessment. In particular, it is noted that the standard use of "median household income" (MHI) can be misleading because an area can have a reasonably high MHI and still have a very large percentage of people who are at the lower end of income percentiles.

The 2014 LTCP Update noted that the City is facing affordability concerns with the implementation of the LTCP, and it is likely that expensive modifications to the City's two regional wastewater treatment plants will also be required. The Update concluded that the City will need to work with NDEQ and others to address these concerns, and it may be necessary to adapt the schedule for LTCP implementation to factor in other elements. As noted elsewhere in this paper, the City and PMT are actively looking for ways to reduce the cost of the CSO Program and are developing strategies to reduce impacts on ratepayers, particularly in the Communities of Concern.

Environmental Challenges

As LTCP projects have progressed from design to construction, numerous environmental challenges have been encountered. These are the result of working in the older parts of the City, which are the areas that have CSOs. Some of these challenges have included:

- Demolition of a pesticide contaminated building that was on the EPA Superfund list (Economy Products, Nicholas Street Phase 1 Sewer Separation).
- Construction through a remediated Comprehensive Environmental Response, Compensation, and Liability Act (CERCLA) site and construction in a remediated smelter site (Lewis and Clark Landing/Heartland of America Park, SIFM North Segment).
- Construction through a solid waste landfill that is currently a botanical garden (Lauritzen Gardens, Martha Street Sewer Separation).
- Construction in wetland areas requiring Clean Water Act Section 404 permits (numerous locations; largest is at the MRWWTP).
- Migratory birds nesting during construction (robins, owls, bald eagles, Canadian geese, barn swallows).

The City and PMT have worked through issues at these and numerous other difficult environmental sites, in addition to obtaining routine permits and approvals for projects. The City has been able to resolve these items because tools and resources were developed to address environmental concerns early on.

It is the City's goal to complete the implementation of the LTCP without any enforcement actions on LTCP projects. To help in the achievement of this goal, the City and PMT developed an Environmental Plan as a tool for staff, as well as designers and contractors. The Plan is based on Environmental Management Principles and has received significant coordination, involvement, and endorsement from City leadership. The Plan provides guidance and resources to ensure that all projects completed as part of the LTCP comply with all applicable environmental and permitting laws and regulations. The Environmental Plan includes guidance and work flows to address environmental issues such as:

- Permitting Processes
- Cultural Resources
- Wetlands and Waters of the U.S.
- Endangered Species Act (ESA)
- Migratory Bird Treaty Act (MBTA)
- Contaminated Soils/Lead
- Materials Management
- Environmental Due Diligence
- Material Spill
- Construction (specifications, permitting, compliance)
- Construction Dewatering

As a companion to the Environmental Plan, an Environmental Field Guide (see Figure 8) was developed that provides information taken from the Plan, but focuses on specific issues that can come up in the field. The Guide also includes names and phone numbers of resources that can be reached by individuals working in the field.

Figure 8. Environmental Field Guide for CSO Program

In addition to the Environmental Plan, the PMT has developed several other tools. This included working with the NDEQ on the development of a Materials Management Plan that provides guidance on what sampling is needed and when materials need to be handled in a special manner if determined to be above certain threshold concentrations. NDEQ's approval of this Plan means that additional approvals are not needed as long as the Plan guidelines are followed. In addition, it became apparent early on that several of the construction sites would need to discharge contaminated groundwater. Rather than having to obtain individual permits for each site, the

PMT worked with NDEQ to develop a general permit that covers the dewatering of contaminated groundwater into the Missouri River. This not only greatly reduced the time needed to cover these discharges, but provided a certainty regarding the treatment requirements needed.

To ensure that the procedures in the Environmental Plan and permit requirements are followed, the City and PMT developed two specifications that are required for every construction site. These include a Permitting Specification that requires compliance with permits obtained for the project and a Temporary Environmental Controls Specification. The latter not only requires compliance with the appropriate elements of the Environmental Plan, but requires that the contractor assign an Environmental Control Officer with responsibility for ensuring compliance with environmental requirements.

Green Infrastructure

As part of the development of the LTCP from 2006 to 2009, a process was developed to identify potential Green Infrastructure sites within the combined area for inclusion in the LTCP. The Basin Consultants implemented the process and developed possible Green Infrastructure projects for each Basin. These projects were evaluated and developed into a list of sites that were included in the LTCP as recommendations to evaluate in greater detail. No percentage of wet weather volume capture was assumed for Green Infrastructure in achieving the LTCP's level of control target. As the LTCP has been implemented, there has been an increasing emphasis on Green Infrastructure.

During LTCP implementation, a procedure was developed whereby each sewer separation project team is required to perform a Grey to Green Infrastructure evaluation for the project area to identify potential Green Infrastructure elements that can be incorporated cost effectively. The procedure was developed such that a good faith effort to incorporate Green Infrastructure is a part of every sewer separation project and is considered from the outset of the project rather than at the end. The procedure requires that the designer show there are physical, engineering, or financial reasons for not including Green Infrastructure components in each project. This effort has resulted in many of the Green Infrastructure elements of projects noted previously and illustrated on Figure 9.

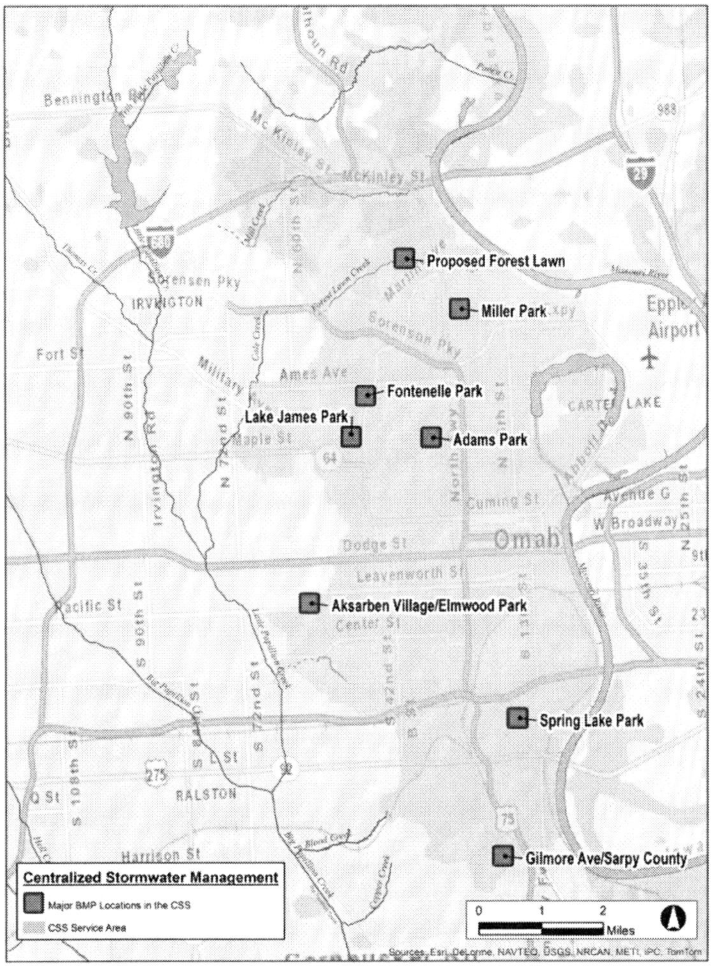

Figure 9. Locations of Major Green Infrastructure Sites

A process has also been developed to evaluate additional opportunities for the incorporation of Green Infrastructure into the LTCP in areas where there are no CSO projects. As part of this evaluation, the PMT developed a geographic information system (GIS) model, and a site suitability evaluation was performed to determine if there are Green Infrastructure projects that can be implemented in the CSO area to assist in the reduction of CSO events and costs. The City hired a consultant to evaluate some of these areas in further detail and to identify specific projects for potential implementation. This effort also identified areas where establishment of stakeholder relationships could result in further reduction of stormwater entering the combined system.

As part of the overall adaptive management strategy, the PMT will continue to assist the City with the development, implementation, and documentation of an overall Green Infrastructure Program, including develop of a business case that describes the benefits of Green Infrastructure Program recommendations and outlines the reasons for implementation of the Program to promote endorsement by the general public and others. Incorporation of an incentive program may be a tool that is used to promote Green Infrastructure.

Cost Escalation

As implementation of the LTCP has progressed, the contracting market in the Omaha Metro area has become challenging, with many contractors and subcontractors engaged in multiple projects, which minimizes their ability to pursue new bids. The strong contracting market has also impacted the labor market. It has been stated by local contractors that labor rates are 10 to 20 percent higher than typical because of increased demand. Contractors have recently expressed some difficulty in putting a sufficient labor force together for projects, and are finding it necessary to supplement their local labor force with individuals from outside of the area.

Examples of large, local projects being contracted in the near-term in addition to CSO projects include those for the Council Bluffs Interstate System, Metropolitan Utilities District, Nebraska Department of Roads, Omaha Airport Authority, Omaha Public Schools, University of Nebraska at Omaha, and University of Nebraska Medical Center. To help maximize the number of contractors bidding on CSO projects, the PMT communicates and coordinates with agencies and municipalities in the Omaha Metro area regarding projects that are proposed to bid within the next several months. Where possible, bidding periods for CSO projects are adjusted to avoid direct conflicts with other projects. This approach recognizes the fact that contractors are limited not only by how much work they can perform at any given time, but also by how many projects they can bid.

In addition to forward planning, summary information on recent bid openings are documented to communicate the number of contractors bidding, bid prices in relation to the engineer's opinion of probable construction cost (OPCC), and any specific issues related to the bids. With the assistance of the City and PMT, CSO design teams take the competitive market into account when preparing OPCCs. For example, consideration is given to adding a premium to labor rates as noted above. Another approach to encouraging competitiveness on bids for CSO projects is for design consultants to obtain input from various contractors during the design process, focusing on constructability, risk allocation, and other factors that could affect cost and willingness to bid.

Thus far, the CSO Program has been positive for local area contractors. Out of the $388 million in construction awarded through February 2017, approximately 84 percent has been successfully won by local general contractors. Because the CSO Program is being paid for by local ratepayers, it is considered positive that most of the contracts are being awarded to local contracting firms.

Saddle Creek RTB Bid

A significant project called for in the LTCP is the construction of an RTB with a design flow of 1,192,000 m^3/day (315 mgd) to serve the Saddle Creek Basin at CSO 205. This RTB was designed and bid, with the bid opening taking place in August 2015. Unfortunately, a single bid of $128.7 million was received, significantly higher than the engineer's OPCC of $91 million. The City rejected this bid. A value engineering effort was completed to identify reasons for the high bid and to find ways to reduce the cost of the project as designed. The result of this effort was a recommended budget of $125 million for the RTB. The City determined that the project as originally designed was too expensive and that a lower-cost, reconfigured facility to address the combined flows from CSO 205 needed to be developed.

The City and PMT worked with the designer to develop revised alternatives that would achieve compliance with the EPA CSO Control Policy at a more affordable cost. A list of twelve alternatives including both RTBs and storage tanks was developed and evaluated. The City established a construction budget goal for the revised control of $85 million by escalating the original budget of the RTB in the 2009 LTCP to the present. The alternatives were evaluated based on capital cost, operation and maintenance (O&M) costs and complexity, volume capture of flows during wet weather in the Saddle Creek Basin, number of partially treated combined sewage overflows remaining, aesthetics, risks, and potential for future expansion.

The evaluation produced a short-list of four viable alternatives that were examined in more detail: a 435,300 m^3/day (115 mgd) RTB, a 605,700 m^3/day (160 mgd) RTB, a 53,000 m^3 (14-million-gallon [MG]) storage tank, and a 75,700 m^3 (20 MG) storage tank. The storage tanks would be fed by a 567,800 m^3/day (150 mgd) lift station. It was determined that the storage tanks would be advantageous because of the lower level of O&M required and the lack of any effluent limits on the discharge. Disadvantages of the tanks included the need to dewater them back into the interceptor that leads to the City's Papillion Creek Wastewater Treatment Plant (PCWWTP), which would take up capacity available for growth, create uncertainty associated with costs, and possibly require additional work upstream in the future to achieve adequate water quality improvement in Little Papillion Creek. The RTBs were advantageous because they would get the water out of the system and not take up interceptor capacity reserved for growth, produce better water quality in the receiving stream, and NDEQ was more familiar with the RTB technology. Disadvantages of the RTBs were the level of staff needed to operate the facility and higher O&M costs.

A decision was made in February 2017 to move forward with the 605,700 m^3/day (160 mgd) RTB. Construction of this RTB would provide an estimated 89 percent volume capture for the Saddle Creek Basin in the representative year. Further, the decision was made to design the RTB with the ability to accept and disinfect peak flows in excess of 605,700 m^3/day, up to a maximum of 1,211,000 m^3/day (320 mgd). This approach is referred to as a "hybrid" RTB approach. If it can be demonstrated that equivalent to primary treatment can be accomplished at flows between 605,700 and 1,211,000 m^3/day, this would increase the volume capture in the Saddle Creek Basin from 89 to 94 percent. Regardless, the hybrid approach will provide the City with the opportunity to disinfect as much combined sewage as with the originally designed RTB at a much lower cost. This would be beneficial to water quality in Little Papillion Creek and downstream. Another advantage to the hybrid approach is the minimal amount of re-design that will need to take place.

With an estimated construction cost of $85 million, this approach is within the budget allocated in the 2009 LTCP escalated to current dollars, and it will save over $40 million from the actual bid received for the original RTB. The City and PMT met with NDEQ in March 2017 to discuss the revised RTB concept and received concurrence. The facility will be re-designed in the near future from the original RTB to the hybrid RTB by reducing the original size of the settling basin by 50 percent.

Minne Lusa Conveyance Sewer

Another significant project planned for in the LTCP was the Minne Lusa Stormwater Conveyance Sewer. In 2016, while nearing the completion of the design for his project and the associated Storz Detention Basin downstream, the City decided to place it on hold, pending a re-evaluation of the Minne Lusa Basin. This decision was based on several factors, including the following:

- During design, the estimated cost of the conveyance sewer and detention basin increased to more than $100 million, or more than double the inflation-adjusted cost used in the planning level estimate developed during the 2009 LTCP.

- Other sewer separation projects executed by the City have performed better than expected in reducing basement backups in the Minne Lusa Basin. This, combined with the higher costs, meant that the conveyance sewer project could no longer be justified based on the cost/benefits assumed for the 2009 LTCP. (The conveyance sewer and the sewer separation projects upstream that connect to the sewer were originally planned to address basement backups and street flooding while achieving a reduction in CSO volume from an area that covers approximately 1.2 hectares (3 square miles) of the 3.6-hectare (9-square-mile) Minne Lusa Basin. Starting in the early 1990s, efforts by the City to address basement backups from this area through a series of partial sewer separation – local separation with the stormwater released back into the combined system – have proven to be very successful in addressing many of these problems.)

- Operational and permitting issues, which were not realized until the final design of the project, were of increasing concern. The City's operations staff expressed concerns related to the ability to operate control gates that were included in the project as a requirement of the City's flood control system. In addition, permitting issues arose at the end of the project that included complications in meeting competing and overlapping regulations and jurisdictions between U.S. Army Corps of Engineers (USACE) Regulatory, USACE staff responsible for reviewing flood control, and the Nebraska Department of Natural Resources.

- The City needs time to fully realize the impacts of work completed to date or already committed to. The City has committed to approximately $90 million of work with the Minne Lusa Basin under the CSO Program. This includes projects that have already been completed and projects currently under design that will be going to construction in the next year or so. The City needs a better understanding of the effectiveness of this work on CSO control. The dollars invested by the City are based heavily on the results of its InfoWorks model of the collection system. Although confidence is high in the model, time to monitor and evaluate the impacts of work as it is completed, as well as further development of model detail, will provide information that can be used to better determine controls necessary to meet the required level of control for the Minne Lusa Basin.

The current path forward for the Minne Lusa Basin is to complete a re-evaluation to determine the most cost-effective way to reduce the volume of combined sewage that overflows at CSO 105. As stated previously, the Minne Lusa Basin covers about 3.6 hectares (9 square miles), or approximately 20 percent of the entire 18-hectare (45-square-mile) combined sewer system. An evaluation is currently underway by the City and PMT as part of the Technical Assessments for Cost Savings Task. The task team is evaluating over 20 alternatives. The current schedule calls for this evaluation to be complete by December 31, 2017. Based on the evaluation, City leadership will decide the best approach to move forward. It will then be necessary to evaluate how the approach fits into the overall LTCP schedule before developing a revised schedule for the next LTCP Update, to be submitted in 2019.

Other Project Challenges

As noted previously, the SIFM is an early action LTCP project that replaces an existing force main that has been in service since the 1960s and provides for increased wet weather flow rates to be delivered to the MRWWTP for treatment. Due to soil conditions, environmental contamination, space limitations, and other factors, the decision was made during design to tunnel a portion of the South Interceptor Force Main Replacement through rock. As of May 25, 2016, the tunnel boring machine (TBM) was within 274 meters (900 feet) of the final receiving shaft. On that day, a subsurface geologic anomaly, consisting of a break in competent rock, was encountered that caused tunneling progress to stop. The TBM being used was not designed to progress through the sands, gravels, and excessive flow of groundwater that was encountered in the anomaly. Initial efforts to resolve this issue involved the installation of dewatering wells to lower the groundwater to reduce the pressure on the TBM. However, once this was accomplished, large boulders and cobbles were discovered, which the TBM could not adequately address.

After a thorough evaluation of alternatives in terms of cost and risk, the City decided to proceed with a rock tunneling approach from the receiving shaft using a more robust micro-tunneling machine (MTBM). Based on the evaluation, this approach is believed to be able to complete the SIFM construction in the shortest schedule and with the lowest overall risk to the City. A rescue shaft to retrieve the existing TBM also will be constructed. There is every expectation that this approach will be successful; however, the issue caused a significant delay and increased costs for the project. It is anticipated that the SIFM project will be completed in early 2018.

Competing Demands of the City's Collection and Treatment System

Over the last 10 years, the City's sewer rates have been devoted heavily to implementing the LTCP, with wastewater treatment and collection system needs being secondary. However, over the next several years, new regulatory requirements will need to be met, and it is becoming increasingly apparent that a higher level of investment in the existing infrastructure will be needed to keep the existing system functioning. The desire is to invest in the existing infrastructure in the spirit of being a well-managed utility rather than having additional mandatory requirements or Consent Decrees imposed on the City by regulatory agencies. As noted elsewhere under affordability, the City will need to work with NDEQ to determine and accommodate any resulting impacts on the schedule to complete the LTCP.

PATH FORWARD

The City has been implementing the LTCP at a very fast pace since 2009, and a tremendous amount of progress has been made in spite of a number of challenges and changed conditions. The City is committed to moving ahead with the Saddle Creek RTB and other LTCP projects to achieve compliance with the EPA CSO Control Policy. However, at this stage of implementation, it has become necessary for the City to monitor and assess the effectiveness and performance of what has been accomplished and, while continuing to make progress, determine the most cost-effective approach to meeting regulatory requirements. This includes evaluations such as the TACS Task that is focused on reducing the ultimate cost of the CSO Program and the burden on ratepayers. The City and PMT will describe the path forward for the CSO Program in the next LTCP Update submission to NDEQ, due in 2019.

The path forward for the CSO Program will take a more comprehensive perspective, reflecting the needs of the City's overall sewer collection and treatment system, not just CSO. The City must devote sufficient attention and resources to other aspects of the sewer treatment and collection system to ensure proper O&M and longevity. This could include a re-assessment and prioritization in the following areas:

- Review of wastewater treatment facilities to determine their ability to meet future effluent limits such as for ammonia and nutrients.
- Assessments of existing wastewater treatment and other facilities such as lift stations to determine remaining service lives and to allow development of plans for replacement and repair.
- Examination of the collection system to determine condition and to develop plans for future replacement and repair.

Overall, the results of this work will allow the City to meet its regulatory requirements now and in the future, while ensuring that existing infrastructure is maintained so it does not become a source of regulatory noncompliance.

SUMMARY, CONCLUSIONS, AND LESSONS LEARNED

The Omaha CSO Program is being accomplished by the City of Omaha due to the regulatory requirement to meet the EPA CSO Control Policy. Development of an LTCP began in 2006, and implementation of LTCP projects began in 2009. The City's administrative Consent Order with the State of Nebraska currently calls for completion of LTCP implementation in 2027. The City has been implementing the LTCP at a very fast pace since 2009, and a tremendous amount has been accomplished. While significant progress will continue, the CSO Program is currently in a period of performance assessment, evaluation, and adaptive management to determine the optimum path forward in completing the LTCP in a manner that does not overly burden the ratepayers in the community and that accommodates other needs of the wastewater treatment and collection system.

Following are just a few of the lessons learned from the initial years of CSO Program implementation:

- The flexible Consent Order and a good working relationship with regulators have proven to be even more important than originally anticipated.

- Conditions and requirements do indeed change, and the need for adaptive management is continuous, not just when the LTCP is updated.

- A consistent consultant team has proven to be valuable to supplement City staff during critical periods such as the 2011 Missouri River Flood.

- It is valuable to have individuals on the team with varying backgrounds to help address needs and challenges as they occur.

REFERENCES

City of Omaha (2014), Update to the Long Term Control Plan for the Omaha Combined Sewer Overflow Program, Omaha, NE.

City of Omaha (2009), Long Term Control Plan for the Omaha Combined Sewer Overflow Program, Omaha, NE.

TAPPING INTO COMMUNITY PARKS WITH GREEN INFRASTRUCTURE

Andy Szatko, City of Omaha Stormwater Program

ABSTRACT

Green infrastructure (GI) is being utilized by numerous communities to minimize the effects of stormwater runoff on their environment, reduce combined sewer overflows (CSO), improve their landscape by embracing water instead of treating it as problem to get rid of, or any other number of reasons. GI is the preserving, connecting, and mimicking of natural processes that manages stormwater close to where it lands. GI practices can take on a variety of forms depending on the scale, but common practices used at a site level commonly include bioretention systems, bioswales, wetlands, and permeable pavement. The City utilizes and promotes the use of GI practices across the City to help remove pollution, protect the environment, and provide an amenity to the community. The City of Omaha (City) also utilizes GI practices to enhance the implementation of the CSO Program's Long Term Control Plan (LTCP) by reducing the magnitude, duration, and frequency of CSOs as well as reduce the overall costs with the implementation of the LTCP.

The City's combined sewer system (CSS) is generally located between 72nd Street and the Missouri River and has a generally hilly topography. The historic City of Omaha's Park and Boulevard system was designed to make use of the existing hilly topography and are generally situated where water naturally flows. These spaces provide a natural opportunity to manage stormwater through the integration of GI practices at these parks and improve the overall implementation of the CSO Program's LTCP. Collaboration across departments has been critical to the success of completed projects, especially with the Parks Department and the MS4 program (Omaha Stormwater Program). Lessons learned from each project are adapted to the next to improve overall implementation and performance.

KEYWORDS

Green infrastructure, stormwater, MS4, CSO, low impact development, LID, parks

INTRODUCTION

Stormwater runoff has significant impacts on the water quality of local streams and rivers in the United States (US EPA, 1996; US EPA, 2009). In an effort to mitigate these damaging effects, local governments (City of Omaha) have incorporated the use of green infrastructure (GI) as a best management practice (BMP) within their stormwater management programs. These GI practices significantly reduce stormwater volumes, peak flows, pathogens, sediment loading, erosion, and nutrient concentrations through infiltration into the soil subsurface (Winogradoff, 2002; Dietz and Clausen, 2005; Collins et al., 2010; Beutel and Larson, 2015).

Green infrastructure is a powerful and beneficial approach to stormwater management that aims to preserve natural areas and mimic natural processes through the use of local soils, native vegetation, and land features to capture and manage rain close to where it falls. GI practices can range in scale from regional (floodplains and wetlands) to local (urban and residential) landscapes. Examples of more common GI practices include green roofs, bioswales, rain harvesting systems (rain barrels), rain gardens, and bioretention systems.

DISCUSSION

Omaha has both a municipal separate storm sewer system (MS4) and a combined sewer system (CSS). The MS4 program (Omaha Stormwater Program) promotes the use of GI practices as a means to reduce erosion, protect streams, improve water quality, promote infiltration, improve the aesthetics of the community, and utilize it as a resource for schools. Demonstration projects have been constructed across the City at maintenance yards, schools, streetscapes, and parks to demonstrate these benefits. These projects often are utilized by the Omaha Stormwater Program in their education and monitoring programs to continually assess their performance, create awareness of them, and encourage their use.

The CSO Program utilizes GI practices in much the same way but also utilizes it to improve the implementation of the LTCP. Each project in the CSO program, as part of the design process, is evaluated to see how GI could be used to reduce pipe sizes, enhance the community, or provide other project-specific benefits. The historic Omaha Parks and Boulevard System winds its way through eastern Omaha and are generally located in the lower areas where water naturally drains towards, providing an excellent opportunity to manage stormwater. While the park sites themselves provide a great opportunity to manage stormwater, they are parks first and foremost and embraced by the community. They could not simply be removed or significantly altered.

Collaboration between all involved City Departments, especially the Parks Department, and the community has been a significant part of the success with completed GI practices to date. Here is a summary of completed and in-progress park GI projects and a map of their location:

- Elmwood Park - 2012
- Spring Lake Park - 2016
- Adams Park - 2016
- Field Club Trail, Vinton - 2016
- Albright Park – In construction
- Fontanelle Park – In construction
- Hitchcock Park – Construction to begin May 2017
- Hanscom Park – 95% design
- Forest Lawn park – 60% design

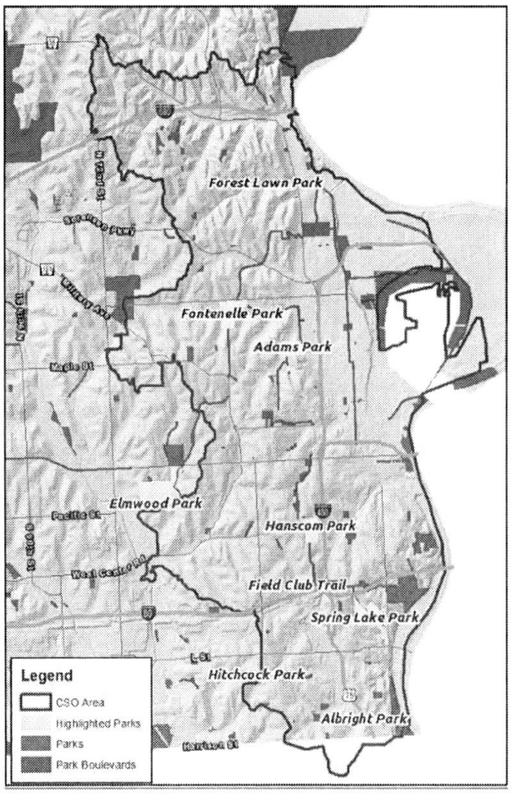

Figure 1 Map of Omaha CSO Completed CSO Green Infrastructure Projects

Descriptions of completed projects in parks and those implemented by the Omaha Stormwater Program are included below to gain a better understanding of how the City of Omaha has utilized green infrastructure in its parks and the benefits of embracing stormwater as a resource and not as a waste product.

Elmwood Park Diversion

The Elmwood Park Diversion was part of the Aksarben Village Sewer Separation Project.

The diversion was identified during the screening process for green infrastructure within the combined sewer system service area. The design diverts 29 acres in the upper reaches of the watershed into the adjacent watershed that contains Elmwood Park. The diversion allowed for the construction of smaller pipes at the bottom of the Aksarben Village watershed. The diversion is estimated to have saved $550,000 on the project overall, which includes the costs associated with building the green infrastructure project.

The diverted stormwater runoff passes through a trash screening structure to minimize debris from entering the diversion. Access hatches and a driveway to the structure were also constructed to allow for easy removal of the built up debris. From the screening structure it then flows to a splitter structure. The structure directs a certain amount of flow, up to approximately 23 cfs, through a bioswale and three bioretention gardens, beginning at the top of the park ravine. Flows exceeding the capacity of this pipe are diverted into a 42" concrete bypass pipe that extends to the bottom of the ravine.

Seven slotted-weir structures were constructed within the ravine to reduce stormwater velocity and lower erosion potential. The weir structures have vertical drops that vary from one to three feet. Runoff flows through a native grass bioswale, separated from the bioretention gardens with a berm. The bioretention gardens are located above three of the weir structures and are designed as an offline system, meaning that stormwater does not actively flow through them. Flows encounter the weir structure first, with some flowing through the low-flow notch while the remainder backs up into the bioretention garden. This offline design decreases shear stress on the garden soils and plantings. Once the garden is full, water will overtop the high-flow weir as it continues flowing down the ravine. At the base of the ravine, a dry detention basin collects flows into an outlet structure that connects with the bypass pipe and flows to Elmwood Creek.

Figure 2 Photos of the Elmwood Park Diversion from construction through establishment

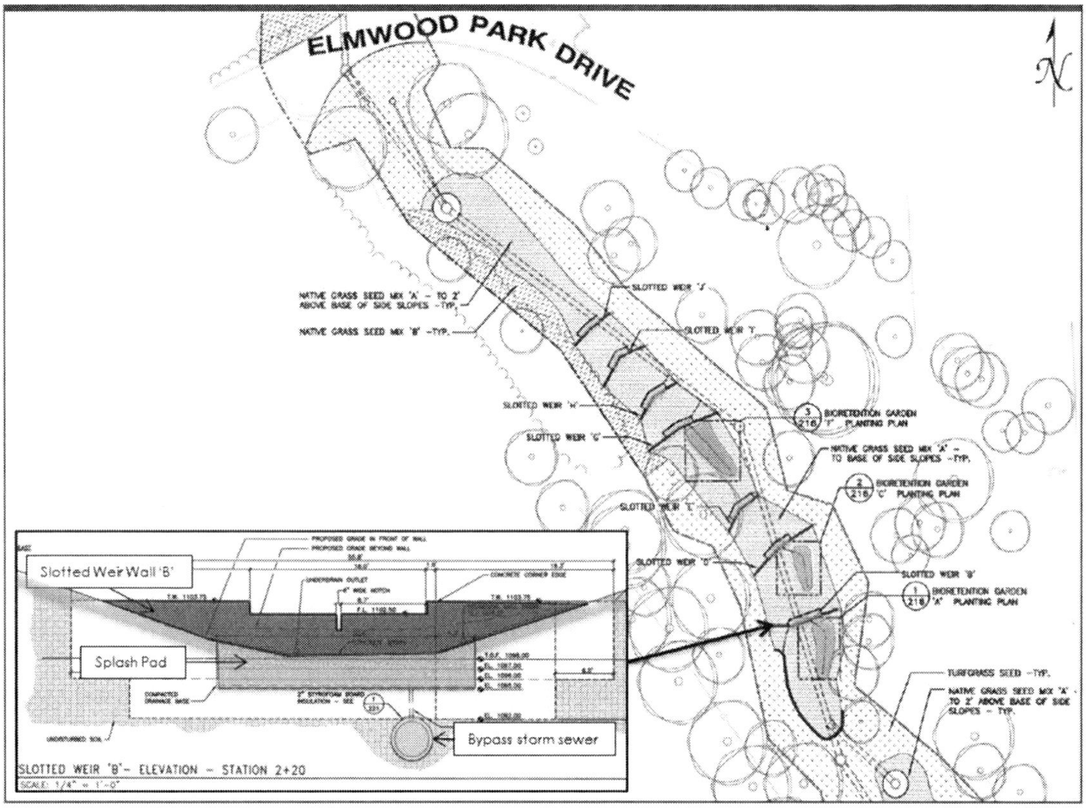

Figure 3 Project layout

Adams Park Wetlands

The John Creighton Boulevard (JCB) Stormwater Conveyance Sewer Project is part of the City of Omaha's Combined Sewer Overflow (CSO) Long Term Control Plan (LTCP). The Project consists of sewer separation and wetlands designed to reduce CSO volume, sewer backups, improve water quality, and provide a neighborhood amenity. The sewer separation project area encompasses the JCB and Miami subareas and consists of 273 acres of predominantly residential property generally bounded by Hamilton Street on the south, Maple Street on the north, 33rd Street on the east and 38th Street on the west. Following completion of the new storm sewer system, the existing combined sewer was converted into a sanitary sewer that will complete sewer separation for the sub-basin. Approximately 27,000 linear feet of storm sewer was constructed in the sewer separation project and consists of pipes as large as 84 inches in diameter to as small as 8 inches in diameter.

Figure 4 Adams Park Wetlands looking south in Fall 2016

The Adams Park wetlands and detention area is designed to provide the maximum practical stormwater storage to reduce the size of the detention basin outlet pipe known as the JCB Stormwater Conveyance Sewer, which will be designed and constructed in a subsequent CSO project. The wetlands consist of an area covering approximately 14 acres, separated by an embankment for a future park road. A low–hazard dam with emergency spillway was constructed on the north side of the park, near Bedford Avenue, to provide detention volume of up to approximately 77 acre-feet, which will detain the 100-year storm event. The wetland facility is designed to provide water quality benefits and community enhancement elements.

Figure 5 Layout of Adams Park Wetlands Project

The sewer separation project will re-connect the detention basin outlet pipe to the existing combined sewer, until the permanent outlet conveyance pipe (JCB Stormwater Conveyance Sewer) is designed and constructed. The Adams Park wetlands will allow for the reduction in the size of the basin outlet pipe size from 108-inch diameter (no detention) to 60-inch diameter, which results in an estimated project cost savings of approximately $4.5 million. The JCB Stormwater Conveyance Sewer will extend from Adams Park to JCB and Paxton Boulevard, where it will connect to the future tunnel drop shaft of the Minne Lusa Stormwater Conveyance Sewer, which will convey the stormwater to the Missouri River.

Construction of the project in Adams Park began on September 3, 2014 and was completed in June 2016. Sewer separation work began in March 2015 and was completed in early 2017. There is a separate construction contract for the installation of trees, wetland plantings and landscape features which includes a 2-year maintenance agreement to ensure the plantings are well-established before the project is considered complete.

Spring Lake Park
Spring Lake Park was established during the 1870's as an urban park. The dramatic topography of the park includes several forested steep slopes and valleys speckled with springs. The park originally contained three small ponds, but they were drained in the 1930's.

The City of Omaha CSO Program identified the opportunity to reintroduce the lake back into the park as part of phase one of a sewer separation project for the area (OPW 51997). The sewer separation project site is a 416-acre watershed which is approximately 50% residential and 50% within Spring Lake Park, which includes a nine-hole golf course. This project was a collaborative effort with members from the Spring Lake Neighborhood Association, Spring Lake Park Habitat Restoration and Preservation Team, CSO Program, Nebraska Environmental Trust, Papio-Missouri River NRD, and Keep Omaha Beautiful.

Green Infrastructure (GI) components within phase one of the project saved an estimated $5,225,000 from the final project cost of $10,400,000 by eliminating or reducing the extent and size of "gray" components. In addition, the City of Omaha received $1.3 million in grants from Nebraska Environmental Trust for design and construction of GI infrastructure within the park. The project is being built in two phases. Phase one commenced on May 2014 and was substantially completed in late 2016. Phase two of the project is scheduled to begin in 2017.

Phase one of this project incorporates several significant GI components, including a fishing lake, constructed wetland, submerged gravel wetland, 750 linear feet of constructed stream, 7 dry detention basins, 1 infiltration basin, and 1 bioretention systems. This project will provide many community benefits, recreational opportunities, and increased wildlife diversity and habitat, as well as improve GI utilization well into the future.

DESCRIPTION	LAKE	CONSTRUCTED WETLAND
System Footprint	1.55 acres	0.45 acres
Under-Drain	None	None
Pre-Treatment System	5 pre-treatment basins, detention basin, constructed stream, plunge pool, submerged gravel wetland, bioretention, constructed wetland	2 pre-treatment basins, detention basin, constructed stream, plunge pool, submerged gravel wetland, bioretention
Outlet Control	Stop logs, gate at lake bottom, 10' diameter cone grated overflow	Stop logs
Contributing Area	134.61 acres	98.51 acres
Predominant Land Use	Residential/Park	Residential/Park
Percent Impervious (%)	37%	35%
Predominate Soil Types	Pohocco-Judson silt loams	Pohocco-Judson silt loams

DESCRIPTION	INFILTRATION BASIN	SUBMERGED GRAVEL WETLAND	BIORETENTION SYSTEM
System Footprint	1,325 ft^2	5,200 ft^2	2,000 ft^2
Under-Drain	None	6" Perforated PVC	4" Perforated PVC
Pre-Treatment System	Pre-treatment basin	2 pre-treatment basins, detention basin, constructed stream, plunge pool	Off-line system
Outlet Control	Concrete overflow ribbon	Underdrain with elbow	Beehive grate
Contributing Area	3.85 acres	45.27 acres	2.67 acres
Predominant Land Use	Residential	Residential/Park	Park
Percent Impervious (%)	30%	35%	30%
Predominate Soil Types	Contrary silty clay loam	Pohocco-Judson silt loams	Contrary silty clay loam

Figure 6 Overview of Spring Lake Park Green Infrastructure

REFERENCES

Beutel, M.W. and L. Larson. 2015. Pathogen removal from urban pond outflow using rock biofilters. *Ecological Engineering*, 78, 72-78. Doi:10.1016/j.ecoleng.2014.05.001

Collins, K.A., T.J. Lawrence, E.K. Stander, R.J. Jontos, S.S. Kaushal, T.A. Newcomer, N. B. Grimm, and M.L. Cole Ekberg, 2010. Opportunities and challenges for managing nitrogen in urban stormwater: A review and synthesis. *Ecological Engineering*, 36, 1507-1519. Doi:10.1016/j.ecoleng.2010.03.015

Dietz, M.E. and J.C. Clausen, 2005. A field evaluation of rain garden flow and pollutant treatment. Water, Air, and Soil Pollution. 167:123-138.

US EPA (United States Environmental Protection Agency), 1996. National Summary of Water Quality Conditions. https://www.epa.gov/sites/production/files/2015-09/documents/1996_national_water_quality_inventory_report_to_congress.pdf , *accessed* December 2016.

US EPA (United States Environmental Protection Agency), 2009. National Water Quality Inventory: Report to Congress, 2004 Reporting Cycle, EPA 841-R-08-001, Washington, D.C. https://www.epa.gov/sites/production/files/2015-09/documents/ 2009_01_22_305b_2004report_2004_305breport.pdf , *accessed* December 2016.

Winogradoff, Derek A, 2002. Bioretention Manual. Programs and Planning Division, Department of Environmental Resources, Prince George's County, Maryland.

Enhanced Affordability Analysis of Combined Sewer Overflow Long-Term Control Plans – Omaha's CSO! Program

Jason Mumm, Stantec, James Theiler, City of Omaha, Andy Baker, Stantec, and Carol Malesky, Stantec

ABSTRACT

The City of Omaha has an Administrative Order of Consent with the State of Nebraska, which has been negotiated with the Nebraska Department of Environmental Quality (NDEQ). The Order requires the City to meet the requirements of the Combined Sewer Overflow (CSO) Control Policy of the Clean Water Act by 2027. More than 90 projects identified in the City's Long Term Control Plan (LTCP) are scheduled to improve water quality in area rivers and streams at a cost of approximately $2.0 billion, with over $400 million spent since the LTCP was implemented, beginning in 2010.

For local communities engaged in consent decree negotiations with federal or state regulators related to compliance with the Clean Water Act and/or Safe Drinking Water Act, completion of a Financial Capability Assessment (FCA) is standard requirement (USEPA, 1997). The FCA is both a quantitative and qualitative benchmark that is intended to provide a basis of comparison between communities. The components of the FCA are defined by USEPA guidance promulgated twenty years ago; comprising a variety of demographic and economic metrics, and a single key quantitative metric: the Residential Indicator. The Residential Indicator compares an imputed cost per household for sewer service to the community's median household income. Since promulgation of these requirements, based on input from the US Conference of Mayors, National Association of Clean Water Agencies (NACWA), and various other parties, USEPA has advised consent decree communities to submit additional information to enhance the understanding of financial issues. The FCA is extremely important in USEPA's determination of the compliance schedule imposed in these cases.

The results of an enhanced affordability assessment approach applied in the City of Omaha, called the Weighted Average Residential Index or WARi™ are intended to demonstrate the financial burden of the City's LTCP on its residents, as well as fairly reflect the benefits of the City's existing rate assistance programs. Results of this analysis better characterize the financial burdens of Order on Consent compliance for the City. WARi provides the City with a valuable analytical tool to precisely identify the neighborhoods where financial burdens are most pronounced, and to focus its limited financial assistance resources in ways that can achieve measurable results that further reduce financial burdens of the LTCP.

KEYWORDS

Affordability, Financial Capability Analysis, Consent Decree, rates, financial burden

INTRODUCTION

Communities relying solely on the USEPA-required components of the FCA have often found that the financial measures that indicate whether a proposed control plan is a low or high financial burden are very limited and do not portray a complete or even an accurate picture.

These limitations often result in inaccurate assessments where a control plan is deemed to be of low- or medium- burden when a closer inspection of the details would suggest otherwise. The City of Omaha and its consultant, Hawksley Consulting (now part of Stantec), have found that the key limitations of the USEPA methodology are that: a) it uses one imputed bill to assess household burden rather than actual bills for service; b) it considers only median household income for the entire service area as a measure of household income rather than considering the full distribution of income levels; and c) it ignores the concentration of impacts in neighborhoods as well as the distribution of population in those neighborhoods.

In response to USEPA's Financial Capability Assessment Framework for Municipal Clean Water Act Requirements released in 2014 (USEPA, 2014), Stantec developed a detailed affordability approach that provides a more complete and accurate picture of the City's financial capability. USEPA's new financial capability assessment framework builds on EPA guidance by encouraging communities to provide additional information that better characterizes residential impacts and financial strength. This paper summarizes Stantec's affordability approach and its application in assessing the financial burden of the City of Omaha's capital improvement program.

METHODOLOGY

Because of the limitations and shortcomings of the USEPA-required methodology, Stantec developed an enhanced method of assessing affordability called the Weighted Average Residential index (WARi™). In addition to the required FCA criteria, the WARi™ methodology provides additional insight by considering income levels and financial burdens for each census tract: 53 data points for every tract in the City's service area. The financial impacts to the households in each tract are portrayed graphically by linking it to the City's geographical information system (GIS), and these financial impacts are projected into the future based on the costs of the proposed control plan. As a result, we can demonstrate affordability metrics with high resolution across income levels, service levels, and geography.

The methodology is relatively simple, and it offers a more detailed and more accurate assessment of actual financial burdens in communities with varying distributions of household income. For each census tract in the service area:

1. Determine the average residential bill per household by cross-referencing billing data to the appropriate census tract by geocode data.
2. Calculate the financial burden for the average bill in the tract for each of the 16 income bins (standardized) tracked by the US Census Bureau. Financial burden is estimated by dividing the average bill by the midpoint of the income bin.
3. From the US Census data for the census tract, determine the number of households for each income bin.
4. Calculate the weighted average financial burden by multiplying the burden in each income bin by the percentage of population in each bin.

Once the calculation has been completed for each tract, the weighted average impact of the entire service area can be determined by weighting the result from each tract by the total household populations in those tracts.

As an additional measure, the methodology is also readily adapted to determine the EPA Residential Indicator for each census tract, and for the entire service area, thus meeting the current regulatory standards but also allowing for much greater levels of detail. In some cases, the WARi values contradict the financial burden indicated by the EPA Residential Indicator, suggesting either higher or lower threshold limits should be considered.

RESULTS

Given approximately $2.0 billion of capital spending anticipated over the course of the City's LTCP, sewer rates have increased significantly since the Order on Consent was negotiated in 2010. With ten years remaining on the term of the City's LTCP, the enhanced affordability analysis provides the information the City needs about the financial burden at present and how it will change as the City continues to raise rates to finance the improvements necessary for compliance with the LTCP.

Projections of future rate increases rely on numerous factors. The project team prepared a long-term financial planning model to analyze these factors and evaluate alternative scenarios for use in the affordability assessment. Annual forecasts of potential rate increases are applied to the average bills by census tract to project annual changes in WARi throughout the period of the LTCP.

Key indicators from the analysis include quantifying the degree of income skew from the national average, which is considered the "normal" income distribution in the United States; evaluated for the City's entire service area, and presented in Figure 1 for the whole City. The figure shows that the City's income distribution approximates that for the United States, which is not always the case in CSO communities. In fact, we find many communities under consent orders are skewed from the national norm, with greater proportions of households in the lower income bins and fewer in the higher income bins.

In Figure 2, a baseline map shows the variation in median household income among census tracts in the City's service area. Darker red census tracts indicates lower MHI and predicts areas that may have affordability concerns.

The WARi™ analysis was performed in conjunction with an update to the City's conventional FCA, which includes anticipated full costs for the overall management of the sanitary and combined collection system, including both future repair and replacement needs and improvements associated with future regulatory impacts. The WARi™ methodology improves upon USEPA's approach by accounting for income skew and mapping a year-by-year projection of residential financial burdens which typically can be used to help negotiate the most cost-effective water quality solutions at the right time. In the City's case, this enhanced view of affordability provides more useful data regarding the use of ratepayer assistance funds and multifamily household affordability.

DISCUSSION

After several years of rate increases at two to three times the rate of underlying inflation, the City's residents and its political leaders had become concerned that the financial burdens of its CSO program were too great. A previous study based on the EPA guideline approach had indicated that the financial burdens were moderate (medium burden) but these findings did not appear to assuage concerns. For its part, the City began to investigate ways that it might successfully reduce the cost of the LTCP without the need for a time extension or reduction in control level. Proposed refinements to the current CSO program include technical engineering savings and more optimized project schedules. These efforts resulted in proposals that may save the City approximately 16 percent in total costs if approved by the State of Nebraska.

The WARi-based study recently completed found that the current financial burdens for the full system were relatively low and would remain low to moderate through 2027 without any changes to existing LTCP. Additional scenario analysis using the WARi approach helped to illustrate the effects that the City's proposed 16-percent cost reduction may have on residents should the proposals be approved by regulators. With the savings in place, the financial burden would likely remain at a "low" level through the 2027 timeframe.

An important finding, however, was that the number of census tracts rated as "high burden" was substantially reduced as a result of the savings. While the reduced costs would not eliminate all such "high burden" neighborhoods, WARi allowed the City to see exactly where those tracts were located, which ones were of highest concern, and which would benefit most immediately from the proposed savings. In turn, this will enable the City to focus its financial assistance programs more precisely to further improve the financial burdens with measurable results.

Figure 3 presents two scenarios of potential results under two CIPs. WARi projected for 2027 for the City's wastewater service area under the current draft CIP is presented on the left. Red census tracts identify "high burden" neighborhoods and produce a systemwide WARi value of 2.41%. This compares favorably to a "high burden" threshold of 3.9% (the equivalent of 2% of MHI – the USEPA's high burden threshold). Pictured on the right is the outlook for the City's financial burdens assuming regulators will approve its proposals resulting in a 16% reduction in total cost and a WARi value of 2.02%.

In reviewing the City's billing data and demographic data, it was observed that many of the lowest-income census tracts within the City's service area are also tracts with a high percentage of multifamily households. This can be observed in comparing Figure 4 with Figure 2. Multifamily accounts in the City of Omaha's service area are billed as general commercial customers – a different rate schedule than for single family accounts. Because of this different rate structure and because of typically lower usage, on a per household basis, individual multifamily households have a lower burden. Additionally, the City of Omaha provides Sewer Rate Assistance funding (SRAF) to low income households, administered through the Nebraska Low Income Home Energy Assistance Program (LIHEAP). These two factors combine to make a significant difference in the overall affordability of the system. Figure 5 presents a comparison of the projected WARi affordability in 2027 with and without the rate structure benefits and

SRAF program. Without these two factors, the systemwide WARi would increase from 2.02% to 2.72%, a 34.7% increase. The impact of the SRAF program and multifamily rate structure are so significant because they mitigate the rate burden on the customers at the lower extreme of income distribution.

CONCLUSIONS

As part of its negotiations with NDEQ in the past, the City has presented its Financial Capability Assessment using standard EPA guidance on the issue. The results of this current study are intended to expand the analysis of the financial burdens to City residents, as well as fairly reflect the benefits of the City's existing rate assistance programs. Results of this analysis better characterize the financial burdens of Order on Consent compliance for the City. Finally, WARi provides the City with a valuable analytical tool to precisely identify the neighborhoods where financial burdens are most pronounced, and to focus its limited financial assistance resources in ways that can achieve measurable results that further reduce financial burdens of the LTCP.

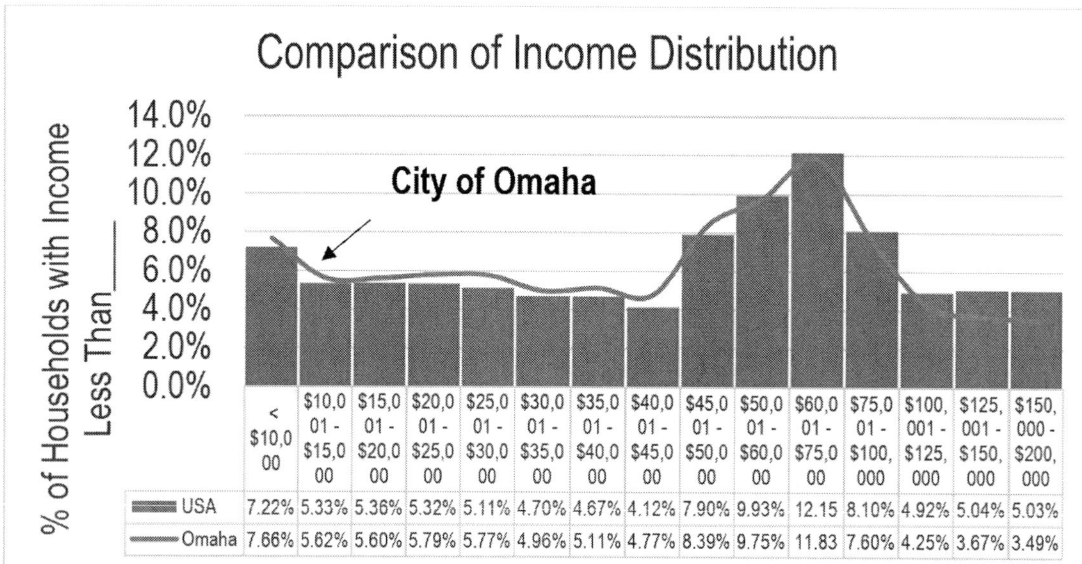

Figure 1. Comparison of City of Omaha Income Distribution with the National Distribution

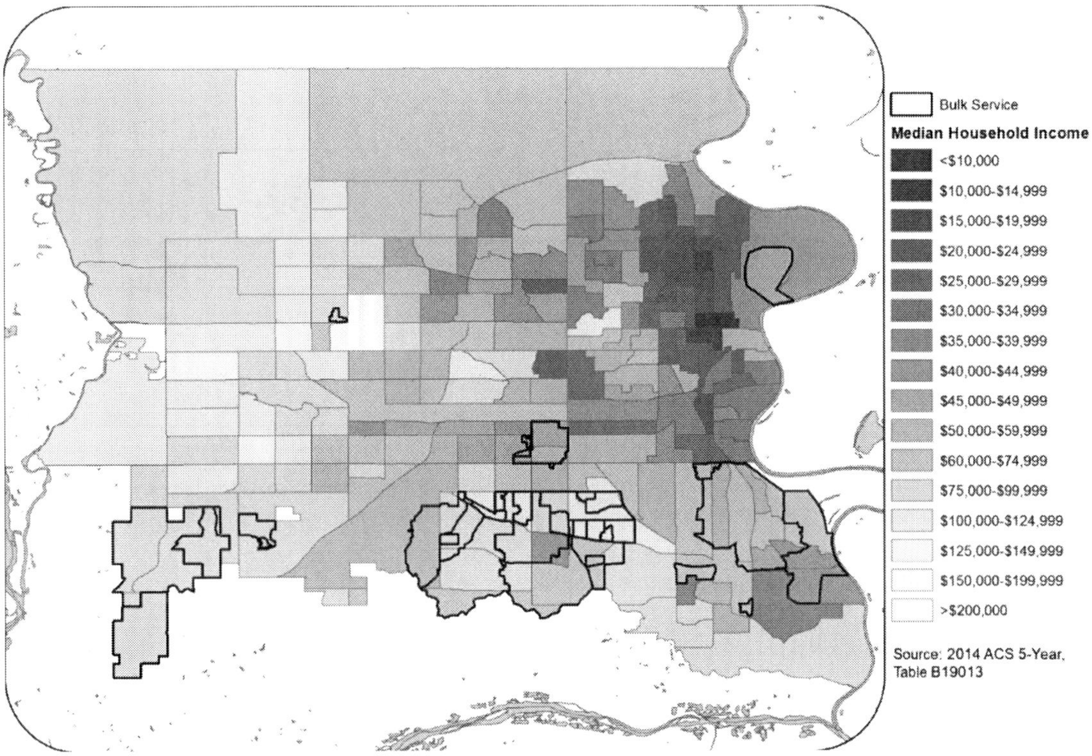

Figure 2. Map of Median Household Income for City of Omaha Service Area

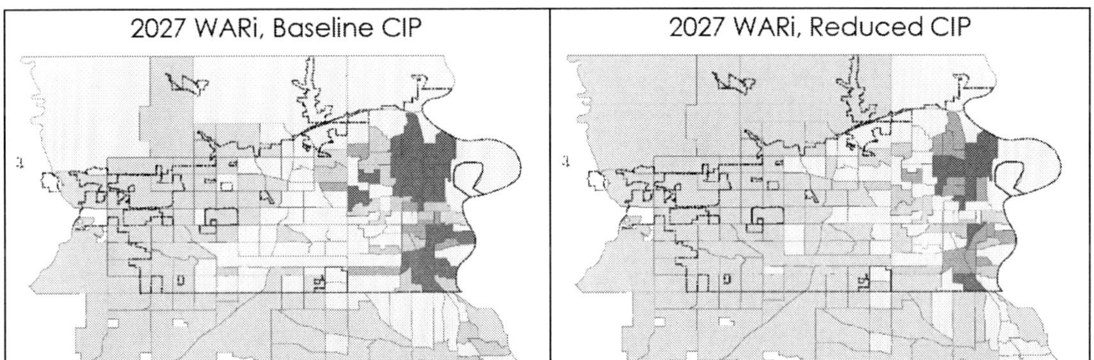

Figure 3. Projected Weighted Average Residential Index (WARi™) in 2027 Under Two CSO CIP Scenarios

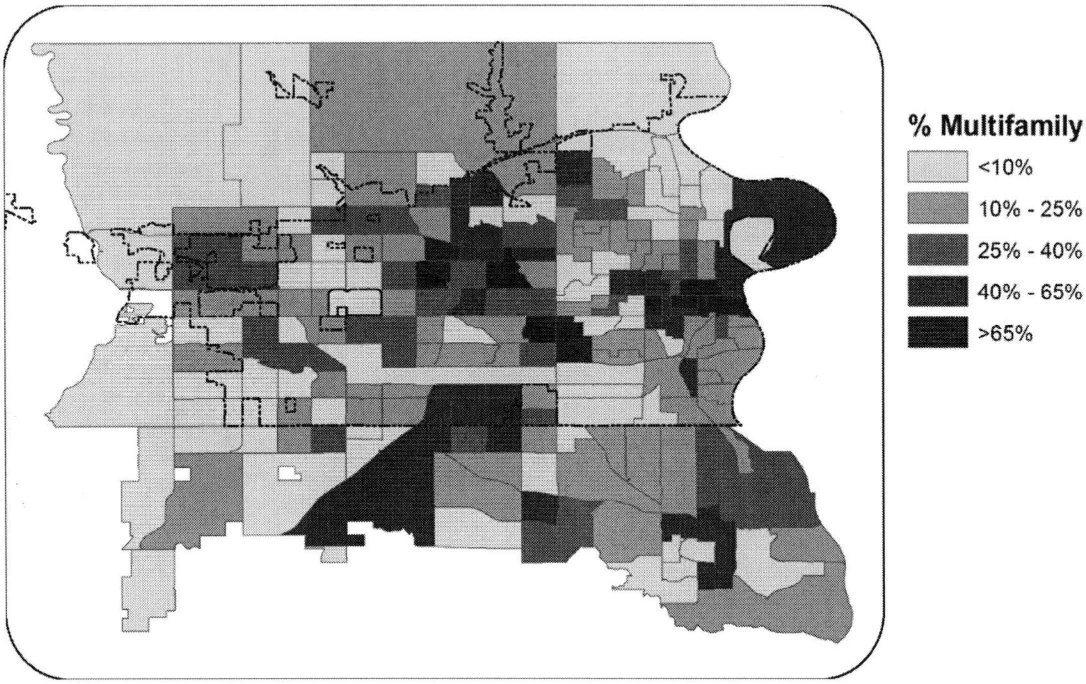

Figure 4. Percentage of Multifamily Households for City of Omaha Service Area

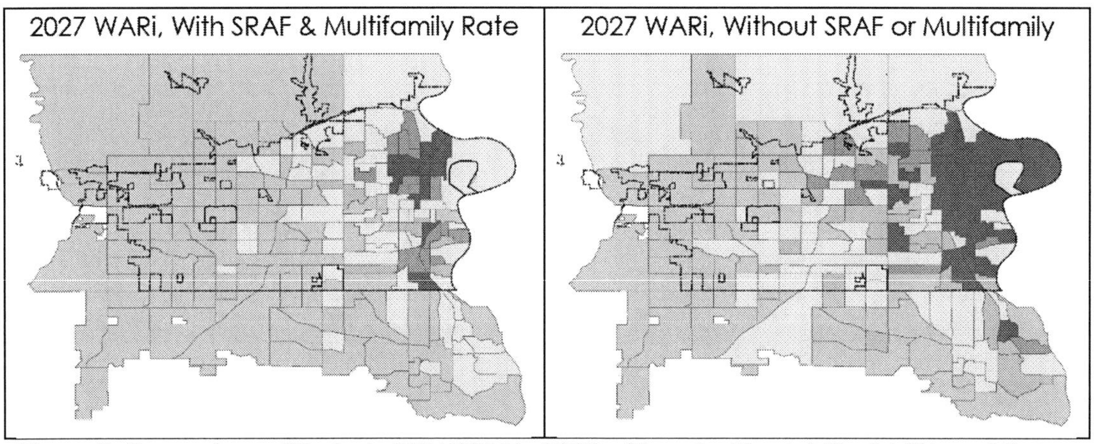

Figure 5. Comparison of Projected WARi™ in 2027 with and without Affordability and Rate Structure Benefits

REFERENCES

USEPA. March 1997. Office of Water. CSO Guidance for Financial Capability Assessment and Schedule Development, EPA 832-B-97-004.

USEPA. November 2014. K. Kopocis and C. Giles. Memorandum: Financial Capability Assessment Framework for Municipal Clean Water Act Requirements. Washington, D.C.

Fort Madison CSO Program Combines Non-Traditional Separation Implementation with Innovative CSO Disinfection

Jay Brady, P.E., Stanley Consultants

ABSTRACT

The City of Fort Madison (City), Iowa is required by a United States Environmental Protection Agency (USEPA) consent order to mitigate its combined sewer overflows (CSOs). This small community of 11,000 embarked on a journey of adaptive management to minimize cost while meeting the CSO mitigation requirements. The current mitigation approach utilizes a combination of sewer separation and overflow disinfection using peracetic acid to reduce overflows and the levels of bacteria entering the Mississippi River. This paper presents an overview of the combined sewer system, the planned separation work, and the evolution of the CSO disinfection system from bench through pilot testing to a fully implemented and effective system. A summary of bench and pilot test results will be provided along with operating results showing system effectiveness. Modifications to the disinfection system during and after pilot testing will be shared to illustrate assorted issues and solutions. The approach the community takes to implement the sewer separation projects including investigation, design, and utilization of city workforce will also be presented.

The chemical disinfection system for the overflows is required to reduce *Escherichia Coli* (*E. Coli*) bacteria to 1,073 most probable number per 100 milliliters (MPN/100 mL). The consent order required a full-scale pilot CSO disinfection system be constructed and operated to demonstrate the feasibility of disinfecting highly variable CSO discharges. As part of the design process, a bench study of peracetic acid (PAA) and sodium hypochlorite (SH) disinfectants was performed by the University of Iowa on behalf of Stanley Consultants, the City's wastewater engineer, to establish potential dose rates and contact times. The bench study utilized jar testing to examine dose concentration-time relationships and impact of suspended solids on disinfection effectiveness. The bench study results showed PAA dosages with 3 to 5 minutes of contact time could achieve E. Coli reductions to 126 mpn/100 mL.

A full-scale pilot chemical disinfection system consisting of chemical feed systems and disinfection contact pipe was designed and constructed. The pilot study results suggested both SH and PAA can be an effective disinfectants for meeting CSO *E. Coli* reductions to the 1,073 mpn/100 mL limit. The data provides evidence that PAA doses that provide residuals of 0.5 mg/L or greater reduce *E. Coli* to down to approximately 100 mpn/100 mL.

While the pilot system demonstrated the technical feasibility of chemically disinfecting combined sewer overflows, the logistics of maintaining and operating several chemical disinfection systems by the small city wastewater staff were determined to be unrealistic. The CSO mitigation plan was modified to continue utilizing the existing chemical disinfection facility at 6th Street with routing of 5th Street overflow to the facility and separation of the other overflow basins.

A project was undertaken by the City to enhance the pilot system to improve operation and maintenance and overall performance as a long-term system and to provide the pipe connection from 5th Street to 6th Street. An overview of this project including performance effectiveness and CSO volume reductions due to enhancements is presented.

The City also embarked on a multiple phase sewer separation project for the 13th Street basin that allowed city crews to self-perform the work. This approach has proven effective at minimizing cash outlays from sewer revenues and minimize additional debt.

KEYWORDS

Peracetic Acid Disinfection, Sodium Hypochlorite Disinfection, Combined Sewer Overflows

INTRODUCTION

The City of Fort Madison, Iowa, a community of 11,000, resides on the Mississippi River in southeastern Iowa. The City entered into a consent agreement with the USEPA to mitigate the impact of its CSOs on the Mississippi River, namely bacteria as evidenced by *E. Coli.* The initial mitigation approach was to disinfect the overflows to reduce the levels of *E. Coli* bacteria entering the river. The consent agreement required a full-scale pilot demonstration of the CSO disinfection system to determine the impacts of CSO flow, duration, and pollutant variability, and the overall feasibility of concept prior to implementation at the other 7 CSO locations.

The City's combined sewers are confined to the older eastern portion of the community. The combined sewer system shown in Figure 1 consists of a series of trunk sewers that run from the toe of the upland bluffs at the northern fringe of the city to its Mississippi riverfront. Historically, these trunk sewers discharged directly into the river. In 1966, a riverfront interceptor was constructed to collect the dry weather flows from the trunk sewers and convey them via the 10th Street Pump Station to the wastewater treatment plant at 20th Street. Leaping weir control structures were installed at each of the trunk–interceptor sewer interface locations to control the amount of flow entering the interceptor during higher wet weather flows. The flow that exceeds the volume allowed by the weirs is discharged to the river via the original trunk sewer outfalls.

Figure 1 Fort Madison Combined Sewer System

Flow monitoring, automated sampling, and floatable control structures were added to the each of the CSO locations to meet the City's nine minimum CSO control obligations. Monitoring and study work including hydraulic modeling performed in conjunction with preparation of the City's CSO long term control plan (LTCP) indicated that the environmental impacts of the CSOs on the Mississippi River are negligible due to low pollutant concentrations and the further dilution of the already dilute CSO water by the much larger Mississippi River. However, *E. Coli* levels in the CSO discharges generally exceeded regulatory limits.

The City decided to approach CSO mitigation by disinfection due to lower initial capital cost. Disinfecting CSOs is challenging due to intermittent flows, highly variable flow rates, and the overall variability of the CSO water quality. The historic CSO flowrates at 6th Street vary between less than 0.0283 cubic meters per second (cm/s) (1 cubic feet per second (cfs)) to a peak of 0.354 cm/s (12.5 cfs).

The CSO water quality also varies. Table 1 data illustrates the variability inherent to Fort Madison's CSOs.

Table 1 CSO Water Quality

Parameter	Minimum mg/L	Median mg/L	Maximum mg/L
BOD_5	2	20	145
TSS	1	135	1480
NH3N	<0.1		1.4
pH	6.9 S.U	7.7 S.U.	8.4 S.U.

BOD5 = 5-Day Biochemical Oxygen Demand
TSS = Total Suspended Solids
NH3N = Ammonia Nitrogen

USEPA, noting these types of challenges for CSO disinfection, required in the 2009 Consent Order that the City perform a full-scale pilot system to demonstrate that the CSOs could be consistently disinfected to meet permit limits for *E. Coli.*

An engineering feasibility study was performed by Stanley Consultants to evaluate potential disinfection technologies appropriate for CSO disinfection with minimal pretreatment upstream of the disinfection process. Chemical disinfection using sodium hypochlorite (SH) or a newly emerging wastewater disinfectant, peracetic acid (PAA), showed the most promise for success. PAA, a very strong oxidant, is a solution of hydrogen peroxide and acetic acid in equilibrium in water. The chemical formula of PAA is CH_3C_3H

Typical solution strength is 15 percent PAA and 22 percent hydrogen peroxide. PAA has been used in previously in the hospital and food/beverage industries as a biocide. PAA offers several advantages over SH including being a faster acting oxidant requiring lower doses and producing no residual byproduct. PAA also has a longer shelf life of six months to a year. However, PAA is expensive relative to SH on a unit cost basis, has limited availability, and is new to regulators.

Bench Study

Additional investigation was necessary to determine the probable values of critical design parameters including chemical dose and contact time needed with anticipated suspended solids concentrations. The University of Iowa Department of Civil and Environmental Engineering was retained to perform bench testing to gain more insight into these design parameters.

Jar testing was performed on simulated CSO effluent to develop chemical disinfection dose rates at varying suspended solids concentrations representative of typical Fort Madison CSO effluent. The simulated CSO effluent consisted of primary clarifier effluent with fine sand addition to attain target total suspended solids (TSS) concentrations of 50, 125, and 200 mg/L. Sodium hypochlorite dose was varied from 10 to 40 mg/l as available chlorine and the PAA dose was varied from 10 to 20 mg/L (Parkin, 2009). The testing was performed to determine doses needed to reduce *E. Coli* to a target of 126 mpn/100 mL in about 5 minutes to minimize contact volume requirements. The target *E. Coli* number was the preliminary discharge limit identified by regulators.

After preparation of the simulated CSO effluent, an initial sample was taken for *E. Coli* and TSS testing. The chemical dose was then applied. The test jars were mixed at 175 rpm to maintain all solids in suspension during the testing. For a given trial, samples were taken from the jar at 10 seconds, 2, 5, 10, 20, and 30 minutes for *E. Coli* testing to develop a dose – time response curve. The samples were immediately dosed with an excess of sodium thiosulfate to quench the oxidation reactions. *E. Coli* was measured using Standard Method 9223B (Parkin, 2009).

Figure 2 presents the results obtained with SH doses as available chlorine at 10, 20, 30, and 40 mg/L with a high TSS concentration of approximately 200 mg/L. The data indicates that a SH dose of 20 mg/L or higher should reduce *E. Coli* below a target of 126 mpn/100 mL within 5 minutes in a well-mixed environment. A SH dose of 10 mg/L will require approximately 10 minutes of contact time to achieve the same target *E. Coli* concentration (Parkin, 2009).

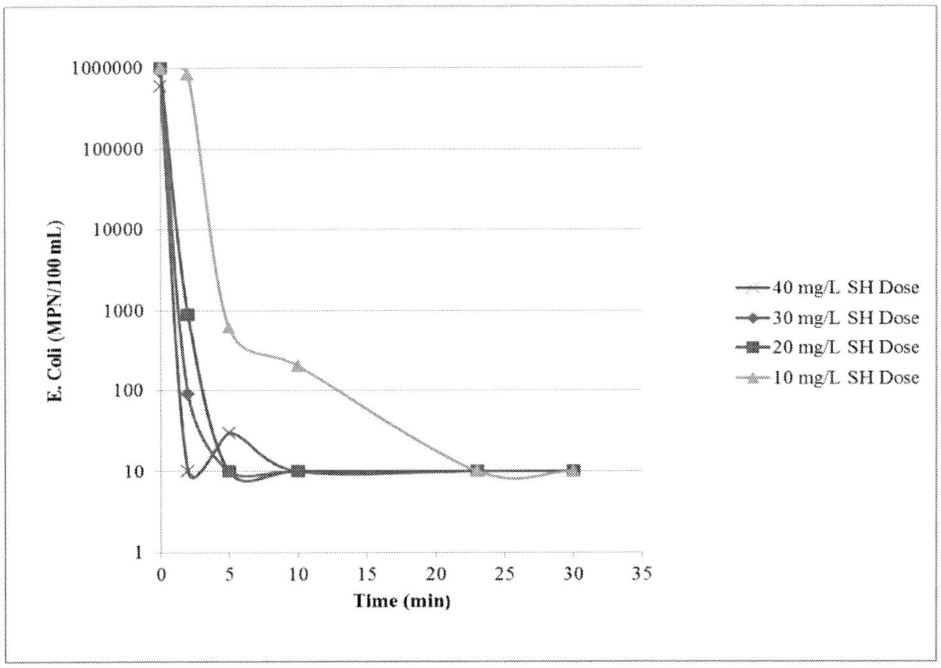

Figure 2 – Results of Variable SH Doses at High TSS Concentration

Figure 3 presents the results obtained with PAA doses at 10, 15, and 20 mg/L with a high TSS concentration of approximately 200 mg/L. The data indicates that a PAA dose of 15 mg/L or higher should reduce *E. Coli* below a target of 126 mpn/100 mL within 3 minutes in a well-mixed environment. A PAA dose of 10 mg/L will require approximately 5 minutes of contact time to achieve the same target *E. Coli* concentration. It should be noted that the 10 mg/L PAA dose consistently achieved near the target bacteria level in 5 minutes, but for some experiments, bacteria levels were somewhat above this target level (Parkin, 2009).

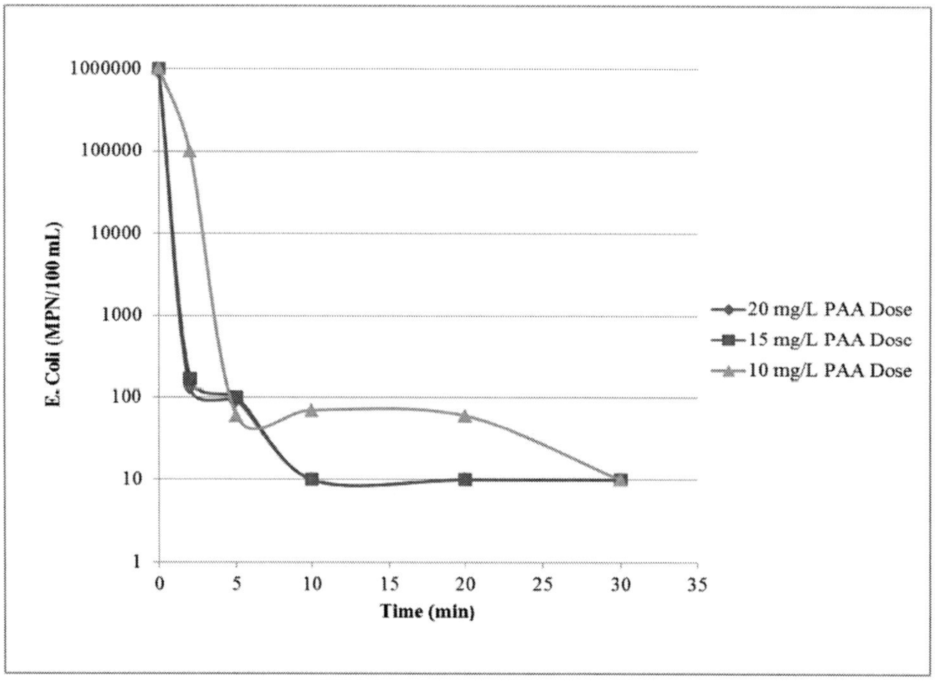

Figure 3 – Results of Variable PAA Doses at High TSS Concentration

Bench study work under ideal and well mixed conditions suggest that a SH dose of 20 mg/L with 5 minutes of contact time or a dose of 10 mg/L with 10 minutes of contact time should reduce E. Coli below 126 mpn/100 mL.

Bench study work under ideal and well mixed conditions suggest a PAA dose of 15 mg/L or higher with 3 minutes of contact time or a dose of 10 mg/L with slightly greater than 5 minutes should reduce *E. Coli* below a target of 126 mpn/100 mL.

Pilot System

The full scale pilot system, consisting of a chemical feed building, SH, PAA, and sodium bisulfite chemical feed systems, debris pretreatment, and disinfection contact chamber, was designed and constructed in 2009 and 2010. Field study work commenced in the fall of 2010 and continued through December 2012. Field study work was hampered by several construction-related issues including flow short circuiting, and drought conditions requiring the extended pilot time period.

The pilot system is located in the City's Riverfront Park on the 6th Street CSO. A contact chamber was provided by constructing a 305 meter (1000 foot) long, 900 mm (36 inch) and 1500 mm (60 inch) diameter polypropylene contact pipe to provide 18 minutes of contact time at 0.354 cm/s (17.4 cfs) as shown in Figure 4.

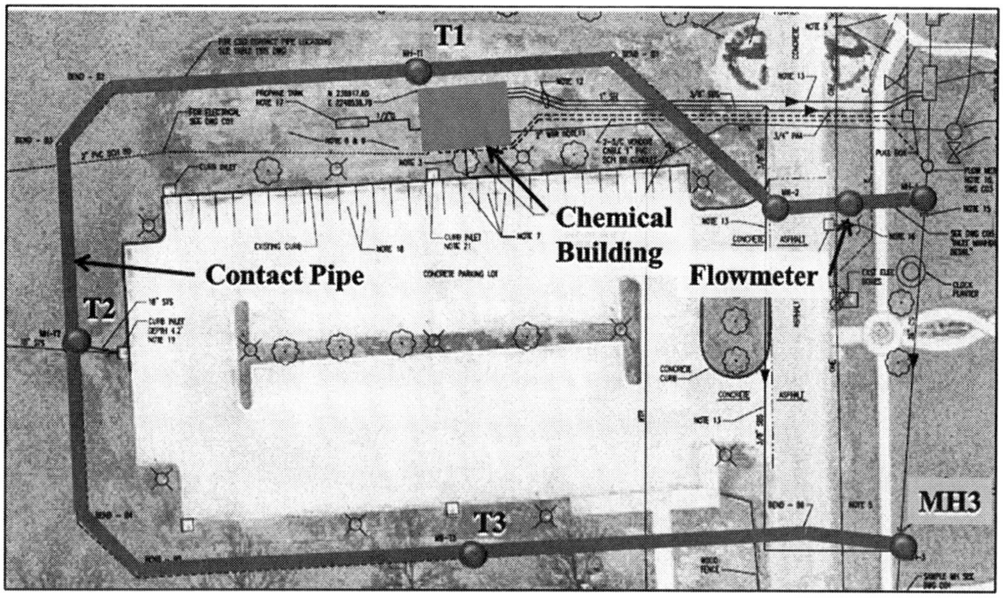

Figure 4 – Pilot Disinfection System Contact Pipe Layout

The contact pipe is equipped with an outlet weir to separate river backwater from the pipe contents and to maintain contact volume and time when the river is low. The pipe is equipped with five manholes for access for cleaning and sampling. Flow is pretreated by a baffle chamber equipped with coarse static screen system for debris and floatables reduction and the flow rate measured with an area velocity flow meter.

The chemical feed building consists of a flood-proofed masonry structure housing SH storage tanks and feed pumps. The PAA feed system consisting of totes and feed pumps are located within the building in a bay separated from the SH system. The building features chemical containment in case of spills or leaks. A sodium bisulfite feed system consisting of totes and feed pumps is also provided to quench the action of the SH or PAA. Separate chemical feed pipes are routed from the building to the diffuser feed point to allow either SH or PAA to be fed. No supplemental mechanical mixing is provided for dispersing the chemicals. Sodium bisulfite piping is routed to the discharge weir. The chemical feed rate is flow-paced based on the upstream flow meter. A data logger records flows and the chemical feed systems are equipped with chemical totalizers. Figure 5 presents the system schematic. Figures 6 – 8 show photographs of various system components.

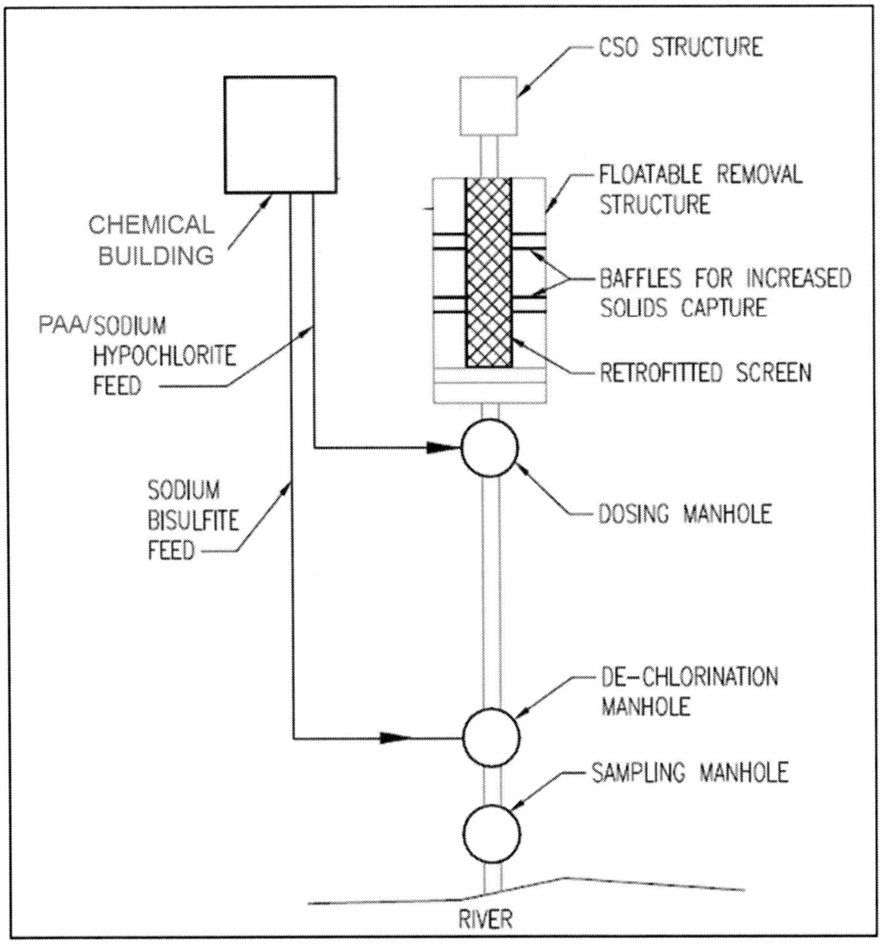

Figure 5- System Schematic

Figure 6 – Contact Pipe Installation

Figure 7 – Chemical Feed Building

Figure 8 – PAA Feed System

The chemical feed system is triggered upon the flow meter sensing CSO flow. The chemical feed pump initially runs at 100 percent capacity for a short period of time to charge the chemical feed piping. The pump then switches to preset flow-paced dosing which continues until CSO flow stops.

Grab samples are obtained upstream of the contact system at the overflow structure (river gate location), the flow meter manhole, T1, T2, T3, and MH-3 and are tested for *E. Coli.* The City strives to start sampling within 30 minutes of the CSO event start. Laboratory analysis of the samples is performed by the City of Fort Madison.

Pilot Data

Field study work was hampered by several construction issues, initial operational challenges, and drought conditions. However, these unintended situations have provided additional insight into the performance of the system and provide real world implications for other systems configured for disinfecting marginally clean wet weather flows.

Thirty-two CSO events occurred at the 6th Street CSO for the period of the pilot study from June 2010 to December 2012. The quality of the overflow water during the pilot study consisted of 5 to 529 mg/L TSS, 3 to 36 mg/L of BOD_5, and 0.1 to 1.5 mg/L of NH3N. The pH varied from 6.2 to 8.4.

As shown in Figure 9, PAA results were variable with some events being effectively disinfected and others having variation in *E. Coli* concentrations along the length of the contact pipe. Some of the variability is due to the overall performance of the chemical feed pumping system. The chemical feed pumps for the PAA are diaphragm pumps which, unfortunately, were not equipped with degassing heads. At times, the pumps became vapor-locked causing the pumps to not start chemical feed or not deliver the full dose expected. The apparent PAA dose delivered was highly variable ranging from 0.9 to 37 mg/L and when combined with variable CSO water qualities produced a range of residual PAA concentrations.

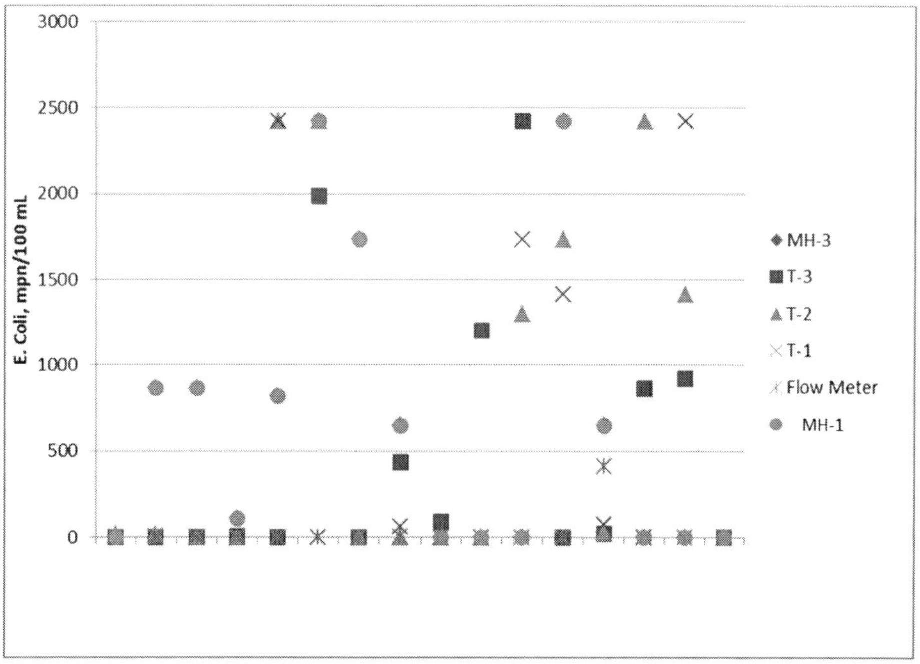

Figure 9 – July 19, 2010 CSO Event Flow

However, the data presented in Figures 10 and 11 does suggest PAA is effective when a residual of 0.5 mg/L or greater can be achieved. Figure 10 shows *E. Coli* concentrations compared to PAA residual. This figure generally indicates that a PAA residual of 0.5 mg/L or more reduces *E. Coli* to 100 mpn/100 L or less. Figure 11 shows *E. Coli* concentrations compared to PAA residual for E. Coli samples below 1000 mpn/100 mL. This figure shows increasing *E. Coli* as PAA residual drops below about 0.5 mg/L.

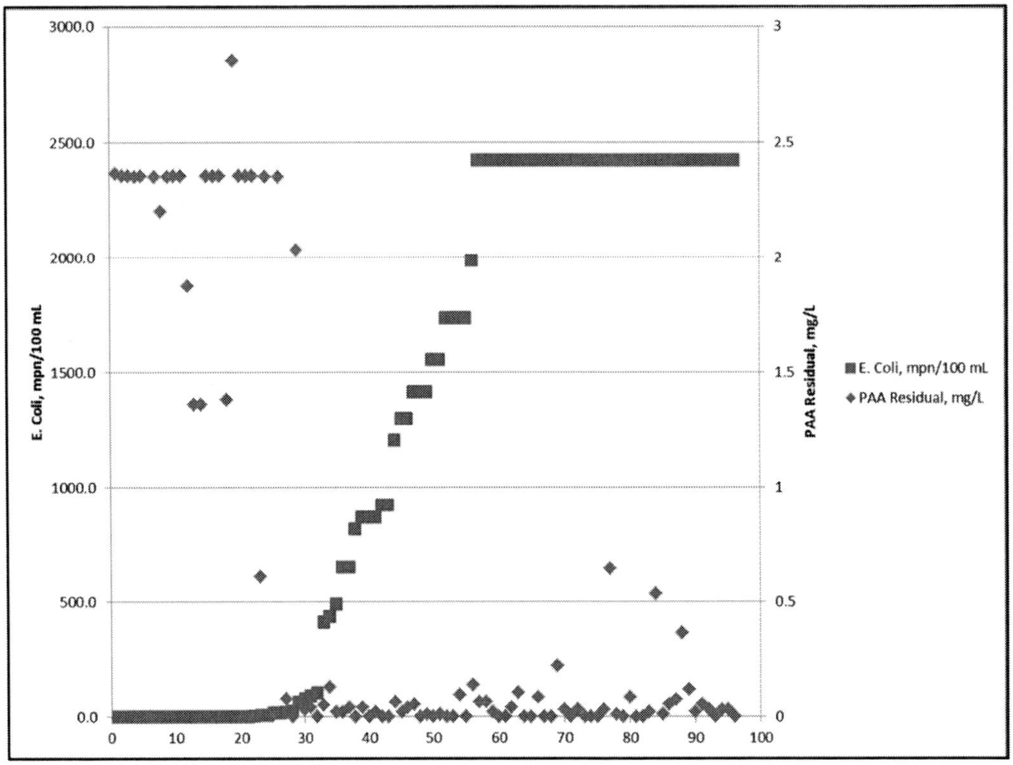

Figure 10 – *E. Coli* Compared to PAA Residual

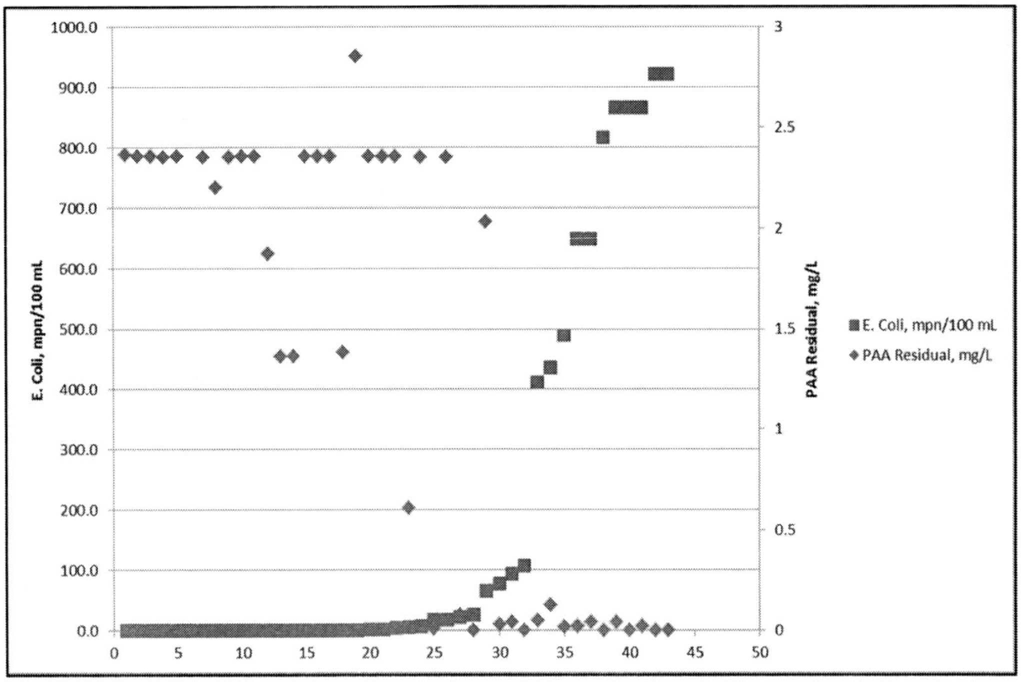

Figure 11 – *E. Coli* Concentrations Below 1000 mpn/100 mL Versus PAA Residual

Figure 12 shows a regression for 2012 and late fall 2011 CSO events indicating that a PAA residual of about 0.5 mg/L is necessary for effective reduction of *E. Coli.*

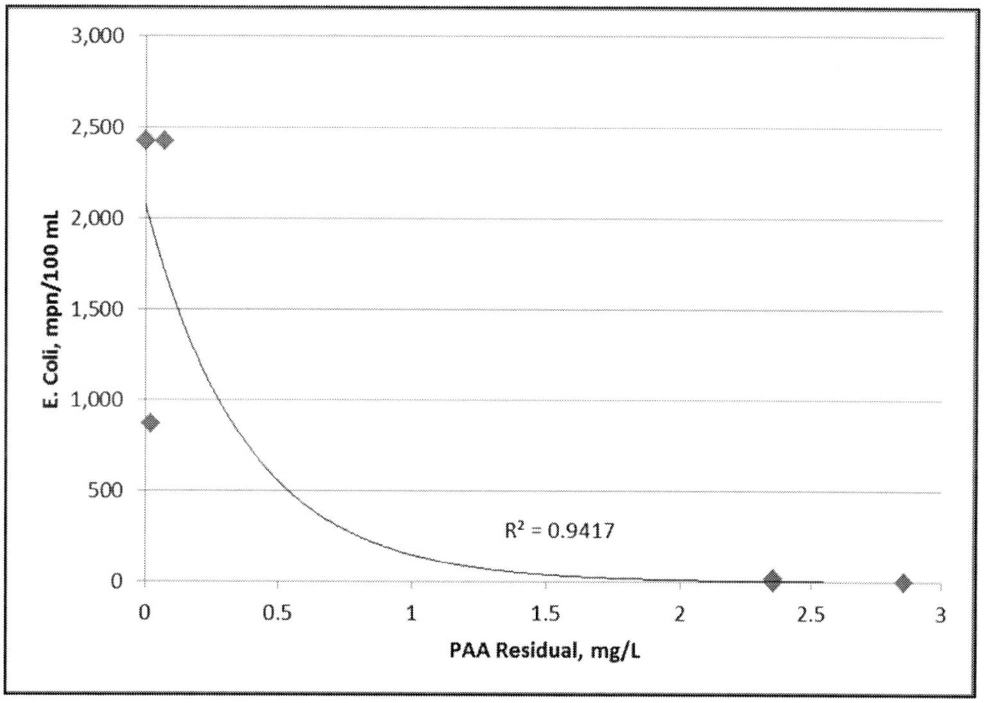

Figure 12 – *E. Coli* vs PAA Residual for 2012 CSO Events

Other factors aside from the chemical pumping issue contribute to the overall variability observed in the PAA results.

The large contact pipe and the feed configuration contribute to the variability. The large contact pipe sized for an extreme flow event holds approximately 379 cubic meters (100,000 gallons). Often the CSO events are a fraction of that volume resulting in the freshly disinfected water traveling only partially down the length of the pipe. Over time the disinfectant is consumed by organics in the water potentially allowing regrowth of bacteria. Some solids settling occurs producing a thin film of bottom solids that can also act as a source for bacteria.

Another issue is the location of the flow meter downstream of the chemical feed point and associated chemical pump control. The chemical pumping is initiated when the flow meter senses flow. A relatively small volume (11.4 cubic meters) of water enters into the contact pipe before the chemical feed starts resulting in non-disinfected water entering the pipe with each CSO event.

Another contributing factor until spring of 2012 was the parking lot storm drain pipe was connected into the contact pipe at T-2 location. The extraneous flows from the parking lot storm water runoff disrupted the hydraulics of the contact pipe and provided another potential source of bacteria.

Preliminary pilot study results suggest PAA can be an effective disinfectant for meeting CSO *E. Coli* reductions. The data provides evidence that doses that provide residuals of 0.5 mg/L or greater reduces E. Coli to down to approximately 100 mpn/100 mL.

Inadequate chemical dispersion especially at low, less turbulent flows may be contributing to the observed variable performance. Consistent chemical disinfectant feed with complete dispersion into the water so the disinfectant contacts all of the water is very important in the overall effectiveness of the disinfection process. Addition of mechanical mixing at the point of application should be considered to improve chemical dispersion.

Bacteria regrowth in the contact pipe and associated sediment may be a factor in the inconsistent results. Draining the pipe between CSO events may reduce this potential source and also allow storage of some CSO water reducing total CSO volumes discharged.

Continuous chemical dose feedback via oxidation reduction potential sensor or similar sensor technology would enhance overall dose control and system performance.

PAA shows promise as a disinfectant agent in wastewater especially in non-ideal wet weather systems. The chemical has a long shelf life of greater than six months. The PAA being used in Fort Madison is from the initial purchase over two years ago at the beginning of the pilot demonstration. Periodic testing of the solution strength indicates that it has only declined by about a percent. This type of shelf life is very advantageous for intermittent wet weather flow applications where a system may be go for a month or more without being used.

Mitigation Plan Revision

City staff recognized the challenge of operating the disinfection system with their limited wastewater staff. Bacteria sampling is very restrictive with need to perform grab samples during the discharge event with limited time to initiate the laboratory analysis. These requirements mean that staff would need to be available during wet weather events when overflows are likely to occur regardless of the time of day or day of the week. Often other issues in the collection system, at the pump stations and at the treatment plant would arise during the wet weather events spreading staff thin. The other recognition was the high cost of implementing additional disinfection facilities in the City's riverfront park and overall public acceptance of multiple facilities. The City determined that multiple disinfection facilities on the riverfront was not a viable alternative.

The City and Stanley Consultants worked with USEPA to amend the Consent Order in 2013 to reflect sewer separation instead of disinfection for the CSO mitigation approach and update the program schedule. Disinfection of 6th Street would continue with extension of piping from 5th Street to 6th Street to facilitate disinfection of 5th street overflows by October 2014. Time was also allotted for making improvements to the 6th Street disinfection system to improve reliability, performance, dose control, and maintenance.

6th Street CSO Disinfection System Improvements

A project was undertaken to make various improvements to the CSO disinfection system for improved operation, maintenance, and reliability. The project also installed pipe between the 5th Street overflow and the 6th Street CSO disinfection system to convey 5th Street overflows to the disinfection system. Table 2 summarizes the major improvements and the anticipated benefit.

Table 2 – 6th Street CSO Disinfection System Enhancements

Improvement	Benefit
Replace PAA pump with peristaltic	Prevent vapor lock by PAA, improving pump dosing reliability
Add mechanical mixing at PAA feed location	Rapid PAA dispersal improving effectiveness
Add TSS meter to flow into system	Facilitate PAA dose control with TSS feedback
Add water level sensing at discharge weir	Accurately determine when flow goes to river vs being retained in contact pipe
Add contact pipe drain to sanitary	Allows pipe to be drained between overflows minimizing regrowth and reducing overflow volume
Exhaust system modifications	Improved building interior atmosphere by exhausting PAA totes
Control System modifications	Improved pump and chemical dose control and monitoring
Sampling location access improvements	Facilitate sampling
Static trash screen at 5th Street Overflow	Reduce debris entering disinfection system
New overhead power to building	Improve power reliability especially during flood events

Figure 13 provides an overview of the project and the pipe conveyance from 5th Street to 6th Street. Figure 14 provides an enlargement of the CSO disinfection system features.

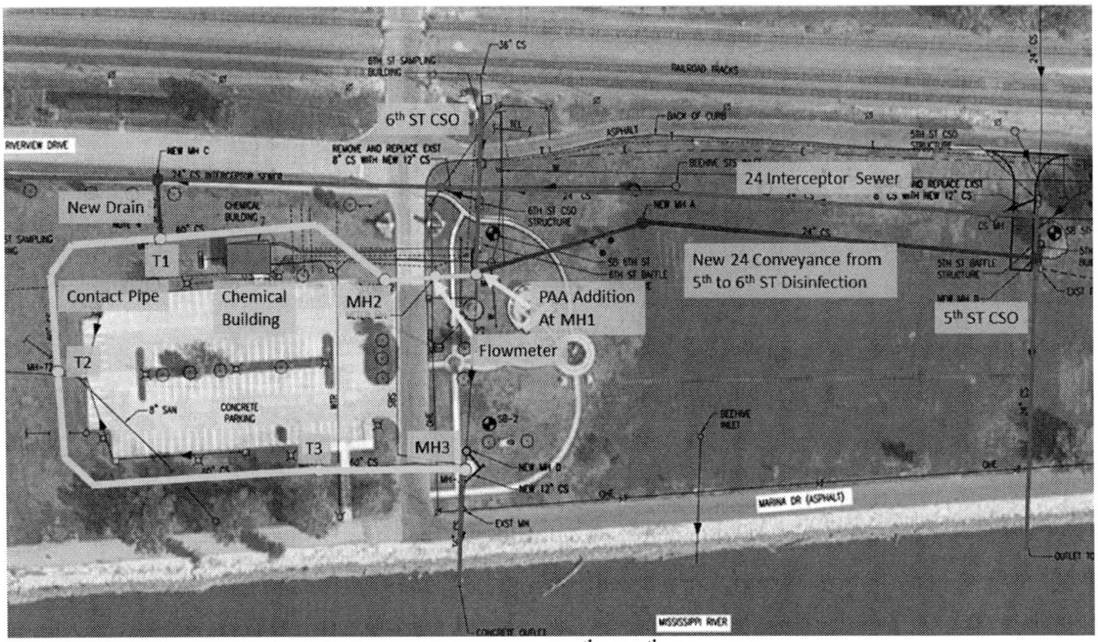

Figure 13 – Overview of Disinfection System and 5th to 6th Street Conveyance

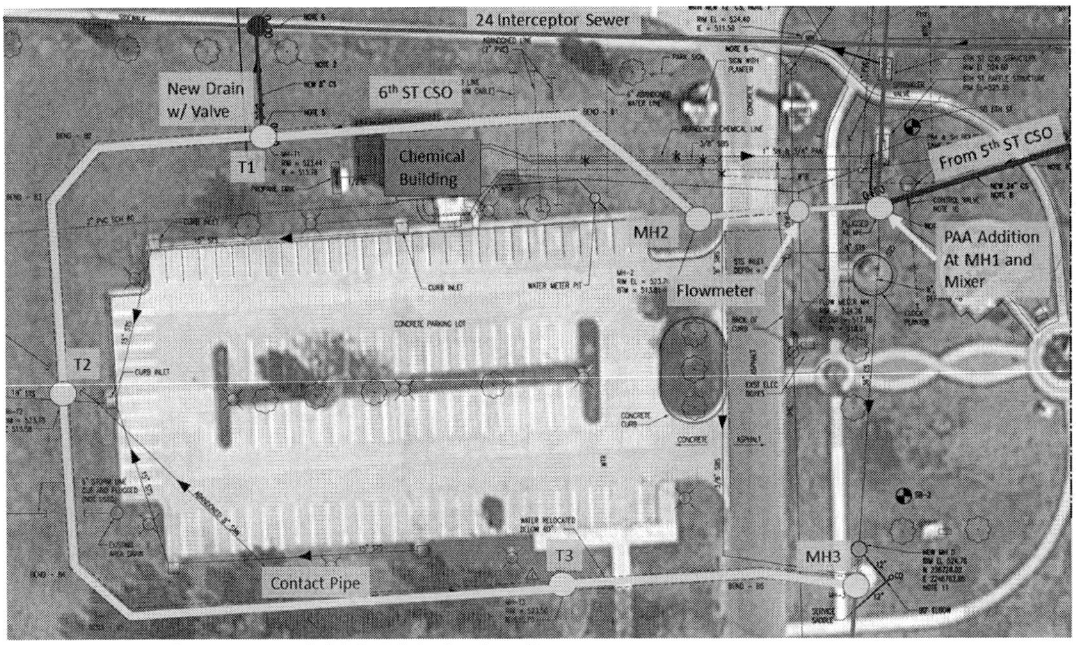

Figure 14 – Enlargement of CSO Disinfection System

Seven CSO events have occurred since the CSO disinfection system enhancements. The system has performed well with E. Coli levels generally reduced to 1 mpn/100 mL. Table 3 presents the CSO disinfection system discharge quality for these CSO events.

Table 3 – CSO Disinfection System Discharge Quality

Event Date	Total Event Flow Gallons	Influent E. Coli mpn/100 mL	Effluent E. Coli mpn/100 mL	Residual PAA mg/L	TSS mg/L	$CBOD_5$ mg/L	NH3N mg/L
7/6/2016	160,270	>2420	2	0.107	81	45	3
7/13/2016	251,160	>2420	1	1.56	150	25	0.11
7/22/2016	303,830	>2420	1	1.56	34	18	0.7
8/5/2016	201,850	>2420	299	0.35	165	17	0.05
8/11/2016	113,540	>2420	1	1.56	106	40	0.95
8/12/2016	366,520	>2420	1	1.027	34	18	0.16
11/2/2016	161,550	>2420	1	1.56	125	40	0.2

Further optimization of the dose criteria and controls is necessary to optimize PAA feed to consistently achieve discharge residual concentrations in the 0.5 to 1.0 mg/L range.

In addition, the City has been able to eliminate multiple overflows associated with smaller events by draining the contact tank down between CSO events and at times even during wet weather events depending on event characteristics.

13th Street Basin Separation Approach

The 13th Street basin is one of the City's larger combined sewer basins with significant volumes of water discharged to French Creek near its confluence with the Mississippi River. However, the actual number of sanitary connections to the combined system pipe network is relatively small.

The project consists of extending a new 10" sanitary sewer nine blocks along the west side of 13th Street to facilitate connection of the scattered households still connected to the combined pipe network. After elimination of sanitary connections to the combined piping and a monitoring period to confirm no sanitary sewage, the combined piping will be declared storm piping. The consent order deadline for completion of the separation is November 15, 2017.

The project was designed in collaboration with City staff to facilitate permitting of the construction to be performed by City staff. City performed multiple small scale investigations to verify piping and house connections through televising and dye testing.

The design was broken into two packages to facilitate construction phasing. Ultimately the City performed the construction work in three phases.

The excavation and backfill work was hired out to a local contractor. City staff installed the pipe and facilitated paving restoration.

The City performed the work for approximately 50 to 60% of the cost of contracting the project out to contractors. City staff feel that this use of city labor versus other maintenance activities was a good trade off compared to cash outlays to hire the work out.

RESULTS AND DISCUSSION

DISCUSSION AND CONCLUSIONS

Bench study work under ideal and well mixed conditions suggest that a SH dose of 20 mg/L with 5 minutes of contact time or a dose of 10 mg/L with 10 minutes of contact time should reduce E. Coli below 126 mpn/100 mL.

Bench study work under ideal and well mixed conditions suggest a PAA dose of 15 mg/L or higher with 3 minutes of contact time or a dose of 10 mg/L with slightly greater than 5 minutes should reduce *E. Coli* below a target of 126 mpn/100 mL.

Pilot study results indicated PAA can be an effective disinfectant for meeting CSO *E. Coli* reductions. The data provides evidence that doses that provide residuals of 0.5 mg/L or greater reduces E. Coli to down to approximately 100 mpn/100 mL.

Several pilot study system issues were identified that hampered the consistent performance of the system including:

- Inadequate chemical dispersion especially at low, less turbulent flows may be contributing to the observed variable performance. Consistent chemical disinfectant feed with complete dispersion into the water so the disinfectant contacts all of the water is very important in the overall effectiveness of the disinfection process. Addition of mechanical mixing at the point of application should be considered to improve chemical dispersion.
- Bacteria regrowth in the contact pipe and associated sediment may be a factor in the inconsistent results. Draining the pipe between CSO events may reduce this potential source and also allow storage of some CSO water reducing total CSO volumes discharged.
- Continuous chemical dose feedback via oxidation reduction potential sensor or similar sensor technology would enhance overall dose control and system performance.

PAA shows promise as a disinfectant agent in wastewater especially in non-ideal wet weather systems. The chemical has a long shelf life of greater than six months. The PAA being used in Fort Madison is from the initial purchase over two years ago at the beginning of the pilot demonstration. Periodic testing of the solution strength indicates that it has only declined by about a percent. This type of shelf life is very advantageous for intermittent wet weather flow applications where a system may be go for a month or more without being used.

Improvements to the 6th Street CSO Disinfection System appear to have improved system performance as demonstrated in 7 CSO events since the improvements. These improvements include:

- Replacement of PAA feed pump with peristaltic pump that doesn't vapor lock.
- Addition of mechanical mixing to improve PAA dispersion.
- Addition of contact pipe drain that allows accumulated water to be drained back into the sanitary system. Draining the pipe minimizes regrowth issues and minimizes overflows by retaining smaller storm overflows.

- Improved control and feedback including total suspended solids meter to give ability to optimize PAA dose.

Finally City has been able to extend its cash resources by self-implementing sewer separation in the 13th Street basin trading city labor to prevent cash outlay.

ACKNOWLEGEMENTS

1. City of Fort Madison, Iowa especially Jeff Helling, Wastewater Superintendent; the dedicated laboratory staff; and the operations staff who have to go out and sample, Rick Kuhljuergen, Retired Wastewater Superintendent, and Larry Driscoll, Public Works Director.
2. University of Iowa, Professor Gene Parkin, Graduate Students, Jessica Beswick, Matt Reusswig.

REFERENCES

Enviro Tech Chemicals, Inc, 2009, *Peracetic Acid*

Kitis, M., 2003, *Disinfection of Wastewater with Peracetic Acid: A Review*

Meakim, J.; Rolak Jr., J.; Howarth, J., 2008, Point Source Disinfection of New Jersey Wet Weather Events: Stormwater and Combined Sewer Overflows (CSO's).

Parkin, G.F.; Beswick, Jessica; Reusswig, Matt, 2009, *Peracetic Acid Disinfection of Simulated Combined Sewer Overflow* Prepared by the University of Iowa for Stanley Consultants (Unpublished).

Stanley Consultants, 2009, *Disinfection Engineering Study, Combined Sewer Overflows* Prepared for the City of Fort Madison, Iowa.

Stanley Consultants, 2006, *Final Combined Sewer Overflow Long Term Control Plan* Prepared for the City of Fort Madison, Iowa.

United States Environmental Protection Agency, 1999, *Combined Sewer Overflow Technology Fact Sheet Alternative Disinfection Methods*, EPA 832-F-99-033, Washington D.C.

Parallel Master Planning and Implementation to Address Significant Stormwater Needs

Michael Butterfield, P.E.[1*], David Wallace, P.E.[2], Michael Schubert, P.E.[1]

[1] HDR
[2] City of Cedar Rapids
[*] Email: Michael.Butterfield@hdrinc.com

ABSTRACT

In June 2014, torrential rains of nearly 8 inches per hour fell in Cedar Rapids, leading to flash flooding across the city. The localized flooding that occurred during the event was significant and widespread throughout the city, causing substantial damage to private property, including massive sewer backups and washed-out roads.

The magnitude of the damages and increasing frequency of these extreme events prompted the City to review and reemphasize the capacity and condition of its stormwater infrastructure. However, time was of the essence and the needs were so significant that they felt they couldn't sacrifice construction time while waiting to shift money from the capital improvements plan (CIP) to a master plan. Therefore, the City decided that parallel master planning and implementation was the most efficient way forward. Through this approach, the City has simultaneously continued essential improvements and refocused their long-term CIP.

KEYWORDS: stormwater, master planning, sewer modeling

INTRODUCTION

It has been nearly 20 years since the City of Cedar Rapids last completed a Stormwater Master Plan. At that time in the mid-1990s, the estimated $14.4 million for recommended improvements was beyond funding capabilities and numerous recommendations were never implemented.

Fast forward to June 29 and 30, 2014 when 3.5 to 5.5 inches of rain fell on the City of Cedar Rapids in less than 6 hours. Rainfall intensities peaked above 8 inches per hour for short periods in some locations. The localized flooding that occurred during the June 2014 event was significant and widespread throughout the City, causing substantial damage to private property, including massive sewer backups and washed-out roads.

The magnitude of the damages and increasing frequency of these extreme events prompted the City to review and reemphasize the capacity and condition of its stormwater infrastructure. However, time was of the essence and the needs were so significant that they felt they couldn't sacrifice construction time while waiting to shift money from the capital improvements plan (CIP) to a master plan. Therefore, the City decided that parallel master planning and implementation was the most efficient way forward. Through this approach, the City has simultaneously continued essential improvements and refocused their long-term CIP.

Over the course of nine months in 2015 and 2016, Cedar Rapids and HDR developed a hydraulic model of the City's stormwater system and refocused the CIP program. The first step of the process was a prioritization of known stormwater needs. The City's intent for that first step was to move forward with the highest priority – and most readily implemented – stormwater needs while taking the initial steps to produce a living Master Plan that will become more comprehensive through additional effort over the next several years.

METHODOLOGY

The approach taken by Cedar Rapids was to develop a "living document" to be built over several years and regularly revisited. The intent is to identify and move forward with capital improvements to address the highest priority stormwater needs while producing a living Master Plan that will become more comprehensive through additional modeling and master planning effort over the course of several years.

The goals for the master plan are intertwined, yet divergent. Each of these goals is kept in mind as priorities are set, evaluation criteria are defined, and potential projects are developed.

1) *Address flash flooding*: As the impetus for renewing a Stormwater Master Plan, this is a critical goal. The hydraulic model was the essential element to understand which bottlenecks were causing surface flooding, ideal locations for detention, etc.
2) *Reduce pumping capacity demands on the Cedar River flood control system*: The large drainage basins behind the City's proposed flood control system necessitate large pump stations to handle interior drainage during high river events. The City's intent is to use the Master Plan to identify detention throughout the watershed that can help reduce the peak flow rates at the pump stations.
3) *Enable growth that improves stormwater quality & quantity concerns*: As the city continues to grow, the Master Plan will be used to establish policy to enable development that achieves water quality goals and reduces downstream flood risks.

The initial phases of the master planning effort were a collaborative effort of City and consultant staff to bring a fresh perspective to accomplish the following specific objectives:

1. Reflect the City's vision for the future as presented in the City's comprehensive plan (EnvisionCR).
2. Develop a hydraulic model that can serve as a platform to better define needs and consider cost-effective solutions.
3. Reflect City staff's efforts to maintain and preserve the existing stormwater system through asset management.
4. Prioritize and recommend projects for the capital improvements plan.
5. Identify financial needs and potential funding to close the gap between revenues and needs.
6. Consider potential policy solutions to address fundamental issues that contribute to current flooding problems.
7. Establish the foundation for annual updates to expand and improve upon the initial plan.

Phases 1 & 2 (2016)

The first phase of the Master Plan, completed in 2016, was intended to identify and move forward with capital improvements to address the highest priority stormwater needs while producing the living Master Plan document.

As a result of the 2008 Cedar River flooding and June 2014 flash-flooding event, considerable response has been provided to the City. Disaster assistance was provided to replace a major storm sewer that was 110 years old and to perform repairs on several detention basins. City funds provided for the replacement of three large culverts, installation of safety inlet improvements on large storm sewer openings, regular maintenance repairs to numerous pipes and structures, and initiate construction of regional detention basins.

Priority & Readiness

With limited resources available for stormwater improvements, the City's first goal for the Master Plan was to target those financial resources on projects that will provide the greater system performance benefit to residents in an objective, systematic, and open manner. Identifying those critical capital needs was based on two philosophies – priorities and readiness.

Determining the project's priority was done by objectively scoring a project against a number of weighted criteria, including:

- Health and Safety,
- Cost–Benefit (Cost of Potential Damage / Cost of Project),
- Current Design Capacity,
- Asset Functionality,
- Water Quality,
- Associated Considerations,
- Sanitary Sewer Inflow Conveyance,
- Future Growth, and
- Easements.

Criteria scores were weighted and tallied into a project prioritization ranking score. The prioritization ranking scores for all projects were sorted from highest (most severe) to lowest (least severe). The result was a defendable process for evaluating and scoring each of the City's storm water projects using a consistent set of pre-determined criteria.

During the project prioritization process, it became apparent there was another factor that affected fiscal year CIP selection: project readiness. To immediately move forward with addressing the needs of the system, projects had to be ready to construct. So, as the prioritized projects were slated into the CIP, a second evaluation was used to determine the project's readiness by accounting for factors such as: land acquisition, previous City commitments, and funding sources. This two scale approach will ensure that critical issues within a system are being addressed while allowing the City flexibility to allocate funds to specific projects as warranted by readiness factors.

Macro-Model & First Basin Model

A fundamental objective of the Master Plan was development of a model of the City's stormwater system as a tool to evaluate existing stormwater infrastructure, identify capital improvement projects, analyze conceptual designs of these projects, and update the entire system model following project implementation. The modeling framework utilizes a multi-step, multi-scale approach to evaluating storm water infrastructure performance throughout the city.

The first step focused on a citywide "macro-scale model" that incorporated large pipes, open channels, and major detention facilities of City's storm water conveyance system. The macro-scale model provided a broad overview of the performance of major components of the storm water system and allowed for prioritization of succeeding analyses. The resulting macro-scale model also provided a foundation for the more detailed "basin-scale models" to be developed in subsequent phases. The basin-scale models include a much more extensive pipe network and simulate both the one-dimensional (1D) pipe and channel flow and two-dimensional (2D) overland flow.

The storm water models were developed with InfoWorks Integrated Catchment Model (ICM) software by Innovyze. The model dynamically simulates rainfall runoff, pipe flow, open channel flow, storage, routing, pumping, and overland flow to evaluate the performance of the City's storm water infrastructure. InfoWorks ICM uses Geographic Information System (GIS) based inputs from the City's database, which included topography, soil type, land use, pipe network, and detention pond data. Other data (such as as-built plan sets) were used to fill-in or infer missing information in the City's GIS database. Significant gaps in the database were identified and survey data were collected to improve the storm sewer GIS database and develop reliable model inputs.

The macro-model was qualitatively validated using the City's observations and radar rainfall reconstruction of the June 29-30, 2014 event. This event had rainfall depths varying from 3.5 to 5.5 inches across Cedar Rapids, most of which fell in one hour. 5-minute rainfall intensities developed from the rainfall radar data for the event peaked above 8 inches per hour (8 in/hr over a 5-minute period) in some locations. Based on the peak 1-hour rainfall depth, this event was similar in magnitude to a 500-year 1-hour duration rainfall event. Model-predicted results were reviewed and discussed with City staff and found to generally coincide with anecdotal flooding information from that event including flooding call logs and other data gathered after the storm. A sample of the results from the macro-model is shown below in Figure 1; the E Avenue NW basin is depicted.

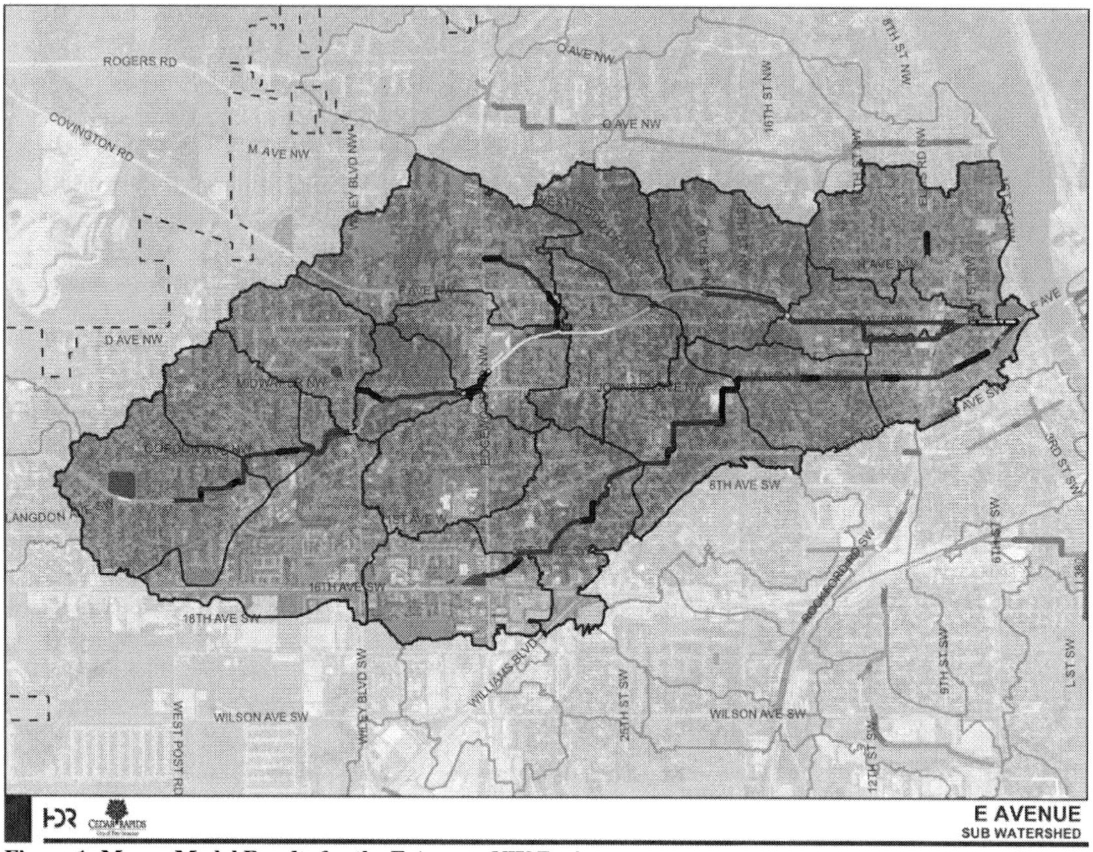

Figure 1: Macro-Model Results for the E Avenue NW Basin

The macro-model was used to simulate the 5- and 100- year 24-hour nested storm events, which incorporates the intensities for storm durations ranging from 5 minutes to 24 hours in a single 24-hr temporal rainfall distribution. The system performance analysis assumed uniform rainfall distribution over the entire City. Both the nested and uniform distribution are conservative when considering the entire basin in that they likely overestimate rainfall and runoff from any given 5- or 100-year storm event that is varies more spatially and temporally. However, the uniformly distributed nested storm is a useful tool in evaluating the level of service provided in the system because it produces the same flow return interval throughout all points of the system, independent of catchment size or time of concentration. By evaluating the system under these assumptions, specific elements within the system were evaluated for a return interval with one simulation, and performance was compared to the Cedar Rapids Metro Area standards criteria for the storm water infrastructure.

Basin Models

An essential benefit of the macro-model was helping to identify problem areas in the system. It showed bottleneck pipes, surcharged areas that resulted from downstream bottlenecks, as well as flows and volumes in the system. The big-picture views that the macro-model afforded enabled the City to target and prioritize the 23 watersheds and sub-watersheds in the city for basin-level modeling. In conjunction with other factors influencing prioritization, the capital improvements plan assembled during the initial Master Plan laid out an order for completing individual basin studies.

The detailed basin-level models add 2D overland flow and additional detail to the 1D macro-model. The basin models simulate ponding and overland flow for storm water flows in areas with collection systems exceeding the existing capacity of the storm sewer system. As additional basin models are developed over the coming years, the results will be used to inform and improve the capital improvements plan (CIP), develop basin-wide mitigation alternatives, and evaluate individual projects. This multi-year strategy will also allow the macro-model to be improved and updated as more data becomes available.

The same 5-year and 100-year 24-hour nested storm events used in macro modeling, uniformly distributed over the basins, were simulated for the basin-scale models. The results of the basin-scale models are being used to consider flood mitigation strategies. The potential mitigation strategies include green infrastructure retrofits, distributed detention, regional storage, and conveyance improvements. This process, which utilizes modeling results and available GIS data, prioritizes strategies that reduce the amount and intensity of runoff using storage and/or green infrastructure, considers the effectiveness in reducing storm water management issues downstream, and then mitigates remaining issues through conveyance improvements. In other words, it categorizes mitigation strategies as green infrastructure, local distributed detention, regional detention, and conveyance improvements with an emphasis on the first three.

To-date, three basin-scale models have been developed: Kenwood, "O" Avenue NW, and "E" Avenue NW.

- *Kenwood*: The Kenwood sub watershed was selected as the initial basin-scale model for several reasons. Most notably, Kenwood was selected because in June 2014 it had the highest reported densities of flooding incidents and the highest number of calls to dispose of flood damaged materials as solid waste. The Kenwood basin also had a flooding related loss-of-life. Furthermore, the Kenwood area is nearly fully developed, among the older neighborhoods, and issues and potential solutions are not well understood. The Kenwood basin model was used to consider flooding mitigation strategies within the basin. The potential mitigation strategies include green infrastructure retrofits, distributed detention, regional storage, and conveyance improvements. An example of the basin-scale model results is shown below in Figure 2, depicting the 5-year storm results for the Kenwood basin.
- *"O" Avenue NW*: One of the highest priority projects identified project prioritization projects was an improvement to the Harrison Basin, a large detention basin in the "O" Avenue NW sub watershed. This was a problem area in the June 2014 flash flood and the basin has experienced overtopping, leading to overland flow and damages on private property. Additionally, this sub watershed leads to a large stormwater pump station proposed for the City's flood control system, so significant reductions in peak flows were desired to address both flash-flooding and interior drainage. Two detention basin projects were developed using modeling results to mitigate downstream flooding; these projects have been added to the CIP for implementation.
- *"E" Avenue NW*: The "E" Avenue NW watershed is one of the most complex basins in Cedar Rapids, covering more than 2,900 acres on the west side of the Cedar River. Two

sub-watersheds within the overall basin converge just upstream of the outfall to the river, both of which are fully enclosed in box-culverts for at least 1-mile upstream of the outfall. The northern branch, called Vinton Ditch, is a regulatory floodplain with numerous residences located in the 100- and 500-year floodplains. As a result of these two branches outfalling to the river at a single location, managing storm drainage in the context of the Cedar River flood control system requires a strategy beyond pump stations. The "E" Avenue study, currently in progress, is using the model to identify distributed detention, regional storage, green infrastructure retrofits, and – if necessary to address undersized storm sewers – conveyance improvements.

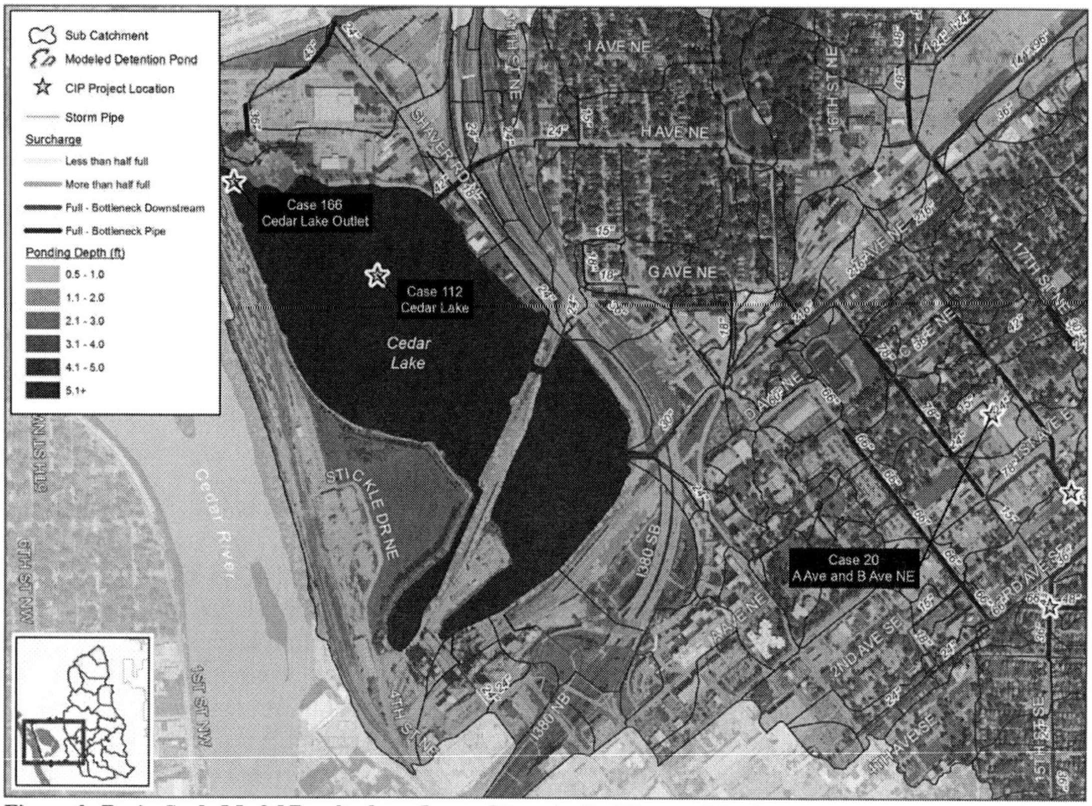

Figure 2: Basin-Scale Model Results for a 5-year Storm in the Kenwood Basin

Living Document

Work on the first phases Stormwater Master Plan came to a close in mid-2016 and the document was adopted by the Cedar Rapids City Council. The council did so knowing that only the first steps had been taken. The Master Plan was designed to be a living document, receiving regular attention, refinements, and further development.

The focus of phases 1 and 2 was the hydraulic modeling and prioritizing immediate needs. However, several other sections were initially laid out for future work:

- *Growth Areas*

- *Asset Management*
- *Financial Needs & Potential Funding*
- *Policy Issues*

Growth Areas

Continued growth in Cedar Rapids is a reflection of the City's resiliency. EnvisionCR, the City's Comprehensive Plan, identifies the City's vision for the future, targeting growth areas to the West, Southwest, South, North, and Northwest. EnvisionCR also includes several specific stormwater-related initiatives including watershed planning, capital improvements planning, policy refinement, best management practices, and sustainability. With a few exceptions, the City's CIP does not include expenditures to address growth related stormwater infrastructure. Given the magnitude of funding needs, it is critical that growth areas be served by stormwater infrastructure paid for by the associated development(s). This will require conceptual planning to identify stormwater needs in growth areas. It will also require due diligence to assure that growth related stormwater infrastructure is adequate for both initial development and future growth in the surrounding watershed. In addressing stormwater needs of planned growth areas, the City will be able to get ahead of future development to address downstream stormwater issues or, at the least, keep them from getting worse.

Asset Management

While much of the public focus has been on identifying and implementing capital improvements to address flooding problems, the Public Works Department is also charged with operating and maintaining the existing stormwater system. The Department is in the process of significantly overhauling practices, procedures, and policies for maintaining and preserving the City's existing stormwater infrastructure. Significant operational changes will be implemented to inventory assets, assess the condition, perform maintenance, and refine management policies. Additional recommendations have been provided to further enhance management of existing stormwater assets through standardizing maintenance activities using the Iowa Stormwater Management Manual and benchmarking of other utilities to refine existing O&M standards.

Financial Needs & Potential Funding:

Over the past eight years, stormwater utility fees have generated approximately $1.5 to $3 million per year for stormwater capital improvements. CDBG funding has provided approximately $9 million of additional funding over the last several years. With the aforementioned capital needs of $50 to $100 million or more, there is a significant gap between available funding and capital needs.

Other stormwater utilities share many of the same needs and challenges as Cedar Rapids. National survey results indicate that most utilities fund stormwater management in whole or in part through stormwater user fees that account for more than all other funding sources combined. Survey results also indicate that nearly all stormwater utilities use cash in lieu of debt financing, and that the majority indicate they do not have adequate funding to meet their total needs.

The current stormwater financial plan includes a 3.0 percent increase in stormwater utility fees in FY 2017 and a 5.0 percent increase in annual fees thereafter. It does not include borrowing. With the increase in fees, forecast funding for capital needs increases from the current $1.3 million in fiscal year 2016 to approximately $2.0 million in fiscal year 2017 and $2.4 million per year by fiscal year 2021. The refinement to incorporate impervious area into the fee structure is forecast to generate an additional $0.5 million in fiscal year 2017 increasing to an additional $2.0 million by fiscal year 2021, subject to actual implementation timing and phasing.

Policy Issues

In the past, existing regulatory standards and policies have focused upon utilizing gray infrastructure construction (concrete basins, inlets, piping, etc.) to meet water quality requirements and to control and convey runoff. As development continues to create more impervious surfaces, coupled with changing weather patterns and higher rainfall intensities, many communities are discovering their existing stormwater systems are unable to manage the increased runoff intensities. Rather than continuing to construct gray infrastructure, communities are incorporating "green infrastructure" into their existing stormwater management policies.

Increasing efforts to incorporate "green infrastructure" into existing stormwater management and other infrastructure policies tops the list of policy recommendations in the Master Plan. This is consistent with the guidelines of EnvisionCR and is consistent with what other communities are doing regionally and nationally. Green infrastructure provides a more holistic stormwater management approach that mimics aspects of the natural hydrological cycle, including retention, infiltration, and evapotranspiration. Promoting stormwater infiltration and runoff reduction postpones indefinitely or delays the need for future gray infrastructure projects.

Other policy related issues for consideration to enhance the City's existing policies follow. These recommendations typically originate as Public Works staff respond to citizen requests and target the safety and welfare of residents.

- Develop a private cross-connect program that considers the increased stormwater flows.
- Implement a post construction grading survey for all development sites to assure that planned grading is achieved.
- Implement a "Soil Quality Plan" to reduce runoff to each area of infiltration post development (initial plan adopted in November 2016).
- Incentivize green practices (the City initiated a cost-sharing program to assist property owners in implementing these).
- Develop a funding policy to clarify liability for non-City stormwater property losses.
- Require low-level openings and lot corner grades on site development plans that reflect the associated drainage plans.
- Move towards regional rather than individual style detention basins for economies of scale and advance planning ahead of new development.
- Consider targeted buy-outs in lieu of more expensive mitigation for flood prone properties.
- Increase educational program awareness in collaboration with the school districts.

RESULTS

The initial Stormwater Master Plan was adopted by the Cedar Rapids City Council in late 2016. Despite the document outlining potential capital needs of twice the cost that was estimated two years prior, the "hard truth" was understood and the Master Plan was well-received. The concept of the living document made it clear that the Cedar Rapids Public Works Department had established objective priorities and was working to address the most important issues.

As the City continues to build on the document, the GIS database has grown, coordination with on-going street improvement programs has become more frequent, and priority policy changes are being discussed and addressed by city leadership. While the Master Plan remains in development, the essential framework is complete and the City has a clear path forward.

September 2016 Flood

In September 2016, heavy rains north of Cedar Rapids swelled the Cedar River, leading to the second largest flood on record in Cedar Rapids. The City's efforts on the Stormwater Master Plan indirectly contributed to part of the City's success in fighting that flood. The City's Flood Response Manual only identified the use of manhole risers to place around inlets to the storm sewers; but the large drainage area and the shear number of manholes made this an impractical approach.

As the flood threat became clear, attention was immediately turned towards finding a more effective use of resources in the limited time available. The knowledge of the system gained through the master planning process enabled the City to target pumping locations that would be most effective, which sewers to plug, and which manholes to raise. Further, while the chances for precipitation during the high river event were slim, data from the stormwater modeling could have been used to identify needed pumping capacities.

In turn, the September 2016 flood has impacted planned work for the Stormwater Master Plan. The flood highlighted vulnerabilities in the storm sewer system, making it a learning opportunity that may better prepare the City for future flood fights. Plugging sewers was one of the most effective tools during the flood fight. However, plugs can be difficult to remove after the flood. A lesson learned, then, is to have secure, reliable closure gates to isolate the sewers from the river. As gates are added, they will aid future flood responses by the City and provide a long-term benefit as the permanent system is constructed.

CONCLUSIONS

Approaching the Stormwater Master Plan by creating a living document has enabled the City to establish priorities, build a framework for modeling, and build consensus with City leadership all while continuing to make progress with capital improvements to the storm sewer system.

Through this process, there are several key lessons learned:

1. The success of a living document and future modeling both depend on a sound framework at the beginning. A little extra time organizing an outline can pay dividends as the work progresses as it will be clear as to the workflow to be followed, the work

remaining, and the work that is complete. For modeling, the framework will establish consistency in results as the work progresses over the years.

2. Clear priorities are essential for all aspects of the Master Plan. From prioritizing modeling work to construction projects to sections of the Master Plan, a living document will be successful if the priorities are clearly established and communicated.
3. Policymakers are assets for building consensus, gain support for funding, and communicating with the public. By understanding their priorities and putting that in a broader context, their issues can get addressed in manner consistent with the overall priorities of the system.
4. Cities have numerous activities going on simultaneously, so working collaboratively across City departments is essential to success. Knowing the other projects in an area can influence priorities. For example, if an urgent streets project is being designed in the location of a potential stormwater need, accelerating the stormwater project will avoid reconstructing a brand new street. With limited funds, holistic views of infrastructure throughout a community become essential to prioritize capital improvements across multiple needs.

ACKNOWLEDGEMENTS

The authors gratefully acknowledge the cooperation of staff at the City of Cedar Rapids Public Works Department and their assistance in developing the Cedar Rapids Stormwater Master Plan.

Davenport, Iowa 14th District Diversion Study

Karol Giokas, P.E., RJN Group
Tom Leabhart, P.E., Davenport, IA

ABSTRACT

The 14th District sanitary sewer basin in Davenport, Iowa has high levels of inflow and infiltration (I/I), resulting in sanitary sewer overflows and basement backups. When the West Side Tunnel came online in 2013 much of the flow that previously was tributary to the south Duck Creek interceptor was diverted to the tunnel. This resulted in the south Duck Creek interceptor having additional capacity. The purpose of this study was to evaluate ways to reduce the occurrence of overflows and basement backups in the 14th District area by utilizing that potential excess capacity in the south Duck Creek interceptor. Different diversion options were evaluated along with different levels of flow reduction to come up with a recommended option for the city to reduce basement backups and overflows.

KEYWORDS

Basement backups, sanitary sewer overflows, flow metering, flow monitoring, flow diversion, hydraulic modeling.

INTRODUCTION

The City of Davenport, Iowa (City) has had problems with sanitary sewer overflows (SSOs) and basement backups in the sewershed known as the 14th District. The area served by the 14th District is one of the City's older areas, situated on a bluff northeast of downtown Davenport, with homes dating from the 1870s. The basement backups in the area have resulted in several lawsuits being filed against the City. The City set a goal to eliminate SSOs and basement backups in wet weather conditions up to a 5-year storm recurrence interval. RJN Group, Inc. (RJN) had been working with the City performing flow monitoring and sanitary sewer evaluation services (SSES) in multiple location in the City including the 14th District. RJN was also selected to evaluate options for diverting flow from the 14th District to the south Duck Creek interceptor.

FLOW MONITORING

Over a period of 4 years 25 flow meters and multiple rain gauges would be installed in and around the 14th District basin to understand the system response to rainfall. The original metered area is shown as the small area outlined in yellow as 14th District Study area in Figure 1. This metering in 2011 indicated that a restriction was present downstream (out of the study area) and was restricting the flow in the 14th District interceptor. In 2012, the metering was expanded to include the downstream sewers that connected to the south Duck Creek interceptor. Unfortunately, there was insufficient rain during the metering period in 2012, and the flow meters were reinstalled in 2013.

FIGURE 1
2014 DIVERSION ANALYSIS STUDY AREA

The West Side Tunnel (WST), installed to allow for continued development in the northwest portion of the City, was completed in 2013. The WST was also designated to take the flow from the western portions of the north and south Duck Creek interceptors. This resulted in available capacity in the two Duck Creek interceptors, also shown in Figure 1. The three periods of flow metering provided a good baseline for flows in the 14th District interceptor.

However, no metering had been performed on the north and south Duck Creek interceptors. Therefore, in 2014 RJN installed meters along both Duck Creek interceptors, as well as the 14th District interceptor. This data would provide the information needed to calibrate a model of all three interceptors. The meters used were ADS Triton meters on telemetry, with dual depth sensors and a Doppler velocity probe. The meters were installed, calibrated, and maintained to obtain high-quality data to use in a hydraulic model.

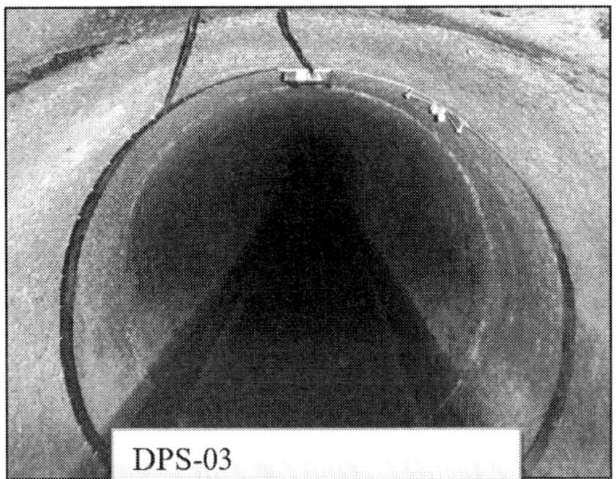

DPS-03
Meter Installation

In this case, where the City is considering committing millions of dollars for construction it is important to get good rain data, both dry days and multiple storms of varying sizes to be confident in the model calibration. For this reason the metering period was designated as 90 days, with an option to extend if that would become necessary. In this case that was not required, as Figure 2 shows the 1-hour rain events that occurred during the study.

FIGURE 2
1-HR RAIN EVENTS

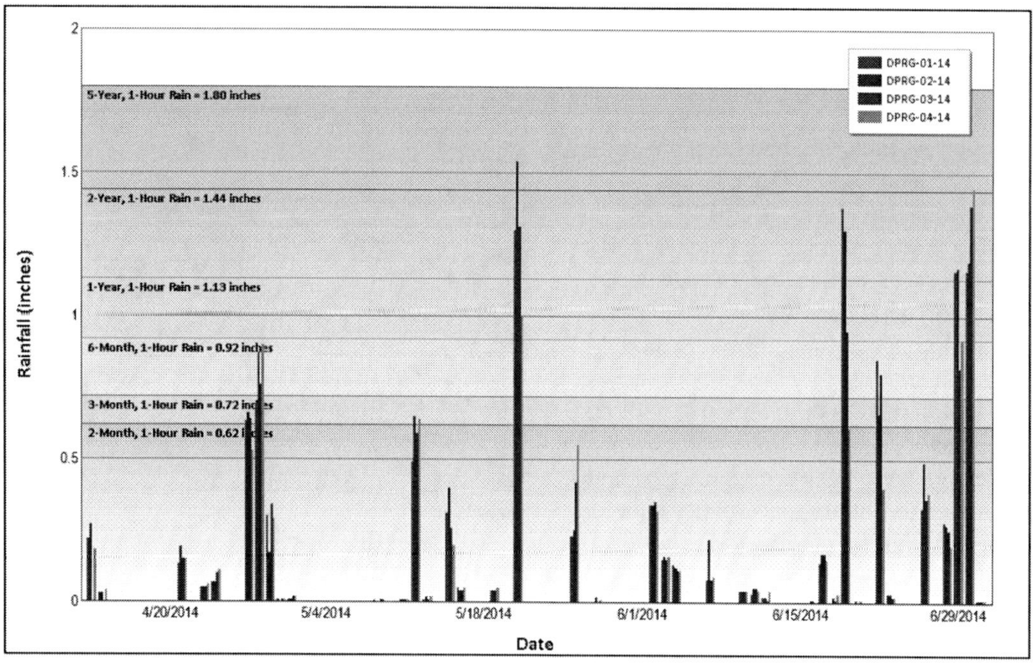

MANHOLE INVESTIGATION AND SURVEY

For the accurate modeling of the sewers in Davenport, GPS sub centimeter rim elevations were collected for manholes along the model route. Many of these manholes were located along Duck Creek which is heavily wooded in some areas, resulting in some challenges to the survey crew.

Collecting Manhole Inspection Data at MH DN-40

Invert depths were collected on the structures to be incorporated into the model during inspections of the manholes. Because of the critical nature of the model and the importance of getting the pipe diameters correct, descent inspections were performed on structures where pipe diameter transitions occurred. Manhole inspections also documented structural and I/I defects present in the manholes. Recommendations for remediation of those defects was provided to the City.

Map changes, resulting from the field investigations were also communicated back to City staff so that the GIS asset inventory could be updated to more closely match the field configuration.

These included incorrect connectivity, new structures, non-existent structures, and pipe diameter changes.

MODEL DEVELOPMENT

The skeletal model, which had been developed previously to include all sewers 12" and larger for the 14th District interceptor, was now expanded to include the north and south Duck Creek interceptors. The collected flow data was used to calibrate this larger model to match the system response to various wet-weather events.

Not only was the model used to evaluate the effect of changes on the 14th District, but also on the south and north Duck Creek interceptors. It is important that options for relief in the 14th District do not simply move the problem to another area and, in this case, negatively impact the south Duck Creek interceptor.

CAPACITY OF THE INTERCEPTORS

To understand the effect of the extra flow from the diversion area on the existing Duck Creek Interceptor flow, it is first important to understand the current capacity of the interceptors. This can be assessed by computing the Sustainable Peaking Factor (SPF), or the ratio of full pipe flow to dry-weather flow for each segment along the sewer.

Flow data collected during dry-weather, low-groundwater periods was analyzed to determine the average dry-weather flow (ADWF) for each basin in the study area. A summary of the ADWF for each basin is given in Table 1. The ADWF was then used to model the current condition of flow during dry weather conditions.

TABLE 1
ADWF SUMMARY

Meter	ADWF (mgd)
DP-03-14	0.58
DP-12-14	1.06
DPN-01	0.60
DPN-02	1.42
DPN-03	0.00
DPS-01	0.17
DPS-02	0.26
DPS-03	2.90
DPS-04	5.29

The calculated capacity, based on pipe diameter, slope, and material, for each pipe segment was compared to the ADWF of that segment to get the segment SPF, and is shown in Figure 3.

Sustainable Peaking Factors can be categorized as follows:

- **Less than 4** – The sewer capacity is lower than design standards for new pipe dry weather flow and capacity improvements are necessary. I/I reduction and storage are also likely components of a full solution.
- **Between 4 and 8** – Sewer has sufficient capacity for dry weather flow, but wet weather capacity is limited. Moderate inflow and infiltration (I/I) during major rain events may surcharge the system. Capacity improvements, storage, and I/I reduction options should be evaluated.
- **Between 8 and 15** – Sewer has sufficient capacity for dry weather flow and should have sufficient wet weather capacity up to a 5-year storm for wet weather flow with moderate I/I. However, systems with significant I/I may still surcharge the system in major rain events. Capacity improvements are generally not justified, but I/I reduction

improvements are typically utilized with storage included in lieu of addressing expensive private sector sources.

- **Greater than 15** – Sewer has sufficient capacity for dry weather flow and should have sufficient wet weather capacity through a 10-year storm event unless I/I is severe.

Figure 4 shows the SPF of all the sewers modeled for this study. As can be seen, the south Duck Creek interceptor has greater than 15 SPF through the upper reaches of its span. This extra capacity was created when all of the tributary flow west of DS-53 near Wilkes Ave, was diverted to the West Side Tunnel in 2013. Also seen in Figure 4 is the limited capacity in the north Duck Creek interceptor, the 14th District interceptor, and the stretch of sewer after the two Duck Creek interceptors merge together south of Duck Creek and east of the railroad tracks.

A new sanitary system is typically designed to convey 3-4 times the assumed dry-weather flow. Therefore, a peaking factor less than 4 for any specific basin is generally not considered to be an I/I issue as the sewer should have capacity to handle the excess flow. A system hydraulic model is calibrated according to the meter data gathered and standard rain distributions are applied to the study area in order to pinpoint areas with surcharges and/or backups.

In smaller wet weather events, such as a 1-Year storm, the north and south Duck Creek interceptors currently have large amounts of freeboard, or vertical distance between the ground elevation and the sanitary flow hydraulic grade line. When freeboard gets to zero, the model is predicting a sanitary sewer overflow. Freeboard is also an important tool used to indicate the risk of basement backups in the area. Because of the variability of basement depths relative to sewer depths the allowable freeboard for each house will be different, but it is clear that as freeboard decreases the risk of a basement backup increases.

FIGURE 3
CURRENT CONDITION SUSTAINABLE PEAKING FACTORS

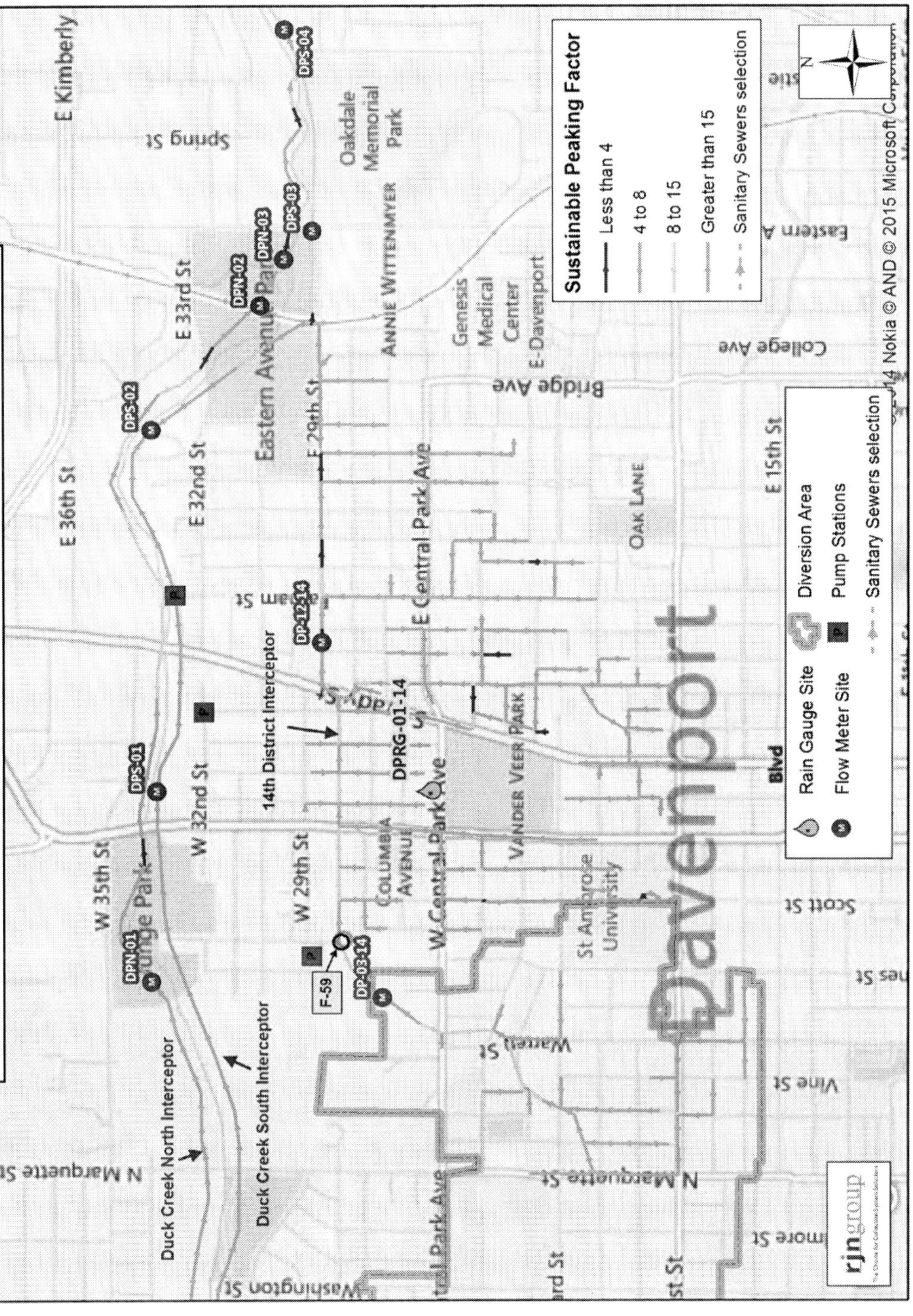

For larger rain events, such as a 5-year or 10-year rain event, the hydraulic model predicted that the Duck Creek interceptors begin to take on significant amounts of I/I causing backups where the two interceptors come together. The volume of excess flow in the Duck Creek interceptors results in loss in freeboard, as can be seen in Figures 4-6.

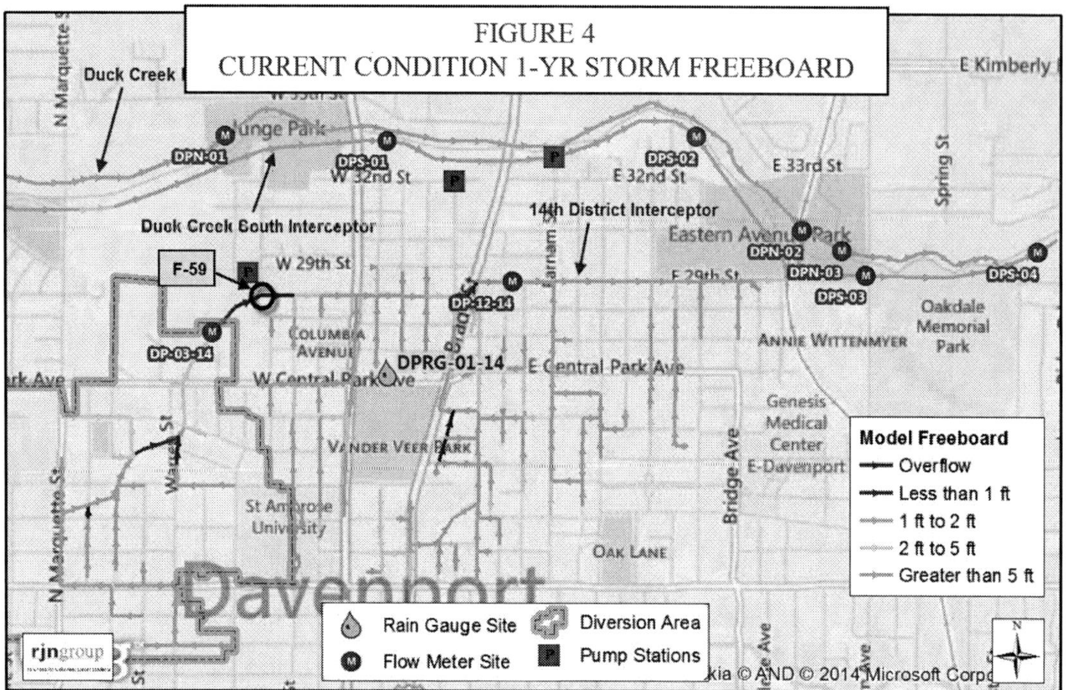

FIGURE 4
CURRENT CONDITION 1-YR STORM FREEBOARD

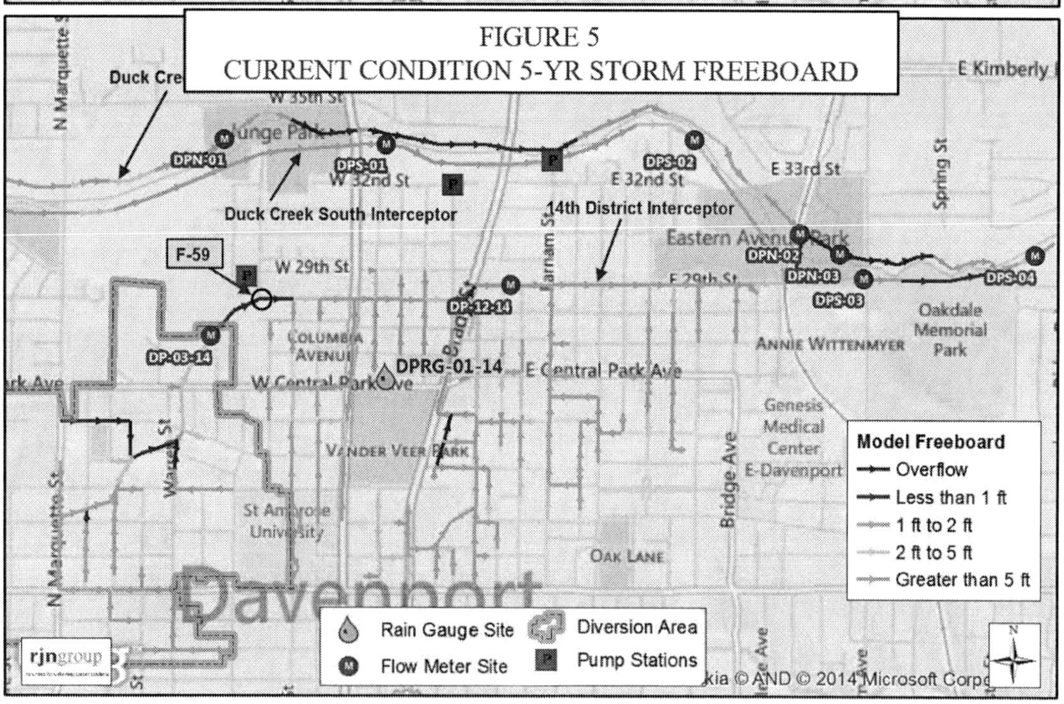

FIGURE 5
CURRENT CONDITION 5-YR STORM FREEBOARD

FIGURE 6
CURRENT CONDITION 10-YR STORM

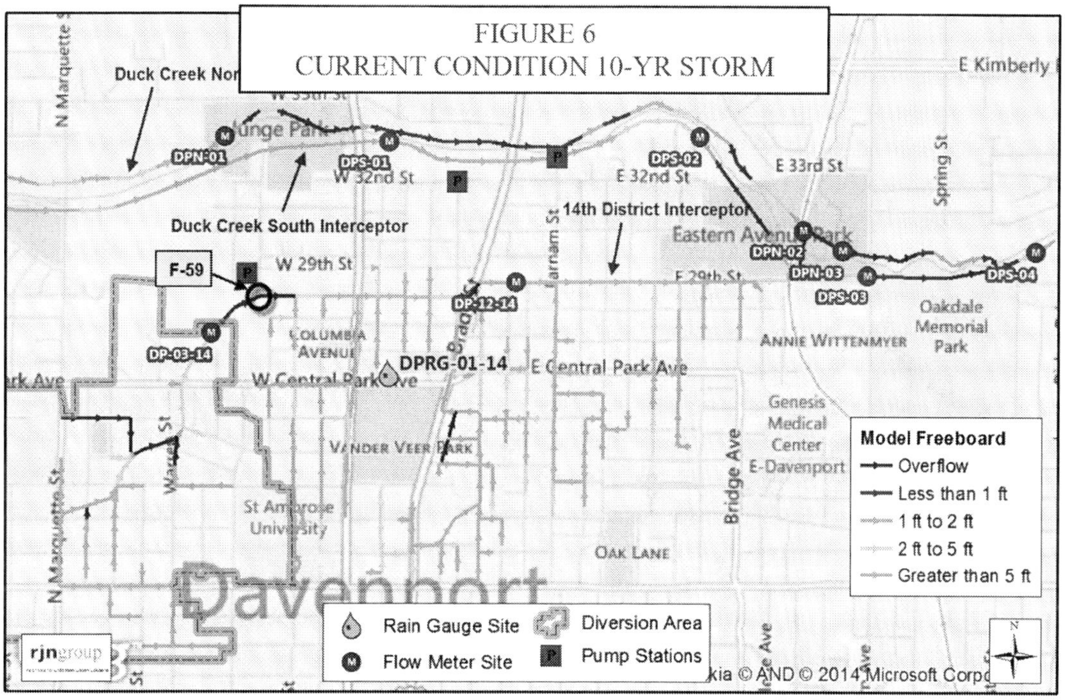

OPTIONS EVALUATED

The limitations in the 14th District interceptor can be seen in the preceding Figures 4 through 6 at manhole F-59 just downstream of meter location DP-03-14. F-59 is a site where the combination of high levels of excess flows and the manhole depth of only approximately 4 feet resulted in frequent overflows. The location that was assessed for possible diversion of flow north to the south Duck Creek interceptor was just upstream of this manhole at manhole F-61.

The hydraulic model indicated that, while flow reduction was necessary, it alone would not be sufficient to meet the City's goal of reducing the occurrence of basement backups and the SSO at F-59. Four scenarios for flow storage and/or flow diversion utilizing the excess capacity in the south Duck Creek interceptor resulting from the recently installed the West Side Tunnel were evaluated by RJN using the enhanced system model. Minimum freeboard in the modeled system was the criterion used to determine how effectively the various options resolved the excess flow issues in the 14th District.

The four scenarios evaluated were:

- Scenario 1 - Diversion of all of the flow from the 14th District interceptor at manhole F-61 to the south Duck Creek interceptor
- Scenario 2 - Partial diversion of flow from the 14th District interceptor at manhole F-61 to the south Duck Creek interceptor
- Scenario 1A - Diversion of all of the flow from the 14th District interceptor at manhole F-61 to the south Duck Creek interceptor and incorporating storage
- Scenario 2 - Partial diversion of flow from the 14th District interceptor at manhole F-61 to the south Duck Creek interceptor and incorporating storage

Each scenario was also modeled with no flow reduction or 20% flow reduction in either the entire 14th District study area or in just the area that was being diverted to the south Duck Creek interceptor. Concurrent with the flow monitoring and hydraulic modeling study, the City initiated a comprehensive sanitary sewer evaluation survey (SSES) with RJN. The work included public system inspections including manhole inspections and survey, closed-circuit televising (CCTV), and smoke testing. In addition, targeted private property inspections were also completed. Upon completion of the SSES inspections, a flow balance was completed to estimate inflow and infiltration (I/I) into the system and to provide the City with recommended flow reduction options.

The results of the modeling indicated that the best solution for the excess flow issues was a combination of diversion of all flow from the 14th District interceptor at F-61to the south Duck Creek interceptor, 20% flow reduction in the 14th District, parallel relief piping in select localized areas, and inclusion of a storage tank. This is basically Scenario 1A with the additional local sewers.

Figures 7 and 8 show the resulting modeled freeboard for Scenario 1A for a 1-year rain event and a 5-year rain event. The other scenarios modeled did not resolve the excess flow issues for up to a 5-year rain event. Also note that the model indicated that for all scenarios the north Duck Creek interceptor had very low levels of available freeboard, indicating high levels of surcharge in the sewer. This issue could not be resolved by any of the options being considered in the study.

FIGURE 7
SCENARIO 1A, 20% FLOW REDUCTION – 1-YEAR RAIN

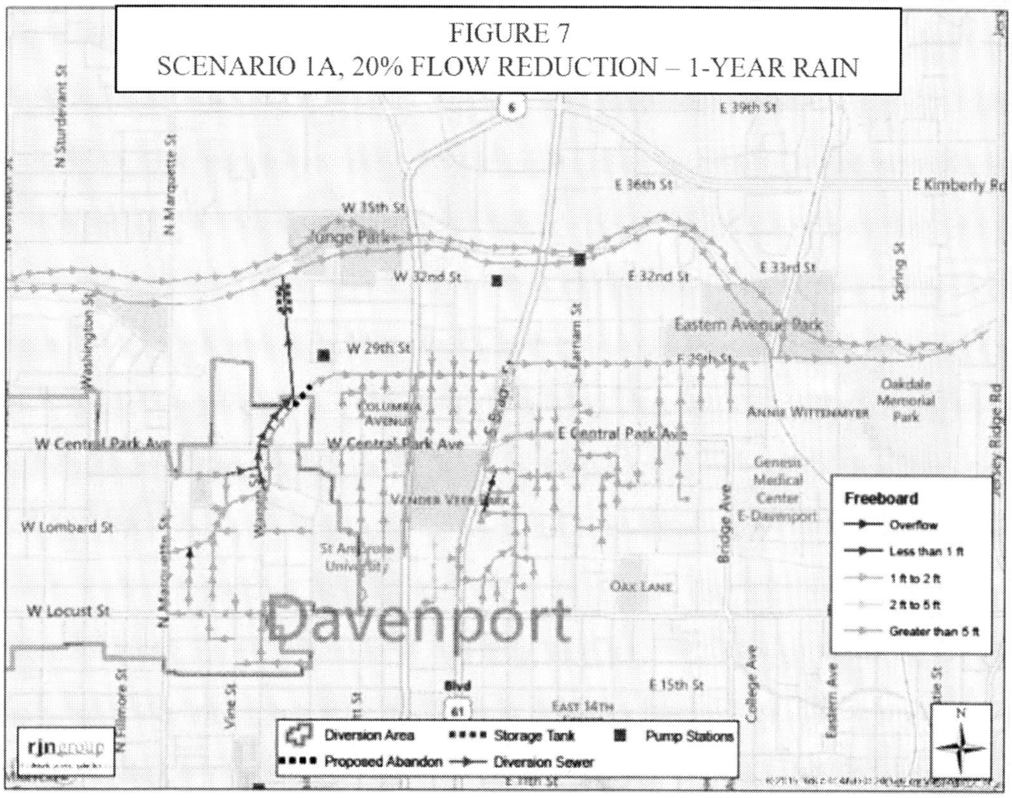

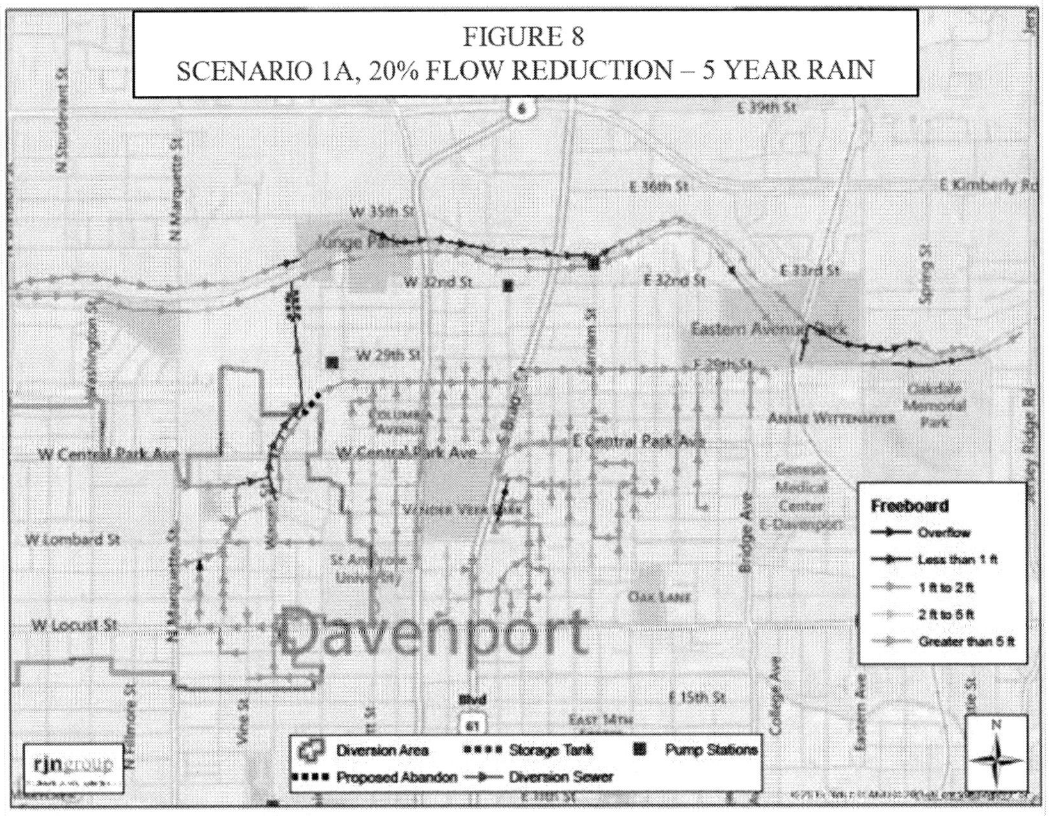

FIGURE 8
SCENARIO 1A, 20% FLOW REDUCTION – 5 YEAR RAIN

RECOMMENDATIONS

The recommendations from the study included:

- Continuation of system rehabilitation and clear water source removal in the 14th District with a goal of at least 20% reduction of wet weather flows through manhole and mainline rehabilitation and "quick fix" private sector defects;
- Complete diversion of flow from the 14th District to the south Duck Creek Interceptor;
- Incorporation of parallel relief sewers in a local area;
- Design considerations to allow for a future 1.5 MGD storage facility; and
- Continuing flow monitoring to measure the change in system response during wet weather events resulting from improvements.

It was recommended that the City implement these recommendations in a phased approach, using flow metering as system changes were completed to confirm the resulting flows in the system matched those predicted in the model. This implementation would involve starting with a modified Scenario 2, using a variable flow diversion structure at F-61, to start with partial flow diversion. The ultimate solution for the area will require the incorporation of the storage and parallel sewer shown in Scenario 1A. This will eventually result in full diversion of all flow from the 14th District interceptor at F-61.

Flow reduction would need to continue by addressing public sector sources, and some "easy" private sources such as downspouts or cleanouts. With the goal of no basement backups or SSOs are experienced during a 5-year storm event, the level of flow reduction achieved by the City could impact the size of the excess flow storage tank that will be needed. This is the major reason for the phased approach recommended. Flow metering data collected and analyzed following the construction of the diversion structure and implementation of flow reduction would result in more accurate sizing of the storage structure and the parallel sewers needed for the resulting system.

SUMMARY AND CONCLUSIONS

In order to relieve surcharging and the resulting overflows and potential backups in the 14th District study area, the City of Davenport needs to reduce flows in the areas tributary to the 14th District interceptor. However, based on system models, flow reduction through I/I removal alone is not sufficient for system wide relief. RJN evaluated potential options for diversion of flow to the south Duck Creek interceptor. It is important that options for relief in the 14th District do not simply move the problem to other areas and, in this case, negatively impact the area tributary to the Duck Creek interceptor. Available freeboard was the criterion used for determining how well a scenario addresses the needs in both the 14th District and the south Duck Creek area.

Only Scenario 1A, with flow reduction, will reduce the risk of overflows in the 14th District at all locations without increasing the potential of overflow or basement backups in the south Duck Creek interceptor area. However, a partial Scenario 2 implementation could be utilized as an initial step to provide some immediate benefit without significantly impacting the Duck Creek interceptor. Implementing Option 2 initially with flow reduction can reduce the cost of implementing scenario 1A by reducing the size of the wet-weather storage needed or it can provide an extra measure of storm protection over the 5-year storm summarized here.

The City completed the implementation of the variable diversion structure in 2017. There have not been significant rain events to date, nor are there flow meters installed, to provide any data indicating the impact on flow levels of the approach.

Water Environment Federation (WEF)
601 Wythe Street
Alexandria, Virginia 22314
USA

ISBN 978-1-5108-6316-3

9 781510 863163

CURRAN ASSOCIATES INC.
proceedings.com